About the Author

Tony Benson lives in Kent, England with his wife Margo. He grew up in a Kent village, and began his career in the 1970s in an optics laboratory at the Royal Aircraft Establishment in Farnborough. He went on to become a Chartered Engineer and, after a successful career spanning more than three decades in research and development in the aerospace and transit industries, he left the corporate world behind and began a new career making stringed instruments and writing.

Website: **tonybenson.org**
Email: **TonyMBenson@yahoo.com**
Blog: **TonyMBenson.blogspot.com**
Facebook: **TonyBensonAuthor**
Twitter: **@TonyMBenson**

Brass and Glass

Optical Instruments and Their Makers

by Tony Benson

First published in 2021 by Tony M. Benson
Second edition 2023
Copyright © 2017-2023 Tony M. Benson
All rights reserved. No part of this book may be reproduced or transmitted in any form or by any means, electronic or mechanical, including photocopying, recording, or by any information storage and retrieval system without the written permission of the author, except where permitted by law.

The publisher expressly prohibits the use of the contents of this book in any manner for purposes of training artificial intelligence technologies to create text or any other content, including without limitation, technologies that are capable of generating works in the same style or genre as this content.

Cover design by J.D. Smith Design Ltd.
Cover images: Photographs of optical instruments reproduced with the kind permission of Charles Miller Ltd., auctioneers.

02P240306

ISBN: 978-0-9576527-7-4

Front cover pictures, left to right, top to bottom:

1: Dissecting microscope by E. Leitz, Wetzlar, Germany c1920
2: Leitz compound monocular tripod microscope, c1920
3: 10-inch theodolite by Adie & Wedderburn, Edinburgh, Scotland c1890
4: 2-inch refracting telescope by Dollond, London, England, c1860
5: C19[th] brass and leather binocular telescope signed G.&S. Merz in München, Germany.
6: Surveying level by A. Meissner, Berlin, Germany, c1830
7: 6½-inch. radius vernier sextant by Crichton, London, England, c1920
8: Culpeper-type microscope by E.J. Elliott, London, England, C19[th]

Back cover pictures, left to right:

1: Late C19[th] lacquered brass spectrometer by Elliott Bros., London, England
2: Three pairs of Kriegsmarine binoculars: Zeiss 7x50, Zeiss 8x60, Leitz 7x50
3: 2½ -inch reflecting telescope by John Cuthbert, London, c1852

Also By Tony Benson

Fiction

An Accident of Birth
Galactic Alliance: Betrayal

Dedicated with love to my wife

Margo

Acknowledgements

My thanks to the following: Dominic de Mattos, Steve Tonkin, William Cook, Lisa Ashby, David Silver, and Manish Patel for sharing their specialist expertise and knowledge and, in doing so, contributing to make this a better book; To Steve Gill of the Quekett Microscopical Club for generously sharing his knowledge and research notes; To Jenny Wetton; Christine Chappelle of the Manchester Literary and Philosophical Society; To Mike Mopar for his help with interpreting German texts; To Katie Abrahams for permission to use the **JB codes** list, published by her father Peter Abrahams on his website *The History of the Telescope and the Binocular*; To Ebay user *Dolemole* for permission to use his photos; To Charles Miller for his encouragement, and Charles Miller Ltd. for permission to use the photos in the cover design.

The writing of this book has involved a huge amount of effort over several years. I could not have done it without the help, support, patience, and wisdom of my wife, Margo, to whom I am eternally grateful.

Forward

The inspiration for this book came from my interests in stargazing, photography, microscopy and bird-watching. Soon I began to take an interest in classic and vintage telescopes, microscopes, binoculars, cameras, and optical instruments in general. I found the lack of reference material frustrating - I would see an interesting looking vintage telescope at auction, but to find out more about the maker I would often have to refer to extraordinarily expensive reference books, many of which are out of print. Further, the information I sought was frequently scattered around multiple sources, and time-consuming to collate. I wanted an affordable, simple, encyclopaedic reference in which I could find the name of the maker or optical instrument, and a concise but informative article about them. This book is my attempt to remedy that gap in the literature.

This book is presented in encyclopaedic form, with alphabetically ordered entries, each entry being a short but informative article about the given subject. The articles contain cross-references to each other, and references to outside resources, with a numbered list of external references at the back of the book. Cross-references within the text are highlighted in non-italicised bold text. These mostly refer to the names of makers included in the book. Names of optical instruments included in the book are not similarly highlighted, due to the high number of occurrences. Numbered external references are enclosed in square brackets.

What this book is

This book is an alphabetical reference to optical instruments and their makers, from the earliest to the end of the 20th century. The main focus is on consumer instruments such as telescopes, binoculars, microscopes, cameras, and other related instruments, but there are also articles on navigation and surveying instruments, laboratory instruments, military instruments, and other instruments that are not primarily for the consumer market. There are appendices containing information on selected related subjects such as optical glass and eyepiece designs.

An optical instrument is an instrument that uses one or more lenses, mirrors, prisms, or some combination of them, to enable the user to view an image of an object. The scope of this book sometimes strays beyond that to include interesting or important related instruments, in order to provide context. For example, the astrolabe is included, even though it uses no mirrors, lenses, prisms to manipulate the light.

Since mankind began to create and use optical instruments there have been countless makers of every kind of instrument. It is not possible to include an entry on every one of them, and it is not the author's intention to do so. However, it is intended that makers of note should be included, as well as most of the makers the reader might encounter when perusing an auction or boot fair.

Some optical instruments are marked with the name of the retailer rather than the maker. In many cases it is simply not known whether the retailer also made the instrument, or purchased it wholesale and re-branded it for sale. It should be noted that if an instrument appears at auction bearing a name not known as a maker, no assumptions can be made regarding the original maker without concrete evidence.

Throughout the history of optical instrument makers, those who make the instruments have frequently augmented their business by retailing instruments made by others. In many cases these are marked with the name of the instrument maker who retailed them rather than the original maker. This can make the attribution of an instrument difficult. Moreover, many instrument makers manufactured parts of their instruments and had other parts made for them, such as optical components or brass-work. These factors should always be considered when attributing an instrument to a maker.

Note: Where a person or company is listed in the following sections as a maker of optical instruments, the reader should not take this as meaning they manufactured all of the constituent parts or, indeed, all of the instruments they sell or sold. Many makers, both now and historically, have some parts made for them, often by overseas manufacturers. Most makers of optical instruments do not manufacture the optical glass used in their instruments.

This book is intended to be worldwide in scope. Some entries are more detailed than others and, in some cases, this reflects the level of available information. Future editions of this book will expand on some entries, and add new ones, as well as correcting any errors that have come to the author's attention.

What this book is not

This book is not a guide to choosing, or assessing the value of an optical instrument. It contains entries for historical instrument makers, but it is not an historical treatise on instrument makers or the development of their craft. It does not contain tutorials on related subjects.

While this book gives definitions of some terms used in optics, such as polarization, types of aberration, etc., it is intended to convey the concepts to a non-technical reader, and is not a rigorous mathematical treatment.

Sources

The information in this book is gathered from a multitude of sources. These include: birth/marriage/death records, census records, newspapers, advertisements, journals, catalogues, trade cards and trade labels, museums, trade directories, postal directories, extant instruments, auctions, auction catalogues, and more. Many books have been written about mathematical, scientific, surveying, nautical, and optical instrument makers and the instruments themselves. The authors of those books include subject matter experts such as museum curators and academic researchers, and such books, many of which are out of print, have also provided valuable information in the preparation of the present volume.

Disclaimer

The information provided in this book is for general informational purposes only. All information in this book is provided in good faith, however the author makes no representation or warranty of any kind, express or implied, regarding the accuracy, adequacy, validity, reliability, or completeness of any information in this book. The author assumes or undertakes no liability for any loss or damage suffered as a result of the use, misuse or reliance on the information and content of this book.

Viewing the sun - WARNING

WARNING: Never view the sun directly with the naked eye. Never view the sun with the aid of an optical instrument unless it is fitted with a suitable solar filter. Always follow the advice and instructions of the manufacturer of the solar filter. If you are not certain of the effectiveness of a solar filter, do not use it. Avoid second-hand solar filters unless they have been professionally tested. Never use a vintage solar filter.

Failure to follow these warnings may result in permanent damage to the eyesight, or blindness.

Differentiating names

Where confusion may arise due to two or more instrument makers sharing the same name they are differentiated in this text with the numbers (I), (II), etc. For example, John Smith (I), John Smith (II). Some referenced books use a similar scheme, and the reader should note that the numbering in other texts may not correspond to that used here.

A Note on I and J

The letter j did not exist in the English alphabet until the late-middle-ages. At that time, it was customary to lengthen an i when it was in a position such as an initial, in which case the tail of the i was lengthened. The j form evolved to represent the sound of i as a consonant, and was used interchangeably with i. It wasn't until the C17th that the differentiation between the two forms of the letter became more common, and the practise of interchanging i and j was still used by some people into the C18th and C19th.

As an example, the reader will see that **John Preston Cutts** (1787-1858) sometimes signed his instruments J.P. Cutts, and sometimes I.P. Cutts.

A Note on Addresses

Addresses shown for the instrument makers in this volume are drawn from various sources, including business directories, census data, books and directories about instrument makers, advertisements, catalogues, and other sources. In general, these are the business addresses for the instrument maker. However, in some cases they may be residential addresses, or a business address that is also a residential address. The dates given for the instrument maker's occupation of an address are those drawn from records, and do not necessarily represent the full duration of the instrument maker's occupation of the address.

Omissions and Inaccuracies

If you find that I have omitted an entry, or included inaccurate information, please let me know by emailing me at **tonymbenson@yahoo.com**. When you do so, please try to provide me with verifiable information, and include any available references. If I include your suggested update in a future edition, you will receive a mention in the acknowledgements page.

Reviews
Positive reviews are the life-blood of an author. If you enjoy this book or find value in its contents, please consider leaving a review.

Contents

A-Z Alphabetical Entries	1
Glossary	419
Abbreviations	422
Appendix I: Barr & Stroud Binocular Models	424
Appendix II: Eyepieces	426
Appendix III: German Wartime Makers' Ordnance Codes for Optical Products	428
Appendix IV: Japanese Manufacturers' Codes for Optical Products	431
Appendix V: Russian and Soviet Makers' Marks	435
Appendix VI: Optical Glass	437
Appendix VII: Charts and Timelines	439
Appendix VIII: The Electromagnetic Spectrum	445
Appendix IX: The London Craft Guilds	447
Appendix X: The Dollond Patent Dispute and the Petition of 1764	454
Appendix XI: Scale Dividers' Marks	458
Bibliography and References	461

A

Abbe Apertometer
See apertometer.

Abbe, Ernst
Ernst Abbe, (1840-1905) [45] [142] [155], was a German physicist, whose theoretical and technical innovations in optical theory led to great improvements in optical design. In 1863 Abbe joined the University of Jena, rising to professor of physics and mathematics in 1870, and director of the astronomical and meteorological observatories in 1878.

In 1866 Abbe was hired by **Carl Zeiss** as research director of the Zeiss optical works. In 1876 Zeiss made Abbe his partner and successor. In 1884 **Otto Schott**, Ernst Abbe, and Carl Zeiss founded a glass making company, **Schott and Associates Glass Technology Laboratory**, in Jena.

Abbe developed his theory of image formation in the microscope, and the sine condition that is named after him, and these form the basis of high-performance optics. In 1886 Abbe invented the apochromatic lens for microscope objectives. The design of the orthoscopic eyepiece is also credited to Ernst Abbe.

Abbe became head of the Zeiss company after Zeiss's death in 1888, and in 1889 he set up the Carl-Zeiss-Stiftung (Carl Zeiss Foundation) to run the two companies, Zeiss and Schott.

Abbe-König Prism
See roof prism.

Abbe Refractometer
See refractometer.

Abbe Spherometer
See spherometer.

Aberrations, Optical
All optical instruments that use mirrors and/or lenses to focus an image suffer from optical aberrations to some degree [7] [63]. Minimising optical aberrations is expensive, and is a factor that drives the high cost of fine instruments. A description of some of the main types of aberration can be found in the following entries: astigmatism, chromatic aberration, coma, distortion, field curvature, spherical aberration.

Two kinds of optical aberration that arise in poorly collimated optics are: (a) when the optical axes of two or more optical elements (lenses or mirrors) are misaligned such that their optical axes are parallel but not coaxial. e.g. a lens or mirror is shifted sideways; (b) when one or more optical elements is rotated in an axis perpendicular to the optical axis - e.g. a lens or mirror is tilted out of alignment. See collimation.

Abney Clinometer (Abney Level)
See clinometer.

Abraham, Abraham
Abraham Abraham (c1799-1863) [3] [8] [18] [49] [50], was a maker of microscopes, surveying instruments, and other instruments of Liverpool, England. Adresses: 8 Lord St. (1818-1820); 10 Lord St. (1821-1823); 7 Lord St. (1822-1828); 6 Lord St. (1827); 9 Lord St. (1829-1834); 76 Lord St. (1835-1836); 78 and 84 Lord St. (1837-1839); 20 Lord St. (1839-1850).

He was born in Exeter, Devon, England, the son of optician, **Jacob Abraham**. He started business in Lord Street, Liverpool in 1817 as a maker of microscopes, surveying instruments, etc. He was one of the first to make and sell inexpensive achromatic microscopes outside of London. He made them with lenses by **Nachet** of Paris. In about 1833 **Charles West (II)** began to work for Abraham, and for 24 years acted as his manager.

In 1838 Abraham opened a branch of his business in Glasgow, A. Abraham & Company, run by **Simeon Phineas Cohen**. Addresses: 8 Exchange Square, Glasgow (1838-1840), and 82 Queen St., Glasgow (1841-1843). Cohen took over the branch in 1843, but this closed in 1853 when Cohen was declared bankrupt.

In 1841 Abraham opened a branch of his business, **Abraham and Dancer**, in Manchester, in partnership with **John Benjamin Dancer**. They continued with this partnership until 1844, after which Dancer acted as sole manager of the firm, trading under his own name.

In c1849 Abraham formed a partnership in Lord Street, Liverpool with his manager, Charles West (II), and **George Smart Wood**. The partnership traded as Abraham & Co., and was terminated in 1855 [35 #21840 p216]. George Wood continued the business as Abraham & Co. He signed his instruments *Wood Late Abraham*, and in the early 1880s he traded as Wood, George S. (late Abraham & Co.)

Abraham & Dancer
Abraham and Dancer, [18] [49] [50], was a partnership, formed in 1841 between **Abraham Abraham** and **John Benjamin Dancer**, trading at 13 Cross Street, King Street, Manchester.

Their trade card of 1842 [53] reads: "Abraham & Dancer, 13 Cross Street, King Street, Manchester, manufacturers of Optical, Mathematical and Philosophical instruments, Achromatic microscopes, New and improved self-acting electro-magnetic machines, for medical purposes; Electrotype apparatus, Materials of every description; Moulds, plaster casts &c., Gas microscopes, Polariscopes, Lanterns for dissolving views, Achromatic object glasses, adapted to microscopes and instruments, or models of every description for public institutions and lectures. Instruments carefully repaired."

The partnership continued until 1844, when Dancer bought out Abraham's interest and continued the business under his own name. The termination of the partnership was formalised in 1845, [35 #20458 p1027].

Abraham, Jacob

Jacob Abraham, (fl.1809-1842 d.1845) [3] [18] [53] [276 #1068] [319], was an optician and mathematical instrument maker from Exeter, England. By 1809, he had moved to Bath, where he traded at St. Andrew's Terrace. In 1819 he was trading at 1 Bartlett St., Bath and, from 1830 until 1842, at 7 Bartlett St. **John Davis (I)** served an apprenticeship with him. He had two daughters and two sons, one of whom was the notable instrument maker **Abraham Abraham**.

In 1830 he opened an additional branch in Cheltenham, where he traded at an address next to Montpelier Rotunda. The address is also referred to in some directories as adjoining Mr. Thompson's Pump Room. According to Jacob Abraham's trade card, dated c1837, "J. Abraham, Optician and Mathematical Instrument Maker to his R.H. The Duke of Gloucester, and His Grace The Duke of Wellington. 7 Bartlett St. Bath and at his shop adjoining Mr. Thompson's Pump Room, Cheltenham."

Among the instruments illustrated on his trade card are a telescope, octant, microscope, spyglass, spectacles, and a globe. He is also known to have sold sundials and drawing instruments. A table orrery by Jacob Abraham is in the History of Science Museum, Oxford, [270]. He was retired by the time of his death in 1845.

Abraham, Robert

Robert Abraham, [61] [306], of 81 Aldersgate Street, London, England, was a photographic dealer, and agent for several manufacturers of cameras. The firm was founded in 1886 by Robert Abraham. The company is also recorded as trading at 22 Charterhouse Buildings, London. One of his suppliers was **Newman & Simpson**. Robert Abraham, died in 1887, after which the firm continued in business until 1890, at which time **A. Adams & Co.** began to trade at 81 Aldersgate Street.

Robert Abraham advertised the *Abrahams Detective Camera*. However, it is not known whether it was manufactured by Abraham. The London Post Office Directory listed the company as apparatus manufacturers from 1888 until 1890.

Achromat, Achromatic

An achromat, or achromatic lens, normally consists of two or more lens elements, so designed to bring a beam of light to focus perfectly in two frequencies of light, normally red and blue. As a result, the chromatic aberration over the full visible spectrum is limited. The achromatic lens was controversially patented by **John Dollond (I)**, but the invention is usually attributed to **Chester Moor Hall** (see Appendix X). A telescope or other instrument with an achromatic objective lens is sometimes referred to as an achromat.

Acme Optical Works

See **John W. Sidle & Company**.

Acuter

Acuter is the terrestrial optics brand of **Synta Technology Corporation**. They produce spotting scopes and binoculars as well as telescopes and mounts.

Adams & Co., A.

A. Adams & Co., [306], of London, England, was established during the 1880s as a photographic dealer, trading from 1900 at 81 Aldgate Street, formerly the address of **Robert Abraham**. By the 1890s they were designing and producing high quality cameras. In 1892 the company relocated to Charing Cross Road. From 1922 the company traded as A. Adams & Co. Ltd. In 1932 they moved to 122 Wigmore Street, and shortly afterwards to 53 Wigmore Street. There is no known connection with **R.T. Adams & Co.**

Adams, George (I)

George Adams (I), (1704-1772) [8] [18] [44] [46] [90] [264] [397] [441] [442] [445], was a maker of scientific instruments and globes, of Fleet Street, London, England. Addresses: 4 doors east of Shoe Lane, Fleet St. (1734); Near the Castle Tavern, Fleet St. (1735); *Tycho Brahe's Head*, Corner of Racquet Court, Fleet St. (1738-1757); 60 Fleet St. (1767).

He was the father of **George Adams (II)** and **Dudley Adams**. In 1724 George (I) was apprenticed in the **Grocers' Company** to James Parker, mathematical instrument maker. Following Parker's death in 1726, he completed his apprenticeship with **Thomas Heath (I)**, being freed in 1733. In 1734 he started his own business as a mathematical instrument maker in Fleet Street, and the business continued until 1817. Among his apprentices were George Adams (II), and **Robert Tangate**, who was turned over to him by **John Morgan** in 1758. One of his employees was **John Miller**. George (I) was appointed instrument maker to His Majesty's Office of Ordnance. He was notable for his globes, quadrants, astronomical orreries, microscopes and telescopes, as well as drawing, gunnery, and surveying instruments. In 1756 he was appointed mathematical instrument maker to the Prince of Wales, and retained the position when the prince became King George III. Upon the death of George Adams (I) his ordnance business went to **Jeremiah Sisson**. Some of his instruments can be seen in the King George III collection at the London Science Museum.

Adams, George (II)
George Adams (II), (1750-1795) [8] [18] [46] [90] [264] [397] [445], of *Tycho Brahe's Head*, 60 Fleet Street, London, England, was the son of **George Adams (I)**. He served his apprenticeship with his father, beginning in 1765, and was freed of the **Grocers' Company** in 1772 by Ann Adams, his father's widow, only a few weeks after his father's death. On George (I)'s death George (II) succeeded his father, and carried on the business together with his wife **Hannah Adams**. By 1777, he had renewed his family's appointment as instrument maker to His Majesty's Office of Ordnance. Among his apprentices was his brother, **Dudley Adams**. He succeeded his father in his appointment to King George III, and in 1787 George Adams (II) became optician to the Prince of Wales (the future King George IV). He developed a substantial ordnance trade in the build-up to the Napoleonic wars. He wrote various illustrated text-books, and pre-deceased his mother Ann. Upon his death in 1795, he was succeeded in business by his wife, Hannah.

Adams, Dudley
Dudley Adams (1762-1830), [8] [18] [46] [90] [397] [445], was an optician, globe maker, and mathematical and philosophical instrument maker, of Charing Cross, London, England. Addresses: 53 Charing Cross (1788-1796); 60 Fleet St. (1796-1826); 6 Jewry St., Aldgate (1800-1822); 19 Charles St., St. James St. (1819); 10 Waterloo Place, near Carlton House (1819); 42 St. Paul's Churchyard (1821); 22 Ludgate St. (1823-1827).

He was the son of **George Adams (I)**. He served his apprenticeship as a mathematical instrument maker with his brother **George Adams (II)**, beginning in 1777, and was freed of the **Grocers' Company** in 1788. His mother gave him the globe plates and tools that had belonged to his father, and in mid-1788 he opened a shop at 53 Charing Cross. He concentrated on globe making but also sold all types of mathematical instruments. In 1794 he was appointed globe maker to King George III. On the death of his brother in August 1795 Dudley quickly secured the ordnance appointment for himself. However, in 1806 he lost the ordnance trade to **Matthew Berge**. In 1810 he reissued his father's obsolete Treatise on Globes, calling it the "30th" edition. Dudley's son, George Adams (III) served as his apprentice beginning in 1811, but never took the freedom. In May 1817 Dudley was declared bankrupt, and the Adams instrument business came to an end.

Adams, Hannah
Hannah Adams, (fl.1795-1796 d.1810) [18] [445], was a mathematical instrument maker of 60 Fleet St., London, England. She succeeded her husband, **George Adams (II)**, in business when he died in 1795. She wrote to His Majesty's Office of Ordnance, hoping to take over her husband's appointment, but her brother-in-law **Dudley Adams** had reached them first, and was given the appointment. In 1796, she sold the company's stock and the rights to her husband's books to **W.&S. Jones**. The shop premises were taken over by Dudley Adams. Evidence suggests that Dudley Adams may have attempted to force her out of business. She died in 1810. A barometer bearing her name is recorded, but its whereabouts is currently unknown.

Adams, Michael
Michael Adams, (fl.1829-1842) [18], was an optician, and optical, mathematical, and philosophical instrument maker of London, England. Addresses: 15 Tavistock Row, Covent Garden (1829-1830); 16 Wilderness Row, Goswell St. (1829-1830); 8 Tavistock Row (1831); 51 Fleet St. (1841); 3 White Hart Yard, Holborn (1842). He served his apprenticeship with the optician and spectacle maker, John Byard, beginning in 1818, and was a Freeman of the **Blacksmiths' Company**.

Adams, Nathaniel
Nathaniel Adams, (1708-1743) [8] [18] [38] [256] [273] [274], of *The Golden Spectacles*, Charing Cross, London, England, was an optical instrument maker. As well as spectacles he is reported to have made telescopes, prisms, scioptic balls, barometers, lanterns, opera glasses, microscopes, air pumps, and other instruments. He served his apprenticeship with **Edward Scarlett (I)**, beginning in 1722, and was freed of the **Spectacle Makers' Company** in 1730. It is considered unlikely that he was related to the London family of George Adams (I) et al. Among his apprentices were **Francis Watkins (I)**, and **John Margas**, who was turned over to **John Cox (II)** in 1741 or 1742.

Adams & Co., R.T.
R.T. Adams & Co., [306], were camera case manufacturers, and camera makers of London, England. The company appeared in the directories during the 1890s in Hatton Garden (1892), St. Ann's Road, Stamford Hill (1893), and Seven Sisters Road (1898). There is no known connection with **A. Adams & Co.**

Addison & Co., John

John Addison & Co., (fl.1820-1838) [18] [276 #1472] [322], were globe makers, and optical, mathematical, and philosophical instrument makers, of London, England. Addresses: 9 Skinner St., Snow Hill (1820-1821); 116 Regent St. (1822-1825); 50 London St., Fitzroy Square (1825-1826); 7 Hampstead Rd., Tottenham Court Rd. (1827-1828); 275 Strand (1829-1830). They were awarded a royal appointment as globe makers to King George IV.

John Addison was a globe maker who began his business in London in c1800, and continued to conduct his business under his own name until the formation of John Addison & Co. in 1820.

Adie and Son

Adie & Son, [8] [18] [263] [271], of Edinburgh, Scotland, were optical and scientific instrument makers. Addresses: 58 Princes St. (1835-1843); 50 Princes St. (1844-1876); 37 Hanover St. (1877-1880).

The company was formed when **Alexander James Adie** went into partnership with his eldest son **John Adie**. The business traded as Adie & Son from 1835 until 1881. **Angus Henderson** worked for Adie & Son prior to setting up his own business in 1861. It is not clear when Alexander retired from the business, but John continued the business until his death in 1857, at which time John's brother **Richard Adie** took over. Richard ran the company until his death in 1881 when the firm of Adie & Son ceased business.

Adie and Wedderburn

Adie and Wedderburn, [263] [169] [271], were opticians, photographic dealers, and mathematical, optical, philosophical, and surveying instrument makers, of Edinburgh, Scotland. Addresses: 33 Hanover Street (1881-1882) and 17 Hanover St. (1883-1899). The firm was a partnership between **Richard Adie** and **Thomas Wedderburn**. This partnership was formed during the last years of Richard Adie's life, prior to the closure of the business **Adie & Son**. **George Hutchinson** was an assistant in Adie and Wedderburn and, when they ceased business, he transferred the company's stock to his own business. The company Adie and Wedderburn was succeeded in 1913 by **Richardson, Adie & Co. Ltd.**

Adie, Alexander James

Alexander James Adie, (1775-1858) [8] [18] [46] [263] [271] [276], was an optical and scientific instrument maker from Edinburgh, Scotland. Addresses: 15 Nicolson St. (1823-1829) and 58 Princes St. (1830-1834).

After the death of his parents when he was young, he was adopted by his maternal uncle, the optician **John Miller (II)**. He served an apprenticeship with his uncle, and from 1803 he became a partner in his uncle's firm, which traded as **Miller & Adie**. He continued to trade under this name after his uncle's death until 1822 when the firm became Alexander Adie. During the mid-1820s to the mid-1830s Alexander made jewel lenses, [280] [395], which he supplied to **Sir David Brewster**. Alexander had three sons, **John Adie**, **Richard Adie** and **Patrick Adie**. He later partnered with his eldest son John, and from 1835 the business traded as **Adie & Son**.

Adie, John

John Adie, (1805-1857) [8] [18] [46] [263], was an optical and scientific instrument maker of Edinburgh, Scotland, and the eldest son of **Alexander James Adie**. He is believed to have learned his trade from his father, and spent some time working with **Troughton & Simms** in London. He went into partnership with his father and, from 1835, they traded as **Adie & Son**. It is not clear when his father retired from the business, but John continued to run it until his death in 1857, at which time his brother, **Richard Adie**, took over.

Adie, Patrick

Patrick Adie, (1821-1886) [8] [18] [44] [263], was a mathematical, optical, surveying, and philosophical instrument maker from Edinburgh, Scotland. He was **Alexander James Adie**'s youngest son. He worked in his father's workshop, and at Sir Thomas Makdougall Brisbane's observatory [51], near Kelso, Roxburghshire, Scotland, prior setting up business in 1847 as an optician and surveying-instrument maker in London. Addresses: 395 Strand, London, England (1847-1868) – the address which had earlier been used by **John Bennett (II)**; 15 Pall Mall (1869-1885); 29 Regent St., Westminster (1869-1870); Tothill St. (1875).

He designed and made surveying instruments that were used in the Great Trigonometrical Survey of India, which finished in 1852, and for the construction of railways in the UK and abroad. He was awarded medals at the Great Exhibition in London in 1851, the Exposition Universelle in Paris in 1855, and the Great London Exposition in 1862. He invented the first single-operator optical rangefinder.

Following his death in 1886, the firm continued to operate under his name until 1942, trading at Broadway Works, Westminster (1890-1901).

Adie, Richard

Richard Adie, (1810-1881) [3] [8] [18] [46] [263], was an optician, and mathematical and philosophical instrument maker of Liverpool, England. Addresses: 28 Bold St.

(1835-1837); 26 Bold St. (1835); 55 Bold St. (1839-1865); 4 Exchange Buildings (1868-1870); 5 Edmund St. (1870); 5 Harrington St. (1872-1876).

He was the son of **Alexander James Adie** of Edinburgh, Scotland, and began his own scientific instrument making business in Liverpool in 1835. His workshop manager was **Thomas Wedderburn**. Upon the death of his brother, **John Adie**, in 1857 he took over the family business of **Adie and Son** in Edinburgh, while continuing his business in Liverpool. He entered a partnership with Thomas Wedderburn, trading as **Adie and Wedderburn**. Upon his death in 1881 the firm Adie and Son ceased business.

ADOX Kamerawerk GmbH

In 1938, **Dr. C. Schleussner Fotowerke GmbH** assumed control of the **Wirgin** factory in Wiesbaden, Germany. Based on the acquired knowledge and technology, they began camera production in 1939 at the former Wirgin factory, naming the firm ADOX Kamerawerk GmbH, and using the *ADOX* brand-name. During WWII the ADOX Kemerawerk military products were marked with the **ordnance code** dpg. In 1948, the Wirgin factory was sold back to Henry Wirgin, and the ADOX Kemerawerk was relocated to another factory in Wiesbaden. Camera and film production continued until 1962, when the company was acquired by DuPont, and subsequently the brand was sold to **AGFA**. Following the closure of the AGFA imaging branch in 2005, the ADOX brand changed hands again, culminating in the use of the brand-name by the film company Fotoimpex, who went on to produce camera film as ADOX Fotowerke GmbH.

AE Optics

AE Optics, [**199**], was a telescope making company initially in Luton, Bedfordshire, England, and later in the Cambridge area. It was formed in about 1977 when **Jim Hysom** and his brother Robert split **Astronomical Equipment** to form AE Mechanics, run by Robert Hysom, and AE Optics, run by Jim Hysom. During the late-1960s, in addition to telescopes, AE Optics sold Russian-made binoculars. The company continued in business until about 1993 when Jim Hysom went on to form the company **Hytel Optics**.

Aerial Telescope

The aerial telescope was invented by **Christian Huygens**, a design which he first published in 1684. It consists of a very long focal length refracting telescope with no tube. The objective is mounted separately from the eyepiece, and may be positioned at the top of a pole or other high object. The objective and the eyepiece are joined together using a rope which, held taut, enables pointing the telescope. The objective may be mounted in a scioptic ball. Variants on this design followed, but few examples exist that were made later than the C17th, since it is awkward and inconvenient to use.

Aeronautical & General Instruments Ltd., (AGI)

See **Agilux Ltd.**

Aeroscope

The Aeroscope was a portable hand-held motion picture camera powered by compressed air. It was the first hand-held movie camera, and was invented by the Polish inventor **Kazimierz Prószyński**, in collaboration with **Arthur S. Newman**. It went into production in 1912 in the UK, manufactured by **Newman & Sinclair**.

AGC

Alfred Gauthier of Calmbach (AGC) was a company founded by Alfred Gauthier in 1902 in Calmbach, Germany. They specialised in camera shutters, and used brand-names *Prontor, Pronto, Vario,* and *Ibso*. Their logo, shown here, depicts a part-open leaf-shutter with the company initials AGC. In 1910 **Zeiss** acquired a stake in the company, and in 1931 they took ownership. During the 1960s the company branched out into producing optical production machinery and medical technology. At some point the company was renamed Prontor AG, and in 2014 it was taken over by Hitech, and became Hitech Prontor GmbH.

AGFA

The company was founded in 1867 in Berlin, Germany, as a dye factory, and renamed Aktien-Gesellschaft für Anilin-Fabrikation (AGFA) in 1873. They began to produce developing materials for colour photography in 1916. In 1925 AGFA became a subsidiary of **IG Farben**, and as part of the move IG Farben acquired the **Rietzschel** camera works in Berlin from Bayer. Initially they re-badged Rietzschel cameras with their name, but they began manufacturing their own cameras in 1926. In 1952, when IG Farben began their liquidation process, AGFA once again became an independent company. In 1960 AGFA acquired **Iloca**. After the DuPont acquisition of **ADOX** in the late-1960s ADOX was subsequently acquired by AGFA.

In 1964 AGFA merged with Gevaert Photo Producten N.V. to become Gevaert-Agfa N.V. in Mortsel, Belgium, and Agfa-Gevaert AG in Leverkusen, Germany. In 1981 the Agfa-Gevaert group became wholly owned by Bayer. They continued camera production until 1983. Subsequent camera products were made for them by other companies and badged with the AGFA brand. During the 1990s they began to diversify, and they discontinued all their digital camera products in 2001. In 2004, Agfa-Gevaert divested all its photographic activities to the new, independent company AgfaPhoto, but in 2005 AgfaPhoto filed for insolvency.

Agilux Ltd., AGI

Agilux Ltd., [306], of Purley Way, Croydon, Surrey, England, was founded in 1946 by its parent company, Aeronautical & General Instruments Ltd., (AGI).

AGI was founded in 1915, and manufactures instruments and systems for the defence and civilian markets. They produced an aerial camera, used during WWII, and after the war they set up Agilux Ltd. to produce cameras for the civilian market. Agilux produced a number of camera models over the years, including the *Agiflash*, *Agimatic*, *Agifold*, and *Agiflex*. During the 1960s Agilux made inexpensive plastic-bodied cameras for **Ilford** and others to sell under their own brand-names. In 1969 Agilux ceased production of cameras. AGI, however, continued to produce specialised cameras for aerial and military use.

Ahrens, Carston Diederich

Carston Diederich Ahrens, (1837-1918) [53] [68] [264] [319] [320] [402], was a German born optician and optical glass prism worker, who moved to London, England as an infant. Addresses: 5 Baker Street, Lloyd Square (1873); 373 Liverpool Road (1874-1880); Sudal Road, Norbiton; 34 King's Road, Kingston. He produced fine quality prisms, including double-image prisms made from Iceland spar, and his name appears on various mono and binocular microscopes and polariscopes. Many of his prisms are in binocular microscopes made to the design of **Francis Herbert Wenham**. An early pair of prismatic binoculars by Ahrens from c1880s is in the Louwman Collection of Historic Telescopes in the Netherlands. In 1884 Ahrens was granted an English patent for "Improvements in the construction of erecting binocular prisms".

A.I.C. Phototechnik GmbH

See **Soligor**.

Ainsley, T.L.

T.L. Ainsley, [44] [276 #1072] [322], of Cardiff and Barry Dock, Wales, and Mill Dam, South Shields, Tyne and Wear, England, was a compass maker and adjuster, chart and navigation warehouse, and chronometer maker to the admiralty. The company was founded by Thomas L. Ainsley (b.1825 fl.1858-1886), who was a nautical publisher, optician, and nautical instrument maker. Instruments bearing this name include barometers; boxed sextants; octant in ebony, ivory and brass; ship's brass bulkhead clock; marine compass; single draw brass and leather telescope, with tapered barrel, 25" closed, 31¼" extended, objective 1⅞", engraved with a military crow's foot followed by *M 462, T.L. Ainsley. South Shields. S1*. The company continued in business until at least 1951.

Ainsworth & Son, William

William Ainsworth, (1850-1917) [58], was born in England, and emigrated to the USA in 1853. He became a master watchmaker with the Elgin Watch Company, and in 1880 he began his own instrument making firm in Colorado - the William Ainsworth Company, making spring balances. He went on to produce surveying instruments, and in the mid-1890s he became the sole manufacturer of the Brunton pocket transit. In 1905 the company became William Ainsworth & Son, and continued in business until 1938.

Aitchison and Co.

James Aitchison, (1860-1911) [1] [29], served an apprenticeship with a firm of opticians in High Holborn, London, England. In 1889, he set up his own practice at 47 Fleet Street, to establish what he considered more scientific methods of assessing eyesight. He began to work in partnership with **Alexander Thomas Clarkson**, trading as Aitchison & Co. at 47 Fleet St. and 42 Bishopsgate St., until the partnership was terminated in 1891 [35 #26140 p1223]. He then continued the business himself, trading under the same name.

In 1899, just before the second Boer War, Aitchison brought out his New Range Finding Field Glass. In 1904, at the beginning of the Russo-Japanese War the Imperial Russian Government placed a large order for his field glasses. Instruments bearing the name Aitchison & Co. include: microscopes; Galilean binoculars; prism binoculars; opera glasses; hand-held telescopes; tripod-mounted brass astronomical telescopes; lightweight aluminium folding field binoculars, as issued to officers during the second Boer War.

In 1908 Aitchison rescued the **Wray Optical Works Ltd.** by merging with them. In 1911 James Aitchison died, and his son Irvine took over the business. He placed the emphasis on retailing rather than manufacturing optical instruments, although he retained ownership of Wray Optical Works Ltd., of Ashgrove Road, Bromley Hill, Kent, until the 1970s. In 1927 Aitchison & Co. acquired the goodwill, leases and stock of **Dollond & Co.**, and began to trade under the name **Dollond & Aitchison**. Upon the sale of the Wray Optical Works Ltd., Dollond & Aitchison were allocated shares in **Hilger & Watts**. Their investment in Hilger & Watts was not particularly successful, and was sold when they shed their photographic business, shortly before **Rank** purchased it at a better price.

Aitchison, James
See **Aitchison & Co.**

AJACK
See **A. Jackenkroll Optische Anstalt GmbH**.

Akron
Akron was a department store based on Los Angeles, USA. From the mid-1950s until the 1970s they sold a 2.4-inch refractor imported from Japan using their own brand-name. The telescope was mounted on a combination equatorial/alt-az mount.

Aktien-Gesellschaft für Anilin-Fabrikation
See **AGFA**.

Alderblick
Alderblick is a brand of binoculars manufactured by **Carton Optical Industries Ltd.** in Japan, and imported and re-branded by **Valley Microscope** of Canada and **Vickers** of the UK. Alderblick binoculars by Carton were also marketed as *Alderblick Fernglasser*, **Celestron** *Ultima*, and **Swift** *Ultra-lite*, among others.

Aldis Bros.
Aldis Bros., [20] [44], of Sare Hole Road, Sparkhill, Birmingham, England, were makers of lenses for cameras and projectors, epidiascopes, strip projectors, visual aids for educational purposes, photographic enlargers, and daylight and masthead signalling lamps. The company was formed in 1901 when Hugh Lancelot Aldis left **Dallmeyer**, and partnered with his brother, Arthur C. Aldis, who had also worked briefly for Dallmeyer.

A 1912 advertisement shows a selection of lenses. During WWI they dedicated their production capacity to producing optical munitions, including riflescopes. An offshoot of this work was their Daylight Signalling Lamp, or *Aldis Lamp*. They also produced lenses for high altitude aerial photography. In 1919 the firm Aldis Brothers was listed as member of the **British Optical Instrument Manufacturers' Association**.

They exhibited in the 1947 British Industries Fair, with their catalogue entry, [450], listing an epidiascope and dual-purpose projector, and listing the company as manufacturers of: Epidiascopes; Strip projectors; Educational visual aids; Photographic enlargers; Daylight and masthead signalling lamps.

In 1957 **R.B. Pullin and Co.** acquired Aldis Brothers.

Aldis Optical Level
An Aldis optical level is a form of bubble level in which the bubble acts as a negative lens, and magnifies the view of a fixed mark.

Aldis Spherometer
See spherometer.

Alexander
Alexander, (fl.c1790), [16] [18] [276 #910], of Yarmouth, England: An octant bearing this name has its scale divided by **Spencer, Browning & Rust** (see Appendix XI). No other instruments are known to be signed with this name, and hence it is not clear whether Alexander was the owner, retailer, or maker of the octant.

Alhazen
See **Ibn al-Haytham, Al-Hassan**.

Alidade
An alidade is a sighting device which is pivoted so that it can be rotated against a circular scale marked in degrees. It is used to measure the angle to a distant object. An alidade may form part of an astrolabe, or a surveyor's plane table, or may perform the same function on other instruments.

The term alidade is also used to describe the upper, rotating part of a theodolite, comprising the telescope and associated parts.

See also Reeve's distance finder.

Alinari
Alinari, of Via Giusti 4, Torino (Turin), Italy, was a telescope manufacturer during the 1960s. Their catalogue shows reflecting and refracting telescopes, mounts and accessories aimed at the amateur market.

Allan, James (I)
James Allan (I), sometimes spelled Allen, (fl.1776-1820 d.1821) [2] [44], was an optical instrument maker of 12 Blewitt's Buildings, Fetter Lane, London, England. He may have been an employee of **Jesse Ramsden** until about a year prior to Ramsden's death, when he started his own instrument making business.

He built a self-correcting dividing engine, incorporating improvements of his own design. These improvements, for which the Society of Arts awarded him a gold medal, were described in a letter to them published in 1810, [427]. The letter was accompanied by testimonies from **John Stancliffe** and **Matthew Berge**.

In a letter to the Society of Arts in 1811, [428], concerning his improvements to the reflecting circle, he wrote, "… I have been induced to make several circular instruments of reflection in various ways, but none upon so simple a construction, or so cheap, as the present, nor so well calculated to prove any untruth, as my improvement upon Borda's; and I believe it will now be generally adopted for use. There have been great numbers of Borda's circles made; I myself assisted about twenty-five years ago to make many, also since I have been in business for the last twelve years on my own account, but I never found any of them to give satisfaction till I invented the present improvement."

In 1816, he was awarded by the Society of Arts a gold medal for an improved theodolite, and a silver medal for a method, and machine, for cutting screws [429].

A letter to James Allan (I) from the Board of Longitude dated 1820 [430], stated, "I have the pleasure to inform you that the Board of Longitude has this day been pleased to grant you the sum of one hundred pounds, as a reward for your ingenuity, displayed in the improvement of the dividing engine." According to the Board of Longitude sub-committee, which included **William Cary** and **William Hyde Wollaston**, Allan's dividing engine, when compared with those of Jesse Ramsden and John Stancliffe, was the least worn. A letter in the Board of Longitude's files, written by Wollaston, states that Allan's circle was more accurate than Ramsden's.

James Allan (I)'s sons, **James Allan (II)**, and **John Allan** were also notable instrument makers. Note that [276 #912], and some other sources, appear to confuse James Allan (I) with James Allan (II). Note also that [2] gives James Allan (I)'s date of death as 1816, whereas [44] has the full text of an obituary dated 1821, from the *Inverness Courier* [431].

Allan, James (II)

James Allan (II), sometimes spelled Allen, (fl.1797-1832) [2], instrument maker of 12 and 13 Blewitt's Buildings, Fetter Lane, London, England, was the son of **James Allan (I)**. He was a shopman in **Jesse Ramsden**'s business for three years prior to Ramsden's death in 1800, after which he stayed with the business to work under **Matthew Berge**.

James Allan (II) formed a partnership with **Nathaniel Worthington**, trading as **Worthington & Allan** at 196 Piccadilly, London. The partnership appeared in the trade directories from 1822 until 1832.

James Allen, of 196 Piccadilly, London, attended the London Mechanic's Institute from 1825-1826.

Note that [276 #912], and some other sources, appear to confuse James Allan (I) with James Allan (II).

Allan, John

John Allan, sometimes spelled Allen, sometimes known as Jonathan, (b.1786 fl.1810-1825) [2] [58] [322], was an optician, and mathematical instrument maker, and the son of **James Allan (I)**. He emigrated to the USA in 1807, and settled in Baltimore, Maryland, where he set up in business as an instrument maker. Addresses: Fell St. (1810); 86 Bond St. (18114-1815); 1 Fell St. (1819). He advertised, "As all instruments sold by me are graduated by my father's improved self-correcting engine, for which the Society of Arts voted him their Gold Medal, on the 19th of May, 1810—they are warranted and will be kept in repair one year, gratis."

Note that according to [322], John Allen worked at 12 Blewitt's Buildings, Fetter Lane, London, England after 1825. According to [18], he was also at this address from 1790-1794. This is the same address as occupied by James Allen (I) and (II).

Alldridge Ltd., R.S.
See **Chapman & Alldridge**.

Allied Impex Corporation

Allied Impex Corporation (AIC) was an American import/export business, which used the brand-name *Soligor* for cameras and lenses imported from Japan. They imported *Miranda* cameras, and took control of the **Miranda Camera Company** in the late 1960s. In 1968 they formed a German subsidiary AIC Phototechnik GmbH, which was renamed **Soligor GmbH** in 1993.

Alment, John

John Alment, (c1740-1787) [34] [215] [304] [322], of 34 Mary's Abbey, Dublin, Ireland, was an optician, and maker of optical, philosophical and mathematical instruments. He worked as foreman for **John Margas** prior to setting up his own business. According to an advertisement in the *Dublin Mercury* in 1767: "John Alment, Mathematical instrument maker, Next door to the *Sign of the White Hart*, Mary's-Abbey, who for several years worked for Mr Margas of Capel Street, but now in his own account, makes the following articles viz. Cases of drawing instruments, Theodolites, Sircumfronters [sic], Sundials, Electrical machines, Weather glasses, and various other instruments too tedious to insert, all of which he is determined to sell as low as possible, and hopes that, and the goodness of his work, will recommend him. N.B. As said Alment is a new beginner, he hopes for encouragement, as it will always be his study to excel in the workmanship of his instruments".

Instruments bearing his name include: surveying compasses; garden sundial; Gregorian telescope of length 356mm, with a removable eyepiece and red glass accessory. The telescope is in an oak case bearing a label which reads: "John Alment Optician at *ye Sign of ye Spectacles* in Mary's Abbey Dublin. Makes Optical

Philosophical & Mathematical instruments. Viz Spectacles, Concave glass, Telescopes, Microscopes, &c Reading & opera glasses, Air pumps, Electrical machines, Barometers, Thermometers, with variety of Drawing & surveying instruments."

Almucantar, Almacantar
An almucantar, otherwise known as an almacantar, is:

(1) A circle of the celestial sphere parallel to the horizon. A line of constant altitude.

(2) An instrument used to measure changes in latitude, or to determine the latitude. This is achieved by measuring the transit of stars across an almucantar (see 1). Conversely, if the latitude is known the measurements may be used to determine the time. The almucantar was invented by **Seth Chandler**, and the first one was built by **John Clacey**.

Alpa
Alpa is a camera brand that was created by **Pignons SA**, a Swiss company, founded in 1918, making parts for the watch-making industry. In 1933 they began a collaboration with **Jacques Bogopolsky**, in which he produced a camera design, which was launched in 1942 as the *Alpa-Reflex* – the first of the cameras which were to become the Pignons' *Alpa* brand. They did not produce lenses, instead having them made by other makers including **Asahi**, **Yashica**, and **Kern & Co.**

In 1976 and 1980, Pignons introduced Alpa branded cameras, the *Si2000* and *Si3000*, manufactured in Japan by **Chinon**. In 1990 Pignons SA declared bankruptcy, and in 1996 the Alpa brand was purchased by Capaul & Weber Ltd., creating the company ALPA Capaul & Weber, which was incorporated as ALPA Capaul & Weber Ltd. in 2007, to produce medium format cameras.

Alpen Optics
Alpen Optics was a US company, founded in 1996, selling imported binoculars, spotting scopes, riflescopes, and hunting accessories, using their own Alpen brand. In 2018 the company was taken over by **Bresser**, and became Bresser's *Alpen* brand.

Altiscope
An altiscope is an arrangement of lenses and mirrors or prisms which enables a person to view an object in spite of obstructions in the direct line of vision. See also Periscope.

Altitude-Azimuth (Alt-Az) Mount
An altitude-azimuth, or alt-az mount, used to mount an instrument such as a telescope, is one which allows rotation in two axes, the *altitude* axis and the *azimuth* axis.

The altitude axis is horizontal, allowing rotation to direct the telescope higher or lower in the sky. The azimuth axis is vertical, allowing rotation to any point on the compass. The most common types of alt-az mount are tripod or pillar mounted, and floor mounted as in a Dobsonian mount. If an alt-az mount is on a wedge, so angled to hold the azimuth axis parallel to the Earth's axis (i.e. at the observer's latitude angle), the altitude axis can then act as a declination axis, and the mount operates as an equatorial mount.

Alt, Robert
Robert Alt, [1], was leader of a group of spectacle-makers who founded the **Spectacle Makers' Company** of London in 1629.

Amadio, Joseph Philip
Joseph Philip Amadio, (1812-1892) [18] [50], of 118 St. John's Street Rd, London, England, was an optician and mathematical instrument maker, and a bookseller. He was a microscope retailer, and the son of a barometer and thermometer maker. From 1840 until 1853 he traded at 6 Shorters Court, Throgmorton Street, London. He was freed of the **Spectacle Makers' Company** by purchase in 1843. He sold microscopes and slides, and some of the instruments he retailed bore his name. It is believed that he bought in the metal parts and may have made the lenses himself. During the mid-1860s, for about three years, he worked in partnership with his brother, Francis (d.1866), trading as F.&J. Amadio.

Amalgamated Photographic Equipment Manufacturers Ltd. (APEM)
Amalgamated Photographic Equipment Manufacturers Ltd., (APEM), [61] [306], formed in 1929, comprised those member companies of **APM** that specialised in sensitised materials for photography. These were the Rotary Photographic Co., **Marion & Co.**, Paget Prize Plates Co. Ltd., and **Rajar Ltd.** In 1929, shortly after APEM was formed, it became a part of the **Ilford** group of companies.

APEM is not to be confused with the trademark that was used by **APM** for their cameras, which was APEM, and sometimes APeM.

Amalgamated Photographic Manufacturers Ltd. (APM)
In 1920 the Selo company was formed in Essex, England, by the companies **Ilford**, the Imperial Dry Plate Co., and the Gem Dry Plate Co. In 1921 it became the Amalgamated Photographic Manufacturers Ltd. (Initially "APM 1921", then APM), [44] [61] [306], including a

consortium of seven more British companies. These were **Kershaw Optical Co.**, **A. Kershaw & Sons**, **Marion & Foulger**, Rotary Photographic Co., **Rajar Ltd.**, Paget Prize Plate Co., and **Marion & Co.**

The trademark used by APM for their cameras was APeM, and sometimes APeM. This is not to be confused with **APEM**.

APEM was formed in 1929, and APM was renamed **Soho Ltd.**, as the sales division of A. Kershaw and Sons. In about 1946, prior to becoming part of **Rank**, it was renamed **Kershaw-Soho Ltd.**

Ambrotype

An ambrotype, [70], is a positive image produced with a plate exposed using the wet collodion process. The exposed plate, which is the collodion negative, bears the image in blackened silver iodide on an otherwise transparent glass plate. The silver iodide may be bleached with nitric acid or bichloride of mercury, yielding white metallic silver. If the reverse of the plate is then blackened with black varnish, paper or velvet, the image may be viewed as a positive image. This is an *ambrotype*, introduced in 1852 by Peter W. Fry of the Calotype Club. The result is a single image, and no further prints may be produced from it.

American Camera Co., The

The American Camera Co., [61] [306], of 399 Edgeware Road, London, England, was a manufacturer of large format mahogany and brass cameras. By 1888 they were trading at the Edgeware Road address. In 1889 they introduced the smaller *Demon Detective Camera*. In 1892 the company was incorporated as The American Camera Co. Ltd., and at about that time they briefly traded at 124 Old Broad Street, London. In 1894 they ceased trading.

American Optical Company

The American Optical Company, of Southbridge, Massachusetts, USA, was founded in 1826 by William Beecher. The company's core product throughout its history has been spectacles and sunglasses. In 1867 **Charles B. Boyle** was appointed director of overseas manufacturing. In 1878 they began a collaboration with the **Blair Camera Co.** to produce a portable wet-plate camera system. Microscope production began in 1935 with the acquisition of the **Spencer Lens Company**. In 1956 **John A. Brashear Co. Ltd.** became a subsidiary, but was later sold. In 1963 American Optical acquired an interest in **C. Reichert, Optische Werke**, and in 1972 they took full ownership. In 1986 the American Optical Company was acquired in a merger with **Cambridge Instrument Co. Ltd.**

Amici, Giovani Battista

Giovani Battista Amici, (1786-1863) [24] [45] [142] [264] [268], was an Italian mathematician, optical instrument maker, astronomer and natural scientist. He graduated in engineering and architecture at the University of Bologna in 1807, having studied under the mathematician Paulo Ruffini.

He began to design and make scientific instruments, and attempted to solve the problem of spherical aberration in achromatic microscope objectives. He turned his attention from lenses to mirrors, which do not suffer from chromatic aberration, and in 1813 he devised the reflecting microscope, using an elliptical mirror. The object to be viewed is placed at the near focus of the elliptical mirror, and an enlarged image is viewed through an interchangeable eyepiece at the far focus. Amici mounted the microscope tube horizontally for easy viewing. The reflecting microscope was further developed by **John Cuthbert**, with others produced by instrument makers such as **S.J. Rienks**.

Amici was professor of mathematics at the University of Modena from 1815-1825, and in 1831 he was appointed astronomer to the Grand Duke of Tuscany, serving as director of the observatory at the Museo di Fisica e Storia Naturale (museum of physics and natural history) in Florence. He continued in this position until 1859, after which he was appointed director of microscopical research at the Museum.

Amici established a reputation for the high quality of the optical instruments he made. He made advances in microscope optics and spectroscopy, and is known to have made about 300 microscopes, as well as telescopes, telescope mirrors, some up to 12 inches, spectrometers, micrometers, camera lucidas, and other optical instruments. Amici is credited with the introduction of the oil immersion objective in microscopy in 1840, and the water immersion objective soon after.

Amici Roof Prism

See roof prism.

Anallatic Lens

An anallatic lens is one placed in the focal plane of the objective of a surveyor's telescope to enable it to act as an anallatic telescope.

Anallatic Telescope

An anallatic telescope is one used by surveyors that has an additional lens in the focal plane of the objective, so designed to enable tacheometrical measurements to be made without the need for a correcting constant.

Anamorphic Lens
See lens.

Anastigmat, Anastigmatic
An anastigmat is an anastigmatic lens. A lens that is not astigmatic; i.e., it is free from astigmatism. The term is also applied to a lens that corrects astigmatism.

Anderson, Henry
Henry Anderson, (c1817-1904) [50], was an optical instrument maker specialising in microscopes, born in London, England. He is believed to have learned his trade from **Andrew Ross** in Clerkenwell, London, with whom he worked from the late-1830s until the 1860s or 1870s. At that time, he set up his own business in Islington, London, working with his two sons, Henry James Anderson (1846-1910) and William Alfred Anderson (1848-c1920). By 1884 they were trading as H. Anderson and Sons, [264], and continued to do so until at least 1891.

Anderson and Sons, H.
See **Henry Anderson**.

Angioli, Flaminio de
Flaminio de Angioli, (fl.second quarter of C18th) [374], was a surveying and physical instrument maker of Milan, Italy. He specialised in making surveying instruments, such as theodolites, levels, and graphometers, as well as pantographs, protractors, and compasses. He is also known to have made mechanical models.

Angle Mirror
Another name for an optical square.

Angular Aperture
The angular aperture of a lens is its angular size as seen from a point at the focus of the lens. Thus, using simple trigonometry, if the diameter of the lens is D, and the focal length is f, the angular aperture θ is given by:
$$\theta = 2 \arctan(D/(2f))$$
See also numerical aperture.

Angular Resolution
The angular resolution (minimum resolvable angular separation in radians, θ_{min}) for an optical system, such as a telescope, microscope, camera, &c., with a circular aperture, is given by the *Rayleigh Criterion* [7]. Since the angles treated by this calculation are small, the *small angle approximation*, $\sin\theta \approx \theta$, applies:
$$\sin\theta_{min} \approx \theta_{min} = 1.22(\lambda/D)$$
Where D = diameter of the aperture, and λ = wavelength of the light. For visible light $\lambda \approx 3.9 \times 10^{-7}$m to 7.0×10^{-7}m

Hence, for example, the angular resolution for an 80mm telescope $\theta_{min} = 5.9475 \times 10^{-6}$ radians (violet) to 1.0675×10^{-5} radians (red) = 1.23" (violet) to 2.2" (red). Thus, in this example, the average angular resolution $\theta_{min} \approx 1.7$"

Angus, Herbert Francis
Herbert Francis Angus, (1872-1948) [50] [320], of London, England, was a manufacturer and retailer of scientific instruments. By 1897 his occupation was recorded as optician, and in 1899 he was freed of the **Spectacle Makers' Company**. In c1901 he became manager of **C. Baker**, and he continued in that role until his resignation in about 1909.

In 1909 Herbert Angus entered a partnership with **Mansell James Swift** of **James Swift & Son**, and began trading as H.F. Angus & Co., at 83 Wigmore Street, London. They advertised as manufacturers, importers, and exporters of scientific instruments, especially microscopes, prepared slides, and accessories. Examples of optical instruments can be seen bearing the name H.F. Angus & Co.; however, it is not clear to what extent these were made by his company.

In 1913 his partnership with Swift was dissolved, [35 #28735 p4910]. This notice is informative, since it refers to M.J. Swift as a manufacturer of scientific and optical instruments, and H.F. Angus as a dealer in scientific instruments. Following the dissolution of the partnership, Herbert Angus continued to trade as H.F. Angus & Co. He was succeeded in business by **Hawksley & Sons** in 1920.

Anone, Frans
Frans Anone, also known as Francis, or Frs., (fl.1802-1808) [18] [322], was an optical and philosophical instrument maker of London, England. Addresses: 2 Holborn; 82, and 242, High Holborn; 51 Fetter Lane (dates uncertain); 26 High Holborn (1802-1808). He is known to have made: telescopes; barometers; thermometers. He is also known to have sold: hygrometer.

Anra Manufacturing Engineers
Anra Manufacturing Engineers of Northridge, California, USA, was founded in 1958. They manufactured and sold telescopes, binoculars, and telescope mounts and accessories. Their 1960s catalogue shows reflecting telescopes from 4½ inches to 16 inches, Newtonian-Cassegrain telescopes from 6 inches to 16 inches, as well as equatorial mounts, 7x50 wide angle binoculars, and parts and accessories for telescope makers. The company changed its name to **Optical Craftsmen**, and continued in business until at least the 1980s.

Ansco
Ansco is a Chinese brand of photographic equipment and binoculars, manufactured by **W. Haking (Xinhui) Optical Ltd.**

Apertometer
An apertometer, [287], is a device used to measure the numerical aperture of a microscope objective.

Abbe apertometer: This design, introduced by **Ernst Abbe**, and produced by **Zeiss**, first appeared in their catalogue in 1889. Its construction, use, and history are described in [287].

Beck apertometer: [53] [57] Made by **R.&J. Beck**, this is a glass block, measuring 3x1 inch by ⅜ inch thick. On its lower surface is a scale, and on its top surface is a microscopic object upon which to focus. Once focused the block is viewed through the microscope without an eyepiece, (or with a low power eyepiece), to reveal how much of the scale can be seen. This design is difficult to use with an immersion objective.

Cheshire apertometer: The Cheshire apertometer, invented by **Frederic J. Cheshire**, was also made by **R.&J. Beck**. It consists of a glass disc, mounted in a brass case with a removable top. The disc has a scale printed on its underside, and a focus point on the top surface. A special telescope eyepiece us used to view the scale. It has the advantage that it may be used with an immersion objective.

Metz apertometer: The Metz apertometer is a simple cardboard scale that cannot be used with immersion objectives.

Aperture Stop
The aperture stop, [7], of an optical system is any element in the system that limits the amount of light reaching the image formed by the system. This may be, for example, the rim of a lens, or the iris diaphragm that determines the focal ratio of a camera. The physical limitation presented by the aperture stop determines which of the rays of light reaching the instrument will ultimately form an image. The aperture stop determines the light-gathering capability of the optical system. An increase in the aperture stop increases the amount of light reaching each image point.

As an example, a pair of binoculars may or may not have an aperture stop as an internal element. Where no such internal element exists, the aperture stop of the binoculars is the rim of the objective lens. Where an internal aperture stop element is included, this reduces the size of the aperture. Hence, in the case of a pair of binoculars with 60mm objectives, with an internal aperture stop, the binoculars do not benefit from the full light-gathering capability of the 60mm lenses.

The aperture stop is one of the factors which determines the entrance pupil and exit pupil of the optical system. The aperture stop does not stop highly oblique rays of light from entering the system – see field stop.

Aplanat, Aplanatic
An aplanat, or aplanatic lens, is one which is free from spherical aberration and coma. Such a lens satisfies the sine condition.

The aplanatic telescope was invented by **Robert Blair**.

APEM
See **Amalgamated Photographic Equipment Manufacturers Ltd.**

APL
See **Apollo Business and Industry**.

APM
See **Amalgamated Photographic Manufacturers Ltd.**

Apochromat, Apochromatic
An apochromat, or apochromatic lens, normally consists of three or more lens elements, so designed to bring a beam of light to focus perfectly in three frequencies of light, normally red, green, and blue. Thus, the chromatic aberration is considerably less than for an achromat. It was invented by **Ernst Abbe** in 1868. A telescope with an apochromatic objective lens is sometimes referred to as an *APO*.

Apollo Labs; Apollo Business and Industry (APL)
Apollo Labs/Apollo Business and Industry was a Japanese company manufacturing telescopes. They used the maker's mark APL (see Appendix IV). Their telescopes were sold under various brand-names including **Monolux**, **Selsi**, **Bushnell** and **Mayflower**. Research suggests that APL was taken over by **Koyu** some time prior to 1970 when they changed their name to **Vixen**. Following their acquisition by Koyu, APL continued to mark their products with APL as well as, later, *Vixen*.

Apomecometer
An apomecometer is an instrument for measuring the approximate height of buildings, trees, etc. This is similar to an optical square, but reflects at 45° rather than at 90°. A sight is taken to a spot near the foot of the object whose height is to be measured, level with the observer's eye. Then, by moving closer to, or further away from the object, the position is noted where the top of the object as seen by reflection in the apomecometer corresponds with the spot on the object. The height of the object is then

equal to the distance of the observer from it, plus the height of the sighting spot above the ground.

Apps, Alfred

Alfred Apps, (1839-1913) [8] [361], was an optical, mathematical and philosophical instrument maker of 433 Strand, London, England. Sources differ on when his business started and ended. [53] states it was 1866-1911, and [264] states it was 1869-1901.

According to his trade card [270]: "Alfred Apps, Optical, Mathematical & Philosophical instrument maker, 433 Strand, London W.C. Seller of Spectacles, Eye preservers, Telescopes, Microscopes, Astronomical, surveying & drawing instruments, Self-registering barometers, Thermometers, Anemometers, Magnetographs."

His obituary by the Institute of Electrical Engineers, [44], states that he retired in 1911. The relationship between Alfred Apps and Apps & Co, of 455 Strand, [44], is unclear.

Aquatic Microscope

See microscope.

Arax

Arax was a camera brand of the **Arsenal Factory**. Production began sometime after WWII, and continued until 2009. The stock was then purchased by Gevorg Vartanyan, who continued the business, selling refurbished and new cameras, lenses, and accessories.

Archbutt, J.&W.E.

John Archbutt, (fl.1838-1864) [18], was a mathematical instrument maker and pawnbroker, who traded at 20 Bridge Road, Lambeth, London, England, from 1838 until 1864. At some point the company became John Archbutt & Sons, and by 1866 their advertisements show them as J.&W.E. Archbutt, trading at 201 Westminster Bridge Road, Lambeth, [44], as manufacturers of scientific measuring instruments, such as levels and theodolites etc, and mathematical drawing instruments.

Archer, Frederick Scott

Frederick Scott Archer, (1813-1857) [46] [70], of Hertford, England, was the inventor of the wet collodion process of photography. He served his apprenticeship with a silversmith and bullion dealer, after which he started his own business as a sculptor. Wishing to produce images of his work, he developed an interest in photography. In 1851, four years after the invention of collodion, Frederick Scott Archer devised the collodion process which, being faster than the albumen process of **Niépce**, dominated photography over the following decades. He did not, however, patent the process. He continued as a photographer and inventor, and died poor in 1857.

Archer, William

William Archer, (fl.1771-1786) [18], was an optical and mathematical instrument maker of, London, England. Addresses: Giltspur St. (1771); Fleet St. (1774); 2 Johnson's Court, Fleet St. (1777-1780); Fetter Lane (1783). He served his apprenticeship with the mathematical instrument maker, John Bush, beginning in 1753, and was freed of the **Stationers' Company** in 1763. He had four apprentices, the first being **William Price**, beginning in 1771, and the last being **John Johnson Evans**, who was turned over to William Price in 1786.

Argus

Argus Cameras, Inc., of Ann Arbor, Michigan, USA, began in 1929 as a radio manufacturer named International Research Corporation, founded by Charles A. Verschoor and William E. Brown Jr. The company was renamed International Radio Corporation and, in 1936, produced their first camera, the 35mm *Argus A*, aimed at the mass-market. In the same year, the company was renamed to become Argus. Argus grew to become a successful camera brand, producing a large number of models until the company was sold in 1957 to Sylvania Electric Products Co. However, with sales declining due to competition from Japanese imports, the company underwent further changes of ownership. By the early 1960s camera manufacturing in Ann Arbour had ceased, and the company began to sell imported cameras under the Argus brand-name, some of which were made by **Mamiya**. Camera production had largely ceased by the late 1970s.

Armourers and Braziers' Company

See Appendix IX: The London Guilds.

Armstrong, Thomas & Brother Ltd.

Thomas Armstrong & Brother Ltd., [53] [138] [139] [260]. In 1825, Joseph Armstrong (d.1851) set up business as jeweller and silversmith at 261 Deansgate, Manchester, England. The shop was known locally as the Old Deanery, and was later renumbered 88. In 1828 he married Sarah Booth and had three sons: Thomas, George and Alfred. By the time he died in 1851, Joseph had brought his elder sons into the business. Thomas, who had already been managing the firm for some years, and may have served an apprenticeship with an optician, expanded the business to include the manufacture of spectacles and optical instruments. In 1868, Thomas took his brother, George, into partnership and changed the company's name to Thomas Armstrong & Brother.

In 1965, the company was taken over by **Harrisons Opticians**. By this time, the firm had branches in both Oxford Street and Nelson Street in Manchester, on the Downs in Altrincham, and on Bold Street in Liverpool. In 1968, The Harrison group was taken over by **Dollond & Aitchison**.

Arnold & Richter Cine Technik
See **ARRI**.

Aronsberg

Maurice Aronsberg (1835-1911) and his brother Woolf, (or Wolf), usually known as William (1833-1908), [**53**] [**319**] [**379**], were from Courland, now a part of Latvia. In about 1850, they emigrated to Scotland, and by 1855 were running a company of glass merchants and glaziers in Glasgow. In c1858 they moved to Liverpool, and following this they both set up business as opticians; Maurice in Liverpool, and William in Manchester. Their name is sometimes spelled Aronsburg.

Maurice set up the company M. Aronsberg & Co. in Liverpool, England. In 1858 he married Eva Prag, (c1839-1899) from Germany. Maurice is listed in the census as an optician in Manchester Street, Liverpool in 1861. Advertisements for M. Aronsberg & Co. at the sign of the *Golden Spectacles*, (and later *Blue Spectacles*), 4 Manchester Street began to appear in 1862. They traded at that address until at least 1866, advertising spectacles and eye-glasses, opera and field glasses, stereoscopes, stereoscopic slides, microscopes (from the best English and French makers), mathematical drawing instruments, micro-photographs, and magic lanterns. By 1870 they were listed as mathematical instrument makers and opticians, trading at 39 Castle Street, Liverpool. Subsequent Liverpool addresses were: 66 Falkner Street (1871); 167 Chatham Street, Mount Pleasant (1881); 40 Manchester Street (1891); 38 & 40 Manchester Street (1901). Maurice worked in partnership with Gabriel Phillips, (start date unknown), trading as **P. Gabriel & Co.**, watchmakers, opticians, and jewellers of Manchester Street, until 1878.

William Aronsberg, having moved to Liverpool, married Ernestine Prag (c1838-1890) in 1858. Liverpool addresses: 66 Bamber St., West Derby (1861); 36 Great Oxford St. (1863). He set up business as an optician, and dealer in stereoscopic and photographic portraits, but in 1863 he was declared bankrupt [**35** #22785 p5221]. By 1871 he was living at 1 South Bank, Broughton, Preston, and in 1864 he started a new business as W. Aronsberg & Co., optician, and mathematical and philosophical instrument maker, opposite the Infirmary at 3 Lever St., off Piccadilly, Manchester. His subsequent Manchester addresses were: 12 Victoria St. (1868-1923); 103 Market St. (1878-1891); 471 Oxford St., Chorlton-on-Medlock (1881); 68 Plymouth St., Chorlton-on-Medlock (1891). He made and sold astronomical, nautical and surveying instruments, microscopes, barometers, and mathematical and drawing instruments. His business was successful, and he was generous with his contributions to charities and education. His sons, Aron, sometimes spelled Aaron (c1860-1928), and Ralph (b.c1861 d.possibly 1884), both worked with him as opticians.

From 1890, William worked in partnership with Aaron, trading as William Aronsberg & Son until the partnership was dissolved in 1892 [**35** #26320 p4910], after which William continued the business. In 1893, he was once again declared bankrupt [**35** #26427 p4325]. In 1896 Aaron took over the company, and ran it under his own name. Aaron continued to run the company until his retirement in 1923, when his business premises at 12 Victoria Street, Manchester were taken over by **A. Franks**.

Instruments bearing their name include: Galilean field binoculars and opera glasses; Galilean binoculars manufactured by **Lemaire**, bearing the Aronsberg name; binocular telescope; 8x25 and 8x27 prismatic binoculars; one, three and four-draw telescopes in brass and leather; ornate oak cased aneroid barometer, thermometer and timepiece; holosteric barometer. Instruments are sometimes marked *Paris* and *Liverpool*. Their advertisements additionally listed model steam engines, magic lanterns, and philosophical instruments.

ARRI

ARRI was founded in 1917 in Munich, Germany as Arnold & Richter Cine Technik, by August Arnold and Robert Richter, when they began to sell film printing machines of their own design. In 1918 they began film production, and were producing their own independent films by 1920. ARRI introduced the first film camera of their own design in 1924, which was a 35mm camera, and at about this time they began to produce lighting equipment, and they began renting out equipment to other film-makers. They introduced their first 16mm film camera in 1928.

In 1937 ARRI introduced the first reflex mirror shutter film camera, the ARRIFLEX 35, so that the viewfinder image is seen through the lens being used for filming. In the same year they produced their first Fresnel lamp-heads. In 1944 the ARRI factory in Munich was destroyed by Allied bombing. Production was relocated for the duration of the war, and the factory was rebuilt after the war. Sales of the ARRIFLEX 35 flourished after the war with exports contributing to its success. Their film-making and lighting products, as well as their rental business and their film production services, kept pace with new technological developments over the next decades, with ARRI continuing to grow into an international company, with subsidiaries in the US, UK, Italy, Australia, and Canada.

Arsat
Arsat was a camera lens brand of the **Arsenal Factory**, used for **Zeiss** copy lenses.

Arsenal Factory
Arsenal is a manufacturer of optical and optoelectronic devices for space, aviation and ground equipment for military and civil purposes. The Arsenal Factory, now known as Arsenal Special Device Production State Enterprise, was established in 1764 in Kiev, Ukraine. Following WWII they manufactured cameras, many of which were based on designs from makers and brands such as **Hasselblad**, **Zeiss Ikon**, **Contax**, **Nikon**, and **Pentacon**. The Arsenal Factory camera brands were **Kiev** (Киев), **Salyut**, and **Arax**. They also produced camera lenses using the brand-named **Arsat** and **Mir**. For Russian and Soviet makers' marks, see Appendix V.

Asahi Optical Company
Founded in 1919 in Tokyo, Japan, Asahi Kogaku Kogyo Co. Ltd. established a reputation for lens manufacturing and polishing. They supplied lenses to Japanese camera and optical instrument manufacturers. In 1932 they started making military binoculars and optical rangefinders. In 1952 they produced Japan's first SLR camera, the *Asahiflex I*. In 1957 Asahi purchased the **Pentax** brand-name from **Zeiss Ikon**. The name is derived from the words *pentaprism* and *Contax* (a Zeiss Ikon camera model). In 1958, Asahi Optical Company products were first sold in the US under the brand-name *Pentax*. From 1959, for about three decades, Asahi used the **JB code** JB6. In 1967 the Asahi Optical Company took over the **Fuji Surveying Instrument Company Ltd.**, which became an Asahi brand. In 1976 they changed the *Fuji* surveying instrument brand-name to *Pentax*. In 2002 the Asahi Optical Company was renamed **Pentax Corporation**.

Ashmore, William
William Ashmore, (fl.1825-1850) [**18**] [**276** #1767], was an optician and optical instrument maker of Sheffield, England. From 1825 until 1837 he traded in partnership with **Thomas P.G. Osborne** under the name **Ashmore & Osborne** at 42 Burgess Street. In 1837 he began to trade under his own name at 17 Fitzwilliam Street. He subsequently moved to 103 Fargate (1839); 101 Fargate (1841).

According to an advertisement dated 1847 [**44**]: "William Ashmore. Manufacturer of Spectacles, Telescopes and all kinds of Optical instruments. Every variety of optical glasses, ground by steam power. Optical works, 104 Fargate, nr. Burgess St., Sheffield. The advertiser has been established in the above business upwards of twenty-five years, it being wholly superintended by himself. Foreign orders executed to any extent, on the shortest notice."

William Ashmore continued in business until 1850 or 1851, and was succeeded by **Leedham & Robinson**.

Ashmore & Osborne
Ashmore & Osborne, [**18**] [**276** #1767], was a partnership between **William Ashmore** and the optician **Thomas P.G. Osborne**. The partnership traded at 42 Burgess Street, Sheffield, England, from 1825 until 1837.

According to an advertisement dated 1828 [**44**]: "Ashmore & Osborne, 42 Burgess Street, Sheffield. Manufacturers of fine steel and all other kinds of Spectacles, Reading glasses, Optical instruments, and grinders of optical glasses. Day and night telescopes."

Instruments bearing their name include: brass three-draw telescope with mahogany; 1¼-inch single draw brass telescope 20½ inches closed, 36½ inches fully extended.

The signatures on some instruments indicate that, as with Thomas P.G. Osborne, they had a branch in London, England. Examples include: brass four-draw telescope with mahogany barrel, signed *Ashmore & Osborne Day or Night London*; brass three-draw telescope with mahogany barrel by Ashmore & Osborne, London.

ASKANIA Mikroskop Technik Rathenow GmbH
ASKANIA Mikroskop Technik Rathenow GmbH is a microscope manufacturer based in Rathenow, Germany. The company was established in 1991 when the microscope technology division of **Rathenower Optische Werke GmbH** was split off to form the new company.

Askania Werke AG
Askania was founded in 1871 by **Karl Bamberg**, in Berlin, Germany, as Werkstätten für Präzisions Mechanik und Optik (Workshop for precision mechanics and optics). They began to produce marine compasses, chronometers, naval instruments and astronomical equipment for observatories. Through a series of acquisitions, the company became Askania Werke AG in 1921. In 1923 they acquired the firm of **Hans Heele**. The company continued to expand, producing film cameras and surveying equipment. During WWII they produced gunsights, gyroscopes and periscopes, and developed the flight controls for the V1 and V2 rockets. During wartime

they marked their products with the **ordnance codes** bxx, kjj or ppx. After the war they re-established their business, and expanded into foreign markets, collaborating with **Keuffel & Esser** in the USA market. In the 1960s they were taken over by the Dutch firm Oldelft and, following a series of take-overs, the last by Siemens, the company ceased to exist. The company name re-emerged in 2006 for a firm making wrist-watches. The brand-name was also re-used by the company **ASKANIA Mikroskop Technik Rathenow GmbH**, formed in 1991 from the microscope technology division of **Rathenower Optische Werke GmbH**.

Astigmatism

Astigmatism in an optical system, [7], is an aberration that occurs when a point on an object is offset from the optical axis, with the result that the cone of light from the object point strikes the lens asymmetrically. In the resulting image different parts of the light cone focus differently, resulting in a spread rather than a single point, and yielding an asymmetrical blur. If the object point is close to the optical axis the effect is negligible, but the further it is from the optical axis the more significant the effect.

Astigmatism occurs in well-formed optical elements such as spherically symmetrical lenses and concave paraboloidal or spherical mirrors. A lens or mirror which is designed to be free of astigmatism is referred to as an anastigmat.

Astigmatism in vision is an inability of the eye to bring rays of light to a common focus on the retina, and is caused by a structural defect in the eye. Vision defects are beyond the scope of this book.

Astrograph

An astrograph is a telescope designed for the purpose of astrophotography.

Astrola

Astrola was a trade-name used by **Cave Optical Co.**, and then by **Hardin Optical Company**.

Astrolabe

The astrolabe, [278], is an astronomical computer, designed to solve problems relating to the positions of the Sun and the stars in the sky, and measurements of time. The principle of the astrolabe has been known for over two thousand years, and the first known astrolabe was made in about 400 AD. It was highly developed in the Islamic world prior to its introduction into Europe in the C12[th].

There are various kinds of astrolabe, the most common form being the *planispheric* astrolabe, in which the markings on the disk represent the celestial sphere projected onto the equator. The astrolabe remained the most popular astronomical instrument until about 1650 when more specialised and accurate instruments began to replace it. See also Prismatic Astrolabe, Danjon Prismatic Astrolabe.

Astro-Mechanik

Astro-Mechanik of Mannheim, Germany, made the *Purus*, a clockwork sidereal tracking device, used to mount a camera for the purpose of astrophotography.

Astronomical Equipment

Astronomical Equipment, [199], was a telescope making company formed in Luton, Bedfordshire, England, after **Jim Hysom** left **Optical Surfaces** in 1966, in order to work for Leicester Astronomical Centre Ltd., which was the company formed by **Cliff Shuttlewood**.

By 1967 Leicester Astronomical Centre Ltd. had been renamed Astronomical Equipment, with Jim Hysom, his younger brother, Robert, a machinist, and with Cliff Shuttlewood designing and engineering the telescopes. They produced Newtonians, Cassegrains, Newtonian-Cassegrains, and Maksutovs, including 3½-inch Maksutovs with primary mirrors by **David Hinds**. The company continued in business until about 1977, when it was split into AE Mechanics, run by Robert Hysom, and **AE Optics**, run by Jim Hysom.

Astronomical Interferometer

See interferometer.

Astronomik

Astronomik was founded in 2000 by Eric Vesting and Gerd Neumann of Hamburg, Germany. The company produces optical filters, principally for astronomy and astrophotography.

Astro Optical Industries Co. Ltd.

Astro Optical Ind. Co. Ltd. of Tokyo was a Japanese maker of high-quality telescopes during the late-1950s through to the 1980s. They sold telescopes under their own name as well as under their brand, **Royal**. Additionally, their telescopes were sold under the brands **Monolux**, **Tasco**, **Sears**, and others. They used several maker's marks (see Appendix IV) during their years of making telescopes.

Astro-Optische Werkstaetten Georg Tremel

See **Georg Tremel**.

Astroscope

The Astroscope was a marine quadrant, using glass prisms instead of mirrors, invented by **Caleb Smith** in about 1734.

Astroscopics

Astroscopics was a US maker of medium to large mirrors, mounts and telescope systems, based in Montclair, California. They traded during the 1960s to the 1980s.

AstroSystems Ltd.

AstroSystems Ltd., [199], was a company formed in the late-1970s by Rob Miller, an ex-employee of **Fullerscopes**, with co-founders and directors **Peter Drew** and **David Hinds**, in Luton, Bedfordshire, England. The company was incorporated in 1976. They made and sold Newtonian and Cassegrain telescopes and mounts. Their telescopes had fine quality mirrors by David Hinds, and ranged from 4 inches to 24 inches. The aim of AstroSystems was to produce uncomplicated, high-quality telescopes at highly competitive prices. At the time they were also the main agent for Celestron and Vixen in the UK. They set up a sales subsidiary, Astro Promotions, which continued in business after AstroSystems was wound up.

ATIK Cameras

ATIK Cameras, of Norwich, England, is a manufacturer of specialised CMOS and CCD cameras, filters, and accessories for astrophotography, fluorescence microscopy, scientific imaging, and other specialist applications.

Atkey & Sons, Pascall

Pascall Atkey & Sons, [323], were chandlers and yacht equipment manufacturers of 29/30 High Street, Cowes, Isle of Wight, UK. They sold books, charts, sailing gear, and Pansy Heaters bearing their name.

According to an advertisement dated 1899 [44]: "Pascall Atkey & Son. Contractors to English and foreign governments. Manufacturers of ships and yachts' cooking stoves. Capstans. Compasses, Winches, &c. and fittings. Cowes, I. of W."

According to an advertisement (undated): "Pascall Atkey & Son Ltd. Cowes. Established 1799. By appointment to Her Majesty Queen Elizabeth II, Suppliers of Chandlery and Yachting equipment. Yacht equipment manufacturers."

Instruments bearing their name include telescopes and sextants. It is not clear to what extent they made the optical instruments themselves. One auction listed a two-draw telescope as "made by **Ross** for Pascall Atkey & Sons". There are, however, other examples of nautical instruments marked as manufactured by Pascall Atkey & Sons.

Atkinson, John J.

John J. Atkinson, (fl.1845-1891 d.1898) [18] [306], was a philosophical and photographic instrument maker, and theatrical ornament maker, of 33 and 37 Manchester St., Liverpool, England. His company traded as a wholesaler and retailer, and among their photographic products were cameras bearing their name. These included the *Atkinson's Portable Camera* (1857), *Woodward's Patent Solar Camera* (1859) and the *Eclipse* (1891), as well as mahogany and brass stereo cameras.

Following his death in 1898, John J. Atkinson was succeeded by his son Frederick Atkinson, and by 1902 the company was trading at 66 Victoria Street, Liverpool, as Atkinson Bros (late J.J. Atkinson). They continued to use this name, advertising "photo requisites", until at least 1904.

Auriscope

An auriscope is a medical instrument used for inspection of the ear. It provides a view of the external ear canal and tympanic membrane, or eardrum. Also known as an otoscope.

Autochrome

The Autochrome process, [70] [71], was an early three-colour photographic process invented by the **Lumière** brothers in 1903, and patented by them in 1904. The process used a glass plate coated with microscopic grains of potato starch, dyed red, green and blue, over which was a panchromatic emulsion. The plate was exposed through the glass side so that the light traversed the starch layer before reaching the emulsion, with the red, green and blue starch particles acting as colour filters. After development, a *reversal* process was used, in which the plate was re-exposed and redeveloped, resulting in a transparency in which tiny points of the three primary colours combined to form a colour image. The **Lumière** brothers did not begin to manufacture Autochrome plates commercially until 1907, when large-scale production with a suitable panchromatic emulsion was possible.

Automatic Level; Auto Level

An automatic level, otherwise known as an auto level or self-levelling level, is a kind of level which has a compensating mechanism so that, as long as the base is close to level, the instrument automatically removes any error resulting from any variation from level. Thus, the auto level is quicker to set up than a dumpy level, and easier to use.

Avimo

Avimo, **[19] [44] [232]**, of Taunton, Somerset, England, was formed in the late-1930s to manufacture equipment under license for the French company Bronzavia. The Avimo advertisement in the 1947 British Industries Fair catalogue, **[450]**, included cameras and equipment for research work. It listed the company as manufacturers of: Optical, Mechanical, and Electrical instruments; Aeronautical instruments; High speed continuous film cameras; Cinematograph cameras; Film assessors and projectors; Cathode ray recording equipment; Film scanning microscopes; Lenses; Prisms; Reflectors; Scientific research instruments.

Avimo set up a workshop in which they carried out refurbishment and reconditioning of Navy issue **Barr & Stroud** binoculars. In 1979, in competition with Barr & Stroud, they submitted a bid to supply both the British Royal Navy and Royal Air Force with hand-held binoculars, to a single unified design. This, with suitable modifications, would serve the various requirements of troops on the ground, seamen aboard their various craft, and airmen in helicopters or aeroplanes. The Avimo fixed focus *Binocular Prismatic, General Purpose, 7x42*, was selected, and Avimo became the prime supplier of British military binoculars.

In 2001 **Thales** merged with Avimo Optical Imaging and Avimo Thin Film Technologies, and re-branded as Thales Optronics. In 2005 Thales Optronics was sold, and became **Qioptiq**, and in 2013 Qioptic was acquired by the US company Excelitas Technologies Corp.

Avizard, René and Charles

The brothers René Gabriel Avizard, (b.1853), and Charles Xavier Avizard, (b.1855) became the proprietors of the business **Maison de l'Ingénieur Chevallier** when they purchased it in 1883 **[50]**.

Aylward, H.P.

Henry Prior Aylward, (fl.1867-1901) **[53] [264] [319] [320]**, was a microscope and accessory maker of Manchester, England, best known for his microscope slides, and for his 1881 design of a concentric dissecting and mounting turntable. From 1867 until c1885 his address was 15 Cotham Street, Manchester. An undated advertisement lists microscope slides and accessories as "Sold only by H.P. Aylward, Microscopist, Maker of Microscopes and All Accessories, 164 Oxford St., Manchester". He traded at this address from 1886 until 1901. He also retailed microscopes by other makers of the day, including **Ross**, **R.&J. Beck**, **C. Reichert**, and **Leitz**.

Ayscough, James

James Ayscough, (1718-1759) **[8] [18] [276 #226]**, was an optical, mathematical, and philosophical instrument maker of London, England. Addresses: *Great Golden Spectacles* in Ludgate St., near St. Paul's (1749); *Golden Spectacles & Quadrant*, Ludgate St. (1751); *Great Golden Spectacles & Quadrant*, Ludgate St., near St. Paul's (1759); *Sir Isaac Newton's Head*, Ludgate St.

He served his apprenticeship with **James Mann (II)** from 1732 until c1739, and was freed of the **Spectacle Makers' Company** in 1740. He worked in partnership with James Mann (II), trading as **James Mann & James Ayscough** from 1743 until 1747, after which he continued to trade under his own name. He became Master of the Spectacle Makers' Company in 1752.

According to his trade card **[53] [361]**: "Optician, At the *Great Golden Spectacles*, in Ludgate-Street, near St. Paul's, London. Makes and sells (wholesale and retail) Spectacles and reading-glasses, either of Brazil-pebbles, white, green or blue glass, ground after the truest method, set in neat and commodious frames. Concaves for short-sighted persons. Reflecting and refracting telescopes of various lengths, (some of which are peculiarly adapted to use at sea); Double and single microscopes, with the latest improvements; Prisms; Camera obscuras; Concave and convex speculums; Magick lanthorns; Opera glasses; Barometers and thermometers; Speaking and hearing trumpets; With all sorts of optical as well as mathematical and philosophical instruments. Together with a variety of maps, and globes of all sizes."

James Ayscough supplied metals to **J. Dollond & Son** **[1 p24]**. **Chester Moor Hall**'s achromat was said to be in Ayscough's shop window prior to **John Dollond (I)**'s work on achromatism. **[3] [8]**. Upon James's death in 1759 he was succeeded by his wife, **Martha Ayscough**.

Ayscough, Martha

Martha Ayscough, (fl.1759-1767) **[18]**, was an optician of London, England. Addresses: *Great Golden Spectacles and Quadrant*, Ludgate-Street (1759); 33 Ludgate St. (1767). Upon the death of her husband, **James Ayscough** in 1759, she succeeded him in business, and continued to trade under his name until 1767, at which time the business was taken over by **Joseph Linnell**.

B

Baader, Michael
See **Optische Fabrik M. Baader**.

Baader Planetarium
Baader Planetarium is an optical instrument and accessories company in Mammendorf, Germany. The company was founded in 1966 by Claus Baader, and produces telescopes, eyepieces, spectrographs, solar observing filters and Herschel prisms, astronomy and astrophotography filters, and other accessories.

The corporate or family connection between Baader Planetarium and **Optische Fabrik M. Baader** is unclear [**174**].

Backstaff
A backstaff is a device that was used in marine navigation to determine the noon altitude of the sun. In its simplest form it consisted of a staff with a sliding transom. At the far end of the staff was a sighting vane. The user stood with their back to the sun, sighted along the staff, and adjusted the position of the transom until the shadow cast by its top coincided with the sighting vane. The position of the transom on the staff was then used to calculate the altitude of the sun. The backstaff was further developed by the naval captain and navigator, John Davis (c1550-1605), into the instrument known as Davis's Quadrant.

Baigish
Baigish is a trade-name of the Russian company **KOMZ**, and is used for a range of binoculars, monoculars, and night vision binoculars, entry level and above. Some modern Baigish optical instruments are reputedly made in China.

Bailey, Henry Page
Henry Page Bailey, (1883-1962) [**24** p358] [**185**], was born in Wisconsin, USA and, after graduating, moved to California. He was a professional dentist, and an amateur astronomer, mirror maker and telescope maker, and is believed to have made the first Schmidt camera outside of Germany. This was an f/2.4 system attached to a 15-inch Cassegrain reflector. He also designed a series of modified horseshoe mounts (see equatorial mount) allowing the telescope to point to the pole star.

Bailey, John William
John William Bailey, (c1822-1900) [**18**] [**264**] [**319**] [**320**], was a mathematical instrument maker and manufacturing optician of 162 Fenchurch Street, London, England, trading at that address from 1859 until 1871. He served his apprenticeship with the mathematical and philosophical instrument maker James Gardner, beginning in 1837, and was freed of the **Grocers' Company** in 1844. According to his obituary in the journal of the Quekett Microscopical Club, he was a mathematical instrument maker and optician who "although more concerned with theodolites and sextants than microscopes, yet he originated a very convenient portable stand with folding tripod foot, packing into a very small compass, which he exhibited at the earlier soirées at University College." He is known to have sold sextants and microscope accessories, and his signature appears on some microscopes.

Bailey, Robert
Robert Bailey, (fl.1873-1904) [**319**], of 14 Bennetts Hill, Birmingham, England, was an optician, and an optical, philosophical and mathematical instrument maker. From 1875 until 1877 the *Post Office Directories* list him as an optician at 15 Bennetts Hill and, in 1878 and 1879, as "Bailey, R. Optician &c. to the Birmingham Eye Hospital (for many years with Messrs. **Field & Son**, of New Street), manufacturer of Microscopes, Telescopes, Barometers, Thermometers &c.; Drawing instruments of the best quality; Sole agent for Bourdons own make pressure & vacuum gauges. 14 Bennetts Hill." The *Kelly's Directories* from 1880 until 1904 list him as "Robert Bailey, Optician, 14 Bennetts Hill." An advertisement dated 1889 includes meteorological and photographic instruments.

Baillou, Francois de
Francois de Baillou, (c1700-1774 fl.1734-1764) [**48**] [**53**] [**270**] [**322**] [**371**], was a highly-regarded French optician and optical instrument maker who relocated to Milan, Italy before the beginning of his career. He described himself as "Professor of optics in Milan, maker and seller", and is reputed to have pursued his own research interests in a variety of scientific areas. In 1750 he was appointed Imperial Optician to Empress Maria Theresa of Austria.

Instruments bearing his name include: nine-draw astronomical and terrestrial telescope, 148 inches length, diameter 3¼ inches, dated 1734; drum microscope in wood and gilt-tooled leather with cylindrical leather-bound case, dated 1738 (Bonhams, 2014); compound microscope, dated 1755; two 1¾ inch diameter plano-convex lenses, dated 1755; small one-draw spyglass in wood and fish-skin, with tooled-leather draw tube, dated 1759; telescope, length 57 inches, dated 1762; wooden bodied monocular opera glass with tortoise shell and guilt decoration, 2¼-inch diameter, in leather case, dated 1763 (Christies, 2008); two-draw opera glass, 3¼ inches closed, diameter 1¼ inches, dated 1764; large seven-draw telescope, dated 1764; two screw-barrel microscopes.

According to his published literature, he also made: binocular telescopes; binocular opera glasses camera obscuras; magic lanterns; anamorphic lenses.

Baird, Alfred W.
Alfred W. Baird, (c1863-1928) [44] [263], began working as a compass adjuster for **Kelvin and James White** in 1884. He then became personal assistant to Kelvin, and in 1914 he was made a director of the company. After the deaths of both Lord Kelvin and James White, the company name was changed, including his name and that of **James Thomson Bottomley**, to **Kelvin, Bottomley and Baird**.

Baird, Andrew H.
Andrew H. Baird, (fl.c1889-1940s) [44] [53] [61] [122] [306], of 33-39 Lothian St., Edinburgh, Scotland, was a scientific instrument maker and chemical dealer. The business began at 15 Lothian Street, then moved to 37-39 Lothian Street in 1895, and subsequently expanded into the neighbouring addresses. Andrew H. Baird retailed photographic supplies, some of which he manufactured himself. His own products included the *Todd-Forrett* magnesium flash lamp, various cameras, and his *Lothian* branded instruments including a stereoscope, a cyclist camera, and a quick-level top for cameras. He adopted a triangular trade-mark containing the initials A.H.B. The symbol represented a top view of a tripod.

Baird & Sons, T.
See **Thomas Baird**.

Baird, Thomas
Thomas Baird, [61], was an optician of 34 Queen Street Glasgow, Scotland. In 1904 he began to trade, primarily as a dispensing optician, as T. Baird & Sons. When Thomas retired in 1933 the business continued under the same name, run by John Baird and William Torrence Baird. There was also a branch at 54 St. Enoch Sq., Glasgow, which continued under the management of Alexander Baird from 1931. In 1970 T. Baird & Sons was acquired by House of Fraser. Instruments bearing his name include a fine brass 4-inch refracting telescope, 50 inches long, with tripod and eyepieces, in a wooden box.

Baird & Tatlock
Baird & Tatlock, [263] [320], were laboratory equipment suppliers, and mathematical and philosophical instrument makers. The firm was established by Hugh Harper Baird and John Tatlock, and first appears in the directories in 1881, trading at 100 Sauchiehall Street, Glasgow, Scotland. Prior to entering this partnership, John Tatlock had been an assistant to **William Thomson**, beginning c1862, and had developed an interest in the use of scientific and philosophical instruments in education.

In 1889 the firm moved to 40 Renfrew Street, Glasgow, and Hugh Harper Baird moved to London, where he opened a branch at 14 Cross Street, Hatton Garden. By 1896 the company had expanded into 10 Drummond Street, Edinburgh, and in that year the company was divided into two independent concerns. Hugh Harper Baird continued to trade as Baird & Tatlock in London, and John Tatlock continued to trade under the same name in Glasgow and Edinburgh, [35 #26814 p274].

The London firm became a private company, Baird & Tatlock (London), in 1903, and in 1978 was acquired by E. Merck.

The Glasgow and Edinburgh firm expanded into Liverpool in 1904, and in 1911 they opened a branch in Manchester. In 1915 they became Baird & Tatlock Ltd. In 1925 they merged with **John J. Griffin & Sons Ltd.**, and in 1929 the merged company changed their name to become **Griffin & Tatlock**. Examples of instruments by Baird & Tatlock include a Nörremberg polariscope.

Baker, Charles
Charles Baker of London, England, [3] [44] [319] [320], was a firm of mathematical, optical, philosophical and surgical instrument makers, of London, England. Addresses: 244 High Holborn (1851-1930); 243 High Holborn (1855-1887, 1894-1912), 244a High Holborn (1859-1878), and 245 High Holborn (1880).

According to their advertisements, the company was established in 1765. Charles Baker (1820-1894), after whom the company was named, ran the company until his death. They began making microscopes in about 1851. In 1867 they began to market the travelling microscope designed by **William Moginie**. When Charles Baker died, he was succeeded in his role by Thomas Curties, who died in 1896. He in turn was succeeded by his sons Charles and Thomas, and the management of the company remained with the Curties family until its acquisition in 1959 by Vickers (see below).

In 1901 **Herbert F. Angus** began to work for C. Baker as a manager. He continued in this role until he resigned in 1909. In 1904 **Charles Perry**, who had been foreman for **Powell & Lealand**, began to work for C. Baker, making microscopes for Thomas Powell. He continued to do this until Powell & Lealand ceased trading as a company in about 1914, after which he left Baker to start his own company.

In 1919 C. Baker was listed as a member of the **British Optical Instrument Manufacturers' Association**.

In the 1929 British Industries Fair Catalogue [**448**], their advertisement listed their exhibit as optical, scientific and photographic, and listed the company as manufacturers of: Microscopes and accessories; Astronomical instruments; Lenses and prisms (not photographic); Optical benches and accessories; Laboratory instruments and equipment; Scientific models; Surveying instruments; Aeronautical instruments; Industrial testing apparatus.

The company set up works in Balham, South London, retaining the Holborn address for instrument sales and repair. In 1936 the company became C. Baker & Co., and in 1940 it was incorporated as C. Baker of Holborn Ltd. Their instrument works were in Balham until, in 1945 the works moved to expanded premises at the Metron Works, Purley Way, Croydon.

In the 1947 British Industries Fair, their catalogue entry, [**450**], listed them as a manufacturer of: Microscopes and related apparatus for research, medical, education and industry; Photo-micro, micro-projection, and illuminating apparatus; Diascopes; Epidiascopes; Cathetometers; Measuring instruments and laboratory equipment; Lenses; Prisms and mirrors.

In 1959, **Vickers** acquired C. Baker of Holborn Ltd., which was renamed C. Baker Instruments Ltd., This excluded the instrument sales and repair part of the business which remained under the control of Michael Curties, who renamed it Rekab Ltd., retaining the High Holborn premises. In 1962 C. Baker Instruments Ltd., together with **Cooke, Troughton & Simms**, **McArthur Microscopes Ltd.** and Casella (Electronics) Ltd., an offshoot of **Casella**, was renamed Vickers Instruments Ltd.

No record exists of Charles Baker serial numbers between the start of production in 1851 and the Vickers acquisition in 1959. The last serial number they produced, however, was around 40,000, and the majority of Baker models were made towards the end of that period.

Balbreck, Maximilien

Maximilien Balbreck, (1827-1902) [**58**], was a manufacturer of precision mathematical, optical, marine and physical instruments of Paris, France. He began his business in 1854. In 1878 he began trading as Balbreck Aîné, which translates to Balbreck the elder, and in the same year he won a gold medal in the Paris exhibition. He won a gold medal again at the exhibition in Melbourne in 1880. He was made a Knight of the Legion of Honour in 1883, and by 1893 he had entered a partnership with his son, and was trading as Balbreck Aîné et Fils - Balbreck the elder and son.

Balda

In 1908, Max Baldeweg founded a factory in Laubegast near Dresden, Germany, making camera parts. In 1913 the company was renamed Balda-Werk Max Baldeweg, and they introduced the *Balda* brand-name. They introduced their first camera in 1925, and their first 35mm camera in 1935. The factory was heavily damaged during WWII, and at the end of the war, in 1946, the original factory was nationalised. The nationalised company continued to produce cameras under the *Balda* brand until it was renamed Belca-Werk in 1951. The company continued to produce cameras until they were absorbed into VEB Kamera-Werke Niedersedlitz (see **Kamera Werkstätten**) in 1956.

Meanwhile, at the end of the war, Max Baldeweg moved to West Germany to re-found his company as Balda-Werk Bünde. They produced new models in addition to some of their pre-war models, using the *Balda* brand-name, until the 1980s.

Bamberg, Karl

Johann Karl William Anton Bamberg, usually known as Karl, (1847-1892) [**47**] [**374**], served his apprenticeship with **Carl Zeiss**, in Jena, and at the same time he attended lectures in astronomy, optics and physics at the University of Jena. In 1869 he went to work for **Pistor & Martins** in Berlin, where he continued similar studies at the University of Berlin. He started his own company, Werkstätten für Präzisions Mechanik und Optik (Workshop for precision mechanics and optics) in 1871, producing marine compasses, chronometers, naval and surveying instruments, astronomical equipment for observatories, and naval and surveying instruments. He supplied a universal transit instrument to the Berlin Observatory, and established a reputation for both linear and circular graduated instruments. Through a series of acquisitions, the company became **Askania Werke AG** in 1921. In 1923 they acquired the firm of **Hans Heele**.

Bancks, Robert

Robert Bancks, otherwise known as Banks, (1765-1841) [**18**] [**50**] [**53**] [**235**], was a mathematical and optical instrument maker of London, England. Addresses: 25 Piccadilly (1792); 440 Strand (1795-1804); 441 Strand (1805-1829); 119 New Bond Street (after 1829).

He was renowned for having made microscopes for Charles Darwin, and for the Scottish botanist Robert Brown, who discovered *Brownian motion*. He made microscopes by royal appointment to the Prince of Wales, who became Prince Regent in 1811 and was crowned King George IV in 1821. During the 1820s Robert's son, also Robert Bancks, (1799-1830), joined him in partnership as

Bancks & Son. The elder Robert Bancks continued to trade under this name after his son's death in 1830, until he retired four or five years later. Instruments are signed either *Bancks* or *Banks*.

Bancks & Son
See **Robert Bancks**.

Banks, William
William Banks, [**199**], of 40 Newport Street, Bolton, Lancashire, England, was a maker of equatorial mounts. He also provided a service refiguring and re-silvering mirrors for Newtonian telescopes. He wrote *Telescopes: Their Construction, Adjustment & Use* (1920).

In the late-1891 William Banks of 32 Corporation Street, Bolton, advertised equatorially mounted astronomical telescopes with clock drive [**8**].

In 1896 William Banks of 30 Corporation St., Bolton, was elected a Fellow of the Royal Astronomical Society [**120**].

Barbier, Bénard, et Turenne (BBT)
The French firm Barbier et Fenestre was founded in 1862 by Frederic Barbier and Stanislas Fenestre. In 1887 the company was renamed Barbier et Cie, and in 1889 it became Barbier et Bénard. In the early C20th the company became Barbier, Bénard, et Turenne, and became a public limited company in 1919.

BBT specialised in producing equipment for lighthouse beacons, including Fresnel lenses. They expanded their scope to other areas including producing microscopes under license. In 1935 BBT merged with **Société des Etablissements Krauss** to form **BBT Krauss**.

Barbon, Peter
Peter Barbon, (fl.1809-1812) [**18**] [**262**] [**271**] [**274**] [**276** #1076] [**358**], was an optical instrument maker of Edinburgh, Scotland. Addresses: 4 Lothian St. (1809); 18 Nicholson St. (1810); 77 Princes St. (1811-1812). A microscope bearing his name, with the 77 Princes Street address, is in the National Museum of Scotland.

Barclay, Adam
Adam Barclay, (d.1753) [**18**] [**271**], was a mathematical and optical instrument maker of Edinburgh, Scotland.

Barclay, Andrew
Andrew Barclay, (1814-1900) [**263**] [**322**], was an engineer, and an amateur astronomer and amateur telescope maker, of Kilmarnock, Scotland.

In 1840 he formed a partnership with Thomas McCulloch, producing mill shafting and calico printing machinery. This partnership lasted two years, after which he started his own business, building machinery and engines for various industrial applications. He was a fine engineer, but a poor businessman, and consequently suffered frequent financial troubles. By 1859 he was producing railway locomotives, and by 1870 his business was said to be the largest in Kilmarnock. In 1893, following a spate of financial difficulties, he was removed from the company by the shareholders.

His interest in astronomy led him to make a number of telescopes, including refractors, and Gregorian reflectors with metal speculum mirrors he had produced himself. He also made at least one micrometer eyepiece. During the 1850s he was elected a Fellow of the Royal Astronomical Society. He contributed many articles on his astronomical observations to the *English Mechanic* magazine, some of which were of a highly dubious nature, such as his observation of mountains on Jupiter. He was removed as a Fellow of the RAS in 1896 due to the non-payment of his fees.

Barclay, Hugh
Hugh Barclay, (fl.1727 d.1749) [**18**] [**271**], was a watchmaker and optical instrument maker of Edinburgh, Scotland. He was the brother of **William Barclay**.

Barclay, William
William Barclay, (fl.1731 d.1758) [**18**] [**271**], was an optical instrument maker of Edinburgh, Scotland. He was the brother of **Hugh Barclay**.

Bardou
Maison Bardou, [**8**] [**72**] [**197**], was founded in 1819 by D.F. Bardou, making terrestrial and astronomical telescopes, microscopes, and binoculars for field, marine and theatre. The company, at 55 rue de Chabrol, Paris, France, passed into the hands of his son, Pierrre Gabriel Bardou. Records show them trading as Bardou & Fils à Paris in 1855 and 1884. In 1865 the firm passed into the control of Pierre's son Albert Denis Bardou (1841-1893). In 1876, they exhibited as Bardou & Sons at the Centennial Exhibition in Philadelphia, USA. Subsequently they traded as Bardou & Son.

In 1896 **Jules Vial** assumed control of the company, and continued to use the Bardou name. Vial produced a catalogue showing up to 6-inch pier mounted telescopes. By 1899 the company had moved to 59 rue Caulaincourt, Paris, and was no longer selling equatorial telescopes. A 1911 catalogue for Bardou and Son, [**237**], shows refracting telescopes ranging from 2¼ inches to 4¼ inches aperture, as well as accessories. Bardou made lenses for other telescope suppliers, including **J.W. Queen**, **Thomas H. McAllister**, **Benjamin Pike**, Sussfield, and Lorsch.

Barker and Co., John

John Barker and Co. of Kensington, London, was a major UK national department store. They sold instruments such as telescopes and barometers bearing their name, but manufactured for them by the instrument makers of the day. Instruments bearing their name include: ship's bulkhead barometer (1920s); inlaid mahogany barometer (c1900); marine barometer (c1915) "*manufactured by Joseph Hicks and retailed by John Barker*"; late-Victorian binnacle aneroid barometer; 3-draw brass and leather telescope.

The department store John Barker and Co. operated from 1894-1988, and was acquired by House of Fraser in 1957. For a history of John Barker and Co., see [281].

Barlow Lens

The Barlow lens, invented by **Peter Barlow**, is a diverging lens, or arrangement of lens elements, that increases the effective focal length of an optical instrument as seen by an eyepiece or imaging sensor. This has the effect of magnifying the image and reducing the field of view. A Barlow lens may be used in conjunction with an eyepiece in astronomy to increase the effective focal length of a telescope. In photography the teleconverter, works on the same principle, increasing the effective focal length of the attached camera lens. In microscopy a Barlow lens may be used to increase the effective magnification of the objective, and it reduces the working distance between the objective and the sample being observed.

See also focal reducer.

Barlow, Peter

Peter Barlow, (1776-1862) [24] [37] [46], was a mathematician, physicist, and optician, born in Norwich, England. In 1801 he attained the post of assistant mathematics master at the Royal Military Academy, and was subsequently promoted to professor. He resigned in 1847 but, in recognition of his public services, continued to receive his salary. He devised a method of correcting the deviation of compasses in iron ships, and his publications included *An Elementary Investigation of the Theory of Numbers* (1811), *A New Mathematical and Philosophical Dictionary* (1814), and *New Mathematical Tables* (1814). He solved problems in the areas of electromagnetism, marine engineering, railway engineering, and civil engineering, and in 1820 was made an honorary member of the Institution of Civil Engineers.

In 1827 he became interested in solving the problems of aberrations in telescope optics. He developed a number of experimental telescopes using fluid lenses. He proposed a lens that could be used to increase the power of eyepieces, which was first made by **George Dollond (I)** in 1833, and became known as a Barlow lens.

Barnet Ensign

In 1930 **Houghton-Butcher** set up a sales subsidiary **Ensign Ltd.**, but in 1940 the Ensign headquarters was destroyed by enemy action. The assets were sold, but Houghton-Butcher kept the name, and used it for their *Ensign* cameras. In 1945 the company, together with **Elliott and Sons**, who made the *Barnet* brand of film, formed the company Barnet Ensign [306].

In 1947 Barnet Ensign exhibited at the British Industries Fair [450]. They were listed as manufacturers of: *Ensign* cameras; Roll films; Cine and photographic apparatus; Photographic plates; Flat films and papers with the *Barnet* trade-mark.

In 1948 Barnet Ensign merged with **Ross Ltd** to form **Barnet Ensign Ross**, which was renamed **Ross-Ensign Ltd.** in 1954.

Barnet Ensign Ross

See **Barnet Ensign**, Ross Ltd.

Barnett, Thomas

Thomas Barnett, (1768-1816) [18] [53] [274] [361] [401], was an optical, mathematical, and philosophical instrument maker of London, England. Addresses: 61 Tower St. (1789); 61 Great Tower Street (1790-1794); 6 Tower Street (1795-1796); 21 East Street, Lambeth (1799-1802); East St., East Place (1804); 4 Mores Yard, Old Fish Street, nr. Doctors' Commons.

He began an apprenticeship in 1782. In 1789 he was freed by purchase of the **Spectacle Makers' Company**, and began his business at 61 Tower Street. His trade card lists him as optical, mathematical, and philosophical instrument maker to His Majesty's Hon[ble] Boards of Customs & Excise.

An advertisement, listing his address as 61 Great Tower Street, states that he "Makes the following articles, with a variety of others, which are sold, wholesale and retail, at the lowest prices." The list includes Spectacles; Magnifiers; Reflecting and refracting telescopes, both celestial and terrestrial; Micrometers; Opera glasses; Solar microscopes; Aquatic, botanic and other microscopes; Magic lanterns and slides; Camera obscuras; Theodolites, Circumferentors, Plane tables, Level telescopes and other Surveying instruments; Mathematical instruments; Sextants and Hadley's quadrants; Gunner's quadrants; And many other Mathematical, Philosophical and Optical instruments.

In 1816, he is recorded as being in the workhouse.

Barr, Archibald

Archibald Barr, (1855-1931) [26] [46], was a co-founder of **Barr and Stroud**.

Archibald Barr was born in Glenfield, near Paisley in Scotland, and served his apprenticeship with A.F. Craig and Co., boilermakers and engineers. He studied for an engineering degree at Glasgow University during his apprenticeship, and his professor was sufficiently impressed with his performance that he offered Barr a position in his department when he graduated in engineering science in 1876. For eight years Barr was assistant to Professor James Thomson, brother of **William Thomson**, some of whose lectures he attended, and with whom he collaborated in research.

In 1884, having obtained his Doctorate in Science, Barr left Glasgow University to become Professor of Engineering at the Yorkshire College of Science in Leeds. There he set about raising funds to build a new laboratory for his department, and in 1885 he met **William Stroud**. In 1887 Dr. Barr began his first collaboration with Stroud when they devised a lantern-slide camera, which they patented in 1889. In 1889 Dr. Barr was appointed to the Chair of Civil Engineering and Mechanics at Glasgow University, succeeding James Thomson.

In 1888 the War Office advertised for an infantry rangefinder. This was another opportunity for collaboration between Dr. Barr and Dr. Stroud, and they produced a design which they submitted to the War Office. By 1895 they had opened a workshop producing the rangefinders. This success gave birth to what was to become Barr & Stroud, in Anniesland, Glasgow, and a long history of producing optical instruments, largely for the military.

In 1912 Dr. Barr resigned his Chair at the university to concentrate on his business. In 1915 he played a prominent role in setting up an optical glass foundry for Barr & Stroud. He continued in his role in the company, protecting, expanding and diversifying the company's interests, until his death in 1931.

Barr and Stroud

Barr and Stroud, [19] [20] [26] [444], (B&S), were makers of optical instruments, of Anniesland, Glasgow, Scotland. Their London Office, (in 1922), was at 15 Victoria Street, Westminster, SW1. The company was formed by **Archibald Barr** and **William Stroud**. Their collaboration began in 1887 when they devised a lantern-slide camera, which they patented in 1889.

Rangefinders: In 1888 the War Office advertised in the *Engineering* magazine for an infantry rangefinder. They required an instrument that could withstand rough usage in the field, could measure a range of 1,000 yards to within 4%, and with which a man of average intelligence could take four readings a minute. This was another opportunity for collaboration between Dr. Barr and Dr. Stroud, and they produced a design, which they submitted to the War Office. In 1892 they conducted a successful trial of their rangefinder design on *HMS Arathusa*, and by 1895 they had opened a workshop producing the rangefinders. They did not produce the optical parts themselves. They had their prisms manufactured by **A. Hilger & Co.**, and their spherical lenses and plane glass panels manufactured by **Chadburn Brothers**. Despite suffering quality problems with the prisms supplied by Hilger, this arrangement continued until Barr & Stroud began to make their own glass and optical parts in 1916 (see below). The success of their rangefinders gave birth to what was to become Barr & Stroud, in Anniesland, Glasgow. In 1965 they developed, for the Chieftain Tank, the first commercially available laser rangefinder.

Growth and Diversification: During the early years Barr & Stroud worked to build up their sales base at home and abroad, with sales and licensing initiatives in such places as Austria-Hungary, USA and Russia. In 1903 Barr & Stroud were asked by the Holland Torpedo Boat Company to design a rangefinder for submarine periscopes, but they did not begin to manufacture submarine periscope rangefinders until they received an order for their first six instruments in 1916. During WWI they supplied prisms for other optical instrument makers including **J. Brimfield & Co.** and **Kershaw**.

In 1909 **Keuffel & Esser** were granted a license to manufacture Barr & Stroud range and order indicators, and in 1910 they were given permission to begin manufacturing Barr & Stroud rangefinder stands, as the two firms worked to compete with the **Zeiss** – **Bausch and Lomb** alliance [26 p67], known as the **Triple Alliance**. Over the years, Barr & Stroud manufactured periscopes for different applications, including tanks, submarines and nuclear reactors. The latter were produced for the Atomic Energy Authority, for whom Barr & Stroud also developed a high-speed camera capable of 50 thousand - 8 million frames per second. After WWI they began to produce prismatic binoculars (see below), and Cinematographs. They also produced a golf practice device called an *Impactor*, motor cycle engines, and an *Optophone* to convert printed words into sounds for the blind. However, their core business was, and always remained, optical munitions for the military.

Glass-Making: In 1914 Royal Proclamation prohibited firms such as Barr & Stroud from exporting instruments of war. France, however, was exempted from this ban in relation to rangefinders in exchange for a continued supply of glass from **Parra Mantois et Cie**. Despite this, not only were supplies of optical glass disrupted by the outbreak of war, but **Chance Brothers & Co.** of Birmingham, Barr and Stroud's main supplier, could no longer keep up with demand. As a result, Barr and Stroud set up their own facility for producing optical glass, and by the end of 1915 they were producing their own glass. The Ministry of Munitions did not classify Barr and Stroud as glassmakers, and so they received no technical or financial assistance in developing their optical glass-making plant. During WWII Barr and Stroud supplied the Air Ministry, War Office and Admiralty with nearly half of their requirements for optical glass every year. The glass foundry remained in use until 1971 when, in need of modernisation, it was closed, and civilian binocular manufacture was discontinued. [**26** p78, p150, p199].

Binoculars: Barr & Stroud produced both Galilean and prism binoculars, with their first binoculars available in 1919, and civilian binocular production continuing until 1971. During the inter-war period Barr & Stroud began to supply binoculars to the British Admiralty. During WWII and the subsequent years, they were the Admiralty's main supplier of binoculars, and this continued until 1979. At that time, in competition with **Avimo**, they submitted a bid to supply both the British Royal Navy and Royal Air Force with a binocular, with a single unified design. This, with suitable modifications, was to serve the various requirements of troops on the ground, seamen aboard their various craft, and airmen in helicopters or aeroplanes. The Avimo bid was selected, and Avimo became the prime supplier of British military binoculars. For details of Barr & Stroud binocular models, dates of manufacture etc., see Appendix I.

Modern History: In 1997 Barr & Stroud merged with **Pilkington**. In 2000 the merged company became a subsidiary of the French company **Thales Group**, and in 2001 they were re-branded as Thales Optronics Ltd. In 2005 Thales Optronics was sold, and became **Qioptiq**.

The Barr and Stroud brand-name returned to the UK when in 2008 the Barr and Stroud trademark was re-registered, and it is now owned by **Optical Vision Ltd.** (OVL). They are an importer and distributor, and modern Barr & Stroud optical instruments are made abroad for OVL, and imported for distribution.

Barrett, Robert Montague

Robert Montague Barrett, (fl.1849-1875) [**18**] [**322**], was an optical and nautical instrument maker of London, England. Addresses: 4 Jamaica Terrace, Limehouse (1849-1851); 4 Jamaica Terrace, West India Dock Road (1855-1870), and 80 West India Dock Road (1875).

He exhibited in the Great Exhibition in London in 1851 [**328**], for which his catalogue entry no. 349, stated, "Barrett, Robert M., 4 Jamaica Terrace, Limehouse—Manufacturer. Improved lunar sextants. Their object is increased facility in reading off by night. Plain sextant. Improved brass quadrant, divided to half minutes."

Two octants bearing his name are in the National Maritime Museum, Greenwich [**16**] [**54**].

Barrow, Henry

Henry Barrow, (1790-1870) [**3**] [**18**] [**53**] [**91**] [**228**] [**276** #2069], was a mathematical and optical instrument maker of London, England. Early in his career he worked as a journeyman for **Edward Troughton**. From about 1824, having set up is own workshop at 18 Crown Court, Soho, he produced work for leading instrument makers, including **George Dollond (I)**.

In 1830 George Everest of the East India Company took Henry Barrow to India for the Great Trigonometrical Survey of India. Everest had been disappointed with the quality of the surveying instruments he had thus far used. According to [**149** p167], whilst in England Everest had personally superintended the construction of the instruments made for him. In 1830, upon his return to India, he took Henry Barrow with him to start a mathematical instrument factory in Calcutta. A notable achievement of Barrow's was the complete renovation of a 36-inch theodolite. The instrument was originally made by **William Cary**, but was subsequently referred to as the Barrow theodolite due to the extent of the renovation. Henry Barrow returned to England in 1839, and was succeeded in this role by a skilled instrument maker, Syud Mohsin, of Arcot, India.

In 1842 Henry Barrow took over the business of the recently deceased **Thomas Charles Robinson**, and traded as **Robinson & Barrow** at 28 Devonshire Street, Portland Place, then from 1843 until 1845 at 26 Oxenden St., Haymarket.

After 1845 he began to trade under his own name, appearing in the directories as Henry Barrow, and from 1849 until 1851 as **Henry Barrow & Co.**, at 26 Oxenden St., Haymarket. He continued in business until 1864. By 1865 the address was occupied by T. Owen, who appears to have traded as Barrow & Owen.

On the popular auction sites there is a flood of cheap imported modern instruments such as telescopes, sextants, pocket sextants, and compasses bearing the name Henry Barrow or Henry Barrow & Co, and it is difficult to find an example of an instrument that is genuine.

Barrow & Co., Henry

Henry Barrow & Co., [**18**] [**53**], were mathematical and optical instrument makers. **Henry Barrow** was trading under this name by 1849, after having worked in partnership with **Thomas Charles Robinson**, trading as **Robinson & Barrow** until at least 1845. Records show the company located at 26 Oxenden Street, Haymarket, London, England from 1849 until 1851. A trade label stated: "Henry Barrow & Co., successors to the late T.C. Robinson, 38 Devonshire Street. Mathematical & Optical instrument makers to the Lords Commissioners of the Admiralty. 26 Oxenden Street, Haymarket, London." Henry Barrow & Co. are known to have sold telescopes, sextants, magnetic compasses, protractors, and thermometers.

On the popular auction sites there is a flood of cheap imported modern instruments such as telescopes, sextants, pocket sextants, and compasses bearing the name Henry Barrow or Henry Barrow & Co., and it is difficult to find an example of an instrument that is genuine.

Barry, Richard

Richard Barry, (1799-1811) [16] [18] [276 #925] [361], was a mathematical instrument maker and stationer, who ran a Navigation Warehouse. He traded at 290 Wapping St., London, England from 1800 until he relocated in 1801 to 106 Minories, London. The instruments he sold bearing his name would have been supplied wholesale by the makers of the time. Instruments he is known to have sold include: telescope; sextant; octant; compass. Instruments bearing his name include: octant with scale marked with a foul anchor scale divider's mark (see Appendix XI). He was succeeded in business by M. Barry, who continued trading as a Navigation Warehouse at 106 Minories until 1828.

Berthélemy, A.

A. Berthélemy, [322] [360], was a C19th French optical and mathematical instrument maker. Instruments bearing his name include a telescopic level. In 1895 the firm was acquired by **Ponthus & Therrode**. See **Albert Lepetit**.

Barton, John Henry

John Henry Barton, (1851-after 1919) [29] [56], of Cheshire, England, was an innovator in the field of optical instruments. During the period 1897-1919 he applied for several patents including a moveable hood for use with magic lanterns and, in 1897, a design for prismatic binoculars which was produced by **Ross Ltd.** as the first prismatic binocular to be manufactured in Britain. Barton worked for Ross, presumably from its incorporation in 1897, until 1905 when he set up his own business as a manufacturing optician at 196 Clapham Park Rd., London. He patented the *minim* binocular which was exclusively sold by **Negretti & Zambra**. In 1912 Barton was in debt, and he was rescued by a cash investment and the creation of a limited company, **Barton Linnard Ltd.**

Barton Linnard Ltd.

Barton Linnard, [29] [56] [110], of 196 Clapham Park Road, London, England, was incorporated in 1912 as a limited company with **John Henry Barton** as Managing Director. During WWI the company mainly made prismatic binoculars, box sextants, riflescopes and prismatic compasses. They continued to produce prismatic binoculars after the war and, following a resolution in Dec. 1923, the company was liquidated in Jan. 1924, [35 #32895 p208]. Instruments bearing their name include: binoculars; military gun sights; telescopes; sextants.

Baserga, Alfred Jean François

Alfred Jean François Baserga, (1833-1901) [50], was the son-in-law of **Pierre Queslin**, married in 1860 to **Adélaïde Queslin**. Following the death of her father, they continued Pierre Queslin's business in the name **Maison Chevallier**.

Bass, George

George Bass, (1692-1768) [1] [2] [8] [18] [44] [276 #6], was an optical instrument maker, and optical grinder and polisher, of Bridewell Precinct, London, England (1733), and later of Fleet Ditch, London (1764). He served his apprenticeship with the optical instrument maker **Ralph Sterrop**, beginning in 1706, and was freed of the **Spectacle Makers' Company** in 1716/17. He became Master of the Spectacle Makers' Company in 1747, and from 1754-1757. Among his apprentices was **David Drakeford**. He was subcontracted by **Edward Scarlett (I)** and **James Mann (II)** to make both optical elements of the achromat invented by **Chester Moore Hall**. In 1764 he was one of the petitioners who attempted to revoke the patent obtained by **John Dollond (I)** for the achromatic lens, and which was enforced after Dollond's death by his son, Peter (See Appendix X).

Basset & Gowin

Bassett & Gowin, [8] [199], of Saltdean, Sussex, England were, according to their 1960s catalogue, "Makers of high quality astronomical telescopes and accessories". Their catalogue shows: Tools and accessories for amateur telescope making; Mirrors from 4 inches to 16 inches; Secondary mirrors; Cassegrain mirrors from 6¼ inches to 16 inches; Mirror blanks; Newtonian and Cassegrain telescopes up to 16 inches; German and fork equatorial mounts. Their advertisements appeared during the 1960s and 1970s.

Bassnett & Co., Thomas
Thomas Bassnett & Co., [379], nautical, mathematical, and optical instrument makers of 10 Bath St., Liverpool, England, are recorded as trading during 1887-1888. It is not clear whether **Thomas Bassnett** was still working in the business in Liverpool at that time. The firm was the Liverpool branch of **M. Walker & Son**.

Bassnett & Son, James
James Bassnett & Son [322] [379] [382] [438], were optical, mathematical, and philosophical instrument makers of Liverpool, England. Addresses: 58 Robert St. North (1855); 8 Robert St. The firm was a partnership between **James Bassnett** and his son **Thomas Bassnett**, formed in 1855. The partners were jointly declared bankrupt in 1857 [35 #21969 p1857], but the partnership appears to have continued in business until 1865. Instruments bearing their name include: telescope; marine compass; wheel barometer; stick barometer.

Bassnett, James
James Bassnett, sometimes spelled Basnett or Basnet, (fl.1829-1865) [3] [16] [18] [276 #2070] [379] [438], was an optician, and optical and mathematical instrument maker, clock maker, and chronometer maker of Liverpool, England. Addresses: 4 Barnes Court, Shaws Brow (1829); 13 Robert St. North (1834); 1 Robert St. North (1834-1853); 16 Robert St. North (1835); 19 Robert St. North (1837); 58 Robert St. (1857); 8 Robert St. (1857-1860).

He added the second "s" to his name in 1841. In 1855 he began to work in partnership with his son, **Thomas Bassnett**, trading as **James Bassnett & Son**. They were jointly declared bankrupt in 1857. Instruments bearing his name include: large single-draw marine telescope; ebony framed octant with brass fittings; blackened brass triangle pattern sextant; marine stick barometer; marine chronometer. He is also known to have sold: marine barometer; thermometer.

Bassnett, P.
P. Bassnett, (fl.mid-late C19th), was an optician of Liverpool, England. Instruments bearing his name include: single draw rope-decorated marine telescope, 55mm objective, length 75cm, 104 cm fully extended, with the *Try Me* logo indicating that it was manufactured by **J.P. Cutts** (two examples).

Bassnett, Thomas
Thomas Bassnett, (fl.1855-1865) [379], was a nautical, mathematical, and optical instrument maker of Liverpool, England. He was the son of **James Bassnett**, and worked in partnership with his father, trading as **James Bassnett & Son** beginning in 1855. They were jointly declared bankrupt in 1857, but the partnership appears to have continued until 1865, after which Thomas Bassnett continued to work under his own name. Instruments bearing his name include: brass field glasses; marine chronometer; mahogany cased brass marine sextant; patent sounder.

See **Thomas Bassnett & Co.**

Bate, John
John Bate, (1809-1840) [18] [276 #1773] [355], was an optical instrument maker of 17 Poultry, London, England. He was the son of **Robert Brettell Bate**, and worked in the family business. From 1824, his address was 20 & 21 Poultry. He never served an apprenticeship, but he described himself as an optical instrument maker, and helped with his father's workload, as well as carrying out work on his own account. In 1832, he patented an *anaglyptograph*, a machine for engraving medals. He died of consumption, aged 30.

Bate, Robert Brettell
Robert Brettell Bate, (1782-1847) [3] [18] [46] [50] [276 #1079] [355], was a mathematical, scientific, nautical and optical instrument maker of London, England. From 1807 he traded at 17 Poultry, London. In 1816 he collaborated with **Sir David Brewster** on the development of the Kaleidoscope. He moved to larger premises at 20-21 Poultry in 1824, where he employed about twenty people. He was freed of the **Spectacle Makers' Company** in 1822, and became Master in 1828-1829. He took on **Alfred Chislett** as an apprentice in 1825, but turned him over to **William Gilbert (II)** in the same year. His father-in-law (who was also his uncle) was Benjamin Sikes, the excise collector who invented an improved hydrometer that bears his name. Robert Bate was awarded the right to make the hydrometer, and appointed mathematical instrument maker to HM Excise. In 1819, upon the death of **Matthew Berge**, Bate took over Berge's position as mathematical instrument maker to the Board of Ordnance.

He had royal appointments as optician to Kings George IV and William IV, and Queen Victoria, and was appointed sole Admiralty Chart Agent in 1830. He made bullion balances for, among others, the Bank of England and the East India Company. He had two sons, Bartholomew and John. Bartholomew who was deaf, served an apprenticeship with his father, beginning in 1822, and was freed of the Spectacle Makers' Company in 1829. He did not, however, work with his father. **John Bate** did not serve an apprenticeship, but worked with his father in the business until his untimely death in 1840.

Following Robert's death in 1847 his wife, Anna Maria, continued to run the business until she closed it down in 1850, the year before she died. She sold the rights to publication of the Admiralty charts to Robert's ex-employee, **John Dennett Potter**.

Baudry et Lorieux
Baudry et Lorieux, [**360**], a firm of nautical instrument makers of Paris, France, was a partnership, formed in 1855, which succeeded **Schwartz dit Lenoir** in business. The partnership was terminated about a year later, following which **E. Lorieux** continued the business under his own name.

Baumann & Co., Alfred
Optische Werke Alfred Baumann & Co. KG were optical instrument makers of Cassel, Germany, [**94**]. During the 1920s they manufactured the **Fata Morgana** miniature binoculars invented by Alfred Baumann.

Bausch & Lomb
Bausch & Lomb, [**201**], is an American firm of opticians, formed in 1853 when John Jacob Bausch, a German immigrant, set up an optical goods shop in Rochester, New York. To fund the ongoing business, he borrowed from his friend, Henry Lomb, also originally from Germany and, as the business grew, the partnership was formed. In the early years, Bausch & Lomb manufactured rubber eyeglass frames as well as a variety of precision optical products. Wishing to enter the microscope business they employed **Ernst Gundlach** in 1876 to head their microscope department. Gundlach, however, was not financially astute, and his employment with Bausch & Lomb ended in 1878. They continued to produce and improve their microscopes and other products. By 1900 microscopes were probably their main product and, by 1903, the firm had been issued patents for microscopes, binoculars, and a camera shutter based on the eye's reaction to light. For details about Bausch & Lomb microscope serial numbers and dates, refer to [**320**].

In 1905 **Fauth & Co.**, by then renamed **Geo. N. Saegmuller & Co.**, merged with Bausch & Lomb, and became Bausch, Lomb, Saegmuller Co. In 1907 the name was changed back to Bausch & Lomb. In 1908 **Zeiss** formed a commercial agreement with Bausch & Lomb and, shortly after this, the Bausch & Lomb, Zeiss, and Saegmuller subsidiaries merged to form the Bausch & Lomb Optical Company, which became known as the **Triple Alliance**, and lasted until WWI.

In 1936 Bausch & Lomb created the *Ray Ban* brand of sunglasses. They continued to produce sunglasses under this brand, including the iconic *Wayfarer* and *Aviator* models, until they sold the brand in 1999 to the Italian Luxottica Group.

During WWII Bausch and Lomb binoculars were made for military use. Binoculars were also made to Bausch and Lomb patterns by companies such as **Westinghouse**, **Nash-Kelvinator**, and **Universal Camera Corp**. These Bausch & Lomb pattern binoculars carried US Military designations such as M3, M8, M9, and M13 (all 6x30), and others.

In 1987 the Optical Systems Division of Bausch & Lomb was acquired by the **Cambridge Instrument Co. Ltd.**, and eventually become part of **Leica Microsystems**. In 2007 Bausch & Lomb was acquired by a private equity company, and in 2013 it was sold to the Canadian company Valeant Pharmaceuticals. Their core business became eye health care products and equipment.

BBT Krauss
The French company BBT Krauss, [**106**], was formed in 1935 by a merger between **Barbier, Bénard, et Turenne** and **Société des Etablissements Krauss**. Instruments bearing their name include prismatic binoculars and photographic camera lenses. Some binoculars bearing the name BBT Krauss also bear the name **Huet**. This was either a collaboration or some other arrangement between the two companies.

B.C. & Co.
See **Broadhurst, Clarkson and Co.**

Beale, John
John Beale, (fl.1805-at least 1830s) [**18**] [**53**], was a mathematical philosophical and optical instrument maker of London, England. Addresses: 76 Maid Lane, Southwark (1805); 17 Alfred Place, Camberwell (1811); Southampton Street, Clerkenwell (1822). His trade card, bearing the Maid Lane address, shows illustrations of his products, including spectacles, a telescope on a library stand, and an electrical machine. **James Prentice** served his apprenticeship with Beale, probably from the late-1820s to the early-1830s, prior to emigrating to the USA in 1842 and setting up his own business.

Beard, R.R.
Robert Royou Beard, sometimes known as Robert Royon Beard, (1856-1932) [**62**] [**109**] [**319**] [**320**] [**335**], was a scientific instrument maker and lantern accessory manufacturer of London, England. Addresses: 62 Alscot Road, Bermondsey (1891-1893), and 10 Trafalgar Road (1894-1932). He served his apprenticeship with a company of brass-finishers, Oakley's, who produced parts for magic lanterns. He began his business as a lantern accessory manufacturer in 1882, and in 1886 he was awarded a patent for his *Eclipse* universal self-centring

slide-carrier. In 1888 he was awarded a patent for a gas regulator for limelight lamps.

In 1895 he published a catalogue, entitled *An Illustrated Catalogue of Scientific Instruments and Optical Lanterns for Scientific and General Purposes, Oxy-hydrogen Apparatus, Patent Microscopes for Lantern Projection and Table Use*. Soon after the **Lumière** brothers introduced their Cinematograph in 1896, Beard began production of the Beard Cinematograph. The Census entries list his profession as: electrical engineer (1881); engineer (1891); scientific instrument maker and employer (1901); scientific instrument maker, cinematographs and accessories (1911).

His company R.R. Beard Ltd. continued after his death, manufacturing projection equipment, and stage and film lighting equipment. The company merged with Photon Lighting Ltd and, in 1988, was incorporated as Photon Beard Ltd. They continued in business until 2020 when they ceased operations due to a downturn in business as a result of the Covid-19 pandemic.

Beavess, Edward
Edward Beavess, otherwise known as Beaviss, (fl.1729-1762) [18] [322], was an optical, mathematical, and philosophical instrument maker of London, England. He traded two doors down from the Brown Bear in Seacoal Lane, Snow Hill. He was also an engine maker, clock maker, and lock maker, and is known to have sold: orrery; balance.

Beck & Söhne, Christoph (CBS)
Georg Christoph Beck founded Christoph Beck & Söhne, in Kassel, Germany, in 1892 producing opera glasses, binoculars, magnifiers and microscopes. Their *Tordalk* model binoculars, 11x80 (introduced in 1963) and 22x80, were popular, premium quality products. Their binocular bodies were made from an aluminium/magnesium alloy, and were particularly lightweight. Other binocular models included 8x32, 7x50, 10x56, 25x80. The company continued in business until 1985.

Beck Apertometer
See apertometer.

Beck, Conrad
Conrad Beck, (1864-1944) [18] [285], was the son of **Joseph Beck**, and served an apprenticeship with him beginning in 1879. He served as President of the **Scientific Instrument Manufacturers' Association of Great Britain** from its inauguration in 1915 until 1922. He was elected to the Quekett Microscopical Club in 1884, and as a fellow of the Royal Microscopical Society in 1885. He was the author of *The Microscope, A Simple Handbook*, published in 1921, *The Microscope, An Advanced Handbook*, published in 1924, and *The Microscope, Theory and Practice*, published in 1938 – all published by **R.&J. Beck**.

Beck, Joseph
Joseph Beck, (1821-1891) [18] [264], of Stamford Hill, London, England, was the son of Richard Low Beck, the business partner and nephew of **Joseph Jackson Lister**. He served an apprenticeship with **William Simms (II)** beginning in 1846, and was freed of the **Goldsmiths' Company** in 1853. He was a microscope maker, and the younger brother of **Richard Beck** of **Smith & Beck**. In 1857 Joseph joined the partnership, which became **Smith, Beck & Beck**. This partnership continued until 1866 when **James Smith (I)** left. The two Beck brothers then reorganised and continued to trade together as **R.&J. Beck**. Joseph Beck's son, **Conrad Beck**, served an apprenticeship with him beginning in 1879. **Thomas Smithies Taylor** also served an apprenticeship with Joseph Beck, beginning in 1879.

Beck, R.&J.
R.&J. Beck, London Ltd., [8] [29] [44] [264] [306], of 69 Mortimer Street, London, England, and later, 68 Cornhill Street, were opticians and optical instrument makers. Their manufactory was Lister Works, Holloway Road, London. The partnership was founded in 1866 by the brothers **Richard Beck** and **Joseph Beck**, following the departure of **James Smith (I)** from their previous partnership, **Smith, Beck & Beck**. Richard Beck, however, died later that year. Following his brother's death Joseph formed a new partnership with **Charles Coppock** and Robert Kemp. The new partnership continued under the existing name of R.&J. Beck. Over the next decades they established a strong reputation for their microscopes, telescopes, cameras and lenses. Some cameras bearing the **Ensign** brand have lenses marked *Ensign Beck*, and some **Thornton Pickard** *symmetrical* lenses are marked *Thornton Pickard-Beck*. T.E. Lawrence (1888-1935), otherwise known as Lawrence of Arabia, used a Beck plate camera.

During WWI, R.&J. Beck's primary contribution to the war effort was gun sights, but they also produced microscopes, telescopes, army trench periscopes, eye test glasses for opticians, camera lenses, cameras, and other instruments. In 1915 they received an order for 2500 pairs of x6 prismatic military binoculars, but only a tiny number were delivered. In 1919 the company was listed as a member of the **British Optical Instrument Manufacturers' Association**. In 1947 Beck exhibited at the British Industries Fair. Their catalogue entry, [450], listed them as manufacturers of: Microscopes and accessories; Spectroscopes; Photographic lenses; Opaque projectors; Sound recording apparatus; Optical units; Lenses; Prisms; Flats in glass, quartz, Iceland spar and fluorspar; Magnifiers; Specialised optical instruments.

In 1968 the company became a subsidiary of **Ealing Corporation** of USA. Subsequently, through a series of changes of ownership and acquisitions, the company finally became privately owned UK company Beck Optronic Solutions, specialising in precision optical solutions in infrared imaging, security, defence, manufacturing & industrial, medical & scientific, and simulation.

For R.&J. Beck catalogues, see [57] and [334]. For details of microscope serial numbers and dates, refer to [320].

Beck, Richard

Richard Beck, (1827-1866) [18] [264], of 6 Coleman St., London, England, was the son of Richard Low Beck, the business partner and nephew of **Joseph Jackson Lister**. He was a microscope maker, and served his apprenticeship with **James Smith (I)**. In 1847, at about the time he would have completed his apprenticeship, he entered a partnership with James Smith (I), trading as **Smith & Beck**. In 1857 Richard's younger brother **Joseph Beck** joined the partnership, which became **Smith, Beck & Beck**. This partnership continued until 1866 when Smith left. The two Beck brothers then reorganised and continued to trade together as **R.&J. Beck**, but Richard Beck died later that year, aged only 39.

Becker & Co., F.E.

F.E. Becker & Co., [44] [53], of Hatton Wall, Hatton Garden, London, England, were makers of scientific instruments and laboratory equipment. Addresses: 33-37 Hatton Wall (1900-c1910); 17, 19, 21, 23, 25 and 27 Hatton Wall (1912-1920s); Nivoc House, 17-29 Hatton Wall (1930s).

The company was founded in 1872, and in 1897 was taken over by **W.&J. George Ltd.** of Birmingham. F.E. Becker continued to trade under their own name until the 1950s when the companies became **W.&J. George and Becker**.

They exhibited at the 1937 British Industries Fair. Their catalogue entry, [449], listed them as offering laboratory equipment for all branches of industrial research and routine tests, and as makers of: Scientific instruments, including *Nivoc* brand analytical balances; Lamp-blown glassware; Thermometers; Apparatus for gas, tar, and oil testing.

Instruments bearing their name include: brass microscope with serial number, and engraved *F.E. Becker & Co.*; Laboratory sextant in case with label that reads *F.E. Becker & Co. W.&J. George (London) Ltd. Proprietors. Hatton Wall London*. A similar label, with the same wording, is in the case of a measuring (travelling) microscope at the Science Museum, London.

Becker, August

August Becker, (fl.1874-1906) [374], was an astronomical, surveying, and physical instrument maker, and microtome maker of Göttingen, Germany. August Becker purchased the workshop of **Moritz Meyerstein** in 1874, and continued the business, also supplying instruments to the observatory at Göttingen. The company was purchased in 1906 by the precision balance manufacturer, **Florenz Sartorius**.

Beier; Beirette

Beier, [347], was founded in 1923 as Freitaler Kameraindustrie Beier & CO., in Freital, near Dresden, Germany. Initially the company produced wooden plate cameras, and in 1931 they introduced their first 35mm camera. In 1934 the company became Kamera Fabrik Woldemar Beier Freital II. The first *Beirette* model of camera was introduced in 1939. During WWII Beier devoted its production to the war effort, and at the end of the war, the Soviet Union dismantled and removed the assets of the manufacturing plant. Despite this, the factory resumed work in 1945, with its first post-war camera model introduced in 1949. In 1972 the company was nationalised to become VEB Kamerafabrik Freital. In 1976 the company became a part of VEB **Pentacon**, which, in 1985, became a part of VEB **Carl Zeiss** Jena.

Beilby, Richard and Charles

The brothers Charles and Richard Beilby, [3] [18] [319], began trading in 1808 as scientific, optical, mathematical, and philosophical instrument makers at 2 Clare Street, Bristol, England, succeeding **Joshua Springer** at that address.

In an advertisement in the *Bristol Weekly Intelligencer* dated 1808 they stated: "R.&C. Beilby (Successors to Mr Springer) respectfully inform the proprietors and captains of ships, that they have a large and well assorted stock of Compasses, Quadrants, Telescopes, and other Musical instruments, which may be depended upon as correct and good. N.B. Any of the above instruments repaired with accuracy." (Note the curious, but correctly transcribed, reference to musical instruments).

One of their products was the kaleidoscope invented by **Sir David Brewster**, which they claimed to have an exclusive licence to manufacture.

The firm continued as a partnership until 1813. Charles Beilby then ran the firm under his own name from 1814 until 1819. In 1818 he advertised in **William Withering**'s botanical textbook, "The Botanical Microscope invented by Dr. Withering, which is more portable and convenient than any other, is now manufactured by Mr. Beilby, optician, Clare Street, Bristol."

By 1820 the firm was operating under the name of Richard Beilby. **John King (I)**, who had been their foreman, took over the premises at 2 Clare St. upon their retirement in 1821.

Beilby, Thomas

Thomas Beilby, (c1787-1860) [3], of York, England, became a partner in the firm of Charles and Luke Proctor. The firm is recorded as **Proctor and Beilby** in 1788 in Birmingham, and in 1800 in Sheffield. He married Deborah Proctor in 1811, and the marriage certificate gives his age as 24.

Belca

See **Balda**.

Bell and Howell

Bell and Howell was founded in 1907 in Chicago, Illinois, USA, manufacturing motion picture projectors. In 1909 they began producing 35mm movie cameras, and by 1934 they were producing 8mm movie cameras for the amateur market. This business continued alongside their automated mailing machine products, and in 1961 they partnered with Canon to enter the still photography market, selling Canon manufactured cameras under their own brand-name. They also produced a number of other models of still camera, including stereo cameras made in Germany. In 1970 they ceased production of movie cameras. They had an Electronics and Instrumentation Division in Basingstoke, England.

They also produced microfilm products, consumer binoculars, and telescopes for gun-sights and tank-sights. The production dates for their optical instruments are unclear. Bell and Howell movie cameras and projectors were marketed in the UK under the brands **G.B. Bell and Howell** and **Bell and Howell Gaumont**.

In 2000 Bell and Howell began to focus on the information technology industry, and there followed sales of their non-core businesses, and various changes in organisation and ownership.

Bell and Howell Gaumont

Bell and Howell Gaumont was a subsidiary of the **Gaumont British Picture Corporation**, distributing **Bell and Howell** movie cameras and projectors in the UK under its own brand-name.

Bellingham and Stanley

Bellingham and Stanley, [29], of 71 Hornsey Rise, London, England, was founded in 1914 by Leonard Bellingham and Frank Stanley for the development of high-quality optical instruments. In September 1915 the Ministry of Munitions ordered a number of prismatic binoculars, of which a small number were delivered prior to cancellation of the order.

At the 1947 British Industries Fair their catalogue entry, [450], listed them as manufacturers of: Sugar and oil testing saccharimeters; Polarimeters and refractometers; Polarimeter tubes; Colour testing apparatus; Spectrographs for analysis; Spectrometers; Photometers; Research apparatus; Prisms and lenses in glass and other materials.

In 2007 the company was acquired by Nova Analytics who were, in turn, acquired by ITT Corporation in 2010. In 2011 ITT Corporation was divided into three companies, with Bellingham and Stanley in Tunbridge Wells, Kent, England, producing polarimeters and refractometers as a subsidiary of Xylem Inc.

Beltex Optics

Beltex Optics is a company in Belarus, founded in 1991, that makes optical instruments marketed under the name **Yukon**.

Belthle, Christian Friedrich

Christian Friedrich Belthle, (1829-1869) [47] [374], was an employee of **Carl Kellner**. Upon Kelner's death in 1855, the company was run by **Louis Engelbert** until 1856, when C.F. Belthle, together with Kellner's widow, began to run the company. In 1865 Ernst Leitz became a partner, and in 1869, upon the death of Belthle, Leitz took control of the business, running it under the name **Leitz**.

Bencini, Antonio

Antonio Bencini, [348], opened a woodworking shop after WWI in Florence, Italy, where he made cameras. In 1920 he entered a partnership, forming the company **FIAMMA** (Fabbrica Italiana Apparecchi Macchine Materiali Accessori), making box cameras and professional wooden cameras. FIAMMA was acquired by **Ferrania** in c1935, and Bencini moved to Turin. There he entered a new partnership, forming the company **Filma**, making cameras. In 1937, Filma was also taken over by Ferrania.

Antonio Bencini then founded the company ICAF, in Milan, making simple and affordable cameras, with their first cameras produced c1939. In a period of successful growth, the company name was changed several times, to CMF (late 1930s), CMF Bencini (1946), and Bencini SpA (early 1950s). In the mid-1980s the company's product distribution was taken over by Cafer Ltd., and they continued production of Bencini cameras until the late-1980s.

Bénèche, Charles Louis

Charles Louis Bénèche, (1826-1901) [**172**], of 55 Grossbeerenstrasse, Berlin, Germany, was a microscope maker. In about 1850 he entered a partnership with **Rudolph Wasserlein**, trading as **Bénèche & Wasserlein**. The partnership continued until about 1860, after which Bénèche continued his business alone.

Bénèche & Wasserlein

Bénèche & Wasserlein, of Berlin, Germany, was a partnership between the two microscope makers, **Charles Louis Bénèche** and **Rudolph Wasserlein**. The partnership was formed in about 1850 and terminated in about 1860, after which both microscope makers continued their respective businesses.

Bennett, John (I)

John Bennett (I), sometimes spelled Bennet, (fl.1735-1770 d.1770) [**18**] [**276** #332] [**358**], was a mathematical, philosophical, and optical instrument maker of *the Globe* in Crown Court between St. Ann's, Soho and Golden Square, London, England. Sometimes this address appears as Crown Court, Little Pulteney St. He is also recorded as trading at *the Queen's Head*, in Crown Court, Knaves Acre (1746-1747). He served his apprenticeship with the mathematical instrument maker Thomas Franklin, beginning in 1723, and was freed of the **Stationers' Company** in 1731. Among his apprentices was **James Simons**, beginning in 1757. Another was the mathematical instrument maker **James Search**, who began his apprenticeship in 1764, and was turned over to the tin plate worker Peter Balchin upon John Bennet (I)'s death in 1770.

In 1764 John Bennett (I) was one of the petitioners who attempted to revoke the patent obtained by **John Dollond (I)** for the achromatic lens, and which was enforced after Dollond's death by his son, Peter (See Appendix X).

John Bennett (I) was awarded royal appointments as instrument maker to the Duke of Gloucester, and the Duke of Cumberland. He appears in three trade cards in the Science Museum collection, [**53**] [**361**] [**383**]:

(1) Card dated c1735-1745: John Bennett, at *the Globe* in Crown Court between St. Ann's, Soho and Golden Square, London. The card illustrates instruments including an octant, sundials, telescope, barometer, anchor, universal ring dial, terrestrial globe, spectacles, pocket globe, rules, and dividers.

(2) Card dated c1753-1770: John Bennet, at *the Globe* in Crown Court, St Ann's, Soho, London. "Mathematical, Philosophical and Optical Instruments made and repaired in the best manner. Instrument Maker To their Royal Highnesses The Duke of Gloucester and Duke of Cumberland. Also the new Nautical Almanack by authority of ye Commissioners of Longitude." The card illustrates instruments including a theodolite, microscope, globe, telescopes, barometers, thermometers, and a plane table with alidade.

(3) Undated card: John Bennett, at *the Globe* in Crown Court, between St. Ann's Soho & Golden Square, London. "Makes and sells all sorts of mathematical, philosophical and optical instruments, of various materials, according to the newest improvements..." The card illustrates instruments including a theodolite, terrestrial globe, orrery, microscope, telescope, waywiser, sectors, rules, and camera.

Instruments bearing his name include: Gregorian telescope on pillar and tripod; telescope with double prism eyepiece; sextant, 16-in radius; theodolite; pantograph; pair of globes; circumferentors; stick barometer. He was succeeded in business by James Search.

Bennett, John (II)

John Bennett (II), (fl.1745-1768, d.1770) [**8**], was an instrument maker, at 28 Medway, and 395 Strand, London, England. (The Strand address was later used by **Patrick Adie**). It is possible that he is the same person as **John Bennett (I)**.

Bennett, Leonard

Leonard Bennett, (fl.1799-1826) [**18**] [**276** #1277], was a mathematical and optical instrument maker of London, England. Addresses: 2 Queen's Square, Aldersgate St. (1799); 26 Charles St., Hatton Garden (1802-1826).

Reference [**2**] notes a possible connection between a Leonard Bennett and **Jesse Ramsden**, and [**373**] indicates that someone named Bennett was a pupil of Jesse Ramsden, and helped to train **William Simms (II)**. [**234**], similarly, states that William Simms (II) "was apprenticed to a mathematical instrument maker called Bennett (who had been one of Ramsden's workmen)".

Bennett, Thomas

Thomas Bennett, (fl.1810-1867) [**34**] [**215**] [**304**], was an optician and manufacturer of mathematical, optical and philosophical instruments of Cork, Ireland. Addresses: Patrick St. (1810-1817); 2 Patrick St. (1820); 65 Patrick St. (1824); 45 Patrick St. (1826-1828); 124 Patrick St. (1844-1867). Examples of his work include: three-draw mahogany-tubed telescope, 170mm closed, 370mm extended; marine and domestic barometers; dumpy levels; ships' binnacles; compasses; thermometers; vacuum gauges; steam gauges; electrical apparatus. Other instruments he is known to have sold include: octant; quintant; sextant; Sikes hydrometer; clinometer. Thomas Bennett was succeeded in his business by **Reynolds & Wiggins** at the same address.

Benoist Berthiot
In 1836 Louis Berthiot, [192], started an optical glasswork in the Champagne region of France. In the mid C19th F. Benoist joined the business, which was renamed Benoist Berthiot. As well as optical glass the company made projection lenses, specialised camera lenses, and eyeglasses. The company became a subsidiary of the eyeglass manufacturer Essilor. Benoist Berthiot has no relationship with **Claude Berthiot** or his companies.

Berge, John
John Berge, (1751-1808) [1] [2] [18] [319], was an optician of London, England. Addresses: At Mr. Dollond's, St. Paul's Chruchyard (1773); Johnson's Court, Fleet Street (1790-1793); 3 Crane Court, Fleet St. (1794-1803); Crane Court, Fleet Street (1796); 26 Lower Eaton St., Pimlico (1804-1808).

He was the older brother of **Matthew Berge**. He served his apprenticeship with **Peter Dollond**, beginning in 1765, and was freed of the **Spectacle Makers' Company** in 1773. He was the first in the records of **Dollond & Co.** to appear with the title *Optician*, and he continued to work for the company until 1791. Note that [276 #571] appears to confuse John Berge with Matthew Berge [276 #1083].

Berge, Matthew
Matthew Berge, (c1753-1819) [2] [18], was an optician, and optical and mathematical instrument maker of 199 Piccadilly, London, England. He was **Jesse Ramsden**'s foreman, ensuring that any instruments not built by Ramsden himself were satisfactorily completed, and repairs and refurbishments were carried out. Ramsden taught Matthew Berge his method of hand-dividing circles, a method he kept largely to himself. Matthew was the younger brother of **John Berge**.

Berge took control of the business and premises, and presumably the workforce, following Ramsden's death in 1800. He completed the instruments that Ramsden had personally been building, and began to trade as "Matthew Berge successor to the late J. Ramsden, 199 Piccadilly, London". Shortly after this the street was re-numbered, and the address became 196 Piccadilly. He produced a one-page catalogue listing a large number of optical, mathematical and philosophical instruments made and sold by his firm. Berge had only one registered apprentice, **Nathaniel Worthington**. Among his employees was also **James Allan (II)**.

In 1806 Berge took over from **Dudley Adams** as the sole supplier of mathematical instruments to the Board of Ordnance. The optical instruments, levels, sextants, and other instruments produced under Berge's management were usually signed *Berge Late Ramsden*. There are also instruments signed *Berge London Late Ramsden*, *Berge*, *M Berge London* or *Matthew Berge London*. Matthew Berge ran the workshop in Piccadilly until his death in 1819. Upon his death **Robert Brettell Bate** took over Berge's position as mathematical instrument maker to the Board of Ordnance. Matthew Berge was succeeded in business at 196 Piccadilly by Nathaniel Worthington who, shortly afterwards, entered a partnership with James Allan (II), trading as **Worthington & Allan**.

Berger, C.L.
Christian Louis Berger, (1842-1922) [58] [210], of Stuttgart, Germany, served an apprenticeship with a local surveying instrument maker, Christian Saeger, after which he worked in instrument shops in England and Germany before emigrating to the USA in 1866. There he worked for E.S. Ritchie & Son in Boston, makers of compasses for marine navigation, and for John Upham, before entering a partnership with **George Louis Buff** in 1871, and trading as **Buff and Berger**. The partnership continued until 1898, following which Christian Berger traded as C.L. Berger & Sons, producing instruments for engineers and surveyors, as well as astronomical telescopes and transits. The business began to decline in the mid-C20th, and in 1995 it was acquired by the Chicago Steel Tape Company.

Berry & Mackay
Berry & Mackay, [262] [263], were nautical instrument makers of 59½ Marischal St., Aberdeen, Scotland. **James Berry**, a clockmaker and optical instrument maker, started his business in 1835, and Alexander Spence Mackay (d.1914) became a partner in 1879, forming Berry & Mackay. In 1880 the company relocated to 65 Marischal St. The company remained in business until 1975.

According to a 1951 advertisement [44]: "Berry and Mackay. Marine Opticians and M.O.T. certified compass adjusters. Chronometers and clocks repaired, cleaned and rated, Chart agents, Sextants, Binoculars, Barometers, Gauges and patent logs repaired and tested. Prompt attention."

Berry, James
James Berry, (1808-1890 fl.1835-1878) [18] [263] [271] [322], was an optical and nautical instrument maker, and chronometer and watch maker, of Aberdeen, Scotland. Addresses: 52 Castle St. (1835-1852); 53 Marischal St. (1852); 88 Union St. (1853-1856); 59½ Marischal St.

(1866-1878). He served his apprenticeship with William Spark, a watch and clock maker and, in 1837, became a Freeman of the craft guild, the Hammermen Incorporation, one of the incorporated trades of Aberdeen.

Following the completion of his apprenticeship, he started his own business as a watchmaker and jeweller in Kincardineshire. In 1835 he moved his business to Aberdeen. By 1843 he was also making nautical instruments. By 1866 he was listed as a chronometer maker, nautical instrument maker, and optician. He also stocked chronometers and other nautical instruments by other makers, and it is not clear to what extent he made the instruments he sold.

James Berry worked in partnership with his son, George Allan Berry, until 1865, trading as **James Berry & Son** then, in 1879, he entered a new partnership, this time with Alexander Spence Mackay, trading as **Berry & Mackay**. The company continued under the same name after James Berry's death in 1890.

Berry, James & Son

James Berry & Son was a partnership between **James Berry**, and his son, George Allan Berry (b.c1833). It is not clear when James began to work in partnership with his son but, in 1857, their partnership, trading as James Berry & Son at 88 Union Street, Aberdeen, Scotland, was terminated [**170** #6743 p926]. It appears that in the same year they formed a new partnership, also trading as James Berry & Son. The business moved to 59½ Marischal St., and George took on the additional premises at 29 St. Nicholas St. This partnership continued until 1865 [**170** #7584 p1311].

Bertele, Ludwig Jakob

Ludwig Jakob Bertele, (1900-1985) [**40**], was a German designer of optics. In 1919 he started working for the **Ernemann** company where his innovative lens designs earned him a strong reputation. In 1926, when the Ernemann company was taken over by **Zeiss Ikon**, he continued his work with Zeiss.

Berthiot, Claude

Claude Berthiot, (1821-1896/7) [**192**], a French optical instrument maker, began his business in 1857 making photographic lenses. In 1884 he was joined in his business by his nephew Eugène Lacour, and the name was changed to **Lacour-Berthiot**. After Claude Berthiot died in 1896 or 1897, Eugène Lacour continued to run the business. Claude Berthiot has no connection with **Benoist Berthiot**.

Berthiot, Louis

See **Benoist Berthiot**.

Bertram, Ernst & Wilhelm

Ernst & Wilhelm Bertram, of Munich, Germany, was founded in 1919 by Ludwig Leiner and Ernst Bertram as Leiner and Bertram. They began with making camera shutters, and in 1928 Leiner left the company, at which time they began to make light meters. They became E.&W. Bertram some time before WWII. During WWII, the products they made for the military were marked with the **ordnance code** kln.

Bertrand Lens

A Bertrand lens is a specialised converging lens, used in a microscope, which acts as a relay lens to bring the back focal plane of the objective into focus at the eyepiece. This is generally built into the microscope eyepiece tube, and allows the objective back focal plane to be observed through the microscope eyepiece. It may be used in one of two applications:

In a polarising microscope, it may be positioned between the analyser and eyepiece, and brings an interference pattern, formed at the back focal plane of the objective, into focus at the image plane of the microscope.

In a phase contrast microscope, a Bertrand lens may be positioned between the objective and eyepiece, and used to facilitate the accurate alignment of the condenser annulus with the phase plate in the objective. Where a Bertrand lens is not used for this purpose, the same task may be achieved by means of a phase telescope.

The lens was first introduced in 1878 by the French mineralogist Emile Bertrand (1844-1909), and was an adaptation of an earlier design by **Giovani Battista Amici**, introduced in 1844.

Bézu, Hausser, and Company

Messrs. Bézu and Hausser, [**50**] [**172**], both worked in the role of foreman for **Hartnack and Prazmowski** at 1 Rue Bonaparte, Paris, France. They took over the company from Adam Prazmowski in 1883, and produced microscopes under their own name until 1896, when the company was acquired by **Nachet et Fils**.

Bidstrup, Jesper

Jesper Bidstrup, (1763-1802) [**2**] [**3**], was a Danish optical, mathematical and philosophical instrument maker. He served his apprenticeship with the Swedish immigrant, Johannes Ahl (1729-1795), in Denmark and was awarded a university grant, together with a royal grant, to travel to London and learn the instrument trade. This was with the objective of establishing an indigenous instrument trade in Denmark. Unable to obtain a position with **Jesse Ramsden** or any other premier instrument maker he took a position with "Mr. White". In 1789 he went on from there to work for a workshop run by a former apprentice of Ramsden, Mr. Higgins, who was a subcontractor for **Nairne & Blunt** and **George Adams (II)**. By the time he left the employment of Higgins in 1790 he had a good collection of tools, models and machines, but was in debt as a result. He asked his Danish sponsors for more money to enable him to acquire the remaining tools, models and machines he needed in order to return to Denmark. In 1793 he set up his business in St. Martin St., Leicester Square, London, and published his eight-page *Catalogue of Optical, Mathematical & Philosophical Instruments* [200]. Most of these instruments, however, were almost certainly not made by him.

In 1798, heavily in debt to the Danish government, he smuggled his equipment back to Denmark and set up his workshop there, only to die four years later. His workshops, tools and belongings were declared to be state property, and the business placed in the hands of his fellow Dane **Jeppe Smith**. Smith continued the business successfully, but without the first-hand knowledge gained by Bidstrup.

Bildt Jr., Johannes van der
Johannes van der Bildt Junior, (1736-1780) [256] [378], was a telescope maker of Franeker, in Friesland, Netherlands. He was the eldest son of **Jan Pietersz van der Bildt**, and served his apprenticeship with his father. He moved a short distance away to Bolsward in 1764, and returned to Franeker in 1767. He made principally Gregorian telescopes.

Bildt, Bauke Eisma van der
Bauke Eisma van der Bildt, (1753-1831) [256] [378], was a telescope maker of Franeker, in Friesland, Netherlands. He was the grandson of **Jan Pietersz van der Bildt**, from whom he learned his telescope making skills. As with the other telescope makers in the family, he made principally Gregorian telescopes. In 1787 he moved to Buiksloot, near Amsterdam where he made telescopes and spectacles. In 1806 he returned to Franeker for a position at the university, where he worked for the rest of his life.

Bildt, Jan Pietersz van der
Jan Pietersz van der Bildt, (1709-1791) [256] [378], was a scientific instrument maker and telescope maker of Franeker, in Friesland, Netherlands. He trained as a carpenter, and went on to give physics lectures at the court of Prince William IV of Orange. In the mid-1740s he began to make scientific instruments and telescopes, and established a reputation as an eminent telescope maker, training a number of apprentices. He mostly built Gregorian telescopes. A telescope, made by him, which can be used as either a Gregorian or Cassegrain telescope, is described in [387]. His sons **Johannes van der Bildt Junior**, and **Lubbertus van der Bildt** both served their apprenticeships with him, as did his grandson, **Bauke Eisma van der Bildt**, and continued the business after his death.

Bildt, Lubbertus van der
Lubbertus van der Bildt, (1738-1778) [256], was a telescope maker of Franeker, in Friesland, Netherlands. He was the second son of **Jan Pietersz van der Bildt**, and served his apprenticeship with his father. He made principally Georgian telescopes, and signed his speculum mirrors LVDB on the reverse side. [378] indicates that he also made microscopes.

Billcliff Camera Works; Billcliff, Joshua
Joshua Billcliff (1820-1899), [44] [61] [306], was a cabinet-maker who established his camera-making business in Manchester, England, in 1860. Addressess: 56 Stretford Rd. (1861-1872); 62 Devonshire St., Hulme, with works at 1 Perry St. (1873-1874); 93 Coupland St. with works at 1 Perry St. (1875-1881); 27 Richmond St, Boundary Lane (from 1882).

In the last two decades of the C19[th] he advertised as "wholesale and retail manufacturer of all types of photographic apparatus: cameras – including copying, enlarging and process types – plate boxes, tripods, stands, etc." The company made cameras for companies in the Manchester area, including the **Thornton-Pickard Manufacturing Co. Ltd.**, **William Chadwick**, and **J.T. Chapman**.

Following the death of Joshua Billcliff, the company was operated by his sons, initially under the name Executors of J. Billcliff, and changing the name to Billcliff Camera Works in 1907. The firm continued to operate until at least 1943.

Bilora
Bilora was a brand-name used by **Kürbi & Niggeloh** for their camera products, (*Bilora Bella*, *Bilora Boy*, *Bilora Bonita*, etc.), between 1935 and 1975. In 1994 Kürbi & Niggeloh split into two independent companies, Kürbi & Niggeloh BILORA GmbH, producing plastic products, and Kürbi Otto Toennes GmbH, producing photographic accessories which they continued to sell under the *Bilora* brand-name. In 2016, Kürbi & Niggeloh BILORA was renamed BILORA Kunstofftechnik GmbH, whose core business is tool making, plastic injection moulding, and finishing.

Binocular
Binocular: Of or relating to seeing with both eyes. This can refer to a binocular microscope, telescope or other optical instrument.

Binoculars

The term *binoculars* generally refers to an instrument used for far viewing with two optical paths, one for each eye, giving binocular vision. Different kinds of binoculars include: Galilean; Porro prism; reverse Porro prism; roof prism. Some designs include image stabilisation, night-vision or image intensifiers. [43] [64] [65] [68] [94] [147].

Binocular Telescope: A binocular telescope is a pair of binoculars in which each optical path is constructed as a telescope. The term is often used for large binoculars, sometimes constructed using two reflecting or refracting telescopes. The term is also sometimes used to refer to an early kind of *field glass*.

Field glass originally referred to low magnification non-prismatic binoculars that produced an erect image using an intermediate relay lens, in the same manner as hand-held telescopes or spyglasses. This design results in limited magnification, since higher magnifications require longer tube length. The term *field glass* is, however, sometimes more widely used to describe any binocular or monocular of sturdy design for outdoor use.

Galilean binoculars consist of two Galilean telescopes, joined in such a way that the user may view through one with each eye. Some Galilean binoculars have a central hinge, allowing adjustment of interpupillary distance, and some do not. Opera or theatre glasses are frequently made to this design and, historically, so were field glasses, marine glasses, etc. Porro prism or roof prism binoculars have replaced Galilean binoculars for most modern applications.

Porro prism binoculars use a double Porro prism in each optical tube to provide an erect image to the viewer. The orientation of the prisms is generally such that the objectives are spaced further apart than the oculars, thus increasing the stereo vision effect.

Reverse Porro prism binoculars are more compact since the orientation of the prisms is such that the objectives are closer together than the oculars.

Roof prism binoculars use a roof prism in each optical tube to provide an erect image to the viewer. The roof prisms are generally of the Schmidt-Pechan or Abbe-König design, in which the exit beam is coaxial with the entry beam. Thus, the objectives are at the same separation as the oculars. This restricts the maximum objective size to less than about 60mm, being the distance between the user's pupils. Binoculars with larger objectives are usually of the Porro prism design.

Compact binoculars are generally either reverse Porro or roof-prism design, with an objective size of 25mm or less.

Binocular Microscope

See microscope.

Binocular Telescope

See binoculars.

Biolam

Biolam is a brand of biological microscope produced by the **Leningrad Optical Mechanical Association** (LOMO).

Bio-Pleograph

The Bio-pleograph was an early kind of motion picture projector, devised by the Polish inventor **Kazimierz Prószyński**, [335], in about 1898, shortly after his invention in 1894 of the Pleograph. Like the Bioscop the Bio-pleograph was a double-film projector, alternating between films to increase the frame rate. It was cumbersome, however, and was not a commercial success.

Bioscop (Bioskop)

The Bioscop is an early type of motion picture projector invented in 1895 by the German film-makers, brothers Max and Emil Skladanowsky [335]. The Bioscop was a double-film-band projector system with two lamps and two lenses. Each of these optical paths projected frames from a 54mm film, using a synchronising mechanism to alternate between them, giving the necessary 16 frames per second.

Bioscope

The Bioscope, [160], was co-invented by the American film producer **Charles Urban** and the engineer Walter Isaacs in 1896. It was designed to overcome the perceived problems with the Vitascope marketed by Edison. The Bioscope had no shutter, and used an eccentric-beater movement to provide the stop-go motion of the film. The resulting projected image was smoother than that of the Vitascope, with no flicker. However, the un-shuttered movement of the film resulted in a "rain" effect which, while not desirable, was considered less of a problem than the pronounced flicker. The Bioscope was operated by a handle, so that it did not need an electrical power supply which, at the time, was not reliably available.

Charles Urban moved to the UK to join the Warwick Trading Company in 1897, and his Bioscope was marketed by them as the *Warwick Bioscope*. Urban left the Warwick Trading Company in 1903 to form his own company, and continued to market his invention as the *Urban Bioscope*. The Bioscope was further improved by Cecil Hepworth and Alfred Darling in the UK, and became a popular form of projector for many years.

Bird, John

John Bird, (1709-1776) [2] [3] [8] [16] [18] [24] [44] [46] [256] [276 #232] [303] [396] [397] [443], was a maker of scientific instruments, born in Durham, Ireland. In 1740 he began to work for **Jonathan Sisson** in London, England, dividing standard measures. Sisson was an ex-employee of **George Graham**, for whom John Bird also worked prior to setting up his own business.

By 1745 he was established in his own workshop at the *Sea Quadrant*, Court Gardens, Strand, London, England, making machine tools and small mathematical instruments. This address is also recorded as the *Sea Quadrant* near the New Exchange Buildings in the Strand. He had a workshop in Berwick St. (1771), and records list him at Little Marybone St. (sic) (1772), and 29 Little Marylebone St., Marylebone (1776).

He made a dividing engine that is believed to be the first to compensate for changes in its operating temperature. He was highly skilled at dividing scales, and was soon in demand making large astronomical instruments for international observatories. In the 1750s he became the first to use a tangent screw in the adjustment of the index arm of a navigational instrument, first using it on a reflecting circle. He made equipment for the Royal Observatory at Greenwich, including an all-brass 8-foot mural quadrant, and similarly supplied equipment for observatories in St Petersburg, Cadiz, Stockholm, and Göttingen as well as others. He worked with **John Stancliffe** producing instruments for the Radcliffe Observatory at the University of Oxford.

Together with Captain John Campbell he devised the marine sextant as an improvement on **John Hadley**'s quadrant.

In 1758 he made the standard yard to be kept in the House of Commons, and he made a duplicate in 1760. Both were lost when the Houses of Parliament were destroyed by fire in 1834. In 1763 the surveyors, Charles Mason and Jeremiah Dixon, used Bird's instruments when surveying the border between Pennsylvania and Maryland in the USA, later to become known as the Mason-Dixon line.

In 1764 he was one of the petitioners who attempted to revoke the patent obtained by **John Dollond (I)** for the achromatic lens, and which was enforced after Dollond's death by his son, Peter (See Appendix X).

In 1767 Bird received £500 in remuneration from the Board of Longitude for training an apprentice and to write an account of his dividing methods. It is not known whether he trained an apprentice, but he wrote of his methods in two publications. *The Method of Dividing Astronomical Instruments* in 1767, and *The method of constructing mural quadrants. Exemplified by a description of the brass mural quadrant in the Royal Observatory at Greenwich* in 1768. In 1772 he moved to larger premises in Little Marylebone Street, where he continued to work until his death in 1776.

Bird-Jones Telescope

See telescope.

Birefringence

Birefringence is a property of certain crystalline materials, such as calcite or Iceland spar, in which the refractive index differs for different crystallographic directions, resulting in two different indices of refraction. This property can be used to produce a polarizing prism such as the Wollaston prism or the Nicol prism.

Bithray & Steane

Bithray & Steane, [50], opticians, mathematical and philosophical instrument makers of London, England, was a partnership between **William Bithray** and Thomas Steane. Both partners had been employed by **Thos. Rubergall** and, upon Rubergall's death in 1854, they took over his business, trading as Bithray & Steane.

According to their trade card, [145]: "Bithray & Steane, many years with & successors to Thos. Rubergall, Optician to the Queen, Mathematical & Philosophical Instrument Maker, 27 Coventry Street, London."

The partnership Bithray & Steane was short-lived, since William Bithray died in 1855.

Bithray, Stephen

Stephen Bithray, (c1789-1858) [18] [50] [276 #1491] [322], was an optician, and mathematical and philosophical instrument maker of the Royal Exchange, London, England. He was freed of the **Wheelwrights' Company** in 1826, by which time, he had been working for **Joseph Smith (II)** for some time, possibly since soon after Smith started his business in 1811.

In 1826, following the death of Joseph Smith (II) the previous year, he took over his former employer's premises at 4 N Piazza, Royal Exchange [305], where he ran his own business until 1838. Then, while the Royal Exchange was rebuilt following a fire, he traded at 6 Spread Eagle Court, Finch Lane, Cornhill, and from 1844 he was at 29 Royal Exchange. His nephew, **William Bithray**, served an apprenticeship with him prior to working for **Thomas Rubergall**. Among the instruments Stephen Bithray made were thermometers, pocket thermometers, barometers, sextants, microscopes, telescopes and sundials. After Stephen Bithray's retirement, the premises at 29 Royal Exchange were taken over by **Henry Macrae**.

Bithray, William
William Bithray, (1816-1855) [50], was a nephew of **Stephen Bithray**, with whom he served his apprenticeship as an optician, and mathematical and philosophical instrument maker. He worked for **Thomas Rubergall** until Rubergall's death in 1854. William Bithray, together with Thomas Steane, who was another of Rubergall's employees, then took over the business and operated it in partnership as **Bithray & Steane**. William Bithray, however, died the following year.

Blachford & Imray
Blachford & Imray, [18], were optical, philosophical, mathematical, and nautical instrument makers, and chart sellers, of 116 Minories, Tower Hill, London, England. The company was a Nautical Warehouse, which was a partnership between **James Imray** and **Robert Blachford**, and operated between 1836 and 1845. They were joined in business by Michael Blachford in 1839. Instruments bearing their name include: octant; large single-draw brass and leather *day & night* telescope; barometers. Upon the termination of the partnership James Imray continued in business at the same address.

Blachford, Robert
Robert Blachford, (fl.1804-1840) [18] [276 #1086], was a chart seller who operated a Navigation Warehouse at 114 Minories, Tower Hill, London, England (1804-1817). He also traded at 137 Minories, Tower Hill (1805). He was the son-in-law of **John Hamilton Moore**, whom he succeeded in business. In 1817 he began to trade as Robert Blachford & Co., at 114 Minories, Tower Hill (1817-1818), then 1 Little Tower St. (1819-1820), and 79 Leadenhall St. (1821-1828). Instruments bearing his name include: telescope; sextant. In 1828 he began to work in partnership with William Blachford, trading as Robert & William Blachford at 116 Minories, Tower Hill. **James Imray** joined the company in 1836, and a new partnership was formed, trading as **Blachford & Imray**.

Blackie, William
William Blackie, (1808-1838) [169] [171] [395], a British optician, was raised by his maternal grandfather, a gardener, whose name he adopted. He began his career as a gardener, with a brief but unfulfilling foray into merchant seafaring. During his work he found a bottle-bottom, and its optical properties inspired in him an interest in lenses. He taught himself to grind lenses, and went on to study at the Leith Mechanics' Institute. He became highly skilled at making both glass lenses and jewel lenses, including diamond and garnet. His skill brought him to the attention of many eminent people, including **Sir David Brewster**, to whom he supplied jewel lenses. Although William Blackie died relatively young, he established a strong reputation for his lens making skill, particularly for microscope lenses.

Blacksmiths' Company
See Appendix IX: The London Guilds.

Blair & Co., H.G.
H.G. Blair & Co., of Cardiff, Wales. According to their 1860 trade card [53] [361]: "H.G. Blair & Co., 95 Bute Street, Cardiff. From Bristol, established 1829. Chronometer makers, Opticians and compass adjusters. Chronometers lent on hire, re-sprung, adjusted and rated. Electric time signals received daily from the Royal Observatory, Greenwich. Admiralty Chart Agents."

Instruments bearing their name include: sextants; marine chronometers. A sextant by **Heath & Co.** has the inscription *Made for H.G. Blair and Co., Cardiff & Barry*.

Blair, Archibald
Archibald Blair, (fl.1827) [18] [271] [276 #802], was an optical instrument maker of 16 Broughton Place, Edinburgh, Scotland, and later in London, England. He was the son of **Robert Blair**, and worked with his father, producing aplanatic telescopes, invented by his father.

Blair Camera Co.
The Blair Camera Company of Boston, Massachusetts, USA, was founded as the Blair Tourograph Company in 1878 by Thomas Henry Blair, collaborating with **American Optical Company** to produce a portable wet-plate camera system. In 1881 the name was changed to Blair Tourograph & Dry Plate Company. In 1886 the name changed again to Blair Camera Company. In the late 1880s they began marketing cameras for the **Boston Camera Company**. They traded in competition with **Kodak** until, in 1899, Kodak acquired the company and moved the manufacturing to Rochester, New York. They continued to produce cameras under their own name until 1907, after which they traded as the Blair Camera Division of Eastman Kodak.

Blair, Robert
Robert Blair, (1748-1828) [46] [276 #802] [388], was a surgeon, inventor, and optical instrument maker of Garvald, Haddingtonshire, Scotland. He served as a surgeon in the Royal Navy, an experience which inspired an interest in navigation and navigation instruments. He devised an improvement to Hadley's quadrant, for which he was rewarded financially by the commissioners of longitude. In 1875 he was appointed regius chair of practical astronomy at the University of Edinburgh. He received a salary for this position, but in his forty-two

years of tenure he gave no lectures, and did little for the University. In 1876 he was elected to the Royal Society of Edinburgh.

His son, **Archibald Blair**, was an optical instrument maker in Edinburgh, and later in London. Robert Blair worked to improve on the achromatic telescope lens, resulting in his design for a triplet of two crown lenses separated by a solution of either antimony or mercury in hydrochloric acid. He worked with his son in London, producing several telescopes with his triplet lens. This invention he referred to as an "aplanat", or aplanatic telescope.

Robert Blair had a change of career in 1793 when he was appointed to the Admiralty's board in a role overseeing the care and custody of prisoners of war.

Blakeney, John W.
John W. Blakeney, (fl.1853-1873) [263] [271], was a nautical instrument maker of Glasgow, Scotland. From 1853 until 1860 he worked in partnership with **Paul Cameron**, trading as **Cameron & Blakeney**. Following this, from 1861, he traded under his own name, retaining the address at 94-96 Jamaica Street (1861-1863). He also took on premises at 25 Turner's Court (1861-1873), which he initially used as a workshop, before consolidating his business there in 1864.

Blazed Grating
Another name for an echelette grating.

Bleuler, John
John Bleuler, (c1756-1829) [18] [50] [53] [276 #687] [383], was an optical, philosophical, and mathematical instrument maker of 27 Ludgate Street, near St. Paul's Churchyard, London, England. He served his apprenticeship with **Henry Raynes Shuttleworth**, being freed of the **Spectacle Makers' Company** in 1779, and serving as Master in 1794, and from 1809-1810. Evidence suggests that he worked for Shuttleworth at 23 Ludgate Street until opening his own shop at 27 Ludgate Street, where he succeeded **Thomas Whitford** in business in 1790. He married Shuttleworth's daughter, Elizabeth, in 1787.

Blunt, Edward
Edward Blunt (fl.1823-1826) [18] [401], of 22 Cornhill, London, England, was the son of **Thomas Blunt (I)**. He served an apprenticeship with his father, beginning in 1813, and was freed by patrimony of the **Spectacle Makers' Compay** in 1825. He worked in partnership with his father, and succeeded him in his business in 1823, trading under his own name. He was succeeded in business at this address by **Thomas Harris & Son**.

Blunt, E.&G.W.
E.&G.W. Blunt, [58], was a New York shop offering nautical books, charts, and instruments, sextants, quadrants, etc.

Edmund and George William Blunt were American specialists in nautical instruments during the mid-19th century. After initially trading in nautical books, charts and instruments, the brothers began to focus on the design of scientific instruments in the mid-1850s. Their early instruments were imported, and at least some came from England.

Blunt, Thomas
Thomas Blunt (I), (fl.1760 d.1823) [8] [18] [276 #577] [383] [397] [401], of 22 Cornhill, London, England, was a mathematical and optical instrument maker. He served his apprenticeship with **Edward Nairne**, beginning in 1760, and was freed of the **Spectacle Makers' Company** in 1771. In 1774, they formed the partnership, trading as **Nairne and Blunt**. The partnership continued until 1793, after which Thomas Blunt traded under his own name.

His trade card, [3], dating around 1800, states: "Thomas Blunt, Mathematical instrument maker to His Majesty, No. 22 Cornhill, London. Telescopes for sea and land, Sextants, Hadley's quadrants, Marine barometers, Azimuth & other compass's, Spectacles & reading glasses, Opera glasses, Theodolites & other surveying instruments, Microscopes, Globes, Thermometers, Electrical machines, Sun dials, Cases of drawing instruments, Rules, Pencils, &c &c."

He was Master of the Spectacle Makers' Company from 1792-1793, and from 1815-1816, and was awarded a royal appointment to King George III in 1785. Thomas Blunt (I) had three sons, all of whom served their apprenticeships with him: Thomas (II), freed 1811, William, freed 1825, and **Edward Blunt**.

Thomas (I) formed various partnerships with his sons, trading under the names T. Blunt, and Blunt and Son. When partnered with his son Thomas (II), they signed their instruments *T.&T. Blunt*. After Thomas (I) died, his son Edward continued the business, trading as E. Blunt.

Boddy, Thomas
Thomas Boddy, (fl.1849) [18], was a mathematical instrument maker of 523 Oxford St., and 3 Great Carter Lane, both in London, England. He served his apprenticeship with **William Peter Piggott** of the **Merchant Taylors' Company**, beginning in 1840, but there is no record of his freedom. He worked in partnership with William Peter Piggott, trading as Piggott & Boddy, opticians and mathematical instrument makers. The partnership was terminated in 1849 [35 #20593 p769].

Boecker, W. Emil
W. Emil Boecker, (1851-1945) [50], was a German microscope maker who worked for the **Leitz** microscope company in Wetzlar from 1869 to 1871. In 1879 he started his own business, and from 1883 to 1885 he worked in partnership with **Gottlieb Fecker**. After this he returned to work for Leitz until his retirement in 1931.

Bogopolsky, Jacques
Jacques Bogopolsky, (1895-1962), was a Ukrainian engineer and camera designer. He began to produce a 35mm combined cine camera and projector in 1934 in Geneva, Switzerland, and created the **Bolex** brand in 1928. In 1930 he sold his company to Paillard, who made it their **Paillard Bolex** division, and he continued to work for the company for about five years. By the late 1930s he was working with the Swiss watch-part manufacturer **Pignons SA**. He designed the *Bolka Reflex* 35mm SLR camera, which was later named the *Alpa-Reflex*, and after which Pignons named their camera brand **Alpa**.

Bogopolsky, at various times in his career, also went by the names Bolsky, Boolsky and Bolsey. After his work with Pignons, he emigrated to the USA. By 1947 he was going by the name Jacques Bolsey, and he founded the **Bolsey** camera company in New York, producing mainly rangefinder and TLR cameras.

BOLCo
See **British Optical Lens Company**.

Bolex
Bolex, [346], was a company established in Geneva, Switzerland by the Ukrainian engineer **Jacques Bogopolsky**. In 1924 he began to produce a 35mm combined cine camera and projector for the amateur market. In 1928 he began to use the Bolex name, with the introduction of a 16mm camera, and a 16mm projector, with further 16mm camera and projector models introduced in 1929 and 1930. In 1930 he sold the company to Paillard, to form their **Paillard Bolex** division.

Bollemeijer & Brans
Bollemeijer Optician is a photographic supplies company in Rotterdam, the Netherlands. A brass and leather three-draw 53mm telescope bearing the name Bollemeijer & Brans was either made by them or retailed by them, perhaps in partnership.

Boller & Chivens
Boller & Chivens, [193], was a manufacturer of high quality professional astronomical telescopes and other optical and mechanical instruments. The company was formed in Pasadena, California, USA in 1946 by Harry Boller (1915-1997) and Clyde Chivens (1915-2008). It began to grow, manufacturing and selling oil well deep hole research instruments, motion picture editing instruments, aerial photo reconnaissance instruments, and other scientific devices.

The company began their work in astronomical instruments in 1956 when they manufactured the slow-motion right ascension and declination drives for the 120-inch telescope at the Lick Observatory. Soon after this they built a 36-inch fork telescope mount and telescope. They continued building large telescopes and mounts until 1965 when the company was taken over by **Perkin Elmer**, and became their Boller & Chivens division. The Perkin Elmer Applied Optics Division ground and polished all of the Boller & Chivens mirrors up to 40 inches.

In the early 1980s Perkin Elmer's Boller & Chivens division relocated to Costa Mesa, California, to merge with their Applied Optics Division. This merged division continued to produce optical instruments including large telescopes and mounts, and finally ceased production of telescopes in the late-1980s.

Bolsey
The Bolsey company, of New York, USA, was a camera company founded in 1947 by Jacques Bolsey, formerly known as **Jacques Bogopolsky** of Ukraine. They mainly produced 35mm rangefinder and TLR cameras, and cine cameras, designed by Bolsey, until his death in 1962. Bolsey cameras were initially manufactured for them by **Pignons SA**, and subsequently by Obex Corporation of Long Island, NY.

Bonanni, Filippo
Filippo Bonanni, (1638-1723) [274] [322], was a Jesuit scholar with a keen interest in microscopy, which included the making of innovative compound microscopes. He pioneered the use of a spring stage to hold the sample slide in position, a concept later adopted by **Nicolaas Hartsoeker** in his screw-barrel microscope. In 1691 he published *Observationes circa Viventia, quae in Rebus non Viventibus Reperiuntur. Cum Micrographia Curiosa*, generally referred to as *Micrographia Curiosa*, detailing his methods, microscopes, and observations, complete with illustrative engravings. He also published a number of works on other topics including, in 1722, *Gabinetto Armonico Pieno d'Istromenti Sonori*, including 152 engravings showing musical instruments from around the world.

Bontemps, Georges
Georges Bontemps, [24] [156], was a French optical glass maker. He was director of the glassworks in **Choisy-le-Roi**, when it was founded in 1828 by **Henry Guinand**.

Bontemps learned his optical-glass-making skills from Guinand, and continued to work in the company, developing his own skills and knowledge, until 1848. With the advent in 1848 of the second French revolution, Bontemps fled to England and took up employment with **Chance Brothers**, Birmingham. At that time the Guinand optical glass was superior to the English glass, and Bontemps brought the necessary skills to Chance Brothers to equal, and ultimately exceed, the quality of the Guinand glass.

Boots

Boots is a UK chain of chemists and pharmacists, founded by John Boot in 1849 as an herbal medicine shop in Nottingham, England. In 1987 the company formed Boots Opticians Ltd., with the acquisition of **Curry & Paxton**, and the **Clement Clarke** chain of optometrists. Optical instruments bearing the Boots brand-name include cameras, photographic light-meters, and binoculars. Cameras, retailed by Boots, and bearing their brand-name, were manufactured by makers including **Braun**, **Beier**, **Franka**, **Houghtons Ltd.**, **King**, and **Bencini**.

Boots merged with **Dollond & Aitchison** in 2009. The name Dollond & Aitchison disappeared, and the major shareholder in the new company was Alliance Boots. In 2012 the American pharmaceutical company Walgreens became a major shareholder, and in 2014 Walgreens completed the acquisition, with Boots becoming a subsidiary of the American multinational holding company, Walgreen Boots Alliance.

Borda Repeating Circle

See repeating circle.

Bostok, Joshua

Joshua Bostock, (fl.1758-1780) [18] [322], was a mathematical and optical instrument maker of St. Mary le Strand, London, England (1758-1766), and Feathers Court, Drury Lane, London (1774-1780). In 1764 he was one of the petitioners who attempted to revoke the patent obtained by **John Dollond (I)** for the achromatic lens, and which was enforced after Dollond's death by his son, Peter (See Appendix X). Instruments bearing his name include a microscope.

Boston Camera Co.

The Boston Camera Company was a camera manufacturer, founded in Boston, Massachusetts, USA, in 1884, by Samuel N. Turner. In the late 1880s they entered an agreement with the **Blair Camera Company**, under which the Blair Camera Co. marketed their products, bringing them to the European market. In 1895 the company was acquired by **Kodak**, and became part of the Eastman Kodak Company.

Boston Optical Works

See **Robert B. Tolles**.

Bottomley, James Thomson

James Thomson Bottomley, (1845-1926) [44] [263], was the nephew of **William Thomson**. He was born in Belfast, Ireland, and his mother was William Thomson's sister. He graduated from Queen's College, Belfast, and Trinity College, Dublin. In 1890, after working as an assistant and demonstrator at the university in Belfast, he moved to Glasgow University as the Arnott and Thomson demonstrator in the Department of Natural Philosophy, then run by his uncle, William Thomson. He held the position whilst carrying out his own research, and was increasingly delegated to lecture William Thomson's students until 1899 when William Thomson resigned his professorship.

In 1900, when William Thomson formed **Kelvin and James White Ltd.**, James Bottomley joined the firm, and **Alfred W. Baird** was a director of the company. In 1907, upon the death of William Thomson, James Bottomley was appointed chairman of the company. In 1913 the organisation of the company changed, and it was renamed **Kelvin, Bottomley and Baird**. He continued in his role as chairman until his death.

Bouguer, Pierre

Pierre Bouguer, (1698-1758) [24] [45] [155], was a French physicist and mathematician who achieved an appointment as professor of hydrology by the age of fifteen. He was interested in solving problems relating to navigation, and joined an expedition to Peru to determine the distance associated with one degree of latitude on a meridian line near the equator. He invented a double image micrometer which was a variation on the invention of **Servington Savery**.

Bowman, Robert

Robert Bowman, (fl.1802) [18] [271], was an optical instrument maker of Calton Hill Observatory, Edinburgh, Scotland.

Box Camera

See camera.

Boyle, Charles B.

Charles B. Boyle, [185], was a maker of binocular telescopes, working in New York, USA, in the 1860s and 1870s. In 1867 he was appointed director of overseas manufacturing for the eyewear makers **American Optical**

Company. In an article in Scientific American in 1880 he described a binocular telescope for medical purposes that could focus on a subject 1 to 10 feet from the instrument.

Brachymedial Telescope
The Brachymedial telescope is a kind of catadioptric telescope that uses a Mangin mirror.

Brachyte
The Brachyte, or Brachy telescope, was a kind of tilted component telescope conceived by **Karl Fritsch** and **J. Forster**. They referred to it as an *excentric Cassegrain*, using a tilted primary and secondary mirror to avoid the need for a hole in the primary mirror. It is similar to the more modern Schiefspiegler telescope.

Bradford, George
George Bradford, (1817-1851) [18] [54] [276 #1284] [322] [361], was a mathematical, optical, and navigational instrument maker of 99 Minories, London, England. According to his trade card, "Real maker of Mathematical instruments, 99 Minories, London. Sextants, quadrants, & telescopes, warranted of the best workmanship. Likewise clean'd & repaired with accuracy & dispatch. Charts & navigation books of the latest publication. Captains, merchants, and shopkeepers supplied on moderate terms. NB The first opticians in the Minories."

Instruments bearing his name include: telescopes; octants; sextants; mariner's compass; dry card compass and binnacle; circumferentor.

See **Isaac Bradford**.

Bradford, Isaac
Isaac Bradford, (fl.1794-1825) [18] [276 #929] [322], was a mathematical instrument maker and optician of London, England. Addresses: Wapping Old Stairs; 87 Bell Dock, Wapping (1794-1801); 69 Bell Dock (1802-1807); 136 Minories, Tower Hill (1808-1824). Isaac worked with his brother John, trading as Isaac and John Bradford, during the period from 1795 until 1822. By 1823 he was trading as Isaac Bradford & Co., optician, and mathematical and optical instrument maker. He continued in business under this name until at least 1824.

According to his trade card, "Isaac Bradford & Co. Mathematical and Optical instrument makers. Nº. 136 Minories, London. Make in the most accurate manner all sorts of sextants, quadrants, telescopes, and compasses, with every description of mathematical & optical instruments, of the best and latest improvements and inventions. Sea charts and navigation books of the latest publications. Captains, merchants, & shopkeepers supplied on the most moderate terms."

Isaac began to trade as Isaac Bradford & Son in 1825, and the business continued under this name until 1836. Addresses: 136 Minories, Tower Hill (1825-1836); 125 Minories (1833). Reference [35 #18277 p2020] records the dissolution in 1826 of a partnership between Isaac Bradford and George Adams Bradford, trading as Isaac Bradford & Son at 136 Minories. An unverified source states that Isaac Bradford's son was **George Bradford** of 99 Minories.

Braham Brothers (I)
Braham Brothers (I), (fl.1872) [319], manufacturing opticians of 6 George Street, Bath, England, was a partnership between **David Braham** and **Philip Braham**, sons of **George Braham**.

Braham Brothers (II)
Braham Brothers (II), (fl.1852) [319], opticians and mathematical instrument makers, of St. Augustine's Back, Bristol, England, was a partnership between Henry J. Braham and Lewis Braham, sons of **John Braham**. The partnership was dissolved by mutual consent in 1852, [35 #21332 p1831], after which Lewis alone continued the business.

Braham, David
David Braham, (b.1844 fl.1865-1867) [18] [319], was an optician of 6 George Street, Bath, England (1859), and the son of **George Braham**. He succeeded his father in business and, subsequently worked in partnership with his brother, **Philip Braham**, trading as **Braham Brothers (I)**.

Braham, George
George Braham, (c1812-1865) [18] [319], of Bath, England, was an optical, mathematical and philosophical instrument maker, and the brother of **John Braham**. From 1842 his address was 6 George Street. This appears to have been his principal address, but he was also recorded at York Buildings, St. Michael, in 1841, and at 9 George Street in 1852. Two of his sons were **David Braham** and **Philip Braham**.

George was succeeded in business by his son, David, who continued at 6 George Street until at least 1867. David Braham also worked in partnership with Philip, trading as **Braham Brothers (I)**.

Braham, John
John Braham, (1799-1864) [3] [18] [50] [53] [319], was an optical, mathematical, and philosophical instrument maker, born in Plymouth, England, to a Jewish family, and the brother of **George Braham**. He began his *mathematical instrument warehouse* at 12 Clare Street, Bristol, England, in 1828. Within a year he had moved to

42 College Street, and in the early 1830s he moved again to 10 St. Augustine's Parade, corner of Hanover St. He also traded at number 17.

He described himself as "Optician, Mathematical, Philosophical and Nautical instrument maker; Agent for the sale of the charts and plans published by the Lords Commissioners of the Admiralty" and, according to an advertisement dated 1855, "Optical, Mathematical and Philosophical instruments of every description made and repaired." He also retailed optical instruments.

He exhibited at the Great Exhibition in London in 1851, but this was mostly spectacles, and not considered innovative. The business traded as John Braham & Co. during the years 1839-1851. Two of John's sons, Lewis Braham (b.c1830), and Henry J. Braham (b.c1829), traded in partnership as **Braham Brothers (II)**. Following the death of John, his company was run by another son, Joseph Braham (b.c1836), who was declared bankrupt in 1866, [35 #23185 p6145]. The company continued to trade until 1874, when it was taken over by an ex-employee and former apprentice of John Braham, Matthew W. Dunscombe (1841-1918).

Braham, Philip

Philip Braham, (b.c1843) [319], was an optician, and nautical, mathematical, and philosophical instrument maker of 6 George Street, Bath, England (1859), and the son of **George Braham**. At age 17 he is listed in the Census as an optician. An instrument label bears the words "Philip Braham, Optician, Bath. Nautical, Mathematical and Philosophical instruments of every description made and repaired."

In 1859, someone named Philip Braham is listed as an agent for **Field & Co.** If this refers to the same person, he was aged about 15 at the time. In 1872, he worked in partnership with his brother, **David Braham**, trading as **Braham Brothers (I)**.

Brander & Höschel

Brander and Höschel, [256] [322] [374], was a partnership between **George Friedrich Brander** and his son-in-law, **Christof Kaspar Höschel**. The partnership was formed in 1774, and continued until 1783. Instruments bearing their name include: microscope; sextants; theodolite; artillery level; reflecting quadrant; alidade; compass; meridian compasses; meridian sundial; equatorial sundials; silver mathematical instrument set; protractors.

Brander, Georg Friedrich

Georg Friedrich Brander, (1713-1783) [8] [53] [256] [274] [374], was an optical, mathematical, philosophical, surveying, and nautical instrument maker of Augsburg, Germany. He studied in Nuremberg under the eminent mathematician and astronomer, Johann Gabriel Doppelmeyr (1677-1750), and went on to set up his business in Augsburg in 1737. In 1760, his son-in-law, **Christof Kaspar Höschel**, joined the firm. They began to work in partnership in 1774, trading as **Brander and Höschel** until Brander's death in 1783. He is known to have made: telescopes; astronomical sectors; microscopes; octants; sextants. Instruments bearing his name include: simple microscope; screw-barrel microscope; box microscope; modified Cuff model microscope; microscope compendium; 16-inch Gregorian reflector; equatorial telescope; portable camera obscura; declination compass.

Brandreth, Timothy

Timothy Brandreth, (b.1679) [18] [274] [275 #517] [319] [322] [442], was an optical instrument maker of London, England. He served his apprenticeship with **Ralph Sterrop** of **Sterrop & Yarwell**, beginning in 1693, and was freed of the **Spectacle Makers' Company** in 1701. Timothy Brandreth worked as a journeyman instrument maker for **John Marshall**, after which he began to work with **George Willdey**, an ex-apprentice of **John Yarwell**. They worked in partnership from c1706 until c1713, trading as **Willdey & Brandreth**. Timothy Brandreth took on two apprentices; Nathaniel Cook in 1713/14, and John Ward in 1714.

Brandt Lens Works, R.E.

R.E. Brandt Lens Works of Prescott, Arizona, USA, was the company of Richard E. Brandt. In his early career he worked for **Optical Craftsmen**. Having started his own company, he worked with the optical engineer Dr. P.J. Peters producing objective lenses and achromatic doublets of 6-inch f/13.3, 8-inch f/13.3, 10-inch f/16.6 and 12-inch f/16.6. He also produced optical flats for use in folded tube designs of refracting telescope, for which he also produced complete tube assemblies.

Brashear Co. Ltd., John A.

John Alfred Brashear, (1840-1920) [8] [24] [59], of Pittsburgh, Pennsylvania, USA, was a maker of fine telescopes, lenses and mirrors. Brashear worked in the steel mills in Pittsburgh, and pursued his interests in astronomy in his spare time. With no technical education or training he taught himself to make lenses and mirrors, and devised his own improved method for silvering mirrors, now referred to as the *Brashear process*. This process became the widely preferred method for silvering mirrors until the vacuum aluminizing process was developed in the 1930s. He quickly established a reputation for lens-making and, with financial help from the philanthropist William Thaw, established his own optical workshop, and began a full-time career in making optical instruments. Soon after this, Thaw helped him set up a new and bigger workshop near the Allegheny Observatory. Brashear became Acting Director of the Allegheny Observatory in 1898, and held the post until 1900. In 1901 he became Acting Chancellor of the Western University of Pennsylvania, a position he held until 1904.

John Brashear remained active in education and in the Allegheny Observatory while producing high quality optical instruments, telescopes and spectrographs. He produced telescopes for professional observatories and the amateur market. Some of Brashear's large reflectors include: the 37-inch reflectors for the Lick Observatory's station in Chile and the University of Michigan Observatory, 30-inch reflectors for the Allegheny Observatory and the University of Illinois, and a 36-inch reflector for the Steward observatory at Tucson, Arizona. Brashear's son-in-law J.B. McDowell was an expert in glass production, and eventually took over his business.

In 1926, after the death of J.B. McDowell, **J.W. Fecker** took over the Brashear business. The company continued after Fecker's death, and in 1956 it became a subsidiary of the **American Optical Company**. After several more changes of ownership, in 2000, the company became Brashear LP, and subsequently **L3 Brashear**.

Braun AG

Braun AG, a consumer products manufacturer, was founded in 1921 by the mechanical engineer Max Braun, in Frankfurt, Germany. The company is best known for its range of shavers, which they began to produce in the 1950s. They began to produce slide projectors in 1954, and in 1962 they acquired the movie camera manufacturer **Niezoldi & Krämer**, and began to sell their products using the *Nizo* brand-name. They also distributed Japanese manufactured SLR cameras. A majority share in Braun AG was acquired by Gillette in 1967. In 1980 the film and photo section was sold to Robert Bosch GmbH, manufacturers of the well-known Bosch power tools. Production of movie cameras ceased in 1985.

Braun Camera-Werk, Carl

Braun was founded in 1915 as Karl Braun KG in Nuremberg, Germany, making binoculars and components for the radio industry, as well as semi-finished precision-engineered products. During WWII they marked their instruments with the **ordnance code hkm**. In 1948 the name was changed to Carl Braun Camera-Werk Nürnberg, and in 1950 they began to produce box-cameras. They expanded their product range, including the introduction of their 35mm *Paxette* branded camera models. In 1955 they began production of slide projectors. They ceased manufacturing cameras in the late 1960s. Their production of slide projectors grew successfully until the advent of digital photography. The company was declared bankrupt in 2000, and in 2004 it re-emerged as Braun Photo Technick GmbH selling binoculars, monoculars, spotting scopes, digital cameras, slide projectors, digital photo frames, film scanners, and accessories.

Bray, Max

Max Bray, (1912-2000) [**185**], was an optical instrument maker from Oregon, USA, who began making telescopes at the age of 19. From 1933 he worked in optical engineering for the **Tinsley Laboratories** in Berkeley, California, USA, and after WWII he founded the Bray Optical Company in Los Angeles. He produced military and aerospace optics as well as consumer telescopes. He contributed to the photographic optics for the Mariner 4 mission to Mars which, on July 15th 1965, photographed Mars, being the first photos ever taken of another planet from space. After his retirement in 1982 Max Bray continued to produce telescopes from his home until his death in 2000.

Breithaupt

Breithaupt, [**374**] is a precision mechanical-optical instrument maker, which was established in Kassel, Germany, in 1762 by Christian Breithaupt (1736-1800). From 1799, the company was run by Christian's son, Friedrich Wilhelm Breithaupt (1780-1855), and H.C.W. Breithaupt, and continued to produce astronomical, geodetic, topographical and physical instruments. In 1831 Georg Breithaupt joined the company, which he renamed F.W. Breithaupt & Sohn, and in 1836 they began to produce instruments for mining and surveying. Among Breithaupt's apprentices were **F.G. Voigt**, **J.C. Dennert**, **Otto Fennel** with Georg Breithaupt from 1841-1848, and **Rudolf Winkel,** with Georg Breithaupt from 1848-1857. In 1864 the company entered its fourth generation, with Freidrich Breithaupt, who had worked with **Steinheil** in Munich, and **Repsold** in Hamburg, and Dr. Wilhelm Breithaupt, who had worked with **Secrétan** in Paris.

During WWII, F.W. Breithaupt & Sohn products made for the war effort were marked with the **ordnance code** clk. In 1943, then in its sixth generation with the Breithaupt family, the company's factory was largely destroyed, and in 1945, near the end of WWII, the company buildings were completely destroyed. They restarted production in 1946 in Kassel, and went on to make high quality optical instruments for the surveying, geological, mining and metrological industries.

Bresser

Bresser is a German company, founded before 1970. From 1999-2009 they were part of **Meade Instruments**, after which they were owner-operated by three major shareholders: Rolf Bresser – son of the founder, main supplier **Jinghua Optical & Electronics Co. Ltd.**, and CEO Helmut Ebbert (since 2002). Their brands include *Bresser*, **National Geographic**, **Explore Scientific**, **Alpen**, and **Lunt**.

Brewster Angle

When a ray of light in a medium of refractive index n_1 strikes the optical surface of a second medium of refractive index n_2, some light is reflected and some is refracted. When the light strikes the optical surface such that there is a 90° angle between the reflected and refracted rays, this results in reflected light that is linearly polarized parallel to the plane of the interface, the remainder being refracted.

This occurs when the incident angle θ_B is such that its tangent is the ratio of the two refractive indices. This angle, which can be simply derived from Snell's law, is known as the *Brewster angle* [7].

$$\tan \theta_B = n_2/n_1$$

This phenomenon of polarization by reflection was first studied by the French physicist, engineer and mathematician, Étienne Malus (1775-1812), in 1808 while studying double refraction (birefringence), but at that time polarization was not fully understood in terms of wave theory. However, soon afterwards, the work of **Thomas Young** and others changed this. **Sir David Brewster** conducted research into, among other things, the use of polarization and refraction in crystallography, and discovered this rule in 1813, known as *Brewster's law*.

Brewster, Sir David

Sir David Brewster, of Edinburgh, Scotland, (1781-1868) [8] [24] [37] [46] [70] [**169**], was a scientist with an interest in optics. His father was a rector, and David was expected to follow in his father's footsteps by going into ministry. However, he developed an interest in optics. When he was just ten years old, he made his own telescope under the guidance of **James Veitch**, who became his mentor in both scientific and religious matters. In 1813 he invented the kaleidoscope, and in 1816 he worked in collaboration with **Robert Bate** on its development.

In 1812 Brewster began work on a compound lens that was to later become known as the Fresnel Lens. By 1820 he was already working with the Northern Lighthouse Board to introduce this kind of lens. This led to a priority dispute over the invention with **Augustin Jean Fresnel**, who's contemporary, but independent invention of the lens was also being adopted in lighthouses.

In 1820 he published his design for what subsequently became known as the Coddington Lens. In 1844 he devised the lenticular stereoscope, an improved form of Wheatstone's reflecting stereoscope, that used lenses instead of mirrors. **George Lowdon** claimed to have been the first to make them, but the design was popularised by the French instrument maker **Jules Duboscq**. Brewster was supplied with some of his photographic apparatus by **Thomas Davidson**. In 1849 he proposed the stereo camera, which was first built by **J.B. Dancer** as a prototype in 1853, thus allowing the stereo images to be taken simultaneously.

Brewster was a prolific writer, and much of his energy was spent on journalism and popularising science. However, this did not stop him in his research into optics, and his particular areas of interest included the use of polarization and refraction in crystallography. He empirically discovered Brewster's Law, defining the *Brewster angle* (qv).

Brewster's interests in crystallography and microscopy led him to experiment with jewel lenses with a view to increasing the resolving power of the microscope, [**162** Ch.5] [**280**] [**395**]. These lenses were supplied by local opticians such as **Alexander Adie**, **Peter Hill**, **James Veitch** and **William Blackie**. Brewster's experiments met with success, but the jewel lenses were expensive, and soon superseded by the Wollaston doublet, and subsequently by the achromatic lens.

Brightness, Image

Image brightness is proportional to $(A/M)^2$, where A is the effective objective aperture diameter, and M is the magnification of the optical system. The constant of proportionality depends largely on the transmission properties of the optics - the higher the transmission, the closer to maximum brightness.

A factor in *perceived* image brightness is its dependence on the observer's pupil diameter. A young, healthy adult will have a dilated pupil diameter of about 7mm, while an older person may have a dilated pupil diameter closer to 5mm. The smaller pupil diameter admits less light, and hence reduces the image brightness. If the exit pupil of the eyepiece is larger than the observer's pupil, a loss of image brightness will be experienced. Thus, an older adult with a dilated pupil of 5mm, using 7x50 binoculars with an exit pupil of about 7mm will experience nearly 50% loss of brightness (the difference between the area of the exit pupil and the area of the observer's pupil).

For the example of 7x binoculars, a practical limit for image brightness is achieved with an aperture of 49mm for a young adult. A larger objective aperture will not enhance perceived image brightness for any observer. According to the same principles, however, an optical instrument with higher magnification will benefit from a larger objective diameter to offer more perceived image brightness.

Brimfield and Co., J.

The firm of J. Brimfield & Co., [20] [29], was a British state-sponsored company established with the objective of making up for the shortfall in optical munitions in production in the early stages of WWI. The company was set up by the Ministry of Munitions, and the state bore the costs of setting up and equipping the factory. Brimfield were paid for the cost of labour and overhead expenses, plus a defined profit margin. They were contracted to produce 5000 military binoculars, with prisms supplied by **Barr & Stroud**, and lenses supplied by the **Guaranteed Lens Co.** Due to a shortage of supplies of lenses, and poor mechanical quality, the rejection rate was high, and the company never achieved its production targets. The company was shut down at the end of the war.

Briois, A.

A. Briois, [53], of 4 Rue de la Douane, Paris, France, was a chemist, and a member of the Société Française de Photographie. The company produced a limited number of cameras on a model designed by an Englishman named Mr. Thompson in 1862. The camera was reputedly inspired by a Colt pistol, and was a pistol-grip camera with a **Petzval** lens, which could expose pictures on four small glass photographic plates.

Britannia Works Company

See **Ilford**.

Britannic

See **Broadhurst, Clarkson and Co.**

Britex

Britex was a trade-name used by **W. Ottway & Co. Ltd.** for some of their products.

Britex (Scientific Instruments), of 523-4 Bank Chambers, 329 High Holborn, London, WC1. was a listed exhibitor in the 1947 British Industries Fair. Their catalogue entry, [450], listed them as manufacturers of: the *Britex Minor* microscope; the *Britex* mechanical stage; the *Britex Spotter* telescope; Drawing lamp; Bench lamp; Small bench lamp; Colorimeter; Mesoscope; Paper testing machine; Vertical illuminator; Drawing eyepiece.

Some Britex (Scientific Instruments) microscopes were marketed by Pyser-Britex (Sales) Ltd., (see **Pyser-Britex Group**). Some of these were marketed for use by children. Some toy-grade instruments can be found bearing the *Britex* name.

British Acoustic Films

British Acoustic Films Ltd. of London, England, was founded in 1925 jointly by Gaumont & Co. and the Electrical Fono Film Company of Denmark. Addresses: Denmark St. (1926-1929); Palladium Buildings, Argyll St., Islington (1930); Woodger Road, Shepherds Bush (from soon after 1930). Factory in Mitcheldean (from 1941). They produced 35mm and 16mm sound recording and reproduction equipment, 16mm silent cameras and projectors, and 8mm cameras and projectors. They sold their 35mm equipment via **GB Kalee Ltd.**, and sold their 8mm and 16mm equipment via **GB Equipments**. By the mid-1930s the company was controlled by **J. Arthur Rank**. In 1948 Rank merged British Acoustic Films with **Taylor, Taylor & Hobson Ltd.**, **A. Kershaw and Sons Ltd.**, **G.B. Kalee Ltd.**, and Gaumont-Kalee Seating Ltd., to form their subsidiary **British Optical and Precision Engineers**.

British Crown Glass Co.

British Crown Glass Co., [46], of Spon Lane, Smethwick, Birmingham, England, was formed in 1816 by Thomas Shutt, making crown window glass. The company was acquired in 1824 by Robert Chance, and became **Chance Brothers and Co.** in 1836.

British Optical Instrument Manufacturers' Association (BOIMA)

The British Optical Instrument Manufacturers' Association, founded in 1916, was a body that endeavoured to represent the most important manufacturers, and was devoted to furthering the interests of the Trade. Its address was Duke Street, London, England. The *Dictionary of British Scientific Instruments*, [30], compiled in 1919, was published under its auspices in 1921. The 28 key members of the Association are listed on pages 3-4 of the Dictionary.

The "key members" in 1919 were: **Aldis Brothers, Charles Baker, R.&J. Beck, Cambridge Scientific Instrument Co. Ltd., Joseph Casartelli, CF Casella, Chance Brothers, JH Dallmeyer, Dollond & Co., Endacott Scientific Instrument Co., Sir Howard Grubb & Sons, Ltd., United Kingdom Optical Co. Ltd., Heath & Co. Ltd., Adam Hilger Ltd., Henry Hughes & Son Ltd., Kelvin, Bottomley & Baird, Ltd., Negretti & Zambra, Ottway & Co. Ltd., W.G. Pye & Co., Ross Ltd., W.F. Stanley & Co. Ltd., James Swift & Sons Ltd., Taylor, Taylor & Hobson, AG Thornton Ltd., Troughton and Simms Ltd., W Watson & Sons Ltd., ER Watts & Son, John Lilley & Sons, Ltd.**

The historical relationship between BOIMA and the **Scientific Instrument Manufacturers' Association of**

Great Britain (SIMA) is unclear. According to [331], SIMA was formed as the successor to BOIMA in 1953. However, the SIMA Bulletin of September 1948, [330], predates this, and lists the past presidents of SIMA dating back to 1915/1916. It also lists 95 SIMA member companies, including most of those listed as BOIMA members in 1919. An unconfirmed source, [332], states that SIMA was an alternative name for the BOIMA.

British Optical Lens Company (BOLCo)

The British Optical Lens Co., [433], was founded in 1911 as a manufacturer of spectacle lenses in Key Hill Drive, Sheffield, England. **Edwin Elliott** (1878-1968) acted as one of their agents, and purchased the company in 1911. In 1912 they were listed as optical glass makers, by which time they had relocated to Birmingham, addresses: Regent Street, Smethwick (1912-1915); 56 & 57 Frederick Street (1914-1915); Warstone Lane (1916-1926); Brearley St. (1927-1930s); 126 & 128 Brearley St. (1930s-1940); Victoria Works, 315 Summer Lane (1936-1968); Bescot Crescent, Walsall (c1938-1983).

During WWI they produced polished green glass for goggles for use by British troops in the Egyptian desert. During the 1920s they began to produce Bakelite mouldings for radio parts, and for a diverse range of other products. During the 1930s they produced body mouldings and lenses for the **Coronet Camera Co.**, and in 1935 they began to produce the *VP Twin* camera, under their own name, for the high street chain, Woolworths. This used a shutter produced by the Coronet Camera Co.

During WWII they produced anti-gas eye-shields and periscopes and, in 1940, their optical department in Brearley St. was destroyed by incendiary bombs, precipitating a relocation to Bescot Crescent. In 1952 they began to make lenses for **K.G. Corfield**. By the time Edwin Elliott retired in 1966, he had six factories producing plastic and optical items. In 1982, receivers were appointed, and in 1983 the company was taken over by **Reed International Group**.

British Optical and Precision Engineers

British Optical and Precision Engineers, [44], was a private company formed in 1947 as part of **Rank**, including the acquisition of **Taylor, Taylor & Hobson Ltd.** [135], which was renamed Rank Taylor Hobson. In 1948, [134], the company went public and acquired **A Kershaw & Sons**, **British Acoustic Films**, **G.B. Kalee Ltd.**, and Gaumont-Kalee Seating Ltd. In 1956 British Optical and Precision Engineers Ltd. was renamed Rank Precision Industries. In 1997, Rank Taylor Hobson was sold to Schroders Ventures, trading as Taylor Hobson.

British Photographic Industries Ltd.

British Photographic Industries Ltd., [61] [306], was a joint venture company, set up in 1915, holding majority shares in **Houghtons Ltd.**, **W Butcher & Sons**, Butchers Film Services, **Houghton-Butcher Manufacturing Co.**, Austin Edwards, and Fordham & Co. The company's objective was to "organise, promote, deal with and protect the trade interests of the British Photographic Trade generally". The company went public in 1920, and continued to be listed in the directories until 1941.

Broadhurst, Clarkson and Co.

Broadhurst, Clarkson and Co., [8] [319] [406] [407], was a telescope maker and optical instrument retailer of Telescope House, 63 Farringdon Road, London, England. The business was founded by Robert Henry Broadhurst, following the dissolution of his partnership with **Alexander Thomas Clarkson** in 1908, and retaining the telescope tube-drawing machine from Clarkson's business. Alexander Clarkson was not a part of the Broadhurst, Clarkson and Co. business.

The company quickly gained a reputation for producing fine quality telescopes. During WWI, they took on new premises, at 69 Fenchurch St., and in Watford, for manufacturing telescopes and producing lenses. This was where they produced telescopes for the military. Their lens-grinding method was based on that used by **John Marshall** more than two hundred years earlier, and their telescope tube manufacturing was based on the method used by **Joshua Lover Martin** in 1782.

Some of their telescopes are marked *Britannic B.C. & Co. Ltd.* Sometimes the mark bears only the word *Britannic*, and sometimes the name is shown only as *B.C. & Co. Ltd.*

In 1947 they exhibited at the British Industries Fair. Their catalogue entry, [450], listed them as manufacturers of: Telescopes - astronomical, naval, and marine; Deerstalker; Portable and rifle range models; Astronomical accessories; Starfinders; Sun and star diagonals; Eyepieces; Comet eyepieces; Barlow lenses; Tripod stands; Brass tubing; Prisms; Lenses, etc.

During the 1950s, Leslie Broadhurst became the last of the Broadhurst family to run the business. At around this time, the business began to suffer as a result of competition from cheap imports of Japanese telescopes. By 1970, all the company's factories had closed except for the manufacturing facilities at Telescope House, 63 Farringdon Road. In 1973 **Dudley Fuller** bought the Broadhurst and Clarkson business, renaming it Broadhurst, Clarkson and Fuller. Broadhurst, Clarkson & Co. was liquidated in 1975 [35 #46618 p8217]. The telescope tube-drawing equipment was decommissioned in the early 1990s, and is now in a private collection. In 2018 the company passed into the ownership of one of their ex-employees, Steve Collingwood, who went on to operate under the name of SC Telescopes.

Brock, G
G. Brock, (fl.c1775-c1800) [142] [274] [322] [369 p23], of London, England, was a maker of microscopes. His name appears on a solar microscope dated c1775. A brass monocular microscope in the Billings collection, dated c1800, bears the inscription *Brock Invenit et Fecit, London*. Note that [276 #443] refers to an optical instrument maker B. Brock whose name appears on a solar microscope dated c1773. This may result from a confusion of the initial.

Broderers' Company
See Appendix IX: The London Guilds.

Brooks, Alfred
Alfred Brooks, (1809-1865) [18] [287] [322], was an optician, and mathematical and philosophical instrument maker of 41 Ludgate Street, London, England. He began his own business in 1845, and claimed in his advertisements that he had more than 20 years' experience working for **Dollond**.

According to his advertisement in 1851 [301], "Brooks (From Dollond's) Optician and Mathematical instrument maker, respectfully solicits from the public a continuance of their patronage. Having had upwards of 20 years' experience in Dollond's, they may rely on all articles submitted by him are of the most perfect character, and at moderate prices. Spectacles, Telescopes, Microscopes, Opera and race glasses, Sextants, Quadrants, Compasses, Barometers, Surveying instruments, Rules, scales, &c., of every description. 41 Ludgate Street, St. Paul's."

He filed for bankruptcy in 1860. Records show his business continuing until at least 1861, when the bankruptcy proceedings concluded, [35 #22494 p1308].

Brown, Benjamin
Benjamin Brown, sometimes spelled Browne, (fl.1797-1831) [3] [276 #930a; 1497] [322], was a mathematical, nautical, and optical instrument maker of Bristol, England. Addresses: Quay (1792-1824); Wilson St. (1814-1815); 14 Wilson St. (1816-1819); 2 Old Park Hill (1825-1826); *opposite the Crane*, 7 Quay (1825-1831). Instruments bearing his name include: octant, 1792, [262]; octant, c1831, [16], with scale marked with anchor symbol (see Appendix XI).

Brown, James
James Brown, (1836-1913) [263], was an optical instrument maker of 76 St. Vincent Street, Glasgow, Scotland. He was born in Jamaica to a missionary family from Scotland, to which country he returned in 1837. He worked for **M. Gardner & Co.** for nine years prior to setting up his own business as an optical, mathematical and philosophical instrument maker in 1871.

According to his advertisement of 1876: "Brown's Optical & Philosophical instrument depot, 76 St. Vincent Street, Glasgow. Surveying instruments, Drawing instruments and material. Spectacles and eye glasses in gold, silver, steel, and tortoise-shell frames. Microscopes by the best French and English makers. Aneroid barometers from pocket size upwards, with and without scale for measuring heights. Opera and field glasses, achromatic, in all the various qualities. Pedometers, Radiometers, Thermometers, Stereoscopes, Graphoscopes, Magic lanterns. Photographs of Scottish, Swiss, Italian, and other scenery. N.B. State of sight ascertained by improved optometer."

James Brown was a prominent ophthalmic optician, and continued in business until shortly before his death in 1913. The company also began to sell electrical and medical instruments, and to produce sundials. The company continued until 1928, after which the premises were acquired by **Dollond & Aitchison**.

Brown and Co., Neville
Neville Brown and Co., [44], of 44 Berners Street, London, England, were makers of photographic equipment, including a range of cameras, and projection equipment. They collaborated with Johnson & Sons of Hendon, London for the manufacture and marketing of some of their products. They were distributors for some **Aldis Bros.** products. The company was acquired in 1959 by **R.B. Pullin and Co.**

Browne & Son, Henry
Henry Browne & Son of Barking, Essex, England, were makers of nautical and aeronautical instruments. The company exhibited at the 1947 British Industries Fair. Their catalogue entry, [450], listed them as manufacturers of: Nautical instruments; Compasses; Clocks; Binoculars; Telescopes; Thermometers; Sounding apparatus; Aircraft instruments; Signalling equipment; Meteorological instruments; Barographs; Thermographs; Aneroid barometers; Sextants; Yacht equipment; Navigational instruction equipment.

Brownell, Frank A.
Frank Alexander Brownell, (1859-1939) [122], was a Canadian born cabinet maker who moved to Rochester,

New York, USA, in the 1870s. He served his apprenticeship with **Yawman & Erbe**, during which time he developed an interest in camera-making.

He set up his company, the Brownell Manufacturing Company, and began to make cameras in the early 1880s. In 1885 he entered a collaboration with **George Eastman**, who asked him to make a wooden roll-film holder. Brownell's company grew to become a major producer of cameras. Frank Brownell contributed to the design and manufacture of the early cameras sold by the **Eastman Kodak Company**, some of which were assembled by Yawman & Erbe. One of Frank Brownell's designs was the *Brownie* box camera, introduced in 1900, and initially manufactured by his company. He continued to design and build cameras for Kodak until they took over his company in 1902. Frank Brownell stayed with Kodak until 1906 when he left to work in the motor industry.

Brownie Camera

The **Kodak** Brownie camera, introduced in 1900 at the price of $1, was a low-cost mass-produced camera which brought photography to the popular market. It was designed by **Frank Brownell**, and initially manufactured for the **Eastman Kodak Company** by Brownell's company. Kodak continued to produce Brownie box cameras in various models until at least the 1960s. They also produced bellows cameras and solid body roll film cameras branded as Brownies, the last being made in the 1980s.

Browning & Co., John

John Browning & Co., [18], was the name used by **John Browning (III)** after he stopped trading as **Spencer, Browning & Co.** in 1870. In 1900, John Browning & Co. was taken over by **W. Watson & Sons Ltd**, but continued to trade under the same name until 1945.

Browning, John (I)

John Browning (I), (fl.1782-1806) [18] [90], of London, England, was a mathematical instrument maker. He was the brother of **Samuel Browning (I)**, and the son of Samuel, a husbandman. He served his apprenticeship with **Richard Rust**, beginning in 1768 and was freed of the **Grocers' Company** in 1782.

Addresses: 49 Virginia Street; Pennington Street, Ratcliff Highway (1793); Prince's Square, Ratcliff Highway (1797); 67 Ratcliff Highway (1799-1811); 25 Prince's Square, Ratcliff Highway (1800); 17 Wellclose Place, New Road, St. George in the East (1803); 117 Ratcliff Highway (1805-1809); Wellclose Place (1806). He last appears in the directories in 1806.

See Appendix VII.

Browning, John (II)

John Browning (II), [18] [90], of London, England, was a mathematical instrument maker. He was the son of **Samuel Browning (I)**, and served as his apprenticeship with his father beginning in 1795. He was freed of the **Grocers' Company** in 1803. See Appendix VII.

Browning, John (III)

John Browning (III), (1830/31-1925) [18] [46] [90], of London, England, optical and mathematical instrument maker, was the son of **William Browning**, and the brother of **Samuel John Browning** and William Spencer Browning. He served his apprenticeship with his father, and went on to work with him at 111 Minories, London. It is not clear when John (III) became a partner in **Spencer, Browning & Co.**, but in 1855 the partnership, then consisting of himself, his father William Browning, and his brother Samuel John Browning, was dissolved [35 #21831 p4843]. From this date Samuel John traded at 52 High Street, Portsmouth, while William and John (III) each worked "on their own account" at 111 Minories. The name of Spencer, Browning & Co., however, continued to be used until 1870, after which John (III) traded as **John Browning & Co.**

In 1872 John (III) moved from 111 Minories to 63 Strand, London. He was for some years the leading English designer and maker of spectroscopes, though he also sold a wide range of other optical equipment. In 1870 **Adam Hilger** worked for him, having arrived in England from Paris, France, upon the outbreak of the Franco-Prussian war. Hilger rose to the position of foreman before leaving to start his own business. John (III) is known to have mounted many of the mirrors made by **George Henry With**. From 1895 to 1900 he served as the first president of the British Optical Association. In 1900, John Browning & Co. was taken over by **W. Watson & Sons Ltd**, but continued to trade under the same name until 1945.

See Appendix VII.

Browning, Richard

Richard Browning, [18] [90], of 66 Wapping, London, England, was a mathematical instrument maker. He was the son of **Samuel Browning (I)**, and the brother of Samuel (II), John (II), and **William Browning**. He was freed of the **Grocers' Company** in 1818 by patronage.

Richard and William both inherited shares in the company **Spencer, Browning & Rust** from their father, and they continued to run the company, together with **Ebenezer Rust (II)**. Ebenezer Rust (II) died in 1838 and, in 1840, they changed the name to **Spencer, Browning**

& Co. By the time the partnership was dissolved in 1855, [**35** #21831 p4843], Richard was no longer a partner.

See Appendix VII.

Browning, Samuel

Samuel Browning (I), (c1752-1819) [**18**] [**90**], was an English mathematical instrument maker. Addresses: 327 Wapping St., London (1782-1783); Eaton Socon, Bedfordshire (1786-1792); Wapping, London (1795-1818). He was the son of Samuel, a husbandman, and the brother of **John Browning (I)**. He served his apprenticeship with **Richard Rust**, and was freed of the **Grocers' Company** in 1782. Among his apprentices was **Joseph Fairey (I)**. His sister married **William Spencer (I)**, and in c1781 Samuel (I) entered a partnership with William Spencer (I), trading with his brother-in-law as **Spencer & Browning**. In 1784 **Ebenezer Rust (I)** joined the partnership, and they began to trade as **Spencer Browning and Rust**.

Samuel (I) had four sons: Samuel (II), **John (II)**, **Richard**, and **William**. Samuel (II) and John (II) both served their apprenticeships with Samuel (I), and Richard was freed of the Grocers' Company by patronage in 1818. Samuel (II) did not take his freedom. Samuel (I) bequeathed shares in Spencer, Browning & Rust to his sons Richard and William.

See Appendix VII.

Browning, Samuel John

Samuel John Browning, (fl.1840-at least 1862) [**18**] [**90**], optician, was the son of **William Browning**, and the brother of William Spencer Browning and **John Browning (III)**. He served his apprenticeship with his father, beginning in 1840, but did not take his freedom of the **Grocers' Company**. He worked in Minories, London, England, with his father and with John (III), in **Spencer, Browning & Co.** until their partnership was dissolved in 1855, [**35** #21831 p4843]. From this date Samuel John traded at 52 High St., Portsmouth. He is also recorded as trading at 66 High St., Portsmouth. He is known to have sold octants, theodolites and telescopes.

See Appendix VII.

Browning, William

William Browning, (d.1862) [**18**] [**46**] [**90**], was an optician of 66 Wapping, London, England (1838), and Minories, London (1840-1851). He was the son of **Samuel Browning (I)**, and the brother of Samuel (II), John (II) and Richard. He was freed by patrimony of the **Grocers' Company** in 1840. His sons, **Samuel John Browning**, William Spencer Browning, and **John Browning (III)** served their apprenticeships with him. William inherited a share of the company **Spencer Browning and Rust** from his father, as did his brother Richard. They continued to run the company with **Ebenezer Rust (II)**. In 1840, following the death of Ebenezer Rust (II) in 1838, they changed the name to **Spencer, Browning & Co.** In 1855 the partnership, then consisting of William Browning and two of his sons, Samuel John, and John (III), was dissolved [**35** #21831 p4843]. From this date Samuel John traded at 52 High Street, Portsmouth, while William and John (III) each worked "on their own account" at 111 Minories. It is not clear when William retired, but the name of Spencer, Browning & Co. continued to be used after his death, until 1870, after which John (III) traded as **John Browning & Co.**

See Appendix VII.

Brownscope

The Brownscope Manufacturing Company of New York, USA, was a manufacturer and supplier of telescopes, microscopes and binoculars. Their advertisements appear in *Popular Science* as early as 1933, and an example of a *Brownscope Monoscope telescope*, which appears to be a 4-draw spyglass, states that it is dated 1962 and manufactured by **Criterion**. Their 1951 catalogue shows spyglasses, Galilean and prismatic binoculars, and microscopes, as well as a Brownscope telescope making kit. Their products appear to have been aimed at the budget end of the market.

Bruce, William

William Bruce, (fl.1795-1863) [**8**] [**18**] [**276** #1092], was an optical instrument maker, brass turner, and optical turner of King's Head Court, Shoe Lane, London, England. He was the son of a brass founder, Joseph Bruce, of London. He served his apprenticeship with **John Mason**, beginning in 1788, and was freed of the **Drapers' Company** in 1795. He collaborated with **Charles Robert West**, and together they were awarded a patent for more easily portable telescopes.

Bruce & Sons, Wilson & Gillie

Wilson & Gillie Bruce & Sons, were nautical instrument makers of Cardiff, Newport and Barry Docks, Wales. Instruments bearing their name include: sextants dated C19th; sextant with calibration date 1917; leather covered brass hand-held telescope; boxed marine chronometer.

Brugger & Breitsamer

Brugger & Breitsamer, [**374**], were optical and mechanical instrument makers of Munich, Germany. The company was a partnership between Otto Brugger and B. Breitsamer. Both had served their apprenticeships with **Steinheil & Söhn**. Brugger from 1857 to 1861, and

Breitsamer from 1861 to 1877. The company was founded sometime after 1877.

Brunner Frères

Brunner Frères, [154] [197], was a partnership between the brothers Emile Brunner (1834-1895) and Léon Brunner (1840-1894), who were the sons of **Johann Josef Brunner**. Both brothers worked with their father, and they took over his business when he died in 1862. They were highly regarded French optical instrument makers specialising in geodetic, surveying and astronomical instruments. Like their father they exhibited at several of the national exhibitions. They had a strong reputation for their astronomical instruments, some of which were installed in observatories in Paris, Nice, Toulouse, Lisbon, Cairo, Lyon, Algiers, and others. Léon died in 1894, and the company closed upon the death of Emile in 1895.

Brunner, Johann Josef (Jean)

Johann Josef Brunner, (1804-1862) [154] [197], was a highly regarded French optical instrument maker specialising in geodetic, surveying and astronomical instruments, as well as making microscopes, equatorial telescopes, clock drives and meridian circles. He learned his instrument making skills in Vienna and, in 1828, moved to Paris where he began working under the name Jean Brunner. He worked with the instrument maker Frédéric Hutzinger, and subsequently with **Vincent Chevalier**. He opened a workshop at 34 Rue des Bernardin.

In about 1845 the Bureau des Longitudes asked **Henry Gambey** to produce a large equatorial telescope for the Paris Observatory. Gambey died in 1847 and was thus unable to fulfil this request, so Brunner built the telescope. In the 1850s Brunner moved his business to larger premises at 183 Rue de Vaugirard. He was a frequent exhibitor at the national exhibitions, and in 1855 he was on the jury of the scientific instruments class in the Paris Universal Exhibition. He also exhibited in the London Exhibition in 1862, shortly before his death. When he died his sons Emile and Léon continued the business as **Brunner Frères** until 1895.

Brunnings

Brunnings was a photographic, optical, and scientific instruments shop at 135 High Holborn, London, England. The family run business was established in 1921, incorporated in 1942, and dissolved in 2002. The business was run in its later years by David Brunning (d.2006). Their 1980 astronomical telescope accessories list shows telescope mirrors, mirror kits and blanks, mirror making supplies, a variety of eyepieces, finders, books and maps.

Brunson

Brunson was founded in 1927 by A.N. Brunson in Kansas City, Missouri, USA. The company provides precision tooling and equipment for surveying and various other industries. In the early 1980s they acquired **Cubic Precision**, formerly the optical tooling division of **Keuffel & Esser**. Their optical products include laser trackers, alignment telescopes, transits, precision levels, collimators, micrometers, eyepieces, camera kits, and accessories, as well as test and calibration equipment.

Brunton Outdoor

Brunton Outdoor is a US company which supplies navigational instruments (mainly compasses), and binoculars. In 1992, Brunton introduced a line of binoculars and other optical equipment aimed at the hunting and outdoor recreation market. The optics line was discontinued in 2014.

Bryceson, William

William Bryceson, (fl.1824-1839) [18], was an optical, mathematical, and philosophical instrument maker, and binnacle maker, of Commercial Road, London, England, where he operated a navigation warehouse. Addresses: King's Arms Place (1824-1826); 5 Union Terrace (1826-1827); 2 Union Terrace (1831-1835); 4 Union Terrace (1837-1839).

Bryson, James Mackay

James Mackay Bryson, (1824-1894) [8] [53] [263] [271], was a scientific and optical instrument maker from a family of renowned watch and clock makers in Edinburgh, Scotland. His father, Robert Bryson, (1778-1852), founded the watch and clockmaking business in 1810, and was one of the co-founders of the Edinburgh School of Arts, later known as the Heriot-Watt University. James served an apprenticeship, finishing around 1843, but it is not known with whom he served his apprenticeship. He then went to Hamburg in Germany, where he studied with and worked for **Adolf and Georg Repsold**. From there he went to Munich where he studied under **Georg Merz**, learning how to make lenses for astronomical instruments, and other related skills. James returned to Edinburgh in 1850, and set up his optician's business. Addresses: 65 Princes Street (1850-1853); 24 Princes Street (1853-1855); 60 Princes Street (1855-1866). After this his address was variously 60 and 60A Princes Street until 1893. His business was a dealership, selling instruments by other makers, as well as selling instruments he made himself. **Angus Henderson** worked for James Bryson prior to setting up his own business in 1861. In 1894 James Bryson was succeeded at 60 Princes Street by **Turnbull & Co.**

B^tee Fr. & Etr.

B^tee Fr. & Etr., is a mark sometimes found on French instruments. It is short for Brevetée France & Étranger, which translates to Patented in France and Overseas.

Buchner, Nicolaus

Nicolaus Buchner, sometimes Nikolaus, (1813-1879) [374] [437], was a spectacle maker and optical instrument maker of Frauenplatz 10, Munich, Germany. He advertised as Court and University Optician. The workshop was established in 1801 by his father, Franz Xaver Buchner (b.c1780), who succeeded one of his relatives, the spectacle maker Julius Paulus Hirn (d.1799), in business, having worked for him since the age of 14. Nicolaus succeeded his father in 1851, and ran the company under his own name until his death in 1879. He was succeeded by his son-in-law, Franz Xaver Brantl, who continued to operate the company under the name of Nicolaus Buchner. The business continued to operate with this name until at least the mid C20th. Instruments bearing his name include: Galilean binoculars; barometer; barograph; spectacles, horn rimmed monocle; 4-draw black painted aluminium telescope in leather case, 53mm objective, opening to 95cm length, with the inscription *Nicolaus Buchner. K.b. Hof und Universitäts Optiker. München, Frauenplatz 10*.

Buff and Berger

Buff and Berger, [58], of Boston, Massachusetts, USA, was a partnership between **George L. Buff** and **Christian Louis Berger**. They made surveying, astronomical, mathematical and philosophical instruments. The partnership was formed in 1871, and continued until 1898 when the firm was succeeded by C.L. Berger, and **Buff and Buff**.

Buff & Buff

See **George Louis Buff**.

Buff, George Louis

George Louis Buff, (1837-1923) [58], of Giessen, Germany, began to work for **A. Repsold & Söhne** in Hamburg at age 17. In 1858 he went to work for **Thomas Cooke & Sons** in England then, in 1864, he emigrated to the USA to work for **Stackpole & Brother** in New York. In 1869 he started a business making his own design of surveyor's transit, and in 1871 he moved to Boston, and entered a partnership with **Christian Louis Berger**, trading as **Buff & Berger** until 1898. After this, George Buff continued the business with his sons, trading as Buff & Buff. The company continued successfully until it began to decline in the mid-C20th, and they ceased trading in the 1980s.

Builder's Level

Another name for a dumpy level.

Buist & Co., James

James Buist & Co., (fl.1859-1886) [271], were spirit level makers and optical instrument makers of Edinburgh, Scotland. They succeeded **John J. Liddell** in business. Addresses: 91 South Bridge (1859); 5 Nicholson St. (1860-1884); 67 South Bridge (1885-1886). They were succeeded in business by **James Buist & Sons**.

Buist & Sons, James

James Buist & Sons, (fl.1887- at least 1900) [271], were spirit level makers and optical instrument makers of 67 South Bridge, Edinburgh, Scotland. They succeeded **James Buist & Co.** in business.

Bulloch, Walter H.

Walter H. Bulloch, (1835-1891) [201], of Glasgow, Scotland, emigrated to the USA in 1852, and learned his trade as a tailor in New York. In a change of career, he then served an apprenticeship with **Benjamin Pike & Son**, working his way up to foreman of the company. He left Pike to start his own company. He entered a partnership with **William Wales** in about 1864, in which Bulloch made microscope stands and Wales made the optics. This partnership continued until 1867. After this, Bulloch sold objectives made by Wales. Bulloch's own business first appeared in the directories from 1868 in Philadelphia, continuing until his death. He advertised variously as making optical, mathematical and surveying instruments, as well as being a model maker. However, his only known instruments are his microscopes, microtomes, and microscope accessories.

Bunders, Jan

Jan Bunders, (fl.1807) [220] [256], of Amsterdam, Netherlands was a telescope maker who succeeded **Harmanus van Deijl** in his business. Unlike van Deijl, Bunders is believed to have made telescopes only, and not microscopes. Bunders may have advertised his business as the successor to van Deijl, since there are examples of telescopes signed *Jan van Deijl Bunders* and *Jan van Deijl Bunders en Zoon* (en Zoon is Dutch for *and Son*).

Bunsen's Photometer

See photometer.

Bunsen's Mirror Photometer

See photometer.

Bureau, Thomas

Thomas Bureau, (fl.1756-1760) [18] [276 #452] [322], was an English microscope maker. In 1756 he made a compound microscope to a design by Dr. Demainbray, Superintendent of the King's Observatory for George III. The microscope was given to George III by Dr. Demainbray, and is now in the collection of the History of Science Museum in Oxford [270].

Burlini, Biagio

Biagio Burlini, (1709-1771) [73], was an optician and optical instrument maker with an optical laboratory and shop at *The Sign of Archimedes*, Fondamenta del Rosmarino, Venice, Italy. Examples of his work include simple and compound microscopes and hand-held telescopes, with tubes of fish-skin covered cardboard, and mounts of ivory, horn or brass. A six-draw telescope, opening to eight feet in length, is in the Science Museum collection. Based on his published booklets he is believed to have also made spectacles, opera and spy glasses, magic lanterns, watch glasses, and other optical instruments. Burlini's instruments were well made with higher quality optical components than later examples by makers such as **Leonardo Semitecolo**.

Burnet & Son, W.

W. Burnet & Son, [199], were optical instrument makers of Boston, Lincolnshire, England. According to a July 1969 advertisement, "Makers of astronomical telescope eyepieces and components, Ramsden eyepieces ¼-inch to 1¼-inch."

Burrow, W.&J.

W.&J. Burrow, [43] [110], of Malvern, Worcestershire, England, were producers of spring waters and associated products. They exhibited in the 1862 Great Exhibition, [258], showing the *Malvern Landscape Glasses* model of Galilean binoculars, and the *Burrows Target Telescope*. These models were presumably made for them and branded with their name.

Burton, Charles Edward

Charles Edward Burton, (1846-1882) [34] [216] [304], from near Dublin, Ireland, was a mirror maker who learned his speculum mirror grinding and polishing skills from **William Parsons (I)**. He worked for the Dunsink Observatory, the Greenwich Observatory, and for William Parsons (I). He made high quality silvered glass mirrors ranging from six to fifteen inches, including eight and twelve-inch mirrors for his own observatory. He collaborated with **Sir Howard Grubb** in the design of an improved ghost micrometer, and designed a binocular spectroscope for observing zodiacal light. His binocular spectroscope was built by **John Spencer & Son**, and parts for it were made by Howard Grubb.

Burton, Edward

Edward Burton was an optical, mathematical, and philosophical instrument maker of 47 Church Street, Minories, London, England.

According to his trade advertisement in 1860, [292]: "E. Burton, Optician, Manufacturer of Telescopes, Microscopes, Opera glasses, Barometers, Thermometers, Magic lanterns, Photographic apparatus, and every description of Optical, Mathematical, and Philosophical instruments. Spectacles to suit all sights. Experimental work to order. An illustrated catalogue free for two stamps, or upon application at the Manufactory, 47 Church Street, Minories, London, E."

Edward Burton exhibited at the International Exhibition of 1862 [258]. The following year he had an entry in the *Ninth Londoniad*, [293], as follows: "The Optical Poem. International Exhibition – E. Burton and Co., 2869, Class 13, Optical and Mathematical Instruments. E. Burton & Co., Opticians, manufacturers of Telescopes, Microscopes, Camera lenses, and every description of Optical instruments, 47 Church Street, Minories, London, E. Wholesale and for Exportation." This is followed by the poem:

The light that streamed in floods from ancient days
Descends upon the heroes of my lays;
On whom the intellectual sons of Science call
For instruments required Optical.
Of best material made they can procure,
For length o' time in any clime to endure.
The rare quality and finish of the
Workmanship are hailed o'er the Western Sea.
In journeying thro' th' world remember this
That quality the test of cheapness is.
Th' palm t' Edward Burton and Co. I've given,
Of Church Street, Minories, and 47.
And, too, the eagle eye of Science kens
Our famous heroes' Carte de Visite Lens,
Which from the rising to the setting day,
Have caus'd men of science all others by to lay.

Burton, George

George Burton, (fl.1772-1815) [8] [18] [276 #694], was a philosophical and optical instrument maker of 136 Borough, Southwark, London, England, and the son of **Mark Burton**. He supplied instruments to the mathematician and astronomer William Wales (1734-1798), for his use on James Cook's second voyage. These instruments included portable barometers, thermometers, theodolite, level, and Gunter's chain. He supplied instruments used on various other ships' voyages, including those used by Mr. W. Gooch on Captain Vancouver's voyage to survey the north-west coast of America aboard the *Discovery*.

Burton, James Haly
James Haly Burton, also known as Halyburton, (fl.1767-1774) **[18]**, was a mathematical instrument maker of Johnson's Court, Fleet Street, London, England. He was freed of the **Turners' Company** in 1761. One of his employees was named John Clack.

He may have been one of the petitioners who, in 1764, attempted to revoke the patent obtained by **John Dollond (I)** for the achromatic lens, and which was enforced after Dollond's death by his son, Peter (See Appendix X). Notably, an instrument maker named **J. Clack** was also one of the petitioners.

Burton, John
John Burton, (fl.1758-1775) **[18] [276 #453]**, was an optical instrument maker of Johnson's Court, Fleet Street, London, England. In 1764 he was one of the petitioners who attempted to revoke the patent obtained by **John Dollond (I)** for the achromatic lens, and which was enforced after Dollond's death by his son, Peter (See Appendix X).

Burton, Mark
Mark Burton, (fl.mid-to-late C18th) **[2] [18] [276 #454]**, of Denmark Court, Strand, London, England, was a renowned mathematical instrument maker. In 1756 **Jesse Ramsden** bound himself apprentice to Burton for four years. Among his other apprentices was **Thomas Clarke**. Burton's trade card, dated after 1760, showed him as a mathematical, philosophical and optical instrument maker at *The Euclid's Head*, Near St. Mary's Church, Strand. His son, **George Burton**, was also an instrument maker.

Busch, Emil
Emil Busch, (1820-1888) **[47]**, was born in Berlin, Germany, and was a maker of cameras, photographic equipment and binoculars, in Rathenow, Germany.

Emil Busch's great uncle, the pastor **Johann Duncker** of Rathenow, began to make optical instruments in 1792 and, in 1801, formed his own company. Johann Duncker's son Eduard took over the company in 1819, and Eduard's nephew, Emil Busch, joined the company in 1840. Since Eduard and his wife had no children, he sent Busch for an apprenticeship in optics in his birth city of Berlin. In 1845 Busch took over the Duncker's company, using the name Optische Industrie-Anstalt, Rathenow. He developed steam driven machines for making spectacle lenses and lenses for optical instruments. He earned a strong reputation for his Galilean binoculars and opera glasses. In 1852 he began to produce photographic cameras, and went on to develop innovations in optical lens design. In 1865 he developed his first prismatic field glasses. In 1872 the company was incorporated as Emil Busch AG. For at least part of their history, the company procured its optical glass from Zeiss. In 1908, the firm was renamed to become Emil Busch AG Optische Industrie. In 1927 **Carl Zeiss** became the majority shareholder, after which the Busch company no longer made their own lenses, instead using Zeiss lenses.

During WWII the company, assisted by the use of forced labour, manufactured military optical ordnance, including optical rangefinders for artillery, and large binoculars (10x80) for aircraft detection. They marked their instruments with **ordnance codes** cxn, czn and krq. Before the end of the war the production facilities were largely destroyed.

After the war, with the partitioning of Germany, the company branched into two companies, one in East Germany and one in West Germany.

In Rathenow, East Germany, Emil Busch AG Optische Industrie was nationalised under Soviet occupation, becoming VEB Rathenower Optische Werke, also incorporating the company formerly known as **Nitsche & Günther Optische Werke KG**. In 1966, VEB Rathenower Optische Werke became a part of VEB Carl Zeiss Jena. In 1990, following the reunification of Germany, the company was separated from Zeiss, and privatised to become **Rathenower Optische Werke GmbH**.

In Götingen, West Germany, in 1949, a new headquarters was established to form Emil Busch AG. In 1953 the West German company was renamed Emil Busch GmbH, Göttingen, which later became part of Carl Zeiss AG. See Appendix VII.

Bush, William
William Bush, (fl.1748-1763) **[18] [322]**, was a mathematical and optical instrument maker of Parish of St. Sepulchre, London, England (1748). He was freed by patrimony of the **Stationers' Company** in 1747/8. By 1752 he was trading at Amen Corner, London. He was the son of mathematical instrument maker Joseph Bush (fl.1703 d.1746), and the brother of John Bush (fl.1748-1763) and Thomas Bush (fl.1745-1766), both of whom were freed by patrimony of the Stationers' Company, and were mathematical instrument makers. His wife, Margaret Bush, was also a mathematical instrument maker of the Stationers' Company. Among his apprentices were **John Munford** and **Samuel Street**, as well as Joseph Jackson

and John Nott, both of whom were turned over to Margaret Bush in 1763.

Bushnell
Bushnell is a US brand of binoculars, spotting scopes and optical shooting accessories. The Bushnell brand is part of the **Vista Outdoor Inc.** group.

Buss, Thomas O'Dempsey
Thomas O'Dempsey Lebert Buss, sometimes spelled Odempsey, Odemcy, or Odency, (c1829-1881) [**17**] [**319**] [**322**] [**379**], of London, England, was a mathematical and philosophical instrument maker best known for making hydrometers. Addresses: 3 Upper East Smithfield (1863-1868); 33 Hatton Garden (1866-1895); 48 Hatton Garden (1896-at least 1900).

His company advertised as successors to **RB Bate**, but evidence to support the claim has not been forthcoming, and he is not mentioned in [**355**]. In 1858 he was awarded, jointly with John Adkins of Islington, a patent for 'certain improvements in ships' compasses' [**35** #22174 p3845]. The patent expired due to non-payment in 1861. The London Gazette carried a notice to creditors for demands against the estate of the deceased following his death in 1881 [**35** #25054 p6959].

Following his death, at age 52, the company continued to trade under the name Thomas Odempsey Buss. In 1891, Ellen Buss, his widow aged 36, is listed as a hydrometer manufacturer at 33 Hatton Garden, living with her two sons, Thomas Odempsey Bower Buss (1880-1939), and Odemcy Lebert Buss (b.1881). Thomas O.B. Buss may also have run the business, and Odemcy L. Buss is listed as an instrument maker in 1939.

Instruments bearing the name Thomas Odempsey Buss include: Sikes hydrometers; saccharometer; desk thermometer; microscope; slide rule.

Butcher & Sons, W.
W. Butcher & Sons, [**44**] [**62**] [**306**], of Camera House, Farringdon Avenue, London, England, was founded as W. Butcher & Son in 1866 by William Butcher (I), and his son, William (II). They initially carried out their business as a pharmacist and homeopathic chemist. In 1889 they set up a camera factory, and operated as a separate business selling their own cameras, and as a wholesaler for cameras from other makers. In 1902 the firm was renamed W. Butcher & Sons. In 1915 they were one of the companies who formed **British Photographic Industries Ltd.** In the same year, together with **Houghtons Ltd.**, they formed a joint venture, Houghton-Butcher Manufacturing Co. in order to share manufacturing space. The two companies finally merged in 1926 to form Houghton-Butcher (Great Britain) Ltd.

Butchers' Company
See Appendix IX: The London Guilds.

Butterfield, Michael
Michael Butterfield, (c1635-1724) [**8**] [**53**] [**256**] [**275** #369] [**276** #22], was an English mathematical and optical instrument maker who spent his career working in Paris, France. He traded at *Aux Armes d'Angleterre*, Quai des Morfonedes, Paris. He had a reputation for making a kind of sundial, adjustable for differing latitudes, which bears his name. He was also innovative in microscope-making, and made other instruments including astronomical rings, graphometers, brass quadrants, levels with sighting telescopes, squares, graduated rules and sectors, proportional compasses, and others. He was appointed Engineer to King Louis XIV.

Butyro Refractometer
See refractometer.

Byers Co.
Edward R. Byers Co., of 29001 West Highway 58, Barstow, California, USA, are makers of large precision telescope mounts, drive systems, heliostats and research-grade telescopes. The company was established by Edward R. Byers in the 1950s.

Byrne, John
John Byrne, (fl.1847) [**184**] [**186**] [**188**] [**189**], was a telescope maker in New York, USA. He served his apprenticeship with **Henry Fitz**, beginning in 1847, and continued to work for him until Fitz' death in 1863. Following the death of Fitz, John Byrne began to build telescopes using his own name. By 1892 he was making objectives for **Gall & Lembke** telescopes. A catalogue of Byrne's telescopes shows portable telescopes from 3 inches to 5½ inches aperture, and observatory telescopes from 6 inches to 9 inches.

Bywater & Co.
Bywater & Co., (fl.1821) [**3**] [**18**], were opticians, stationers, and map and chart sellers of 20 Pool Lane, Liverpool, England. The company was formed by **John Bywater**, and was succeed by **John Bywater & Co.**

Bywater & Co., John
John Bywater & Co., (fl.1822-1831) [**3**] [**18**], were opticians, mathematical instrument makers, stationers, and Navigation Warehouse of Liverpool, England. Addresses: 20 Pool Lane (1822-1831); 18 Pool Lane (1824); 19 Pool Lane (1825-1827); 42 Seymore St. (1825-1827). The company was formed by **John Bywater**, succeeding

Bywater & Co. in business. They were succeeded by **Bywater, Dawson & Co.**

Bywater, Dawson & Co.

Bywater, Dawson & Co., [3] [18] [322], were optical, mathematical, and nautical instrument makers, and map sellers, of 20 Pool Lane, Liverpool, England. The company was a partnership, formed in 1829, between **John Bywater**, Samuel Dawson, and Edward Melling, and continued after John Bywater's death, until 1836.

The notice of termination of the partnership, [35 #19482 p941], stated: "We the undersigned, Samuel Dawson and Edward Melling, of Liverpool, in the county, of Lancaster, Opticians and Stationers, surviving partners of the late John Bywater, trading under the firm of Bywater, Dawson, and Co. under articles of partnership, for the term of seven years, ceased and expired on the 1st day of September last: As witness our hands, 31st day of March 1837. Samuel Dawson, Edward Melling."

Instruments bearing their name include: octant; portable compass; marine compass.

The company was succeeded by **Dawson & Melling**, optician, stationer, and navigation warehouse, trading in South Castle Street, Liverpool.

Bywater, John

John Bywater, (fl.1813-1835) [3] [18] [276 #1288] [322], was an optician and stationer of Liverpool, England. Addresses: 9 Mount Pleasant (1813-1814); 49 Gloucester St. (1816); 52 Gloucester St. (1818-1824); 42 Seymour St. (1832-1835); 44 Seymour St. (1832-1835). By 1821 he was trading as **Bywater & Co.**, and by 1822, **John Bywater & Co.**, opticians, mathematical instrument makers, and navigation and stationary warehouse, in Pool Lane. He then traded in partnership as **Bywater, Dawson & Co.**, from 1829 until his death in c1835.

C

Cahill, Patrick
The Cahill family business, (fl.1876-1922) [34] [215] [304], was a firm of opticians and mathematical instrument makers of 13 and 17½ Wellington Quay, Dublin, Ireland. The Cahill business shared these two addresses with the business of **P.J. Keary**, and additionally traded at 20 Wellington Quay around 1883, and number 15 from 1915-1922.

Patrick Cahill founded the business in 1876 and, in 1887, began to trade as Patrick Cahill & Co., then from 1888 to 1891 as Patrick Cahill & Son. The business also variously traded as Patrick Cahill Junior, K. Patrick Cahill, Patrick Cahill Ltd. "Opticians to his Holiness Pope Pius X", and finally Patrick Cahill from 1919 until 1922.

A three-draw spyglass dated c1900 bears the name Patrick Cahill.

Cail, John
John Cail, (fl.1825-1865) [3] [16] [18] [276 #1500], was a mathematical, philosophical, nautical, and optical instrument maker of 2 New Bridge St., Newcastle upon Tyne, England.

John Cail worked for **Edward Troughton** for a period which, according to [44], was the early 1820s. He went on to start his own business in Newcastle. In 1838 he moved to 61 Pilgrim St., and by 1844 he had additionally taken on premises at 45 Quayside. He is known to have acquired agricultural levels from **Thomas Cooke**, which he engraved with his own name, and re-sold.

His advertisement of 1844, [53], reads "John Cail, Mathematical and Philosophical instrument maker, Optician, &c. 61 Pilgrim Street, and 45 Quayside, Newcastle. Surveying & drawing instruments, Chronometers, Sea charts for all parts of the world, Ivory scales, Sextants, Quadrants, Telescopes, Compasses, Barometers, Thermometers, Spectacles, &c. &c. Ships' binnacles of every description made to order, and on hand."

From 1844 until 1855 he worked in partnership with **Septimus Anthony Cail** who, according to [44], was his brother, trading as **J.&S.A. Cail**. Following the termination of the partnership John Cail continued to work under his own name, trading at 21 Grey St., Newcastle upon Tyne until 1865. This is the same address as the business of **T.B. Winter**, who succeeded him in business at that address. A sextant bearing his name is in the National Maritime Museum, [54].

Cail, J.&S.A.
J.&S.A. Cail, [18], were opticians, and mathematical, nautical, and philosophical instrument makers of 45 Quayside, Newcastle upon Tyne, England. This was a partnership between **John Cail** and **Septimus Anthony Cail**, trading from 1848 until 1855.

Cail, Septimus Anthony
Septimus Anthony Cail, (fl.c1840-1855) [16] [18] [276 #1095], was an optician, and mathematical, nautical, and philosophical instrument maker of Newcastle upon Tyne, England. According to [44] he was the brother of **John Cail**, with whom he worked in partnership as **J.&S.A. Cail** from 1848 until 1855. An octant bearing his name is in the National Maritime Museum, [54].

Cairns, Alexander
Alexander Cairns, (fl.late-C19th) [58] [322] [379], was a nautical, optical, and philosophical instrument maker of Waterloo Road, Liverpool, England. At various times he traded at numbers 12, 13, and 32 Waterloo Rd. In 1867 he applied for a patent for "improvements in the construction of liquid compasses" [35 #23211 p380]. Instruments bearing his name include: single draw telescope in brass with leather cover; sextant; sextant in mahogany case; octant; quadrant; marine chronometer; compass binnacle; wheel barometer; marine barometer with a sympiesometer, and incorporating an aneroid barometer.

Callaghan & Co.
See **William Edmund Callaghan**.

Callaghan, William
William Callaghan, (b.c1817-1874) [18] [319] [320] [322], was an optician of London, England. He worked for **Thomas P. Harris**, and the 1841 Census lists him as an Optician's Apprentice in Harris' household at 51 Great Russell Street, Bloomsbury. By 1845 he was trading at 45 Great Russell Street, Bloomsbury. He is known to have sold spectacles, octants, telescopes, perspective glasses, and microscopes, as well as opera, marine, field and race glasses.

In the catalogue of the Great Exhibition of 1851, W. Callaghan of 45 Great Russell Street, is listed as a manufacturer, showing an improved deer-stalking telescope and a pair of portable steel spectacles [328]. A bellows camera based on the design by **C.G.H. Kinnear** bears his signature. He advertised as the sole agent for **Voigtländer** opera glasses and race glasses and, during the 1860s, as agent for Voigtländer portrait and landscape photographic lenses. In 1855 he announced in *The Times* that he had relocated to 23a New Bond Street, where he traded until his death in 1874. The 1861 Census also shows him located at 32 Conduit Street, Mayfair. He was succeeded in business by his son, **William Edmund Callaghan**.

Callaghan, William Edmund
William Edmund Callaghan (c1853-1890) [**18**] [**319**], was an optician of 23a New Bond Street, London, England. He was the son of **William Calaghan**, and was freed of the **Spectacle Makers' Company** by purchase in 1875. He continued the business of his father, and by 1877 the company was advertising as Callaghan & Co. His son, Edmund Ford Callaghan (c1880-1946) was also a manufacturer and dealer in optical instruments. William Edmund's death was announced in the London Gazette in 1890 [**35** #26087 p4960], after which Edmund Ford Callaghan continued the business at 23a New Bond Street until at least 1909.

Calotype
Calotype, [**70**], is a photographic technique developed by **William Henry Fox Talbot** which uses a sheet of paper sensitised with silver iodide. The sheet is exposed to light, producing a negative image. The negative can be used to produce multiple positives, unlike the Daguerreotype, which can only produce one positive image. A Calotype positive image is reproduced by means of contact printing, and fixed by washing with a fixing liquid.

Calver, George
George Calver, [**8**] [**76**], was a notable mirror maker from near Chelmsford, England, working around 1870. He made, or re-figured, over 2000 mirrors, perhaps as many as 4000, and supplied mirrors to **Ottway**.

Cambridge & Paul Instrument Company
See **Cambridge Scientific Instrument Company**.

Cambridge Instrument Company Ltd.
The Cambridge Instrument Company Ltd., [**44**] [**320**] [**444**], was a manufacturer of scientific, medical, and optical instruments, with their head office at 13 Grosvenor Place, London, England. The company was established in 1924 when the Cambridge & Paul Instrument Company, (formally the **Cambridge Scientific Instrument Company**), became a public limited company, and was renamed to become the Cambridge Instrument Company Ltd. They made scientific and medical instruments, including electron microscopes. In 1934 the foreman, S.W.J. Stubbens, left to form **Unicam Instruments**.

Cambridge Medical Instruments Ltd., (CMI), was incorporated in 1935, and in 1990 changed its name to Cambridge Instruments Ltd., [**315**]. It is believed to have been the medical instruments division of the Cambridge Instrument Co. Ltd.

In 1968 the company was taken over by the George Kent Group, and there followed a series of mergers and acquisitions, including in 1986 a merger with **American Optical Company**, who already owned **C. Reichert, Optische Werke**. In 1987 they acquired the Optical Systems Division of **Bausch & Lomb**. Microscopes and eyepieces, made in the USA, can be found bearing the name *Cambridge Instruments*. In 1990 the company merged with the Wild Leitz AG Group, resulting in the formation of **Leica Heerbrugg AG**.

Cambridge Optronics Ltd.
Cambridge Optronics Ltd., [**315**], of 10 Albany Road, Bedford, Bedfordshire, England, was formed in 2007 by the optical designer, Keith Dunning, and the industrial designer, Richard Dickinson (who was formerly Head of Design for Sinclair UK), by renaming **Science of Cambridge Ltd.** Dunning and Dickinson designed several folded optics microscopes, including the **Lensman** which they produced themselves, the **Meade** *Readview*, and both the *Micron* and *Trekker* made by **Enhelion**. They also produced the **Newton NM1** [**314**], which was a folded optics microscope, in several variants, for the Millennium Health Foundation charity. The NM1 was manufactured in China, and retailed at a significantly higher price than many folded optics microscopes of the time. The company was dissolved in 2017.

Cambridge Scientific Instrument Co. Ltd.
The Cambridge Scientific Instrument Company, [**44**] [**320**] [**444**], was established in 1881 by **Horace Darwin**, and Albert George Dew-Smith (a lens grinder from the observatory at Cambridge University), with William T. Pye (father of **William G. Pye**) as foreman. In 1919 the company was listed as a member of the **British Optical Instrument Manufacturers' Association**. In 1919/20 they undertook a merger with the **Robert W. Paul Instrument Co.**, and changed their name to the Cambridge & Paul Instrument Company. In 1924, upon becoming a public limited company, they changed their name again to become the **Cambridge Instrument Company Ltd.**

Camera
A camera is a device for taking photographs, moving pictures, or video signals. It uses an aperture and/or a lens to focus an image on a light-sensitive film or sensor.

Box camera: A simple form of camera consisting of a rigid light-proof box with an objective lens and shutter at one end, and film at the other end, which is at the focus of the objective lens.

Cine camera: A camera used to record moving pictures.

Folding camera: A camera which may be folded when not in use. The objective lens is normally mounted on a plate at one end of collapsible bellows. The other end of the bellows is fixed to the camera body, which contains the film. When the bellows are fully extended the film is at the focus of the objective lens.

Single-lens reflex (SLR): An SLR uses a *reflex mirror* behind the objective lens, usually together with a pentaprism, to enable the user, when looking through the viewfinder, to view through the same lens that is used to form the camera image. This means that the reflex mirror is in the path from the objective lens to the film or sensor. In order to take a photograph, either the mirror is raised when the shutter is triggered, or some other method such as a partially-reflecting mirror is used.

Twin-lens reflex (TLR): A TLR has two objective lenses. One to form the image for the photograph, and one for the viewfinder. The two lenses are usually adjacent and close together, and a *reflex mirror* is used to direct the viewfinder image to a convenient angle. Since the viewfinder objective is close to the photographic objective, parallax is kept to a minimum.

Rangefinder camera: A rangefinder camera has a small, built-in coincidence rangefinder (see rangefinder). In some cases, this is integral with the viewfinder. The user views directly through the rangefinder window to the subject, which is brought into coincidence with an image formed, using mirrors and/or prisms, through a second, separate window in the camera body. The rangefinder is adjusted until the two images coincide, and the adjustment indicates the distance to the subject. In some cases, this is mechanically coupled to the focus mechanism of the objective lens. Since the baseline of the rangefinder (the distance between the two windows) is constrained by the size of the camera, accuracy is limited.

Video camera: A video camera is one that is used to view, and sometimes to record, moving images either digitally or on tape.

Viewfinder camera: A camera in which the viewfinder is separate from the image-forming optics of the camera. This arrangement suffers with parallax, which is more of an issue with close subjects than with distant ones.

See also ambrotype, Daguerreotype, ferrotype, interference colour photography, reflex mirror, tintype. The French website *Collection Appareils*, [272], shows more than 10,000 camera models for a large number of different brands. See also [70] [71] [349] [350] [351].

Camera Lucida
A camera lucida is an optical instrument consisting of a prism or an arrangement of mirrors, which allows the user to view an image, such as a microscope image, simultaneously with a surface, such as paper or canvas. The image appears superimposed on the surface, enabling its outline to be traced. The camera lucida was patented in 1806 by **William Hyde Wollaston**. See also optigraph.

Camera Obscura
In its earliest form, the camera obscura operated as a darkened room or chamber into which light was introduced via a small hole in one wall. This produced an inverted image on the opposite wall [45] [70]. The principle was known to Aristotle (384-322 BCE), and it was described in around 1000 AD by the notable Islamic physicist **Al Hasan Ibn al-Haytham**. A lensless camera of this kind is sometimes referred to as a pinhole camera, and may be used to form an image on a photographic medium such as photographic paper.

The term camera obscura also later became used for an optical device, based on the original concept that, by means of a lens, prism, or mirror, creates an image on a white surface within a darkened chamber. This was used for observation, or as an aid to sketching.

Camerawerk Sontheim
See **Nettel Camera-werke**.

Cameron & Blakeney
Cameron & Blakeney [18] [263] [271], were mathematical, nautical, and optical instrument makers of Glasgow, Scotland. The partnership was formed in 1853 between **Paul Cameron** and **John Blakeney**, and continued until 1860. Addresses; 76 Great Clyde St. (1853-1855); 102 Great Clyde St. (1856-1859); and finally at 94-96 Jamaica St. When the partnership was dissolved, both went on to work under their own names.

An octant by Cameron & Blakeney is in the National Maritime Museum, Greenwich [16] [54]. An azimuth dial, and a 1½ inch 3-draw telescope in brass and leather, are in the National Museums, Scotland [262].

Cameron, Paul
Paul Cameron, (fl.1851-1869) [263] [271], was an optical, mathematical, and philosophical instrument maker of 87 London St., Glasgow, Scotland. He succeeded **Robert Finlay** in business at this address.

He exhibited at the Great Exhibition in 1851 in London [294] [328]. His catalogue entry no. 356 stated: "Cameron, Paul, 87 London Street, Glasgow—Inventor. Azimuth compass, adapted to solve various problems in nautical astronomy, practical navigation, and civil engineering.

Engineer's improved indicating level. Mathematical and nautical slide rule, for the use of engineers and naval officers. Improved thermometer, steam and vacuum gauge."

In 1853 he entered a partnership with **John Blakeney**, trading as **Cameron & Blakeney**. This partnership was dissolved in 1860, after which Paul Cameron traded as Paul Cameron & Co. until 1864, at 11 & 19 Howard Street. In the 1861 Census, he was listed as a "Mathematical and nautical instrument maker employing 12 men and 5 boys."

In 1864 he was subject to sequestration proceedings, brought against him by brass-founder and gas-fitter Hugh Buchan. This was dismissed in September 1864 [**170** #7468 p1811], but the next day new similar proceedings were brought by the Sheriff of Lanarkshire [**170** #7469 p1193], and were not concluded until 1865.

During 1865 he traded as **Houston & Cameron** for a short period, after which he reverted to Paul Cameron & Co. at 25 Howard St. (1866-1867); 2 York Place (1868); 178 Broomielaw (1869).

Campana, Guiseppe
Guiseppe Campana, [**256**], was a contemporary of **Eustachio Divini** and **Guiseppe Campani**, and was an Italian optical instrument maker. The **Nachet** collection of microscopes includes a microscope signed by Guiseppe Campana.

Campani, Guiseppe
Guiseppe Campani, (c1635-c1715) [**8**] [**53**] [**256**] [**388**], was an Italian astronomer and optical instrument maker. He learned his lens-grinding skills in Rome, and specialised in grinding large lenses. He worked both in Rome and Bologna. His lenses were made from Venetian glass, and were made using a lens-grinding lathe of his own design. He established his reputation making telescopes and microscopes, and is believed to have originated the concept of a screw-barrel for microscope focussing. He was a contemporary, and rival of **Eustachio Divini**, who had a similar business, and both of whom supplied lenses to the astronomer Giovanni Domenico Cassini (1625-1712).

Campani published the results of his astronomical observations. In 1664-1665 he observed with his own instruments the moons of Jupiter, discovered by Galileo only 55 years earlier, and the rings of Saturn.

According to [**274**], all signed microscopes by Campani are made of cardboard or wood. Campani was the son-in-law of the German optician, **Johann Wiesel**.

Christian Huygens wrote about one of Campani's telescopes saying, "The beauty of the Campani telescope of the Abbé Charles (de Bryas) is that it is without any iris coloration and that there are no visible defects in the eyepieces; the aperture is fairly large, without making the objects appear curved, and finally it gives a very clear image because of the quality of the lenses."

Following Campani's death, his collection of optical instruments was acquired by Pope Benedict XIV for the Instituto di Bologna. Some of his instruments and lenses, including an aerial telescope, are at the Paris Observatory, and more are at the Conservatoire National des Arts et Métiers in Paris.

Campbell, W.
W. Campbell, (fl.1820-1831) [**18**], was an optician who worked in partnership with **William Harris (I)**, trading as opticians, and mathematical and optical instrument makers **William Harris & Co.** at 50 High Holborn, London, England, and as W. Campbell & Co. at Bey dem Rathause no. 26, Hamburg, Germany. The firm is known to have sold globes, compasses, balances, and microscopes.

Campbell-Stokes Recorder
A Campbell-Stokes sunshine recorder is a device used for measuring the hours and intensity of sunshine. It comprises a glass sphere which focuses the sun's light onto a card upon which the focused beam burns a mark. The burn progresses across the card as the sun moves across the sky. Cloudy periods appear as gaps in the burn, and the brighter the sunlight the more intense the burn.

Canzius, J.H. Onderdewijngaart
Jacob Hendrik Onderdewijngaart Canzius, (1771-1838) [**256**] [**264**] [**274**] [**374**] [**367**], was a lawyer of Delft, Netherlands. When the French invaded in 1796, his legal practise failed because of his allegiance to the House of Orange. He underwent a change of career and, in 1797, he set up a workshop in Delft manufacturing scientific instruments. Some of the instruments were his own inventions, but he did not make the instruments himself – he employed craftsmen, recruited both locally and from Germany, to do the manufacturing. By 1800 his business employed 40 people, and was reputed to have seventeen departments, with specialities ranging from lens and mirror-polishing and brass-casting to wire-drawing and carpentry.

In 1803, he expanded his business by opening a second factory in Delft. His catalogue of 1804 offered about 650 items, including: Mathematical, physical, and chemical instruments; Constructional and mechanical instruments; Surgical tools; Musical instruments; Anatomical preparations; and more. Many of the chemical and electrical instruments in the catalogue, as well as a portable fire-engine, were the inventions of Martinus van

Marum [362]. Canzius's company supplied many instruments to Van Marum, and was the manufacturer of at least some of the microscopes bearing the name of **Louis François Dellebarre**.

Canzuis received a gold medal for his exhibit at the Utrecht exhibition of national industry in 1808, by which time a period of rapid expansion of the business had left him in considerable debt. As a consequence, the business closed in 1810 and Canzius's houses, factories, and their contents were sold at public auction. His subsequent career was varied, including a short spell as a museum director in Brussels.

Carena
Carena was a brand-name used by the German camera retailer **Porst** for their re-branded cameras.

Carlaw, David
David Carlaw, (1832-1907) [262] [263], was a mathematical instrument maker and engineer of Glasgow, Scotland. He began a seven-year apprenticeship with **Thomas Rankine Gardner** in 1846, after which he worked for "Messrs. Hommersley" in London before returning to Glasgow (this may be a misspelled reference to **J. Hammersley**). Upon his return, he worked for **James White**, rising to the position of manager, and worked on the instruments used in the laboratory of **William Thomson** of Glasgow University. In 1860 he set up his own business as an instrument and model maker, producing such instruments as theodolites and other surveying instruments, compasses, and spirit levels. His model making included working models of steam engines as well as other engineering models and machines.

Glasgow addresses: South Portland St. (1860); Sydney Court, 62 Argyle St. (1865-1871); Havelock Buildings, 75 East Howard St. (1872-1874); Ropewalk Lane (1875-1897); 81 Dunlop St. (1892-1897).

In 1894 the company was renamed David Carlaw and Sons, in which he worked together with his three sons until his death in 1907. The firm continued to prosper, diversifying into other areas including motor cars.

Carlill, John
John Carlill, (fl.1834-1837) [18] [276 #2093], was a mathematical, optical, and nautical instrument maker and ship chandler, of Northside Old Dock, Hull, England.

Carpenter, Philip
Philip Carpenter, (1776-1833) [3] [18] [53] [264], was a mathematical, optical and philosophical instrument maker of Birmingham, England. Philip had a keen interest in microscopy, and made microscopes, kaleidoscopes, projectors, and magic lanterns and slides, as well as other optical instruments. His nephew was the famous naturalist William Benjamin Carpenter, who published *The Microscope and its Revelations*. Philip traded at Inge Street from 1808-1812, then from 1815-1828 at Bath Row. He took on additional premises at 111 New Street in 1823, and in 1829 he took on premises at 33 Navigation Street. In 1830 the address at 33 Navigation Street was taken over by **Robert Field**, who had been his foreman at the factory in Bath Row. Carpenter continued his business at 111 New Street until his death in 1833, and was succeeded by his sister, Mary. In 1835 she entered a partnership with William Westley, and the business became **Carpenter & Westley**.

Carpenter & Westley
Carpenter & Westley, [8] [18] [53] [264], were globe makers, and mathematical, optical and philosophical instrument makers of 111 New Street, Birmingham, and 24 Regent Street, London, England. The company succeeded **Philip Carpenter** in 1835 following his death, and was a partnership between his sister, Mary, and William Westley. The company continued in business until 1914.

Carpentier, Jules
Jules Carpentier, (1851-1921) [53] [335] [410], was an inventor and scientific and electrical instrument maker, based in Paris, France. He studied at the École Polytechnique, after which he served a brief apprenticeship as a fitter in the workshops of a railway company. Following the death in 1877 of the physicist Heinrich Daniel Ruhmkorff, Carpentier purchased his workshops in ruc Campollion, Paris at auction. He restructured and modernised Ruhmkorff's company, introducing new methods and tools, and began to manufacture electrical instruments. He named the firm Ateliers Ruhmkorff-J.Carpentier, Ingénieur Constructeur. In c1878 he relocated the firm to 20 rue Delambre, and soon expanded into numbers 14, 16, 18, and 22. He also took on premises at 98 Boulevard Montparnasse. The firm was, and continued to be, primarily a manufacturer of high-quality electrical equipment.

He began to make cameras in the 1890s, being noted for his *photo-jumelle a répétition* models. The name means binocular, but although the *photo-jumelle* had two lenses, it was not a stereo camera, one of the lenses being used for the viewfinder. These were marketed in England by the **London Stereoscopic Company**. He also introduced the *focomètre* and the *focograde*, used to measure the characteristics of eyepieces.

In 1895 Carpentier began to manufacture the *Cinématographe* for the **Lumière** brothers, of which he went on to manufacture hundreds. In 1909 he collaborated with Lumière to produce the *Cinématolable* camera, using 35mm film, which he went on to manufacture. Jules Carpentier is thought to have been the first to patent, in 1896, the Geneva drive, otherwise known as a Maltese cross, used to advance the film in a projector. He also designed submarine periscopes and trench periscopes.

He invented and produced two devices for use with a piano. One attached to the piano and recorded the keystrokes onto a paper roll, and the other attached to the piano and replayed the keys from the recorded paper roll.

Jules Carpentier's death in 1921 was the result of a car accident. The company was renamed Societé Anonyme des Ateliers Jules Carpentier, with his son, Jean, as the general director. They merged with Société Anonyme des Industrie Radio-Électrique in 1939, and shortly after this the company was sold, and the name Carpentier ceased to be used.

Carrew, John William

John William Carrew, otherwise known as I.W. Carew, (fl.1835-1866) [**18**] [**276** #2094], was a nautical, mathematical, and philosophical instrument maker of 18 Wapping Wall, London, England (1835-1838), and 13 Wapping Wall (1839-1866). He is known to have sold: sextant.

A possibly related notice appeared in the London Gazette on 7th December 1858 [**35** #22207 p5369], which read, "NOTICE is hereby given, that the Partnership heretofore subsisting between us the undersigned John William Carrew and James Rule, as Watch and Chronometer Makers, at No. 15, Fenchurch-street, in the city of London was on the 29th day of September now last past, dissolved by mutual consent; and all debts due to the said partnership will be received by the said John William Carrew.—Dated this 4th day of December, 1858. Jno. Wm. Carrew. James Rule."

Carroll, George A.

George A. Carroll, (1902-1987) [**185**] [**202**], of Texas, USA, began life in aviation, flying and building aircraft. His interest in astronomy stemmed from seeing Halley's comet when he was 8 years old, and he built his first telescope in 1934. In 1940 he joined the Vega Aircraft Corporation as Staff Engineer-Head of Service Design, and soon after that the company merged with the Lockheed-California Company. Carroll joined the staff of the Lockheed Solar Observatory and built telescopes for use there. In 1947, together with others, he formed an amateur astronomy association that was to become, in 1957, the Stony Ridge Observatory. He designed and built the 30-inch reflecting telescope for the observatory, and procured funding from Lockheed for the observatory in exchange for telescope time for its lunar mapping project, taking high resolution images of the moon's surface to survey for landing sites for the Apollo programme. Asteroid *144633 Georgecarroll*, discovered by members of the Stony Ridge Observatory, was named after him.

Carte du Ciel

The Carte du Ciel [**352**] was a project, started by Ernest Mouchez (1821-1892), Director of the Paris Observatory, in 1887. The project, inspired by the sky-charts and astrograph design of **Paul & Prosper Henry**, ultimately involved 22 observatories around the world using specially designed astrographs. The objective of the *Carte du Ciel* project was to produce a complete astrographic catalogue of the sky and, from it, produce a full set of sky-charts.

The astrographs for all of the French observatories, as well as some others, were supplied by Paul & Prosper Henry, with mounts supplied by **Paul Gautier**. **Howard Grubb** supplied the objectives for all of the astrographs used in observatories in countries of the British Empire, as well as supplying most of the mounts for them. The mount for the astrograph used at the Vatican observatory was supplied by **Henry Guinand**.

By the time the last results were published in 1962 the astrographic catalogue was complete, but the sky-charts were only half completed. The project was terminated in 1970.

Carton Optical Industries Ltd.

In 1930 Carton Optical Industries, Ltd. was founded as a wholesaler of spectacle frames in Tokyo, Japan. In 1951 they began importing and exporting various kinds of optical goods as a trading company. In 1964 they established their own factory in Saitama to manufacture optical goods such as microscopes, binoculars, telescopes, magnifiers and ophthalmic instruments under a registered *Carton* brand. Instruments made by Carton Optical Industries Ltd. may bear the maker's mark COC (see Appendix IV).

Carton Optical Canada Inc. (Valley Microscope)

Carton Optical Canada Inc., operating as Valley Microscope, is the exclusive Canadian wholesale & retail distributor of *Carton* brand microscopes, magnifiers & binoculars made by **Carton Optical Industries Ltd.** Their binocular range includes **Alderblick** binoculars imported from Japan, bearing the name *Carton Alderblick*.

Cary, George

George Cary, (c1788-1859) [**18**] [**276** #1290], of 181 Strand, London, England, was the nephew of **William Cary**, and the son of **John Cary (I)**. He served his apprenticeship with his father, beginning in 1802, and was freed by patrimony of the **Goldsmiths' Company** in 1827. He worked in partnership with his brother, **John Cary (II)**, trading as George & John Cary [**276** #1290]. Together they also continued William's business after his death, trading as William Cary.

Cary, John (I)
John Cary (I), (1754-1835) [**18**] [**276** #695], was a cartographer of London, England, and the brother of **William Cary**. He served his apprenticeship with the engraver William Palmer of London, beginning in 1770, and was freed of the **Goldsmiths' Company** in 1778. **George Cary** and **John Cary (II)** were his sons. As well as carrying on his own business as an engraver and seller of maps, charts and globes, John (I) worked in partnership with his brother, trading as John & William Cary. He was succeeded in business by his sons George & John Cary.

Cary, John (II)
John Cary (II), (1791-1852) [**18**], of 181 Strand, London, England, was the nephew of **William Cary**, and the son of **John Cary (I)**. He was freed of the **Goldsmiths' Company** by patrimony in 1840. He worked in partnership with his brother, **George Cary**, trading as George & John Cary [**276** #1290]. Together they also continued William's business after his death, trading as William Cary.

Cary, William
William Cary, (1759-1825) [**18**] [**46**] [**50**] [**264**] [**276** #810], was a scientific and optical instrument maker in London, England, who served his apprenticeship with **Jesse Ramsden**. His brother, **John Cary (I)** was a cartographer. William and John (I) worked together making globes, compasses and other instruments, trading as John & William Cary.

William started his own business in about 1785, in the Strand, London, making scientific instruments. He made sextants, microscopes, theodolites, and reflecting and refracting telescopes. **Henry Porter** served an apprenticeship with him. William Cary equipped **Francis Wollaston**, as well as observatories including Zürich and Königsberg, with circles and telescopes. Records show that he marketed about three quarters of the malleable platinum produced by **William Hyde Wollaston** [**128**]. William Cary also invented the box-mounted portable microscope, announced in 1826, after his death, by **Charles Gould**. Charles Gould is believed to have been the manager and head machinist for the Cary business, and **Henry Gould** is believed to have been Cary's foreman.

After William's death the business continued under the name of William Cary until about 1853, run by his nephews **George Cary** and **John Cary (II)**, sons of his brother John (I). How and when the business transitioned into the hands of the Gould family is unclear, but by 1856 **Charlotte Gould**, following the death of her husband Henry, continued the business as Gould late Cary. In 1863 Henry Porter acquired an interest in the business, and sometime between 1874 and 1876 he is believed to have become the sole owner. Henry Porter continued to use the name Cary, and is known to have been trading as Cary & Co. at 7 Pall Mall, London, until at least 1898. Upon Henry's death in 1902, his sons Sydney and Clement inherited the business of Cary Porter, Ltd., and continued to maintain the reputation of their predecessors. By 1931, however, the business was no longer in operation.

George Cary and John Cary (II) also continued the cartography business of John Cary (I) after his death, trading as G.&J. Cary, although they ceased map-making in about 1846, and ceased trading as G.&J. Cary in 1850.

Casartelli
The Casartelli family, [**3**] [**18**] [**50**] [**126**] [**384**] [**438**], originally from Tavernerio, Italy, were scientific and optical instrument makers of Liverpool and Manchester, England. The family is known to have intermarried with the **Ronchetti** family, also from Tavernerio, in at least two generations.

Louis Casartelli, (fl.1821-1848), also known as Lewis, or Luigi Antonio, was an optician and philosophical instrument maker. He took over the Manchester business of Baptist Ronchetti upon his retirement in c1810. In c1815 he exchanged his business with that of Charles Joshua Ronchetti, and moved to Liverpool, trading in King Street, and later in Duke Street. He was succeeded by the partnership Anthony and Joseph Casartelli, (fl.1845-1849), at 20 Duke Street. The Joseph in this partnership was Joseph Louis Casartelli.

Joseph Louis Casartelli, (1823-1900), also known as Giuseppe Luigi, was the most renowned member of the family. He emigrated from Tavernerio to join his family's business making barometers and thermometers in Liverpool. He continued the business at 20 Duke St. in Liverpool, and in 1851 he married Jane Harriet Ronchetti. In 1852 he succeeded Charles Joshua Ronchetti in his business at 43 Market Street, Manchester, when Charles Ronchetti retired to Italy. Joseph Casartelli's company produced optical instruments, including microscopes, telescopes, cameras, and microphotograph slides. They also sold instruments made by other makers, but bearing their name.

Joseph Louis Casartelli's son, Joseph Henry, joined the business in 1882. In 1896 Joseph Henry was made a partner, and the business began to trade as Joseph

Casartelli & Son. The company continued to trade under that name, and in 1919 Joseph Casartelli & Son of 43 Market St., Manchester was listed as a member of the **British Optical Instrument Manufacturers' Association**. In 1922 the company moved to 18 Brown Street, Manchester. Following the depression of the 1930s the Casartelli name continued as three separately owned companies, two in Manchester, and one in Liverpool, which subsequently became Casartelli Instruments Ltd.

Cassegrain Telescope
See telescope.

Casella, Charles Frederic
Charles Frederic Casella, (1852-1916) **[44] [46] [444]**, a scientific instrument maker of London, England, was a son of **Louis Pascal Casella** and Maria Tagliabue. Upon the death of his father in 1897, Charles took control of the business **Louis Casella & Co.**, and renamed it C.F. Casella. Though the company name was changed they continued for some time to produce catalogues and instruments under the name of Louis Casella.

Charles was a poor businessman, and despite his interest in the instruments themselves, the company went into decline. Recognising this, he employed Rowland Miall as business manager, and Robert Abraham, an engineer with capital to invest. Between the three of them they revitalised the company, and in 1910 it was incorporated as C.F. Casella & Co. In 1911 they published their catalogue of surveying and drawing instruments **[238]**. In 1919 the firm was listed as a member of the **British Optical Instrument Manufacturers' Association**. Their optical instruments included sextants, theodolites and telescopes.

In 1947 they exhibited at the British Industries Fair. Their catalogue entry, **[450]**, listed them as manufacturers of: Meteorological, Scientific, Industrial, Astronomical, Laboratory, Photogrammetric, and Lighthouse instruments; Air filters; Bacteria counters; Cathetometers; Dust counters; Hydrometers; Hygrometers; Manometers; Pitot tubes; Stereoscopes; Thermometers; Water samplers.

Casella, Louis Marino
Louis Marino Casella, (1842-1923) **[44] [46]**, a scientific instrument maker of London, England, was the first son of **Louis Pascal Casella** and Maria Tagliabue. He joined his father's company as a young man, and retained an interest in it while pursuing a career elsewhere. He became chairman of D. Gilson & Co., whose business was "Screws and terminals in brass or steel and all small repetition work". After WWI, D. Gilson & Co. partly amalgamated with **C.F. Casella & Co.**, who absorbed some of their production capacity.

Casella, Louis Pascal
Louis Pascal Casella, (1812-1897) **[18] [44] [46] [438]**, was born and raised in Edinburgh, Scotland, but moved to London, England, to take employment with the instrument maker **Caesar Tagliabue**. In 1838 he married Caesar's daughter, Marie Louisa Tagliabue, and entered a partnership with Caesar as **Tagliabue & Casella**. Louis and Marie's two sons, **Charles Frederic Casella** and **Louis Marino Casella** both maintained an interest in the business.

Following Caesar Tagliabue's death in 1844, Louis Casella bought out his sister in law's interest in the firm, and in 1848 he renamed the firm to Louis Casella & Co. Louis continued as a maker of a range of scientific instruments, including thermometers, hydrometers, and drawing and surveying instruments, meteorological instruments, and photographic accessories. When he died in 1897 his son Charles Frederic took over, and renamed the firm to C.F. Casella & Co.

Casson, George
George Casson, (fl.1814-1824) **[18] [276 #1291]**, was an optical, mathematical, and philosophical instrument maker of 2 George Terrace, Commercial Rd., London, England. He was freed by patrimony of the **Girdlers' Company** in 1814. He is recorded as working for "Troughton" and, although no forename is given, the dates suggest that it was **Edward Troughton**. He is known to have sold: telescope.

Catadioptric Telescope
See telescope.

Cathetometer
A cathetometer is an instrument used to take measurements of differences in vertical height at a distance from the object being measured. It consists of a horizontal telescope on a vertical mount. The telescope is moved up and down the mount, maintaining its horizontal orientation, to take measurements which are read off on a vertical scale.

Catoptric
A catoptric instrument is one which uses the principle of reflection, particularly relating to the use of mirrors.

Catoptric Micrometer
See micrometer.

Cauchoix, Robert Algaé
Robert Algaé Cauchoix, (1776-1845) **[197]**, was a French scientific and optical instrument maker near Pont Royal, Paris, France. In 1800 he married the daughter of **Jean-**

Charles Gonichon and **Paule Gonichon**. He made barometers, spherometers, micrometers, and telescope optics. He sometimes used quartz for the fabrication of his achromatic objectives, and he used glass from **Guinand** for his large telescope objectives. In 1829 he made an 11¾-inch objective with a focal length of about 19 feet, at the time the largest in the world. In 1831 he completed a 13.3-inch objective with a focal length of 25 feet for the astronomer Edward Joshua Cooper, once again the largest of its time. He also made the optics for the Northumberland Equatorial telescope built by **Troughton & Simms** for Cambridge University in 1832, a guiding refractor used on the 30-inch **Grubb** reflector at Greenwich, and many other instruments. He was succeeded in his business by his nephew M. Rossin.

Cavalieri, Bonaventura
Bonaventura Cavalieri, (1598-1647) [8] [155] [256], was an Italian mathematician, and a priest of the Jesuate order. He studied the works of Euclid, and developed his "method of indivisibles", a theory which paved the way for the later development of integral calculus. He experimented with the use of mirrors in telescopes and, although the degree of success of this work is not clear, it preceded the development by **Isaac Newton** of the reflecting telescope.

Cave Optical Co.
Thomas R. Cave, (1923-2003), of California, USA, served in the army during WWII, and afterwards worked for Herron Optical Co. He started his own business, Cave Optical Co., in 1950 in Long Beach, California, working on government optical contracts. Cave started re-figuring mirrors for amateur astronomers on a spare-time basis, and soon began to make telescopes, specialising in long-focus Newtonians which were ideal for planetary viewing. He sold his telescopes under the *Astrola* brand-name. Cave had several mirror-makers, one being **Alika Herring**, who left the company at the end of the 1950s. Tom Cave was a good friend of John Krewalk Sr. of **Criterion Mfg., Inc.** and, rather than competing with each other they would help each other out when the work-load was too high.

Cave Optical built thousands of telescopes, and an estimated 83,000 primary mirrors, many of which were made for the aerospace industry. Cave's biggest mirror was 36", built for the Mexican Defence Ministry. They also made about 200 refractors. They made the 6-inch lenses themselves, and had their 4-inch lenses made by **A. Jaegers** of Lynbrook, New York.

Tom Cave sold the business in 1979 for reasons of health, and it ceased trading soon after. In 1999 he transferred the trade-name *Astrola* to one of his ex-employees, **Larry Hardin**.

Cavers, William
William Cavers, (fl.1811) [18], was an optical instrument maker of 11 Mount Pleasant, Gray's Inn Lane, London, England.

C&D (Scientific Instruments) Ltd.
C&D (Scientific Instruments) Ltd., [319] [320], was an optical instrument manufacturer of Hemel Hempstead, Hertfordshire, England. The company was formed in 1942 by two optical engineers, named Chandler and Dicker. Initially they traded as Chandler and Dicker, and worked on Ministry of Defence contracts. At some point, they changed their name to C&D (Scientific Instruments) Ltd. During the 1960s they entered the education optics market, producing magnifiers, microprojectors, microscopes, microtomes, lamps, and projection microscopes, as well as a specialised microscope for inspection of record player styluses. During the 1970s the company faced competition from low-cost Japanese imports. To face this challenge, they began to produce industrial optical instruments. In 1985 the decision was made to enter voluntary liquidation, which was completed in 1986 [35 #50699 p14010]. Instruments bearing their name include: cast metal and brass theodolite with nameplate which reads *Chandler and Dicker, London*; 1941 WW2 spirit level marked *Level spirit Nº2 Mk1* by Chandler and Dicker; small field microscope in conical metal case, marked *The Field C&D London*.

CDGM
CDGM is a US optical glass manufacturer.

Celestron
Celestron is a **Synta** brand of telescopes and related equipment [136]. The company was founded by the American Tom Johnson in the late-1950s as Valor Electronics to manufacture electronic components for the aerospace market. The movement from electronics to telescopes was prompted by Tom's interest in making a six-inch reflecting telescope, but one that might be more compact and suitable for his two young sons. At first a hobby, this developed into a business that underwent several transformations before Celestron emerged.

In 1980, Celestron was acquired by Diethelm Keller Holding, based in Zurich, Switzerland. It was then acquired by **Tasco** in 1998, and subsequently by **Synta** in 2005.

Some Celestron branded binoculars are manufactured by the **Kunming United Optics Corporation** of China.

Chadburn
The Chadburn family, [3] [18] [44] [53], of Sheffield and Liverpool, England, were opticians and optical instrument

makers for several generations, who traded under various names and partnerships. William Chadburn began his business as an optician in 1816, and two years later he began to work in partnership with **David Wright**, trading as **Chadburn and Wright** until 1825, manufacturing optical goods and dealing in a variety of hardware. William Chadburn continued in business until at least 1834, when his business was at Albion Works, Nursery Street, Sheffield, and was succeeded there by Chadburn Brothers.

John Chadburn began his business in 1821 at 3 Mulberry Street, Sheffield. From 1830 to 1833 he traded as Chadburn & Co., initially in Albion Works, then in Mulberry Street.

By 1837 Chadburn Brothers were trading at Albion Works, Sheffield. The partnership was between Alfred Chadburn and Francis Wright Chadburn, and they were joined in 1841 by Charles Henry Chadburn. An 1841 advertisement stated "Opticians, &c. Manufacturers of all kinds of Spectacles, Telescopes, Microscopes, Opera glasses, Optical lenses and every variety of Optical instruments; Improved garden syringes, &c.; Pneumatical, Electrical, Magneto electric, and all kinds of Philosophical instruments; Working models of steam engines &c. to order. Also improved ship deck and berth or side illuminators, or ventilators, air valves, &c. ... Every variety of spectacles to suit the various defects of vision." Chadburn Brothers were granted Prince Albert's Royal Warrant, and by 1847 they were advertising as "Opticians &c. to HRH Prince Albert".

Chadburn Brothers exhibited at the 1851 Great Exhibition in London, [328]. The partnership was dissolved in 1865 [35 #22938 p659], but was re-formed, probably without Charles Henry Chadburn who already had a wholesale and retail business in Liverpool. By 1893 Chadburn Brothers was run by Mr. W.T. Morgan, who was a nephew of the Chadburn brothers, and had by then been associated with the company for more than forty years. Beginning in the 1890s Chadburn Brothers supplied spherical lenses and plane glass panels to **Barr & Stroud** for their rangefinders [20]. This continued until 1916 when Barr & Stroud began to make optical glass, and make their own optical components.

In 1845 Charles Henry Chadburn started a branch of the company in Liverpool, where he continued to trade until sometime after 1851. His card read: "Chas Hy Chadburn, Optician & Mathematical instrument maker to His Royal Highness Prince Albert. Wholesale and Retail. 71 Lord Street, Liverpool. Manufactory Sheffield. All kinds of Optical, Mathematical, Philosophical & Nautical instruments made, repaired & accurately adjusted." This business was succeeded in Liverpool by Chadburn and Sons, makers of ships' telegraphs. They were succeeded in 1898 by Chadburn's (Ship) Telegraph Co., also of Liverpool. There are examples of telescopes bearing this name.

In 1944 the Chadburn's (Ship) Telegraph Co. was renamed Chadburn's. A 1951 advertisement shows addresses in Glasgow, Liverpool, London, Newcastle-On-Tyne, and Belfast. In 1961 Chadburn's was listed as general engineers, and manufacturers of marine equipment, including nautical instruments, speed indicators, counters and manual steering gear, as well as automatic soft drink vending machines.

Chadburn & Wright
Chadburn & Wright, (fl.1818-1825) [3] [18] [276 #1509], were opticians of Sheffield, England. Addresses: 81 Ladies' Bridge (1818); 85 Wicker (1821-1822); 40 Nursery St. (1825). The company was a partnership between **William Chadburn** and **David Wright**. In the directory of 1825, they advertised as "opticians and manufacturers of spectacles, reading and opera glasses, telescopes, etc., also dealers in all kinds of hardware". They also advertised: magic lanterns; diagonal mirrors; pocket camera obscuras; single and compound microscopes. They were succeeded in business by the opticians Wright & Sykes.

Chadburn, William
See **Chadburn**.

Chadwick, William Isaac
William Isaac Chadwick, (1848-1913) [53] [61] [70] [109] [306], was an optician and optical instrument maker of 2 St. Mary's St., Deansgate, Manchester, England and, from 1889, 10 St. Mary's St., Deansgate. He had a magic lantern and slide supply business. He had a keen interest in magic lanterns and stereoscopic photography, and designed stereo cameras. He had produced his first stereo camera design by 1895, initially manufactured by **Billcliff Camera Works**. Chadwick not only supplied instruments, but he taught photography, published a manual on stereo photography, and stimulated a renewed growth in interest in stereo photography. He also produced the *Leach's lantern microscope*, on which he presented a paper to the London Microscopical Society in 1891. The article was reprinted in the *Optical Magic Lantern Journal*, which also published several other articles by him on the subject of magic lanterns and slides.

Chamberlain, James Bradley
James Bradley Chamberlain, (fl.1827-c1872) [18] [53], was an English optician, manufacturer of spectacles, and "Manufacturer of improved barometers, ...achromatic

telescopes, and every description of drawing, and mathematical instruments at the lowest remunerating prices." He traded from 1827 until 1848 at 37 Broad Street, Bloomsbury, London, and from c1845 until at least 1872 at 203 High Holborn, London. In 1849 he began to trade as J.B. Chamberlain & Son. Instruments bearing their name include: mahogany barrelled brass three-draw spyglass signed *Chamberlain & Son, 203 High Holborn, London*.

Chamberlain & Son
See **James Bradley Chamberlain**.

Champneys, James
James Champneys, sometimes known as Champness, (fl.1760-1777) [18] [276 #583] [322] [366], was an optical, mathematical, and philosophical instrument maker of Near the Royal Exchange, Cornhill, London, England. He served his apprenticeship with the mathematical instrument maker Richard Winn, beginning in 1752, and was freed of the **Stationers' Company** in 1760. Among his apprentices, beginning in 1761, was **John Cuthbertson**, who was to become his son-in-law.

In 1764 he was one of the petitioners who attempted to revoke the patent obtained by **John Dollond (I)** for the achromatic lens, and which was enforced after Dollond's death by his son, Peter (See Appendix X). It was in the same year that he began to make achromatic telescopes, which he continued, despite the failure of the petition, until 1765, when he was successfully sued by **Peter Dollond** for infringement of the patent. He was ordered by the court to cease production of achromatic telescopes, and to pay Dollond £150 compensation, and £100 expenses.

Perhaps unable or unwilling to discharge his debt to Peter Dollond, in 1768 he went to the Netherlands, and settled in Amsterdam. There he worked in partnership with John Cuthbertson, trading as **Cuthbertson & Champneys**. It appears, however, that they were unable to compete effectively with the already established firm of **Harmanus van Deijl** and **Jan van Deijl**.

While John Cuthbertson stayed in Amsterdam, James Champneys returned to London in c1771, and there he had to face his debt to Peter Dollond. In 1772 he was declared bankrupt [35 #11221 p2], in proceedings that were finally settled in 1775 [35 #11558 p4]. Following this, he once again began to trade as an optician in Cornhill, where he is last recorded in 1777.

Chance Brothers and Co.
Chance Brothers & Co., [46], was established in 1824 when Robert Lucas Chance, (1782-1865), bought the glassworks of the **British Crown Glass Co.** in Spon Lane, Smethwick, Birmingham, England. In 1836 the company was renamed Chance Brothers & Co. They made blown window glass, and began to manufacture optical glass in Birmingham in 1838. With the advent in 1848 of the second French revolution, **Georges Bontemps**, director of the glassworks founded by **Henry Guinand** in **Choisy-le-Roi**, fled to England and took up employment with **Chance Brothers**. At that time the Guinand optical glass was superior to the English glass, and Bontemps brought the necessary skills to Chance Brothers to equal, and ultimately exceed, the quality of the Guinand glass.

At the time of WWI, the Russian Company for Optical and Mechanical Production was trying to master the secrets of optical glass smelting. They had little success, and eventually managed to persuade Chance to sell the technology for 600,000 gold rubles. Their production of optical glass was then organised at the **Petrograd Porcelain Works** [27].

In 1919 Chance was listed as a member of the **British Optical Instrument Manufacturers' Association**. They began making glass fibres at Firhill, Glasgow in the late-1920s, and **Pilkington** acquired an interest in the company in 1938. Prior to WWII a "shadow" optical factory was erected by Pilkington at St. Helens in case of war damage to the Chance Birmingham plant.

By 1945 Pilkington had acquired a 50 per cent shareholding in Chance Brothers, and in 1952 Pilkington took full control of the company. In 1957, the optical side of both companies were combined in the new **Chance-Pilkington Optical Works** at St. Asaph, North Wales.

In 1997 Pilkington merged with **Barr and Stroud**. In 2000 the merged company became a subsidiary of the French company **Thales Group**, and in 2001 they were re-branded as Thales Optronics Ltd. In 2005 Thales Optronics was sold, and became **Qioptiq**. In 2006 Pilkington was acquired by the NSG Group, and became an NSG brand.

Chance-Pilkington Optical Works
The Chance-Pilkington Optical Works in St. Asaph, North Wales, was set up in 1957 when the **Pilkington** glass company, consolidated their optical glass making with that of **Chance Brothers & Co.** which, by then, was a wholly owned subsidiary.

Chandler and Dicker
See **C&D (Scientific Instruments) Ltd.**

Chandler, Seth Carlo
Seth Carlo Chandler, (1846-1913) [45] [203] [205], of Boston, Massachusetts, USA, was a theoretical and practical astronomer and inventor. He worked for the US Coastal Survey until 1870, then worked as an actuary until he joined the Harvard University Observatory in 1881. In

1896, following the death of the previous editor, he became chief editor of the *Astronomical Journal*, a role he kept until 1908. He was a friend of the optician **John Clacey** and, when Chandler invented the chronodeik for determining true local time, Clacey made both the prototype and the first models to be sold. Chandler is most remembered for his work on variations in latitude. He invented the almucantar, for detecting changes in latitude, the first of which was also made by John Clacey. With this invention Chandler discovered a periodic wobble in the rotation axis of the Earth, now known as the *Chandler Wobble*.

Chapman & Alldridge
Chapman & Alldridge, [320] [344], of 39 Mortimer Street, London, England, was a manufacturer of microscopes and accessories. They made microscopes during the 1920s, but the company was liquidated in 1928. R.S. Alldridge acquired the assets of the company and continued as R.S. Alldridge Ltd. The company foundered after the death of Alldridge, and in 1949 it was taken over to become **Sartory Instruments Ltd.**, with a completely new range of products.

Chapman, J.T.
J.T. Chapman, (1843-1907) [53] [61] [122] [306], was a camera maker of Manchester, England. The company was founded in 1874 by Josiah Thomas Chapman, the nephew of **J.T. Slugg**. He served his apprenticeship as a chemist and druggist with J.T. Slugg, from whom he also learned about optics. In 1871, he began to work in partnership with **J.B. Payne**, trading as **Payne & Chapman**. After the partnership was dissolved in 1874, Chapman started his own business at 162 Deansgate. He relocated to 168 Deansgate then, in 1884, to 7 Albert Square. As well as pioneering work in dry plates and emulsions, he began to produce cameras. Initially these were made for him by companies such as **Billcliff Camera Works** and **Lejeune & Perken**, but he went on to set up his own factory. Following Chapman's death, the company continued making cameras until the 1920s. The company was run briefly by an employee, William Hughes, and from 1917 it was again run by the Chapman family until 1968, when it merged with Frederick Foxall Ltd. to become Foxall & Chapman.

Chapman, James
James Chapman (I), (fl.1774-1804) [18] [276 #812] [322], was an optician, and optical and mathematical instrument maker, and a ship chandler, of London, England. Addresses: St. Catherine's near the Tower (1774-1796); *Hadley's Quadrant* opposite the King's Store House, St. Catherine's (1776); 41 St. Catherine's (1793); 5 St. Catherine's (1794). He served his apprenticeship with **Christopher Stedman**, beginning in 1760, and was freed of the **Stationers' Company** in 1769. Among his apprentices were his sons, James Chapman (II) and Josiah Chapman. Instruments bearing his name include: single-draw refracting telescopes [358]; octant [262]. He is also known to have sold: sextant.

James Chapman (II), (fl.1819-1827), was an optician, and mathematical instrument maker of 8 Upper East Smithfield, Tower Hill, London, England. He served his apprenticeship with his father, beginning in 1792. He is known to have sold: telescope.

It is not clear whether James (II) made and sold instruments while still working at his father's address. Hence it is not certain that instruments bearing the name James Chapman may be attributed to father or son by means of the address, if present on the instrument.

Chérubin
Chérubin d'Orleans, (1613-1697) [256] [274] [288], was a French Capuchin monk with an interest in mathematics and optics, and studied the works of natural philosophers such as **Johannes Kepler** and **Galileo Galilei**. Chérubin made binocular telescopes, and is reputed to have been the first to design a binocular microscope – a claim which was disputed by **Daniel Chorez**. Instruments by Chérubin were owned by members of European royalty, including King Charles II of England and King Louis XIV of France, the latter whom, among others, provided valuable patronage for his work.

Chérubin's first published work on optics was *La Dioptrique Oculaire*, Paris, 1671 [317], which became a successful and popular reference on practical optics. He also published *La Vision Parfaite*, Paris, 1677 [317], in which he described a binocular microscope. A second volume to this work was entitled *La Vue Distincte*. In this book he describes a microscope with multiple objectives mounted on a disc, multiple sample holders mounted on another disc, and both discs centred on a common axis parallel to the optical axis of the microscope. This appears to be the first museum microscope.

Cheshire Apertometer
See apertometer.

Cheshire Collimator
A Cheshire collimator is a simple device used to align the optical axis of a reflecting telescope to achieve collimation. It is a device mounted in place of the eyepiece in the telescope focuser. It uses a sighting hole, an angled mirror, and a cross-hair to sight along the optical axis and evaluate any necessary corrections to the alignment. It was invented by **Frederic J. Cheshire**.

Cheshire, Frederic J.

Frederic J. Cheshire was a lecturer in physics at the Birkbeck College, London, England. He is known for his inventions of the Cheshire apertometer and the Cheshire collimator. He also co-translated *The Zeiss works and the Carl-Zeiss stiftung in Jena; Their scientific, technical and sociological development and importance* [33].

Cheshire Optical Instrument Co.

The Cheshire Optical Instrument Co. was a manufacturer of telescopes and mounts, of Mableton, Georgia, USA. According to their 1985 catalogue they sold "quality portable refracting telescope systems". Products in the catalogue include a portable pier, equatorial mounts, and refracting telescopes ranging from a 3.2-inch f/3.5 to a 6-inch f/15. The catalogue also includes a *Cheshire-Byers 58 Equatorial Mount*. Note that the *Byers 58* mount was produced by **Byers Co.**

Chester, Edward

Edward Chester, (fl.1782-1785) [18] [322], was an optician and mathematical instrument maker of 8 King St., Tower Hill, London, England and, from 1784, 64 Upper East Smithfield, Tower Hill, London. He served his apprenticeship with **Benjamin Cole (II)**, beginning in 1771, and was freed of the **Merchant Taylors' Company** in 1778. Among his apprentices was **John Snart**.

Chevalier, A.

A. Chevalier, (fl.1770-1791) [16] [18] [276 #698], was a nautical instrument maker of Guernsey, Channel Islands. Instruments bearing his name include an octant, dated c1780, with scale divided by **Edward Pritchard** (see Appendix XI). He also retailed books and charts produced by **John Hamilton Moore**.

Chevalier, Paris

Three members of the French family of Chevalier were notable instrument makers in Paris. Vincent Chevalier (1770-1841), his son, Charles Chevalier (1804-1859), and grandson, Arthur Chevalier (1830-1872). Vincent and Charles are known to have worked together [37] [70] [256] [264] [320] [408].

The family firm began with Vincent's father, Louis Vincent Chevalier (1743-1800), who started his business in 1765, trading at 31 Quai de l'Horloge, Paris. He made lenses, telescopes, and mirrors. His three sons, Louis, Nicholas, and Jacques Louis Vincent (known as Vincent), all worked for him but, by the time of his death the firm was not widely known, and Vincent had already left to join the army. The firm closed with Louis Vincent's death in 1800.

Vincent returned to Paris in 1803, and set up his own workshop in Rue de Grenelle St. Germain, where he soon began to manufacture telescope eyepieces and objectives. He expanded into 21 bis rue de l'Horloge, then relocated to number 67 in 1810, and 69 in 1818. He began to work in partnership with his son, Charles, and together they made improvements to the camera obscura, and developed a successful trade making microscopes. Charles made instruments for **Giovani Battista Amici**, including a polarizing microscope. He made various models of microscope, including student models.

In 1826 **Joseph Niépce** purchased, through his cousin, **Claude Niépce de St Victor**, his first professionally made camera obscura from Charles and Vincent Chevalier. Chevalier was fascinated by Niépce's work, particularly when shown an example of a heliograph. When Chevalier subsequently met **Louis Daguerre**, he advised Daguerre to contact Joseph Niépce, ultimately resulting in a partnership between the two.

Vincent and Charles Chevalier supplied Niépce with an achromatic meniscus lens designed by Vincent for photographic purposes. This was one of the first lenses to be designed specifically for photography, and became a source of renown for the Chevaliers. This was not, however, the only kind of lens they made. **W.H. Wollaston** had designed in 1812 the periscopic camera obscura, featuring the periscopic meniscus lens. The design was only widely taken up when the Chevaliers began to make them in 1829.

Daguerre bought his experimental cameras from Charles and Vincent Chevalier until 1839 when he began to purchase them exclusively from **Alphonse Giroux**. These latter instruments, however, included the Chevalier achromatic meniscus lens.

The Chevalier achromatic meniscus lens, although successful at the time, had a small aperture, and hence was a slow lens. This limitation was solved by **Josef Petzval** who created the successful Petzval-Voigtländer lens, which made portrait photography a more realistic endeavour.

From 1832, Vincent ran the business at 69 Quai de l'Horloge while Charles relocated to run his own business at 163 Palais Royal, Galerie de Valois, with workshops at 1 Rue Neuve des Bon Enfants until c1845, after which the workshops were at 1 bis Cour des Fountaines. In 1846, he opened a photographic laboratory at 2 Rue de Valois.

Charles' son, Louis Marie Arthur Chevalier (known as Arthur) began to work with him in 1848. Following the death of Charles in 1859, Arthur took over the firm, then trading at 158 Palais Royal, Galerie de Valois. He continued to expand and improve the company's products. By 1860 the firm was advertising many different instruments, including: theodolites; microscopes; telescopes; micrometers; protractors; globes and spheres; barometers; magnetic compasses; spectacles; drawing instruments, and much more.

Although the date the company finally closed is not known, the last known catalogue published by the Chevalier firm was in 1885, thirteen years after Arthur's death, and there is little evidence of the firm's activity after this date.

Chevallier, Jean Gabriel Augustin

Jean Gabriel Augustin Chevallier, (1778-1848) [50] [265] [266] [320], the grandson of Paris optician **François Trochon**, was a renowned optical instrument maker of 1 Quai de l'Horloge, Paris, France. The family business was started in 1740, and J.G.A. Chevallier began his business producing microscopes, telescopes and other optical instruments in 1796, working under the name l'Ingénieur Chevallier. He was knighted with la Légion-d'Honneur, which allowed him to use the title *le Chevalier* (the Knight). His use of this title can cause confusion between Jean Chevallier and the **Chevalier** family of optical instrument makers. The Chevallier business moved to 15 Place du Pont Neuf, sometime between 1839 and 1845.

L'Ingénieur Chevallier's entry in the 1884 edition of *Galignani's New Paris Guide*, [267], states: "Optician, 15 Place Du Pont Neuf, Paris (Facing the statue of Henri IV.) L'Ingénieur Chevallier, Optician and maker of Scientific instruments. Opera and field glasses, Sporting and astronomical telescopes, Microscopes, Spectroscopes, Surveying and levelling instruments, Theodolites, Sextants, Compasses, Metrological instruments, Barometers, Mountain barometers, Thermometers, Eyeglasses and spectacles, Photographic apparatus and all kinds of Philosophical instruments. New patented *Photo-Field Glass*, the most instantaneous and portable apparatus. This old and celebrated establishment, founded in 1740, exists only at Place Du Pont Neuf 15, Paris. English spoken."

Jean's son **Pierre Chevallier** started his own business in 1835. Jean's daughter **Marie Chevallier** married **Alexandre Ducray** and in 1842, at around the same time the business moved to 15 Rue de Pont-Neuf, J.G.A Chevallier sold his business to Marie and Alexandre. Jean retained an advisory role in the company until his death in 1848, after which his daughter and son-in-law began to trade under the name **Maison de l'Ingénieur Chevallier**.

Chevallier, Marie Louise Mélanie

Marie Louise Mélanie Chevallier, (1815-1897) [50], was the daughter of **Jean Gabriel Augustin Chevallier**. Together with her husband, **Alexandre Ducray-Chevallier**, she purchased her father's business in 1842. Upon her father's death in 1848 they changed the name of the business to **Maison de l'Ingénieur Chevallier**. Her husband Alexandre died in 1879, and she continued to run the business until 1883 when she sold it to the brothers **Charles and René Avizard**.

Chevallier, Pierre Marcel Augustin

Pierre Marcel Augustin Chevallier, (1797-1841) [50], was the son of **Jean Gabriel Augustin Chevallier**, and started his own business in about 1835. He sometimes signed his instruments *A. Chevallier* for his name Augustin, and sometimes *Chevallier Ingéneur Opticien*. Since his business was at 1 Rue de la Bourse, Paris, France, his work can be differentiated from that of his father by the address. In 1842, following the death of Pierre Chevallier, the business was taken over by **Pierre Queslin**.

Chinon

Chinon Industries Inc. was founded in 1948 in Japan by Chino Hiroshi as Sanshin Seisakusho. In 1962 the company name was changed to Sanshin Optics Industrial Co. Ltd., and in 1972 the name was changed again to Chinon Industries Inc. The company produced various models of binoculars and cameras. During the late 1970s and 1980s Chinon produced cameras for **Pignons SA** to be sold under their **Alpa** brand. In 1985 they began producing cameras for **Kodak**, and in 1997 Kodak Japan Ltd., a subsidiary of the Eastman Kodak Company, became the majority shareholder.

Chislett & McAll

Chislett & McAll, (fl.c1850), nautical and philosophical instrument makers of 8 Postern Row, Tower Hill, London, England was, according to [322], a partnership between **Alfred Chislett** and **John McAll**. The address was previously occupied by John McAll until at least 1834. It was also occupied by Alfred Chislett from 1849 until 1855, which may have been the period during which the partnership Chislett & McAll conducted their business. Instruments bearing their name include: wheel barometer.

Chislett, Alfred

Alfred Chislett, (fl.1834-1855) [18], was an optical, nautical, mathematical, and philosophical instrument maker of 27 Greenfield St., Whitechapel, London, England (1837-1845), and 8 Postern Row, Tower Hill, London (1849-1855). He is also associated with the addresses 7 Budge Row, Cannon St., and Gloucester St., Commercial Road. He began his apprenticeship with **Robert Brettel Bate** in 1825, but was turned over in the same year to **William Gilbert (II)**. He was freed of the **Spectacle Makers' Company** in 1837. He took on at least seven apprentices, four of whom were turned over to other apprentice masters. According to [322], he worked in partnership with **John McAll**, c1850, trading as **Chislett & McAll**.

Choisy-le-Roi Glass-Works

The glass-works in Choisy-le-Roi, near Paris, France, [**24**, p252] [**156**], was founded by the optical glass maker **Henry Guinand**. The glass-works was founded in 1828 with **Georges Bontemps** as its director. In 1848 Bontemps fled the second French Revolution, and went to England to work for **Chance Brothers**. Upon Guinand's death in 1852 his firm passed into the hands of his grandson, **Charles Feil**, and was renamed Feil & Co. Feil operated the company until 1887, when the firm passed to his son-in-law, **Edouard Mantois**, and was renamed **Parra-Mantois** in about 1900.

Chorez, Daniel

Daniel Chorez, (fl.1616-1625) [**256**] [**322**], was a mathematical and optical instrument maker of Paris, France. Addresses: rue de Périgueux de Marais; and later, Isle Nôtre-Dame, at the Sign *Au Compas*. A sector, made in 1616, is his earliest known work. He is known to have made telescopes and microscopes. According to a pamphlet published in 1625, he began to make field-glasses in 1620. He made telescopes, microscopes, binocular telescopes, and binocular microscopes. He also made instruments which he claimed could be used as both telescope and microscope. Chorez disputed the claim by **Chérubin** to have made the first binocular microscope.

Chrétien, Henri Jacques

Henri Jacques Chrétien, (1879-1956) [**166**], was an astronomer and inventor from Paris, France. After studying engineering, he went to work for the Nice Observatory. He co-invented the Ritchey-Chrétien telescope with the American astronomer **George Willis Ritchey**. He is also known for his innovations in optics for widescreen cinematography, and as a co-founder of the *Institut d'Optique Théorique et Appliquée*.

Christian, George

George Christian, (fl.1825-1867) [**16**] [**18**] [**358**], was an optician, and mathematical and nautical instrument maker of Liverpool, England. Addresses: 16 Regent St. (1825); 16 Strand St. (1827-1829); 20 Strand St. (1831-1832); 11 Strand (1834-1853); 113 Duke St. (1855); 11 Crooked Lane (1857); 133 Duke St. (1859-1867); 9 Canning Place (1864-1867). He is known to have made: sextant; octant. Octants bearing his name are in the Manx Museum, Isle of Man, and the Town Docks Museum, Hull.

Christie & Wilson

Christie & Wilson, [**44**] [**263**], was a firm of nautical opticians and compass adjusters in Glasgow Scotland. The company was set up in 1916 as a partnership between **Andrew Christie** and James Wilson, who had previously worked for **Whyte, Thomson & Co**. An advertisement of 1951 describes them as certified B.O.T. compass adjusters of 90 and 130 Broomielaw, Glasgow, makers of C.&W. compasses, Practical nautical opticians, Chronometers, Clocks, Watches. Agents for Admiralty charts and publications.

Christie, Andrew

Andrew Christie, (b.1856) [**263**] [**271**], of Glasgow, Scotland was a compass adjuster and maker of marine navigation instruments. He worked for **Duncan McGregor** prior to setting up his own business in 1890 at 27 Clyde Place, Glasgow. There he made nautical and mathematical instruments, chronometers, clocks and watches, and worked as a compass adjuster and an optician. By 1896 his business had grown, necessitating an expansion into 28 Clyde Place.

After a year-long partnership with the shipbroker James Gilchrist Lee, beginning in 1901, Andrew Christie worked independently, trading at the Glasgow addresses: 192 Hyndland Road (1902-1907); 34 Robertson St. (1908-1910); 54 Broomielaw from 1911. In 1916, he entered a partnership with James Wilson, trading as **Christie & Wilson**, nautical opticians and compass adjusters.

Christie, George

George Christie, (fl.1794-1837) [**18**] [**276** #940], was an English optical and mathematical instrument maker of Leigh on Mendip, Frome, Somerset, and 11 Strand, Liverpool. In 1794 he published a pamphlet entitled *Construction of a Sea Telescope*, and sent it to the Board of Longitude.

Chromatic Aberration

Chromatic aberration is an optical aberration which results in colour fringing, most noticeable in higher contrast parts of an image. It is the result of dispersion, (see *Properties of Optical Glass* in Appendix VI), a property of the optical material (e.g. glass), and can be to some extent corrected by such devices as an achromatic or apochromatic lens arrangement. There are two forms of chromatic aberration: longitudinal chromatic aberration, which is an error of focus, and chromatic difference of magnification, which is an error of magnification.

Longitudinal chromatic aberration: This occurs when different frequencies of light are brought to focus at different distances from the lens. Hence, there is no one image plane in which all frequencies of light are in perfect focus. This is also known as *axial chromatic aberration* because the different colours on an image come to focus on image planes at different distances along the optical axis. This results in undesirable colour fringing over the entire image.

Chromatic difference of magnification: This is also known as *lateral chromatic aberration*, and occurs when the magnification of an image point varies with the wavelength of light. The effect occurs with off-axis rays of light, and is hence more prevalent towards the image edges in an uncorrected system, where the point images of details vary in size for different wavelengths of light, resulting in colour fringing.

Chronodeik

A chronodeik, [204], is an instrument used to measure local time to within a second. This is accomplished by taking two measurements, one in the morning and one in the afternoon. The instrument consists of a small vertically mounted telescope, with a suitable solar filter, with a mirror mounted such that it may be tilted to direct the sun into the telescope. By viewing through the telescope, the observer is able to measure the time at which the sun crosses a line of altitude, indicated by a horizontal wire. The instrument is set during the morning measurement and, as the sun crosses the same altitude in the afternoon, the second measurement is taken. The mean of the two measurements corresponds to true noon.

The chronodeik was invented by **Seth Chandler**, and the first chronodeiks were built by **John Clacey**.

Chronophotography, Chronophotographic Gun

The chronophotographic gun was invented by **Étienne-Jules Marey** in 1882 for his study of the movement of birds. It is shaped like a rifle, and takes photographic images at 12 per second. The sequence of images is all taken on a single frame, thus giving a vivid sense of motion in a single image. This technique is referred to as *chronophotography*.

Ciceri & Co.

Several instrument makers named Ciceri had businesses in London, Edinburgh, and Paris during the C19[th], many of them being principally carvers, guilders, and barometer makers. The extent to which they were able to make optical instruments is unclear, but it was not uncommon for such companies to sell instruments bearing their name, having purchased them wholesale. Often the relationships between the proprietors of the various businesses are not clear, but a family link is likely, as is a connection with the Italian immigrant community. The following businesses may be related:

Ciceri & Co., London, England. Instruments bearing this name include: brass three-draw telescope, 30mm objective, 25½ inches fully extended, missing leather cover. Brass three-draw telescope with leather cover, 1½-inch objective, 25½ inches fully extended.

Ciceri & Co., were wire workers of 4 Eyre Street Hill, Leather Lane, Holborn, London, England. This partnership between Joseph Ciceri and John Chaplin May was dissolved in 1878 [**35** #24569 p2413].

Ciceri & Co., (fl.1875) [**379**], were carvers and guilders of Frederick St., Edinburgh, Scotland. Reference [**263**] shows evidence that this firm had no connection with the firm of **Ciceri & Pini** of Edinburgh.

Ciceri et Cie, (fl.c1800) [**256**], were philosophical instrument makers of rue Saint-Honoré and rue du Faubourg-Saint-Martin, both in Paris, France. They are known to have sold instruments by English makers.

Many other Ciceri family connections were involved with the barometer and thermometer manufacturing trade. These include Francis Ciceri, who worked in partnership with **Caesar Tagliabue**'s son Antoni Tagliabue, trading as Tagliabue & Ciceri at 31 Brook St., Holborn, London. This partnership was terminated in 1842 [**35** #20612 p2150], but continued to be listed in the directories until 1849. He also worked with **Henry Negretti**, trading at 31 Brook St. and 19 Leather Lane, Holborn. The partnership was terminated 1846 [**35** #20628 p2816], after which Francis Ciceri continued in business at 31 Brook St., and Henry Negretti at 19 Leather Lane.

Ciceri & Pini

Ciceri & Pini, [**18**] [**263**] [**379**] [**438**], was a firm of carvers and gilders, located in London, England (1825-1845). In 1842 they took over the premises of the barometer makers, Battistessa & Co., at 8 & 9 Calton St., Edinburgh, Scotland. They advertised as picture frame makers, barometer and thermometer makers, and telescope makers. In 1852 they took on additional premises at 81 Leith St., Edinburgh, and continued in business until the partnership between John Ciceri and Joseph Pini was dissolved in 1858 [**170** #6809 p1099]. They were succeeded in business by a new partnership, trading at the Leith St. address as Ciceri, Mantica & Torre. This partnership was between John Ciceri, John Ciceri Jr., Anthony Mantica, and Paul Della Torre.

Cine

Cine refers to instruments such as movie cameras and projectors that are used in association with moving pictures.

Cinématograph

A Cinématograph, sometimes called a Kinematograph, is a motion picture camera, projector and printer combined into one unit. The device was invented in the 1890s by **Auguste and Louis Lumière**, who named it the *Cinématographe Lumière*. Unlike the Kinetoscope the Cinématograph could project an image onto a screen, for viewing by an audience of more than one person.

Cintel

Cintel was a British company, founded in 1927 by John Logie Baird. The company was bought by Blackmagicdesign in 2014, and they retain Cintel as a brand-name. They specialise in cameras, scanners and digital imaging of film for broadcast and post-production. During the early years of the company, Cintel's optics were designed and manufactured by **Wray Optical Works Ltd.**

Circular Micrometer

See micrometer.

Clacey, John

John Clacey, (1857-1931) [**185**] [**203**], was an optician from Maryland, USA, and was highly renowned for the lenses he ground and figured. His interest began with amateur astronomy and, being unable to afford a telescope, he built one. He learned to grind and figure lenses from the works of **Fraunhofer**, **Gauss**, **Steinheil** and others, and finished his lenses using "removal of zones by local correction", with frequent tests against an artificial star. Before he had finished building his first telescope, he was asked by **Seth Chandler** to build another like it. Chandler was impressed, and asked him to help with the optics on his experimental almucantar for detecting changes in latitude, and for his first chronodeik for determining true local time. A subsequent order twelve chronodeiks was the beginning of Clacey's professional career as an optician. He made the optical parts himself and had local machine shops make the metal parts.

Clacey made lenses for four and five-inch telescopes, as well as making smaller telescopes, some with a shorter focal length than was common at the time. He moved to a workshop in Cambridgeport, Massachusetts, and made lenses up to 12 inches in diameter. Among his customers were the Coast and Geodetic Survey, the Blue Hills Meteorological Observatory, Harvard College Observatory, William Brooks's Smith Observatory, the Japanese Navy, Georgetown University, Amherst College, the Vatican Observatory, and others. Wishing to work only on lenses, rather than the other parts of telescopes he took a position on the staff of the National Bureau of Standards. He continued to produce fine optical components for them for the rest of his career.

Clack, J.

J. Clack, (fl.1764) [**18**], was a mathematical instrument maker of Saffron Hill, London, England. He may be the John Clack who was employed by **James Haly Burton**. In 1764 he was one of the petitioners who attempted to revoke the patent obtained by **John Dollond (I)** for the achromatic lens, and which was enforced after Dollond's death by his son, Peter (See Appendix X).

Clark & Sons, Alvan

Alvan Clark, (1804-1887) [**8**] [**24**], of Massachusetts, USA, went into business with his sons, Alvan Graham Clark and George Bassett Clark in 1846. They formed the company Alvan Clark & Sons, and gained a reputation for making fine telescopes. They ground their own lenses, and built spherometers to aid in the process.

In 1865 their 18½-inch refracting telescope, then the largest in the world, was installed in Dearborn Observatory in Chicago. They manufactured the objective lenses for the James Lick 36-inch refractor at the Lick Observatory in California, the Yerkes 40-inch refractor at the Yerkes Observatory in Wisconsin and, in 1893, the 24-inch Bruce doublet astrograph for Harvard College University.

Carl Lundin began to work fof the firm in 1874, and rose to become a senior optician. His son, **Robert Lundin** joined the firm in 1896, and rose to become head of the optical department, where he stayed until he left to start his own company in 1929.

Alvan Clark and Sons were supplied with optical glass by **Feil & Co.** at the **Choisy-le-Roi** glassworks, Paris, France, for many of the telescopes they made. In 1882-1885 Feil & Co. supplied them with the crown and flint glass blanks for the 36-inch James Lick refractor. In 1895, when the glassworks were under the direction of **Edouard Mantois**, they supplied Alvan Clark & Sons with the glass blanks for the 40-inch Yerkes refractor.

Alvan Clark & Sons also made large telescopes for other major observatories as well as smaller instruments for education and amateur astronomy.

Clark, John

John Clark, (fl.1749-1796) [**271**], of Edinburgh, Scotland, was a goldsmith, jeweller, and optical instrument maker. Addresses: Luckenbooths (1749); Parliament Close (1751-1755); Opposite the Guard (1733-1782); At the Cross (1786-1788); 13 Parliament Close (1793-1796).

Clark, Thomas
Thomas Clark, (fl.1733, d.c1743) [18], of the Golden Head near Arundel Street, Strand, London, England, was an optical instrument maker. He served his apprenticeship with **George Willdey**, beginning in 1724, and was freed to the **Spectacle Makers' Company** in 1733.

Clarke & Son
See **William Clarke (III)**.

Clarke, Edward
Edward Clarke, (fl.1804-1846, d.1859) [18] [34] [46] [304], was an optical, philosophical and mathematical instrument maker from Dublin, Ireland. By 1810 he was working as an optician in Lower Sackville Street, and from 1815 until 1817 he worked in partnership with **Richard Spear**, trading as Spear & Clarke. Following this, although missing from the directories of 1813, 1814, 1818 and 1822, Clarke is otherwise listed as an optician in Dublin until 1832, trading as E. Clarke & Co. from 1823.

By 1833 he had moved to London, England, and is believed to be the Edward Marmaduke Clarke, ([276 #1099] refers to him as Edward Montague Clarke), who then began working for **Watkins & Hill**. Soon after this he started his own instrument making business in Strand, London, England. He appears to have largely made electrical instruments, and found himself in dispute with the American instrument maker Joseph Saxton over the originality of the magneto-electric machines he made. There is a record of a mounted refracting telescope, 72 inches in length, bearing his signature, made when he was in Dublin [215].

Clarke, Thomas
Thomas Clarke, (fl.1805) [18], was an optician of 13 Wapping Wall, Shadwell, London, England. This is the same address as concurrently occupied by **William Clarke (III)**. He served his apprenticeship with **Mark Burton**, beginning in 1759.

Clarke, William (I)
William Clarke (I), (c1810-1873) [3] [50], optician and scientific instrument maker, was a senior journeyman for **Thomas Davies King**. Upon King's retirement in 1858, Clarke entered a partnership with **Henry Husbands**, another of King's senior journeymen. They traded at 1 Denmark Street, Bristol, England, as **Husbands and Clarke**. Clarke continued in this partnership until his retirement in 1870.

Clarke, William (II)
William Clarke (II), (fl.1722-1743) [18], was an optical instrument maker of Near Union Stairs in Wapping, London, England. He advertised Hadley's quadrants for sale, and is known to have sold backstaffs.

Clarke, William (III)
William Clarke (III), (fl.1797-1822) [18] [276 #942], was a nautical and mathematical instrument maker of 13 Wapping Wall, Shadwell, London, England. In 1805, **Thomas Clarke** occupied this same address. William (III) was the son of Samuel Clarke, a glazier of Aldgate. He served his apprenticeship with jeweller John Wood, beginning in 1776 and, in 1779, was turned over to Samuel Clark to complete his apprenticeship. He was freed of the **Farriers' Company** in 1786. He is known to have made: quadrant; compass. His business continued as Clarke & Son, trading at the same address from 1821 until 1834.

Clarkson, Alexander T.
Alexander Thomas Clarkson, (c1848-1918) [44] [319] [407] [409], was a telescope maker of London, England. According to [406], he took over the business of **Robert Mills** in 1873. In fact, this would have been after **Harriett Mills** succeeded Robert, and was herself succeeded by Alfred Mills after 1869. Nonetheless, with this acquisition, he also took ownership of the telescope tube-drawing machine which originated with **Joshua Lover Martin**.

Alexander Clarkson began to work in partnership with James Aitchison, who had started his company in 1889, trading as **Aitchison & Co.** at 47 Fleet St., and 42 Bishopsgate St. The partnership was terminated in 1891 [35 #26140 p1223], after which Aitchison continued the business himself, trading under the same name, and Alexander Clarkson ran his own company.

During the 1890s Robert Henry Broadhurst joined him, and they began to work in partnership as A. Clarkson and Co., trading as scientific instrument makers and gas compressors. In the 1895 trades directory, he is listed at 28 Bartlett's Buildings, Holborn Circus, as "wholesale manufacturer of monocular and binocular microscopes, second-hand microscopes by best makers in stock". By the time the partnership was terminated in 1908, [35 #28231 p1897], the firm was a partnership between Clarkson, Broadhurst, and William Henry Towns, and was already trading at: 63 Farringdon Road; 63 Charterhouse Street; Bowen's Mills, Phoenix-place; 338 High Holborn; and 2 Staple Inn, all in London. Following the termination of the partnership, the firm wis divided into two. Alexander Clarkson carried on the business with William Henry Towns, at 338 High Holborn, and 2 Staple Inn.

Robert Broadhurst carried on the business at: 63 Charterhouse Street; Bowen's Mills, Phoenix-place; and 63 Farringdon Road, where the telescope tube-drawing machine remained. He renamed his business

Broadhurst, Clarkson & Co., even though Alexander Clarkson was not a part of it.

Claudet, Antoine François Jean
Antoine François Jean Claudet, (1797-1867) [45] [46] [70] [71], was a French photographer and inventor. After his education he became co-director of the glassworks of **Choisy-le-Roi**, near Paris, France, and in 1829 he opened a warehouse in London, England, as an outlet for the firm's specialities of glass shades, painted glass, and sheet glass. He settled in London, and in 1833 he invented a successful machine for cutting glass cylinders. In 1834 he entered a partnership with **George Houghton** trading as **Claudet and Houghton**. Following a recommendation from his friend, **Noël Marie Paymal Lerebours**, he received personal instruction from **Louis Daguerre** in 1839, and acquired a license for the process, setting up his own daguerreotype studio, and making his own innovations in the daguerreotype process.

His extensive entry in the 1851 Great Exhibition catalogue, [328], lists his exhibits including: Multiplying camera obscura; Photographometer; Dynactinometer; Focimeter; Screens; Patent photographic camera obscura; Dark boxes; Brass frames; Mercury box; Apparatus for cleaning and finishing a daguerreotype plate; Daguerreotype pictures, and many other items.

He was elected a fellow of the Royal Society in 1853.

Claudet and Houghton
In 1834 **Antoine François Jean Claudet** entered a partnership with **George Houghton**, trading as Claudet and Houghton, [44], with premises at 89 High Holborn, London, England. The company sold glass products, including optical glass for photographic and microscope applications. Sometime between 1864 and 1868 they became Claudet, Houghton & Son. In 1876 they became George Houghton & Son, and in 1892 they became George Houghton & Sons.

Clavé
The company Clavé, [226], was founded in 1937 by Serge-René Clavé (d.1988) in Paris, France. Initially the company focused on producing optical instruments and systems for the military, including periscopes, sighting systems and other optical systems. During the early 1950s Clavé began to offer telescope accessories. Among these were eyepieces, including a Plössl type eyepiece (see Appendix II) designed by **Jean Texereau**. Clavé expanded their offerings to include eyepieces, prisms, lenses, mirrors, Barlow lenses, filters, and more. Clavé initially used optical glass supplied by **Parra Mantois et Cie**, then later by **Schott AG**, and during the 1980s they used American glass.

Following the death of Clavé in 1988 the company was taken over by Kinoptik. Clavé eyepieces were produced until 1999, and the Clavé stock was sold until 2005.

Cleare, J.
James Cleare, (fl.1763-1764) [18], was a spectacle maker and telescope maker of Mitre Court, Fleet Street, London, England. He served his apprenticeship with **John Cuff**, beginning in 1745/6, and was freed of the **Spectacle Makers' Company** in 1753.

As detailed in Appendix X, two petitioners named J. Cleare of Fleet Street, were listed as signatories in the attempt to revoke the patent obtained by **John Dollond (I)** for the achromatic lens, and which was enforced after Dollond's death by his son, Peter. The father of James Cleare was John Cleare, for whom no occupation is listed. It is possible that the two entries in the list for J. Cleare may have been father and son.

Clement Clarke
Clement Clarke was founded in 1917, making ophthalmic instruments, and subsequently they opened a chain of optometrists. The chain of optometrists was acquired by **Boots** in 1987, along with **Curry & Paxton**, to form Boots Opticians Ltd. The ophthalmic instrument making part of Clement Clarke was acquired by **Haag-Streit** in 1989.

Clinometer
A clinometer is an instrument that measures angles of inclination. This is usually achieved using a graduated scale, and a plumb line or bubble level, or some other method of ascertaining true vertical, or true horizontal reference.

Clinometer gun-sight: A gun-sight which usually uses a bubble level as a reference, and allows a sighting to be taken along a prescribed angle with respect to level, determined by its angle adjustment.

Abney Clinometer: Otherwise known as an *Abney level*, it is a hand-held instrument, with a sighting tube through which the user sights a line. A mirror allows simultaneous viewing of a bubble level, which may be adjusted to show the angle of inclination on a graduated arc. The user may then use trigonometry to calculate the height of an object.

Watkin Clinometer: Designed to measure the inclination of a surface, for example on a ship. This clinometer has a sturdy frame, usually brass, in which a level is mounted, pivoted at one end. The elevation of the other end is adjusted by rotating a cylinder whose rim is marked in degrees and minutes.

Watkin Mirror Clinometer: A portable clinometer designed for surveying and field work. It consists of a circular case containing a curved angle scale attached to a weighted pendulum. The line is sighted through an aperture in the rim, so that the scale may be read, magnified by a concave mirror.

Clockmakers' Company
See **Appendix IX: The London Guilds**.

Clothworkers' Company
See **Appendix IX: The London Guilds**.

CMF
See **Antonio Bencini**.

Coachmakers' Company
See **Appendix IX: The London Guilds**.

Cock, Christopher
Christopher Cock (sometimes Cocks, Cocke, Cockes, or Cox), (fl.1669-1696) [8] [18] [24] [53] [256] [274] [275 #285] [388], was a lens grinder and optical instrument maker of London, England. Addresses: *Long Acre* (1693); *The Two Twisted Posts* in Long Acre (1696); *The Blue Spectacles* near St. Anne's Church (1696). He began his apprenticeship in 1657 with John Stonehall, and was turned over to **John Reeves**. He was freed of the **Turners' Company** in 1669, and subsequently freed of the **Spectacle Makers' Company** in 1680/81. One of his apprentices was **Edward Scarlett (I)**, beginning in 1691.

He produced lenses, telescopes, microscopes, and dioptric lanterns, among other instruments, for eminent members of the Royal Society, including **Robert Hooke**, **Sir Isaac Newton**, John Flamsteed [275 #318], and **James Gregory**. He made microscopes to the design of Robert Hooke, both for the Royal Society, and for other customers, and he supplied Sir Isaac Newton with a 14-foot telescope. In 1663 he worked with **Richard Reeves** to make a speculum for James Gregory, but the result was considered unsatisfactory.

Coddington Lens
A Coddington lens is a spherical lens with a deep groove cut around its rim or equator. The axis of the circular groove is the axis of the lens, and the groove limits the spherical aberration. This type of lens is used in small magnifiers [7] [30]. The design was first published by **Sir David Brewster** in 1820, and subsequently by Henry Coddington in 1845. Despite Brewster's earlier claim, the lens was named after Coddington [264].

Coelostat
A coelostat, [152], is a siderostat with its mirror mounted in a specific configuration determined by **Gabriel Lippmann** in 1895. This configuration is the only one in which a siderostat will yield an image that does not suffer from field rotation.

In Lippmann's configuration the plane mirror is mounted so that a line element of its surface (usually a diameter) is coincident with the polar axis. The mirror is rotated about the polar axis once every 48 sidereal hours to follow the apparent motion of the stars. Since the reflected beam moves through twice the angle moved by the mirror, this results in a stationary view.

The view may be advanced or retarded in right ascension by moving the mirror ahead of or behind the clock drive. Changes in declination may be achieved in one of two manners. Either the telescope can be moved to view the mirror at a different declination, or a second mirror may be used. When a second mirror is used for this purpose it introduces another optical surface, and hence more optical loss. Thus, this arrangement is generally only used for solar viewing, where the light is abundant and, in this context, it is a form of heliostat.

Cohen, S.P.
Simeon Phineas Cohen, (1805-1894) [3] [8] [18], formed a partnership with **Abraham Abraham** in 1838, and traded as A. Abraham & Co., a branch of the A. Abraham business. Their address was 8 Exchange Square, Glasgow, Scotland until 1841, when they moved to 82 Queen Street. Cohen took the branch over in 1843. From 1845 he traded in Buchanan Street, and later in St. Vincent Street. He was subject to sequestration proceedings, and declared bankrupt in 1853, mostly due to his debts to Abraham, [35 #21496 p3325] [170 #6340 p984].

Cole, Benjamin (I)
Benjamin Cole (I), (1695-1766) [3] [18] [234] [276 #26], was a mathematical, philosophical, and optical instrument maker, born in Oxford, England. He served his apprenticeship with William Cade, beginning in 1710, and was freed of the **Merchant Taylors' Company** in 1719. He worked for **Thomas Wright (I)** at the *Orrery and Globe, next the Globe & Marlborough Head Tavern* in Fleet Street, London, until at least 1733. His son, **Benjamin Cole (II)**, served as his apprentice beginning in 1739, by which time he was trading at Poppins' Court, Fleet Street. In 1748 Benjamin Cole (I) took over the business of Thomas Wright (I), his trade address then known as *The Orrery, next the Globe Tavern*, Fleet Street. Benjamin (I) worked in partnership with Benjamin (II), trading as Benjamin Cole & Son from 1750 until 1766. **John Wright** served as Benjamin Cole (I)'s apprentice from 1750 to 1760. Upon his death, Cole was succeeded in his business by his son, Benjamin (II).

Cole, Benjamin (II)
Benjamin Cole (II), (1725-1813) [18] [234] [276 #336], an English mathematical, philosophical, and optical instrument maker, was the son of **Benjamin Cole (I)**. He served as his father's apprentice from 1739, and was freed of the **Merchant Taylors' Company** in 1746. He was elected to the Livery of the Merchant Taylors' Company in 1763. Upon the death of his father in 1766 he succeeded him in business at *The Orrery, next the Globe Tavern*, 136 Fleet Street, London. Among his apprentices was **Edward Chester**, beginning in 1771. In 1782 his business was purchased by **John Troughton (II)**.

Cole, Humfrey
Humfrey Cole (sometimes Humphrey or Humfray), (c1525-1591) [18] [46] [275 #21] [336], was a goldsmith, engraver, mathematical instrument maker, and die sinker (a person who engraves the dies used for minting coins, medals, etc.). In 1563 he was working as a die sinker at the Royal Mint in the Tower of London, England, a position which provided for lodgings in the Tower. He continued in this role for at least fifteen years, during which time he had already started making mathematical instruments. Of the instruments made by Cole during his career twenty-six have survived, the earliest of which was made in 1568, and most of which are now in museums, [53] [54] [145] [270]. These include: compendiums, gunner's folding rules, a nocturnal, altitude dials, theodolites, astrolabes, surveyor's folding rules, horizontal sundials, a nautical hemisphere, and a plane-table alidade. In 1572 he engraved a map of the Holy Land for Richard Jugge's *Bishop's Bible*, on which the signature panel reads: *GRAVEN BI HUMFRAY COLE GOLDSMITH A ENGLISH MAN BORN IN YE NORTH AND PERTAYNING TO YE MINT IN THE TOWER · 1572*. Cole supplied instruments for Martin Frobisher's 1576 voyage of exploration attempting to find a north-west passage to the Far East. In 1582 Cole's address is recorded as *neere unto the North dore of Paules'*, London.

Cole, Thomas
Thomas Cole, (fl.1839-1869) [18], was a nautical, mathematical, and philosophical instrument maker of 21 Hannibal Rd., Mile End, Stepney, London, England. He served an apprenticeship with **William Cook** of the **Masons' Company**, beginning in 1812. He is known to have sold: magnetic compass; course corrector.

Cole, William
William Cole, (fl.1762-1764 d.1799) [18], was a mathematical and optical instrument maker of Lambeth, London, England (1762), and Strand, London (1764). In 1764 he was one of the petitioners who attempted to revoke the patent obtained by **John Dollond (I)** for the achromatic lens, and which was enforced after Dollond's death by his son, Peter (See Appendix X).

Collimation, Collimate
Collimation is the process of aligning the various optical elements of an instrument to ensure that all the elements in an optical path share the same optical axis. Where an optical instrument has more than one optical path, as in binoculars, collimation additionally ensures that the optical axes are parallel to each other and to the hinge if one is present.

Two kinds of optical aberration that arise in poorly collimated optics are: (a) when the optical axes of two or more optical elements (lenses or mirrors) are misaligned such that their optical axes are parallel but not coaxial. e.g. a lens or mirror is shifted sideways; (b) when one or more optical elements is rotated in an axis perpendicular to the optical axis - e.g. a lens or mirror is tilted out of alignment.

See also collimator, Cheshire collimator.

Collimator; Collimated
A collimator is:

(1) An optical device which forms a parallel beam of light from a diverging source. The parallel beam of light is referred to as *collimated*. Light from a far distant source, such as a star, is sufficiently parallel to be considered collimated.

(2) An instrument used to assist in the collimation of an optical device. See collimation, Cheshire collimator.

Collins, Charles
Charles Collins, (1837-c1915) [50] [264] [319] [320], was a mathematical, philosophical, and optical instrument maker of 77 Great Titchfield Street, London, England. He was the son of Henry Collins, a map and globe maker. He began his business, with his first advertisement appearing in 1863, and specialised in producing microscopes, and retailing these and prepared slides. In 1871 he began to retail his instruments from nearby 157 Great Portland Street, while still making them at his original address. He continued his business until at least 1911, and lived until at least 1915. His nephew, also named Charles Collins, was a maker of prepared microscope slides.

Collodion Process
Guncotton, otherwise known as nitrocellulose, was discovered in 1846 by Professor C.F. Schönbein of Basel, Switzerland. In 1847 Flores Domonte and Louis Ménard discovered that, by dissolving guncotton in ether, a clear gelatinous liquid resulted. This was named collodion, and soon became important for medical dressings.

In 1851 **Frederick Scott Archer**, [70], used collodion instead of albumen for glass plate photography, improving on the process devised by **Niépce**. The resulting *wet collodion process* involves adding iodide or bromide to collodion, and wetting the glass plate with it. The plate is then sensitised (in the dark) by immersion in silver nitrate then, while still moist, exposed in the camera, and fixed with potassium cyanide or sodium thiosulphate. This produces a negative image that can be used to produce multiple prints. Alternatively, the exposed plate may be turned into an ambrotype to produce a positive image.

The wet collodion process resulted in the fastest exposures possible at the time, and dominated photography over the subsequent decades.

If, instead of using a glass plate for the wet collodion plate, a thin black-enamelled metal sheet is used, the result is called a tintype, and cannot be used for the production of multiple prints.

Colmont

Colmont of Paris, France, was a maker of Galilean and Porro prism binoculars and opera glasses, probably during the late-C19th and early C20th. Instruments bearing their name include WWI military binoculars, some with the mark *MG*, which stands for *Ministère de la Guerre*, which was the French Ministry of War until shortly after WWII.

Colonial Optical Company

Colonial Optical Company, [94], of Los Angeles, CA., USA was an importer of Japanese telescopes, binoculars, riflescopes and microscopes. Evidence suggests that they operated from the 1950s on, and that they used **Mayflower** as a brand-name for their imported optics. See their 1962 catalogue [239].

Colorimeter; Colorimetry

Sometimes also spelled colourimeter, colourimetry.

A colorimeter, also known as a tintometer, is an instrument that measures the quality of a colour by means of comparison with standard colours, or combinations of them. This may be used to determine the concentration of a solution of a coloured substance by comparison with a standard solution, or standard colour slides. The design of a colorimeter may utilise a Lummer-Brodhun cube.

Colorimetry is the use of a colorimeter for these purposes.

Colposcope

A colposcope is an instrument for medical examination of the cervix.

Coma

Coma (comatic aberration) is an optical aberration that is dependent on lens or mirror shape, and is present when off-axis incoming rays of light magnify differently depending on which part of the lens or mirror they traverse. The result in, for example, an image of a field of stars is that the stars become more distorted further from the centre of the image, and the distortion blurs a point, giving it a comet-like tail (hence the name coma).

Coma occurs because the off-axis rays of light from a distant point are all at the same angle to the optical axis but, because of the curve of the optical surface, they strike it at different angles at different points on the surface. This results in different refraction or reflection, depending on where they strike the optical surface.

Coma Corrector

A coma corrector is a specially shaped optical lens, so designed to correct coma in an optical instrument. A coma corrector for a reflecting telescope is usually mounted on or near the focusing tube.

Combined Optical Industries Ltd. (COIL)

Combined Optical Industries Ltd., of Aylesbury, Buckinghamshire, England, otherwise known as COIL, is a manufacturer of plastic optics. The company was established in 1936 in Slough, England, by Arthur Kingston. The company pioneered plastic moulding techniques for optics using injection and compression moulding. They also have overseas offices in Latrobe, PA, USA, and Shanghai, China.

Combs, Oliver

Oliver Combs (sometimes spelled Combes, Coombs, or Coombe), (fl.1691-1750 d.c1752) [18] [53] [319] [322] [361] [401], was an optician and optical instrument maker of London, England. Addresses: Leadenhall St. (1695); *The Spectacles* in St Martin's Court near Leicester Fields (until 1747); Within two Doors of Essex St., in the Strand by Temple Bar (from 1747); *The Spectacles* ye Second House from Essex Street Near Temple Bar. He was the son of William Combs, a tailor in Somerset. He served an apprenticeship with Johatnan Payton, beginning in 1680, and was freed of the **Needlemakers' Company** in 1688. He served as Master of the Needlemakers' Company in 1724-1725. He worked as a journeyman for the spectacle-maker **William Radford**, and for **Edward Scarlett (I)**, and was freed of the **Spectacle Makers' Company** as a Foreign Brother in 1727/8. In 1742, he was summoned by the Spectacle Makers' Company for arrears of his quarterage payments.

According to his trade card, "Oliver Combs, (from Mr Scarlet) Optician, at the Spectacles, ye Second House from

Essex Street near Temple Bar, London. Grinds optick glasses, of all lengths, & spectacles after the new method marking ye focus on the frame, approved by the most learn'd in opticks, as the exactest way of fitting different eyes: Reading glasses in rock crystal & white or green glass. He suits the most short sighted. Concave & convex mirrors, Magick lanthorns, Camera obscuras, Skye opticks, Prisms for Sr Isaac Newton's experiments in light & colours. Sells barometers, thermometers, reflecting telescopes & perspectives of all lengths. Variety of single and double microscopes, reflecting telescopes Gregorians and Newtonians of all lengths, at reasonable rates."

Commodore

Commodore was a brand of Japanese made binoculars. Not to be confused with the **Swift** *Commodore* model of binoculars.

Common Main Objective Stereo Microscope

See microscope.

Compass Microscope

See microscope.

Comptoir Général de Photographie

Comptoir Général de Photographie was a French photographic and optical equipment maker, owned by **Félix-Max Richard**. In 1895, following a legal battle with his brother, **Jules Richard**, Felix-Max Richard sold the company to one of his directors, **Léon Gaumont**, who purchased it jointly with the engineer who built the Eiffel Tower, Gustave Eiffel, the astronomer and naturalist, Joseph Vallot, and financier Alfred Besnier. The new owners changed the name to **L. Gaumont et Cie**.

Comyns, Henry

Henry Comyns, (fl.1826-1867) [**18**] [**276** #1810], was an optical, mathematical, and philosophical instrument maker of Chelsea, London, England. Addresses: 15 Asylum Terrace, King's Road (1826); 17 King's Road East (1839-1845); 5 Hereford Terrace, King's Road (1846-1855).

Concave Diffraction Grating

A concave diffraction grating is a reflection grating consisting of a concave spherical surface with lines ruled as a projection of equidistant parallel lines on an imaginary plane surface. This produces sharp spectral lines without the need for lenses or mirrors. One advantage of this is that it may be used in those parts of the ultraviolet and infrared spectrum that are absorbed by glass. The concave diffraction grating was invented by **Henry Augustus Rowland**.

Concave Grating Spectrograph

See spectrograph.

Condenser Lens

A condenser lens is an optical lens that takes diverging light from a point source and renders it into a parallel or converging beam for illumination of an object. This is used in applications such as microscopes, enlargers and projectors.

A condenser for use in a microscope produces a light cone compatible with the numerical aperture of the objective in use. This is achieved by means of focusing the condenser lens and adjusting its aperture until the illuminating light is focused into a converging cone onto the specimen, and diverges from there to exactly fill the objective lens. The light cone reaching the objective lens should match the numeric aperture of the objective for optimum illumination, contrast and depth of field.

Conrady, Alexander Eugen

Alexander Eugen Conrady, (1866-1944) [**319**] [**321**], was an optical designer and lecturer, born in Burscheid, near Cologne in Germany. He studied mathematics, physics and chemistry at university. He developed an interest in optical design for telescopes in the early 1890s. He emigrated to England and, after moving to London in 1896, he began designing and making microscopes. In 1902 he was elected a Fellow of the Royal Astronomical Society, and in the same year he was engaged by **W. Watson & Sons** as a scientific advisor and optical designer. This work was mainly concerned with microscopes, but he also worked on other instruments, including photographic objectives, telescope objectives, and instruments for the armed services, such as submarine periscopes. Watson instruments with one or more optical component designed by Conrady were sometimes marked **Watson-Conrady**. In 1903 he naturalised as a British citizen.

In 1917 the Imperial College of Science and Technology in London set up their Technical Optics Department. Conrady left Watson, and joined Imperial College as Professor of Optical Design. Among his students were his daughter Hilda Conrady Kingslake and her husband Rudolf Kingslake. In 1927 he published *Applied Optics & Optical Design*. He began making notes for a second volume, but did not complete it prior to his death in 1944. The second volume was completed by his daughter and her husband, and published posthumously in 1960.

Constantine, Pickering & Co.

Constantine, Pickering & Co., (fl.c.late-C19th-early-C20th) of South Shields, England, & Cardiff, Wales, produced nautical instruments. Examples of instruments bearing their name include sextants, a box for a quintant, a dry card compass, and a brass telescope. These may be related to the Constantine and Pickering Steamship Co. Hence the instruments may have been made for them, and marked with their name, rather than made by them.

Contax

Contax was a **Zeiss Ikon** brand-name used from 1932 for their Contax Rangefinder camera and, from 1947, for a single lens reflex camera. In 1973 Zeiss Ikon licensed the name to **Yashica** who used it for their Contax range of cameras with interchangeable Contax/Yashica or Zeiss lenses. In 1982 Yashica was acquired by **Kyocera**, who continued to produce Contax cameras and lenses until 2005.

Contessa Camera-werke Drexler & Nagel

Contessa Camera-werke Drexler & Nagel was founded initially as **Drexler & Nagel** by the camera designer **Dr. August Nagel** in Stuttgart, Germany, in 1908. In 1919 Dr. Nagel bought the **Nettel Camera-werke** company, and renamed the combined company to **Contessa-Nettel**.

Contessa-Nettel

Contessa-Nettel was a camera maker based in Stuttgart, Germany. The company was formed in 1919 by the merger of **Contessa Camera-werke Drexler & Nagel** with **Nettel Camera-werke**. In 1926 Contessa-Nettel merged with **Ernemann**, **Goerz** and **ICA**, with financing from **Zeiss**, to form the company **Zeiss Ikon**.

Cooberow, Matthew

Matthew Cooberow, also spelled Cowperow, Cowperoe, Cooperoe, or Cooberro, (fl.1690-1720) [**18**] [**401**], was a spectacle maker of Angel St., London, England (1695). He was freed of the **Spectacle Makers' Company** in 1690. Among his apprentices were **James Mann (II)**, beginning in 1699, **Thomas Lincoln**, beginning in 1708, and **William Radford**, who was freed in 1720. **Damaine Middlefell** may also have served an apprenticeship with him.

Cook & Son, William

William Cook & Son, [**18**], quadrant maker, was a partnership which succeeded **William Cook** in business at 178 High St., Shadwell, London, England. The partnership began in 1820, and lasted only until 1821, after which **George Cook** continued the business at the same address.

Cook, George

George Cook, (fl.1821-1836) [**18**], was an optical, mathematical, and philosophical instrument maker of 178 High St., Shadwell, London, England. He succeeded **William Cook & Son** at this address in 1822, and may have been the son of **William Cook**. In 1834 the business relocated to 44 Ratcliff Highway, London, and George Cook continued to trade there until 1836.

Cook, William

William George Cook, (fl.1799-1820) [**18**] [**276** #1104], was a mathematical instrument maker and a maker of quadrants and compasses, of London, England. He traded in Cork St., St. George in the East, from 1799 until 1800, after which he traded at 178 High St., Shadwell, until his company became **William Cook & Son** in 1820. He served his apprenticeship with the mathematical instrument maker William Bailey, beginning in 1782, and was freed of the **Masons' Company** in 1794. Among his apprentices was **Thomas Cole**. He was subject to bankruptcy proceedings in 1820 [**35** #17626 p1618], and again in 1828 [**35** #18524 p2129].

George Cook is not listed as one of William's apprentices, but may have been his son, since William Cook & Son was succeeded in business by George Cook at the same address.

Cooke, John

John Cooke, (fl.1761-1767) [**18**], was an optical instrument maker of Cock Lane, Snow Hill, London, England. He served his apprenticeship with the mathematical instrument maker, John Bush, beginning in 1748/9, and was freed of the **Stationers' Guild** in 1760. During the 1760s he collaborated with **William Eastland (I)**, making microscopes bearing the name Eastland & Cooke. In 1764 he was one of the petitioners who attempted to revoke the patent obtained by **John Dollond (I)** for the achromatic lens, and which was enforced after Dollond's death by his son, Peter (See Appendix X).

Cooke Optics Ltd.

Cooke Optics Ltd. are manufacturers of cinematic camera lenses based in Leicester, England [**446**].

In 1997, Taylor Hobson, formally **Taylor, Taylor & Hobson**, was acquired by Schroders Ventures. In 1998 the optics division was sold to the American distributor, Len Zellan, who began to operate it as Cook Optics Ltd., producing cinematic camera lenses.

The objective of the company was to continue the optical tradition started in 1893 by **H. Dennis Taylor** with the design of the *Cooke Triplet* photographic lens. This was produced under license for **T. Cooke & Sons** by

Taylor, Taylor & Hobson, and was the first of a series of photographic lenses, marketed by them as *Taylor-Hobson Cooke Lenses*, and including cinematic camera and projector lenses.

Cooke's Reversible Level

The Cooke's reversible level is similar to a dumpy level, but with the difference that the telescope may be reversed, end-to-end, to enable the correction of collimation errors. This reversal is similar to that of the Y-level, but with the telescope held in two rigid sockets.

Cooke, Thomas

Thomas Cooke, (1807-1868) [3] [24] [228] [234], was born in Yorkshire, England. After studying navigation, he had wanted to go to sea, but changed his mind. He went into teaching, and studied mathematics and optics. He began making telescopes, and in 1837 he opened a small optical business in York. This was the company that was to become **Thomas Cooke and Sons** in 1868. One of his products was agricultural levels, some of which he sold to **John Cail**, who re-sold them bearing his own name. After a few years he moved to larger premises and began constructing equatorially mounted telescopes, building bigger and bigger telescopes as time passed. In 1855 he began work on a telescope factory, the Buckingham Works in Yorkshire. He continued building large telescopes as well as pursuing his interests in steam engines and surveying instruments. He acquired his large optical glass discs from the newly started **Chance Brothers** in Birmingham. Cooke's largest telescope was the 25-inch Newall refractor which was, at the time, the largest telescope in the world. It took seven years to build, and Cooke died a year before its completion. Thomas Cooke also constructed a fine self-acting dividing engine, which he finished building shortly before he died in 1868.

See Appendix VII.

Cooke and Sons, Thomas

Thomas Cooke and Sons, (fl.1868-1922) [3] [20] [24] [228] [234], were makers of scientific and optical instruments of Buckingham Works, York, England. The firm was founded in 1868, the year of **Thomas Cooke**'s death, and manufactured large observatory telescopes, telescopes for the amateur market, spectacles, opera glasses, as well as steam engines, turret clocks, drainage levels, surveying instruments and machine tools. In about 1860 they had begun to build a telescope with a 25-inch lens for the industrialist Robert Newall. The telescope took many years to complete, and proved to be a financial strain on the company. It was still not completed by the time of Thomas's death. Upon Thomas's death his two sons, Thomas (II) (1839-1919) and Charles Frederick (1836-1898), took over running the company. In 1871 they completed the Newall telescope which was, for just one year, the largest refracting telescope in the world. Newall, however, was unsatisfied, and tried to force them into liquidation. The business was rescued, and purchased by James Wigglesworth (1825-1888), and continued with Thomas and Charles in their roles in the company. They began to make observatory domes, one being the dome for the Royal Observatory's 28-inch telescope in Greenwich. In 1894, Charles Frederick Cooke retired.

From the beginning of the C20th they began manufacturing optical munitions, including depression rangefinders, gun and rifle sights, and surveying and engineering instruments tailored to military needs. By 1912 Thomas Cooke & Sons had developed an optical rangefinder on a gyro-stabilised mount, with a fire control system designed by the inventor Arthur Pollen, to compete with **Barr & Stroud**'s rangefinder offerings. It used new glass developed by **Schott**, had a brighter, higher contrast image, and was easier to use. It was, however, more complex, the Admiralty were not inclined to purchase products using imported glass, and Thomas Cooke & Sons had a less constructive relationship with the Admiralty than Barr & Stroud. Further, it failed to live up to expectations during trials, and by 1914 the Thomas Cooke & Sons rangefinder had been rejected by the Admiralty. It did, however sell well abroad. During WWI, in addition to their own wartime production, Thomas Cooke & Sons supplied lenses to **Kershaw** for their binocular production.

In the early 1880s **Harold Dennis Taylor** began to work for the company. He was an innovative optical designer who designed the *Cooke Photo Visual* telescope objective. This was a triplet apochromat that used innovative glass from Schott. This was a highly influential design, being probably the first triplet apochromat telescope objective. In 1893 he was made Optical Manager, and in the same year he designed the *Cooke triplet* photographic lens which achieved a remarkably flat field and aberration free image. Thomas Cooke & Sons did not manufacture the lens themselves, since they did not at that time manufacture photographic lenses. Instead, Taylor licensed the production to **Taylor, Taylor & Hobson** (with whom he was not related).

In 1897 the firm was incorporated as a limited company. In 1908 H.D. Taylor's son **Wilfred Taylor** joined the firm as an apprentice. In 1915 **Vickers** acquired 70% of shares in the company and, in the same year, Thomas Cooke (II), the last of the Cooke family in the firm, retired. In 1916 Thomas Cooke & Sons purchased a controlling interest in **Adam Hilger & Co.**, but in 1926 Hilgers purchased the shares back. In 1922 the firm merged with **Troughton and Simms** to become **Cooke, Troughton and Simms**.

See Appendix VII.

Cooke Troughton and Simms

In 1922 **Troughton and Simms** merged with **T. Cooke and Sons** to form Cooke, Troughton and Simms, of which **Vickers** held a majority shareholding [228] [234] [320]. By this time, the Simms family were the only one of the three founding families to have members still working in the company.

They continued to produce a range of instruments, including telescopes, microscopes, levels, theodolites and tacheometers, on which they offered discounts to counter the depressed post-war market. Having failed to achieve profitability, Vickers liquidated Cooke, Troughton & Simms in 1924, and re-floated it under their direct control. In 1926 their optical designer **Wilfred Taylor** played a major role in the development of the Tavistock Theodolite, which went into production in 1930. They also developed the *Heape-Grylls Cinema Machine*, shooting 500 to 5000 frames per second for use in ballistics experiments. In 1932 they began production of the *Vickers Projection Microscope* for projecting images from metal specimens, and they expanded their microscope offerings for the medical, educational and industrial markets. In 1938 **Grubb Parsons** took over the astronomical instrument making part of the business.

During the 1930s ordnance orders began to boost business, and during WWII they manufactured a large range of products for the Admiralty, the Ministry of Supply and the Air Ministry. This included clinometers, flash spotting instruments, dial sights, sighting telescopes, predictor telescopes, identification telescopes, stereoscopic telescopes, long internal base naval rangefinders, Tavistock theodolites, optical scale theodolites, tank periscopes, tail drift sights, telescopic sights, and tank traverse indicators. As the war neared its end their sales were once again dominated by civilian orders and, among other products, they increased their range of microscopes.

In 1947 Cooke, Troughton and Simms were listed as an exhibitor at the British Industries Fair. Their entry in the catalogue, [450], listed them as manufacturers of surveying instruments, theodolites, levels, etc. The entry went on to list: Microscopes for all purposes, biology, pathology, zoology, botany, physics, chemistry, petrology, metallurgy; Stereoscopic microscopes; Vickers projection microscope; Optical measuring instruments. By this time, however, lower priced imports from Europe were threatening business. During the early 1950s this problem was compounded by the rise in Japanese imports.

By 1956, **James Simms Wilson** and Wilfred Taylor were joint managing directors. In that year, they both retired, as did Arthur Simms who was sales manager. This marked the end of an era, since James Simms Wilson and Arthur Simms were the last of the Simms family in the business.

In 1959 the company acquired **McArthur Microscopes Ltd.**, **C. Baker Ltd.**, and Casella (Electronics) Ltd., an offshoot of **Casella**. The expanded business was then renamed in 1962 to become Vickers Instruments Ltd.

See Appendix VII.

Coombes, J.

J. Coombes, (fl.c1889-1930) [16], was an optician, admiralty agent, and nautical instrument maker of 87 Fore Street, Devonport, and The Observatory, Fore Street, Devonport, England. He also offered test certificates for sextants, showing index errors. He claimed that his business was founded in 1806 – however, this was the date when the business of **William Charles Cox** was founded. J. Coombes joined the Cox business, trading in partnership as **Cox & Coombes**. By 1889, the partnership of Cox & Coombes was no longer trading, and the business continued under the name of John Coombes (who may have been the same J. Coombes). The Coombes business continued until at least 1930.

Instruments bearing his name include: boxed drawing instrument set; telescopes; sextants; naval barometer; surveying barometer; pocket barometer; Galilean field binoculars; a sextant by J. Coombes is at the National Maritime Museum, Greenwich [54].

Cooper, Michael A.

Michael Andrew Cooper, (fl.1851-1880) [18] [409], was an optician of 7 St. James's Walk, Clerkenwell Green (1851-1859), and 25 St. James's Walk, Clerkenwell Green (1865-1880), both in London England. His daughter, Matilda Sarah Ada Cooper (b.1850), was married to the microscope maker **Edward John Weeden** in 1877.

Coopers' Company

See Appendix IX: The London Guilds.

Coppock, Charles

Charles Coppock, (1837-1900) [50], of London, England, graduated in civil engineering in 1856. Shortly after graduation he joined the firm **Smith, Beck & Beck** where he became a microscope maker. He was still with the firm in 1866 when the departure of Smith resulted in the formation of **R.&J. Beck**. When **Richard Beck** died later that year Charles Coppock became a partner in the company which did not, however, change its name. In the early 1880s Coppock left R.&J. Beck and began his own company at 100 Bond Street, London, where his business continued to produce microscopes, slides and accessories until his death in 1900.

Coppock, James T.
James T. Coppock, of London, England, and Paris, France. Instruments bearing their name include: Coppock, London, telescopes; Coppock, Paris, *Telinor* 8x25 binoculars.

Cordwainers' Company
See Appendix IX: The London Guilds.

Corfield, K.G.
K.G. Corfield, [44] [99] [306], was a company founded in 1950 by Kenneth Corfield as a maker of scientific instruments and cameras in Merridale Works, Wolverhampton, England. He had been making and marketing his *Corfield Lumimeter* at his home since 1948. The company developed their range of photographic products, and in 1952 they began to source lenses from the **British Optical Lens Company**. In 1959 the company moved to Ballymoney, Northern Ireland. In around 1960 they received an injection of capital from Guinness, and began to additionally manufacture brewery equipment. During the 1960s they ceased production of camaras, and in 1967 their factory was taken over by **Smiths Industries**. They continued in business until 1971.

Corner Cube Prism
See prism.

Corning
Corning, [24], is a US optical glass manufacturer. They introduced the **Pyrex** brand in 1915. Since 1998 the Pyrex brand has been licensed by a spinoff company of Corning throughout the world.

Because of its low thermal expansion properties, Pyrex is used as a substrate for astronomical mirrors. In 1933 Corning made the mirror substrate for the 74-inch David Dunlap reflector telescope built by **Grubb-Parsons** for the David Dunlap Observatory in Toronto, Canada. This telescope has two modes of operation. As a Newtonian reflector it operates at f/4.9 with a focal length of 30 feet. As a Cassegrain it operates at f/18 with a focal length of 111 feet. The David Dunlap observatory opened in 1935. In 1948 Corning made the Pyrex glass mirror substrate for the 200-inch f/3.3 reflecting **Hale** Telescope in the Palomar Observatory in California.

Coronet Camera Co.; Coronet Ltd.
The Coronet Camera Co., [44] [306], was founded in 1926 at 48 Great Hampton St., Birmingham, England. They produced a large number of models of low-cost cameras, as well as accessories and films. In 1932 they moved to 310 Summer Lane, Birmingham, and in 1933 the company name was changed to Coronet Ltd. Among their many camera models was the *Midget* miniature camera, using 16mm roll-film, introduced in 1935, which is a popular collectors' item, and the *3-D* stereo camera, using 127 film. During the 1930s the **British Optical Lens Co.** (BOLCo) produced body mouldings and lenses for Coronet and, in 1935, Coronet began to supply shutters for the *VP Twin* camera produced by BOLCo.

In 1946 the company became part of **Dufay-Chromex**. In 1947 they introduced the *Cameo*, a plastic-bodied miniature camera, also using 16mm film, which was a derivative of the *Midget*, but took photos in landscape orientation rather than portrait. In the 1950s Dufay-Chromex changed their name to Dufay Ltd. Coronet continued to produce many models of camera until the 1960s, marking them variously with the name Coronet Ltd, Dufay-Chromex Ltd., or Dufay Ltd.

Correction Collar (Microscope Objective)
A microscope objective correction collar is a feature of some microscope objective lenses which facilitates the use of the objective with cover slips of differing thickness. Most microscope objectives are designed to be used with cover slips of a standard thickness and refractive index. An objective with high numerical aperture, however, is sensitive to the smallest variations in these factors, so that a difference of just a few micrometres in thickness can result in severe image degradation. A correction collar provides for the adjustment of the central lens group in the objective to compensate for these variations in thickness of the cover slip.

Cosina
Cosina is a high-end Japanese optical glass and camera manufacturer founded in 1959. They have produced cameras under their own Cosina brand-name, as well as cameras re-branded for sale by many other companies worldwide. In 1999 they began to use the **Voigtländer** brand-name under license from **Ringfoto**. In the early C21[st], **Zeiss** revived the **Zeiss Ikon** name with cameras manufactured by Cosina. Production of these *Zeiss Ikon* cameras ceased in 2012.

Coudé System
A coudé system, [151], is used in a reflecting telescope, usually of Cassegrain design, and is a means of bringing the image to focus at a fixed eyepiece location that does

not move with the motion of the telescope. This is usually achieved by the use of a flat mirror that redirects the light path from the secondary mirror, at right angles along the declination axis. It may then be redirected by another flat mirror to be coaxial with the polar axis. Depending on the physical arrangement and the design of the mount the coudé system may utilise one, two or three or more mirrors to bring the light path to its coudé focus.

The coudé system was first proposed by the astronomer **Maurice Loewy** in 1871 for a telescope referred to as an *equatorial coudé*, but it was not until 1882 that the first example was built by **Paul Gautier**, sponsored by the philanthropist Raphael Bischoffsheim.

Couldrey, Joseph
Joseph Couldrey, (fl.1818-1851) [18] [276 #1297], was an optical and mathematical instrument maker of London, England. Addresses: 4 Church Yard Passage, Tooley St., (1822); 26 St. Thomas' St. East, Borough (1839-1851); 26 New St., Dean St., Borough. He is known to have made hydrometers and saccharometers.

Coulter Optical Co.
Coulter Optics of Idyllwild, California, USA, was founded in 1967 by Jim Jacobsen and continued in business until the company went bankrupt as a result of Jacobsen's illness in 1995. They were best known for their Dobsonian telescopes, but also made refracting telescopes, and mirrors for amateur telescope makers. In 1996 the company was sold to Murnaghan Instruments of Florida, who continued to produce telescopes using the Coulter name until 2001. They went on to produce mirrors and telescope kits for online sale.

Cousens & Son Ltd., B.C.
B.C. Cousens & Son Ltd., [44], were chronometer, watch, and nautical instrument makers, and compass adjusters, of Commercial Chambers, Arwenack Street, Falmouth, Cornwall, England. According to a 1951 advertisement, the company was founded in 1850, and were "by appointment, agents for Admiralty charts and publications". Instruments bearing their name include a single draw brass and leather marine telescope with 2¼" objective, length 24½" when closed, and 33" when extended; single draw brass and leather marine telescope, length 31" when extended; cased brass and black lacquered sextant; binnacle compass in brass casing with spirit burner lamp; ship's binnacle compass; brass ship's clock.

Cowley Level
A Cowley level is a device used for site levelling, and is accurate up to distances of about 30m. It uses prisms and mirrors, and has two optical paths from the objective to the eyepiece. One optical path is aligned level with the instrument, and the other is levelled by means of a pendulum. The user views through the eyepiece, and sees a split view. The relative orientation of the two parts of the view indicate level both from front-to-back, and side-to-side. The Cowley level is mounted on a special tripod with a vertical rod which, when the level is placed on it, releases the pendulum for levelling. Thus, the delicate pendulum mechanism is stowed safely when not in use.

Cox & Coombes
Cox & Coombes were opticians and nautical instrument makers of Fore Street, Devonport, England. The company was founded as a partnership between **William Charles Cox** and **J. Coombes**. W.C. Cox traded under his own name until 1857, so the partnership was likely to have been formed shortly after that date. After W.C. Cox died in 1874, the partnership continued, with William Joseph Cox in partnership with John Coombes (who may have been the same J. Coombes). After William Joseph Cox's death in 1888 [35 #25839 p3972], John Coombes continued the business under his own name.

Cox & Coombes are known to have made: telescopes; binoculars; spectacles. They also manufactured sextants for the Royal Navy.

Cox & Son
Cox & Son, [18], were opticians of 3 Barbican, London, England. The partnership, between **William Cox (I)** and, presumably, his son **James Cox**, began in 1788, and continued until 1796.

Cox & Son, J.
J. Cox & Son of London, England, were C19th telescope makers. They may have been **James Cox**, and his son **Frederick Cox**, who traded at 5 Barbican, London in the early 1830s.

Instruments bearing this name include: single-draw "Night or Day" telescope in brass with a mahogany barrel, signed *J. Cox & Son London* in cursive script [54].

Cox, Frederick
Frederick Cox, (b.1809 fl.c1833) [18], was an optician of 5 Barbican, London, England. He was the son of **James Cox**, with whom he shared this address. He served his apprenticeship with the cooper and optician, John Herbert Vaughan, beginning in 1824, and was freed of the **Coopers' Company** in 1833. He may have been the son in the partnership **J. Cox & Son**.

Cox, Frederick James

Frederick James Cox, (fl.c1845-1882) [18] [53] [61] [306], was an optician and optical instrument maker of London, England. Addresses: 100 Newgate St. (1845-1870); 22 Skinner St. (1856-1866); 26 Ludgate Hill (1866-1882); 98 Newgate St., (1875-1901).

He is recorded as having been born in 1834/5 and started his career as an optician in 1845. Since he would have been ten at the time it is possible that one of the two dates is incorrect. However, some apprenticeships began at an early age, so he may have begun an apprenticeship at that age. He began making photographic equipment in 1858, at which time he claimed that the business had been established for 130 years, suggesting it was started in 1728, perhaps by one of his forefathers. His products included some cameras and equipment for the collodion process, and some for use with dry plates, as well as stereoscopic cameras and others.

From 1882 the business of F.J. Cox continued at 26 Ludgate Hill, being then run by **H.&E.J. Dale**. Instruments bearing the name of F.J. Cox include: cameras; photographic lenses; telescopes.

Cox, H.W.

H.W. Cox and his brother L.A. Cox, [8] [24] [199], were the first English makers of Schmidt cameras. They made their first Schmidt camera in 1936. It was an f/1 6½-inch x 9-inch, and they used it in conjunction with a 12-inch reflector they had also made. H.W. Cox also made an f/1.5 6½-inch x 10-inch Schmidt together with J.P.M. Prentice in 1939. In 1947 H.W. Cox collaborated with **Frederick J. Hargreaves** and **John Victor Thomson** to form **Cox, Hargreaves & Thomson**. According to Thomson, however, H.W. Cox never joined the firm [221].

Cox, Hargreaves & Thomson

Cox, Hargreaves & Thomson, [9] [24] [199], of Coulsdon, Surrey, England, were makers of optics for large telescopes. The company was established in 1947 by the telescope maker **H.W. Cox**, the astronomer and mirror maker **Frederick J. Hargreaves**, and the mirror maker **John Victor Thomson**. According to John Thomson, having collaborated to form the company, H.W. Cox never actually worked for them. In 1951 the company moved into the Tunnel Shelter underneath Cane Hill Hospital, thus acquiring a vibration free working environment with stable temperature.

In 1951 Cox, Hargreaves & Thomson supplied the tube and optics for a 16-inch x 24-inch f/2.5 Schmidt camera for the Royal Observatory, Edinburgh, Scotland. They also figured the 50-inch mirror for a Gregorian-Schmidt at the Commonwealth Observatory, Canberra, Australia. In 1957 they delivered a Schmidt telescope with a 98cm primary mirror and 65cm correcting lens to the Vatican Observatory in Italy.

In 1961 John Thomson, together with two other members of the company, J. Mortleman and **Jim Hysom**, left Cox, Hargreaves and Thomson to start the company **Optical Surfaces**.

Cox, James

James Cox, (fl.1810-1857) [18] [322], was an optician, and mathematical and philosophical instrument maker of London, England. Addresses: Parish of St. Giles, Cripplegate, (1810); 3 Beach St., Barbican (1816); 5 Barbican, Aldersgate St. (1822-1855); 51 Banner St., St. Lukes (1822-1855); 85 Lombard St. (1839).

He was freed by patrimony of the **Spectacle Makers' Company** in 1843. According to [18], he was the son of **William Cox (I)**. Hence, he may have been the son trading in partnership with his father as **Cox & Son**, although if this was so, his career would have spanned 69 years – possible if he began his apprenticeship at age 14, as was common at the time. Nonetheless, given the dates, William (I) seems unlikely to have been the sponsor of his freedom.

In 1820 he began to work in partnership with **Joseph Cox**, trading as **Joseph & James Cox**, until 1822. His son, **Frederick Cox** was also an optician, and also traded at 5 Barbican during the 1830s. James and Frederick may have been the proprietors of **J. Cox & Son**.

Cox, John (I)

John Cox (I) was a well-regarded C17th spectacle maker in London [24]. He is known to have made mirrors for reflecting telescopes – one for **James Gregory** and one for the Royal Society, but it is said that in neither case did he produce a mirror of sufficient quality.

Cox, John (II)

John Cox (II), (fl.1742-1764) [18], was an optical instrument maker of St. John's Court, Cow Lane, London, England. He served his apprenticeship with **Thomas Lincoln**, beginning in 1724, and was freed of the **Spectacle Makers' Company** in 1732. Among his apprentices was **John Margas**, who was turned over to him by **Nathanial Adams** in 1741 or 1742. In 1764 John Cox (II) was one of the petitioners who attempted to revoke the patent obtained by **John Dollond (I)** for the achromatic lens, and which was enforced after Dollond's death by his son, Peter (See Appendix X).

Cox, Joseph

Joseph Cox, (fl.1789-1822) [18] [276 #816], was an optician, and optical and mathematical instrument maker of London, England. Addresses: 4 Queen's Square, Bartholomew Close, (1789); Queen St., Bartholomew Close (1790); 3 Barbican (1792-1816); 5 Barbican, Aldersgate St. (1817-1822). He was a Freeman of the **Turners' Company**. In 1820 he began to work in partnership with **James Cox**, trading as **Joseph & James Cox**, until 1822.

Cox, Joseph & James
Joseph & James Cox, [18], opticians and mathematical instrument makers of 5 Barbican, London, England, was a partnership between **Joseph Cox** and **James Cox**. The partnership was formed in 1820, and terminated in 1822 [**35** #17797 p410].

Cox, William (I)
William Cox (I), also known as William Cock, (fl.1765-1786) [18] [276 #816], was a perspective turner and optician of London, England. Addresses: Love Lane, Wood St. (1767); Barbican (1767); 3 Barbican (1784-1786). The address at 3 Barbican was later occupied by **Joseph Cox**. William (I) was a Freeman of the **Turners' Company**. He began to work in partnership as **Cox & Son** in 1788, trading at the same address, and the company continued under that name until 1796. The son in this partnership may have been **James Cox** who, according to [18], was William (I)'s son.

Cox, William (II)
William Cox (II), (fl.1777-1780) [18], was an optical instrument maker of 55 Houndsditch, London, England. He served his apprenticeship with **John Cuff**, beginning in 1767. He was turned over to **Charles Lincoln** in 1771, and was freed of the **Spectacle Makers' Company** in 1778.

Cox, William Charles
William Charles Cox, (1786-1874) [16] [18] [81] [276 #1517] [358], of Fore Street, Devonport, England, and Southside Street, Plymouth, was an optician, and mathematical and philosophical instrument maker with a deep interest in astronomy. Fore Street addresses: 93; 86 (1822-1839); 86B (1830); 87 (1851); 89 (1852-1856); 83 (1857). Southside Street addresses: 35; 24 (1856).

His family business was founded in 1806, although whether he was the founder is unclear. He was the Devonport agent in charge of the depot of Admiralty chronometers, and had a close relationship with the Chronometer Department of the Royal Observatory. He was associated with **Robert Brettell Bate** as an agent for Admiralty charts. In 1831 he was elected Fellow of the Royal Astronomical Society.

According to his trade card, "W.C. Cox, Optical & Nautical instrument maker, removed to 93 Fore Street, Devonport, & 35 South-side Street, Plymouth. Sextants, quadrants, compasses, telescopes, barometers, binnacles, &c., Surveying and mining instruments on the most improved principles. Charts by various makers to all parts of the world. Admiralty chronometer agency. A great variety kept ready for sale, new and second-hand, accurately rated by transit observations. Nautical instruments of every description repaired and adjusted at a short notice."

Records show W.C. Cox trading at the Fore St., Devonport and Southside St., Plymouth addresses from 1822 until 1857. After this, he entered a partnership with **J. Coombes**, trading as **Cox & Coombes**. Following his death in 1874, the partnership continued under the proprietorship of William Joseph Cox and John Coombes.

Instruments bearing W.C. Cox's name include: telescopes; sextants; sympiesometer; artificial mercurial horizon.

Craig, Henry W.
Henry W. Craig, (c1817-1863) [50], was an Englishman who emigrated to the USA in 1850. He moved first to New York, then Ohio, settling in Cleveland where he worked as a janitor in a medical college. His knowledge of optics was self-taught, and in 1862 he was granted a patent for his invention of the Craig Microscope. He may have made the early microscopes himself, but he then began to have them manufactured and distributed under license. After Henry Craig's death his wife continued to licence the production of his design.

Craig Microscope
The Craig Microscope, [50], invented by **Henry Craig**, is a simple drum microscope with an illuminating mirror at the bottom and the lens at the top. The lens consists of a small flint glass sphere fused onto a plate of crown glass on its underside. The focus of the lens is the bottom surface of the crown glass plate. The sample, such as a drop of liquid, is placed directly on the underside of the crown glass plate.

Crawford, Henry
Henry Crawford, sometimes spelled Craford, (fl.1737-1761) [90], was a mathematical instrument maker of London, England. Addresses: Denmark Street, Ratcliffe Highway (1737); Ratcliffe Highway (1741). He was the son of Thomas Crawford, a gunmaker of London. He served apprenticeship with **John Gilbert (I)**, beginning in 1724, and was freed of the **Grocers' Company** in 1737. Among his three apprentices was John Robinson, son of Richard, beginning in 1753, and freed in 1761. A single

draw brass & wood telescope signed *Crawford London Day or Night* may be attributed to him.

Crichton Brothers
Crichton Brothers, [**276** #2108] [**319**] [**322**], opticians and nautical instrument makers of London, England, was a partnership between John David Crichton and Charles D. Crichton, the sons of **John Crichton**. The partnership was formed in December 1869 following the termination of the partnership **John Crichton & Son** [**35** #23575 p157].

Crichton, John
John Crichton, (c1806-1870) [**18**] [**276** #1518] [**319**] [**322**], of London, England, was an optical, nautical, mathematical and philosophical instrument maker. Addresses: 32 Fore Street, Limehouse (1831); 112 Leadenhall Street (1834-1865). He served his apprenticeship with **Benjamin Messer**, beginning in 1820, and was freed of the **Worshipful Company of Makers of Playing Cards** in 1834, upon which he moved to the Leadenhall Street address, where he is known to have sold barometers, thermometers, magnetic compasses, drawing instruments, artificial horizons, sextants, microscopes, and telescopes. He also produced a Microscopic Phantasmagoria, which could be used either as a phantasmagoria lantern or a lucernal microscope, and a wide selection of slides for use with it.

His trade card stated "Manufacturer of Mathematical, Optical and Nautical Instruments to the Honorable East India Company and to the Honorable Corporation of the Trinity House." In the Great Exhibition of 1851, he exhibited a number of nautical and optical instruments [**328**]. He was awarded medals for his sextants and drawing instruments.

John Crichton formed a partnership with his son, John David Crichton (b.c1848), trading as **John Crichton and Son**. After the partnership was terminated in December 1869, John David Crichton continued the business with his brother Charles D. Crichton (b.c1851), trading as **Crichton Brothers**.

Crichton, Joseph
Joseph Crichton, (fl.1838-1851) [**276** #2108] [**322**], was a nautical, optical, mathematical, and philosophical instrument maker of 112 Leadenhall Street, London, England. This is the same address as **John Crichton**.

Crichton & Son, John
John Crichton & Son, [**18**] [**319**], opticians and nautical instrument makers of 11 Billiter Street, off Leadenhall Street, London, England, was a partnership between **John Crichton** and his son John David Crichton. The partnership was terminated in December 1869 [**35** #23575 p157], and the business continued as **Crichton Brothers**.

Crickmore, Thomas
Thomas Crickmore, (1781-1822) [**18**] [**274**] [**276** #1109] [**412**], was an optical instrument maker of Suffolk, England. He trained as a weaver, but changed career to become an optician, trading at premises in New Street, Ipswich. According to [**413**]: "Crickmore, Thomas, came from Beccles to Ipswich, in 1809, and died the 6th of January, 1822, about forty years of age. On his first arrival in this town he was a performer in the band of the Suffolk militia, and at the peace he married and settled here as an optician, and had he lived, would probably have risen to eminence in that occupation; for though he was self-taught, his knowledge and excellence of execution was known and appreciated by many men of science."

He collaborated with the engineer and scientist Captain Henry Kater F.R.S., of the Royal Engineers [**276** #999], to conduct a comparison of the light-gathering ability of Gregorian and Cassegrain telescopes. Their conclusions, although controversial, were published in 1813 in the *Philosophical Transactions of the Royal Society of London*.

He is known to have made microscopes, as well as Newtonian, Gregorian, and Cassegrain telescopes, for which he cast, ground, and polished the telescope mirrors. It is believed that he made bespoke instruments to order, rather than keeping a significant stock. According to an advertisement of 1821, "…after a long interval occasioned by severe indisposition, he has re-commenced manufacturing of different kinds of reflecting telescopes, solar and other microscopes, &c., to which he has added, the making, in the best style of workmanship, lathes, for ornamental turning, or for engraving upon wood, ivory, or metal."

Criterion Mfg., Inc.
Criterion Manufacturing Inc., [**136**], was started as the Criterion Co. by John Krewalk Sr. in about 1954 in Connecticut, USA. They began by making small refractors, reflectors, and eyepieces and, as the demand for larger telescopes grew, they built bigger telescopes, up to 16" aperture. John Krewalk's son, John Jr., joined the company in 1974. John Krewalk Sr. was a good friend of Tom Cave of **Cave Optical Co.** and, rather than competing with each other they would help each other out when the work-load was too high.

During the 1970s, competition from **Celestron** and **Meade** began to put pressure on Criterion, and in 1981 the Krewalk family sold Criterion to **Bushnell**, then a wholly owned division of **Bausch & Lomb**. They continued to manufacture telescopes until 1997 when Bausch & Lomb closed Criterion to focus on other products.

Critical Angle
See total internal reflection.

Cross-staff
The cross-staff, sometimes known as a fore-staff, radius astronomicus, groma, Jacob's staff, or baculus Jacob, was a simple tool, used in astronomy, navigation, and surveying to measure angular distance. It consists of a long staff, normally wooden and of square section, with a transom mounted across it, which can be slid back and forth along the staff. The staff is marked along its length with trigonometric graduations. To measure the angular distance between two distant points, the user holds the staff to the eye, and views along it to one point, moving the transom until its end coincides with the second point. The marking on the staff in the position of the transom yields the angular distance between the points. As an example, this may be used to determine latitude, by measuring the angular position of the pole star above the horizon. It was first used for marine navigation in the early C16th. It could be used to measure the altitude of the sun, but this necessitated looking into the sun, and it was superseded for this purpose by the backstaff in the late C16th.

Crouch, (Pearce) Henry
Pearce Henry Crouch, (1838-1916) [50] [61] [306] [320], who usually went by the name Henry Crouch, served his apprenticeship with **Smith, Beck and Beck**. He left them in about 1862 to begin his own business, trading as H.&W. Crouch, in partnership with his younger brother, William Manning Crouch (1840-1910) whose profession was recorded as a civil engineer. They conducted their business initially at an address on Regent's Canal Dock, Commercial Road, London, England, making microscopes, telescopes, opera and race glasses, spectacles, and other instruments. By 1864 they had taken on additional premises at 64A Bishopsgate Street. The partnership lasted until 1866 when William left, after which Henry continued to trade successfully under his own name.

In 1868 he moved his business to 54 London Wall, and in 1883 he moved again to 66 Barbican. With his business thriving, in 1886 he incorporated his business as Henry Crouch Ltd., and at about this time he began to advertise that he was making camera lenses and cameras. By 1892, at about the time his son Pearce H.H. Crouch joined his business, he opened a second shop at 141 Oxford Street. Some of his instruments were retailed through S. Maw, Sons & Thompson, and S. Maw, Son and Sons. In 1907 he sold his business to S. Maw, Son & Sons, but continued to work for them.

Crouch, H.&W.
See **Henry Crouch**.

Crown Glass
See *Types of Optical Glass* in Appendix VI.

Crown Optical Company
The Crown Optical Company, [136], was set up in New York, USA, in 1909 to manufacture optical instruments including photographic lenses and binoculars. (Not to be confused with Japanese made binoculars bearing the name Crown or COC). They began to produce military optics during WWI, with Galilean and Porro prism models being made for the armed forces of foreign countries, including Britain. In 1917, when the US entered WWI, the Crown Optical Company couldn't keep up with domestic demand for military optics, and was nationalised. From that time, they ceased using the name Crown Optical Company, and became known as the **U.S. Naval Gun Factory Optical Shop Annex**.

Cubic Precision
Cubic Precision was formed in the early 1980s when Cubic Corporation of California, USA, acquired the optical tooling division of **Keuffel & Esser**, and renamed it Cubic Precision. The company was acquired from Cubic Corporation in 1997 by **Brunson**.

Cuff, John
John Cuff, (1708-1772) [3] [18] [200] [264], was an optical instrument maker at the *Reflecting Microscopes and Spectacles*, opposite Sergeant's Gate, Fleet Street, London, England. He was at this address from 1737 to 1757 when he moved to an address near Salisbury Court. He started his apprenticeship with **James Mann (II)** in 1722, gained freedom of the **Spectacle Makers' Company** in 1729, and became Master of the Company in 1748. Among his apprentices were **Henry Raynes Shuttleworth**, **James Cleare**, and **William Cox (II)** who was turned over to **Charles Lincoln** in 1771.

He began trading in Fleet Street in 1737 making optical and mathematical instruments, barometers and thermometers, and he devised an improved solar microscope. In 1750 he was declared bankrupt, [35 #9009 p3], but managed to stay in business by selling his personal possessions. Reputedly **Benjamin Martin** took premises adjacent to Cuff's shop and brazenly took business away from Cuff, who, in 1757, moved to a new address at the *Double Microscope, three Pair of Golden Spectacles & Hadley's Quadrant* opposite Salisbury Court in Fleet

Street. In 1764 his address is listed as Strand, London. In the same year, he was one of the petitioners who attempted to revoke the patent obtained by **John Dollond (I)** for the achromatic lens, and which was enforced after Dollond's death by his son, Peter (See Appendix X).

John Cuff's trade card when he was at the *Reflecting Microscope and Spectacles* showed: Spectacles and reading glasses; Refracting telescopes; Newtonian and Gregorian reflecting telescopes; Single and double microscopes; The solar microscope; The aquatic microscope; Culpepper's microscope; Wilson's Pocket Microscope, &tc.; The camera obscura, magic lanterns, convex and concave speculums or looking glasses; Multiplying glasses; Barometers; thermometers; opera glasses, and other instruments.

Culpeper, Edmund
Edmund Culpeper (I), (1670-1737) [**18**] [**46**] [**53**] [**90**] [**264**] [**275** #422] [**396**] [**442**], was an English optical and mathematical instrument maker, born in Dorset, England, who began his career as an engraver in London. Addresses: *Old Mathematical Shop*, Black & White Horse, Middle Moorfields; *Old Mathematical Shop, Cross Daggers*, Middle Moorfields; Under the Piazza at the Royal Exchange; *Cross Daggers*, Moorfields (1700-1731); Near the Royal Exchange, Cornhill (1737).

He was a skilled engraver of scales, dials and sectors. His father was a clergyman in Tarrant Gunville, Dorset. Edmund served an apprenticeship with the mathematical instrument maker Walter Hayes from 1684, and was freed of the **Grocers' Company** in 1713. He succeeded Walter Hayes in his shop. He began to make simple microscopes, but went on to make various kinds including tripod-mounted microscopes, and made improvements to the mechanical and optical designs then current. One result of his endeavours was the eponymous *Culpeper microscope*, although it is not certain that he invented it. He also advertised a "Long Telescope for Celestial Observations." Edmund (I)'s son, Edmund (II) became free of the Grocers' Company by patrimony, and took his own son, John, as his apprentice.

Curry & Paxton
Curry & Paxton of London, England, was an ophthalmic optician who produced ophthalmic instruments and eyewear beginning in the late-C19[th]. During the 1920s-1940s they traded at 120-126 Albert St., London, and in 1928 they were also trading at 22 Wigmore St., London. They were acquired by **Boots** in 1987 along with the **Clement Clarke** chain of optometrists, to form Boots Opticians Ltd. Instruments bearing their name include: Galilean binoculars.

Cushing's Level
A Cushing's level is similar to a Cooke's reversible level, but the telescope cannot be reversed. Instead, the fittings for the objective and eyepiece are both the same, so the objective and eyepiece may be removed and interchanged, thus reversing the level to enable the correction of collimation errors.

Cuthbert, John
John Cuthbert, (1783-1854) [**18**] [**142**] [**264**] [**268**], was an optical, mathematical, and philosophical instrument maker of London England. Addresses: 16 Paradise St., Lambeth; Westminster Rd.; 40 Great Windmill St. (1809); 113 St. Martin's Lane (1816-1817); 5 Bridge Rd., Lambeth (1822); 445 Strand (1822); 22 Bishop's Walk, Lambeth (1826); 5 Purbeck Place, Lambeth, near Church St. (1826); 58 Brook St., West Square, Lambeth (1833).

He made reflecting telescopes, and from 1826 he began to produce reflecting microscopes - notably his horizontal reflecting microscopes, designed using a plan recommended by the microscopist Dr. C.R. Goring, that used a concave elliptical speculum, and was based on the reflecting microscope invented by **Giovani Battista Amici**. Whereas the Amici microscopes had a single elliptical mirror, and relied on changing the eyepiece to vary the magnification, Cuthbert's instruments were equipped with a selection of mirror objectives of different focal lengths. Thus, the user could choose the eyepiece and objective combination most appropriate to the observation being made. John Cuthbert established a strong reputation for the quality of the instruments he produced.

When in 1830 **J.J. Lister** published his design for the achromatic refracting microscope objective lens, the reflecting microscope, being expensive and inconvenient to use, was superseded.

Cuthbert, T.
T. Cuthbert, (fl.1815-1822) [**276** #1303], was an optical and mathematical instrument maker of 113 St. Martin's Lane, Charing Cross, London, England.

Cuthbertson & Champneys
Cuthbertson & Champneys was a partnership between **John Cuthbertson** and his father-in-law, **James Champneys**. The partnership began in 1768, trading in Amsterdam, Netherlands, and continued until c1771, when James Champneys returned to London. Their price list included refracting telescopes, of which none is known to have survived. Instruments bearing their name include a reflecting telescope.

Cuthbertson, John

John Cuthbertson, (1743-1821, fl.1766-1821) [**18**] [**276** #340] [**322**] [**366**], was a mathematical, optical, and philosophical instrument maker of 54 Poland St., Oxford St., Soho, London, England. He served his apprenticeship with **James Champneys**, beginning in 1761, and was freed of the **Stationers' Company** in 1794. When James Champneys, by then his father-in-law, relocated to Amsterdam, Netherlands, in 1768, Cuthbertson accompanied him, and they traded together as **Cuthbertson & Champneys**. This partnership was short-lived, however, and Champneys returned to London in c1771. John Cuthbertson began to specialise in air pumps and electrical machines, and continued to build a successful business, making instruments for, among others, Martinus van Marum, who was amassing a collection of instruments for Teyler's Museum [**362**]. Among his apprentices were **Hartog van Laun** and **Hendrick Hen**. From 1773 until 1777, John Cuthbertson shared an address with **Jacobus van Wijk (II)**. In 1774 his brother **Jonathan Cuthbertson** joined his business. Jonathan, however, subsequently relocated to Rotterdam, and started his own business. In 1793 John Cuthbertson returned to London, where he traded at 53 Poland St., Oxford St. until 1800, after which he traded at 54 Poland St., Broad St. until his death in 1821.

Cuthbertson, Jonathan

Jonathan Cuthbertson, (bap.1744 d.1806) [**18**] [**276** #340a] [**322**] [**362**] [**366**], was a nautical, optical, mathematical, and philosophical instrument maker of Dearham, Cumberland, England. In 1774 he went to Amsterdam, Netherlands, to join the business of his brother, **John Cuthbertson**. He then relocated to Rotterdam, where he began his own successful business. In 1794 he published *Verhandeling over de Verrekykers* (Treatise on Telescopes).

Cutts, J.P., (Cutts, I.P.)

Note: The J. in the makers name, in its various forms, often appears as I.

J.P. Cutts, [**3**] [**18**] [**44**] [**50**] [**258**], optical instrument maker, of Sheffield, London, and New York. Addresses: Near St. Paul's Church, 58 Norfolk St., Sheffield (1822-1825); Division St., Sheffield (1828-1841); 3 Crown Court, Fleet Street, London (1836); 43 Division St., Sheffield (1839); 39 Division St., Sheffield (1852, 1870); 56 Hatton Garden, London (1852). 51, 53, 55 & 57 Division St., Sheffield (1882).

The company was first established by John Preston Cutts (1787-1858), in 1804 in Sheffield, England. J.P. Cutts would have been 17 at this time, and records suggest he served an apprenticeship with **Proctor and Beilby**. He began trading as J.P. Cutts. Two of his sons, William and Henry joined the business, and they began to trade as J.P. Cutts and Sons sometime between 1829 and 1845. In about 1845 John Sutton (c1788-1859) joined the firm, and they began to trade as J.P. Cutts, Sons and Sutton.

In 1834 James Chestermann (1792-1867), began to work with J.P. Cutts. Their association lasted until the death of John P. Cutts, including the partnership Cutts, Chestermann & Co., which was trading by 1855.

An advertisement of 1839 reads: "IP. Cutts. 43 Division Street, Sheffield. Manufacturer of gold, silver, tortoiseshell & steel spectacles. Achromatic, pocket, ship and astronomical telescopes; Single, compound and solar microscopes; Brass and electrum drawing instruments; Sextants and quadrants; Marine, mountain, portable and wheel barometers; Pocket, bath, chemical and brewers' thermometers; Ivory and boxwood rules and scales; Surveyors' measuring tapes and chains &c. &c. Also sole manufacturer of Chesterman's patent lever roasting jacks; Double and single action spring door hinges; Self-acting window blind and map rollers; Spring tape measures &c. &c. Late warehouse in London removed to Sheffield"

The firm advertised from 1839 as "opticians to Her Majesty". J P Cutts' partner, John Sutton died in 1859. Their signature appears in the following forms: *J.P. Cutts*; *J.P. Cutts, Sons & Sutton*; *J.P. Cutts, Sutton & Co.*; *J.P. Cutts & Co.*; *J.P. Cutts, Sutton & Sons*; *J.P. Cutts, Sutton & Son*; *J.P. Cutts & Sons* (in all cases with the J. sometimes showing as an I.) Records of their work continue until 1887.

Reference [**225** p348] shows the logo used by I.P. Cutts Sutton & Son at the time of its publication in 1870. It shows a **foul anchor** with the words *Trade Mark* above, and *TRY ME*. below. The same logo is shown on an advertisement of 1882, along with a different logo for ships' logs and sounding machines. The referenced handbook lists the company as: "Manufacturers of telescopes, microscopes, spectacles, readers, eye-glasses, opera, race and marine glasses. Also manufacturers of every description of dram bottles for the pocket, for rail, race &c. Game bags, pouch and rifle slings and sling cases for use of sportsmen &c. Importers of French and German goods as above."

D

Dach Prism
Dach prism is another term used for roof prism. The word dach is short for *dachkante*, German for *roof-edge*.

Dacora
Dacora was founded in 1946 in Reutlingen, Germany, by Bernhard Dangelmaier, manufacturing low-cost cameras which were sold under their own brand, and re-branded by other companies. They continued camera production until shortly after the company changed hands in 1972.

Daguerre, Louis
Louis Jacques Mandé Daguerre, (1787-1851) [31] [37] [70] [71], of Carmeilles-en-Paris, France, was the inventor of the daguerreotype, the first practicable photographic process. He met **J.N. Niépce** through dealing with the optician **Chevalier**. In 1829 Niépce and Daguerre formed a partnership to develop their mutual interest in capturing camera obscura images using Niépce's heliographic process. They had not made much progress when Niépce died in 1833, and Daguerre continued with his efforts. By exposing a silver plate to iodine, the plate would become photosensitive. Having then exposed the plate in a camera obscura it could be developed using mercury vapour to form a direct positive image. When the process was announced in 1839 the basic process remained the same although the exact formulation had changed. The process was made freely available worldwide, except in England, where Daguerre retained a patent. Daguerre was awarded a pension by the French government for his work. Some of Daguerre's early instruments were made for him by **Antoine Molteni**.

Despite Daguerre's process preceding the work of **William Henry Fox Talbot**, and being made freely available, it was Talbot's technique of developing multiple positive images from a negative that led the way for the future of popular photography.

Daguerreotype
In 1829 **Louis Daguerre** began to work in collaboration with **Joseph Nicéphore Niépce** [31] [37] [70] [71]. Niépce had been performing experiments with the objective of forming enduring images using light sensitive materials, and their work together led to the invention of the daguerreotype – the first viable form of photography.

A silver-coated copper plate is exposed to iodine with the result that the surface of the plate becomes photosensitive silver iodide. The plate is then exposed in a camera obscura, after which it is developed using mercury vapour to form a direct positive image. The developed image is then fixed by exposing it to salt water or some other fixative such as sodium thiosulfate. Finally, the plate is washed in distilled water and mounted.

Early daguerreotypes required lengthy exposures, making portraits challenging, but as the process was refined and developed, the exposure times became shorter.

A collection of daguerreotype images may be seen at the online *Daguerreobase* [240].

Daiichi Kōgaku
Daiichi Kōgaku, was a Japanese camera and telescope maker. The company began business sometime in the 1930s under the name **Okada Kōgaku**, and was renamed Daiichi Kōgaku sometime around 1951. Soon after this they began to use the *Zenobia* brand-name for some of their cameras. They sold telescopes under brands, including **Southern Precision Instruments** in the USA. Having failed to adapt quickly enough to the shift in demand from folding cameras to 35mm cameras, the company closed in 1955. In 1956 the company re-emerged, this time under the name **Zenobia Kōgaku**, and continued in business until 1958.

Dainteth, Thomas
Thomas Dainteth, (fl.1799) [18], was a telescope maker of *By Hick's Hall*, Clerkenwell, London, England. He served his apprenticeship with the mathematical instrument maker Joseph Bush of the **Stationers' Company**, beginning in 1744, and was freed by patrimony of the **Clothworkers' Company** in 1753.

Dale, H.&E.J.
H.&E.J. Dale, [61] [306], of 26 Ludgate Hill, London, England, was also known as the Ludgate Photographic and Scientific Store. In 1883 they acquired the optical instrument makers **F.J. Cox**. By this time, they were additionally trading at 4 Little Britain, and 9 Kirkby Street, London, the latter address being their photographic camera works. Following their acquisition of F.J. Cox, **Arthur S. Newman** worked in their photographic department. They produced pocket cameras, tourist cameras, and studio cameras, and continued to trade until at least 1890.

Dall, Horace E.S.
Horace Edward Stafford Dall, (1901-1986) [53] [320] [342], of Luton, Bedfordshire, England, was an innovative and prolific amateur telescope maker and mirror and lens maker. He began his career in the drawing office of aircraft manufacturers Hewlett & Blondeau. In 1918 he began to work for the manufacturing engineers George Kent Ltd., for whom he designed equipment for fluid flow measurement until his retirement in 1965.

In 1925 he began to make telescope mirrors, and went on to make telescopes and eyepieces. During the 1930s he worked with an architect to design himself a house in which the attic was specially constructed to incorporate a camera obscura capable of projecting a seven-foot image of the sun. He invented, independently of Alan Kirkham, a form of Cassegrain telescope having a prolate ellipsoidal primary mirror and a spherical secondary. This became known as the Dall-Kirkham Cassegrain. He made jewel lenses for simple microscopes, achieving a record numerical aperture of 1.92. During WWII he took on the role of official repairer of Leitz microscopes, which necessitated making hundreds of lenses. Throughout his life he remained a prolific innovator and maker of optics, and an inspiration and mentor to others.

Dall-Kirkham Cassegrain Telescope
A Dall-Kirkham Cassegrain telescope is one with a prolate ellipsoidal primary mirror and a spherical secondary. This was invented independently by **Horace Dall** and the amateur astronomer Alan Kirkham, and named for both inventors.

Dallaway, Joseph James
Joseph James Dallaway, (fl.1799-1809) [18] [53] [90], was an optician and mathematical instrument maker of London, England. Addresses: 4 George Lane, Behind the Monument (1799-1803) and 147 Tottenham Court Road (1803-1809).

He served his apprenticeship with **Edward Troughton**, beginning in 1789, and was freed of the **Grocers' Company** in 1799. Instruments bearing his name include: 3-draw brass and mahogany telescope, 1¾" diameter, 9¾" length, extending to 31".

Dallmeyer, J.H. Ltd.
J.H. Dallmeyer Ltd., [44] [50] [61] [306], was an optical instrument maker of London, England. Addresses: 19 Bloomsbury St. (1860-1887); 25 Newman Street, Oxford St. (1888-1913); 83 Denzil Rd., Neasden (1906-1911); Church End Works, High Road, Willesden (1911-1957); Carlton House, 11d Regent Street, Piccadilly Circus (1920-1925); 31 Mortimer St., Oxford St. (1925-1941).

The company was established by the Anglo-German optician, John Henry Dallmeyer, (1830-1883) [8] [45] [46] [264]. He served his apprenticeship with an optician in Osnabruck, Germany, and in the early 1850s he began to work in London for **Andrew Ross**. After a period of absence from Ross he returned as scientific advisor, and later married Ross's daughter, Hannah. On Ross's death in 1859, Dallmeyer inherited part of his fortune, and the telescope manufacturing portion of his business.

J.H. Dallmeyer produced telescopes and photoheliographs, including those for observatories, and was renowned for the quality of his work. He also produced original, innovative designs for camera and microscope lenses. In 1866 he introduced the *Rapid Rectilinear* lens, similar to the **Steinheil** aplanat, which was introduced at the same time. These lenses were used as a standard for camera objectives for some years. He was advised by his doctors to travel for his health, and died in 1883 while on a cruise off the coast of New Zealand, being lost overboard, presumed drowned.

His second son, Thomas Rudolphus Dallmeyer, (1859-1906) [24] [37] [46], assumed control of the business on the failure of his father's health. In 1891 Thomas Dallmeyer patented and introduced the telephoto lens at the same time as, but independently of, its introduction by Steinheil.

In 1919 the company was listed as a member of the **British Optical Instrument Manufacturers' Association**. In 1947 they advertised a variety of photographic and projection lenses at the British Industries Fair [450]. The business, continued until at least 1957.

Dalton, Charles X.
Charles X. Dalton, [201], was an American microscope maker in the latter half of the C19[th]. He worked for **Charles Spencer** as a brass worker and, when **Robert Tolles** left Spencer to form his own company in Canastota in 1858, Dalton went to work with him. When Tolles moved to Boston, Dalton continued to work with him until Tolles' death in 1883. Dalton then took over the company, and continued to run it until at least 1895 when he advertised as "Charles X. Dalton, successor to the late R.B. Tolles, Boston Optical Works", located at 30 and 48 Hanover Street, Boston.

Dancer, John Benjamin
J.B. Dancer FRAS, (1812-1887) [3] [8] [18] [46] [49] [50] [70] [127] [384], of Liverpool and Manchester, England, was an inventor, optician, and maker of microscopes, barometers, surveyors' compasses, and other scientific instruments. Addresses: 11 Pleasant St., Liverpool (1835); 21 Pleasant St., Liverpool (1839-1845); 13 Cross St., King St., Manchester (1841-1845); 43 Cross St., King St., Manchester (1847-1848).

His father, **Josiah Dancer**, had worked for **Edward Troughton**, but later worked in partnership with his own father, **Michael Dancer**. In 1816 Michael Dancer died, and Josiah moved his business to Liverpool. Here John Dancer learned his trade in his father's workshops, and inherited the business when his father died in 1835.

J.B. Dancer's experimental work led him to several innovations and inventions. He introduced the use of limelight in magic lanterns, and later invented a dissolving mechanism to fade one image into the next. He was the first to use porous, unglazed jars to separate the chemicals in a Voltaic battery, and he discovered a method of copper plating using electrolysis. Upon the announcement of the Daguerreotype in 1839, he began to experiment with the process. He also began to make, and improve the designs of good quality, reasonably priced achromatic microscopes.

In 1841 J.B. Dancer began to work in partnership with **A. Abraham**, trading as **Abraham and Dancer** at 13 Cross Street, King Street, Manchester. The partnership continued until 1844, when Dancer bought out Abraham's interest and continued the business under his own name. The termination of the partnership was formalised in 1845, [35 #20458 p1027]. He began to supply photographic apparatus, and taught the process to customers. He supplied apparatus used by the physicist J.P. Joule in his work on the mechanical equivalence of heat. John Dancer's expertise in both microscopy and photographic equipment led him to develop the first microphotograph in 1852. He went on to produce a large catalogue of microphotographs for viewing with a microscope. In 1853 Dancer made the first prototype stereo camera, a concept proposed by **Sir David Brewster** in 1849.

In the London International Exhibition of 1862, he was awarded a prize medal and an honourable mention for his microscopes, and for the invention of the microphotograph. He had two entries in the catalogue, [258]. The first was entry 2889, "Dancer, J.B., Manchester.—Binocular and monocular microscope; Microscopic photographs; Micrometers; Equatorial telescope; Dissolving view lantern." The second was entry 3070, "Dancer, J.B., 43 Cross Street, Manchester.—Microscopic photographs; Landscapes."

During the 1870s, his business was thriving. However, suffering with diabetes, his eyesight began to fail, and he retired in 1878.

Dancer, Josiah

Josiah Dancer, (1779-1835) [3] [18] [276 #1304], was an optician and mathematical instrument maker of London, and later Liverpool, England. He was the son of **Michael Dancer**, and the father of **John Benjamin Dancer**.

London address: 52 Great Sutton St., Clerkenwell (1817). Liverpool addresses: 20 Grafton (1821); 14 New Quay (1823); 22 Wolfe St. (1831-1832); 4 Chapel Walks (1834); 13 Pleasant St. (1835). He had a residence at 20 Grafton (1823-1829).

He worked for **Edward Troughton**, and later worked in partnership with his father, Michael. He was succeeded in business by his son, John Benjamin Dancer.

Dancer, Michael

Michael Dancer, (fl.1776-1817 d.1817) [3] [18] [274] [276 #947] [385], was a mathematical instrument maker of London, England. He was the father of **Josiah Dancer**, and the grandfather of **John Benjamin Dancer**. He served his apprenticeship with **Robert Tangate**, beginning in 1766, and was freed of the **Joiners' Company** in 1776.

Addresses: Bride Lane, Fleet St. (1776); 4 Bangor Court, Shoe Lane (1778-1786); 11 New Street Square, near Shoe Lane, Fleet St. (1788-1790); Blewitt's Buildings, Fetter Lane, and 32 Rosoman's Row, Clerkenwell (1793); Rosoman St., Clerkenwell (1796-1800); Red Lion St., Clerkenwell (1801); 55 Great Sutton St., Clerkenwell (1804-1805); St. James, Clerkenwell (1808); 52 Great Sutton St., Clerkenwell (1810-1817).

Among his apprentices were **William Green (I)**, **Ebenezer Hoppe**, **John Lilley (I)**, **Thomas Pickering**, and **Benjamin Jasper Wood**. He later worked in partnership with Josiah, who succeed him in business in 1817.

Danjon Prismatic Astrolabe

The Danjon prismatic astrolabe, [236] [277], was an improvement, devised by André-Louis Danjon, to the prismatic astrolabe. Danjon's improvement was to introduce a Wollaston prism at the focus. Because this improved instrument removed the *personal errors* introduced by the observer, it was also known as the impersonal prismatic astrolabe.

Dann, William

William Dann, (fl.1841-1844) [18], was an optician, and optical and mathematical instrument maker of Pelham St., Nottingham, England.

Dark Field Microscopy

Dark field microscopy, otherwise known as *dark ground* microscopy, is a method used to enhance contrast when viewing unstained, transparent specimens such as biological specimens, and is a less complex approach than phase contrast microscopy.

In dark field microscopy an opaque disk is used to block the centre of the light cone from the condenser, leaving a focussed, but hollow cone of light to converge on, and illuminate, the specimen plane, then diverge again towards the objective. With no specimen in place, the diverging hollow cone reaches the objective, but passes around the objective lens – the opaque disk being specially sized to ensure this is so. Thus, without a specimen in place the view is completely dark. When a specimen is in place it scatters some of the light. It is this scattered light that reaches the objective lens, and forms an image.

While dark field microscopy is simpler than phase contrast microscopy, most of the light does not contribute to the image, and so the image is formed with much lower levels of light.

Darton & Co., F.
F. Darton & Co., [44] [319] were wholesale manufacturing opticians and scientific instrument makers, of Clerkenwell Optical Works, 142 St. John Street, and 52 Clerkenwell Road, London, England. The company was established in 1834 by Alfred Oborne and, sometime before 1873, it was named F. Darton & Co.

In 1914 [110] lists **Francis A. Darton** as the principal, with specialities including: Standard meteorological instruments for observatories; Recording meteorological instruments; Spectacles and folders; Opera, field and marine glasses; Prism binoculars; Microscopes and telescopes; Meteorological instruments; Barometers and thermometers; Electro medical apparatus; Electrical accessories; Model motors; Dynamos.

In 1925 the company began the process of voluntary liquidation, [35 #33048 p3399], and the process was completed in 1936, [35 #34283 p3107]. The company was subsequently reformed as the limited liability company F. Darton & Co. Ltd.

In 1951 [327] lists the company as Darton (F.) & Co., Ltd., of Watford Fields, Watford, Hertfordshire, specialising in lenses, magnifiers, hygrometers, barometers, and electrical contact thermometers.

Darton, Francis Arthur
Francis Arthur Darton, (1850-1923) [44] [319], of London, England, was an optical and mathematical instrument maker. He was freed by patrimony of the **Clothworkers' Company** in 1850. In 1872 Charles Nathaniel Burrows began an apprenticeship with Darton, which he completed in 1883. By 1873 Francis Darton was the Principal of **F. Darton & Co.**

Darwin, Sir Horace
Sir Horace Darwin, (1851-1928) [46], born in Downe, Kent, England, was the youngest surviving son of the naturalist, Charles Darwin, and was a civil engineer and scientific instrument maker. He graduated from Trinity College, Cambridge in 1874. He served his apprenticeship with the engineering company Easton and Anderson of Erith, Kent, after which he went to work as a consulting engineer in Cambridge. He soon began to design instruments for use in teaching and research and, in 1878 he founded the **Cambridge Scientific Instrument Company** in partnership with Albert George Dew-Smith, a lens grinder from the observatory at Cambridge University. Darwin was the principal shareholder and, by 1891, he was in full control of the company, working closely with the university. He was a Justice of the Peace, and was mayor of Cambridge in 1896-1897. In 1903 he was elected a Fellow of the Royal Society, and he was knighted in 1918.

David White Company
The David White Company, of Milwaukee, USA, was the manufacturer of the *Stereo Realist* cameras from 1947 until 1971. Among their camera models was the *Realist 45*, which was manufactured for them by **Iloca** of Hamburg, Germany, as a version of their *Stereo Rapid* camera, but without the rangefinder or the self-timer.

Davidson & Co., F.
F. Davidson & Co., [53] [320], was founded in about 1890 by F.W. Davidson, an optician and inventor of ophthalmic instruments. The company traded at 29 Great Portland St., London, England. In 1914 the company introduced their *Davon* Super-Microscope, and *Davon* Micro-Telescope. From 1923 the company address was 143 Great Portland St., where they continued in business until about 1938.

Davidson, James
James Davidson, (fl.1851-1859) [18] [271], was an optical instrument maker of 39 South Bridge, Edinburgh, Scotland.

Davidson, Thomas
Thomas Davidson, (fl.1840-1853) [18] [263] [271], of Edinburgh, Scotland, was a mathematical and optical instrument maker, a maker of photographic apparatus, and a pioneering photographer. Addresses: 12 Royal Exchange (1840-1843); 63 Princes Street (1844); 67 Canongate (1845). 12 Royal Exchange (1848); 187 High Street (1850); 4 Infirmary Street (1852-1853).

He worked for **John Davis (II)** for about two years, beginning in 1836. In 1848 he began to trade as Thomas Davidson & Co. He is known to have produced an improved camera obscura, oxyhydrogen microscope and polariscope, and an improved camera lucida, and he supplied photographic apparatus to **Sir David Brewster**.

Davies, B
B. Davies, (fl.1836-1837) [18], was an optical instrument maker of 101 High St., Marylebone, London, England.

Davies, David
David Davies, (fl.1823-1832) [18] [271], was a mathematical and optical instrument maker of Glasgow, Scotland. Addresses: 110 Nelson Street (1823-1825); 100 Nelson Street (1825); 98 Trongate (1826-1832).

Davies, John
John Davies, (fl.1764) [18], was an optical instrument maker of Charing Cross, London, England. In 1764 he was one of the petitioners who attempted to revoke the patent obtained by **John Dollond (I)** for the achromatic lens, and which was enforced after Dollond's death by his son, Peter (See Appendix X).

Davis & Son, John
John Davis & Son, [357], optical and surveying instrument makers of 21 Irongate, Derby, England, was a partnership between **John Davis (I)** and his son Henry Davis. In about 1840 the company began to produce mining equipment. Following the death of John Davis (I) in 1873, Henry continued the business. In 1875 the company relocated to All Saints Works Amen Alley. A catalogue, dated 1877, shows their products to include turret clocks, weather vanes, surveying instruments, miner's lamps, and electric bells for mining and domestic use. They also supplied and installed industrial electric lighting and pumps. By 1882 they additionally had a site at 118 & 119 Newgate St., London. In 1900 they opened a subsidiary in Baltimore, MD, USA, and in 1912 the subsidiary became an independent spin-off, Davis Instruments. In 1910 Henry's son Wilfred Henry Davis took over running the company until his death in 1917.

In 1925 the products manufactured by John Davis & Son included theodolites, dials, levels, anemometers, water gauges, hygrometers, and a broad range of mining engineering equipment. Precision instrument making ceased during the 1950s. In 1962 the company was finally sold out of the Davis family. This may have been when it was renamed Davis Derby, after which it continued to produce engineering, technology, and safety equipment for the mining, quarrying, and warehousing industries.

Davis Bros.
Davis Brothers, [18] [276 #2113] [319], were opticians and mathematical instrument makers of 1 Lower Terrace, Islington, London, England and 33 New Bond St., London. [322] records them trading from 1820 until 1860, with a manufactory at 209 High Holborn. The participants in the partnership presumably changed over time. **Isaac Davis** and Marcus Davis traded as Davis Brothers in a partnership which was terminated in 1840 [35 #19929 p2977].

Davis Instruments
Davis Instruments is a meteorological instrument manufacturer of California, USA.

During WWII, various American companies made plastic or cardboard sextants to help satisfy the increased demand. These were designed for use in lifeboats, or in the lifesaving rubber dinghies carried by aircraft. After the war, such instruments were still produced for use in training and leisure sailing.

Davis Instrument Corporation, [16] [58], was founded in San Leandro, California, USA by Englishman William A. Davis. Initially they produced a range of inexpensive plastic nautical instruments. These included the plastic marine sextant, first manufactured in 1963, which continued in production until well into the C21st, by which time several variants of differing complexity were available. The company changed hands in 1969, and in 1984 they began to produce a hand-held wind meter for marine use. They subsequently acquired the company Digitar, and began to produce weather stations, which became their principal product line. In 2019 the firm was acquired by Advanced Environmental Monitoring.

Davis, Bernard
Bernard Davis, (fl.1859-1872) [319], of 430 Euston Road, London, England, was an optician and optical instrument maker. In addition to spectacles, he advertised a magic lantern with dissolving views and slides, and a combined microscope and telescope. He also advertised, in 1861, optical instruments at wholesale prices, including pocket telescopes, compound microscopes, and opera and field glasses.

Davis, C.&D.
See **Clara Davis**.

Davis, Clara
Clara Davis, (fl.1829-1834) [18] [276 #1824], of 12 South St., Finsbury, London, England, was an optician, and mathematical and philosophical instrument maker. From 1829 until 1830 she worked in partnership as **C.&D. Davis**, optical, philosophical, and mathematical instrument makers at the same address.

Davis, David
David Davis, (fl.1816-1827) [18] [276 #1529] [322], was an optician, and optical, philosophical, and mathematical instrument maker of 8 Macclesfield St., Soho, London, England. He worked at this address at the same time as **Hannah Davis**, and succeeded her in business. He is believed to have served his apprenticeship with **John Davis (III)** in 1783. From 1825 he additionally traded at

9 Macclesfield St., Soho, where **Elizabeth Davis** traded from 1830.

Davis, Edward
Edward Davis, (b.c1808 fl.1833-1861) [3] [18] [319] [357], was an optical and mathematical instrument maker of Liverpool and Cheltenham, England. He was a nephew of **Gabriel Davis**. He began work selling his uncle's instruments, and went on to establish his own business as an "optician and instrument repository". By 1833 he was trading at 171 High St., Cheltenham. By 1839 he was trading at 65 Bold St., Liverpool together with his brother, **John Davis (I)**. In 1843 he is recorded at 45 Bold St.

By 1852, following his father's death, he was trading at his father's address of 35 Boar Lane, Leeds. He is known to have sold telescopes, barometers, and thermometers. A compound microscope with fitted wooden box bears his signature [358]. The company continued in business until its sale 1923.

Davis, Elizabeth
Elizabeth Davis, (fl.1830-1834) [18] [276 #1529] [322], was an optician, and mathematical and philosophical instrument maker of 9 Macclesfield St., Soho, London, England, an address at which **David Davis** had previously traded.

Davis, Gabriel
Gabriel Davis, (c1790-1851) [3] [18] [319] [357], optician and optical instrument maker, was a Jewish immigrant from Bavaria who settled in Leeds, England. He set up his business as an optical, mathematical, and surveying instrument maker in 1779, trading at 20 Boar Lane. His nephews, **Edward Davis** and **John Davis (I)** worked with him in his company, and began their careers travelling around Leeds, Liverpool, Cheltenham and Derby, selling instruments from their uncle's Leeds workshop. By 1826 Gabriel was working in partnership with Edward, trading as **Gabriel & Edward Davis** until 1831. By 1834 he was trading at 24 Boar Lane, and at Saddler Street in Durham. He continued at the Boar Lane address until he moved to number 35 in 1848.

Davis, Gabriel & Edward; Davis G.&E.
Gabriel & Edward Davis, (fl.1826-1831) [18], opticians and mathematical instrument makers of Boar Lane, Leeds, England, was a partnership between **Gabriel Davis** and his nephew **Edward Davis**. The partnership was terminated in 1831 [35 #18792 p696].

Davis, Hannah
Hannah Davis, (fl.1796-1823) [18] [276 #1309] [322], was an optician, and optical and mathematical instrument maker of 8 Macclesfield St., Soho, London, England, adjacent to the location occupied a few years earlier by **John Davis (III)**. **David Davis** also worked at 8 Macclesfield St. from 1816, and succeeded her in business.

Davis, Isaac
Isaac Davis, (fl.1832-1842) [18] [319], was an optician, and mathematical and philosophical instrument maker of 1 Lower Terrace, Islington, London, England. He traded in partnership with his brother Marcus as **Davis Bros.** until 1840 [35 #19929 p2977], after which he worked under his own name.

Davis, John (I)
John Davis (I), (1810-1873) [3] [18] [319] [357], optical and mathematical instrument maker of Derby, England, was a nephew of **Gabriel Davis**. He served his apprenticeship with **Jacob Abraham** in Bath, after which he began to work for his uncle in Leeds. He toured Liverpool, Cheltenham, and Derby selling his uncle's products in temporary shops at each location. He traded in Rotten Row, Derby (1830-1843), 101 High St., Cheltenham (1843), 14 Iron Gate, Derby (1835), and with his brother **Edward Davis** at 65 Bold St., Liverpool (1839). In 1843 he settled in Derby with his family, and began to trade at 21 Irongate, where he established his workshop for making instruments. He formed a partnership with his son, Henry Davis, trading as **John Davis & Son**, and Henry continued the company following John (I)'s death in 1873 [35 #24127 p4259].

Davis, John (II)
John Davis (II), (fl.1836-1842) [18] [263] [271], was an optician and mathematical, and philosophical instrument maker of 64 Princes St., Edinburgh, Scotland. In 1841 he relocated to 78 Princes St., where he stayed until the close of his business in 1841. Instruments bearing his signature include: portable universal equatorial sundial; 6-inch Gregorian telescope; sympiesometer; brass surveyor's circumferentor. **Thomas Davidson** worked for him for about two years, beginning in 1836, prior to starting his own business. John Davis (II) was declared bankrupt in 1840, and again in 1842. He advertised for sale 'every description of optical, mathematical and philosophical instrument', but the extent to which he made them is unclear.

Davis, John (III)
John Davis (III), (fl.1779-1784) [18] [322], was an optician of London, England. Addresses: At Mr. Scarlett's (1777); Macclesfield St., Soho (1779); 9 Macclesfield St., Soho (1780); Dean St., Soho (1784). He served his

apprenticeship with **Edward Scarlett (II)**, and was freed of the **Spectacle Makers' Company** by purchase in 1777. **David Davis** is believed to have served an apprenticeship with him in 1783.

Davis, Messrs
Messrs Davis, **[18] [319]**, was a trade name used by **Edward Davis** and **John Davis (I)** when trading at 101 High St., Cheltenham, and 65 Bold St., Liverpool around 1842.

Davis's Quadrant
Davis's quadrant, **[16] [256]**, was invented by the English naval captain and navigator, John Davis, (1550-1605) **[275 #46]**, at the end of the C16th. It is a form of backstaff that consists of two arcs mounted on a staff. One, an arc of 60°, has a short radius, and is mounted above the staff. The other, an arc of 30°, has a longer radius, and is mounted below the staff. The two arcs together total 90°, hence the term quadrant. The two arcs have a common centre, at which is placed a sighting vane. The upper arc has a sliding shadow vane, which can be moved to any point on the arc. The lower arc has an eye vane which can also be slid to any point on the arc. In use, the shadow vane on the upper arc is positioned a little below the expected altitude of the sun. The user stands facing away from the sun, and sights through the eye vane to the sighting vane. The position of the eye vane on its arc is adjusted to bring the shadow of the shadow vane, cast by the sun, onto the sighting vane. The altitude of the sun is then calculated as the sum of the angles on the two arcs.

Dawson & Melling
Dawson & Melling, **[3] [18] [276 #2114]**, were opticians, stationers, map and chart sellers, and Navigation Warehouse, of Liverpool, England. They traded at 20 South Castle St. (1837), and 39 South Castle St. (1839). They succeeded **Bywater, Dawson & Co.** in business in 1837, by which time the firm was a partnership between Samuel Dawson and Edward Melling. They acted as chart agents for **Robert Brettell Bate**. They are known to have sold: repeating circle. In 1839 the firm became **Dawson, Melling & Payne**.

Dawson, Melling & Payne
Dawson, Melling & Payne, **[3] [18]**, were opticians, mathematical instrument makers, and Navigation Warehouse, of 39 South Castle St., Liverpool, England. The partnership was formed in 1839, succeeding **Dawson & Melling**, and was terminated in 1842, by which time the firm was a partnership between Henry Melling and George Patmore Payne **[35 #20189 p257]**. Instruments bearing their name include octants. The company was succeeded in business by **Melling & Payne**.

Day, Francis
Francis Day, (fl.1826-1837) **[18]**, of 37 Poultry, London, England, was an optician, and mathematical, philosophical, and optical instrument maker. He was the son of the philosophical instrument maker, James Day. He was freed of the **Spectacle Makers' Company** by redemption in 1826. From 1827 he is recorded at 97 Poultry. He is known to have sold microscopes and sextants. An example of his work is a **Culpeper** style microscope.

Day, John
John Day, (fl.1716-1727) **[18]**, was an optician of London, England. He served his apprenticeship with **Jane Sterrop**, beginning in 1708/9, and was freed of the **Spectacle Makers' Company** in 1716. Among his apprentices was **Peter Eglington**.

Daystar Filters
Daystar Filters of Warrensburg, Missouri, USA, is a company that specialises in solar filters, solar telescopes, mounts, and accessories.

Deane, David
David Deane, (fl.1735-1777) **[18] [276 #161]**, was an optical instrument maker of Smithfield, London, England. He served his apprenticeship with **Matthew Loft**, beginning in 1728, and was freed of the **Spectacle Makers' Company** in 1735. He served as Master of the Spectacle Makers' Company in 1760. In 1764 he was one of the petitioners who attempted to revoke the patent obtained by **John Dollond (I)** for the achromatic lens, and which was enforced after Dollond's death by his son, Peter (See Appendix X).

Deijl, Harmanus van
Harmanus van Deijl, (1738-1809) **[220] [256] [264]**, of Amsterdam, Netherlands, was an optical instrument maker, and the son of **Jan van Deijl**. They had a successful business making and selling achromatic telescopes, and are known to have made a binocular refractor. In 1762 they began to work together to produce achromatic microscopes, but it was not until after Harmanus published their work in 1807 that he established a reputation as the first to commercially produce them. He was succeeded in his business by **Jan Bunders**.

Deijl, Jan van

Jan van Deijl, (1715-1801) [220] [256], of Amsterdam, Netherlands, was an optical instrument maker, and father to **Harmanus van Deijl**.

Deleuil, Louis Joseph

Louis Joseph Deleuil, (1795-1862) [58] [113] [302], was a scientific instrument maker of Paris, France. He produced mathematical, physical, electrical, and chemical apparatus, and had a strong reputation, particularly for balances and for standard weights and measures.

He was the first to manufacture Raspail's simple chemical microscope upon its invention in about 1830.

Louis retired in about 1857, and was succeeded in his business by his son, Jean Adrien Deleuil (1825-1894). The business grew and prospered under Jean until 1893 when, going blind, Jean was succeeded by the engineers Jules Velter and André Pillon. Pillon died in 1899, after which the firm continued as Maison Deleuil Velter & Cie, Successeurs. Velter died in 1903, and in 1911 the company was renamed **Société Industrielle d'Instruments de Précision**.

Following the death of **Félix Pellin** in 1940, the company Philibért and Félix Pellin, successors to **Philibért François Pellin**, merged with Société Industrielle d'Instruments de Précision. They continued in business until the early 1950s.

Dell'Acqua, Carlo

Carlo dell'Acqua, (1806-after 1870) [17] [374], was a physical, optical, astronomical, and electrical instrument maker of Milan, Italy. After his education, he took up a role as a mechanic at the Imperial Regio Ginnasio Liceale di Sant'Allessandro. Following the death of **Carlo Grindel** in 1854, dell'Acqua applied for the newly vacant job at the astronomical observatory at Brera, Milan, but it was awarded to Grindel's son, **Francesco Grindel**. Francesco, however, died in 1859, and dell'Acqua was awarded the job of *macchinista* (machinist), constructing instruments for the observatory, and keeping the instruments in a good state of repair. He also provided precision instruments to schools, universities, and other institutions around Italy. In 1863, he published a catalogue showing more than eight hundred instruments and machines.

In 1864, Carlo dell'Acqua, along with engineer and mathematician Luigi Longoni, and photographer Alessandro Duroni, founded the **Tecnomasio Italiano**, while continuing in his role at the observatory.

Dellebarre, Louis François

Louis François Dellebarre, (1726-1805) [256] [264] [274], of Abbeville, France, is known for his microscopes, many of which were produced while he lived in the Netherlands, initially in Leiden, then in the Hague, and finally Delft, before moving to Paris, France, where he died in 1805.

In 1770 he introduced his *Universal Microscope*. This, he claimed, was achromatic, based on the principles of Euler, with an arrangement of lenses, some crown glass and some flint glass. Nonetheless it was not achromatic and, while the microscope achieved a degree of popularity, it was not considered to be of fine quality. In 1777 he presented a *Universal Microscope*, and a memorandum describing it, to the to the Académie des Sciences in Paris. Some of Dellabarre's microscopes bear his name, and some bear the name of **J.H. Onderdewijngaart Canzius** of Delft, who manufactured them for him.

Dennert & Pape

Dennert & Pape, [114] [217]. In 1857 **Carl Plath** started a business in Hamburg, Germany, making surveying instruments. Johann Christian Dennert, (1829-1920), one of his employees, purchased the company from him in 1862, when Carl Plath took over the business of David Filby. Martin Pape, (1834-1884), became a partner with Dennert in 1863, and the company began to trade as Dennert & Pape. According to their first price list they offered: theodolites, levels, levelling rods, planimeters, drawing instruments and scales with any graduation, made of ivory, brass, nickel-silver, or silver-plated brass.

In 1872 they began to manufacture slide rules, and changed their name to Dennert & Pape, Mechanical-Mathematical Institute. Martin Pape died in 1884, at which time J.C. Dennert became the sole owner until he retired at the age of 80, leaving the business in the hands of the Dennert family. They introduced the trademark *Aristo* for their slide rules in 1936, but continued to use the name Dennert & Pape on their other products. During WWII they manufactured instruments for the German military, and marked them with the **ordnance code** gwr.

In 1952 they began to use *Aristo* as a trademark for all their products, and in 1956 the company was renamed Dennert & Pape Aristo-Werke. In 1978 they discontinued production of slide rules and the Dennert family relinquished the company. It continued as a computer aided design business named Aristo Graphic Systeme.

Dennis, John Charles

John Charles Dennis, (fl.1839-1861) [18] [53] [54] [307], was an optical, scientific, philosophical, and nautical instrument maker of London and Bristol, England. Addresses: 118 Bishopsgate Street, London (1839-1849); 122 Bishopsgate Street, London (1850-1862); Bristol (1845).

Examples of his work include: marine barometer; microscopes; his patented lunar sextant; globes. In 1849 he was elected a fellow of the Royal Astronomical Society. In 1851 he exhibited at the Great Exhibition in London [294], where he was listed as an inventor and manufacturer, and showed an improved astronomical reflecting circle, lunar sextant, and half sextant, theodolite, and drawing instruments.

In 1862 his entire stock was sold at auction [308 p3], consisting of barometers, thermometers, microscopes, telescopes, sextants, quadrants and other nautical instruments, opera and race glasses, drawing instruments, spectacles, reading glasses, electrical machines and apparatus, a quantity of brass and unfinished work, wood models, several useful lathes, vices and tools, nests of drawers, glass cases, &c.

Dent, Edward John
Edward John Dent, (1790-1853) [17] [18] [38] [54], of London, England, was a chronometer, watch, and clock maker. Addresses: 43 King St., Long Acre (1826-1827); 33 Cockspur St., Charing Cross (1846-1851); 34 Royal Exchange (1846-1851); 82 Strand (1846); 61 Strand (1851). He took part in the chronometer trials at the Royal Greenwich Observatory. From 1830 he worked in partnership with the chronometer maker John Roger Arnold, trading as Arnold & Dent. The partnership was terminated in 1840 [35 #19900 p2190]. He formed the company Edward Dent & Co.

Wishing to devise a simple transit measurement method for the correction of timepieces, he worked in collaboration with James Mackenzie Bloxam, a barrister of Chancery Lane, who had invented a meridian instrument. They developed and patented the concept and, in 1843, began to sell it as the dipleidoscope, manufactured by Dent. Dent published *A Description of a new Dipleidoscope, or Double Reflecting Meridian & Altitude Instrument, with plain Instructions for the Method of Using it in the correction of Time-Keepers*. Following Edward Dent's death, his stepson, Frederick William Dent, continued the business, including production of the dipleidoscope.

Depression Rangefinder
A Depression Rangefinder is an early form of optical rangefinder, invented by Major H.S. Watkin of the British Royal Artillery. See rangefinder.

Derby Crown Glass Company
The Derby Crown Glass Company, [44], was a UK optical glass manufacturer. The company was formed in 1913, and began to produce optical glass in 1916. Around 1921 the company was taken over by **Charles Parsons** for the **Parsons Optical Glass Company**.

Deregni (De Regni), Angelo
Angelo Deregni, [75], was a contemporary of **Leonardo Semitecolo**, making telescopes in Venice, Italy, in the late-C18th. His telescopes are similar in construction to those of Semitecolo, being fabricated of cardboard tube, with horn or brass mount.

Derry, Charles
Charles Derry, (fl.1838-1857) [18] [276 #2115], was an optical, philosophical, and mathematical instrument maker of London, England. He attended the London Mechanics Institute from 1824 until 1826.

Addresses: 10 Little Coram St, Brunswick Square (1824-1826); 67 Warren St., Fitzroy Square (1829); 6 Leigh St., Burton Crescent (1838); 19 Compton Place, Brunswick Square (1839-1840); 7 Leigh St., Burton Crescent (1839-1851); 74 Judd St., Brunswick Square, (1855).

Desiccator Pump
A desiccator pump is a manually operated pump used to replace the air within the body of a suitably equipped pair of binoculars with dry air. The desiccator pump attaches to special ports that are integral to the binocular prism housing, and pumps air through a bed of desiccant into the binoculars, thus expelling air that may be humid. The desiccant is dried periodically using heat.

Desilva, William
William Desilva, (fl.1851-1881) [18] [322], was an optician, chronometer maker, and nautical instrument maker of Liverpool, England. Addresses: 175 Great Howard St. (1851); 40 Regent Rd. (1853); 44 Regent Rd. (1855); 38 Regent Rd. (1857-1865); 5 Bath St. (1867); 114 Duke St. West (1868); 112 Duke St. West (1871); 126 Duke St. West (1872-1881); 37 Bath St. (1877-1881); Regent Rd., opposite the Bramley-Moore Dk. Gate.

According to his trade card, "William Desilva, Chronometer maker, Optician, and Nautical instrument maker, Regent R^d. opposite the Bramley-Moore D^k Gate, Liverpool. Sextants, octants, compasses, telescopes, barometers, sympiesometers, thermometers, binacles, watches, &c &c."

He was declared bankrupt in 1856 [35 #21920 p3053], and in 1861 a patent was awarded "To William Desilva and Thomas Fletcher Griffith, of Liverpool, in the county of Lancaster, Opticians, for the invention of 'an improved construction of instrument for taking observations at sea or on land.'" [35 #23245 p2474]. Instruments bearing his name include: telescopes; sextants; octant; compass; barometers.

Deutsche Spiegelglas AG
Spiegelglashütte auf dem Grünen Plan was a glass-making factory founded in 1744 in Grünenplan in Lower Saxony, Germany. In 1871 the factory was acquired by Gebrüder Koch, and the name was changed to Deutsche Spiegelglas AG. In 1907 they expanded their products to include sheet glass, watch glasses, scientific and technical glassware, and optical glass. In 1930 **Schott** acquired an 80% interest in the company. During WWII they used the **ordnance code** doq on their products. In 2002 Schott took 100% ownership of the company, and it became their thin glass manufacturing plant.

De Vere (Kensington) Ltd.; De Vere Ltd.
De Vere (Kensington) Ltd., [44] [306], founded in 1947, was a manufacturer of professional grade photographic equipment, including cameras, and were best known for their photographic enlargers and darkroom equipment. Their manufacturing plant was at Thalia Works, Thayers Farm Road, Beckenham, Kent, England, and their offices were located in Kensington High Street, London. In the 1960s they opened a new manufacturing plant at Seven Brethren Bank, Barnstaple, North Devon. The company name was changed to De Vere Ltd., and they continued in business until 1992.

Diagonal
A diagonal is a device that alters the direction of the optical path, usually by 45° or 90°, by means of a prism or mirror. It is commonly used in telescopes to bring the eyepiece to a more comfortable position, but may also be used in other instruments, such as microscopes.

Dial Sight
A dial sight is an artillery gun sight for use in indirect fire (aiming at a target that the gunner cannot see). It has a graduated circle for reading azimuth angles. During setup, the sight is aligned to a *zero line*, based on a known azimuth direction. When preparing to fire, a direction to the target is then given by the gunnery director, and this is dialled into the sight. The gun is then turned until the dialled angle is aligned with the zero line, with the result that the gun is aimed at the target.

According to [20] the dial sight was "awkward to use, bulky, and relatively fragile". In 1904 the dial sight was improved by **C.P. Goerz** as the *panoramic dial sight*. In 1908 Goerz set up C.P. Goerz Optical Works Ltd. in Britain to license the panoramic dial sight for British manufacture. They were made by **Barr & Stroud**, **R.&J. Beck**, **T. Cooke & Sons**, C.P. Goerz, **Ross Ltd.** and **Vickers**, with the Vickers orders subcontracted to T. Cooke & Sons.

Dialyte Lens
The *separated thin-lens achromat* (dialyte) is a lens arrangement in which the two elements of the achromat are separated by a small distance.

Diaphanometer
An instrument used for measuring transparency, particularly of the atmosphere, or of a liquid.

Diatom Prism
See prism.

Dick, A.B.
Allen B. Dick, microscopist, was the inventor of a mechanism by means of which the polarizer and analyser in a petrological microscope may be rotated either simultaneously or independently. This was achieved by means of a connecting rod and gears built in to the microscope stand. He developed this concept in collaboration with **James Swift & Son**, who consequently made the idea available as the Swift Dick model petrological microscope. The concept was described by A.B. Dick in 1889 in the *Journal of the Royal Microscopical Society*, [439].

Dick, William
William Dick, (fl.1838-1844) [18], was an optical and mathematical instrument maker, watch maker, ironmonger, and ship chandler of Glasgow, Scotland. Addresses: 57 Trongate (1838); 94 Jamaica St. (1839-1841); 96 Jamaica St. (1842-1844).

Dickman, John
John Dickman, (fl.1794-1854) [18] [271] [276 #1829], was a Scottish nautical instrument maker, clock maker, and chronometer maker. Addresses: Kirkgate, Leith (1794-1796); Bernard St., Leith (1797-1813); 33 Shore, Leith (1814-1842); 142 George St., Edinburgh (1841-1842); 91 Princes St., Edinburgh (1843); 6 Charlotte Place, Leith (1844-1848); 4 Charlotte Place, Leith (1848-1854).

In 1835 he received a Royal Appointment to King William IV and, when Queen Victoria ascended the throne in 1837, the Royal Appointment continued.

Dickson, William
William Kennedy-Laurie Dickson, (fl.1883, d.1935) [335], was a British inventor and filmmaker, who was born in France and, in 1879, emigrated to the USA. In 1883 he began to work for **Thomas Edison** and, in 1888, by then a senior associate of Edison's, he began to work on motion pictures. By 1892 Edison and Dickson had developed the Kinetograph motion picture camera, and the Kinetoscope

motion picture viewing device. In 1893 they opened the *Black Maria* film studio, designed by Dickson. It had a hinged roof that could be raised, and the studio pivoted to catch the sun.

William Dickson went on to controversially assist the **Latham Brothers**, while still in Edison's employment, in a project which led to the announcement in 1895 of the Panoptikon projector, which was then in competition with Edison's own product. This perceived disloyalty led to Dickson leaving Edison's employment that same year. He continued to work for the Latham Brothers, but did not like to be associated with their lifestyle. In 1894 he had, together with three friends, Koopman, Marvin, and Casler, set up the KMCD group. Having left the Latham Brothers, he turned his attention back to this endeavour and, after a spell of design work, he became a travelling cameraman, filming in the USA then, from 1897, in England, and in 1899 in South Africa, covering the Boar War.

Didbin's Hand Photometer
See photometer.

Diederichs, Carl
Carl Diederichs, (fl.1875-1898) [374], was an optical, physical, and surveying instrument maker of Göttingen, Germany. He served his apprenticeship with **Moritz Meyerstein** following which, in 1875, he began to work in partnership, trading as Diederichs & Bartels. In 1890 he left this partnership, and established his own company, trading under his own name. The firm continued in business until 1898, when it was taken over by **Spindler & Hoyer**.

Dienstglas
Dienstglas is German for *service glass*, and is a marking sometimes found on German military optical instruments such as binoculars.

Diffraction Grating
A diffraction grating is a series of fine parallel lines, regularly spaced, which produces a spectrum by the interference of light. In its simplest form it is narrow, closely spaced, parallel slits in an otherwise opaque material. A diffraction grating may be either reflective or transmissive, and in the case of the concave diffraction grating, invented by **Henry Augustus Rowland**, has a concave curved surface. It can also take the form of an echelon grating, echelette grating, or echelle grating. See also Thorp's diffraction grating, reflection grating.

Dillon, Mr.
Mr. Dillon, (fl.1722) [18] [53] [361], was an optician and optical instrument maker of Long Acre, next Door to the White Hart, London, England.

According to an advertisement dated 1722 [370]: "New perspective-glasses are to be sold by way of subscription at Mr. Dillon's in Long-Acre, next Door to the White-Hart, where proposals may be had. With the help of those perspective-glasses, any one that looks forwards, may take a view of any object, that is on the right hand, or on the left; and no body can discover what he looks at. Some other uses of the same glasses are described in the said proposals."

DIN
DIN is an abbreviation for Deutsches Institut für Normung, the German national standards institution.

The DIN standard for the distance between the objective flange and the eyepiece flange on a compound microscope is 160mm. This is the most common of the two most widely used standards for this distance. The other is the **JIS** standard of 170mm

Dioptra; Diopter; Dioptre
The dioptra, otherwise known as a *diopter* or *dioptre*, was an early instrument used in surveying and astronomy, and was a precursor of the theodolite. The dioptra was described by the Greek engineer and scientist Hero of Alexandria in his book *Treatise on the Dioptra*, in about AD 100.

It consists of a sighting rod or tube mounted so that it may be turned about both its azimuth axis and its altitude axis. Each axis may have a graduated arc to give a reading of angle. The base is levelled, and angles are measured to a distant terrestrial or astronomical object.

Dioptre
A dioptre is a unit of measurement of the power of a lens. It is the reciprocal of the focal length, expressed in metres.

Dioptric Micrometer
See micrometer.

Dipleidoscope
A dipleidoscope (double image viewer), [17] [38] [54], is an instrument used to determine the time of noon in order to correct a timepiece. It consists of a hollow prism with two internal surfaces mirrored, or can be constructed from three pieces of glass. It is positioned so that the sun reflects from the plain glass face, and a double reflection is produced by the two mirrored faces. Thus, two images of the sun are viewed by the observer, moving in opposite

directions. The two images coincide as the sun's centre transits the local meridian.

The invention was made by a barrister, James Mackenzie Bloxam, who referred to it as a 'meridian instrument', since it could be used to measure the transit time of any celestial object on the ecliptic. He collaborated with the chronometer maker **E.J. Dent**. They patented the device, and in 1843 Dent began to manufacture them.

Dipping Refractometer
See refractometer.

Director
A director is a device or mechanism used to determine the parameters for aiming a gun. In its simplest form it is similar to a gun sight, mounted on a tripod, but without a gun. This is used to take a bearing to the target, which is then communicated to the gunner. In more sophisticated systems, such as on a warship, a director may be a part of the fire control system providing automated aiming parameters, such as bearing and gun elevation, and the impulse to fire, to the gun installation.

Direct Vision Prism
See prism.

Direct Vision Spectroscope
See spectroscope.

Dispersion
See *Properties of Optical Glass* in Appendix VI.

Distortion
Distortion is an optical aberration that arises when the magnification either increases or decreases with distance from the centre of the image. Positive distortion occurs when the magnification increases toward the edge of the field. This is sometimes called *pincushion* distortion. Negative distortion occurs when the magnification decreases toward the edge of the field. This is sometimes called *barrel* distortion.

Divided Object Glass Micrometer
See micrometer.

Dividing Engine
Dividing engine: [2] [3] [228] [256] [443]. During the C18th, considerable advances were made in the capabilities of optical instruments. With these advances came higher expectations in such disciplines as navigation, astronomy, and surveying. Effective measurements could only be made with these instruments if angular directions and lateral movements could be measured with a high degree of accuracy. Thus, the art of engraving divided scales on the arcs of sextants, the axes of telescope mounts, and the arcs of theodolites, required a level of accuracy never before achieved. The solution to this was the dividing engine. Two important types of dividing engine were devised, the circular dividing engine and the linear dividing engine, each subject to sophisticated design and construction, and competitive advances in design.

The use of a circular mechanical device for cutting horological gear-wheels began as early as the C17th, and a significant contribution to this art was made by **Robert Hooke** with his invention of such a device, an example of which is in the Science Museum, London.

This concept was taken to a new level of mechanical precision by the clockmaker **Henry Hindley**, with his invention of a circular dividing engine for use in cutting clock-wheels, using a screw and worm gear to rotate the dividing plate in a full circle. He subsequently adapted this design for dividing scales on astronomical instruments. An employee of Hindley was **John Stancliffe**, who subsequently went to London and took employment with **Jesse Ramsden**. Ramsden further improved on the design, and made two circular dividing engines, the second being an improvement on the first. A sextant divided on Ramsden's second engine was reported to have graduations engraved over the full 120° to an accuracy of one two-thousandth of an inch. This set a new standard for dividing instruments, and Ramsden received a reward from the Board of Longitude on condition that he published his design. He did so in his book, *Description of an Engine for Dividing Mathematical Instruments* [252], in 1777. Subsequently his design was much copied and improved by other instrument makers in the UK and abroad. Notable dividing engines were constructed by instrument makers such as **James Allan (I)**, **John Stancliffe**, **John Troughton (II)**, **Thomas Cooke**, and **William Simms (II)**, the latter who improved the design to include a self-acting mechanism, driven by steam. By about 1815, there were at least 10 dividing engines operated by instrument makers in London, of which most were copies of Ramsden's second engine. With the improvements in toolmaking and precision engineering since Jesse Ramsden's day, William Simms (II)'s dividing engine, completed in 1843, would not only run unattended, but was 100 times more accurate than Ramsden's engine.

See foul anchor, the symbol used by some makers to denote which dividing engine was used to engrave the scale. See also Appendix XI.

The straight-line dividing engine was also improved by Jesse Ramsden, and his improvements were adopted by instrument makers such as **François-Antoine Jecker** in building their own dividing engines.

In 1845 **Friedrich Adolph Nobert** devised a dividing engine that could be used to produce fine ruled parallel lines for use by microscopists as a micrometer. In doing so he introduced the first physical test for the resolution of an optical microscope, with line-sets of varying separations down to those beyond the resolving power of an optical microscope.

Divini, Eustachio
Eustachio Divini (or Divinis), (1620-1695) [8] [256] [274], was an Italian astronomer and optical instrument maker who established his reputation making telescopes, including aerial telescopes, and microscopes. He published the results of his astronomical observations. He specialised in grinding large lenses. He was a contemporary, and rival of **Guiseppe Campani**, who had a similar business, and both makers supplied lenses to the astronomer Giovanni Domenico Cassini (1625-1712). Some of his instruments are in the Firenze Museum (Florence).

Dixey & Son, C.W.
C.W. Dixey & Son, [8] [18] [361] [436], were opticians, and mathematical and philosophical instrument makers, of 3 New Bond Street, London, England. The firm succeeded **C.W. Dixey & Sons** in 1881. According to [379], the son in this partnership was **Adolphus William Dixey**. The firm traded at the 3 New Bond St. address until at least 1930, and continued into the C21st as a purveyor of eyewear.

Dixey & Sons, C.W.
C.W. Dixey & Sons, [18] [436], were opticians, mathematical, and philosophical instrument makers of 3 New Bond St., London, England. The firm was a partnership, formed in 1863 between **Charles Wastell Dixey** and his sons **Charles Anderson Dixey**, and presumably one or both of Albert Alfred Dixey and **Adolphus William Dixey**. They were succeeded by **C.W. Dixey & Son** in 1881.

Dixey, Adolphus William
Adolphus William Dixey, (fl.1885-1901), was an optician of 3 New Bond St., London, England [35 #25571 p1435]. He had royal appointment from 1885 until 1901 [35], presumably to Queen Victoria. According to [379] he was the son of **Charles Wastell Dixey**, and a partner in the firm **C.W. Dixey & Son**.

Dixey, Charles Anderson
Charles Anderson Dixey, (c1824-c1867) [18], of New Bond St, London, England, was the son of **Charles Wastell Dixey**. He served his apprenticeship with his father, beginning in 1838, and was freed of the **Girdlers' Company** in 1845. In 1863 he entered a partnership with his father, and presumably one or both of his brothers, Albert Alfred Dixey and **Adolphus William Dixey**, trading as **C.W. Dixey & Sons**. He served as Master of the Girdlers' Company in 1865.

Dixey, Charles Wastell
Charles Wastell Dixey, (fl.1838-1863) [17] [18] [436], was an optician, and mathematical and philosophical instrument maker of London, England. Addresses: 335 Oxford St.; Old Bond St. (1838); 3 New Bond St. (1838-1862). He was the son of **Edward Dixey**, and served an apprenticeship with his father beginning in 1812. He was freed of the **Girdlers' Company** in 1838. In 1822 he began to trade as **G.&C. Dixey**, in partnership with his uncle, **George Dixey**. When this partnership ended in 1838 Charles continued to trade under his own name. In the same year he took on his son, **Charles Anderson Dixey**, as an apprentice. His son Albert Alfred Dixey was freed in 1865 and, according to [379], **Adolphus William Dixey** was also his son.

According to his trade card (undated): "Established 1777. C.W. Dixey. Optician & Mathematical instrument maker to Her Majesty; the King and Queen of Hanover; their Royal Highnesses the Dukes of Sussex and Cambridge; the Princesses Augusta and Sophia; the Dutchesses of Kent, Cambridge, and Gloucester; the King of Belgium &c &c. No. 3, New Bond Street, London."

In 1863, C.W. Dixey went into partnership with his sons, C.A. Dixey and presumably one or both of A.A. Dixey and A.W. Dixey, trading as **C.W. Dixey & Sons**. This partnership continued until 1880, after which the firm continued as **C.W. Dixey & Son**.

Dixey, Edward
Edward Dixey, (fl.1805-1843) [18], was an optician, and mathematical, philosophical, and optical instrument maker of London, England. Addresses: 10 Air St., Piccadilly; 370 Oxford St.; 17 City Road; Vine St., Piccadilly (1794); Princes St., Leicester Square (1805); 335 Oxford St. (1810-1843).

He was the son of John Dixey, a victualler of Westminster London. He served his apprenticeship with George Black, beginning in 1786, and was freed of the **Girdlers' Company** in 1794. (Note, [276 #703] states that he lived from 1757-1838, and served his apprenticeship with **George Linnell** beginning in 1771. According to [18] this information relates to a different Edward Dixey, who was the son of John Dixey, a fishmonger of Southwark, London).

Edward's son, **Charles Wastell Dixey** served an apprenticeship with him beginning in 1812. Edward later worked for his son's partnership, **G.&C. Dixey**.

Dixey, G.&C.
George & Charles Dixey, [18] [276 #1536], sometimes styled as G.&C. Dixey, was a partnership between **Charles Wastell Dixey** and his uncle, **George Dixey**, trading as opticians, mechanicians, and mathematical, philosophical, and astronomical instrument makers. The partnership was formed in 1822. At some time, Charles's father, **Edward Dixey**, worked for the partnership, as did **John Dixey**. They initially traded at 78 Bond Street, London, England, but sometime between 1823 and 1825 they took over the business of **William Hawkes Grice**, and moved to 3 New Bond Street, London. They received a royal appointment to King George IV in 1824, and to King William IV in 1830. The partnership ended with the death of George Dixey in 1838, following which C.W. Dixey continued to trade under his own name.

Dixey, George
George Dixey, (fl.1802-1838, d.1838) [18], was an optician, optical turner, and telescope maker of 370 Oxford St., London, England. He served his apprenticeship with **George Willson**, beginning in 1798, and was freed of the **Stationers' Company** in 1809. He worked in partnership with George Willson from 1802 until 1809, trading as **Willson & Dixey**. From 1810 until 1822 he traded under his own name at 20 Vine St., Piccadilly, London. His son, **Lewis Dixey**, set up his own business in Brighton. In 1822 George began to work in partnership with his nephew, **Charles Wastell Dixey**, trading as **G.&C. Dixey** until his death in 1838.

Dixey, John
John Dixey, (b.c1810 fl.1834-1845) [18], was an optician, mathematical instrument maker, and telescope maker of London, and later, Norwich, England. He worked for the partnership **G.&C. Dixey**, prior to relocating to Norwich where he set up his own business. Norwich addresses: Market Place (1834-1841); Upper Market (1836); 5 Exchange St. (1845).

Dixey, Lewis
Lewis Dixey, (c1814-1895) [18] [379], was an optician, and mathematical, optical, and philosophical instrument maker of 62 King's Road, Brighton, Sussex, England. He was the son of **George Dixey**, and is known to have sold: microscope; sector. According to his trade card, "Son of Mr G. Dixey, Optician to the Royal Family, Bond Street, London."

Dixons
Dixons was a UK retailer which began retailing imported Japanese cameras, projectors, telescopes and binoculars in the 1950s, using their own brand-name, **Prinz**. In the 1980s they also re-used the **Miranda** brand-name for imported Japanese cameras manufactured by **Cosina**.

Dobbie & Son, Alexander
Alexander Dobbie, (1815-1887) [18] [263] [271], was the son of a watchmaker in Glasgow, Scotland. He began his business as a watch and clock maker in 1841, trading at 20 Clyde Place. He began making marine chronometers, and from 1851 he advertised his business as a nautical instrument maker and chart seller. He moved his business to 24 Clyde Place in 1857, and in 1873 he expanded to include No. 25. In 1886 Alexander Dobbie entered a partnership with his son, John Clark Dobbie, and began trading as Alexander Dobbie & Son.

John continued the business after his father's death, and expanded into various locations in Glasgow, Greenock, Cardiff, and London. The company specialised in producing and retailing nautical instruments as well as making chemical, mathematical, optical, and philosophical instruments. They were also Admiralty chart agents, and chronometer makers to the Admiralty. In 1893 John C. Dobbie purchased the firm of T.S. McInnes, engineering and mathematical instrument maker. Sometime between 1895 and 1902 he also acquired at least a part of **D. McGregor and Co. Ltd.** The businesses continued separately until John merged them in 1903, and began trading as **Dobbie McInnes Ltd.**

Dobbie McInnes Ltd.
Dobbie McInnes Ltd., [263] [271], were nautical instrument makers of Glasgow, Scotland. The company had its origins in 1841 with **Alexander Dobbie** and, in 1903, merged with T.S. McInnes and **D. McGregor and Co. Ltd.** to become Dobbie McInnes Ltd., under Alexander's son, John Clark Dobbie.

In 1914 the *Whitaker's Red Book of Commerce or Who's Who in Business*, [110], listed Dobbie McInnes Ltd. as manufacturers and patentees of nautical and engineering instruments. It listed their specialities as their *McInnes-Dobbie* patent engine indicators; Pressure and explosion recorders and all accessories for indicating purposes; Improved Bourdon pressure, vacuum and hydraulic gauges; Revolution counters; Furnace deformation indicators and high-class engine and boiler fittings; Dobbie's patent compass and sounding machine; Ships' clocks; Aneroids; Binoculars; Telescopes and general navigational instruments; also official chart agents.

According to [44], in 1967 they were acquired by the engineering company G.&J. Weir. According to [263], in 1984, Dobbie McInnes Ltd. became a part of the Cunningham Shearer Group.

Instruments bearing their name include: sextants; marine chronometers; telescopes; Galilean binoculars.

Dobson, John (I)

John Dobson (I), (1915-2014) [206], was born in Beijing, China and moved to San Francisco, USA, with his family while still a child. He studied chemistry at UC Berkeley and, having graduated, worked in several defence related jobs until 1944 when he joined a monastery as a monk of the Ramakrishna Order. During his 23 years as a monk, he became interested in astronomy, and began making telescopes, which became a serious endeavour for him in the late-1950s. Being a monk he had no money to fund his telescope making, so he had to improvise with whatever materials came to hand. Thus, he came up with the design that is now known as a Dobsonian telescope mount. He continued making telescopes and using them for outreach until he was asked to leave the monastery in 1967 because he was perceived to be neglecting his duties.

He went to San Francisco and, with help from his friends, he continued his outreach and began teaching astronomy and telescope making. In 1968 some of his students began a public service organisation called *Sidewalk Astronomers*, which continued into the C21st.

Dobson, John (II)

John Dobson (II), (fl.1826-1866) [18] [276 #1830], was an optician, and mathematical, optical, and philosophical instrument maker, and drawing instrument maker, of London England.

Addresses: 4 Great Suffolk St., Borough; 24 Crown Row, Walworth; 3 Norfolk Place, East St., Walworth (1826); 13 Newington Causeway (1830); 54 Newington Causeway (1838-1846); Camberwell Green (1855); 1 Unison Row, Camberwell Green (1859); 254 Camberwell Rd. (1865).

He served his apprenticeship with the mathematical and philosophical instrument maker Thomas Lefever, beginning in 1813, and was freed of the **Merchant Taylors' Company** in 1822.

Dobsonian Telescope (Mount)

The term *Dobsonian telescope* is generally used to refer to a Newtonian telescope in a floor-standing alt-az mount. It is named after **John Dobson (I)** who originated and popularised the mount design, and is often used as a simple, portable, and effective mount for large aperture Newtonian telescopes.

Docktree, Walter

Walter Docktree, (b.1874) [61] [306], was a camera maker and inventor. He was works manager for **A.C. Jackson**, after which he worked for **Houghtons Ltd.** By 1910 he had his own company, Walter Docktree & Co., of Celsus Works, 33a Kenmore Road, Hackney, London, England, manufacturing the *Celsus Reflex* camera. He also designed the *Britisher Reflex* camera. In 1915 he began to work for **Staley, Shew & Co.**, taking charge of their production.

Dodd, Andrew

Andrew Dodd, (fl.1837-1847) [18] [271], was an optical, mathematical, and philosophical instrument maker of Glasgow, Scotland. Addresses: 70 Hutcheson Street (1837); 36 Glassford Street (1838-1846); 88 Glassford Street (1847).

The register of *Burgesses & Guild Brethren of Glasgow, 1751-1846,* [398], shows: "Dodd, Andrew, merchant, mechanical instrument maker, 36 Glassford Street, B. and G.B, as mar. Elizabeth, I. dau. To Robert Thomson, gardener, B. and G.B. 2 August 1839".

He was succeeded in business at 88 Glassford Street by **Thomas Mathers**.

Dolci Brothers

See **Fratelli Dolci**.

Dollond & Aitchison

Dollond & Aitchison, [1], was formed in 1927 when **Aitchison & Co.** acquired the goodwill, leases and stock of **Dollond & Co.**, and began to trade under the name Dollond & Aitchison. They acquired the premises of **James Brown** following the closure of the business in 1928. In 1950 they took over the company **A. Franks Ltd.** In 1968 they acquired **Harrisons Opticians**. In 1974 they acquired **Filotecnica Salmoiraghi**, but it was subsequently returned to Italian ownership. In 1981 they acquired **Theodore Hamblin Ltd.** Dollond & Aitchison continued in business until 2009, when they merged with **Boots**, following which the name Dollond & Aitchison ceased to be used.

Dollond & Co.

Dollond & Co., [1] [3] [18] [29] [46] [319], were opticians, and optical, mathematical, and scientific instrument makers of London, England. Addresses: 59 St. Pauls Churchyard (up to c1869); 1 Ludgate Hill (c1870-c1890); 62 Old Broad Street (1894); 5 Northumberland Avenue (1894); 35 Ludgate Hill (1909); 211 Oxford Street (1918); 61 Brompton Road (1918); 2 Northumberland Avenue, Charing Cross (1918); 223 Oxford Street.

The company was originally founded by **Peter Dollond** in 1750, and by 1752 he was working in partnership with his father, **John Dollond (I)**, trading as **J. Dollond & Son**. With the death of his father, Peter Dollond traded under his own name until 1766, when he began to work in partnership with his brother, **John Dollond (II)**, trading as **P.&J. Dollond**. John (II) died in 1804 and, in 1805, Peter Dollond began to trade in partnership with his nephew, **George Dollond (I)**, trading as **P.&G. Dollond**. With the retirement of Peter in 1819, George (I) continued the business under his own name until his death in 1852, and was succeeded by his nephew, **George Dollond (II)**, trading under his own name. George (II) was, in turn, succeeded on his death in 1866 by his son **William Dollond**. William ran the company until his retirement in 1871. By that time, trading as Dollond & Co., and the entire company stock and shop fittings were sold in public auction.

A listing in *The Times* newspaper on 27th July 1871 announced the sale of "Messrs. Dollond and Co., the eminent opticians" and "the whole of the optician's stock". The announcement lists many optical and philosophical instruments, as well as the premises at 59 St. Paul's Churchyard, and the shop furniture and fittings. Dollond & Co. was purchased by John Richard Chant (b.1837 Ret.1891), a former employee, in partnership with Tyson Crawford (1839-1930). Although no Dollond family interest remained, they continued the business, trading under the name Dollond & Co.

According to an advertisement of 1872, "Dollond & Co., (Established 1750.) By appointment, Opticians to Her Majesty, to the Hon. Corporation of the Trinity House. To the Royal National Life-Boat Institution. Have removed from 59, St. Paul's Churchyard, to No. 1 Ludgate Hill, Corner of St. Paul's Churchyard. Manufacturers of Microscopes, Telescopes, Opera glasses, Marine binocular glasses, Barometers, Thermometers, Sextants, &c. Dollond's unrivalled one-guinea opera glass, of great power and clear definition. Dollond's binocular 2-guinea field glass (the Gem), of great power. Dollond & Co.'s celebrated gold, silver, steel, and tortoiseshell spectacles."

In 1891 J.R. Chant retired, and the business continued under the sole proprietorship of Tyson Crawford [35 #26259 p943]. In 1894, the brothers Norman Parsons and Harold Parsons both joined the company as apprentices.

By 1900, many military telescopes were being replaced by field glasses. Dollond sold lightweight, inexpensive field glasses to both sides of the Russo-Japanese War. Their two-guinea War Office field glasses were popular at sporting events, as were the Dollond *Lord Reay* telescopes.

In 1903 Tyson Crawford retired, and sold the company to Norman and Harold Parsons. In 1907, the company was incorporated, and began to trade as Dollond & Co. Ltd.

In 1914 the *Whitaker's Red Book of Commerce or Who's Who in Business*, [110], listed Dollond & Co. Ltd. as "Manufacturers, Opticians to the Admiralty and other government departments, Eyesight specialists," and, "Established by John Dollond in 1750. Specialities: all descriptions of Optical instruments, including Prismatic binoculars, Telescopes, Spectacles, &c.; also Scientific instruments in great variety. Inventions: various, in connection with optical instruments. Branches: Eight, in London."

In 1916 Dollond & Co. Ltd. began to supply binoculars to the War Department for the war effort. This continued throughout WWI. In 1919 Dollond & Co. was listed as a member of the **British Optical Instrument Manufacturers' Association**.

In 1927 the firm was acquired by **Aitchison & Co.**, to become **Dollond & Aitchison**, and thereafter concentrated increasingly on prescription spectacles.

Dollond & Son, John

John Dollond & Son, [1] [18], were opticians, and optical and mathematical instrument makers, of *the Golden Spectacles and Sea quadrant*, Near Exeter Exchange, Strand, London, England. The partnership was formed in 1752 between **Peter Dollond** and his father, **John Dollond (I)**. **James Ayscough** is known to have supplied J. Dollond & Son with metals. The partnership continued until the death of John Dollond (I) in 1761, after which Peter Dollond continued the business under his own name.

Dollond, George (I)

Born George Huggins, and changed his name to George Dollond. See **George Huggins (I)**.

Dollond, George (II)

Born George Huggins, and changed his name to George Dollond. See **George Huggins (II)**.

Dollond, John (I)

John Dollond (I), (1707-1761) [1] [18] [388], was an optician and optical instrument maker of Spitalfields, London, England. He was the son of a Huguenot silk-weaver who fled France after Louis XIV revoked the Edict of Nantes in 1685. John Dollond (I) began his career as a silk-weaver, but pursued his interest in mathematics. In 1727 he married Elizabeth Sommelier (1708-1782), and together they had nine children, of whom two sons, **John Dollond (II)** and **Peter Dollond**, and three daughters survived to adulthood. His daughter, Susan (1728-1798), married William Huggins. Their sons, **John Huggins (I)** and **George Huggins (I)** both worked for the Dollond company. Another daughter, Sarah (1743-1796), married **Jesse Ramsden** and, according to some sources, as part of her dowry Ramsden received a share in the Dollond patent for the achromatic lens. However, Sarah was later estranged from Ramsden.

John Dollond (I)'s son, Peter, under his guidance, set up business in 1750 as an optician in Vine Street, Spitalfields. Within two years, John Dollond (I) was able to give up silk-weaving and join his son in business, trading as **J. Dollond & Son**.

In 1757 John Dollond (I) had a conversation with **George Bass** in Bass's workshop, and figured out how **Chester Moor Hall** had used two optical elements, the crown and flint glasses, to overcome chromatic aberration. After some research of his own, he devised the achromat based on the principles Bass had used. He controversially patented the achromat following encouragement from his son, Peter, and with financial support from **Francis Watkins (I)** (See Appendix X). He entered a partnership with Francis Watkins (I) to market refracting telescopes based on the patent.

Upon John (I)'s death in 1761, Peter continued the business, trading under his own name.

Dollond, John (II)
John Dollond (II), (1733-1804) [1] [18], was an optician of St. Paul's Churchyard, London, England. He was the son of **John Dollond (I)**, and the brother of **Peter Dollond**. He was freed by purchase of the **Spectacle Makers' Company** in 1767, and was Master of the Spectacle Makers' Company from 1790-1791. Among his apprentices were **William Moon**, and John Fairbone, the son of **Charles Fairbone (II)**. **Edmond Gabory** also worked for him. Beginning in 1766, he worked in partnership with Peter Dollond, trading as **P.&J. Dollond** until John (II)'s death in 1804.

Dollond, P.&G.
P.&G. Dollond, [1] [18] [319], opticians, and optical and mathematical instrument makers of 59 St. Paul's Churchyard, London, England, was a partnership between **Peter Dollond** and his nephew, **George Dollond (I)**. The partnership was formed following the death of **John Dollond (II)** in 1804, and succeeded **P.&J. Dollond** in business. In 1820 Peter and George (I) were made opticians to King George IV. Peter Dollond, however, had retired at the end of 1819 [35 #17542 p2203], bringing the partnership to an end, with the result that the royal appointment was never carried out. Following Peter's retirement, George (I) continued to trade under his own name.

Dollond, P.&J.
P.&J. Dollond, [1] [18] [319] [434], opticians, and optical and mathematical instrument makers of 59 St. Paul's Churchyard, London, England, was a partnership between **Peter Dollond** and his brother, **John Dollond (II)**, formed in 1766. They also traded at 35 Haymarket from 1780 to 1793. They made and sold mathematical, philosophical and optical instruments, some of which were used on Captain James Cook's second trip to explore the Pacific. The partnership ended with the death of John (II) in 1804.

According to a trade handbill, dated c1780, "Optical, Mathematical, and Philosophical instruments made by P.&J. Dollond, opticians to His Majesty, in St. Paul's Churchyard, London." It lists: Marine, terrestrial, and astronomical telescopes, both refracting and reflecting; Spectacles; Opera glasses; Single and compound microscopes, slides, and accessories; Solar microscopes; Optical machines for viewing perspective pictures; Camera obscuras; Prisms; Concave and convex mirrors; Hadley's octants; Hadley's sextants; Theodolites; Levelling telescopes; Drawing instruments; Sundials; Globes; Electrical machines; Air pumps; Barometers; Thermometers; Hydrometers; Hydrostatic balances; "With all other Optical, Philosophical, and Mathematical instruments".

Dollond, Peter
Peter Dollond, (1731-1820) [1] [18] [319], was an optician and optical instrument maker of London, England. He was the son of **John Dollond (I)**, and the brother of **John Dollond (II)**. Like his father, Peter Dollond was trained as a silk-weaver and pursued an interest in mathematics. Not wishing to make a career as a silk-weaver, he set up business as an optician in Vine Street, Spitalfields, in 1750, with guidance from his father.

Within two years, John Dollond (I) had joined the business, and by 1752 they were trading in partnership as **J. Dollond & Son** at the *Golden Spectacles & Sea Quadrant*, Near Exeter Exchange in the Strand, London.

Peter was freed of the **Spectacle Makers' Company** in 1755 as a Foreign Brother, and was Master of the Company 1774-1881, 1797-1798, and 1801-1802. Among his apprentices were **John Huggins (I)**, **John Berge**, and **William Gilbert (I)** who was turned over to him by **John Gilbert (II)** in 1769. John Berge completed his apprenticeship in 1773, and continued to work for the Dollond business until 1791.

Upon the death of John Dollond (I) in 1761, Peter inherited his father's achromatic lens patent. He bought out **Francis Watkins (I)**' share of the agreement, made previously between Watkins and John (I). Watkins, however, continued to market achromatic telescopes in violation of the patent, and Peter successfully sued him. Peter subsequently enforced the patent with some vigour, leading to a lengthy dispute in which thirty-five other opticians signed a petition to the Privy Council calling for the patent to be revoked. Peter Dollond prevailed, and continued to enforce his patent. See Appendix X.

Following the death of John Dollond (I), Peter Dollond continued the business, trading under his own name. According to his trade card, "Peter Dollond, Optician to his Majesty and to his Royal Highness the Duke of York. At the *Golden Spectacles & Sea Quadrant*, Near Exeter Exchange in the Strand, London."

In 1766, he began to work in partnership with his brother, **John Dollond (II)**. At this time the company moved to 59 St. Paul's Churchyard, London, trading as **P.&J. Dollond**. In 1805, following the death of John Dollond (II) in 1804, Peter Dollond began to work in partnership with his nephew, **George Dollond (I)**, trading as **P.&G. Dollond**.

Peter Dollond retired at the end of 1819 [35 #17542 p2203], and was succeeded in business by George Dollond (I), trading under his own name.

Dollond, William
William Dollond, (c1826-1893) [1] [18] [319], was the last member of the Dollond family to run the family business of opticians, and optical and mathematical instrument makers. He served his apprenticeship with his father, **George Dollond (II) of the Grocers' Company**, beginning in 1847. He succeeded his father in business in 1866. He retired due to ill-health in 1871, at about 44 years of age, at which time he had only one child, Alfred, who was too young to take over the business. With William's retirement, **Dollond & Co.** was sold at public auction, and purchased by ex-employee J.R. Chant. In the 1871 census William Dollond was still listed as an optician, but was recorded as blind.

Double Image Micrometer
A double image micrometer is a device which produces a double image of an object and, by means of a micrometer, enables the measurement of the angular separation between two parts of the image. This was used, for example, as a heliometer to produce a double image of the sun, allowing the measurement of its angular size, or to measure the angular separation of two stars.

The device was first proposed by **Servington Savery** in 1743 in the form of a divided object glass micrometer. This was achieved by cutting out two centre segments from a lens, and bringing the two outer portions together at the cut. The double image thus formed was measured with a micrometer. The instrument was attached to a telescope, and such a device was used by **George Graham** to measure the diameter of the Sun. Savery also proposed two more arrangements of this invention, the second utilising the two centre pieces of the lens attached back-to-back and the third using two lenses of the same focal length arranged side-by-side.

A variation on the concept was proposed by **Pierre Bouguer** in 1748. This was similar to Savery's third idea, with two lenses arranged side-by-side, with the difference that one could be moved on a screw adjustment to alter the separation between the lenses. This adjustment was used to make the measurement, removing the necessity for an auxiliary micrometer.

In 1754, **John Dollond (I)** used a combination of Savery's and Bouguer's ideas to produce his divided object glass micrometer, in which an objective lens is cut into two through its axis, and one half is moved laterally relative to the other, (in a direction perpendicular to the axis of the telescope), by means of a micrometer adjustment. In this arrangement, each half of the lens forms an image of the stars whose separation is to be measured, and the images coincide when the optical centres of the lens halves coincide. He and his son, **Peter Dollond**, produced instruments to this design and variations of it.

See also double refraction micrometer.

Double Image Prism
See prism.

Double Refraction Micrometer
See micrometer.

Doublet
Doublet is the term usually used in optics to refer to a lens arrangement consisting of two lens elements.

Doublet, T.&H.
T.&H. Doublet, [18] [319] [322], were opticians, spectacle makers and scientific, mathematical, and drawing instrument makers of London, England.

Hannah Doublet (fl.1830-1851), otherwise known as Dublott [276 #1833], was married to Thomas Doublet (I) in 1814. By 1830 she was trading as an optical, mathematical, and philosophical instrument maker at 14 Shepperton Place, New North Road.

Thomas Doublet (II) (1815-1878), was the son of Thomas (I) and Hannah Doublet. By 1832 he was trading as an optician at the same address as his mother. He is also recorded at 19 and 25 Windmill St., Finsbury and, by 1842, at 74 Paul St., Finsbury.

Thomas and Henry Doublet were trading at 74 Paul St. by 1842, and continued in business until 1901, trading at the London addresses: 74 Paul St. (1842-1849); 4 City

Road, Finsbury Square (1849-1859); 6 Moorgate Street (1865-1875); 7 City Rd. (1865); 50 Finsbury Square (1875); 11 Moorgate St. (1880-1895); 39 Moorgate St. (1901). They are known to have sold microscopes, rules, and telescopes.

Thomas and Hannah Doublet were trading at 4 City Road around 1851, and are also known to have traded at 6 Moorgate St.

Advertisements under the name T.&H. Doublet appear from about 1850 until 1900, giving their addresses as 4 and 7 City Road, and 39 Moorgate. This suggests that the partnership was Thomas and Henry Doublet, and it seems reasonable to speculate that the T. refers to Thomas (II). Their various advertisements included opera, coursing, racing, field, and yachting glasses, spectacles, microscopes, magic lanterns, telescopes, barometers, and mathematical instruments. In 1868 their entire stock in trade, as well as fixtures and fittings, were sold at auction due to their premises at 7 City Road "having to be rebuilt".

Douglas, James
James Douglas, (fl.1788-1793) [18] [271], was an optical instrument maker of Calton Hill Observatory, Edinburgh, Scotland, and was the grandson of **Thomas Short**, whom he succeeded in business. The Calton Hill Observatory was founded by **James Short (II)**.

Dove Prism
The Dove prism, named after its inventor, the Prussian physicist Heinrich Wilhelm Dove, takes the form of a right-angled prism, truncated by the removal of its triangular apex. Light entering parallel to its longitudinal axis is refracted toward the longest face where it undergoes total internal reflection before emerging from the prism once more on the longitudinal axis.

The Dove prism has some interesting properties. Since the beam of light undergoes a single reflection, the emerging beam is inverted (up becomes down and down becomes up), but not laterally transposed (left stays left and right stays right). If the prism is rotated on its longitudinal axis the emerging beam is rotated by twice the amount that the prism is rotated, and hence the prism may be used as a beam rotator. The Dove prism introduces astigmatism if it is used with converging light, and hence it is generally used with collimated light, such as for astronomy applications and interferometry.

Dowling, William
William Dowling, (fl.1814-1830) [18] [274] [276 #1538] [319] [322], was an optical, mathematical, and philosophical instrument maker of London, England. Addresses: Lincoln's Inn Passage; 114 Great Russel St., Bloomsbury; 121 Great Russel St., Bloomsbury (1814); Serle St., Lincoln's Inn West Gate (1822); West Gateway of Lincoln's Inn, Serle St., Lincolns Inn Fields (1829) – the last two likely being the same address.

He served his apprenticeship with **Thomas Harris (I)**, beginning in 1804, and was freed of the **Spectacle Makers' Company** in 1814.

According to his trade advertisement at this last address: "Manufactures and sells all kinds of Optical, Mathematical, and Philosophical instruments, of the best workmanship, and according to the latest improvements." The following are listed: "Improved opera glasses, both plain and achromatic, so very desirable in a theatre, in various mountings; Refracting telescopes for land or sea; Reflecting telescopes; Perspective glasses; Single and compound microscopes; Camera obscuras; Diagonal optical machines, for viewing prints; Concave and convex mirrors; Prisms; Magic lanterns with a variety of slides; Barometers and thermometers; Quadrants and sextants; Globes in various sizes; Electrical machines; Air pumps, Pentagraphs, Theodolites; Circumferenters; Levels; Measuring chains and tapes; Drawing instruments; Gauging and plotting rules and scales; Protractors and parallel rules, together with every instrument in general request."

Drakeford, David
David Drakeford, (fl.1761-1764), was an optical instrument maker of Fleet Ditch, London, England. He served his apprenticeship with **George Bass**, beginning in 1728, and was freed of the **Spectacle Makers' Company** in 1736. His son, also named David, served as his apprentice. In 1764 he was one of the petitioners who attempted to revoke the patent obtained by **John Dollond (I)** for the achromatic lens, and which was enforced after Dollond's death by his son, Peter (See Appendix X).

Drapers' Company
See Appendix IX: The London Guilds.

Drew, Peter
Peter Drew is a British astronomer, optical designer, and maker of large aperture binocular telescopes. His shop in Luton, Bedfordshire, England, was called Bedford Astronomical Supplies. In the early 1970s he is believed to have been the first maker in the UK to offer large format reflecting binoculars, and these were sold under his **Terrascan** brand-name. He has made many large binocular telescopes up to 200mm aperture with refracting optics and up to 300mm aperture with mirrors. In 1976 he founded **AstroSystems Ltd.** with co-founders and directors Rob Miller and **David Hinds**. He runs the Astronomy Centre and observatory, which he founded in

1982 on the moors of the Southern Pennines, near the border between Yorkshire and Lancashire, England.

Drexler & Nagel
Drexler & Nagel was a company formed by camera designer **Dr. August Nagel** in Stuttgart, Germany, in 1908. Soon after this the company name was changed to **Contessa Camera-werke Drexler & Nagel**.

Dring & Fage
Dring & Fage, [18] [53] [276 #958], were optical, mathematical, and philosophical instrument makers of London, England. The company was formed in 1790 as a partnership between glassworker and hydrometer maker, John Dring, and **William Fage**.

Addresses: 21 Gracechurch St. (1790-1792); 4 Albion Place (1790); 6 Tooley St. (1792-1796); 248 Tooley St. (1796-1804); 8 Crooked Lane (1801); 20 Tooley St. (1804-1844); 109 Upper East Smithfield (1804); 10 Duke St., Tooley St. (1843-1844); 19 & 20 Tooley St. (1845-1882); 145 Strand (1883-1902); 56 Stamford St. (1903-1938).

The partnership between John Dring and William Fage was terminated in 1817 [35 #17210 p114]. At some point William Fage entered a new partnership with William George, still trading as Dring & Fage, and this was terminated in 1834 [35 #19123 p167], after which William Fage continued to trade under his own name, and Dring & Fage continued without him.

When the business of **R.B. Bate** closed in 1850, various instrument makers bid for the lucrative and prestigious Excise appointment. Dring & Fage were successful, and were appointed Hydrometer and Saccharometer Makers to the Board of Inland Revenue. By this time the company was a partnership between the hydrometer and saccharometer makers, Edward Hall and Edward Jenkins [355].

According to their trade card [361], "Dring & Fage, 19 & 20, Tooley Street, London. Manufacturers of Sextants, Quadrants, Compasses, Telescopes, & every description of Nautical instruments. Map & chart agents." Another trade card reads, "Sikes's improved Hydrometer, made by Dring & Fage. Original hydrometer, saccharometer, & gauging ins^t. Manufacturers & upw^{ds}. of 50 years Hydrometer makers to His Majesty's Honble. Board of Excise. 20, Tooley Street, Southwark, London. Mathematical, Philosophical, & Optical instruments of every description. Rules, thermometers & gauging instruments, for distillers, maltsters, brewers, brandy merchants &c. Old instruments repaired and accurately adjusted."

Drum Microscope
A drum microscope, [264], is a simple form of portable compound microscope invented by **Benjamin Martin** in 1738. It is inexpensive, simple, and easy to use. Because of this it became popular, and many makers have produced examples. It consists of a sliding body tube with the eyepiece at the upper end, and objective at the lower end. This slides in a base tube that contains a stage to support the sample. It allows for illumination by light reflecting from the top surface of the specimen, and may also have a substage mirror for illumination from underneath. The objective may be interchangeable, as may the eyepiece.

Ducati
See **Società Scientifica Radio Brevetti Ducati (SSRD)**.

Duboscq, Louis Jules
Louis Jules Duboscq, (1817-1886) [8] [53] [299], usually known as Jules Duboscq, was a scientific and optical instrument maker of Paris, France. He began an apprenticeship with **Jean Baptiste François Soleil** in 1834, and in 1839 he married Jean Baptiste Soleil's daughter Rosalie. In 1850, upon Jean Baptiste Soleil's retirement, Duboscq purchased the part of his business which was the workshop for making scientific instruments. For a short while he traded as Duboscq-Soleil, but quickly went on to use his own name. He soon established himself a reputation as a leading maker of instruments for physical optics. In 1851, at the Great Exhibition in London, he was awarded a medal for his instruments. He began to make, and popularise, **Sir David Brewster**'s lenticular stereoscopes, as well as making stereo pictures. He had a strong interest in both still and moving image photography, and was innovative both in improvements to photographic and projection technology, and as a photographer. In 1870 **Léon Laurent** married Jules Duboscq's daughter Marie.

In 1885 Jules Duboscq formed a partnership with **Philibért François Pellin**, trading as **Duboscq & Pellin**. Following the death of Jules in 1886 the company was continued by Philibért Pellin.

Duboscq & Pellin
Duboscq & Pellin was a partnership between **Jules Louis Duboscq** and **Philibért François Pellin**. The partnership was formed in 1885 and, following the death of Jules Duboscq in 1886, the company was continued by Philibért Pellin.

Ducray-Chevallier, Alexandre Victor
Alexandre Victor Ducray-Chevallier, (1810-1879) [50], was the husband of **Jean Gabriel Augustin Chevallier**'s daughter **Marie Louise Mélanie Chevallier**. He changed

his name to Ducray-Chevallier when he took an interest in the Chevallier business. Together with Marie he purchased Jean Chevallier's business in 1842. Upon Jean Chevallier's death in 1848 they changed the name of the business to **Maison de l'Ingénieur Chevallier**, and he continued in the business until he died in 1879.

Dufay Chromex Ltd.; Dufay Ltd.

Dufay Chromex Ltd., [**44**] [**306**], of 14-16 Cockspur Street, London, England, was founded in 1936 as an amalgamation of Cinecolor Ltd., Dufaycolor Ltd., Dufaycolor Corporation, and Spicer-Dufay. The objective of the company was to develop and market the *Dufaycolor* film process in colaboration with **Ilford**, and the *Cinecolor* film process in collaboration with **Adam Hilger**. In 1936 they acquired the **Coronet Camera Co.** They exhibited at the 1947 British Industries Fair. Their catalogue entry, [**450**], listed them as manufacturers of: Sensitised photographic materials; Cameras; Optical lanterns; Photographic accessories; Photographic filters; Colour vision products; Cellulose acetate film, and other products. During the 1950s the company changed its name to Dufay Ltd.

Dumaige, M.

M. Dumaige, (fl.1876-1900) [**53**] [**113**], of 9 Rue de la Bûcherie, Paris, France, was a maker of microscopes. A catalogue for microscopes and accessories, sold by **A. le Vasseur & Cie**, states that all of the instruments it shows are made by M. Dumaige, supplier to the Faculté de Médecine, Paris. He is known to have made microscopes and objective lenses.

Dumotiez, Frères

Frères Dumotiez, [**256**] [**322**] [**362**] [**374**] [**423**], were highly regarded manufacturers of philosophical, physical, and electrical apparatus, and microscopes, of 2 rue du Jardinet, Paris, France. The firm was a partnership between the brothers Louis Joseph Dumotiez (b.1757) and Pierre François Dumotiez (1743-1817), which began in 1780. **Cornelius van Wijk** was trained in their workshop from 1786 to 1790. By the end of the C18th, their clientele included a number of eminent scientists, both domestic and international, and their catalogues listed hundreds of instruments for sale. Their last published catalogue, dated 1817, listed over 600 items. In 1818 they were succeeded in business by the eminent physical instrument maker, **Nicolas Constant Pixii**, the son-in-law of one of the brothers.

Instruments bearing their name include: inclinable microscope; lucernal microscope; hydrostatic balances; barometer; Hygrometers; thermometers; compression pumps; fire syringes; Guyton de Morveau-type disinfectant bottles. They also made mathematical and surveying instruments, and the necessary equipment for a full fantasmagoria show.

Dumpy Level

A dumpy level, also known as a *builder's level*, is a tool used by surveyors and builders for finding or transferring a level. Traditionally this consists of a precision telescope with cross-hairs, mounted on a base, usually a tripod, that is levelled using bubble levels, and can only be rotated in the horizontal plane. The invention of the dumpy level is credited to the English civil engineer **William Gravatt**, who apparently conceived of the idea while using a Y level. Variants of the Dumpy Level include the automatic level and the tilting level.

Dunbar-Scott Auxiliary Top and Side Telescope

A telescope attachment for a theodolite, used to sight down inclined or vertical shafts, or steep gradients, in situations where the main telescope cannot be used.

Duncan, William

William Duncan, (fl.1841-1849) [**18**] [**271**] [**322**], was an optical, mathematical, and philosophical instrument maker of Aberdeen, Scotland. Addresses: 46 Dee St. (1841), and 92 Union St. (1842-1849). Instruments bearing his name include: microscope; level; stick barometer.

Duncker, Johann Heinrich August

Johann Heinrich August Duncker, (1767-1843) [**47**], was a German optical industrialist. He was the son of a priest, and began studying theology in Halle in 1786. In 1792 he began the construction of microscopes, which established his reputation as an optical instrument maker. Together with his acquaintance, the minister Samuel Christoph Wagener, he appealed to the Prussian King Frederick William III in 1800 to allow them to establish an optical industrial plant. The King agreed, granting royal privilege, and in 1801 they established the Royal Privileged Optical Industry Institute in Rathenow. Their objective was to introduce the production of all kinds of optical instruments into Prussia, based on scientific knowledge, and to employ invalids and poor children. In 1801, they also were granted a patent for a multi-purpose lens-grinding machine that greatly facilitated production of optical lenses, particularly with less experienced workers. Wagener left the company in 1806, but retained his financial interest as co-owner of the patent.

In 1819, as a result of illness, Duncker handed over control of the company to his son, Edward. In 1834 the company relocated and expanded into new premises in

Rathenow, and in 1845 Edward handed over control of the company to his nephew **Emil Busch**. In 1872 Busch renamed the company Emil Busch AG, and again in 1908 to become Emil Busch AG Optische Industrie.

Dunn, John

John Dunn, (c1791-1841) [**18**] [**263**] [**271**] [**274**] [**322**] [**358**], was an optical, mathematical, and philosophical instrument maker of Edinburgh, Scotland.

Addresses: 7 West Bow (1824); 25 Thistle St. (1825-1827); 52 Hanover St. (1828-1831); 50 Hanover St. (1832-1841). He also opened branches in Glasgow, trading at 157 Buchanan St. (1840-1841), and 28 Buchanan St. (1841).

In the winter of 1839/1840, he displayed a number of instruments at an exhibition in Edinburgh. These were: portable transit instrument; altitude & azimuth instrument; quadrant; sextant; theodolite; thermometer. His trade card listed surveying and drawing instruments, philosophical and chemical apparatus, barometers, thermometers, telescopes, microscopes, and spectacles. He was actively involved in science education and, from 1833 until the end of his life, he was curator of the Museum of the Society of Arts, and responsible for demonstrations of their apparatus. Following his death in 1841 he was declared bankrupt, and his estates were subject to sequestration proceedings [**35** #20008 p2117].

Instruments bearing his name include: level; theodolite; graphometer; pyrometer; survey compass; transit for the adjustment of chronometers; pantographs. His younger brother, **Thomas Dunn**, began to work for him in about 1825, and succeeded him in business.

Dunn, Thomas

Thomas Dunn, (c1803-1893) [**18**] [**263**] [**271**], was an optical, mathematical, and philosophical instrument maker of Edinburgh, Scotland. He was the younger brother of **John Dunn**, for whom he worked beginning in 1825, and whom he succeeded in business. Addresses: 50 Hanover St. (1843-1866) and 106 George St. (1867).

He exhibited at the Great Exhibition in 1851 in London [**328**]. His catalogue entry no. 689a stated: "Dunn, T. Edinburgh—Inventor. Electro-magnetic machine." He was declared bankrupt in 1864, and his estates were subject to sequestration proceedings, [**170** #7450 p938], which continued until the year of his retirement, 1868 [**170** #7831 p291].

Dunnell, John (Dunning)

John Dunnell, otherwise known as Jack Dunning, (fl.1673-1688) [**18**] [**275** #373], was a spectacle maker, turner, and telescope tube-maker of Charing Cross, London, England. He was freed by purchase of the **Turners' Company** in 1673, and freed by purchase of the **Spectacle Makers' Company** in 1683. Among his apprentices was **John Marshall**, beginning in 1673. He worked for both Robert Hooke, and the Astronomer Royal, John Flamsteed (1646-1719). He is believed to have worked in partnership with the turner and lens-maker **Smethwick**.

Duren, Henry

Henry Duren, (fl.1852-1861) [**382**], was a nautical instrument maker of New York, USA. According to his trade label, "H. Duren, Nautical instrument manufacturer, 39 Burling Slip, One door from South Street, New York. Chronometers, Sextants, Quadrants, Telescopes, Compasses, Log glass, Barometers cleaned and repaired on reasonable terms. Old instruments bought or exchanged. Second hand quadrants from $5,—warranted correct. Clocks and watches carefully repaired." Another trade label shows him as a dealer in nautical instruments. His name appears on an octant bearing the scale divider's mark for **Spencer, Browning and Rust**, (see Appendix XI).

Dynameter

The dynameter, devised by **Jesse Ramsden** in about 1775 [**2**], is an instrument consisting of a micrometer, mounted on the eyepiece of a telescope, used to measure the exit pupil (otherwise known as the Ramsden disc). Thus, knowing the diameter of the objective or aperture stop, the magnifying power of the telescope may be determined. Ramsden's dynameters may have mostly been made by **Thomas Jones**.

E

Eagle Optics
See **Seeadler Optik**.

Ealing Corporation; Ealing Catalog
The Ealing Corporation was founded in 1961, producing opto-mechanical products. The company began in Cambridge, Massachusetts, USA, and later moved to South Natick. They published a set of teaching catalogues to promote their products, giving rise to the name Ealing Catalog. Their optics products include lenses, mirrors & beam splitters, prisms & polarizers, and filters.

In 2013 the company was acquired by **Hyland Optical Technologies**, and has sold their products as a Hyland Optical Technologies brand since then.

Eastland & Co.
Eastland & Co. [276 #166] [366], optical instrument maker of Rotterdam, Netherlands, was a collaboration between **William Eastland (I)** and his employee, a master silversmith and instrument maker from Delft, named Jonas Regenboog. They produced optical instruments, some of which were supplied to the Dutch East India Company. Following the death of William Eastland (I) in 1787, Regenboog continued to trade alone using this name. Their instruments were marked *Eastland & Co.*, or *Eastland & Comp*. Instruments bearing their name include microscopes.

Eastland & Cooke
Eastland & Cooke [366], of London, England, was a partnership between **William Eastland (I)** and **John Cooke**. The partnership operated during the 1760s, and they collaborated to make microscopes.

Eastland, William (I)
William Eastland (I), (c1704-1787) [18] [276 #166] [322] [366], was an optical instrument maker of Clerkenwell, London, England. He served his apprenticeship with **Thomas Gay**, beginning in 1718, and in 1719 was turned over to **Thomas Lincoln**. He was freed of the **Spectacle Makers' Company** in 1726. One of his apprentices was "William Eastland, son of George", beginning in c1761, and was turned over to **Henry Raynes Shuttleworth** in 1762 (see **William Eastland (II)**).

In 1764 William Eastland (I) was one of the petitioners who attempted to revoke the patent obtained by **John Dollond (I)** for the achromatic lens, and which was enforced after Dollond's death by his son, Peter (See Appendix X). Despite the failure of the petition, Eastland continued to make achromatic telescopes, and in 1765 he was successfully sued by **Peter Dollond** for infringement of the patent.

During the 1760s he made telescopes and microscopes, and collaborated with **John Cooke**, making microscopes bearing the name Eastland & Cooke. Sometime between 1768 and 1770 he moved to the Netherlands, and settled in the Hague, where he established his own successful business specialising in achromatic telescopes. See **Eastland & Co.**

Some sources indicate that he had a son, also named William Eastland, who was also an instrument maker. However, [366] refutes this, stating that William Eastland (I) was not married, and had no legitimate children.

Eastland, William (II)
William Eastland (II), (c1747-1780) [18] [276 #166] [322] [401]. Some sources indicate that **William Eastland (I)** was his father, but [366] refutes this. It seems likely that he was the "William Eastland, son of George", who began an apprenticeship with William Eastland (I), in c1761. He was turned over to **Henry Raynes Shuttleworth** in 1762, and freed of the **Spectacle Makers' Company** in c1768.

Eastman Dry Plate Company
See **Kodak**

Eastman, George
George Eastman (1854-1932) was an innovator, and co-founder of the **Eastman Kodak Company**. His family were from upstate New York, USA, and were left poor by the early death of his father. George became interested in photography at the age of 24, and he bought all the equipment he would need to take photographs, spread photographic emulsion on glass plates before exposing them, and develop the exposed plates before they dried out. His interest grew, and he decided he wanted to simplify the process. By 1880 he had invented a dry-plate formula, and patented a machine for its production. He set up the **Eastman Dry Plate Company** in partnership with Henry A. Strong – a company that eventually became the **Eastman Kodak Company**.

George Eastman became a wealthy industrialist as a result of his work, but he believed that it was important to take good care of his employees, and in 1919 he gave one third of his company stock to his employees. He also set up retirement annuity, life insurance and disability benefits plans for them. His philanthropy included generous support for the arts, education and dental health-care.

Eastman Kodak Company
See **Kodak**.

Ebeling, Fritz
Fritz Ebeling, [50], was a microscope maker in Vienna, Austria. He worked for **C. Reichert** until 1886 when he left to form a partnership with another ex-Reichert employee, **Ludwig Merker**, trading as **Merker & Ebeling**. This partnership continued until c1892, after which Ebeling continued in business alone.

Ebsworth, George Richard
George Richard Ebsworth, (fl.1829-1835 d.1840) [18] [319] [322], was an optician, and mathematical and philosophical instrument maker of 54 Fleet St., London, England. This is the same address as that of **Richard Ebsworth** at the time. He is believed to have died in a "Frightful and fatal accident upon the Eastern Counties Railway" in which he was severely scalded and suffered "concussion of the brain".

Ebsworth, Richard
Richard Ebsworth, (fl.1820-1835) [18] [274] [276 #1320] [319] [322], was an optician, and optical, mathematical, and philosophical instrument maker of London, England. Addresses: 41 Fleet St.; 68 Fleet St. (1826); 54 Fleet St. (1827). Instruments bearing his name include: refracting telescope; reflecting telescope; two-draw telescope with metal tubes and shagreen cover; microscope; miniature astronomical quadrant; floating sundial; drawing instruments.

Some instruments bear inscriptions such as *Ebsworth, 68 Fleet Street*. This should be interpreted with caution, since **Thomas Ebsworth**, optician, was also trading at 68 Fleet St. in 1824. **George Richard Ebsworth** also traded at 54 Fleet St. from 1829 until 1835.

Ebsworth, Thomas
Thomas Ebsworth, (fl.1824) [18], was an optician of 68 Fleet St., London, England. At that time, these premises were also occupied by **Richard Ebsworth**. Hence, attribution of optical instruments marked *Ebsworth 68 Fleet Street* are ambiguous.

Echelette Grating (Blazed Grating)
An echelette grating, also known as a blazed grating, is a kind of diffraction grating consisting of a substrate onto which is scored a number of equally separated parallel grooves. The groove separation is small, and the cross-section is a triangular sawtooth shape. The critical characteristics are the groove separation, the angle of the sawtooth facet with respect to normal – known as the *blaze angle*, and the wavelength at which the grating operates – known as the *blaze wavelength*.

For high order diffraction patterns the grating will have a high blaze angle, and a relatively low groove density. This special kind of echelette grating is known as an echelle grating.

Echelle grating
An echelle grating is a special kind of echelette grating with a high blaze angle and relatively low groove density. This is used for high order diffraction patterns.

Echelon Grating
An echelon grating is a diffraction grating consisting of a series of plane-parallel glass plates of equal thickness, placed in echelon to form a series of equally spaced steps.

Echelon Spectroscope
See spectroscope.

Ecole Supérieure de Optique
See **Institut d'Optique**.

Ede, C.J.
C.J. Ede, (fl.1839-1844) [18], was an optical instrument maker of 100 St. George's Road, Southwark, London, England. He attended the London Mechanics Institute from 1839 to 1844.

Eden & Hollis
See **Alfred Frederick Eden**.

Eden, Alfred Frederick
Alfred Frederick Eden, (fl.c1830-1852) [18] [274] [276 #1838] [319] [320] [322], was an optician of 6 Langham Place, Regent St., London, England (c1830-1849), and 4 Dowgate Hill (1849-1850s). He worked for, and studied under **Andrew Pritchard**, and may have served an apprenticeship with him. He is known to have sold microscopes, and is believed to have made them.

He worked in partnership with Edwin Wise Hollis, trading as opticians and mathematical instrument makers at 4 Fenchurch Buildings, in the City of London. The partnership was dissolved in 1846 [35 #20601 p1655].

In 1847 he was in debtor's prison, [35 #20783 p3737], described as "Alfred Frederick Eden, formerly of Fenchurch-buildings, Fenchurch-street, in the city of London, in partnership with Edwin Wise Hollis, trading under the style and firm of Eden and Hollis, Opticians, then of No. 8, Railway-place, Fenchurch-street, in the city of London, Optician, then of No. 9, Commercial-place, City-road, Middlesex, Optician, and late of Greenhithe, in the county of Kent, out of business or employment."

In 1852, trading as an optician, agent for the Discount of Bills of Exchange, and shipping agent, he was again declared insolvent [35 #21379 p3022].

Eden, William
William Eden, (fl.1818-1827) [18] [276 #1321] [322], was an optician and optical instrument maker of 30 Lower Holborn, London, England. He served his apprenticeship with **Samuel Jones**, beginning in 1811, and was freed of the **Spectacle Makers' Company** in 1818. Addresses: 30 Castle St., Holborn (1818-1827); 30 Lower Holborn (1826-1827). During 1826-1827 he attended the London Mechanics' Institute in Holborn.

Edgecomb, William C.
Born in 1845, William Cary Edgecomb lived in Mystic, Connecticut, USA, where he taught astronomy and spent much of his time in his observatory. Edgecomb was a skilled scientific instrument maker and, although his instruments are rare, he was largely known for manufacturing high quality binocular telescopes.

Edgeworth, Henry
Henry Edgeworth, also known as Edgworth, (fl.1768-1787 d.1790) [3] [16] [18] [54] [322] [386], was a mathematical, philosophical, and optical instrument maker of *Hadley's Quadrant & Spectacles*, opposite the Dial, on the Quay, Bristol, England. The address was also known as 51 on the Quay, Bristol. It is believed that he came from Longford in Ireland, and served an apprenticeship with **John Margas** in Dublin, beginning in 1760.

According to his trade label, "Henry Edgeworth, Mathematical, Philosophical and Optical instrument maker at the Sign of *Hadley's Quadrant and Spectacles*, Opposite the Dial, on the Quay, Bristol, (being the only person in that City who served a regular apprenticeship in the above branches) Makes and sells ..." There follows an extensive list of nautical, surveying, optical, mathematical, and philosophical instruments.

Instruments bearing his name include: octants; compasses. He was succeeded in business in 1790 by the mathematical instrument maker Richard Rowland.

Edison, Thomas Alva
Thomas Alva Edison, (1847-1931) [37] [155] [335], was a prolific inventor, born in Ohio, USA, and famous for his inventions in the fields of telegraphy, power generation, sound recording, and motion pictures. He had little formal education, and began at an early age selling newspapers and candy on railways. During this time, he studied telegraphy, and was soon working as a telegraphist. He began to experiment and develop improvements to the equipment, and this led to a job in the Western Union Telegraph Company where he was paid handsomely for the rights to his duplex telegraph. He set up his own research laboratory and ultimately, during his career, he was awarded 1180 patents. His innovations included the phonograph, and the commercialisation, together with British inventor **Joseph Swan**, of the electric light-bulb. He pioneered commercial electric power generation and distribution, although his system used direct current, and was superseded by the use of alternating current as developed by Nicola Tesla and George Westinghouse. His innovations were also in the fields of ore extraction, cement manufacture, and electric car batteries.

In 1888 Edison patented the Kinetoscope, co-designed by his assistant **William Dickson**, for viewing moving pictures recorded using their Kinetograph motion picture camera. This used 35mm film strips, with two rows of perforations, made to Edison's specifications by the **Eastman Kodak Company**. In 1895 the Panoptikon projector from the **Latham brothers** began to offer competition, controversially developed with the help of William Dickson. Dickson left Edison's employment as a result of this perceived disloyalty, and in 1896 Edison announced the Edison Vitascope, invented by Thomas Armat, in response to this competition.

Edixa
See **Wirgin**.

Edmund Optics
Edmund Optics is a US optical components manufacturer, established by Norman Edmund in 1943 as Edmund Scientific. In 1982, the company split into two divisions: Edmund Scientific and Edmund Industrial Optics. Their products have included imaging lenses, microscopes, cameras, telescopes and telescope eyepieces.

Edu Science
Edu Science is a company established in 1987 by Chinese founder Mr. David Choi. They are makers of telescopes, microscopes and other items intended for use in science education.

Eglington, John
John Eglington, sometimes known as Eglinton or Egglington, (fl.1764-1786) [18], was an optical instrument maker of Hatton Garden, London, England. Addresses: Great Kirby St., Hatton Garden (1772); Cross St., Hatton Garden (1785). He was the son of a hair merchant, also named John Eglington. He served his apprenticeship with **Peter Eglington**, beginning in 1736, and was freed of the **Spectacle Makers' Company** in 1745. In 1764 he was one of the petitioners who attempted to revoke the patent

obtained by **John Dollond (I)** for the achromatic lens, and which was enforced after Dollond's death by his son, Peter (See Appendix X).

Eglington, Peter

Peter Eglington, sometimes known as Eglinton, Eaglinton, or Egglington, (fl.1736-1772) [**18**], was an optical instrument maker of Strand, London, England. He was the son of a mason, James Eglington. He served his apprenticeship with **John Day**, beginning in 1722, and was freed of the **Spectacle Makers' Company** in 1731. Among his apprentices were his son, also named Peter Eglington, and **John Eglington**. He served as Master of the Spectacle Makers' Company from 1761-1762. In 1764 he was one of the petitioners who attempted to revoke the patent obtained by **John Dollond (I)** for the achromatic lens, and which was enforced after Dollond's death by his son, Peter (See Appendix X).

Ehrenreich Photo Optical Industries

Joseph Ehrenreich was the owner and CEO of Ehrenreich Photo Optical Industries in the USA. In 1954 he became the exclusive US importer and distributor of **Nikon** products from the **Nippon Kōgaku Kōgyō Kabushikigaisha** company in Japan. Ehrenreich played a major role in the growth of Nikon in the US. In 1975 Ehrenreich purchased **Unitron**. In 1981, after the death of Joseph Ehrenreich, Ehrenreich Photo Optical Industries was purchased by Nippon Kōgaku Kōgyō Kabushikigaisha, and renamed Nikon Inc.

Eichens, Freidrich Wilhelm

F.W. Eichens, (1818-1884) [**8**] [**24**], was an eminent precision mechanical engineer, born in Berlin, Germany. He naturalised as a French citizen, and worked as head of the **Lerebours & Secretan** workshop, and as director of the company, until 1866, when he left to start his own company. At the same time **Paul Gautier** left the firm of Lerebours & Secretan to work with Eichens. Eichens went on to produce fine instruments for many observatories. In 1868 Gautier and Eichens built a large siderostat to a design by **Foucault**, who died that year. The following year, together with **A. Martin** and Gautier, Eichens built a 47-inch Newtonian for the Paris observatory. It was on a Secrétan mount, and stood unused for many years because the mount was insufficiently stable. In 1880 Eichens retired, and Paul Gautier bought his business.

Eikow Optics

Eikow Optics was a Japanese maker of binoculars, telescopes and microscopes. In 1990 they merged with **Hino Kinzoku Sangyo Co.** to become **Mizar Co. Ltd**.

The maker's mark used by Eikow is illustrated in Appendix IV.

Electrotachyscope

The Electrotachyscope was the invention of German inventor and photographer Ottomar Anschütz (1846-1907) [**335**]. It consists of a vertical spinning glass disk upon which a series of photographic transparencies are placed in sequence around the edge. It is illuminated from the back in a manner that shows only one image at any one time. Thus, when the spinning disk is viewed from the front, the image is presented with the illusion of motion. The backlit image is not enlarged by projection. Hence it can be viewed by only a small number of people at any one time.

Elliott & Sons

(Note: This company is not related to **William Elliott & Sons**)

Elliott & Sons, [**44**] [**61**], were photographic paper and plate makers of Talbot House, Park Road, Barnet, Hertfordshire, England. The company was formed in around 1890 when Joseph John Elliott (1835-1903) purchased the printing works of the photographers Elliott & Fry, and renamed it Elliott & Son. In 1901 the company was incorporated as Elliott & Sons Ltd.

Elliott & Sons entered an optical, scientific and photographic exhibit in the 1929 British Industries Fair. Their catalogue entry, [**448**], listed them as manufacturers of: Photographic sensitised materials, Plates, Papers and Films under the Trade Mark *Barnet*.

In 1945, together with **Ensign Ltd.**, they formed the company **Barnet Ensign**. In 1947 Barnet Ensign exhibited at the British Industries Fair. They were listed in the catalogue, [**450**], as manufacturers of: *Ensign* cameras; Roll films; Cine and photographic apparatus; Photographic plates; Flat films and papers with the *Barnet* trade-mark.

In 1948 Barnet Ensign merged with **Ross Ltd** to form **Barnet Ensign Ross**, which was renamed **Ross-Ensign Ltd**. in 1954.

Elliott & Sons, William

(Note: This company is not related to **Elliott & Sons**)

William Elliott & Sons, [**17**] [**18**] [**319**] [**435**], drawing instrument makers, and optical, mathematical, and philosophical instrument makers, of 56 Strand, London, England, was a partnership between **William Elliott** and his sons **Charles Alfred Elliott** and **Frederick Henry Elliott**. The partnership was formed in 1850, and terminated upon the death of William Elliott in 1853. Charles and Frederick went on to trade as **Elliott Brothers**.

Elliott Brothers

Elliot Brothers, [17] [18] [44] [46] [319] [361] [435] [444], of London, England, was a partnership between **Frederick Henry Elliott** and his brother **Charles Alfred Elliott**, the sons of **William Elliott**. Addresses: 56 Strand (1854-1857); 5 Charing Cross (1856-1857); 30 Strand (1858-1863); 449 Strand (1874-1872); 112 St. Martin's Lane (1864-1872, 1877-1879); 101 St. Martin's Lane (1873-1877, 1880-1900); 102 St. Martin's Lane (1873-1877, 1880-1899). In 1882 the premises at 101 and 102 St. Martin's Lane is listed as a manufacturing facility.

The brothers had been trading in partnership with their father since 1850 as **William Elliott & Sons**. Upon William's death in 1853, the brothers traded briefly as **F.H.&C. Elliott**, but by 1854 they had changed the name of the business to Elliott Brothers.

In 1856 Elliott Brothers took over the business of **Watkins and Hill**. Charles left the partnership in 1870 [35 #23637 p3537], but the firm continued to use the name Elliott Brothers. The firm's customers included **J.C. Maxwell**, **Sir Charles Wheatstone**, J.W. Strutt, and other eminent scientists.

In 1873 Frederick died, and his wife Susan brought in Willoughby Smith, a leading telegraph engineer, as a partner. At about this time **Robert Paul**, the scientific instrument maker and innovative film-maker, learned his instrument making skills working for Elliott Brothers. In 1880 Susan Elliott died, and control of the company passed to Willoughby Smith. This was the end of the Elliott family connection with the business.

In 1900 the manufacturing facility relocated to Century Works, Connington Road, Lewisham, Kent. At this time their products included telegraphy, electrical, engineering, surveying, drawing, meteorological, marine, and other instruments.

In 1902 Elliott Brothers received a Royal Warrant as *Elliott Brothers of London - Opticians*. They continued to expand in the areas of military and electrical products, including analogue computers for naval (gun) fire control. In 1917 the company name was changed to Elliott Brothers (London) Ltd. In 1946 Elliott Brothers were allocated space in the Shorts' factory at Rochester, Kent. During the early 1950s they began to develop digital computers. In 1950 the company Elliott Automation was formed, with Elliott Brothers as a subsidiary. In 1967 Elliott Automation was taken over by the English Electric Co., which was in turn taken over by the General Electric Company in 1968.

Elliott, Charles Alfred

Charles Alfred Elliott, (1822-1877) [18] [46] [435], was a mathematical instrument maker of London, England. Addresses: 21 Great Newport St., St. Martin in the Fields (1823); High Holborn (1837). He served an apprenticeship as mathematical instrument maker with his father, **William Elliott** of the **Coachmakers' Company**, beginning in 1837. He became a partner in his father's company, **William Elliott & Sons**, in 1850. Following William's death in 1853 the brothers traded briefly as **F.H.&C. Elliott**, then changed the name of the business to **Elliott Brothers**.

Elliott, Edwin

Edwin Elliott, (b.1878, ret.1966) [433], started his own business in 1909 as an agent for a manufacturer of spectacle lenses. In 1911 he bought out the manufacturer of spectacle lenses, and developed the company into the **British Optical Lens Co.** In the 1920s the company began to produce Bakelite parts as well as continuing the optical business, and he combined the two disciplines to produce cameras, as well as making lenses for other camera-makers. In 1966 Elliott retired, and the company continued in business until receivers were appointed in 1982. The company was taken over by **Reed International Group** in 1983.

Elliott, E.J.

E.J. Elliott, (fl.C19th), of London, England, was a retailer of microscopes, and may have been a maker. Instruments bearing his name include: Culpeper style microscope in brass, with wooden base.

Elliott, F.H.&C.

F.H.&C. Elliott, (fl.c1853-c1854) [435], of 56 Strand, London, England, was a partnership between **Frederick Henry Elliott** and his brother **Charles Alfred Elliott**. The brothers traded under this name briefly following the death of their father, **William Elliott** in 1853, and by 1854 they had changed the name to **Elliott Brothers**. A rectangular protractor, dated c1853/54, [270], bears the signature *F.H.&C. Elliott*.

Elliott, Frederick Henry

Frederick Henry Elliott, (1819-1873) [18] [46] [435], was an optician and mathematical instrument maker of 30 King St., Holborn, London, England. He practised as a surveying engineer before joining his father, **William Elliott** and brother, **Charles Alfred Elliott**, in business in 1850, trading as **William Elliott & Sons**. He became a fellow of the Royal Society of Arts in 1850 and a fellow of the Royal Astronomical Society in 1859. He was innovative, and obtained many patents. Upon William's death in 1853 the brothers traded briefly as **F.H.&C. Elliott**, then changed the name of the business to **Elliott Brothers**.

Elliott, Thomas
Thomas Elliott, (fl.1809-1840) [18] [319] [379], was an optical and mathematical instrument maker, and drawing instrument maker of London, England. Addresses: 3 Albermarle St., Clerkenwell; 39 Crown St., Finsbury; 21 Tabernacle Walk, Finsbury; 20 Pitfield St., Old St.; 3 Albion Place, Clerkenwell; 20 Haberdashers Walk, Hoxton (c1830-1840).

He advertised drawing instruments, and is known to have sold: microscope; barometers.

Elliott, William
William Elliott, (1780/81-1853) [18] [46] [319] [435], was an optician, drawing instrument maker, and optical, mathematical, and philosophical instrument maker, of London, England. Addresses: 26 Wilderness Row, Goswell St. (1817); 21 Great Newport Street, St. Martin's Lane (1817-1827); 227 High Holborn (1830-1833); 268 High Holborn (1835-1849). (Note: It is not certain that the 26 Wilderness Row address is for the same William Elliott).

His father, also named William Elliott, was a Yeoman of London. He began his apprenticeship with William Backwell, a mathematical instrument maker of Gray's Inn, London, in 1795 and, although he was trained with Backwell, he was immediately rebound to Thomas Collingridge for the purpose of guild membership. He was freed of the **Coachmakers' Company** in 1804. This was probably the year when he started his own business. On completion of his apprenticeship, he bound his own first apprentice. He trained at least nine apprentices, among whom was one of his sons, **Charles Alfred Elliott**.

In 1850, under a deed of partnership, his sons **Frederick Henry Elliott** and Charles Alfred Elliott were taken into their father's business which then carried on as **William Elliott & Sons** at 56 Strand, London. In 1853, when William Elliott died, his sons went on to trade as **Elliott Brothers**.

Ellis Aquatic Microscope
The aquatic microscope was designed by the naturalist Abraham Trembley (1710-1784) some time before 1744, and promoted by the naturalist John Ellis, thus becoming known as the Ellis aquatic microscope. An improvement on this design is Raspail's simple chemical microscope.

Emden, Abraham van
Abraham van Emden, (1794-1860) [322] [362] [374] [377] [378], was a physical, mathematical, and optical instrument maker of Amsterdam, Netherlands. He was the grandson of **Hartog van Laun**, and succeeded his uncles, Abraham (II) and Jacob van Laun, in business. Following the death of Abraham in 1860 the firm continued, initially run by his wife, Sarah and, following her death, by former employee Willem Boosman. The firm continued in business until 1881.

He is known to have made: microscopes; magic lanterns; dry card compass; boxed set of drawing instruments; thermometer; air pumps.

Emmerich Optical Company, John H.
The John H. Emmerich Optical Company, [136], of New York City, NY, USA, was a manufacturer of lenses and prisms, acquired by Square D in 1940. They were subsequently merged with another Square D acquisition, **Kollsman Instrument Company, Inc.** for production of binoculars during WWII under the **Sard** tradename.

Empire Made
Following the introduction of the Safeguarding of Industries Act in the UK in 1921, the UK import duty on optical glass and optical instruments was increased in 1929 to 50% [444]. An exception was made for articles imported from countries that were part of the British Empire if at least 25% of their value was due to work carried out within the Empire. Consequently, some companies imported the components for their products and assembled them within the Empire.

Empire Made was a marking found on such goods. Examples include binoculars by makers such as **Regent**. During the period these were manufactured, there was a strong demand for Japanese made optics, and the inference is that these *Empire Made* binoculars were likely assembled in Hong Kong, from Japanese parts, and intended for sale in the UK.

Enbeeco
Enbeeco was a trade name used by **Newbold and Bulford**.

Endacott Scientific Instrument Co.
After WWI, **Sherwood & Co.** was renamed to become the Endacott Scientific Instrument Co., [29]. In 1919 the company was listed as a member of the **British Optical Instrument Manufacturers' Association**.

Endoscope
An endoscope is a medical instrument used for internal examination of organs such as the bowel, bladder, stomach, or lung. An endoscope may use a relay lens to extend its reach.

Engelbert, Louis
Louis Engelbert, (1814-1877) [47], was a microscope maker, and an employee of **Carl Kellner**. Upon Kellner's death in 1855, he ran the **Optisches Institut** until 1856 when it was taken over by another employee, **Friedrich**

Belthle. He worked as an independent microscope maker until 1861, when he began to work in partnership with **Moritz Hensoldt**, making microscopes under the name **Hensoldt & Engelbert**. This partnership lasted until his death in 1877.

Engineer Chevallier
Engineer Chevallier (l'Ingénieur Chevallier) was the name under which **Jean Gabriel Augustin Chevallier** traded.

Engiscope
An engiscope is a reflecting microscope designed so that the image is viewed through an aperture in the side of the instrument, in a similar manner to the Newtonian telescope.

Enhelion Ltd.
Enhelion Ltd. of Burwash Manor Farm, New Road, Barton, Cambridge, England, was a company, previously named Mikron Ltd., formed by Keith Dunning and Richard Dickinson, (see **Science of Cambridge Ltd.**, **Cambridge Optronics Ltd.**) Enhelion produced the *Micron* and *Trekker*, which were folded optics microscopes using a design that was derived from the **Lensman**. The *Micron*, (magnification x80 and x160), was branded as an Enhelion product, while the *Trekker*, (magnification x35), was marketed under the company's *Looksmall* brand-name. The eyepiece of the *Trekker* could be detached and used as a x10 loupe magnifier. Unlike the other *Lensman* variants, the *Trekker* did not have the fold-up arm for illumination, but had internal LED side-illumination for use when viewing solid objects.

Enhelion also produced the *Readview*, (magnification x80 and x160), a folded optics microscope similar to the *Micron* but with LED illumination in the fold-up arm, which was packaged and marketed by **Meade**.

These Enhelion microscopes were manufactured for Enhelion in China.

Ensign Ltd.
Ensign, [306], was a brand-name used by **Houghton-Butcher**. In 1930 Houghton-Butcher set up a sales subsidiary Ensign Ltd., but in 1940 the Ensign headquarters was destroyed in enemy action. The assets were sold, but Houghton-Butcher kept the name, and used it as a brand-name for their *Ensign* cameras. In 1945 the company, together with **Elliott and Sons**, who made the *Barnet* brand of film, formed the company **Barnet Ensign**. In 1948 Barnet Ensign merged with **Ross Ltd** to form **Barnet Ensign Ross**, which was renamed **Ross-Ensign Ltd.** in 1954.

Entrance Pupil
The entrance pupil, [7], of an optical system is the image of the system's aperture stop, formed at a point on the object viewed on the optical axis of the system through any optical elements that precede the aperture stop, (e.g. eyepiece).

In the example of a pair of binoculars whose aperture stop is the rim of the objective lens, the entrance pupil is the objective lens itself.

See also exit pupil.

Epidiascope
An epidiascope is a projector, similar to an episcope, that is capable of projecting images of transparent objects as well as opaque objects.

Episcope
An episcope is a projector, designed to project an image of an opaque object onto a screen, using reflected light. It can be used to project an image of a flat item, such as a drawing or sketch, postcard, or a 3-D item such as an insect or leaf.

Épry, Charles
Charles Épry, [12] [72], was an optical instrument maker of Paris, France, who took over the **Lerebours & Secretan** business in 1911. In 1913 Épry associated with Gustave Jacquelin (1879-1939), and together they operated their business as the *sole successors to Lerebours & Secretan*. (Note that the é in Secrétan's name is unaccented on their advertisements and cards). In 1934 the firm of **Georges Prin** became a part of the Secrétan business.

Equatorial Mount
An equatorial telescope mount is one that, by rotating on a suitably orientated axis, enables the telescope to follow an observed star's apparent motion caused by the rotation of the Earth. An equatorial mount has two axes of rotation. One is aligned to be parallel with the axis of Earth's rotation, (i.e. points to the north or south celestial pole, which is the centre of the apparent rotation of the sky), and is called the *right ascension*, *polar*, or *equatorial* axis. The second axis is at right angles to it and is called the *declination* axis. Rotation about the right ascension axis, in order to follow a star's apparent motion, may be achieved by means of a motor drive. This requires rotation at a rate of 360° per sidereal day (23 hours and 56 minutes). Such a motor drive is referred to as a *sidereal drive*, a *clock drive*, or a *right ascension drive*. Early clock drives were driven by clockwork.

The alignment of the mount's right ascension axis with the Earth's rotation axis must be accurate. When an equatorial mount is situated in an observatory this alignment is generally carried out once, and only repeated on the rare occasions when it becomes necessary. For a portable mount, however, the alignment must be carried out each time the mount is set up. To facilitate this, some mounts are equipped with an integral polar scope.

Early Equatorial Mounts: Equatorial mounts date back at least to the C18th, with an early example made by the French clockmaker Philippe Vayringe (1684-1746) [2 pp.26-27]. Some early equatorial mounts were referred to as a *portable observatory*. In the base of the portable observatory is a latitude setting arc, above which is a graduated setting circle for the right ascension axis and a graduated setting circle for the declination axis. This type of mount was made by instrument makers such as **Jeremiah Sisson**, **George Adams (II)** and **Jesse Ramsden**.

German Mount: A German Equatorial mount, invented by **Joseph von Fraunhofer**, is attached to a tripod or other suitable portable stand, or a fixed pier. The tilt of the right ascension axis is adjusted so that it is parallel to the Earth's rotation axis. The declination axis is mounted at one end of the right ascension axis. The telescope is mounted at one end of the declination axis, and a counterweight is placed at the other end to maintain balance and reduce stresses on the mount. The need for a counterweight effectively reduces the payload capacity of the mount.

English Mount: The right ascension axis of the English Equatorial mount is supported at each end in a bearing mounted on a pillar, the two mount points being aligned on a line parallel to the Earth's rotation axis. The difference in height between the two pillars is determined by the latitude of the observatory. The telescope is mounted on a pivot centrally on the right ascension axis, and thus can turn in declination. See also parallactic ladder mount.

The English Equatorial mount was first proposed by **Henry Hindley** in 1741, and the design was brought to fruition by **Jonathan Sisson** soon afterwards. The first all-metal English Equatorial mount was constructed in 1859 for a large **Cooke** refractor owned by Isaac Fletcher of Carlisle, England [96].

Fork Mount: The latitude tilt of the right ascension axis of a fork mount is adjusted by appropriate means such as an adjustable wedge. The end of the right ascension axis terminates in a two-pronged fork. The declination axis bearings are mounted at the ends of the fork prongs, and the telescope is supported between them within the fork. If the telescope is longer than the fork prongs, this limits how close to the celestial pole the telescope can view. The payload capacity of a fork mount is limited by the stresses placed on the right ascension axis bearings by the weight of the telescope. One solution for this problem is offered by the horseshoe mount.

Horseshoe Mount: A horseshoe mount is similar in concept to a fork mount, but has in addition a horseshoe-shaped support at the end of the forks. This rests in bearings to accommodate the rotation in right ascension, and provides support for heavy payloads. The **Hale Telescope [97]** is famously situated in a horseshoe mount.

Erecting Prism
See prism.

Erfle, Heinrich Valentin
Heinrich Valentin Erfle, (1884-1923) [47], was a German optician. He studied at the Technical University of Munich, and graduated with a doctorate in 1907. He worked in the workshops of **Steinheil & Söhne** until 1909 when he began to work for **Carl Zeiss**. He took over the management of Zeiss in 1918, where his innovations led to advances in optical design, and publications about his work.

Heinrich Erfle invented and patented the eponymous Erfle eyepiece (see Appendix II) while he was working for Zeiss.

Ernemann
The Ernemann camera company, [122], was based in Dresden, Germany. It began as the Dresdner Photographische Apparate-Fabrik Ernemann & Matthias, founded in 1889 by Johann Heinrich Ernemann (1850-1928) and his partner Wilhelm Franz Matthias. Matthias left the company in 1899, after which Ernemann renamed it Heinrich Ernemann, Aktiengesellschaft für Cameraproduktion, producing cameras and movie projectors. In 1926 Ernemann merged with **Contessa-Nettel**, **Goerz** and **ICA** to form the company **Zeiss Ikon**.

Ertel & Sohn
Traugott Leberecht von Ertel, (1778-1858) [47] [164] [374], was a German instrument maker. After serving an apprenticeship as a blacksmith he went to Vienna where he worked as a surgical instrument maker. In 1806 he went to Munich and began to work for the **Mathematical-Mechanical Institute Reichenbach, Utzschneider and Liebherr**. There he worked with **Reichenbach** making optical instruments and, in 1815, a year after Riechenbach left the Institute, Ertel also left to work with Reichenbach's new enterprise, which was then named **Mathematisch-mechanisches Institut von Reichenbach & Ertel**.

In 1821 Reichenbach, with many other commitments on his time, handed over the running of the company to

Ertel who, in 1834, with his son Georg Ertel (1813-1863), re-named it Ertel & Sohn, adding the subsidiary name Reichenbach'schen Mathematisch-Mechanischen Institute in honour of the company's founder. Ertel's second son, Gustav Ertel (1829-1875) also later joined the company. The company continued to grow, producing fine optical instruments, and specialising in surveying instruments. In 1858, when Traugott Ertel died, the company continued in the hands of Georg until he died in 1863. After this the company was run by Gustav, and then by Gustav's son Georg.

In 1870 August Diez (d.1920), a renowned instrument maker, joined the company, and by 1876 he was the sole director. In 1890 he purchased the company, and broadened the kinds of instruments they made, as well as moving from the construction of individual instruments toward mass production. In 1911, following a period of rapid expansion, the company was incorporated as T. Ertel & Sohn G.m.b.H., Mathematisch-Mechanisches Institut für Geodätische und Militärwissenschaftliche Instrumente. The company was sold in 1916 and, after supplying the military during WWI, the company lost its way until Walter Preyss became director in 1921. The company was renamed Ertel-Werke A.G. für Feinmechanik, and the focus shifted over the next years to fine quality geodetic/surveying instruments. In 1928 Walter Preyss took sole ownership of the company and renamed it Ertel-Werk für Feinmechanik. After supplying military instruments during WWII, the company prospered under the management of Walter Preyss' son Carl Preyss until 1997 when he left the company, and the company ceased to use the name Ertel.

Ertel, Traugott Leberecht von
See **Ertel & Sohn**.

Établissements Lacour Berthiot S.A.
Établissements Lacour Berthiot S.A., [**192**], was a French optical instrument maker. It began as the business of **Claude Berthiot**, and became **Lacour-Berthiot** when Berthiot's nephew Eugène Lacour joined his uncle's firm. Upon incorporation as Établissements Lacour Berthiot S.A., Eugène Lacour was the general director, and Charles Henri Florian was the technical director. In 1913 the company became **Société D'Optique et de Mécanique de Haute Precision (S.O.M)**.

Etalon
In its simplest form, an etalon, or Fabry-Perot interferometer, [**7**], consists of two plane, parallel, highly reflecting (partially transparent) surfaces separated by some distance d. If the gap is mechanically varied, by moving one of the mirrors it is usually referred to as an interferometer. If the mirrors are fixed it is referred to as an etalon, although it is still, of course, an interferometer. The mirrored surfaces may be the two sides of a single plate, or two separate plates. If two separate plates are used, the non-mirrored sides are often made to have a slight wedge shape (a few minutes of arc) to reduce interference resulting from reflections from these sides. The light that enters by way of a partially mirrored surface is reflected multiple times within the gap, and transmitted through the other partially mirrored surface to a focusing lens. Any one ray of light, being coherent with itself, is thus able to interfere with its multiply reflected self.

The interference is constructive when the etalon gap is a multiple of the wavelength of the incident light. In this case the transmission of light is reinforced by the interference. For other wavelengths of light, the interference is destructive, the extent depending on how out of phase the light is after reflection. The result is that the etalon functions as a narrow bandpass filter whose peak transmission is precisely determined by the etalon gap.

An etalon is sometimes used for solar viewing. The sun emits a broad spectrum of light, but certain features may be observed better using a narrow bandpass filter. For example, hydrogen alpha, with a wavelength of 656.28nm, is a frequency of light emitted when an electron moves between the third and second orbits within a hydrogen atom. A narrow bandpass filter at that wavelength enables viewing of surface features of the sun, such as prominences and filaments, that would otherwise be overpowered by other frequencies of light.

When used as a filter in solar viewing it is generally referred to as an etalon even though the gap is tuneable. The gap, d, is tuned to give a narrow bandpass of typically less than ~ 0.1nm (1 Ångstrom) width at the optimum viewing wavelength. The spacing of solar filter etalon plates generally needs to be changed in order to "tune" the etalon, which shifts the etalon peak to account for environmental conditions, or allows better viewing of off-band phenomena. The distance, d, for hydrogen alpha must be a multiple of 656.28nm.

Eumig
Eumig was an Austrian company, founded in 1919 in Vienna, who produced audio and video equipment. They began producing movie cameras and projectors in the 1930s. In 1969 they acquired the **Paillard Bolex** division of Paillard, which they renamed in 1970 to become Bolex International SA. Eumig was declared bankrupt in 1982.

Evans, John Johnson
John Johnson Evans, (fl.1792-1809) [**18**] [**276** #1118] [**379**], was an optical instrument maker of 87 and

88 Bishopsgate Within, London, England. He served his apprenticeship with **William Archer**, beginning in 1783, and was turned over to **William Price** in 1786. He was freed of the **Stationers' Company** in 1792. He is known to have sold: barometer; hygrometer.

Evans, Thomas J.

Thomas J. Evans, of London, England, was a maker of fine optical instruments in the late-19th century. Inferior modern reproductions from India are frequently sold as the genuine article.

Everest Theodolite

See theodolite.

Evershed Spectroscope

See spectroscope.

Exakta

Exakta is a brand-name that was used by the **Ihagee** camera company. The first camera model of this brand used 127 film, and subsequent models used 35mm film. The Exakta brand was subsequently used by **Pentacon**.

Exit Pupil

The exit pupil, [7], of an optical system, also known as the Ramsden disc, is the diameter of the image delivered to the eye. It is the image of the aperture stop, as seen from a point on the image plane which is on the optical axis of the system, viewed through any intervening optical elements.

In the example of a pair of binoculars whose aperture stop is the rim of the objective lens, the exit pupil is calculated as the diameter of the objective lens divided by the magnification. Hence, for a pair of 10x50 binoculars the exit pupil would be 5mm.

The pupil of the human eye in a healthy young adult, when dilated, is normally around 7mm, but this reduces with age. An older person will typically have a dilated pupil diameter of nearer 5mm. If the exit pupil of the optical instrument is larger than the pupil of the observer's eye, a loss of image brightness will be experienced. See also entrance pupil, image brightness.

Explore Scientific

Explore Scientific is a **Bresser** company. The company was founded by Scott Roberts, a former Meade VP, in Laguna Hills, California, USA, in 2008, using **Jinghua Optical & Electronics Co. Ltd.** as their exclusive manufacturer. Their brands include *Explore One*, **National Geographic** and *Discovery*. In 2009 Explore Scientific became the exclusive distributor of Bresser products throughout the Americas, and expanded their range to include Bresser's other products. Explore Scientific became a wholly owned subsidiary of Bresser.

Extinction Wedge

The extinction wedge, [24 p296], devised in about 1881 by **Charles Pritchard**, comprises a thin prism of neutral glass cemented to a wedge of white glass, resulting in a plate of uniform thickness. This is placed in the eyepiece of a telescope, and can slide in a groove. Thus, the part of the wedge through which the star is being observed is adjusted until the star disappears. The position of the wedge, as determined by a calibrated scale, then gives a measure of the brightness of the star.

Eyepiece

See Appendix II.

Eyepiece Micrometer

See micrometer.

Eye Relief

Eye relief in an optical instrument is the maximum distance between the surface of the eyepiece lens and the eye within which the user can view the full viewing angle offered by the instrument. Thus, eyepieces with short eye relief require the user's eye to be close to the lens, which is usually impractical for those who wear spectacles. Examples of the eye relief for various designs of eyepiece are shown in the *Eyepiece Designs* section of Appendix II.

F

Fabry, Charles
Charles Fabry, (1867-1945) [7] [45], was a French physicist who began his career at Marseilles University researching interference of light. He became well known in the fields of optics and spectroscopy, and in 1896 co-invented the Fabry-Perot interferometer with **Jean-Baptiste Alfred Perot**. In collaboration with Henri Buisson, he discovered the ozone layer in the atmosphere, and its effect in filtering ultra-violet light. He was the first director of the **Institut d'Optique**, and he was appointed professor of physics at the Sorbonne in 1921.

Fabry-Perot Interferometer
See interferometer.

Factory of Optical and Precision Devices H. Kolberg & Co.
The company H. Kolberg & Co., Warsaw, Poland, [115], was set up by **Henryk Kolberg** and other investors in 1921. They set up a production line making binoculars, using optical glass from the **Parra Mantois** glass-works in France, and from the **Schott** glass-works in Germany. In 1923 they began to produce military 6x30 binoculars. Production of military 7x50 binoculars proved uneconomical, but they went on to produce an 8x40 model, as well as loupe magnifiers and microscopes for the education and health industries, and rangefinders, cameras, and sights for anti-aircraft installations.

In 1930 Kolberg sold his shares to **Optique et Précision de Levallois (OPL)**, **Krauss**, and **Barbier Bénard et Turenne (BBT)**. This, together with input from other investors, resulted in the company being re-formed as Polskie Zakłady Optyczne, or **Polish Optical Industries (PZO)**.

Following the sale of his shares, Henryk Kolberg formed a new company, reinstating the name of H. Kolberg & Co., manufacturing and selling binoculars in competition with PZO. They continued in business, supplying the German army during the Nazi occupation, and the company finally closed its doors in 1944.

Fage, William
William Fage, (fl.1790-1855) [18] [276 #2123], was a mathematical, optical, and philosophical instrument maker, and hydrometer and saccharometer maker, of London, England. Addresses: 5 King's Row, Walworth (1834-1835); 10 Great Dover St., Borough (1836-1837); 3 Great Dover St., Borough (1838-1840); 62 High St., Borough (1841-1842); 7 Friar St., Blackfriars Rd. (1855).

Beginning in 1790 he worked in partnership with the glassworker and hydrometer maker, John Dring, trading as **Dring & Fage**. He left the partnership in 1834 [35 #19123 p167], and began to work under his own name.

Fairbone, Charles (I)
Charles Fairbone (I), (fl.1753-1800) [18] [90], was a timber merchant, and mathematical and optical instrument maker of London, England. Addresses: Princes Court, Westminster (1765); St. Martin's Lane (1766); New St. (1773-1800); 20 Great New St., Fetter Lane (1780-1794); New St., Shoe Lane (1780-1801).

He was the son of Timothy Fairbone, a blacksmith of St. Margaret's, Westminster. He served his apprenticeship with **Tycho Wing**, beginning in 1753, and was freed of the **Grocers' Company** in 1765. Among his apprentices were his sons **Charles Fairbone (II)**, Timothy and Henry, and his nephew Isaac Fairbone. Of these, only Charles (II) took the freedom of the Grocers' Company. Another of his apprentices was **George Huggins (I)**, grandson of **John Dollond (I)**, who later changed his name to George Dollond (I). George Huggins (I) began his apprenticeship in 1788, and was freed by Charles (II) in 1804, after the death of Charles (I). According to [256], the Fairbone family was related to the Dollond family.

Fairbone, Charles (II)
Charles Fairbone (II), (fl.1781-1812) [2] [18] [90] [276 #832], was a mathematical and optical instrument maker of 20 Great New Street, Shoe Lane, London, England. There were several Fairbones noted at this address, some of whom were timber merchants and some were mathematical instrument makers. Some records refer to a partnership between Charles Fairbone and **Jesse Ramsden** at this address in 1756, but it is not clear which member of the Fairbone family this may refer to. Charles (II) was the son of **Charles Fairbone (I)**, and served his apprenticeship with his father beginning in 1781. He was freed of the **Grocers' Company** in 1802. The son of Charles (II), John Fairbone, served an apprenticeship with **John Dollond (II)**. Charles (II) was the subject of bankruptcy proceedings that began in 1812 [35 #16579 p419], and concluded in 1813 [282].

Fairey, Joseph (I)
Joseph Fairey (I), (c1782-1871) [18] [264] [276 #964] [319] [322] [379], was an optical, mathematical, and philosophical instrument maker of London, England. Addresses: 15 Fair Street, Horsleydown (1803-1805); 150 Tooley St., Borough (1806 until after 1811); 20 Ratcliffe Highway, Upper East Smithfield (1811-1822); and from 1826-1839 at 8 Northumberland Place, Commercial Rd., and 4 Principal Entrance, London Dock.

The premises at 150 Tooley St. appear to have been used by the family business from c1790 until at least 1860.

Joseph Fairey (I) served his apprenticeship with **Samuel Browning (I)** beginning in 1796, and was freed of the **Grocers' Company** in 1803. He took on a number of apprentices, including his son, **Richard Fairey (I)**. A quadrant bearing his name, with a scale engraved by **Spencer, Browning and Rust**, is in the National Maritime Museum, Greenwich [16] [54].

According to his trade card, in the box lid of a compound microscope, "J. Fairey. Real manufacturer of Mathematical and Optical instruments. No.8 Northumberland Place, Commercial Road, near Cannon Street Road Turnpike, and 150 Tooley Street, Boro. Stationary. Sea charts, Navigation books &c. Wholesale and for exportation. Quadrants, compasses & telescopes cleaned & repaired with the greatest care."

In 1839 Joseph entered a partnership with his son, trading as **Joseph Fairey & Son** at 8 Northumberland Place, Commercial Rd.

Fairey, Joseph (II)

Joseph Fairey (II), (fl.1790-1827) [276 #964] [319] [322] [379], was a partner in the firm **Richard & Joseph Fairey**, trading at 150 Tooley St., Borough, London, England, from 1790-1827. Since **Joseph Fairey (I)** was about 8 years old in 1790, and did not begin his apprenticeship until 1796, it is doubtful that he is the same person as Joseph Fairey (II) although, given the coincidence of addresses, a family connection is almost certain.

Fairey, Richard (I)

Richard Fairey (I), (b.c1815 fl.1839-1860) [18] [276 #964] [319] [322], was a mathematical instrument maker, and watch and chronometer maker, of 150 Tooley St., London, England. He served his apprenticeship with his father, **Joseph Fairey (I)**, beginning in 1829, and was freed of the **Grocers' Company** in 1836. In 1839 he entered a partnership, trading with his father as **Joseph Fairey & Son** until 1858. Following the termination of the partnership, Richard continued the business at the same address until at least 1860.

Fairey, Richard (II)

Richard Fairey (II), (fl.1790-1827) [276 #964] [319] [322] [379], was a partner in the firm **Richard & Joseph Fairey**, trading at 150 Tooley St., Borough, London, England, from 1790-1827. Given that **Richard Fairey (I)** was not born until c1815, he cannot be the Richard of this partnership although, given the coincidence of addresses, a family connection is almost certain.

Fairey, Richard & Joseph

Richard and Joseph Fairey, (fl.c1790-1827) [276 #964] [319] [322] [379], were optical, mathematical, and philosophical instrument makers of 150 Tooley St., Borough, London, England. The partnership between **Joseph Fairey (II)** and **Richard Fairey (II)** began in c1790 and continued until 1827.

Fairey & Son, Joseph

Joseph Fairey & Son, (fl.1839-1858) [18] [276 #964] [319] [322] [379], were optical, mathematical, and philosophical instrument makers of 8 Northumberland Place, Commercial Rd., London, England. The partnership between **Joseph Fairey (I)** and his son **Richard Fairey (I)** began in 1839 and continued until 1858.

Fallowfield, Jonathan

The firm of Jonathan Fallowfield, [61] [122] [306], was established in 1856 in Lower Marsh, Lambeth, London, England, by Jonathan Fallowfield (1835-1920) as a chemist and, soon after, began to supply photographic chemicals. In 1888 the company was acquired by F.W. Hindley, who subsequently began to sell *Fallowfield* branded cameras made by other camera manufacturers, including **Henry Park**. They are not known to have manufactured cameras themselves. In 1890 the company moved to 146 Charing Cross Road. The sales of cameras bearing their brand-name was declining by the time of WWI, and eventually ceased. In 1923 the company moved to Newman Street, London.

Fanmakers' Company

See Appendix IX: The London Guilds.

Fantascope

A fantascope is a large projector box, similar to a magic lantern, mounted on a frame with wheels, thus making it mobile. It was used to project onto the reverse side of a transparent screen, and the projectionist could manipulate the image by bringing the projector closer to, or further from the screen, thus varying the size of the projected image. It also had a variable shutter to enable the projectionist to make the image brighter or fainter. If equipped with two objectives – a double lantern – the use of cross-fades was possible. This kind of projector was used in phantasmagoria shows at the end of the C18th. The fantascope was patented by Etienne Gaspard Robert, known as Robertson. It is believed that some of the first fantascopes were made for Robertson by **Joseph Antoine Molteni**.

Farriers' Company
See Appendix IX: The London Guilds.

Fasoldt, Charles
Charles Fasoldt, (1819-1889) [201], was a German born watchmaker who emigrated to the USA in about 1848 or 1849, and settled in Rome, New York. He advertised that he made watches, clocks, chronometers, and small medical instruments as well as other instruments, and repaired jewellery. In 1861 he moved to Albany, NY, where he also built tower clocks and astronomical clocks. By 1878, having developed an interest in microscopy, he was advertising stage micrometers, ocular micrometers, and test rulings which he claimed were from 5000 to a million lines per inch. He also made and sold microscope lamps and nosepieces, and in 1888 he began to advertise his own "patented microscope". This had a built-in vertical illuminator, and a coarse and fine adjustment mechanism which was designed, he claimed, to prevent breaking objectives or objects by accidental movement of the microscope tube.

Fata Morgana
Fata Morgana, [94], was a brand-name used by Alfred Baumann of **Optische Werke Alfred Baumann & Co. KG** for their miniature binoculars, produced during the 1920s.

Fauth, Camill
See **Fauth & Co**.

Fauth & Co.
Fauth & Co., [8] [189] [208], of Washington DC, USA, was a maker of astronomical and surveying instruments. Camill Fauth, (1847-1925), from Karlsruhe, Germany, moved to Washington DC, USA, along with his brother-in-law **George Saegmuller**, on the invitation of **William Würdemann** to work in his workshop. He stayed there for about four years until 1874 when he started his own instrument making business together with George Saegmuller and another brother-in-law, Henry Lockwood. In 1874 they began to trade as Fauth & Co., making geodetic and surveying instruments, transit instruments, telescopes, and mounts for astronomical observatories. They built their own dividing engine, and made all their instruments in their own shop, with the exception of lenses and mirrors. Their lenses were supplied by **Alvan Clark & Sons**, **Bausch & Lomb**, and **Georg Merz**, and their mirrors by **Brashear**. Camill Fauth retired in 1887, and Saegmuller became owner and sole proprietor of the business. Saegmuller continued to use the name Fauth & Co. on his instruments until about 1892, when he changed the name to Geo. N. Saegmuller & Co., and moved to a new location. In 1905 his company merged with **Bausch & Lomb** in Rochester, New York, and became Bausch, Lomb, Saegmuller Co. In 1907 the name was changed back to Bausch & Lomb with Saegmuller as a subsidiary. In 1908 **Zeiss** formed a corporate agreement with Bausch & Lomb and, shortly after this, the Bausch & Lomb, Zeiss, and Saegmuller subsidiaries merged to form the Bausch & Lomb Optical Company, which became known as the **Triple Alliance**, and lasted until WWI.

Fayrer & Sons
See **James Fayrer**.

Fayrer, James
James Fayrer (I), (b.1760 fl.1798-1849) [18] [228] [276 #2125], was a mathematical, optical, and philosophical instrument maker, and clock maker, of White Lion St., Pentonville, London, England. Addresses: 66 White Lion St.; 35 White Lion St. (1805); 40 White Lion St. (1839). He trained as a clock maker, and married **Edward Troughton**'s niece, Nancy Suddard. James (I) and his son, the mathematical instrument maker James Fayrer (II), both worked for Edward Troughton. James (I) kept the second of the Troughton's dividing engines at his premises in Pentonville. Examples of his work include the drive clock for an equatorial telescope by Edward Troughton. By 1844 he was working in partnership with his sons, trading at 66 White Lion St. as **Fayrer & Sons** until 1849. He is known to have sold: zenith instrument; repeating circle; sextant; pantograph.

Featley, Robert (I)
Robert Featley (I), was an optical instrument maker of Fleet Street, London, England. In 1764 he was one of the petitioners who attempted to revoke the patent obtained by **John Dollond (I)** for the achromatic lens, and which was enforced after Dollond's death by his son, Peter (See Appendix X). He may be the same person as **Robert Featley (II)**.

Featley, Robert (II)
Robert Featley (II), (fl.1772-1783) [18], was an optician and optical instrument maker of Crown Court, Cow Lane, Smithfield, London, England (1772), and Crown & Cushion Court, Cow Lane, Smithfield (1783). He may be the same person as **Robert Featley (I)**.

Fecker, Gottlieb L.
Gottlieb L. Fecker, (c1857-1921) [50] [185], was born to a family of instrument makers in Karlsruhe, Germany. His father and grandfather, as well as his mother's family, were instrument makers. The Fecker family had a factory producing precision optical instruments. From 1883 to

1885 Gottlieb worked in partnership with **Emil Boecker**. After this he traded as Fecker & Company until 1887 when he emigrated to the USA.

He worked for **George Saegmuller** in Washington DC, but was accused of copying designs and sending them to **Warner and Swasey**. He was dismissed by Saegumller in 1895, after which he began to work for Warner and Swasey, where he is credited with the designs for many of their fine instruments. He also invented and patented an improvement to the design of prism binoculars, which were marketed as the patented Warner and Swasey binocular. In 1912 his son **James Walter Fecker** began working for Warner and Swasey, learning his optical instrument making skills from his father until 1921 when, upon the death of his father, he left to start his own company.

Fecker, James Walter
James Walter Fecker, (1891-1945) [24] [185] [209], American optical instrument maker, was the son of **Gottlieb L. Fecker**. In 1912, upon graduation, he began to work with his father at **Warner and Swasey**, learning his skills there until he left in 1921 to form his own company, J.W. Fecker Inc., manufacturing small optical instruments. In 1926, after the death of **John Brashear**'s son-in-law, J.B. McDowell, Fecker took over the Brashear business.

Some of Fecker's notable works include: the 69-inch mirror for the Ohio Wesleyan University; a 61-inch reflector for Harvard College Observatory; the 20-inch Ross lens for the Lick Observatory; a 15-inch refractor for the Cook Observatory at the University of Pennsylvania; many other instruments for observatory or other uses around the world. The company continued after Fecker's death, and in 1956 it became a subsidiary of the **American Optical Company**. After several more changes of ownership, in 2000, the company became Brashear LP, and subsequently became **L3 Brashear**.

Fecker Inc., J.W.
See **James Walter Fecker**.

FED
See **Kharkov Machine-Building Plant**.

Feil, Charles; Feil & Co.
Charles Feil, [8] [24] [156], was the grandson of **Henry Guinand**. He was an optical glass maker in Paris, France during the mid-to-late C19th. He learned his glass-making skills from Henry Guinand, and after his father's death he took over supervision of the glassworks in **Choisy-le-Roi**. Charles Feil named the company Feil & Co. The company made the optical glass used in many telescopes by **Alvan Clark and Sons** including, in 1882-1885, the 36-inch crown and flint blanks used for the James Lick telescope at the Lick Observatory in California, USA, and in 1895 the crown and flint blanks for the 40-inch Yerkes refractor at the Yerkes Observatory in Wisconsin, USA. In 1887 his firm passed to his son-in-law, **Edouard Mantois**, and was renamed **Parra-Mantois** in about 1900.

Feinmess Dresden GmbH
See **Gustav Heyde**.

Fennel, Otto
Otto Fennel, (1826-1891) [374] [382], was a surveying instrument maker of Kassel, Germany. He served an apprenticeship with **Breithaupt** from 1841 until 1848. He established his own business in Kassel in 1851 producing surveying instruments. The company name was subsequently changed to Otto Fennel Söhne KG. During WWII they marked their instruments with the **ordnance code crj**. The firm continued as a family business until 1968.

Ferrania
Ferrania, [122] [348] [351], of Milan, Italy, mainly produced film for cameras, however they did produce some cameras. The company began in 1917 in Milan, as FILM (Fabbrica Italiana Lamine Milano). In 1932, following a merger with SA Michele Cappelli, the company was renamed FILM – Fabbriche Riunite Prodotti Fotografici Cappelli e Ferrania. In 1935 they took over the camera maker **FIAMMA**, founded as a partnership by **Antonio Bencini**. In 1937 the company was renamed Ferrania, and in the same year they took over the camera-maker **Filma**, also founded by Bencini. Between 1947 and 1952 Ferrania collaborated with **Officine Galileo**, who manufactured their *Condor* range of cameras. In 1964, Ferrania was taken over by 3M.

Ferrotype
A ferrotype, [70], also known as a *tintype*, is a photographic image produced using the wet collodion process in which, instead of clear glass, the collodion is applied to a thin black-enamelled metal sheet. The result is a single positive photographic image which cannot be used for the production of multiple prints. The tintype was introduced in 1853 by Adolphe Alexandre Martin, a French teacher and amateur photographer.

Ferson, Fred B.
Fred B. Ferson, (1897-1969) [185], of Ohio, USA, moved to Mississippi in 1912. He became a keen amateur telescope maker, and was a major participant in the "Roof prism gang", a group of amateurs who produced optical

components, mainly roof prisms, for the WWII war effort. After WWII he formed Ferson Optics making amateur telescopes and other optics. He shifted production to military optics during the Korean war, and went on to produce military and aerospace optical components. After the Korean war he began to shift production back to commercial work. The company went on to produce movie camera lenses, and they were awarded a contract with **Bell & Howell** in Chicago for production of 35mm high precision movie camera lenses. They also produced large telescopes for the amateur market, spectrographs, photometers and cameras.

Féry Refractometer
See refractometer.

Féry Spectrograph
See spectrograph.

FIAMMA
Fabbrica Italiana Apparecchi Macchine Materiali Accessori (FIAMMA), [348], camera manufacturer of Florence, Italy, was founded as a partnership in 1920 by **Antonio Bencini**. By the early 1930s they employed over 100 staff, and the company was acquired by **Ferrania** in 1935.

Fidler, Robert
Robert Fidler, [18] [53] [256] [276 #1332] [322] [358] (fl.1805-1822), was an optician, and optical, nautical, mathematical, and philosophical instrument maker of 32 Wigmore St. Cavendish Square, London, England (1810), and 30 Foley St., Cavendish Square (1822). Instruments bearing his name include: sextant; barometer; thermometer; balances; orreries; Atwood machine. He is also known to have sold: cometarium.

Field & Co., R.
The business R. Field & Co. had its origins with **Robert Field (I)** in 1830, and became Robert Field & Son in 1845, producing optical instruments. Robert Field (II) finally sold the company, probably in the early 1870s, to the optician John Anderton (1843-1905), who continued the business as R. Field & Co. The business prospered under his tenure, expanding to produce lantern projectors and photographic equipment.

Field Curvature
Field curvature, [7], also known as Petzval curvature, is an optical aberration resulting in a curved image field, so that it is not possible to achieve perfect focus for the entire image. This is because a simple lens or curved mirror focuses a planar object onto a curved image plane that is symmetrical around the optical axis. This curvature in the image plane is known as Petzval curvature after the Hungarian mathematician Josef Max Petzval (1807-1891).

A positive lens or concave mirror produces *positive* field curvature, in which the centre of the image focusses closer to the eye than its edges. A negative lens or convex mirror produces *negative* field curvature, in which the centre of the image focusses further from the eye than its edges. Field curvature may be overcome by a suitable combination of positive and negative lenses.

A certain amount of field curvature can be accommodated by the eye and so, for visual instruments with a limited amount of field curvature, no correction is necessary. However, for photography, field curvature is more problematic, since photographic images are normally formed on a flat sensor or film. Photographic lenses generally consist of multiple lens elements, and correct for field curvature. When using an instrument such as a telescope, designed for visual use, a field flattener may be introduced into the optical path to correct the aberration for use in photography.

Field Flattener
A field flattener is a specially shaped optical lens or combination of lenses, so designed to correct field curvature. A field flattener is generally designed to match a specific objective or mirror configuration. In the case of a Schmidt-Cassegrain, for example, the field flattener is an integral optical element. In the case of a refracting telescope, it is usually detachable, sold separately as an accessory, and is mounted on or near the focusing tube for use in photography.

Field Lens
A field lens, [7] [22], in an optical instrument is a lens that serves to increase the field of view. It is a positive lens positioned at or near the image plane, which converges the image to avoid or reduce vignetting as the image reaches the next optical element in the instrument. As an example, many eyepiece designs include a field lens so that the diameter of the eyepiece does not have to be inconveniently large to achieve a satisfactory field of view. The field lens is also a useful element in the endoscope, which uses a relay lens to extend the optical reach of the instrument. At each relay a field lens can help to maintain a suitably wide field of view.

The field lens may consist of multiple optical elements to avoid introducing optical aberrations, and may correct for aberrations introduced by other optical elements. The field lens, being at or close to the image plane, must be clean and scratch free, since any marks on the lens will be in focus when the observer uses the instrument.

Field Stop

The field stop, [7], of an optical system is the element of the system which limits the angular size of the object which can form an image. It determines the field of view of the optical system. An increase in the field stop allows the optical system to form an image of those parts of the object which were previously beyond the edge of the image. Most microscope eyepieces, (see Appendix II), have an integral field stop.

See also aperture stop.

Field, John

John Field, (c1765-1830) [18] [274] [276 #833] [319] [361], was an optical, mathematical, and philosophical instrument maker of 74 Cornhill, London, England. He was the son of an apothecary, also named John Field. He served his apprenticeship with **Edward Nairne**, beginning in 1779, and was freed of the **Spectacle Makers' Company** in 1787. He succeeded the instrument maker Matthew Field in business at 74 Cornhill. He served as Master of the Spectacle Makers' Company from 1799-1801 and from 1820-1822. He worked at His Majesty's Mint from 1804 until 1830. His trade card, [53], shows illustrations of reflecting and refracting telescopes, an octant, globe, orrery, armillary sphere, and mathematical instruments. He is known to have sold microscopes.

Field, Robert

Robert Field (I), (c1787-1851) [3] [18] [50] [53] [319] [320], was an optical instrument maker of Birmingham, England. He worked as foreman for **Philip Carpenter** at his Bath Row factory. In 1830 he took over Carpenter's premises at 33 Navigation Street where he began to run his own business. In 1839 he took on additional premises at 111 New Street. From 1845 until his death in 1851 he traded in partnership with his son, Robert Field (II), as Robert Field & Son at 113 New Street. A trade card for Robert Field & Son states that they made: Magic lanterns and slides; Opera and field glasses; Achromatic telescopes; Astronomical telescopes; Combined barometer and storm glass.

In 1851, prior to the death of Robert (I), the firm exhibited microscopes and photographic equipment at the Great Exhibition in London, but did not receive acclaim for their instruments. Following Robert (I)'s death Robert (II) continued the business, and relocated it to Suffolk Street. He went on to win the prestigious Society of Arts' prize for best students' microscopes and for best school microscopes in 1855. Microscopes are recorded with serial numbers from 102 to 998. Robert (II) continued to run the business as Robert Field & Son, appearing in the directories until at least 1863. The firm does not appear in the directories in 1867. He sold the firm to the optician John Anderton, although it is not certain when the sale took place – but it was probably in the early 1870s. John Anderton continued the business as **R. Field & Co.**

Field, Robert & Son

Robert Field & Son was a partnership, beginning in 1845, between **Robert Field (I)** and his son, Robert Field (II).

Field, W.E.

W.E. Field, (fl.c1880-1911) [319] [320], of 17 Mount Road, Hastings, Kent, England, were microscope makers and retailers. Their price-list of 1890 lists their student model of microscope, improved histological microscope, as well as accessories and mounting materials, and microscopes by **Zeiss**.

The census record of 1911 shows two microscope makers resident at 17 Mount Road. These were William Field, aged 67, born in Deptford, Kent, and William Edward Field, aged 41, born in Campden Town, London.

Filma

Filma, [348], was a camera manufacturer of Milan, Italy. The company was founded by **Antonio Bencini** in c1936. They produced box cameras, in 4.5x6cm format on 127 film, and 6x9cm format on 120 film. The cameras were moulded plastic, with distinctive rounded corners. The company was acquired by **Ferrania** in 1937.

Not to be confused with the *Filma* model of roll-film camera produced by the **Thornton-Pickard Manufacturing Co. Ltd.** in 1912.

Filotecnica, La

La Filotecnica. See **Ignazio Porro**, **Filotecnica Salmoiraghi**.

Filotecnica Salmoiraghi

La Filotecnica, otherwise known as Officina Filotecnica, [162 p194-197] [374], in Milan, Italy, was a combined training school and production laboratory where students and their mentors produced and sold optical instruments, principally for topographical and geodetic use. It was founded in 1865 by **Ignazio Porro**, who was then 64 years old. **Angelo Salmoiraghi** (1848-1939) graduated from the Politecnico di Milano, the technical university in Milan, in 1866. Porro, who already knew Salmoiraghi, engaged him in the Filotecnica, and soon promoted him. Salmoiraghi's influence in the Filotecnica increased until he took over ownership, and changed its name to Filotecnica Ing. A. Salmoiraghi.

The company expanded, and produced instruments for astronomy and navigation, as well as cameras, including specially designed cameras for aerial photography during

WWI. Between the wars they also manufactured equipment for foreign companies, including the **Houghton Butcher Manufacturing Co.** During this period, they became a part of Istituto per la Ricostruzione Industriale, (the IRI, set up to rescue enterprises that were failing as a result of the great depression). During WWII they once again contributed to the war effort, producing instruments for the military. Sometime after WWII Filotecnica merged with the opticians **Viganò** to become Filotecnica & Viganò. Upon this merger the production of optical instruments other than eyewear ceased. The brand was purchased in 1974 by **Dollond & Aitchison**, but was subsequently sold back into Italian ownership. The company continued as Salmoiraghi and Viganò.

Finder Scope

The term *finder scope* is generally used to refer to a telescope with a wide field of view, aligned in the same direction as, and usually mounted on the side of, the main telescope. Its purpose is to assist the user in finding the area of the sky that they wish to view through the main telescope - the wide field of view allowing the user to see more of the sky than is possible through the main telescope.

Finlay, Robert

Robert Finlay, (fl.1846-1850) [18] [263] [271], was an optical, mathematical, and philosophical instrument maker of Glasgow, Scotland. Addresses: 46 John St, (1846-1847); 225 George St. (1848); 87 London St. (1849-1850). The London St. address had previously been occupied by **William Green (III)**, and Robert Finlay was succeeded at this address by **Paul Cameron**.

Finnie & Liddle

Finnie & Liddle, [18] [271] [379], were optical instrument makers of 3 North Bank St., Edinburgh, Scotland. The firm was a partnership between **Joseph Finnie** and **William Liddle**, which operated from 1826 until 1827, after which William Liddle continued the business in his own name.

Finnie, Joseph

Joseph Finnie, also known as I. Finnie, (fl.1818-1827) [18] [271] [379], was an optical instrument maker of Edinburgh, Scotland. Addresses: 5 Bank St. (1818), and 3 North Bank St. (1819-1825), after which he worked in partnership with **William Liddle**, trading as **Finnie & Liddle** until 1827.

Fire Damp Photometer

See photometer.

Fisher, Samuel

Fisher, 188 and 189 Strand, London was established in 1838 and remained in business until c.1900. They were a retailer of fine goods, known for their fitted travelling bags and boxes. A pair of **Mars** folding opera glasses bears their name, and the markings *Jumelle "Mars" Brevetée France Étranger. Model Militaire.*

Fishmongers' Company

See Appendix IX: The London Guilds.

Fitz, Henry

Henry Fitz, (1808-1863) [8] [24] [184] [187], was a notable American telescope maker. Trained as a locksmith, he became well known as an amateur astronomer and, in 1839, he travelled to Europe to learn his trade in optics for astronomy and photography. He learned his lens-making skills from English and German opticians, and he established contacts in the French glass-making industry. While there, he also learned the new daguerreotype process.

In 1841 Fitz opened a photographic studio in Baltimore, selling an improved daguerreotype camera and, by 1845, he was beginning to produce refracting telescopes with lenses he had made and figured himself. He went on to produce fine telescopes ranging from 6 inches to 16 inches, equipping astronomers and observatories throughout the US. He took as an apprentice **John Byrne**, who went on to become a renowned telescope maker.

Henry Fitz supplied several telescopes to **Lewis Rutherford** and, in 1858, Rutherford began to experiment with astronomical photography. His experiments led him to develop, together with Fitz, a large astronomical telescope purely for the purpose of photography. Fitz died before the telescope was finished, and his son, **Henry Giles Fitz**, assisted Rutherford with its completion.

Henry Giles Fitz learned his craft from his father and, when his father died in 1863 at age only 55, Henry Giles took over his business.

Fitz, Henry Giles

Henry Giles Fitz, (c1847-1939), known as Harry, was the son of **Henry Fitz**. He learned his lens-making and telescope making skills from his father, and took over his father's business after his death in 1863. He assisted **Lewis Rutherford** with the completion of the photographic telescope that Rutherford had developed with his father, and later helped Rutherford with another, larger telescope. He continued his father's business from 1863 until sometime in the 1880s.

Fizeau, Armand Hippolyte Louis

Armand Hippolyte Louis Fizeau, (1819-1896) [37] [45] [155], was a French physicist, famous for making the first accurate determination of the speed of light in air in 1849. In 1850 Fizeau showed that the speed of light is slower in water than in air. He reached this conclusion at the same time as **Foucault** achieved the same result.

Fizeau investigated the shift in wavelength in light coming from stars. He carried out this investigation independently of the work on the subject by Doppler, and determined that the redshift could be used to measure the relative velocities of stars.

During the late-1830s and the early 1840s Fizeau became interested in photography, and worked to make improvements to the daguerreotype process. In 1840 he introduced the practice of gold-toning daguerreotypes, which helped to protect the surface of the plate, and improved the quality of the image. He did not patent the technique, and it was soon widely used. He devised a method of converting daguerreotype plates into printing plates, but the technique was never adopted widely. In 1845, together with Foucault he produced the first daguerreotype of the Sun.

Flatters & Garnett Ltd.

Flatters and Garnett Ltd., [44] [138] [140], of Manchester, England, was formed by Abraham Flatters (1848-1929) and Charles Garnett (1843-1921). Both shared an interest in natural history, and they met at the Manchester Microscopical Society in the late-1880s. In 1895 Flatters started a business in Manchester making lantern slides and microscope slides. In 1901 Garnett helped him financially by joining the business as a partner, and bringing his son John B Garnett into the business. In 1909 the Garnetts bought Flatters out, and the Garnett family continued the business. Flatters started a new business, Flatters, Milbourne and McKechnie, making lantern slides and microscope slides.

The Garnett family went on to expand Flatters and Garnett Ltd., and after employing an instrument maker they embarked on making specimen collecting apparatus, and ultimately microscopes and micro-projectors. The company went into liquidation after financial troubles in 1967.

Flavelle Brothers and Roberts

The company began in 1840 as a partnership between Henry Flavelle and George Brush, opticians and jewellers at 87 King Street, Sydney, Australia. In 1850 Brush left the partnership. At the same time, Henry's brother, John Flavelle, joined, and they began to trade as Flavelle Brothers, [340]. In 1855 they opened a branch of their business in Brisbane. In 1868, by then trading at 354 George Street, Sydney, John Roberts joined the partnership, and they began to trade as Flavelle Brothers & Roberts. The business was mainly a jeweller, but some optical instruments can be seen bearing their name. Examples include a brass telescope and a brass surveyor's theodolite. The company finally closed in 1932.

Fletchers' Company

See Appendix IX: The London Guilds.

Flicker Photometer

See photometer.

Flint Glass

See the *Types of Optical Glass* section in Appendix VI.

Foca

Foca was a brand-name used by the company **Optique et Précision de Levallois** for their rangefinder cameras and other cameras.

Focal

Focal was a brand-name used by US retail company K-Mart for telescopes, cameras, and binoculars. Examples of Focal telescopes bear the maker's marks for **Towa**, and **Tanzutsu**, (See Appendix IV). An example of Focal binoculars bears the **Astro Optical Industries Co. Ltd.** maker's mark, together with the name *Opko*.

Focal Plane

Focal plane is another name for the image plane.

Focal Reducer

A focal reducer is a converging lens, or arrangement of lens elements, that decreases the effective focal length of an optical instrument as seen by an eyepiece or imaging sensor. This has the effect of making the image smaller, and increasing the field of view. In astronomy a focal reducer may be used in conjunction with an eyepiece to decrease the effective focal length of a telescope. In microscopy a focal reducer may be used to decrease the effective magnification of the objective, and increase the working distance between the objective and the sample being observed. See also Barlow lens.

Focostat Lens

A focostat lens is a lens mounted by means of a universal joint to an arm, and may be attached to a dissecting instrument, mapping pen, or other hand-held instrument. The lens magnifies the subject, which may be a dissection, drawing, etc. An example may be seen illustrated on page 111 of the **R.&J. Beck** microscope catalogue [57].

Folded Optics Microscope

A folded optics microscope is a compact, portable microscope for field and educational purposes [313]. The compact design is achieved by the use of prisms and/or mirrors to fold the optical path. The concept was first popularised by Dr. John McArthur (1901-1996) in the 1930s. The **McArthur microscope** was a small, portable microscope that used prisms to fold the optical path. He formed a company which made microscopes to this design, and had others of similar designs produced under license by makers such as **Charles Hearson**, **Nikon**, **Scientific Optics** on behalf of the Open University, and **Swift Optical Instruments**. The *TWX-1* by **TaiYuan Optical Instruments** was also a derivative of the McArthur microscope.

Several derived designs were produced by the optical designer, Keith Dunning, and the industrial designer, Richard Dickinson (who was formerly Head of Design for Sinclair UK). These designs folded the optics in three dimensions, in contrast with the two-dimensional folding of the light path in the McArthur. In 1989, operating as **Science of Cambridge Ltd.**, they began to produce the **lensman**, with optical design by the British designer Eddie Judd. In 2007 they changed the company name to **Cambridge Optronics Ltd.** Dunning and Dickinson developed derivatives of the *Lensman*, including the **Meade** *Readview*, and the *Micron* and *Trekker*, all of which were produced by **Enhelion** and manufactured in China.

Dunning and Dickinson went on to produce the **Newton NM1**, [314], which targeted a more professional audience, and was designed for the Millennium Health Foundation. This was manufactured in China and produced in several variants until Cambridge Optronics closed for business in 2017.

Folding Binoculars

Folding binoculars are generally constructed in a hinged frame so that they can fold flat while not in use. Examples may be found of folding binoculars by **Mars** and **R.&J. Beck**.

A novel design of field binocular by **Aitchison** has a folding aluminium frame, with barrels made from a spiral of aluminium. As the frame folds closed, the barrels collapse into a puck-shape within the frame. These binoculars were issued to officers during the second Boer War.

Folding Camera

See camera.

Folmer & Schwing

Folmer & Schwing, [122], of New York, NY, USA, was founded in 1887 by William F. Folmer and William E. Schwing, initially manufacturing gas lighting fixtures and bicycles. In about 1896 they began to produce bicycle-mounted cameras. By 1898 they were using the *Graflex* brand-name, and offering a much wider range of cameras. From 1905 until 1926 the company was a part of **Eastman Kodak Company**. In 1926 it was separated into an independent company named Folmer Graflex Corporation. In 1945 the name was changed to **Graflex Inc.**

Ford, William

William Ford, (fl.1748-1793) [18], was an optician and optical instrument maker of London, England. Addresses: Cannon Street (1764); 8 Crooked Lane, Fish Street Hill, (1784-1793). He served his apprenticeship with the spectacle maker Mary Brawn, beginning in 1729, and was freed of the **Spectacle Makers' Company** in 1743. In 1764 he was one of the petitioners who attempted to revoke the patent obtained by **John Dollond (I)** for the achromatic lens, and which was enforced after Dollond's death by his son, Peter (See Appendix X).

Forster, J.

J. Forster, [8], was a German theoretical optician with an interest in tilted component telescopes. He collaborated with **Karl Fritsch** in their design for a telescope he referred to as an *excentric Cassegrain* or *Brachyte*, or *Brachy telescope*.

FOS Limited Partnership Society Ginsberg and Co

Alexsander Ginsberg, (1871-1911) [115], from Sosnowiec, Poland, worked with **Krauss** in Paris, France, and **Zeiss** in Jena, Germany. In 1898 he formed "The First Factory of Optical Instruments in the Country FOS" in Warsaw, Poland, and in 1902 changed the name to FOS Limited Partnership Society Ginsberg and Co. The company mainly produced lenses for photographic cameras, but also produced their own camera models. The company remained in business until shortly after Ginsberg's death in 1911.

Foton

FOTON Optoelectronics, of South Africa, was founded in 1998. They produce opto-electronic, laser and precision mechanics products for various industries, including mining, tunnelling and construction, surveying, astronomy and security. They also sell imported products for professional and recreational use, including Surveying

Instruments, thermal imaging and night vision equipment, astronomical telescopes, binoculars, digital and optical microscopes, rifle sights and laser pointing devices.

Foucault, Jean Bernard Léon

Jean Bernard Léon Foucault, (1819-1868) [7] [24] [72] [153] [155], was a physicist, born in Paris, France, and is most famous for his invention of the gyroscope, and his experiments with pendulums. He showed that a pendulum swings in a plane that is stationary in inertial space, irrespective of any movement due to the rotation of the Earth. He also observed that a pair of dark Fraunhofer lines in the spectrum of the Sun corresponded with a pair of bright spectral lines obtained when heating sodium in the laboratory.

He researched the speed of light, and in 1850 determined that the speed of light in water was less than that in air. He achieved this result at the same time as **Fizeau** reached the same conclusion. This result supported the wave theory of light, and verified the theory, first put forward about 900 years earlier by Islamic physicist **Ibn al-Haytham**, that the refraction of light results from the fact that light travelled at different speeds in different materials.

In 1845, together with Fizeau, Foucault made the first daguerreotype of the Sun. In 1855 he joined the Paris Observatory as a physicist. Following **Liebig**'s invention of a process for silvering a glass mirror, Foucault was one of the first to use this technique, applying it to mirrors for telescopes by **Eichens**. He designed a method for figuring a paraboloidal telescope mirror, and for testing the surface at every stage of the process. He took on **Adolphe Martin** as a pupil, and had to persuade Martin not to give away his trade secrets, although Martin did so anyway after Foucault's death. The use of a silvered paraboloidal glass mirror in a reflecting telescope, using Foucault's method for accurate figuring of the mirror, was a significant advance, and Foucault collaborated with **Marc Secrétan** to develop and market this concept [72]. Foucault also designed the first large siderostat, which was built by Eichens and **Gautier** in 1868, the year of Foucault's death.

Foucault Siderostat

See siderostat.

Foul Anchor

 At the end of the C15th the foul anchor symbol began to be used by the Lord High Admiral of Scotland. A century later the symbol began to be used as the seal of the Lord High Admiral of England [310]. A foul anchor is an anchor which has become fouled on the sea-bed, or one which, when weighed, is found to have its cable wound around the anchor's stock or flukes. The use of the foul anchor symbol by the British Admiralty may have been what inspired its use by **Jesse Ramsden** and other instrument makers to mark marine octant, sextant, and quintant scales.

Some scale dividers added initials or other identifying marks to the foul anchor, while others are known to have used an anchor symbol with no identifying marks, which makes it difficult to be certain which maker engraved it, or with which dividing engine. Foul anchor symbols with and without initials, as well as other marks were used by many scale dividers. For further information on scale dividers' marks, see Appendix XI.

 The **TRY ME** logo of **I.P. Cutts, Sutton & Son** also depicted a foul anchor.

Founders' Company

See Appendix IX: The London Guilds.

Fournier, G.

G. Fournier Opticien of Paris, France retailed optical and photographic products, and may have manufactured them. Addresses: 25 Quai de l'horloge; 22 Avenue De L'Opera. Instruments bearing their name include: Porro prism binoculars, 8x26, 8x30; Galilean binoculars; cine lens.

Français, E.

Jean Pierre Émile Français (1830-1890), usually known as Émile Français, of 3 Rue du Chalet, Paris, France, was an optician who began to produce high quality camera lenses during the 1860s. His lenses bear the signature *E. Français Paris*. Later the E. Français workshops began to produce various models of camera. They were among the earliest makers of twin-lens reflex cameras, a popular model being the *Kinegraphe*, introduced in 1886, and credited to Eugène Français. This was available in various sizes, including stereo versions, [61]. The E. Français workshops continued to produce cameras and lenses until the 1890s.

Examples of their work include: a brass mounted lens bearing the inscription *E. Français Paris Rectilinaire, Grand Angle, Serie E No 4*; a brass mounted convertible 300mm f/4 Petzval lens, which can be converted to function as either a portrait or a landscape lens; a *Cosmopolite* camera bearing the inscription *E. Français. Opticien-Constructeur. 3.Rue du Chalet.3 Paris. Appareil Breveté S.G.D.G.*

Francis, George

George Francis, (fl.1826-1844) [**18**] [**276** #1851] [**319**] [**322**], of London, England, was an optical, mathematical, and philosophical instrument maker. Addresses: 101 and 103 Regent St.; 93 and 96 Berwick St., Soho (1826-1829); 92 Berwick St. (1838-1842). He attended the London Mechanics' Institute in Holborn from 1826 until 1829.

Frank Ltd., Charles

Charles Frank Ltd., [**306**], was a scientific and optical instrument company in Glasgow, Scotland. Charles Frank (1865-1959) was born in Lithuania, and migrated to Scotland in the early 1900s. The company was founded in 1907, when Charles Frank began his business in the Saltmarket in Glasgow. They dealt in and repaired scientific, optical, and photographic equipment, including telescopes, and made cameras. During WWII they focused on navigational instruments and binoculars. The company continued in business until 1974. After Charles Frank died in 1959, his son Arthur continued the business. In 1968 Charles Frank Ltd. took over the premises at 18 Forrest Rd., Edinburgh, Scotland, previously occupied by **George Hutchinson**. The company had a reputation for quality, and failed to reduce their costs sufficiently when the 1970s brought an influx of cheap imports.

The original Charles Frank Ltd. ceased trading in 1974, but the rights to the name were acquired by former staff after the business was sold off. Charles Frank Ltd. continued to trade from Edinburgh, and specialised in retail and wholesale binoculars and telescopes. The new company did not make their own instruments. Their *Nipole* binoculars were made in Japan.

Instruments bearing their name include: Russian manufactured binoculars, believed to be 10x50, marked *Charles Frank. Saxmundham*, bearing the maker's mark for **ZOMZ**.

Franka

Franka began as a small camera maker in Stuttgart, Germany, in 1909, founded by the husband and wife, Franz and Leoni Vyskocil. By 1914 the company was operating in Bayreuth and, in that year, it was renamed Franka Kamerawerk. In 1962 the company was taken over by **Wirgin**, and in 1966 camera production ceased.

Franke and Heidecke

Franke and Heidecke was a camera manufacturer established in 1920 by Paul Franke and Reinhold Heidecke in Brunswick, Germany. They first began to use the name *Rollei* in 1926 when they began to produce a stereo camera named the *Rolleidoscop*. Shortly after this they began to produce their first *Rolleiflex* model. They continued to produce new models of camera, expanding their business. During WWII they marked their instruments with the **ordnance code** gxl. Paul Franke died in 1950, and was succeeded by his son Horst Franke. Reinhold Heidecke died in 1960. In 1962 Franke and Heidecke, was renamed to Rollei-Werke Franke & Heidecke, the company that came to be commonly known as **Rollei**.

Franklin-Adams, John

John Franklin-Adams, (1843-1912) [**100**], was born in Peckham, London, England, in 1843. He took to astronomy late in life, and conceived a scheme for a complete photographic survey of the Milky Way. He afterwards extended this scheme to the photographic charting of the whole heavens from North to South. The telescopes he used, an equatorial mount by **T Cooke and Sons**, and his optical and photographic arrangement, are described by Franklin-Adams, **H. Dennis Taylor** and Alfred Taylor in [**77**]. He died in 1912 at the age of 69. The telescope he used for his sky survey was sold in 1912/13 to The Royal Observatory, Greenwich [**101**].

Franks Ltd., A.

See **Aubrey Franks**.

Franks, A.&B.

A.&B. Franks, [**306**] [**319**], were opticians and optical instrument makers of 95 and 97 Deansgate, Manchester, England. The firm was a partnership between **Aubrey Franks** and his brother, **Benjamin Franks**. An 1896 advertisement shows them offering a *Prestograph* half-plate mahogany field camera. The partnership was terminated in 1897 [**35** #26815 p363].

Franks, A.&J.

A.&J. Franks, [**3**] [**319**] [**384**], were opticians, and optical, mathematical, scientific, and philosophical instrument makers of Manchester, England. They traded at 114 Deansgate and 44 Market St. The firm was a partnership, formed in 1848 when **Joseph Franks** joined the partnership already existing between his mother, Amelia Franks, and his half-brother, **Abraham Franks**. In 1851 his brother, **Henry Franks**, joined the partnership, and ran a branch at 60 Cross St. In addition to their main business making and supplying spectacles, the firm made instruments including telescopes, microscopes, opera glasses, compasses, and barometers. By the time of Abraham's death in 1868, Henry had moved to set up business as an optician in Hull. Joseph continued the business, retaining only the premises at 44 Market St, until his death in c1888, upon which his son, **Aubrey Franks**, took over the business.

Franks, Abraham
Abraham Franks, (1781-1868) [18] [276 #1852] [319] [384], was an optician and optical instrument maker, and lecturer on the anatomy and physiology of the human eye. Born in Manchester, England. He was a son of **Jacob Franks**. By 1835 he was trading at 33 Hick Street, Newcastle-under-Lyme. In 1847, following the death of his father, he entered a partnership with his mother Amelia, running Jacob's business in Manchester. In 1848 his half-brother, **Joseph Franks**, joined the partnership, and the firm was renamed **A.&J. Franks**. Abraham continued to work in this partnership until his death, after which it was continued by Joseph. Abraham also worked in partnership with Herrmann Meyer, trading as **Meyer & Co.** until the partnership was dissolved in 1863.

Also in Hick Street, Newcastle-under-Lyme, in 1834, the records show Charles Franks, optician, umbrella maker, and seller of coins and clothes.

Franks, Aubrey
Louis Aubrey Franks, commonly known as Aubrey, (b.c1853 fl.1878-1917) [53] [306] [319] [384], was an optician, and optical and scientific instrument maker of Manchester, England. He started his business in about 1878 as an optician and fine art dealer at 2 and 4 King Street, and 95 Deansgate. In 1879 he was subject to bankruptcy proceedings [35 #24691 p1998]. A few years after starting his business, Aubrey Franks began to make and sell optical instruments. These included microscopes, telescopes and magic lanterns. By 1880, his company was producing cameras, including the very popular, low cost, *Presto* hand-camera, as well as other optical and mathematical instruments.

In c1888, following the death of his father, **Joseph Franks**, he took over the premises of **A.&J. Franks** at 44 Market Street. He worked in partnership with his brother, **Benjamin Francks**, trading as **A.&B. Franks** from 1896 or earlier, until 1897.

At some point, the company incorporated as A. Franks Ltd., and Aubrey Franks' son-in-law, Maurice Saffer, took over the firm of A. Franks Ltd. in about 1917, the year of Aubrey's death. He continued to trade under the same name, and expanded the business to include radio and television. During the 1920s, the company also had branches in Oxford Street, South King Street, and Bradshaw Gate, Bolton. In 1923 the firm took over the premises at 12 Victoria Street from **W. Aronsberg & Co.** Maurice Saffer continued to manage the firm until his death in 1947, after which it was managed by his secretary until 1950 when she sold it to **Dollond & Aitchison**.

Franks, Aubrey & Isaac
Aubrey & Isaac Franks, also known as Francks, (fl.1835-1838) [18], were opticians of 114 Deansgate, Manchester, England. This was a partnership between Aubrey Franks and Isaac Franks, although the relationship with other members of the Franks family, who also traded at this address, is unclear.

According to an advertisement of 1835, [451], "A. Franks of Manchester, Lecturer on the Anatoty (sic) and Physiology of the eye, and I. Franks, Practical Optician, are staying for a few days only, at Mrs. Taylor's, 33, Charlotte Street, George-Street, Hull, where they will meet any lady or gentleman wishing to consult them, with candour, and inform them, gratuitously, that their own glasses are proper, or convince them, by ocular demonstration, that others would be better; in which case I.F. will be able to remove such glasses as have a tendency to injure the eye and replace them by others calculated to preserve and strengthen the sight. Attendance from ten in the morning to six in the evening. Isaac Franks, Licensed hawker, A. No. 4392."

Franks, Benjamin
Benjamin Franks, (b.c1862 fl.1891-1897) [319], was an optician of Manchester, England. He was a son of **Joseph Franks**. At age 18, he was working as a shipper's clerk near Preston, but by 1891 he was working as an optician in Manchester. He worked in partnership with his brother, **Aubrey Franks**, trading as **A.&B. Franks** until the partnership was dissolved in 1897.

Franks, Henry
Henry Franks, (b.c1840 fl.1851-1868) [319] [384], was an optician of Manchester and Hull, England. He was a son of **Jacob Franks**, and was brother of **Joseph Franks**, and half-brother of **Abraham Franks**. He began to work for the partnership **A.&J. Franks** in 1851, and ran a branch of the company at 60 Cross St., Manchester. By the time of Abraham's death in 1868, Henry had set up business as an optician in Hull.

Franks, Jacob
Jacob Franks, also known as Francks, (b.c1785 fl.1805-1846) [3] [18] [319] [384], was an optician and optical instrument maker, of Manchester, England. Addresses: 283 Oldham Rd.; Withy Grove; 378 Oldham Rd. (1824-1825); 24 St. Mary's Gate (1838); 114 Deansgate (1838-1846). From 1840 he was known as Franks, rather than Francks. He inherited the lens-selling business of his father, Isaac Franks (fl.1781-1818), a Jewish immigrant from the Netherlands who settled in Liverpool, England, in the 1760s, then in 1798 had moved to Manchester.

According to his trade handbill, "J. Franks, Optician. Manchester. Makes and repairs all sorts of optic glasses, telescopes, microscopes, reading glasses &c. &c. With a variety of spectacles for all ages, whether concave or convex. Old ones taken in exchange in any of the above articles. Likewise, excellent tooth powder will make the blackest teeth the finest white. Also excellent eye-water has cured many almost blind. Infallible worm powder for destroying worms in human body. N.B. Umbrellas made and neatly mended."

Jacob had a son, **Abraham Franks**, by his first wife, Mary. He also had two sons, **Joseph Franks**, and **Henry Franks**, by his second wife, Amelia. All three became opticians and optical instrument makers. Upon Jacob's death, his wife Amelia ran the business, forming a partnership with Abraham in 1847, and Joseph in 1848, at which time the business took on premises at 44 Market St. In 1849/1850 the firm became **A.&J. Franks**.

Franks, Jacob Henry
Jacob Henry Franks, (b.c1852 fl.1881-1901) [319], was an optician of Manchester, England. Addresses: 36 Chestnut Street, Cheetham (1881); 17 Worcester Street, Broughton (1901). He was a son of **Joseph Franks**, and worked as an optician on his own account.

Franks, Joseph
Joseph Franks, (c1822-1888) [319], was an optician of Manchester, England. Addresses: 114 Deansgate (1841); 45 Morton St., Cheetham (1851); 10 Parks Place, Cheetham (1861); 39 Cheetham Rd., Cheetham (1871); 19 Wellington Street East, Broughton, Salford (1881). He was a son of **Jacob Franks**, and had ten children, of whom **Aubrey Franks**, **Benjamin Franks**, and **Jacob Henry Franks** became opticians. In 1848, joined the existing partnership between his mother, Amelia Franks, and his half-brother, **Abraham Franks**, trading as **A.&J. Franks**. The partnership continued until Abraham's death in 1858, after which Joseph continued the business alone. Upon Joseph's death in 1888, Aubrey took over the business.

Fraser & Son, William
William Fraser & Son, [8] [18], opticians, of *Ferguson's Head*, 3 New Bond St., London, England, was a partnership between **William Fraser** and his son, **Alexander Fraser**. The partnership began in 1805 and ended with the death of William in 1812.

An interesting telescope bearing the maker's name Fraser & Son is in the Maritime Museum collection [102]. This example was once owned by the Emperor of China, and is highly decorated with enamel, gold and pearls. The end-cap is a watch.

Fraser & Co.
See **William Fraser**.

Fraser & Sons
See **Alexander Fraser**.

Fraser, A.&H.
See **Alexander Fraser**.

Fraser, Alexander
Alexander Fraser, (fl.1799-1818) [18] [436], optician, of 3 New Bond St., London, England was the son of **William Fraser**. In 1799 he began to work in partnership with his father, trading as **William Fraser & Son**. The partnership continued until William's death in 1812. In 1812, Alexander received a royal appointment to King George III. Alexander continued the business under his own name until 1818. Also trading at the same address, were the partnerships Fraser & Sons (1814-1818) and A.&H. Fraser (1816). In 1818 the premises were taken over by **William Hawks Grice**.

Fraser, J.
J. Fraser (fl.1826-1828) [18], was an optical, philosophical, and mathematical instrument maker of 95 Wardour St., Soho, London, England.

Fraser, William
William Fraser, (c1720-1812) [18] [276 #835] [319] [436], was an optical and mathematical instrument maker of *Ferguson's Head*, 3 New Bond St., London, England, initially trading under his own name. He was appointed Mathematical Instrument Maker to the Prince of Wales, who was later to become King George IV.

A catalogue dated c1780, believed to be the only surviving copy, is in the Science Museum Library in London. It is entitled: *Catalogue of Optical, Mathematical, and Philosophical Instruments, made and sold by William Fraser, Optical and Mathematical Instrument Maker to his Royal Highness the Prince of Wales.*

One of his employees was **James Smith (II)**. In 1805 he began to work in partnership with his son, **Alexander Fraser**, trading as **William Fraser & Son**. In 1808 the Post Office directory lists the firm of William Fraser & Co., Opticians &c., trading at 3 New Bond St. William continued to work in partnership with Alexander until his death in 1812.

Fratelli Dolci
Fratelli Dolci (Dolci brothers), [75], were contemporaries of **Leonardo Semitecolo**, making telescopes in Venice, Italy, in the mid-to-late C18th. Their telescopes were

similar in construction to those of Semitecolo. An example is a hand-held 5-draw telescope made in pasteboard, vellum, leather and boxwood - this appears to be typical of their work.

Fratelli Koristka

Fratelli Kroistka (Kroistka brothers), [374], was a company formed in Milan, Italy, in 1881 by the optician Francis (Franz) Koristka (1851-1933). Franz Koristka learned optical engineering and design whilst working for **Angelo Salmoiraghi** before forming his own company in 1881. He knew **Ernst Abbe**, and through this relationship he acquired permission to make use of **Zeiss** patents in his construction of microscopes. The company was renowned for making microscopes, and for their binoculars. In 1929 the company became a part of **Officine Galileo**.

Fraunhofer, Joseph Ritter von

Joseph Ritter von Fraunhofer, (1787-1826) [24] [156] [157] [161] [374] [422], born in Staubing, Bavaria, was a physicist and optician, famous for discovering spectral lines now known as *Fraunhofer lines*. He lost both of his parents during his childhood, and served an apprenticeship with a mirror-maker and ornamental glass cutter in Munich.

In 1806 **Joseph von Utzschneider** employed him in a junior post at the **Mathematisch-mechanisches Institut von Utzschneider, Reichenbach & Liebherr** in Munich. There, studying optics from books, and learning from the head optician Joseph Niggl, he learned the skills of grinding and polishing optical surfaces. He also studied the techniques of optical glass making, his teacher being **Pierre Guinand**, with whom he had a difficult relationship. Soon Fraunhofer was in charge of the workshops and apprentices at the glassworks in Benediktbeuern, and in 1809 he was put in sole charge of the glassworks, and made a junior partner in the company, which was renamed **Optisches Institut von Utzschneider, Reichenbach & Fraunhofer in Benediktbeuern**. In 1814, when Reichenbach left to set up his own instrument-making firm, the company became a new partnership, **Optisches Institut von Utzschneider und Fraunhofer**. In 1819 the optical branch of the business moved from Benediktbeuern to Munich, at which time Fraunhofer was made a full partner and director of the Optical Institute. **Martin Wörle** worked for Fraunhofer for many years, before setting up his own company.

Fraunhofer worked to measure the refractive index and dispersion for the kinds of glass produced at the glassworks. Whilst engaged in his research on refraction he noticed that the spectrum of light emitted by the Sun and other stars contained dark lines. **William Hyde Wollaston** was the first to observe these lines in the Sun's spectrum, but believed them to be simply the boundaries of the colours. Fraunhofer studied them in the Sun and other stars, and showed that the position of the lines depended upon which star he was observing. He also showed that the lines occurred whether using a prism or a grating, thus concluding that they were present in the light, and not a result of the experimental optics. He had previously noticed the two bright spectral lines produced by a sodium lamp, and he realised that these corresponded with two dark lines in the Sun's spectrum, but he didn't follow up this line of investigation.

He classified 574 of the dark absorption lines in the visible solar spectrum, and used them as a benchmark to determine the refractive index and dispersion of glass for given wavelengths. With this work he was able to derive formulae for the design of aberration-free optical lenses. He was innovative in his ways of grinding and polishing lenses, and testing lens surfaces, with the invention of new machines, and the use of Newton's rings for optical testing. In his studies of dispersion, he showed that spherical aberration can be fully corrected by means of only two lenses in the objective.

Joseph Fraunhofer invented the German Equatorial mount, and built many telescopes, his masterpiece being the Great Dorpat Refractor. This 24cm aplanatic f/18 refractor on a German equatorial mount was completed in 1824, and he built it for William Struve, director of the Dorpat observatory.

See Appendix VII.

Freiberger

Freiberger is a German brand of surveying and marine instruments, including marine sextants. The company was founded in Freiberg, Saxony, in 1771 by Gottlieb Friedrich Schubert, who was a *Bergmechanikus*, or master mining mechanic. The company produced instruments for mining and metallurgy. In 1791, under new ownership, the company began to introduce new products, including theodolites, and by the 1870s they were manufacturing instruments on a large scale, with 80 employees. At the end of WWII, under Soviet occupation, the factory was much reduced, but continued to produce surveying instruments. In 1950 the company was nationalised, becoming VEB Freiberger Präzisionsmechanik, and expanded, introducing a wider range of products. Following the reunification of Germany, the company was once again privatised in 1993. In 1994 the company became FPM Holding GmbH, producing geodetic and nautical instruments.

Fresnel, Augustin Jean

Augustin Jean Fresnel, (1788-1827) [7] [155] [300], of Broglie, Normandy, France, began his career as an engineer. In 1814 he began to study optics, and became a notable physicist specialising in the field of optics and, like his contemporary, **Thomas Young**, a proponent of the wave theory of light. While researching the wave theory of light, he originated several devices for studying interference, including the Fresnel double mirror, the Fresnel Prism, and the Fresnel double prism. He also developed optical systems for lighthouses, including the Fresnel lens, which he developed initially in collaboration with **François Soleil**. A contemporary, but independent invention of this same design of lens by **Sir David Brewster** led to a priority dispute between Brewster and Fresnel over the invention.

Fresnel Double Mirror

See interferometer.

Fresnel Double Prism

See interferometer.

Fresnel Lamp/Lantern

A Fresnel lantern is a projecting light source with a Fresnel lens, commonly used in stage or filmset lighting. A Fresnel lamp is one in which the light source is enclosed in a hollow, cylindrical Fresnel lens.

Fresnel Lens

The Fresnel lens is a thin optical lens with the properties of a thick, heavy lens. It comprises a central convex lens, surrounded by a series of adjoining concentric annular sections, each with a surface of different curvature, and with a common focus.

It was invented by **Augustin Jean Fresnel** soon after he began his work in optics in 1814, and independently at about the same time by **Sir David Brewster**, leading to a priority dispute between the two inventors.

Some applications of Fresnel lenses are: lighthouses; photographic lighting; stage and filmset lighting.

Fresnel Prism

See Prism.

Frezzolini, James

James Frezzolini was the founder of Frezzolini Electronics Inc., USA, (Frezzi – Broadcast power and lighting). As head of General Research Laboratories, New York City, NY, USA, he was well known for adapting cameras. An example of this is his adapted and upgraded Auricon 16mm news-film cameras. The *Big Bertha* camera was an adapted Graflex camera with a long telephoto lens, designed for sports photography. A shift lever with adjustable stops allows the user to obtain instant focus at preselected distances.

Friese-Greene, William

William Friese-Greene, (1855-1921) [46] [61] [71] [335], was a photographer, inventor, and an innovator in early British moving picture technology. He was born William Edward Greene in Bristol, England, and served an apprenticeship with a photographer. When he was 18, he opened his own photographic portrait studio, and married Mariana Helena Friese, upon which he changed his surname to Friese-Greene. With the success of his photographic business, he opened shops in Bristol, Plymouth, and his new home in Bath. He also worked in partnership with A.A. & J.W. Collings, with addresses in Bath and London, from 1887 until 1888 [35 #25871 p5964]. In 1888 he entered a partnership with F.W. Simpson in Bath. This partnership was terminated in 1890 [35 #26082 p4687], after which they continued as Friese-Greene, Simpson & Co. Ltd. until 1897 when they announced their voluntary liquidation [35 #26855 p2865].

Having settled in Bath, Friese-Greene began to work with **John Rudge**, pursuing his interest in moving pictures. In 1885 he moved to London where, in addition to six photographic studios, he started his own laboratory in which he began to design moving picture cameras.

By 1888 he had designed his first moving picture camera, and he quickly went on to design a stereoscopic film camera. In 1889, having worked with a civil engineer, Mortimer Evans, to improve on his designs, they registered a patent, which was accepted in 1890, for a camera which they claimed could take ten frames per second. This was one of many patents bearing Friese-Greene's name.

He was, however, while being an energetic inventor and patentee, a poor businessman. He was subject to bankruptcy proceedings and imprisoned in 1891, [35 #26171 p3168], and declared bankrupt in 1892. By the mid-1890s the moving picture industry was dominated by **Thomas Edison**, the **Lumière brothers**, and **Robert Paul**, and Friese-Greene's technical contribution was largely forgotten. Nonetheless, in the late 1890s he began to work on a two-colour film and projection system, resulting in his development of his Biocolour system. This led to a patent dispute with Edison, in which Friese-Greene prevailed, and a dispute with **Charles Urban**, which Friese-Greene also won. However, Edison's system remained more successful.

In 1907 he advertised the claim that "William Friese Greene is the Inventor and Patentee of the Master Patent for Animated Photography", referring to his patent

No. 22,954 dated 29th November 1893 [**35** #28047 p5335]. He gave notice that he was petitioning to extend the patent, and set a date for anyone wishing to challenge it. In 1896 he began a collaboration with **John Alfred Prestwich** to produce a projector, designed to overcome the problem of flicker by means of two vertically arranged projecting lenses, with a mechanism to show the film through one while the next frame was readied for the other. By 1910 he was once again bankrupt, [**35** #28353 p2313] and, following WWI, the industry developments again left Friese-Greene behind. He famously died while making an impromptu speech at a meeting of film distributers in London.

Frith & Co., Peter

Peter Frith & Co., (fl.1814-1885) [**3**] [**18**] [**276** #1557a], were optical and mathematical instrument makers, and powder flask and shot belt makers, of Sheffield, England. They also had branches in London and Birmingham.

Sheffield addresses: Arundel St. (1814-1834); 37 Arundel St. (1822-1828); 58 Arundel St. (1825); 32 Arundel St. (1837); 83 Arundel St. (1841-1854); 81 Arundel St. (1841-1857). 105 Arundel St. (1845). London addresses: 8 Cursitor St., Chancery Lane (1822-1826); 4 Bolt Court, Fleet St. (1841); 5 Bartlett's Buildings, Holborn.

According to their advertisement of 1860, "Peter Frith and Co., Wholesale Opticians, 81 Arundel Street, Sheffield, and 5 Bartlett's Buildings, Holborn, London, E.C. Manufacturers of all kinds of military, naval, tourists', & deer-stalking telescopes; Astronomical instruments; Single and double achromatic photographic lenses, Mathematical instruments, Microscopes, Operas, Reading glasses, Stereoscopes, Spectacles, and all descriptions of convex, concave, and meniscus spectacle eyes and pebbles, for home and exportation. Established 1790."

An advertisement of 1865 additionally offered: Achromatic microscope and telescope objectives; Woollen & linen provers; Twin photographic stereoscopic view lenses; Camera lucida; Right angle and compass prisms; Rifleman's telescopes; Astronomical and surveying instruments made to order. As wholesale manufacturers they rarely signed their instruments.

In 1839, the partnership Peter Frith & Co. between Peter Frith, James Frith, Joseph Frith, and Henry Frith, was dissolved [**35** #19759 p1589]. It was presumably then re-formed without James and Henry, who went on to work as **Frith Brothers**. For a brief period in the mid-1840s Peter Frith & Co. shared the address 105 Arundel St. with Frith Brothers.

Frith Brothers

Frith Brothers, (fl.1839-1860) [**3**] [**18**], were optical, philosophical, and mathematical instrument makers, and powder flask makers of Sheffield, England. Addresses: 105 Arundel St., Sheffield (1841-1846); 46 Lisle St., London (1845). The firm was a partnership between James Frith and Henry Frith who had, until 1839, been part of the partnership, **Peter Frith & Co.** They mostly worked at different premises to Peter Frith & Co. but for a brief period in the mid-1840s, they shared the address 105 Arundel St. with them. They also had a branch in London.

Fritsch, Karl

Karl Fritsch, (1855-1926) [**8**] [**219**], was an Austrian optician and optical instrument maker. He worked for **Wenzel Prokesch** in Vienna, and succeeded him in business in c1873, naming his company the Karl Fritsch Opto-Mechanical Workshop, and trading at Gumpendorferstrasse 31. Some of his instruments bear the inscription *K. Fritsch-Prokesch Wien*, or *K. Fritsch vorm Prokesch*, in acknowledgement of his predecessor. He made and sold achromatic doublets up to 264mm, apochromatic triplets up to 157mm, oculars, spyglasses, binoculars and spectroscopes. In the late 1870s, together with **J. Forster**, he developed a kind of telescope he referred to as an *excentric Cassegrain* or *Brachyte*, or *Brachy telescope*. In 1898 the companies **Plössl** and Karl Fritsch were merged to form **Karl Kahles**.

Fromme Brothers

See **Gebrűder Fromme**.

Fuess, Rudolf

Rudolph Fuess, (1838-1917) [**8**] [**47**] [**50**] [**374**], was a precision instrument maker from Hanover, Germany. He learned his mechanical skills in Göttingen, and in Hamburg he was a journeyman for **Heinrich Schröder**, making astronomical and other scientific instruments. In 1860 he moved to Hamburg, and in 1864 he started his own company producing goniometers, microscopes, and other scientific instruments. In 1880 he was involved in starting the *Zeitschrift für Instrumentenkunde* (Journal of Instruments), and in 1881 he was one of the founders of the Physikalisch-Technische Reichsanstalt (the Imperial Physical and Technical Institute). He was also chair of the German Society for Precision Mechanics and Optics. In 1877 he purchased the glass manufacturer Greiner & Geissler, and increased the scope of his products to include meteorological, geodetic, and hydrographic instruments, spectrometers and heliostats, and other scientific instruments. In 1891 he built a factory in Berlin.

In 1913 Rudolf's son Paul Fuess (1867-1944) took over running the company. The factory in Berlin continued after the death of Rudolf Fuess. During WWI they produced riflescopes and other optical munitions as well as aircraft instrumentation. During WWII they made optical munitions for the war effort, marking them with the **ordnance code** cro.

Fujifilm

Fujifilm is a Japanese film and camera manufacturer. The company was established in 1934 as Fuji Photo Film Co. Ltd. as part of a government programme to establish a domestic photographic film manufacturing industry. Through a series of expansions and acquisitions they grew to become a global group of companies.

Their consumer products include digital cameras, films and film cameras, camera lenses, instant photo cameras and film, and binoculars made under the *Fujinon* brand-name. Their business products include medical systems, graphic systems, photofinishing products, motion picture archival film, storage media, and specialised lenses and other optical products.

Fujinon

Fujinon is a **Fujifilm** brand.

Fuji Surveying Instrument Company Ltd.

The Fuji Surveying Instrument Company Ltd., was founded in 1929, [**333**], in Tokyo, Japan. In 1967 it was taken over by the **Asahi Optical Company** who, in 1976 changed the *Fuji* surveying instrument brand-name to **Pentax**. From that time the name Fuji was no longer associated with the company.

Fullerscopes - Dudley Fuller

Dudley Fuller was an eminent British maker and retailer of telescopes of 760 Finchley Rd., London, England, at his shop called Fullerscopes. In the 1973 he took over **Broadhurst Clarkson & Co.** at Telescope House, 63 Farringdon Rd., London. He renamed the firm Broadhurst Clarkson and Fuller and manufactured Cassegrain reflectors [8].

G

Gabory, Edmund
Edmund Gabory, (fl.1790 d.c1813) [2] [18] [298] [391], was born in Strasbourg, France, and trained under both **Jesse Ramsden** and **John Dollond (II)** in London, England. In 1790 he set up his own workshop in Holborn, London and, in 1795, after marrying and having a daughter, he emigrated with his family to Hamburg, Germany. There, in 1796, he set up his own workshop and retail shop, manufacturing and selling optical, mechanical, and electrical instruments. Following Edmund's death, his son, Edmund Nicolas, and daughter, Mary Ann continued his business. In 1823 Mary Ann Gabory married **Andres Krüss**, who continued the business with Edmund Nicholas Gabory.

Gabory, M.
M. Gabory, (fl.1796) [18], was an optical and philosophical instrument maker of 125 Lower Holborn, London, England, and was one of several instrument makers named Gabory working in Holborn at the time. At this same address, J. Gabory is recorded as a barometer and thermometer maker in 1794.

[379] records A barometer maker named Gabory at 125 Holborn, with estimated work dates of 1800-1820, as well as J. Gabory at 123 Holborn, recorded as working in 1794, and with estimated work dates of 1780-1800. These are likely to be the same people as mentioned above.

Edmond Gabory, who trained under Jesse Ramsden, also had a workshop in Holborn c1790-c1796.

Gabriel & Co., P.
P. Gabriel & Co., watchmakers, opticians, and jewellers of Manchester Street, Liverpool, was a partnership between **Maurice Aronsberg** and Gabriel Phillips. The partnership was terminated in 1878 [35 #24612 p4552], after which Gabriel Phillips continued the business alone.

Gaertner Scientific Corporation
See **William Gaertner**.

Gaertner, William
William Gaertner, (1864-1948) [185] [211], was a German born American optical instrument maker. At age 16 he began a four-year apprenticeship to an instrument maker, after which he attended training school for instrument makers in Berlin. After this he took several jobs as an instrument maker with companies including **F.W. Breithaupt & Sohne** in Kassel, and **A. Repsold & Söhne** in Hamburg, as well as working in London and Rotterdam, and at the University of Prague.

He emigrated to the USA in 1889, and began working for **Buff and Berger** in Boston. After about a year he moved on to work for the US Coast and Geodetic Survey. He then had an extended visit to Germany after which he returned to work as an instrument maker in the astrophysical laboratory of the Smithsonian Institution in Washington DC. In 1895 he worked briefly at the Yerkes Observatory before starting his own company in Chicago. Thanks to the many connections he had accrued during his career his new venture quickly grew into a successful optical instrument making concern. He produced many kinds of instruments including surveying and astronomical instruments, and interferometers for **Alfred Michelson**, and his output was prolific.

Some of the optics and components for Gaertner's instruments were provided by **Octave Leon Petitdidier** and, when Petitdidier died in 1918, Gaertner took over his company. His expanded range of products included etalons, echelons, Lummer plates, spectrographs, spectrometers, interferometers, monochromators, photometers, and measuring microscopes as well as other specialised optical instruments.

In 1923 the firm was incorporated and was called the Gaertner Scientific Corporation, with expanded premises where they produced telescopes, coelostats, heliostats, spectrohelioscopes, solar cameras, prism transits, zenith transits, and other specialist custom instruments.

Gailand Optical Co.
The Gailand Optical Company was active in New Mexico, USA, in the 1950s. They imported Japanese telescope eyepieces (which may have been made by **Nikon**), and sold them under their own Gailand Company brand. See Appendix IV for their maker's mark.

Gajdušek, Vilém
Vilém Gajdušek, (1895-1977) [191], was Czech optician and telescope designer. In 1947 he received the František Nušl Award from the Czech Astronomical Society. He was well-known for his collaboration with **František Kozelský**, and asteroid 3603 is named after him. Hvězdárna Ždánice (the Ždánice observatory) was the first major project undertaken by the **Gajdušek-Kozelský** collaboration, and until its closure it had several examples of their Cassegrain and refracting telescopes.

Gajdušek-Kozelský
Gajdušek-Kozelský, [191], was a collaboration between **Vilém Gajdušek** and **František Kozelský**. They made telescopes including Cassegrain and refracting telescopes. Hvězdárna Ždánice (the Ždánice observatory) was the first major project undertaken by them, and until its closure it

had several examples of their Cassegrain and refracting telescopes.

Galilean Telescope
A Galilean telescope, named after its inventor, **Galileo Galilei**, uses a converging lens as its objective, and a diverging lens as its eyepiece. The objective is either planoconvex or biconvex, and the eyepiece lens is either planoconcave or biconcave. Unlike the Keplerian telescope, this provides an upright image with correct left/right image orientation. However, the diverging eyepiece lens results in a relatively narrow field of view, short eye-relief, and a maximum magnification of about 30.

Galilei, Galileo
Galileo Galilei, (1564-1642) [37] [45] [155] [388], was an Italian natural philosopher, astronomer, and mathematician who made fundamental contributions to the sciences of motion, astronomy, and strength of materials, and to the development of the scientific method.

After beginning his studies in medicine, Galileo found that the subject did not interest him enough, so he changed to study mathematics, mechanics and astronomy. During his studies he discovered that a pendulum will swing at the same frequency whatever its amplitude. By the time he was 25 he was a Professor of Mathematics at Pisa, subsequently holding positions in Padua and Florence. He invented the pendulum clock, but was unsuccessful in making one – a feat that was not achieved until **Christian Huygens** improved the design and built one in 1656. He also invented the hydrostatic balance. In around 1590 Galileo experimented with rolling balls down slopes, and devised the laws of motion of a falling body. He also proved that a projectile would move in a parabolic arc. He published his findings in *Dialogues Concerning Two New Sciences*.

Galileo's interest in astronomy led to both innovation and controversy. Having heard of the invention of the "instrument for seeing at a distance" by **Hans Lippershey** Galileo set about improving on the design and built himself a telescope in 1610. His eponymous telescope design, the Galilean telescope, enabled him to observe the heavens. He was already a supporter of the Copernican heliocentric theory, and used his telescope to find the first evidence to support the theory. He discovered the moons of Jupiter, providing further evidence for the theory. He published these and other findings in *Siderius Nuncus* (Starry Messenger) in 1610. His belief in the Copernican theory, and his scientific evidence to support it, brought him into conflict with the Catholic Church. He published his heliocentric theories in *Dialogue Concerning the Two Chief World Systems* in 1632, but was summoned to Rome and forced to denounce the Copernican theory. He did so under threat of torture and being burned at the stake.

Galileo
See **Officine Galileo**.

Gall & Lembke
Gall & Lembke, of 21 Union Square, New York City, USA, [188] [189], were retailers and producers of optical instruments in the late-C19[th]. The company was a partnership, formed in the 1860s, between Julius Gall, (d.1883), and Charles Lembke, (1835-1903). They were distributors for **John Byrne**, and manufactured telescopes using objectives made by him. Only one catalogue from Gall & Lembke is known, and is believed to be from 1892. The catalogue shows exclusively John Byrne's products.

Galletti, Antoni
Antoni Galletti, (fl.1805-1850) [18] [263] [271] [322] [379], of Glasgow, Scotland, traded as a carver and guilder at 10 Nelson Street (1805-1828), and 21 Nelson Street (1826-1828), after which he moved to 24/25 Argyle Arcade where he traded as an optical, mathematical, and philosophical instrument maker. He claimed to have established his business in 1789, and is known to have sold: barometers; thermometer; hydrostatic bubbles. Prior to opening his business in Glasgow, he had worked in Milan, Italy. Instruments bearing his name include telescopes. He was succeeded in business by his son, **John Galletti**.

According to his trade label, "A. Galletti, Optician and Mathematical instrument maker, (agent for Dica's & Sykes' hydrometers, & Allan's sacharometers) No 21, Nelson-Street, Glasgow. A. G. sells all sorts of improved Spectacles, with pebble or glass eyes, in gold, silver, tortoiseshell, and steel frames; Reading-glasses; Telescopes; Opera-glasses; Drawing instruments; Rules and tapes, for measuring and gauging, to the imperial standard; Theodolites; Globes; Instruments for extracting milk from the breast; Spirit proof beads; Improved magic lanterns; Glaziers' patent diamonds; Quadrants; Spirit levels; Sun-dials; Barometers and thermometers, of all sorts on the most improved principles."

Galletti, John
John Galletti, (fl.1851-1894) [18] [263] [271] [322] [379], was an optical, mathematical, and philosophical instrument maker, carver, and guilder, of 24 Argyle Arcade, Glasgow, Scotland. He was the son of **Antoni Galletti**, whom he succeeded in business. He is known to have sold barometer; thermometer. Instruments bearing his name include: telescope; hydrostatic bubbles. He was succeeded in business by A. McKnight.

Gambey, Henry Prudence

Henry Prudence Gambey, (1787-1847) [154] [197] [198], was a highly regarded French optical instrument maker of the early C19th. He was the son of a clock-maker, from whom he learned instrument making, mathematics and natural sciences. He went to work for a little-known instrument maker called Ferrat near Paris, after which he went on to work as a journeyman for **Etienne Lenoir**. He worked with Lenoir for only a short time before Ferrat offered him the job of *chef d'atelier* in his school, and in 1808 Gambey went back to work for him. In 1809 his father died, and using his inheritance he opened his own shop at 52 Rue du Faubourg, St. Denis, Paris. He met the physicist, astronomer and politician Francois Arago, and this association proved helpful to his career.

His career achievements were many, including building a fine circular dividing engine. He devised improvements to machine tools, and a new kind of theodolite, which was used by the Bureau des Longitudes in their surveys. He exhibited in the national exhibitions, showing instruments such as a repeating circle, theodolite, reflection circle, comparator, precision magnetic compass, heliostat, precision declination compass with reading microscope, and an equatorial mount with two divided circles 3 feet in diameter. He was awarded gold medals for his work in 1819, 1824 and 1827. His work is said to closely resemble that of **Reichenbach** and **Repsold**, [24], and he produced instruments for the Paris observatory, including a mural circle and a transit instrument.

In about 1845 the Bureau des Longitudes asked Henry Gambey to produce a large equatorial telescope for the Paris Observatory. Gambey died in 1847, and was hence unable to fulfil this request, so **Johann Josef Brunner** built the telescope.

Gamer Engineering Co. Ltd.

The Gamer Engineering Co. Ltd. of Sherborne, Dorset, and Preston Park Close, Yeovil, England, were manufacturers of photographic enlargers during the 1950s.

Gandolfi & Sons, Louis

Louis Gandolfi, (1864-1932) [61] [122] [306], was an English camera-maker. From age 12 he worked for a firm of cabinet-makers, after which he learned the camera-making trade during five years spent working for **Lejeune & Perken**. Following this, in 1885, he set up his own business at 15a Kensington Palace, Westminster. London. He made and sold cameras of his own patented designs. By 1896 he had moved his business to 752 Old Kent Road, and in 1913 he moved again to 84 Hall Road, Peckham Rye. He specialised in medium and large format wooden cameras, which he made to order or in small production runs.

In 1928 the company moved to 2 Borland Road, Peckham Rye. By the time of this last move Louis' three sons, Thomas (1890-1965), Frederick (1904-1990), and Arthur (1907-1992), were already working for the company, and when Louis Died in 1932, the brothers began to trade as Louis Gandolfi and Sons. The company remained small, with never more than about six employees, and continued to produce small numbers of made-to-order cameras. During WWII they made cameras for **W. Watson & Sons Ltd.** After the death of Thomas in 1965 the two remaining brothers continued the company until 1982. They sold the company, and it relocated to become Gandolfi Ltd., in Andover, Hampshire. The company finally closed its doors in 2017.

Gardam and Sons, William

William Gardam and Sons, [185], was a company making engineering, surveying and astronomical instruments in New York, USA, from the 1870s until 1929. The company comprised of William Gardam and his sons Joseph and Frederick.

Gardiner, James Blake

James Blake Gardiner, (1836-1909) [50], optician, mathematical and nautical instrument maker, was married to **John King (II)**'s daughter Harriet, and began trading from his father-in-law's shop at 2 Clare Street, Bristol, England, in 1863. By 1865 he was running the business alone. Only one surviving microscope by him is known, and he ran the business until 1867 when he filed for bankruptcy, [35 #23307 p5336].

Gardner, Henry

Henry Gardner, (fl.1805-1835) [304], of Belfast, Ireland, was a watchmaker and dentist. From 1805 until 1807 he worked in partnership with **Job Rider**. He subsequently advertised as watch and clockmaker, silversmith and jeweller, and optical and mathematical instrument maker. From 1809 until 1818 he worked in partnership with **Robert Neill**, trading as **Gardner & Neill**.

Gardner, John

John Gardner (I), (1734-1822) [262] [263] [271], was a mathematical, philosophical and optical instrument maker of Bell's Wynd, Glasgow, Scotland. He began his career in instrument making working for **James Watt**, and by 1769 he was Watts' senior journeyman. As James Watts pursued his interests in steam engines and land surveying, Gardner is believed to have run his instrument making workshop. By 1773 he was working independently, employing several of the journeymen from Watts' workshop, and producing a broad range of instruments.

In 1789 Gardner became an assistant to the land surveyor James Barry and, in 1792, upon Barry's death, he took over the position. Also by 1792 his son John Gardner (II) (1765-1818) was working with him, and they began to trade in partnership with James Laurie as Gardner & Laurie. This partnership continued until 1798, and in 1799 John Gardner (I) entered a partnership with his son John (II), trading as J.&J. Gardner. This partnership continued until John (II)'s death in 1818. Following this John Gardner (I) entered a partnership with Robert Jamieson, in which they traded as Gardner & Jamieson until John Gardner (I)'s death in 1822. After his death the company continued as **M. Gardner & Co.** (sometimes trading as M. Gardner & Sons).

Gardner and Co.

Gardner and Co., [9] [53] [262] [263] [271], of Glasgow, Scotland, were opticians and manufacturers of scientific instruments. The firm was established in 1837 following the bankruptcy of **M. Gardner & Co.**, and initially traded at 44 Glassford Street. They produced instruments such as spectacles, telescopes, microscopes, barometers, thermometers, sun dials, sextants, quadrants, drawing instruments, saccharometers, barometers, and hydrometers. They also produced surveying instruments such as clinometers, mining dials, drainage levels, and surveyors' compasses.

In 1829 the company moved to 21 Buchanan Street. In 1839 they took on **James White** as an apprentice. In 1846 **William Gardner** left to start his own business, and in the same year, **David Carlaw** came to the firm as an apprentice, and James Lyle joined the company. In 1849 Margaret Gardner died, after which **Thomas Rankin Gardner (I)** is believed to have been the principal or only partner. In 1852 Thomas was appointed Optician to Queen Victoria in Glasgow.

In 1860 the company moved to 53 Buchanan Street. By 1861 **James Brown** was a salesman for the company. In 1883 Thomas retired, and his son, Thomas Rankin Gardner (II), entered a partnership with James Lyle. The company moved to 53 St. Vincent Street, where they traded as Gardner and Lyle until 1891, when Lyle left the partnership. Lyle continued in business alone until 1893. Following Lyle's departure, Thomas R. Gardner (II) carried on under the name Gardner & Co. The company continued in business after his death in 1884. In 1899 the company moved to 36 and 40 West Nile Street, and continued to trade there until at least 1920.

Gardner & Co., M.

M. Gardner & Co., (sometimes trading as M. Gardner & Sons), [263] [271], was successor to the business of **John Gardner** which, at the time of his death in 1822, was trading as the partnership Gardner & Jamieson. M. Gardner & Co. is believed to have been a partnership between Margaret Rankine Gardner (1772-1849), the widow of John Gardner (II), and their two surviving sons, **Thomas Rankin Gardner (I)** (c1805-1884) and **William Gardner**. They traded at 43 Bell Street, Glasgow, Scotland, from 1823 to 1825, then they moved to 92 Bell Street where they traded from 1826 to 1831. From there they moved to 44 Glassford Street, where they continued to trade until the partnership was terminated in 1836 due to bankruptcy. The company may have continued to do business during the sequestration proceedings which completed in 1841. However, the new family business, **Gardner & Co.**, was formed in 1837.

Gardner & Lyle

Gardner & Lyle, [263] [271], was a partnership between James Lyle and **Thomas Rankin Gardner (II)**. See **Gardner & Co.**

Gardner & Neill

Gardner & Neill, [304], was a partnership between **Henry Gardner** and **Robert Neill** (also sometimes spelled Neal, Neil, or Neale) of Belfast, Ireland. The partnership was formed in 1809, and continued until 1818. Instruments bearing their name include: microscope signed *Gardner & Neal, Belfast*; 10-inch octant signed *Gardner & Neil * Belfast*; watch signed *Gardner and Neill*; 4-draw brass and mahogany telescope signed *Gardner & Neale, Belfast*, with brass and wood tripod.

Gardner, Thomas R.

Thomas Rankine Gardner (I), (c1805-1884), [263] [271], was the grandson of **John Gardner (I)**, and the son of John Gardner (II) and Margaret Gardner, and became a notable mathematical and optical instrument maker. His son, Thomas Rankin Gardner (II) also worked in the family business. See **M. Gardner & Co.**, **Gardner & Co.**, **Gardner & Lyle**.

Gardner, William

William Gardner, (c1809-1875) [263] [271], mathematical and optical instrument maker, was the grandson of **John Gardner (I)**, and the son of John Gardner (II) and Margaret Gardner. In 1823 he entered a partnership with his mother and his brother, **Thomas Rankine Gardner (I)**, trading as **M. Gardner & Co.**, and subsequently as **Gardner & Co.** In 1846 he left to start his own business as an instrument maker. He also sold

Gargory, James
James Gargory, (fl.1833-1862) [3] [18] [44], of Birmingham, England, was an optician, and maker of drawing instruments, scales and rules for engineers, architects and surveyors. Addresses: 4 Bull St. (1833-1847); 5 Bull St. (1849-1860); 41 Bull St. (1862). He also supplied optical, philosophical, and surveying instruments, including: sextant; theodolite; level; stave; pentagraph; magnetic compass; circumferentor; barometer; thermometer; land chains and measuring tapes. Advertisements date to 1835 and 1847. Instruments bearing his name include: telescope, dated by the auction house as 1874.

Garland, John
John Garland, (fl.c1855-c1888) [61] [306], was a camera maker of Pentonville, London, England. Addresses: 5 Weston St. (1867); 15 Weston St. (1868); 30 Rodney St. (1871-1878/9); 4 Rodney St. (1878/9); 32 Hermes Street (from 1879).

He was trained in the workshops of **Thomas Ottewill**. He claimed to have been "late foreman to Ottewill & Rouch", the latter referring to **W.W. Rouch & Co.** He advertised a whole-plate camera called the *Universal Studio camera*, and various sizes of his *Folding Tailboard camera* model.

Gaumont Léon
Léon Gaumont (1864-1946), [53] [335], was a pioneer in the motion picture industry in France. He formed the company L. Gaumont et Cie in 1895 when, along with three others, he purchased **Comptoir Général de Photographie** from **Félix-Max Richard**. Léon Gaumont had been a director of the company, and when it came up for sale, he bought it, along with three co-investors, the engineer, Gustave Eiffel, who built the Eiffel Tower, the astronomer and naturalist, Joseph Vallot, and the financier Alfred Besnier. The company sold camera equipment and film. In 1897 they employed John Le Contour as their British agent, who shortly afterwards incorporated the agency as **L. Gaumont & Co.**

At the same time, in 1897, L. Gaumont et Cie began to produce motion pictures, with Alice Guy as head of their production company. They also started a chain of cinemas. In 1906 Léon Gaumont raised new financing, reorganised, and renamed L. Gaumont et Cie to become the **Société des Etablissements Gaumont**. The company owned a chain of cinemas, and was a major player in French film-making, and a manufacturer of still and motion picture equipment. Léon Gaumont retired in 1930 with the advent of talking movies and, after a merger with Franco-Film-Aubert to form Gaumont-Franco-Film-Aubert, the company went into liquidation in 1934. The company re-emerged in 1938 as Société Nouvelle des Etablissements Gaumont.

Gaumont & Cie., L.
See **Léon Gaumont**.

Gaumont & Co., L.
In 1897, **Léon Gaumont** appointed John Le Coutour (1858-1905) as the British agent for L. Gaumont et Cie motion pictures and camera equipment. In 1898, together with his assistant, Alfred Bromhead, John Le Coutour formed L. Gaumont & Co., [44], in London. The success of the company grew, and they began to handle films for Hepworth, and the French companies Elge and **Lumière**. In 1927 the company was made public as the **Gaumont British Picture Corporation**.

Gaumont British Picture Corporation
The Gaumont British Picture Corporation was formed when **L. Gaumont & Co.** was made public in 1927, and acquired Ideal Films Ltd. and W.&F. Film Service Ltd. One of the directors was Charles Woolf, owner of General Film Distributors, which was, by the mid-1930s, controlled by **J. Arthur Rank**. In 1928 Gaumont British Picture Corporation acquired the General Theatre Corporation and Gainsborough Pictures, and soon they were equipping their studios to cater for the new talking pictures. By 1937 the company had over 100 subsidiaries in a complex corporate structure. With the industry shrinking, Charles Woolf and J. Arthur Rank proposed taking over Gaumont's film production at Rank's Pinewood studios. In 1941 Rank took control of the Gaumont British Picture Corporation, and in 1947 it was incorporated into the new J. Arthur Rank Organisation.

Among the many subsidiary companies, both prior to and during the Rank ownership, were **Gaumont Kalee**, **G.B. Bell & Howell**, **G.B. Equipments**, **G.B. Kalee**, **G.B. Kershaw**, and **Bell and Howell Gaumont**.

Gaumont Kalee
Gaumont Kalee was a subsidiary of the **Gaumont British Picture Corporation**, distributing professional cinema projectors and sound equipment under its own brand-name.

Gaunt, Charles
Charles Gaunt, (fl.1845-1859) [18], was a telescope maker of 34 Meredith St., Clerkenwell, London, England. By 1859 he had relocated to 54 Red Lion St., Clerkenwell.

Gauss, Karl Friedrich

Karl Friedrich Gauss, (1777-1855) [7] [45] [155], was an eminent German mathematician who made significant contributions to the fields of mathematics, geodesy, astronomy, optics, and natural sciences. His ability and talent in mathematics and languages were recognised while he was a child, and he gained sponsorship from the Duke of Brunswick to attend university to study mathematics. Gauss was appointed director of the observatory in Göttingen, a role which he kept for the rest of his life. He made a considerable contribution to the field of mathematics and mathematical physics, even though some of his important results were not published, and were later attributed to others. In 1799 he was awarded a doctorate by the University of Helmstedt for his proof of the fundamental theorem of algebra.

In 1801 the astronomer Guiseppe Piazzi discovered the asteroid Ceres, but had no time to calculate its orbit before it disappeared behind the sun. Gauss correctly predicted its orbit and where and when it would reappear. Gauss made important contributions to the field of optics, including some of the fundamental laws of electromagnetism. He was the first to rigorously describe the formation of images based on *first order* or paraxial optics, sometimes known as *Gaussian optics*, and he derived the *Gaussian lens formula*. He made other contributions to the field of theoretical optics, and was the inventor of the heliotrope, a signalling instrument used in long distance land surveys.

Gautier, Paul

Paul Gautier, (1842-1909) [24] [72] [154], was an instrument maker from Paris, France. He began an apprenticeship at the age of 13 then, at age eighteen, he went to work for **Lerebours & Secretan** under the supervision of **William Eichens**, head of the Lerebours & Secretan workshop. In 1866, at the age of twenty-four he left them to work with Eichens, who was also leaving and setting up his own company. In 1868 Gautier and Eichens built a large siderostat to a design by **Foucault**, who died that year. The following year, together with **A. Martin** and Eichens, he built a 47-inch Newtonian for the Paris observatory. It was on a Secrétan mount, and stood unused for many years because the mount was insufficiently stable.

In 1876 he opened his own shop and, in 1880, upon the retirement of Eichens, he bought Eichens' business. He went on to collaborate with **Paul and Prosper Henry**, contributing the mounts and drives while the Henry brothers contributed the lenses and mirrors for a number of observatory telescopes. Together with the Henry brothers, Gautier continued to build telescopes for French and international observatories for another two decades. Between 1879 and 1892 Gautier and the Henry brothers built the first coudé telescope, a design proposed by **Maurice Loewy**, for the Paris Observatory, followed by others for observatories in Vienna, Lyon, Besançon and Nice. Gautier produced mounts for a number of double equatorial telescopes, with optics by the Henry brothers, used in the *Carte du Ciel* project for a photographic mapping of the sky. They also built the largest double equatorial in Europe for the observatory of Meudon in Paris. This was a double refracting telescope with objectives of 83cm (for visual observation) and 62cm (for photography). The two refractors were mounted together in a rectangular tube on an equatorial mount.

Gautier built a large refracting telescope with an objective of 49 inches for the 1900 Paris great exhibition, using a lens blank cast by **Edouard Mantios**. The telescope, however, was not a success despite his best efforts, and resulted in significant financial loss. After Gautier's death in 1909 his firm was purchased in 1910 by **Georges Prin** and, in 1934, the Prin firm became part of the Secretan business.

Gay, Richard

Richard Gay, [18] [276 #43], was an optical instrument maker of London, England. He was the father of **Thomas Gay**, and is believed to have traded during the period c1620 until c1700. He was succeeded in business by his son, Thomas.

Gay, Thomas

Thomas Gay, (fl.1668-1732) [18] [275 #327] [276 #43], was an instrument maker of the *Golden Spectacles*, by the Sun Tavern, Behind Royal Exchange, London, England. He was the son of **Richard Gay**, whom he succeeded in business. He served his apprenticeship with **Joseph Howe**, being freed of the **Spectacle Makers' Company** in 1689 or 1690, and became its Master from 1718 until 1720. From 1714 until 1723 records show him trading at *Archimedes & Spectacles*, near the Sun Tavern. He is known to have made telescopes, microscopes, reading glasses, magic lanterns, and weather glasses, as well as other instruments. The optical instrument maker **Matthew Loft** served an apprenticeship with him from 1711 to 1720, and succeeded him in his business.

G.B. Bell and Howell

G.B. Bell and Howell was a subsidiary of the **Gaumont British Picture Corporation**, distributing **Bell and Howell** movie cameras and projectors in the UK under its own brand-name.

G.B. Equipments
G.B. Equipments, [44], was a part of **British Optical and Precision Engineers**. G.B. indicates **Gaumont British**, then part of **Rank**. They made and distributed 8 and 16mm cine equipment, as well as acting as an umbrella for distribution of products from other parts of the group, such as those branded **G.B. Kershaw** and **G.B. Bell and Howell**.

G.B. Kalee
G.B. Kalee Ltd, of London, W1., [44], was a sales organisation for 35mm cine equipment. The company was started by **Abram Kershaw**. The company began as **Kalee Ltd.**, and in 1948 it was acquired, along with Kershaw and other companies, by **British Optical and Precision Engineers**. The company became G.B. Kalee Ltd. G.B. indicates **Gaumont British**, then part of **Rank**.

G.B. Kershaw
G.B. Kershaw was a subsidiary of the **Gaumont British Picture Corporation**, distributing **A. Kershaw & Sons Ltd.** movie cameras and projectors, slide projectors, and still cameras under its own brand-name.

Gebrüder Fromme
Gebrüder Fromme, [172], of Hainburgerstrasse 21, Vienna, Austria, were instrument makers, founded by the brothers, Adolf and Karl Fromme in 1884. The company address changed in 1903 to Herbeckstrasse 29, and again in 1906 to Herbeckstrasse 27. In 1913 the company incorporated as Gebrüder Fromme GmbH, by which time Karl Fromme was no longer part of the business. In 1927 the name was changed to Adolf Fromme Mechanische, Optische und Geodätische Instrumenten-Fabrik, and the company remained in business until about 1970.

Instruments bearing their name include: microscopes; drawing and surveying instruments; a fine, late-C19th, japanned brass theodolite in case, bearing the label *Gebrüder Fromme, Mathematisch-Mechanisches Institut, Wien, III/1., Hainburgerstrasse N^R 21. Special-Fabrikation sämmtlicher Vermessungs-Instrumente für den Forst-, Berg- und Hüttenbau, für Eisenbahn- und Strassenanlagen sowie sämmtliche Maassstäbe, Messbänder, Zeichenrequisiten etc. Vertretung der optischen Werkstätte Carl Zeiss, Jena.* (Translation: Special fabrication of all kinds of surveying instruments for forestry, mining and sluice-house construction, for railway and road facilities as well as all kinds of scales, measuring tapes, drawing props etc. Representing the optical workshops Carl Zeiss, Jena).

Gekroonde Lootsman, De
De Gekroonde Lootsman (The Crowned Pilot), [378] [380], was an Amsterdam firm of cartographers, map and chart sellers, and makers of navigation instruments. The company was founded by **Johannes van Keulen (I)**, who was admitted to the guild of book sellers in 1678 as a book seller and cross-staff maker. After the death of **Gerard Hulst van Keulen** in 1801, the firm no longer manufactured navigation instruments, instead retailing those by other makers. The company continued as a family concern until 1823, and ceased business in 1885.

Geneva Optical Co.
Geneva Optical Co., of Geneva, New York, USA, was founded in 1873 by Andrew Smith. They produced lenses that were sold through the **Standard Optical Company**. Geneva Optical and Standard Optical later merged to become **Shuron Optical**, which continued in business until 1963.

George Ltd., W.&J.
W.&J. George Ltd., [44] [53], of Great Charles Street, Birmingham, England, was founded some time prior to 1897, when they took over the laboratory instrument makers **F.E. Becker & Co.** During the 1930s they marketed laboratory apparatus and instruments under their *Nivoc* brand.

In 1937 they exhibited at the British Industries Fair. Their catalogue entry, [449], listed their products as laboratory equipment for all branches of industrial research and routine tests, and stated that they were makers of: Scientific instruments; *Nivoc* analytical balances; Lamp-blown glassware; Thermometers; Apparatus for gas, tar and oil testing, etc.

In the 1950s they consolidated the business by renaming it **W.&J. George and Becker**. Optical instruments bearing their name include: spectrometer.

George and Becker, W.&J.
W.&J. George and Becker, [44] [53], were scientific instrument makers of Nivoc House, 17-19 Hatton Wall, London, England, and 157 Great Charles Street, Birmingham. In 1897 **W.&J. George Ltd.** took over laboratory instrument makers **F.E. Becker & Co.** Becker continued to trade under its own name until the 1950s when the companies became W.&J. George and Becker.

They continued to use the *Nivoc* brand-name, and in 1947 they exhibited at the British Industries Fair. Their catalogue entry, [450], listed them as manufacturers of scientific instruments and laboratory equipment for educational and industrial research, specialising in chemical and physical apparatus, balances and weights.

The company became **Griffin and George Ltd.** when it merged with the laboratory instrument makers **Griffin & Tatlock**.

Optical instruments bearing their name include: telescope; spectrometer; vernier microscope.

Gerhardt Optik GmbH
Gerhardt Optik GmbH was founded by Julius Gerhardt in 1936 in Naumburg, Germany. Initially they supplied optical components to the industry, and after WWII their products included microscopes.

In 1995, they rescued **Hertel and Reuss** from bankruptcy, and acquired the riflescope maker **Nickel**. In 1996 the company name was changed to Gerhardt Optik und Feinmechanik GmbH. Their registered brands for optical instruments are *Hertel and Reuss* and *Nickel*.

German Equatorial Mount
See equatorial mount.

German Makers' Ordnance Codes for Optical Products
During WWII, German military optical instruments, and those made in occupied territories, were not marked with the name of the manufacturer for security reasons. Instead, they were marked with a three-letter code that represented the maker's name. This same system applied to all German military ordnance. For a list of the codes relating to optical instrument makers, and the makers' names they represent, see Appendix III.

GeoMax
GeoMax is a **Hexagon AB** brand surveying instrument manufacturer and service provider. Their products include total stations, laser scanners, global navigation satellite systems, 3D levelling systems, levels, laser rotators, pipe lasers, field controllers, locators, and accessories.

Gerrard, William
William Gerrard, (fl.1862-1890) [16] [54], was a chronometer, patent lever watch, and nautical instrument maker of *Newton's Head*, 35 South Castle St., Opposite Chapel Walks, Liverpool, England. He was sextant maker by appointment to the Royal Navy. His trade label is found in the box lids of various sextants by other makers, which he may be presumed to have either retailed, repaired, adjusted or calibrated. Instruments bearing his name include: sextants; two-draw telescope.

Ghost Micrometer
A ghost micrometer, [216], uses the images of fixed or moveable reticule lines in the measurement of an image formed in the objective of a microscope, telescope or spectroscope. The lines of the reticule are "ghost lines" (images of lines) instead of material lines, thus avoiding the diffraction effects experienced with directly viewing a scribed line or a wire.

The concept had been discussed, but not put into practice until a system was devised by **Karl von Littrow** for use in a meridional telescope, and adapted by Mr. G.P. Bidder for use in a position micrometer on an equatorial mount. The resulting instrument was effective, but liable to large parallax errors in the case of small misalignments of the prism. A further improved version was devised by **Charles Burton** and **Sir Howard Grubb**.

Gibson, John
John Gibson, (d.1795) [18] [263] [271] [276 #973] [322], was an optical instrument maker, and watch and chronometer maker of Kelso, Scotland. Instruments bearing his name include: reflecting telescope; solar microscope.

Gieves Ltd., Gieves & Hawkes Ltd.
Gieves Ltd., of Saville Row, London, England, was a military and gentleman's outfitters who also sold other accessories. They acquired Hawkes & Co. in 1974 to become Gieves & Hawkes Ltd., [54], who were tailors, and supplied items of uniform to the British Army. Telescopes bearing the inscription *Gieves Ltd.* may be found, but Gieves did not make the telescopes themselves. There is an example in the Imperial War Museum, [261], bearing the inscription *Ross, London No 71395 Gieves Ltd.* Another example is in the National Maritime Museum, bearing the inscription '*GIEVES LTD. / No. 8576*'.

Gilbert & Co.
Gilbert & Co. is believed to be another name used by **Gilbert, Wright & Hooke**. Instruments bearing this name include two octants [16] [374], and a solar microscope [264 p236], all dated c1800. They are also known to have sold: microscope; telescope.

Gilbert & Gilkerson
Gilbert & Gilkerson, [18] [276 #974], were opticians, and scientific, mathematical, and optical instrument makers of 8 Postern Row, Tower Hill, London, England. The firm was a partnership between **William Gilbert (I)** and **James Gilkerson**, formed in 1793, and succeeded in 1809 by **Gilkerson & Co.** They are known to have sold: sextant; rule; globe; equinoctial ring dial.

Gilbert & Son, John
John Gilbert & Son, (fl.1749) [18], opticians and mathematical instrument makers of Postern Row, Tower

Hill, London, England, was a partnership between **John Gilbert (I)** and his son, **John Gilbert (II)**. The firm was succeeded by John Gilbert (II).

Gilbert & Son; Gilbert & Sons
Gilbert & Son, sometimes trading as Gilbert & Sons, [**18**], were opticians and mathematical instrument makers of *Navigation Warehouse*, 148 Leadenhall St., London, England. The partnerships were between **William Gilbert (I)** and his sons **William Gilbert (II)** and **Thomas Gilbert**, formed in 1806. They continued in business after the death in 1813 of William (I), until 1816 (Gilbert & Son), and 1819 (Gilbert & Sons).

Gilbert & Wright
Gilbert & Wright, [**18**], of *Navigation Warehouse*, 148 Leadenhall Street, London, England, was a partnership between **William Gilbert (I)** and **Gabriel Wright**. The partnership began trading in 1789, [**35** #13124 p560]. In 1794 **Benjamin Hooke** joined the partnership, and they traded as **Gilbert, Wright & Hooke** until 1801 when Hooke left the partnership. After this they continued to trade as Gilbert & Wright until 1805.

Gilbert, John (I)
John Gilbert (I), (fl.1719-1749 d.1750) [**9**] [**16**] [**18**] [**53**] [**90**], was the first of a family of optical and mathematical instrument makers in C18th and C19th London, England. Addresses: Little Tower Hill (1716); Postern Row, Tower Hill (1718-1750).

In 1709 he began an apprenticeship with the mathematical instrument maker John Johnson, and was freed of the **Grocers' Company** in 1716. He started to work as a mathematical instrument maker and optician in 1719, and among his apprentices were his son, **John Gilbert (II)**, and **Henry Crawford**. In 1749 he was trading in partnership with John Gilbert (II) as **John Gilbert & Son**. John Gilbert (II) succeeded him in his business.

Gilbert, John (II)
John Gilbert (II) (fr.1744, d.1791), [**9**] [**16**] [**18**] [**53**] [**90**], was an optical and mathematical instrument maker of London, England. Addresses: The Mariner, Postern Row, 7 Tower Hill; Postern Row (1745); The Mariner in Postern Row, Tower Hill (1771); 12 Tower Hill (1776-1781); 8 Postern Row (1782-1791).

John Gilbert (II) served his apprenticeship with his father, **John Gilbert (I)**, and was freed of the **Grocers' Company** in 1744. In 1749 he was trading in partnership with his father as **John Gilbert & Son**. After his father's death in 1750, he worked under his own name as a mathematical instrument maker and optician. Among his apprentices were his son **John Gilbert (III)**, who served his apprenticeship from 1764 until 1771, and **Thomas Ripley**. He also took his son, **William Gilbert (I)**, as an apprentice in 1769, but the apprenticeship was turned over to **Peter Dollond** a few weeks later. He was succeeded in his business by William Gilbert (I).

An undated trade card from the mid-C18th [**361**] reads: "John Gilbert at the Mariner, in Postern Row, Tower Hill, London. Makes and sells wholesale and retails all sorts of Mathematical, Optical & Philosophical instruments viz. Hadley's sextants & octants, in wood and brass. Davis's quadrants, Azimuth steerage, amplitude & common compasses. Gunter scales, Sliding D°. Reflecting telescopes of various sizes, Refracting D° for day & night. Theodolites, Circumferenters, Plain tables, Perambulators or measuring wheels. Pantographs for copying drawings. Parallel-rules. Cases of drawing & surveying instruments, Universal & horizontal dials. Globes of all sizes. Spectacles. Concave and convex mirrors. Opera glasses. Reading & burning glasses. Perspective machines, Double or single, Electrical machines, Barrell & hand air pumps. Microscopes, Camera obscuras, Prisms, Barometers, Hygrometers, Hydrometers, Farenheit's thermometers, D°. for distillery, brewery & botany. Gauging instruments, Rules & scales of all sorts in wood, brass & ivory, with various other instruments not mentioned. Books and charts of navigation. NB. Walking canes sold and mounted in the neatest manner." (Note, it is not clear whether this trade card was for John (I) or John (II)).

Gilbert, John (III)
John Gilbert (III), (fr.1771, d.1780) [**18**] [**90**], was an optician and mathematical instrument maker of St. Paul's Churchyard, London, England. He served an apprenticeship with his father, **John Gilbert (II)**, beginning in 1764, and was freed of the **Grocers' Company** in 1771. Upon the death of **Joseph Linnell** in 1775, John (III) succeeded him in his business, trading at 33 Ludgate St., London until his death in 1780.

Gilbert, Thomas
Thomas Gilbert, (fl.1809-1831) [**18**] [**90**], was an optician and mathematical instrument maker of 148 Leadenhall St., London, England. He served his apprenticeship with his father, **William Gilbert (I)**, beginning in 1801, and was freed of the **Grocers' Company** in 1809. Among his apprentices were the mathematical instrument makers **John Thomas Schmalcalder** and George Schmalcalder, the sons of **Charles A. Schmalcalder**. In 1806 he began to work in partnership with his father, and his brother, **William Gilbert (II)**, trading as **Gilbert & Sons**. Following the termination of this partnership, in 1819 he began to work

in partnership with William Gilbert (II), trading as **W.&T. Gilbert** until 1831.

Gilbert, W.&T.
W.&T. Gilbert, [18], of *Navigation Warehouse*, 148 Leadenhall Street, London, England, was a partnership between **William Gilbert (II)** and his brother, **Thomas Gilbert**, formed in 1819. In 1823 **Andrew Ross** began to work for W.&T. Gilbert, and rose to become their works manager. In 1828 the Gilberts were declared bankrupt, [35 #18512 p1860]. Andrew Ross left to start his own company in 1830. The partnership, W.&T. Gilbert, ceased in 1831, and from 1832 William Gilbert (II) continued in business alone. W.&T. Gilbert are known to have sold compasses, gunner's scales, sectors and sextants.

Gilbert, William (I)
William Gilbert (I), (fl.1776 d.1813) [16] [18] [53] [90], was an optical and mathematical instrument maker of London, England. Addresses: 8 Postern Row, Tower Hill (1776); 148 Leadenhall St. (1795-1813); Stepney Green (1805. Possibly a residence). He began his apprenticeship with his father, **John Gilbert (II)** in 1769, but within weeks his apprenticeship was taken over by **Peter Dollond**. He was freed of the **Grocers' Company** in 1776. He was freed by purchase of the **Spectacle Makers' Company** in 1801, and became Master of the Company in 1807-1808. He had four sons, **William Gilbert (II)**, **Thomas Gilbert**, Henry and Charles, and all except Charles served as his apprentices.

William Gilbert (I) engaged in several partnerships which traded at *Navigation Warehouse*, 148 Leadenhall St., London. He joined the partnership **Gregory & Wright** sometime after its formation in about 1782, from which time they traded as **Gregory, Gilbert & Wright** until 1789. Following the dissolution of this partnership, he continued in partnership with **Gabriel Wright**, trading as **Gilbert & Wright** until 1792. During 1793 he was engaged in a partnership with **Henry Gregory**, trading as **Gregory, Gilbert & Co.** In 1794 he formed a partnership with Gabriel Wright and **Benjamin Hooke**, trading as **Gilbert, Wright & Hooke** until 1801, after which they once again traded as Gilbert & Wright until 1805.

William (I) and **James Gilkerson**'s partnership, **Gilbert and Gilkerson**, began in 1793 at 8 Postern Row, Tower Hill, London. They were succeeded by **Gilkerson & Co.** in 1809.

William (I) conducted business with his sons as **Gilbert & Son**, and sometimes **Gilbert and Sons**, from 1806 until his death in 1813. After this, William Gilbert (II) and Thomas Gilbert worked in partnership as **W.&T. Gilbert**.

Gilbert, William (II)
William Gilbert (II), whose full name was William Dormer Gilbert, (fl.1802-1845) [18] [90], was an optician, and mathematical and philosophical instrument maker of London, England. Addresses: Leadenhall St. (1802-1813); workshops in Woodford, Essex (1826); 148 Leadenhall St. (1832); 97 Leadenhall St. (1840); 138 Fenchurch St. (1845).

He served his apprenticeship with his father, William **Gilbert (I)**, beginning in 1795, and was freed of the **Grocers' Company** in 1802. He was freed of the **Spectacle Makers' Company** by purchase in 1813. Among his apprentices was **Alfred Chislett**, who was turned over to him in 1825 by **Robert Brettell Bate**.

In 1806 he began to work in partnership with his father, and his brother, **Thomas Gilbert**, trading as **Gilbert & Sons**. Following the termination of this partnership William Gilbert (II) and Thomas Gilbert worked in partnership as **W.&T. Gilbert**, from 1819 until 1831. In 1828 the Gilberts were declared bankrupt, and from 1832 William Gilbert (II) continued in business alone.

William (II) was appointed compass maker to the Admiralty. He supplied theodolites to the East India Company for the Survey of India. However, in 1838, following a comparison of the accuracy of their instruments, **Troughton & Simms** took over as the supplier.

Gilbert, Wright & Hooke
Gilbert, Wright & Hooke, [18], were mathematical instrument makers of *Navigation Warehouse*, 148 Leadenhall Street, London, England. The firm began as a partnership, formed in 1789, between **William Gilbert (I)** and **Gabriel Wright**, trading as **Gilbert & Wright**. In 1794 **Benjamin Hooke** joined the partnership, and they began to trade as Gilbert, Wright & Hooke. The partnership continued until 1801 when Hooke left the partnership. After this they continued to trade as Gilbert & Wright until 1805.

Gilkerson & Co., James
James Gilkerson & Co., [16] [18] [276 #1130], were opticians, and optical, mathematical, and philosophical instrument makers of 8 Postern Row, Tower Hill, London, England. They are also recorded as trading at the *Navigation Warehouse*, 148 Leadenhall St. The firm, owned by **James Gilkerson**, succeeded **Gilbert & Gilkerson** in 1809, and continued in business until 1825, after which Gilkerson began to work in partnership with **John McAll**, trading as **Gilkerson & McAll**.

Gilkerson & McAll, James & John
Gilkerson & McAll, [18], were mathematical instrument makers of 8 Postern Row, Tower Hill, London, England. The firm was a partnership between **James Gilkerson** and **John McAll**. The partnership began in 1826, and continued until 1830, after which John McAll continued in business at the same address.

Gilkerson, James
James Gilkerson, (fl.1793-1825) [9] [18] [276 #1130, #1340], was an optical and mathematical instrument maker of 8 Postern Row, Tower Hill, London, England. From 1793 until 1809 he traded in partnership with **William Gilbert (I)** under the name Gilbert and Gilkerson. From 1809 he continued his business as **James Gilkerson & Co.**, and was succeeded by **Gilkerson and McAll**.

Gilles, Alexandre François
Alexandre François Gilles was generally known by the name **Selligue**.

Gillett and Sibert
Gillett and Sibert was a Scottish company, formed in 1960 and incorporated in 1974, that made microscopes, including their popular *Conference Research Microscope*. The company became a part of Elcomatic, possibly in 1990, but optical instruments were not Elcomatic's core business. They continued to make a small number of microscopes under the Gillett and Sibert name until 2012.

Ginsberg, Aleksander
Aleksander Ginsberg was the founder of the **FOS Limited Partnership Society Ginsberg and Co.**

Girdlers' Company
See Appendix IX: The London Guilds.

Giroux & Cie, Alphonse
Alphonse Giroux, [70], was a relative of **Louis Daguerre**'s wife. Daguerre bought his experimental cameras from Vincent and Charles **Chevalier** until 1839, after which he exclusively bought them from Alphonse Giroux. Prior to 1841, the Giroux et Cie's Daguerre cameras were fitted with a lens of diameter of 3¼ inches and focal length 16 inches. This was initially a plano-convex lens, and had a fixed stop of 15/16 inch. This resulted in a focal ratio of f/17, which was very slow for portraiture. Subsequently they were supplied fitted with Chevalier's achromatic meniscus lens which, operating at f/14, was still slow.

Glan-Foucault Prism
See prism.

Glan-Thompson Prism
See prism.

Glass, Optical
See Appendix VI.

Glass Perfection
Glass Perfection is a UK based optical glass manufacturer.

Glover, Joseph
Joseph Glover, (fl.1834-1840) [18], was an optical, mathematical, and philosophical instrument maker of Church St, Hackney, London, England.

Goddard, James Thomas
James Thomas Goddard, (1851-1863) [18] [322], was a telescope maker and photographic lens maker of London, England. Addresses: 35 Goswell St. (1851-1853), and Jesse Cottage, Whitton, Middlesex. Instruments bearing his name include an achromatic astronomical telescope.

Godfrey, Thomas
Thomas Godfrey, (1704-1749) [45], of Philadelphia, USA, was an inventor. He began his career as a glazier, but went on to study mathematics and science. He invented the nautical quadrant independently of **John Hadley**. A competing sea quadrant using prisms instead of mirrors was invented by **Caleb Smith**, but it did not prevail.

Goerz, Carl Paul
The company C.P. Goerz, [55] [374], was founded by Carl Paul Goerz, (1854-1923) [47], in 1886 in Berlin, Germany. Goerz had served an apprenticeship with **Emil Busch** in Rathenow, Germany, and was later a partner of **Eugen Krauss** in Paris, France. In 1895 Goerz opened a branch of his company in New York, USA, which started production in 1902, and became the C.P. Goerz American Optical Co. in 1905. In 1903 C.P. Goerz set up a successful department for manufacturing military optics. In 1905 they set up a manufacturing plant in Vienna, Austria, and in 1908 they set up a plant in Pozsony (now Bratislava, Slovakia). The Austrian plant manufactured military optics in Austria during WWI and WWII, as well as making consumer cameras. Examples of C.P. Goerz binoculars made in Vienna and/or Pozsony bear the mark *Wien und Pozsony*. Other branches of C.P. Goerz included Paris, France (est.1893), London, England (est.1899), St. Petersburg, Russia (est.1905). In 1926 the German branch of C.P. Goerz merged with **Contessa-Nettel**,

Ernemann and **ICA** to form the company **Zeiss Ikon**. The Austrian branch of C.P. Goerz supplied 700 CM2 gunsights, designed by **Barr and Stroud**, to the RAF, before Austria, as part of the Third Reich, went to war with Britain in September 1939 [**26**]. Goerz products also included cameras, camera lenses, binoculars, telescopes, periscopes, and meteorological and aeronautical instruments.

Gofton, William
William Gofton, (fl.1843) [**18**], was an optical and surgical instrument maker of 63 Farringdon St., London, England. He was freed by purchase of the **Spectacle Makers' Company** in 1843.

Gogerty, Robert
Robert Gogerty, (1814-c1856) [**18**] [**61**] [**319**] [**322**], of London, England, was an optician, and optical, mathematical, and philosophical instrument maker, as well as a brass turner, and shower bath and pump maker. Addresses: 25 St. John St., Clerkenwell (1834-1835); 19 Great Sutton St. (1839); 32 King St., Smithfield (1845), and 72 Fleet St. (1849-1855). He served his apprenticeship with **John Stanton**, beginning in 1828, and was turned over to Joseph Barker of the Armorers' and Brasiers' Company in 1834. He was freed of the **Drapers' Company** in 1835.

He advertised as a "Photographic chemist, and manufacturer of lenses, cameras, and all kinds of photographic apparatus". His entry in the catalogue for the 1851 Great Exhibition [**328**] lists him as an inventor of 72 Fleet Street, displaying a model of a pair of direct-acting steam-engines, with paddle-wheels, a plate electrical machine to exhibit negative and positive electricity, a double-barrel air-pump with iron plate, and a delicate galvanometer.

Instruments bearing his name include: brass drum microscope; marine barometer; barometer on a stand with a carved oak dancing bear; six-draw brass telescope in fitted case with extra eyepiece and mounting attachment. He also sold mounted stereo photographic prints.

GOI
See **Vavilov State Optical Institute**.

Goldsmiths' Company
See Appendix IX: The London Guilds.

Gonichon & Fils, Veuve
Veuve Gonichon & Fils (Widow Gonichon & Son), [**416**], was a name used in two partnerships:

The first was a partnership between **Jean Charles Gonichon** and his mother, **Marie Michelle Gonichon**. They began to trade together following the death of Marie's husband, **Jean-Baptiste Gonichon** in 1761, and the partnership continued until 1781.

The second was a partnership between **Paule Gonichon**, who was the widow of Jean Charles, and their son, **Pierre Charles Gonichon**. This partnership was formed following the death of Jean Charles in 1800, and continued until 1806.

Gonichon, Jean-Baptiste Charles
Jean-Baptiste Charles Gonichon (1703-1761) [**256**] [**416**], was an optical instrument maker of Rue de Postes, Paris, France. He began his career as an engineer, mapping the French territory of Louisiana and the French quarter of New Orleans in North America. By 1733 he had moved back to Paris, and began to work as an optical instrument maker, in partnership with **Claude Paris**. He supplied many telescopes to the East India Company. He married Claude's sister, Marie Michelle Paris (**Marie Michelle Gonichon**), with whom he had a son, **Jean Charles Gonichon**, and a daughter, Marie Angélique Catherine (**Marie Putois**), who married **Etienne Antoine Putois**.

Gonichon, Jean Charles
Jean Charles Gonichon, (fl.1761-1800) [**416**], was a telescope maker of Rue des Postes, Paris, France. He was the son of **Jean-Baptiste Gonichon** and **Marie Michelle Gonichon**. Following the death of his father, he worked in partnership with his mother, trading as **Veuve Gonichon & Fils**. In 1780 he married Paule Cojette (**Paule Gonichon**). In 1781, the year their son, **Pierre Charles Gonichon**, was born, he stopped working in partnership with his mother, and continued under his own name. In 1800 his daughter married **Robert Aglaé Cauchoix**. Jean Charles presumably died in 1800 because, following this, his widow worked in partnership with Pierre Charles, once again using the name Veuve Gonichon & Fils.

Gonichon, Marie Michelle
Marie Michelle Gonichon, (fl.1761-1781) [**416**], was an optician of Rue de Postes, Paris, France. She was born Marie Michelle Paris, the sister of **Claude Paris**, and married **Jean-Baptiste Gonichon**. She succeeded her husband in his business in 1761, and worked in partnership with their son, **Jean Charles Gonichon**, trading as **Veuve Gonichon & Fils**.

Gonichon, Paule
Paule Gonichon, (fl.1800-1806) [**416**], was an optician of Paris, France. She was born Paule Cojette, and in 1780 she married **Jean Charles Gonichon**. Upon the death of her husband, she continued the business at Rue des Postes, trading in partnership with their son, **Pierre Charles**

Gonichon, as **Veuve Gonichon & Fils**. In 1802 they moved the business to Rue de la Feuillade, 2, at the corner of Place des Victoires, and the partnership traded there until 1806, after which Pierre Charles continued the business under his own name.

Gonichon, Pierre Charles
Pierre Charles Gonichon (1781-1867) [**416**], was a telescope maker of Paris, France. He was the son of **Jean Charles Gonichon** and **Paule Gonichon**. In 1800, he began to work in partnership with his mother, trading as **Veuve Gonichon & Fils** at Rue des Postes. In 1802 they moved the business to Rue de la Feuillade, 2, at the corner of Place des Victoires, and the partnership traded there until 1806. From 1807 until 1833 he continued the business under his own name.

Goniometer
A goniometer is an instrument for measuring angles, and is particularly used in the study of crystals. The contact goniometer is the simplest form, consisting of two rules, pivoted together, with a graduated circle indicating the angle between the two rules.

A more accurate instrument is the reflecting goniometer, in which the crystal surface reflects a light source. It is rotated until another facet reflects the light source to the same place, and the angle of rotation indicates the angle between the facets. An improvement to this makes use of a collimated light source and a viewing telescope, mounted such that the angle between them may be varied on a circle, and read off on a graduated scale. Many variants exist, one of which is the Mitscherlich goniometer, which uses a vertical circle, whereas the more common form uses a horizontal circle.

Goniophotometer
A goniophotometer is an instrument, similar to a goniometer, used to measure the intensity of light with respect to the angle from which the light source is viewed. This may be used, for example, to measure the spread of light from a vehicle LED headlamp.

Gorizont; Horizon
Gorizont, (Горизонт; Horizon), is a name used by **Krasnogorskij Mechanicheskij Zavod** (KMZ), beginning in 1967, for a series of models of swing-lens panoramic camera.

Gosudarstvennyy Opticheskiy Institut (GOI)
See **Vavilov State Optical Institute**.

Gosudarstvennyi Optiko-Mehanicheskiy Zavod (GOMZ)
The Russian State Optical Mechanical Plant (Государственный оптико-механический завод) [**23**], was formed during the 1930s. When the company **LOMO** was formed in 1962, GOMZ were a part of LOMO. Their camera models included the popular *Lubitel* twin-lens reflex. For Russian and Soviet makers' marks, see Appendix V.

Goto Inc.
The Goto Optical Mfg. Co. was founded in 1926 by Seizo Goto in Tokyo, Japan. He began by making a 25mm f/3.2 refractor which quickly became popular. By the 1950s the company was making telescopes and planetarium projectors, and then complete planetariums. They continued to expand their offerings of both telescopes and planetariums, and were still selling telescopes until at least the late-1970s. Following this they shifted their business focus to planetariums only.

Gough, W.
Walter Gough, (fl.1784-1811) [**18**] [**276** #1134] [**319**] [**322**], otherwise known as William Gough, was an optician, and optical, mathematical, and philosophical instrument maker of Holborn, London, England. Addresses: 21 Middle Row (1784-1790); 20 Middle Row (1790-1794); 23 Middle Row (1799-1811). Instruments bearing his name include: microscope; sundials; air pump; mercury barometer with hygrometer and mercury thermometer in mahogany case with marquetry. He advertised a wide range of instruments.

Gould, Charles
Charles Gould, (c1786-1849) [**18**] [**50**] [**264**], was an optical instrument maker of 272 Strand, London, England. He is believed to have been the manager and head machinist for **William Cary**'s business. His son, **Henry Gould** also worked for the business. Charles was a microscope maker, and an innovator in microscope slide mounting techniques. How and when the Cary business transitioned into the hands of the Gould family is unclear, but the box-mounted portable microscope, invented by William Cary, was announced in 1826, after his death, by Charles Gould, and is generally referred to as a Gould-type microscope.

Gould, Charlotte Hyde
Charlotte Hyde Gould, (c1797-1865) [**50**], was an optician, and was married to **Henry Gould**. At some point Charlotte and Henry are believed to have become the proprietors of **William Cary**'s business. After Henry's

death in 1856 she continued the business as Gould, late Cary. In about 1863, **Henry Porter** acquired an interest in the business, after which it operated as Gould and Porter. The business continued as Gould and Porter after Charlotte's death until sometime between 1874 and 1876, when Henry Porter is believed to have become the sole owner.

Gould, Henry
Henry Gould, (c1796-1856) [**18**] [**50**], a London instrument maker, was the son of **Charles Gould**, and married to **Charlotte Hyde Gould**. By 1835 he is believed to have been foreman in **William Cary**'s business.

Gould & Porter
Gould & Porter, [**50**], was a partnership between **Charlotte Hyde Gould** and **Henry Porter**. The partnership began in about 1863, and continued after Charlotte's death until sometime between 1874 and 1876 when Henry Porter is believed to have become the sole owner.

Gowllands Ltd.
Gowllands Ltd., of Croydon, Surrey, England, [**44**], was formed in 1898 by William Gowlland, Egbert Gowlland, and Charles Septimus Gowlland. They became William Gowlland Ltd. in 1908, and in 1931 they became Gowllands Ltd.

The catalogue for the 1929 British Industries Fair, [**448**], listed their entry as an optical, scientific and photographic exhibit, and listed the company as manufacturers of: Ophthalmic uncut lenses; Ophthalmoscopes; Retinoscopes; Brackets; Trial frames; Laryngoscopic lamps; Keratometers; Orthoscopes; Perimeters; Tables; Laryngoscopic, dental, aural instruments; Trial cases and accessories; Test cabinets, with sterilisers; May-Mortons and combined sets.

Their entry in the users' section of the 1951 *Directory for the British Glass Industry*, [**327**], lists Gowllands Ltd., of 176 Morland Road, Croydon, Surrey, established in 1898. It states that they have eight gas and electric furnaces, and produce: Ophthalmic lenses; Instruments for the examination of the ear, eye, nose and throat, including optically worked lenses and mirrors; Dental mirrors; Magnifiers. The joint managing directors were E. Gowlland and G.P. Gowlland.

In 1998 the company was acquired by Medicamenta of the Czech Republic.

Graflex
Graflex Inc., [**122**], was a camera manufacturer which began as **Folmer & Schwing** and, following a period of ownership by the **Eastman Kodak Company**, became an independent company named Folmer Graflex Corporation. In 1945 the company name was changed to Graflex Inc. In 1956 Graflex Inc. was acquired by the General Precision Equipment Corporation, and in 1968 it was acquired by Singer Corporation. Graflex was dissolved in 1973.

Graham, George
George Graham, (1673-1751) [**3**] [**18**] [**24**] [**46**] [**276**] [**443**], originally from Cumberland, England, was a renowned clock, watch, and instrument maker of *Dial and Three Crowns*, Water Lane, London (1713-1720), and next door to *Globe and Marlborough's Head Tavern*, London (1720-1751), and during that same period, at *Dial and One Crown*, Fleet St. London. He served his apprenticeship with the London watchmaker, Henry Aske, and took on apprentices William Dutton and Thomas Mudge. He later worked for Thomas Tompion (1638-1713), the royal clockmaker, whose niece he married.

He employed **John Bird**, Thomas Mudge, **Jonathan Sisson** and J Shelton. He applied his fine skill as a watch and clockmaker, and his knowledge of astronomy, to the production of astronomical instruments, including sidereal clocks, mural quadrants, meridian transit telescopes and zenith sectors, and equipped major observatories including the Royal Observatory at Greenwich. He was succeeded in his business by his ex-apprentice, the chronometer maker Thomas Mudge.

Granger, R.
R. Granger, (fl.1790) [**18**] [**53**] [**264**], of Tettenhall, Staffordshire, England. Instruments bearing this name include a brass three-in-one magnifier, otherwise known as a "flea", or "acorn" microscope.

Grant, Alexander
Alexander Grant, (fl.1795-1822) [**18**] [**276** #1137], was a mathematical and optical instrument maker of 2 Winckworth's Buildings, City Rd., London, England. He was engaged by Captain Gower, [**276** #840], to make the spiral fitting for his patent log, which was sold exclusively by **Gilbert & Son**.

Grant, James
James Grant, (fl.1837-1848) [**16**] [**18**] [**276** #2139], of 40 Queen St., Hull, England, was an optician, retailer of mathematical and nautical instruments, stationer, and seller of maps and charts. Instruments bearing his name include a sextant with scale showing a scale divider's mark in the form of the initials FW (See Appendix (XI).

Grant, John
John Grant, (fl.1863-1865) [**271**], was an optical and nautical instrument maker of 45 Regent Quay, Aberdeen, Scotland. He succeeded **Smith & Ramage** in business.

Grant, Matthew
Matthew Grant, (fl.1830-1839) [**18**], was an optical turner of 15 Kirby St., Hatton Garden, London, England.

Graphic Telescope
The graphic telescope, [**53**] [**388**], invented by **Cornelius Varley**, is an optical aid for artists. It combines the ability of the camera obscura to project a view of a distant object with the ability of the camera lucida to superimpose an image onto a paper or canvas surface, for use in sketching. Varley's lens system allows great versatility in the size of, and distance to the object to be viewed. Depending on the power of the telescope the magnification can be anything from five to sixty times – the lower magnifications, from five to twenty, being considered by Varley to be the most useful. Varley also offered a portable stand to avoid vibration when using the graphic telescope outdoors.

Graphoscope
A graphoscope is an instrument used for magnifying photographs, engravings etc., and usually consists of a frame containing a mount for the photograph and a lens for viewing it. The frame is commonly designed to fold flat when not in use. See also graphostereoscope.

Graphostereoscope
A graphostereoscope (or stereographoscope) is of similar construction to a graphoscope, but combines the function of the graphoscope with that of the stereoscope, generally having a large lens for mono viewing, and two smaller lenses for stereo viewing.

Gravatt, William
William Gravatt, (1806-1866) [**312**], was a civil engineer from Gravesend, Kent, England. He worked with both Isambard Kingdom Brunel and his father, Marc Isambard Brunel, on the Thames tunnel crossing in London, as well as other projects. He was a friend of **W.H. Wollaston**, **William Simms (II)** and **Edward Troughton**. In 1832 he became a Fellow of both the Royal Society and the Royal Astronomical Society. While working on the London and Dover railway scheme he is said to have invented the dumpy level as an improvement on the Y level.

In 1850 Gravatt was chosen by Reverend John Craig to superintend the construction of the Craig Telescope [**312**], which was built on Wandsworth Common in London. One of the glass optical blanks was made by **Chance Brothers**. There were, however, problems with its lens figuring. **William Parsons (I)** gave advice about correcting the optics, but no corrections were made, and the telescope was not considered a success. Gravatt died in 1866 from an overdose of morphine which was accidentally administered by his nurse.

Gray, John (I)
John Gray (I), (fl.1804-1877) [**18**] [**276** #1138] [**322**], was a mathematical, nautical, and optical instrument maker of Liverpool, England. His business was located near the Dry Dock, Liverpool until 1828, when he moved to 25 Bridgewater Street. In 1811 he traded briefly in partnership as Gray & Lessels.

In 1839 he moved to 26 Strand Street, where **Charles Jones** was already working. There, and at 25 Strand Street, he participated in a series of partnerships. These were **Jones & Gray** until 1841, then **Jones, Gray & Keen** until 1842, and finally **Gray & Keen** until 1855, after which he traded under his own name until 1877.

According to [**3**] there is a likely family relationship between John Gray (I) and **John Gray (II)** - possibly John (II) being the father of John (I).

Gray, John (II)
John Gray (II), (fl.1823-1839) [**18**], was an optician, and mathematical and philosophical instrument maker of London, England. Addresses: 36 Nightingale Lane, East Smithfield; 13 Little Hermitage St., Wapping; 41 Nightingale Lane, East Smithfield (1823); 4 Upper East Smithfield (1839). He served his apprenticeship with the mathematical instrument maker John Kitchingman, beginning in 1783, and was a Freeman of the **Turners' Company**.

A trade label reads "John Gray, Manufacturer of Sextants, Quadrants, Compasses, Telescopes, &c., No 13 Little Hermitage Street, Wapping, London."

According to [**3**] there is a likely family relationship between **John Gray (I)** and John Gray (II) - possibly John (II) being the father of John (I).

Gray & Keen
Gray & Keen, [**3**] [**18**] [**276** #1138] [**322**], were opticians and nautical instrument makers of 25 & 26 Strand St., Liverpool, England. Following the dissolution of the partnership **Jones, Gray & Keen** in 1842, [**35** #20078 p641], **John Gray (I)** and **Robert John Keen** continued to work in partnership. In 1851 Gray & Keen exhibited barometers at the Great Exhibition in London, [**294**] [**328**]. Their partnership continued until 1855.

Gray, Stephen
Stephen Gray, (fl.1694-1736) [**276** #50] [**319**] [**322**], was a scientist, and optical, mathematical, and philosophical

instrument maker of Canterbury, Kent, and Charter House, London, England. He was an experimentalist who designed and made electrical machines, and contributed to contemporary understanding of electricity. He was a Fellow of the Royal Society, to which he contributed papers, including on the "drop" microscope, and one entitled *A Method of drawing a true Meridian Line by the Pole Star*, with a description of a telescope he designed for the purpose. He is known to have made: sundials; microscopes; electrical machines.

Green, John
John Green, (fl.1880s) [201], of Brooklyn, New York, USA, was an American maker of microscopes and objectives. He had worked for **Robert B. Tolles** before setting up his own business. He also retailed microscopes by other makers.

Green, William (I)
William Green (I), (fl.1818-1856) [18] [276 #1872] [322], was an optical, mathematical, and philosophical instrument maker of London, England. Addresses: 7 Bartholomew Terrace, City Rd. (1826); Paternoster Row, and 54 Rahere St., Goswell Rd. (1828); 14 Fountain Place, City Rd. (1838-1855); 10 Guildford Place, Farringdon Rd. (1865). He served his apprenticeship with **Michael Dancer**, the grandfather of **John Benjamin Dancer**, beginning in 1808, and was a Freeman of the **Joiners' Company**.

Green, William (II)
William Green (II), (fl.1778-1790) [276 #722], was a surveyor of South Moulton St., Hanover Square, London, England, and 93 High St. Marylebone, London. In 1778 he published *Description & Use of the Improving Reflecting & Refracting Telescopes and Scale for Surveying*. This invention consisted of a telescope which, using the same principle as the megameter, was able to determine the height of a distant object. Two eyepiece points, adjustable with a micrometer screw, allowed the angle subtended by the object to be measured. Knowing the distance to the object, the height could be calculated, or vice versa. The instrument was sold by the firm of **George Adams (II)**.

Green, William (III)
William Green (III), (fl.1844-1848) [18] [263] [271], was a mathematical, optical, and philosophical instrument maker of Glasgow, Scotland. Addresses: 5 Franklin St. (1844-1845) and 87 London St. (1846-1848). He was succeeded in the London St. address by **Robert Finlay**.

Greenough Binocular Microscope
See microscope.

Gregg, William T.
William T. Gregg was a successful American optician in the late-C19[th], located at 104 Fulton Street, New York. The company manufactured cameras, telescopes and lenses, as well as other optical and surveying equipment.

Gregorian Telescope
The Gregorian telescope, the earliest design of reflecting telescope, was invented by the Scottish mathematician and astronomer **James Gregory**, and described in his book *Optica Promota* in 1663. The first example was built by **Robert Hooke** in 1673. The Gregorian telescope has a paraboloidal primary mirror and a concave ellipsoidal secondary mirror, reflecting the image back through a hole in the primary mirror, where it is brought into focus by an eyepiece.

Gregory & Co., W.
William Gregory & Co., [9] [379], were optical, mathematical, and philosophical instrument makers of 51 Strand, London, England, the address previously occupied by **John Pallant**. The company appears in the directories at this address in 1885 and 1890, and they continued there until at least 1900. **William Gregory (I)** is also recorded as trading at the same address from 1866 until 1891. They advertised as instrument makers to "Her Majesty's Government, War Department, London County Council and the National Rifle Association". Some instruments were signed *William Gregory & Son*.

Instruments bearing their name include: Galilean binoculars; 4-draw telescope; 3-draw brass and mahogany telescope; 2-draw telescope; telescope on tripod, 40-inch length and 2¼-inch objective, 25x magnification; single-draw stalking telescope, c1880. Instruments bearing the name William Gregory & Son include barometers.

Gregory & Wright
Gregory & Wright, [16] [18] [54] [276 #842a], were mathematical instrument makers and opticians, trading at *Navigation Warehouse*, 148 Leadenhall St., London, England. This was a partnership, formed in about 1782 between **Henry Gregory** and **Gabriel Wright**. At some point **William Gilbert (I)** joined the partnership, from which time they traded as **Gregory, Gilbert & Wright** until 1789.

Their trade card, [53], reads: "Gregory & Wright. Mathematical and Optical instrument makers. N°. 148 Leadenhall Street, near the India House, London. The patent improvement added to octants & sextants of the common construction. Wright patent quintant to 160 D without a back observation glass. Globes on an improved construction. Patent perpendicular. New equatorial dial for all latitudes. Visual glasses by G. Wright from Mr. Martin's, Fleet Street. Spectacles of all the different kinds."

Gregory, Gilbert & Co.
Gregory, Gilbert & Co., (fl.1793) [18], were opticians of 148 Leadenhall St., London, England. The firm was a partnership between **William Gilbert (I)** and **Henry Gregory**.

Gregory, Gilbert & Wright
Gregory, Gilbert & Wright, [18], were opticians of *Navigation Warehouse*, 148 Leadenhall St., London, England. This was a partnership formed when **William Gilbert (I)** joined the partnership **Gregory & Wright**, and continued until the partnership was terminated in 1789, [35 #13124 p560], after which William Gilbert (I) and Gabriel Wright continued the business as **Gilbert & Wright**, optical and mathematical instrument makers.

Gregory, Henry
Henry Gregory, (fl.1774-1793) [18] [53] [276 #484], was a mathematical instrument maker of London, England. He began his apprenticeship in 1767 with his father, a mathematical instrument maker also named Henry Gregory, of 148 Leadenhall St., London. He was freed of the **Masons' Company** in 1774. By 1776 he was working in partnership with his father as Henry Gregory & Son.

Following this he engaged in several partnerships at *Navigation Warehouse*, 148 Leadenhall St. In about 1782 he entered a partnership with **Gabriel Wright**, trading as **Gregory & Wright**. They were subsequently joined by **William Gilbert (I)**, and traded as **Gregory, Gilbert & Wright** until the partnership was terminated in 1789, [35 #13124 p560]. During 1793 he worked in partnership with William Gilbert (I), trading as **Gregory, Gilbert & Co.**

Gregory, James
James Gregory, (1638-1675) [45] [46] [155] [275 #282] [388], was a Scottish mathematician and astronomer, responsible for the invention of the eponymous Gregorian telescope, the earliest design of reflecting telescope. He never made a telescope to his design, however, and the first to be made was by **Robert Hooke** in 1673.

In 1663 Gregory published his book *Optica Promota* (The Advance of Optics). In it he described both refractive and reflective optics. In this work he recognised that refracting telescopes will always be limited by aberrations of one kind or another, and proposed his design for a reflecting telescope.

Gregory was chair in mathematics at the University of St. Andrews, Scotland, from 1669 to 1674, and the University of Edinburgh from 1674 to 1675.

Gregory, William (I)
William Gregory (I), (fl.1866-1891 d.1904) [379], was a mathematical and optical instrument maker of 51 Strand, London, England. See **W. Gregory & Co.**

Gregory, William (II)
William Gregory (II), (1830-1836) [18] [276 #1874], was a mathematical, optical, and philosophical instrument maker of Clerkenwell, London, England. Addresses: 8 Berry St. (1834); 9 Hooper St. (1835); 2 Francis Court, Berkeley St. (1836). He advertised as a maker of mathematical rules.

Grice, William Hawks
William Hawks Grice, (fl.1815-1825) [9] [18] [276], was an optician and mathematical instrument maker. In 1818 he took over the business of **Alexander Fraser** at *Ferguson's Head*, 3 New Bond St., London, England. He was appointed mechanician and optician to the Royal Family, and traded under his own name until sometime between 1823 and 1825 when the business was taken over by **George and Charles Dixey**.

Instruments bearing his name include: surveyor's perambulator; small six-draw gilt brass spyglass; ebony stick barometer; 28-inch-long 1½-inch gilt brass three-draw spyglass. He is known to have sold microscopes.

Griffin & George
Griffin and George, [44] [53], of Ealing Road, Alperton, Wembley, London, England, was founded in 1954 upon the merger of **W.&J. George & Becker** with the laboratory instrument makers **Griffin & Tatlock**. Griffin & George was a manufacturer and supplier of science teaching apparatus, and they continued to use the *Nivoc* brand-name for lab essentials, as well as using the Griffin & Tatlock *Microid* brand-name for educational apparatus. Instruments bearing their name include: spectrometer; measuring microscope.

Griffin & Sons, John J.
John J. Griffin & Sons, [18] [53] [61] [122] [263] [306], was established in Glasgow, Scotland as the publisher, and bookseller, John Joseph Griffin & Co. Soon after the announcement of the daguerreotype in 1839 they began to offer daguerreotype chemicals and apparatus. In 1848 the chemical and instrument part of the company relocated to London, England, producing photographic chemicals and laboratory equipment. By 1890 they were trading at 22 Garrick Street, London, and in that year they opened a

Photographic Apparatus and Chemical Department, managed for the first two years by Robert C. Murray of **Murray & Heath**. By the early 1900s they were advertising as "Photographic manufacturers and wholesale dealers", and marketed several models of camera. In 1925 the company merged with **Baird & Tatlock**, and in 1929 the merged company changed their name to become **Griffin & Tatlock**.

Griffin & Tatlock

Griffin & Tatlock, [53] [263], were manufacturers and suppliers of scientific instruments and laboratory equipment. The company was formed in 1925 by the merger of the London firm of Chemical apparatus and chemical suppliers, **John J. Griffin**, with the Glasgow firm **Baird & Tatlock Ltd.** The new company began to use the name Griffin & Tatlock in 1929. They used the brand-name *Microid* for some of their products. In 1954 they merged with **W.&J. George & Becker** to become **Griffin and George Ltd.** Examples of their instruments include a spectroscope.

Griffith, Edward

Edward Griffith, (fl.1809) [18] [276 #1139], was an optical instrument maker of Birmingham, England. He is known by his application for a patent for a table-lamp in 1809, in which he described himself as an optical instrument maker.

Griffith, Ezra H.

Ezra Hollace Griffith, (fl.late-C19th) [201] [264], of Fairport, New York, USA, and later of Rochester, NY, was a microscopist, and a designer of microscopes, objectives, and accessories. He is best known as the designer of the *Griffith Club Microscope*. He designed and sold these, but they were made by **Bausch & Lomb**. He donated an 1881 model of the microscope to the Royal Microscopical Society. An 1881 microscope of this design in the Billings Collection, [369 pp83-84], is attributed to John Field of Birmingham, England (Note: it is not clear who this refers to, or whether it bears any relation to **R. Field & Co.**). In 1892 Griffith claimed that he had a new design of microscope which was to be made by the **Gundlach Optical Company** - however, little is known about this, and none are known to have survived.

Griffiths & Co. Ltd., Walter

Walter Griffiths & Co. Ltd., [61] [306], was a photographic equipment maker of Highgate Square, Birmingham, England. The company was incorporated in 1891, although they had already carried on business as Walter Griffiths prior to that date. The company's early business included photo-lithography and printing services and, later, photographic enlargers. During the 1890s they began to manufacture cameras. They produced a number of different models of camera, and were awarded various related patents. From 1892 their address is recorded as 5 Union Passage, New End Street, Birmingham. In 1901 the company was struck from the register, [35 #27336 p4853], and subsequently re-emerged as the Griffiths Camera Co. Ltd. which, due to its liabilities, was wound up in 1905, [35 #27813 p4658].

Grindel, Carlo

Carlo Grindel, (1780-1854) [322] [374], was an engineer, and astronomical instrument maker of Milan, Italy. He worked for the astronomical observatory at Brera, Milan, with the job-title *macchinista* (machinist), constructing instruments for them, and keeping the instruments in a good state of repair. He was succeeded in this role by his son, **Francesco Grindel**, whom he had taught his instrument making skills.

Grindel, Francesco

Francesco Grindel, (d.1859) [322] [374], was an engineer and astronomical instrument maker of Milan, Italy. He was the son of **Carlo Grindel**, from whom he learned his instrument making skills. Following Carlo's death in 1854, took over his father's role as *macchinista* (machinist) at the observatory, constructing instruments for them, and keeping the instruments in a good state of repair. Upon Francesco's death in 1859, the role of macchinista went to **Carlo dell'Acqua**.

Grocers' Company

See Appendix IX: The London Guilds.

Grubb, Sir Howard

Sir Howard Grubb, (1844-1931) [24] [46], of Dublin, Ireland, was a maker of scientific instruments, and was the fifth and youngest son of **Thomas Grubb**. Howard Grubb joined his father's company in 1865, and took control in 1868.

Under his direction the Grubb Telescope Company, founded by his father, gained significant reputation. After a successful commission to produce the 48-inch refractor for the Melbourne Observatory in 1869, widely considered a masterpiece in scientific instrument making, the firm went on to assist in the construction of telescopes for India, Vienna, South Africa, Greenwich and Dublin. They made many of the astrographs and mounts used in the *Carte du Ciel* sky-mapping project. Following the company's success, Howard Grubb was knighted by Queen Victoria in 1887, and in 1912 was the third recipient of Royal Society of Dublin's Boyle Medal.

Sir Howard Grubb & Company began to build and supply periscopes for **Vickers** submarines in 1901 [20]. They were the only British company making submarine periscopes until the advent of WWI, during which both **Kelvin, Bottomley & Baird**, and **Barr & Stroud** began to make them. In 1917 the Admiralty forced the company to relocate from Dublin to St Albans in Hertfordshire, England. This was partly because they wished to reduce the risk of losses to U-boat attack while crossing the Irish Sea, and partly because of fears over the political unrest in Ireland. By the end of WWI, the move was still incomplete, and the British government terminated their periscope contracts. No more periscopes were made by Grubb, and the ongoing restrictions on their business imposed by the Admiralty and the Ministry of Munitions constrained their ability to recover from the financial losses incurred.

In 1919 the company Sir Howard Grubb and Sons was listed as a member of the **British Optical Instrument Manufacturers' Association**. In 1925 the company was bought by **Charles Parsons**. He merged it with the **Parsons Optical Glass Company** to form the company Grubb Parsons. In 1938 Grubb Parsons took over the telescope making part of the **Cooke, Troughton and Simms** business.

Grubb, Sir Howard and Sons
See **Sir Howard Grubb**.

Grubb Parsons
See **Sir Howard Grubb, Charles Parsons**.

Grubb Telescope Company
See **Thomas Grubb, Sir Howard Grubb**.

Grubb, Thomas
Thomas Grubb, (1800-1878) [18] [24] [46], was an engineer and telescope builder in Dublin, Ireland. In 1833 he founded the **Grubb Telescope Company**.

His first major telescope construction was a telescope to mount a 13.3-inch lens made by **R.A. Cauchoix** of Paris. For its construction he made a full-scale model which he went on to make into a 15-inch reflector for the Armagh observatory. In 1838 he made the "sheepshanks Telescope" for the Royal Observatory, Greenwich. In 1853 he made a 22-inch reflector for the University of Glasgow, and in 1868 he made a 12-inch refractor for the Trinity College observatory in Dublin. He did not, however, make the optical components for these telescopes.

Grubb also made magnetometers for a network of magnetic observatories, and he designed and manufactured machines for engraving, printing and numbering banknotes for the Bank of Ireland. In 1864 Thomas Grubb was elected a Fellow of the Royal Society. In 1866 he began work on a Cassegrain reflector for an observatory in Melbourne. It had two interchangeable 4-foot speculum metal mirrors, rather than using silvered glass, which was gaining popularity at the time. The telescope was not considered a success, largely because of the choice of mirror construction.

The Grubb Telescope Company went on to gain significant reputation in the 1860s under the direction of Grubb's son, **Sir Howard Grubb**.

Grunow, J.&W.
The brothers Julius and William Grunow, [201], were German born American microscope makers. In 1849 Julius Grunow emigrated from Berlin, Germany to New Haven, Connecticut, USA, and studied optics. He began making microscopes in 1852 and at about the same time his brother, William Grunow, emigrated from Germany to join him. They conducted their business in New Haven under the various names Grunow Brothers, J.&W. Grunow, and J.&W. Grunow & Company. They were particularly noted for their binocular microscopes and for their inverted microscopes. In 1863 or 1864 the Grunow brothers moved their business to New York. There they continued in partnership until 1874, from which time Julius Grunow continued alone until about 1892.

Guan Sheng Optical (GSO)
Guan Sheng Optical (GSO) is a major Taiwanese manufacturer of telescopes and accessories. They market their products under their own name as well as making products for companies such as **William Optics**.

Guaranteed Lens Co.
The Guaranteed Lens Co., [20] [29], was a UK optical manufacturer that supplied lenses to **J. Brimfield & Co.**, **Theodore Hamblin Ltd./Precision Optical Co. Ltd.**, and **W. Watson & Sons Ltd.** during WWI.

Guinand, Aimé
Aimé Guinand, (1780-1857) [24] [156], of Switzerland was the son of **Pierre Louis Guinand**, and brother of **Henri Guinand**. He took over his father's workshop, and supplied flint glass in Vienna, Paris and London until 1840.

Guinand, Henry
Henry Guinand, (1771-1852) [9] [156], of Switzerland was the son of **Pierre Guinand**. During Pierre's lifetime Henry worked with him, assisting his glass-making. In 1823 he collaborated with the French government in an

effort to give his father an opportunity to set up a glass-works there, but Pierre was too old to make the move, and died the following year. Henry's brother, **Aimé Guinand**, took over Pierre's business, and Henry went to France. He built the mount for the *Carte du Ciel* astrograph in the Vatican Observatory. In 1828, on the instigation of **Noël-Jean Lerebours**, he founded the **Choisy-le-Roi glassworks**, with **Georges Bontemps** as director, to produce large diameter flint glass discs – a venture which met with great success. Upon Henry's death in 1852 his firm passed into the hands of his grandson, **Charles Feil**.

Guinand, Pierre Louis

Pierre Louis Guinand, (1748-1824) [9] [24] [156], was a Swiss craftsman and optical glass maker. With no formal training in glass-making he performed his own meticulous experiments, repeatedly making glass, noting temperatures, tools and conditions, and measuring the quality of the resulting glass. He continued to improve his methods until he was able to produce fine quality optical glass which was then used for large diameter telescope objectives and mirrors. He was the first to rival, then exceed the quality of the optical glass produced by the English glassmakers.

In 1805 Guinand met **Joseph von Utzschneider**, a wealthy lawyer who was financing a company called the **Mathematical-Mechanical Institute Reichenbach, Utzschneider and Liebherr**. The company would be producing high quality surveying instruments for the military, and Utzschneider recognised that he would need good quality optical glass. Guinand moved to Munich, Germany to work for them, and as a part of his terms of employment he was required to train a member of Utzschneider's staff. This student was **Fraunhofer**, then only twenty years old, and shortly after training him, in 1814, Guinand returned to Switzerland.

Once back in Switzerland he worked making refracting telescopes, and is known to have supplied flint glass to **Noël-Jean Lerebours**, who used the largest to make objective lenses for the Paris Observatory. In 1823 King Louis XVIII of France, visiting an exhibition, saw the work of Pierre Guinand. He proposed to Guinand's eldest son, **Henry Guinand**, that Pierre bring his expertise to France at the expense of the French government. This proposal was followed up by the French government with generous terms. While Henry was working on this proposal an attempt was made to bring Guinand to England to start a glass factory, but Pierre was too old to make either move, and died the following year. His younger son, **Aimé Guinand**, took over his workshop after his death, and supplied flint glass in Vienna, Paris and London until 1840.

Gundlach, Ernst

Ernst Gundlach, (b.1834) [201] [210], was a German optical instrument maker, noted for his microscopes and fine objectives, and was awarded patents for his improvements to telescope eyepieces and objectives. He worked for **Kellner** in the 1850s, along with **Wilhelm and Heinrich Seibert**. He set up his own business in 1859, taking with him the Seibert brothers to work in his company. He exhibited his microscopes in Berlin in 1866. In 1872 the Seibert brothers bought his company, and he emigrated to the USA and began to trade in Hackensack, New Jersey, while maintaining a factory near Berlin, Germany. In 1876 he began to work for **Bausch & Lomb** in Rochester, New York, as head of their microscope department, introducing his designs, and innovations. By 1878 he had stopped working with Bausch & Lomb, and started his own business, **Gundlach Optical Co.** in Rochester. During the early 1880s Gundlach produced the optics for microscopes made by **Yawman & Erbe**. Gundlach had stopped working for Gundlach Optical by 1895, from which time he began to produce photographic lenses, trading as **Gundlach Photo-Optical Co.** He returned to Germany in around 1900.

Gundlach-Manhattan Optical Co.
See **Gundlach Optical Co.**

Gundlach Optical Co.
The Gundlach Optical Company of Rochester, New York, USA, was formed by **Ernst Gundlach** in around 1878. Gundlach himself had stopped working for the company by 1895, but they continued to produce Gundlach microscopes into the C20th. In 1902 they acquired the **Manhattan Optical Co.**, and continued to trade as **Gundlach-Manhattan Optical Co.**, producing microscopes.

Gundlach Photo-Optical Co.
See **Ernst Gundlach**.

Gunner's Quadrant
A gunner's quadrant is an instrument with a graduated quarter-circle, sometimes including a level, used to measure the elevation or depression of a weapon.

Gunter's Quadrant
Gunter's quadrant, [16] [38], was invented by the English mathematician and astronomer, Edmund Gunter (1581-1626) [275 #106], who published his design in 1623. It consists of a flat quarter-circle of wood or brass with sighting rings positioned at each end of one straight side. The face is engraved with a projection of the equator, the tropics, the ecliptic, and the horizon, and sometimes

includes positions of some stars and other information. The corner is drilled with a hole to which a plumb line is attached. The curved edge is marked in degrees. It is used to measure the altitude of the sun or a star, and can be used to determine the time, and for other tasks. See also quadrant.

Gurley, W.&L.E.

W.&L.E. Gurley, [58], was a USA-based partnership between the brothers William Gurley (1821-1887) and Lewis Ephraim Gurley (1826-1897). Initially William Gurley entered a partnership with Jonas H. Phelps, and they traded as **Phelps & Gurley**, making surveying instruments. Lewis E. Gurley worked for the firm, and in 1852 the Gurley brothers took over the company and began to trade as W.&L.E. Gurley. They continued to make engineering and surveying instruments, becoming the largest manufacturer of surveying instruments in the USA. They were incorporated in 1900, and continued to produce large numbers of instruments until they were taken over by Teledyne in 1968. They began trading as Teledyne-Gurley, but then quickly phased out production of surveying instruments.

Gustav Heyde Mathematisch-Mechanisches Institut Optische Präzisions-Werkstätten

See **Gustav Heyde**.

GX Microscopes (GT Vision)

GX Microscopes is a brand-name used by GT Vision for the microscopes they manufacture. Their products include many models of benchtop microscopes for professional, educational and amateur use.

Gyro-Theodolite

A gyro-theodolite is a theodolite with an integrated gyro-compass. It is used in situations where a magnetic compass or GPS cannot function properly, such as mine surveying and tunnel engineering. For example, gyro-theodolites were used in the construction of the Channel Tunnel connecting England with France.

H

Haag-Streit
Haag-Streit group is a multinational company, founded by Hermann Studer and Friedrich Hermann, in Berne, Switzerland in 1858. They design, manufacture and sell orthoptic, optometry and ophthalmic equipment. They acquired the ophthalmic instrument making part of **Clement Clarke** in 1989, which is now a part of Haag-Streit UK. In 1999 Haag-Streit acquired **J.D. Möller**, which was renamed Möller-Wedel Optical GmbH.

Haas & Hurter
See **Hurter & Haas**.

Haas, J.B.
See **Hurter & Haas**.

Hadley, John
John Hadley (1682-1744), [3] [9] [18] [45] [46] [276 #56], was an optical instrument maker and mathematician of Middlesex, England. At the age of 35 he became a member of the Royal Society. With the assistance of his brothers, George and Henry, he constructed a 6-inch Newtonian reflector with a focal length of 62 inches. In doing so he revived the reflecting telescope and began the process of improving on the method of grinding mirrors to achieve consistency. He did not publish his mirror grinding and polishing processes, but contributed them to **Robert Smith** who included them in his *Compleat System of Opticks*. John Hadley built probably the first successful commercial paraboloidal reflecting telescope, and from 1719 until 1726 he worked with **Francis Hauksbee** on improvements to reflecting telescopes.

He became well known for inventing the reflecting nautical quadrant, independently of the American mathematician **Thomas Godfrey**. His quadrant used double reflection to measure angles such as the altitude of the Sun or a star above the horizon [236 pp57-60]. A competing sea quadrant using prisms instead of mirrors was invented by **Caleb Smith**, but it did not prevail.

Hadley's Quadrant
Hadley's quadrant is a double-reflecting octant, invented by **John Hadley**. He presented his invention to the Royal Society in 1731, and added a spirit level to the design in 1734. Because of the double-reflecting operation, 1° on the octant's arc represents 2° separation of the observed objects, and hence it is referred to as a quadrant. It came into general use, superseding the earlier Davis's Quadrant, as it was capable of higher accuracy. It was the forerunner of the modern sextant. Hadley's invention was made independently of the American mathematician **Thomas Godfrey** who arrived at the same idea at about the same time.

Haering
According to an undated trade card, [53] [322] [361], Haering was an optician, and mathematical, physical, and optical instrument maker and seller of l'Aigle d'Or, No. 63 Palais du Tribunat, pres le Café de Foi, Paris, France.

Hahn, A.&R.
A.&R. Hahn, [374] [452], was an optical, mathematical, and surveying instrument manufacturer, established in 1870 in Kassel, Germany by the brothers Arwed and Richard Hahn. By 1871 the firm had 38 employees. In 1877 they patented a design for their Artillerie-Entfernungsmesser (Artillery rangefinder), following which they supplied rangefinders to the military. In 1910 they began to work in collaboration with **C.P. Goerz**, producing movie cameras and projectors with Goerz lenses. After WWI, Hahn produced surveying levels and theodolites, some of which were also produced in collaboration with Goerz. They also began to produce security locks and keys. At some point the company name was changed to AG Hahn für Optik und Mechanik, and in 1927 the company was taken over by **Zeiss Ikon**.

Hahn für Optik und Mechanik, AG
AG Hahn für Optik und Mechanik. See **A.&R. Hahn**.

Hails, Thomas
Thomas Hails, also known as Hailes, Hale, Hales, or Hayles, (fl.1687-1699) [9] [18] [256] [274] [275 #454], was a spectacle maker and optical instrument maker of London, England. He was freed of the **Spectacle Makers' Company** in 1687, and may have been an apprentice of **John Yarwell**. He traded at the *Spectacles & Prospective Glass*, corner of Cannon Alley over against the great North Door of St. Paul's Church, and later at the *Telescope & Spectacles*, corner of Cannon Alley over against the great North Door of St. Paul's Church. His last appearance in the attendance records of the Spectacle Makers' Company was in 1699/1700.

According to his trade leaflet, "There are all sorts of instruments of glass, as Telescop's fitted for sea and land, Microscop's single and dubbel, Prospective glasses great and small, Reading glasses of all sizes, Magnifying glasses, Multiplying glasses, Triangular prisms, Speaking trumpets, Spectacles of all sorts fitted up for all ages, to be sold hole sale and retail and all other sorts of glasses both concave and convex made and sold by Thomas Hails."

Haking (Xinhui) Optical Ltd., W.
Founded in 1991, W. Haking (Xinhui) Optical Ltd. is a large manufacturer of photographic equipment and supplies, located in Jiangmen, Guangdong, China. Their brands include **Halina** and **Ansco**.

Halden, Joseph and Company
Joseph Halden and Company, [53] [141], was founded in 1878 when Joseph Halden entered a partnership with **A.G. Thornton**. The partnership lasted about two years after which Joseph Halden set up his own business in Albert Square, Manchester, England. Initially the new company imported drawing materials and made drawing instruments, and soon they had a healthy wholesale and export trade. They produced the popular *Calculex* circular calculator, which was less expensive than Fowler's calculator. They expanded into making surveying instruments, including theodolites, and in the 1930s they began production of photocopying, photo enlarging and developing apparatus. The company stayed in the Halden family until 1965, and was finally sold to the Ozalid Group in 1969, marking the end of the company.

Hale, George Ellery
George Ellery Hale, (1868-1938) [155], was an American astrophysicist. He was born to a wealthy Chicago family, and studied physics at the Massachusetts Institute of Technology. He developed an early interest in designing and building his own instruments, and with funding from his father he built a solar observatory by their Chicago home. He became a professor of astrophysics at the University of Chicago. In 1889 Hale invented the spectroheliograph, and used it to study the sun's surface features and prominences.

He persuaded Chicago businessman Charles T. Yerkes to fund the building the Yerkes Observatory and its 40-inch refractor for the University of Chicago. The 40-inch telescope was, and remains, the largest refracting telescope ever successfully used for astronomy. It became operational in 1897, and Hale was director of the observatory until 1905.

Hale's father acquired a 60-inch mirror, and Hale persuaded the Carnegie Institute in Washington to fund the building of the Mount Wilson Observatory in California to house the new 60-inch telescope, which became operational in 1908. Hale also sought funding for a 100-inch telescope for the Mount Wilson Observatory. A major contribution was made by **John Daggett Hooker**, after whom the telescope was named, and additional funding came from the Carnegie Institute. The Hooker telescope saw first light in 1917. Hale was director of the observatory from 1904 to 1923.

In 1908 Hale discovered that some of the spectral lines in sunspots were double. This was due to the Zeeman effect, and indicated the presence of strong magnetic fields in sunspots. This was the first discovery of an extra-terrestrial magnetic field.

In 1929 Hale persuaded the Rockefeller Foundation to fund building of a 200-inch reflecting telescope. This was a massive undertaking, and was not completed before Hale's death in 1938. The telescope, together with the Palomar Observatory, specially constructed for it, finally became operational in 1948. The Hale Telescope, as it is known, was, at the time, the largest telescope in the world.

Halina
Halina is a brand of photographic equipment and binoculars, produced in China by **W. Haking (Xinhui) Optical Ltd.**

Hall & Brothers Ltd., A.H.
In 1949, three sons of the **Hall brothers** formed A.H. Hall and Brothers Ltd., a small company in Islington, London, England, which specialised in producing optical instruments for construction and planning.

Hall & Watts Ltd.
The company Hall Brothers was founded in 1880. They manufactured and supplied theodolites, levels, artillery directors and weapon sights for the British military. In 1938 the company was incorporated as Hall Bros. (Optical) Ltd., and became a wholly owned subsidiary of Aeronautical & General Instruments & Co. Ltd. In 1979 Hall Brothers acquired **E.R. Watts** from **Rank**, and became known as Hall & Watts Ltd., producing surveying and military instruments. In 1995 they stopped producing surveying instruments and focused on the defence market, and became known as Hall & Watts Defence Optics Ltd.

Hall Brothers
See **Hall & Watts Ltd.**

Hall, Chester Moor
Chester Moor Hall, [1] [9] [46] [388], was a barrister of the Inner Temple, London, England, who is generally credited with inventing the achromat. His hobby was optics, and in 1729-1730, based on his own experiments he devised a new theory: "that of correcting the refrangibility of the rays of light in the construction of object glasses for refracting telescopes." He had the two optical elements, the crown and flint glasses, ground by two different opticians, **Edward Scarlett (I)** and **James Mann (II)**. Both of these opticians subcontracted the work to the same person, **George Bass**, a grinder and polisher in Bridewell. Chester Moor Hall, having solved the

problem of chromatic aberration in a telescope, never publicised his work in this field.

In 1757, **John Dollond (I)** had a conversation with George Bass in his workshop, and figured out how Hall had used two optical elements, the crown and flint glasses, to overcome chromatic aberration. After some research of his own, he devised the achromat based on the principles Bass had used. He controversially patented the achromat following encouragement from his son, Peter, (see Appendix X).

Hamblin Ltd., Theodore

Henry Thomas Hamblin, (1873-1958), [1] [29] [44] [229], trained as a draughtsman, and set up a firm of opticians in Cavendish Square, London, England. His brother joined him in partnership, and in 1909 he sold the business for "a good price". He then pursued his interests in food reform. Soon after this he opened a new firm of opticians in Cavendish Street in London, which he called Theodore Hamblin Ltd. The company was incorporated in 1909, and was very successful.

In 1915 the company approached the War Office, and offered to make binoculars and optical gunsights for the army to War Department specifications. Once awarded the contract they set up a factory where they also later made ophthalmoscopes and other ophthalmic instruments. Their wartime instruments were made using lenses supplied by the **Guaranteed Lens Co.** They also produced optical instruments under the name **Precision Optical Co. Ltd.**, at 5 Wigmore Street, London, and this may have been where their wartime production took place. In 1919 Henry retired to become what he referred to as "a mystic", and wrote numerous books on the subject of "the science of thought". The firm continued in business, acquiring royal warrants to HRH Prince of Wales, His Majesty King George V, and the King's wife, HM Queen Mary. The firm was acquired in 1981 by **Dollond & Aitchison**.

Hammersley, J.

J. Hammersley, (fl.1841-1902) [53] [319] [322] [411], of Half-Moon Crescent, Islington, London, England, was an optician and optical and mathematical instrument maker. Addresses: Half-Moon Crescent (1841); 10 Half-Moon Crescent (1851-1861); 37, 39, 41, and 43 Half-Moon Crescent (1901).

Joseph Hammersley (I) (c1800-1881) first appears in the census as a brass turner in 1841. In 1861 he is listed at 10 Half-Moon Crescent as a telescope maker. Also at the address were Charles G. Hammersley, aged 26, a telescope maker, William Hammersley, aged 17, a telescope maker, Henry Hammersley, aged 14, a spectacle maker, Joseph Hammersley (II), aged 10, a scholar, and another six members of the Hammersley family.

In 1870, J. Hammersley was listed as a microscope maker in *Wright's Improved Handbook of the Principal Manufacturers and Warehousemen of Great Britain*.

It is unclear when the firm of J. Hammersley moved to 37, 39, 41, and 43 Half-Moon Crescent, but in 1881 Joseph Hamersley (I), aged 81, was listed as an optician and telescope manufacturer at 27 Lichfield Grove, Finchley, Middlesex. He died later that year.

In 1901, bankruptcy proceedings were initiated against George Hammersley and William Hammersley, trading as J. Hammersley, telescope, microscope, and mathematical instrument makers at 37, 39, 41, and 43 Half-Moon Crescent, [35 #27353 p6016]. As a result, the company was acquired by **Negretti & Zambra**. By the time of this acquisition, the company occupied three of the houses. "Captain Joseph Hammersley" (presumably Joseph Hammersley (II)), continued to trade under his own name in two of them following the acquisition. This continued until 1912 when Negretti & Zambra demolished the houses to build their Half-Moon works.

Instruments bearing the name of J. Hammersley include: single draw naval telescope; 3-draw aluminium telescope with 2.25" objective in hard leather case; brass and leather 3-draw telescope; brass telescope on wooden tripod; spectroscope slits with micrometers; surveyor's level; telescope used in the British National Antarctic Expedition 1901-1904.

Hampson, Robert

Robert Hampson, (fl.1865-1871) [122], was a chemist and photographic supplier of 63 Piccadilly, Manchester, England. He succeeded **J.J. Pyne** in business in the mid-1860s. By 1865, **J.T. Chapman** was working for him and, shortly afterward, so was **J.B. Payne**. In 1871, Hampson retired, and Payne and Chapman took over the firm, trading as **Payne & Chapman**.

Hanimex

Hanimex was an Australian company, based in Sydney, established in 1947. They sold cameras and photography supplies as well as binoculars. They imported the products or components from China and Japan, and sold their products in Australia and New Zealand. Their manufacturing division was bought by **Reed International Group** in 1986, and they continued in business until they were purchased by the UK firm Gestetner in 1989, which was in turn taken over by Japanese firm **Ricoh** in 1996.

Hapo

Hapo was a brand-name used by the German camera retailer **Porst** for their re-branded cameras.

Hardin Optical Company

The Hardin Optical Company of Oregon, USA, was formed by Larry Hardin. Hardin worked for Tom Cave of the **Cave Optical Co.** in the 1960s and, in 1999, after retiring, Tom Cave agreed to transfer the *Astrola* brand-name to Hardin. The company made telescopes and eyepieces under the *Astrola* brand-name. The brand-name is no longer used by the company, and they no longer make consumer telescopes or eyepieces.

Hardy & Co., E.H.

E.H. Hardy & Co., (fl.1894-1895) [306], were photographic apparatus makers and importers of 16 Commercial St., Sheffield, England. Their registered trademark was *Delograph*, which was used for a half-plate mahogany and brass camera. They also sold a patent exposing and developing easel, and other accessories for photographic processing.

Hare, George

George Hare, (1825-1913) [61] [306], was a camera maker in London, England. Note that his date of birth is variously listed as 1825, 1826 or 1828. He served an apprenticeship with his father in Yorkshire as a joiner, and after working in that trade for some time he joined the firm of **Thomas Ottewill** in London. After working for them for a relatively short while, he left to start his own business sometime around 1855-1857. By 1857 he was trading at 140 Pentonville Road, and in 1864 he took on additional premises at 1 Lower Calthorpe Street, Gray's Inn Road. In 1867 he was subject to bankruptcy proceedings, [35 #23217 p736]. **Henry Park** worked for George Hare some time prior to starting his own company. Hare relinquished the Pentonville Road address in 1870, and by 1877 he was trading at 26 Calthorpe Street. George Hare is believed to have retired some years before his death in 1913. The firm is believed to have continued beyond his retirement, until about 1911.

Hargreaves, Frederick J.

Frederick J. Hargreaves, (1891-1970) [9] [24] [199] [222], was an astronomer and amateur mirror maker of Bradford, England. He began his career as a patent agent, and became active in the British Astronomical Association (BAA), becoming a fellow in 1925. He earned a reputation for his skill in making mirrors, carrying out work for the Royal Greenwich Observatory and working as a consultant to **Grubb Parsons**. He was president of the BAA from 1942 to 1944. In 1947 he co-formed the optical and engineering company **Cox, Hargreaves & Thomson** in Coulsdon, Surrey, with **H.W. Cox** and **J.V. Thomson**.

Harling, W.H.

W.H. Harling Ltd., [44] [53], was a mathematical, drawing, and surveying instrument maker, founded by W.H. Harling in 1851 at 47 Finsbury Pavement, London, England. Their manufacturing facility was at Grosvenor Works, Hackney, London. By 1883 they were additionally advertising their address as 40 Hatton Garden, London. In 1961 they merged with Blundell Rules, and in 1964 the company was renamed Blundell Harling. Examples of instruments by W.H. Harling Ltd. include: drawing instruments; pocket theodolite; alidade; brass surveyor's level.

Harris & Co., William

William Harris & Co., (fl.1813-1839) [18] [276 #1151], were opticians, and mathematical and optical instrument makers. The company was a partnership between **William Harris (I)** and **W. Campbell**, trading in London and Liverpool, England, and in Hamburg, Germany as W. Campbell & Co. Addresses: Bey dem Rathhause no. 26, Hamburg (1820); 50 High Holborn, London (1816-1835); 50 Holborn, London (1816-1839); 35 Crown St., Liverpool (1816-1839). The firm is known to have sold globes, compasses, balances, and microscopes, and was succeeded by **William Harris & Son**.

Harris & Son, H.

H. Harris & Son, (fl.1811) [276 #1352], were optical and mathematical instrument makers of 50 High Holborn, London, England. Note that this is the address occupied by **William Harris (I)** from 1815.

Harris & Son, Thomas

Thomas Harris & Son, [9] [18] [53] [276 #1150], was an optician, mathematical instrument maker, globe maker, and telescope maker of London, England. The firm was also known to have sold microscopes. It was a partnership between **Thomas Harris (I)** and his son **Thomas Harris (II)**, trading at 140 Fleet Street, from 1802 until Thomas (II)'s untimely death in 1808. The firm continued thereafter as a partnership between Thomas (I) and his son **William Harris (II)**, trading at: 20 Duke St., Bloomsbury (1808); 30 Hyde St., Bloomsbury (1810-1817); 52 Great Russel St., Bloomsbury (1816-1885); 9 Cornhill (1822); 22 Cornhill (1831-1835). The premises at 22 Cornhill were previously occupied by **Edward Blunt**.

In 1819 they had a royal appointment to George, Prince Regent, and this continued when the Regent was crowned King George IV. In June 1828 the partnership between Thomas (I) and William (II) was terminated [35 #18480 p1196] (this appears to be a correction to a notice in the same paper three days previously). This was presumably on the retirement of Thomas (I).

The business continued after the death of William (II) in 1843, trading at: 52 Great Russel Street, Bloomsbury until 1885, as well as: 141A Oxford St. (1851); 43 Great Russel Street, Bloomsbury (1859); 32 Gracechurch St. (1890-1901).

Harris & Son, William
William Harris & Son, (fl.1840-1855) [18], was an optician, philosophical instrument maker, and drawing instrument maker of 50 High Holborn, London, England. The firm succeeded **William Harris & Co.**, and was probably a partnership between **William Harris (I)** and his son Richard Joshua Harris.

Harris, George
George Harris, (fl.1733) [18] [256] [274], of England, is listed as an optical instrument maker in [276 #267]. His name appears on a quadrant dated 1733. However, [9] points out that Harris may have been the owner, rather than the maker.

Harris, George and Henry
George and Henry Harris, (fl.1815-1825) [276 #1350, #1351], were optical instrument makers of 6 Curtain Road, London, England.

Harris, John
John Harris, (fl.1816-1840) [18] [276 #1578], was an optical instrument maker, optical brazier, and brass founder of 145 High Holborn, London, England, then from 1816 until 1840, 22 Hyde Street, Bloomsbury, London.

Harris, Philip
Philip Harris is a brand-name used by Findel Education Ltd., of Accrington, Lancashire, England, for laboratory equipment and supplies, aimed largely at the biology, chemistry and physics education market. Their optical products include microscopes, spectrometers and spectroscopes as well as components such as lenses, prisms, gratings, and a range of products for bench optics demonstrations and experiments in the classroom.

Harris, Richard
Richard Harris, (fl.1790-1810) [18] [276 #1149], optical instrument maker of London, England was the father of **William Harris (I)**. Addresses: 22 Lamb's Conduit St., Red Lion St.; 44 Great Russel St., Bloomsbury; 401 Strand; 50 Red Lion St., Clerkenwell; Red Lion St., Clerkenwell (1788); 404 Strand (1790-1810).

Harris, Thomas (I)
Thomas Harris (I), (fl.1790-1828) [9] [18] [276 #1150], of London, England was a mathematical and philosophical instrument maker, optician, and globe and telescope maker. He served his apprenticeship with **George Linnell**, beginning in 1771, and was freed of the **Spectacle Makers' Company** in 1804. He had two sons, **Thomas Harris (II)** and **William Harris (II)**. Among his apprentices was **William Dowling**, beginning in 1804.

Thomas (I) worked with his son Thomas (II), and subsequently William (II), in partnership, trading as **Thomas Harris & Son**. In 1804, during the partnership with Thomas (I), he traded at 140 Fleet Street and 30 Hyde Street, Bloomsbury.

Following the death of Thomas (II), from 1809 Thomas (I) continued the partnership with William (II), trading at 52 Great Russell Street, Bloomsbury, until their partnership was terminated in 1828, presumably on the retirement of Thomas (I).

Harris, Thomas (II)
Thomas Harris (II), (fl.1802, d.1808) [18], was an optician of London, England, and was the son of **Thomas Harris (I)**. He was freed of the **Loriners' Company** by purchase in 1804. He worked in partnership with his father, trading as **Thomas Harris & Son** at 140 Fleet Street, and in 1808 his address is recorded as Hyde Street, Bloomsbury. He was a Sergeant in the Bloomsbury Volunteers, and died in 1808 in a fire at the Covent Garden Theatre.

Harris, Thomas P.
Thomas Phillips Harris, (c1806-1892) [18] [319], was an optician of 52 Great Russel St., Bloomsbury, London, England. This is the same address as **Thomas Harris (I)**, and overlaps with the end of Thomas (I)'s time there. [18] records Thomas P. Harris at this address from 1824 until 1841, and census records show him at the same address from 1851 until 1881, and at 51 Great Russel St. in 1841. The Census record of 1841 also shows that his employee **William Callaghan** was an Optician's Apprentice in his household. In 1851 Thomas P. Harris employed three men and two boys.

Harris, William (I)
William Harris (I), (fl.1799-1841) [18] [276 #1151], was an optical instrument maker of 47 High Holborn, London, England, then from 1815 at No. 50 – note this is the same address as previously occupied by **H. Harris & Son**. William Harris (I) was the son of **Richard Harris**. He served his apprenticeship with the mathematical instrument maker Joseph Robinson, beginning in 1788, and was freed of the **Clockmakers' Company** in 1796.

He was Master of the Clockmakers' Company from 1830-1832. His apprentices included his son, Richard Joshua Harris. William (I) made a goniometer for **David Brewster**, and took out a patent with him for a micrometer telescope in 1811. In 1813 he entered a partnership with the optician **W. Campbell**, trading in London and Liverpool as **William Harris & Co.**, and in Hamburg, Germany, as W. Campbell & Co. In 1840 William (I) entered a partnership, presumably with his son, Richard Joshua, trading as **William Harris & Son**.

Harris, William (II)

William Harris (II), (fl.1818, d.1843) [9] [18] [276 #1353], was an optician, mathematical instrument maker, and globe maker of London, England. He was the son of **Thomas Harris (I)**, and was freed of the **Spectacle Makers' Company** by purchase in 1818 [401]. He worked in partnership with his father, Thomas (I), trading as **Thomas Harris & Son**. They were awarded a royal appointment to the Prince Regent in 1819 and to King George IV in 1820. William (II) was Master of the Spectacle Makers' Company from 1824-1825. He was declared bankrupt in 1830, [35 #18744 p2376] [35 #18793 p726].

In March 1839 he filed for insolvency [35 #19718 p667], and a month later he was in debtor's prison [35 #19724 p814].

Harrisons Opticians

Harrisons Opticians, [1], had works in Solihull, Exeter and Manchester. They had a photographic branch in Birmingham, and their other consumer branches were high-street opticians. They also had a small property company and a shop-fitting operation. In 1965 they took over **Thomas Armstrong & Brother Ltd**. They were themselves taken over by **Dollond & Aitchison** in 1968.

Hart, Joseph

Joseph Hart, (fl.1785-1812) [3] [18], was an optician, and optical, mathematical, and philosophical instrument maker of Birmingham, England. Addresses: 5 Digbeth (1787-1788); Digbeth (1791-1798); 5 Digbeth (1800-1801); 65 Dale End (1803-1808); Dale End (1808-1812). He began his business as an optician and spectacle maker and, in 1803 when he moved to Dale End, he began making optical, mathematical, and philosophical instruments. A theodolite signed "*Hart Fecit BIRMING^{Hm}*" from c1800 may be attributed to him.

Hartley, C.

C. Hartley, [319], were opticians, and manufacturers of optical and mechanical instruments and appliances, of 29 Great Portland Street, London, England. At the time the partnership was dissolved in 1909, it was between Frederick Nathaniel Davidson, Jack Daniel Jacobs, and Charles George Cooper [35 #28257 p4329]. The business continued after this, run by Frederick Nathaniel Davidson and Jack Daniel Jacobs.

Hartmann & Braun

Hartmann & Braun, [374], was a manufacturer of electrical measurement and test instruments and, later, control and automation equipment. The company was founded by Eugen Hartmann, who had served his apprenticeship with **Steinheil & Söhn**. He formed the company in 1879 as optische Anstalt, physikalisch astronomische Werkstätte (optical institute, physical astronomical workshops) in Würzburg, Germany, and moved to Frankfurt in 1884. In 1882 Wunibald Braun joined the firm. In 1968 began a series of changes of ownership, until the company finally became a part of ABB in 1999.

Instruments advertised in their catalogue of 1894 include: Reading telescopes for use in reading the scales on reflecting instruments (such as a mirror galvanometer); Spectrometer; Total reflectometer; Cathetometer; Photometer; Laboratory theodolite.

Hartnack, Edmund

Edmund Hartnack, (1826-1891) [53], was a German microscope maker. In 1854 he went into partnership with his uncle, **Georg Johann Oberhäuser**, and in 1864 he took over Oberhäuser's business, trading at 1 Rue Bonaparte, and Place Dauphine 21, Paris, France. **Constant Verick** worked for Hartnack prior to starting his own business in 1866. In 1870 the business moved to Potsdam, Germany, but the Paris business was retained, and continued to trade as **Hartnack & Prazmowski**. Hartnack traded at 39 Waisenstrasse, Potsdam, Germany, and the company continued after Hartnack's death, until 1906.

Hartnack and Prazmowski

Hartnack and Prazmowski, [50] [53], were microscope makers of Paris, France. The firm was a partnership between **Edmund Hartnack** and Adam Prazmowski (1821-1885), trading at 1 Rue Bonaparte, Paris, France. Adam Prazmowski was a Polish physicist who specialised in astrophysics. The partnership was formed in 1870 when Hartnack left Paris to move to Potsdam, Germany. Prazmowski sold their instruments in Paris under the name Hartnack and Prazmowski. The partnership was dissolved in 1878, and Prazmowski continued the business under his own name until it was taken over in 1883 by two of his foremen, **Bézu and Hausser**.

Hartsoeker, Nicolaas

Nicolaas Hartsoeker, (1656-1725) [9] [24] [256] [274] [322], was a Dutch optician, optical instrument maker, and physicist. He took an interest in optical glass-making, and showed that slow cooling of the molten glass results in fewer imperfections. He developed his own rigorous method for polishing small lenses. He met **Christian Huygens** and his brother Constantijn in 1678, and taught them his method of making glass spheres for microscope objective lenses. He is frequently credited with the invention of the screw-barrel microscope, but never claimed to have invented it. He made several long-focus telescopes for the Paris Observatory, including two aerial telescopes of focal length 155 and 220 feet. He also made a levelling instrument with telescopic sights.

Harvey & Peak

Harvey & Peak, (fl.1884-1899) [53] [320], were scientific instrument and laboratory apparatus makers, of Beak Street, Regent Street, London, England. The company was founded by William George Harvey and W.F. Peak, who purchased the business of **William Ladd** following his retirement in 1882. In 1891 Harvey & Peak moved to 56 Charing Cross Road. The company was dissolved in 1899. Examples of instruments by Harvey & Peak include a polariscope, and educational polarizing light apparatus.

Harvey, Reynolds and Co.

Thomas Harvey, (1812-1884) [131 p319] [132] [306], was a pharmacist and philanthropist from Leeds, England. He set up business in 1839, then in 1856 he took on partner Richard Reynolds, and they continued in business as Harvey and Reynolds. In addition to the patent medicines and remedies sold as a chemist and pharmacist, they specialised in photographic equipment, and sold other scientific instruments under their own brand-name.

Through various partnerships the firm continued in Leeds as Harvey, Reynolds and Fowler (1861); Haw and Reynolds (1864); Haw, Reynolds & Co. (1867); and **Reynolds and Branson** (1886), and remained in business until the 1970s. Instruments bearing their name include: telescope; brass monocular microscope; C19[th] pocket barometer; compass.

Hasselblad

Hasselblad is a Swedish manufacturer of premium quality cameras, lenses, and photographic accessories. The company was established in 1841 as a trading company, F.W. Hasselblad & Co., founded by Fritz Victor Hasselblad in Gothenburg, Sweden. Their first venture into photography came in 1885 when Fritz's son, Arvid Viktor Hasselblad, entered an agreement with **George Eastman** to make F.W. Hasselblad & Co. the sole Swedish distributor of the **Eastman Dry Plate company**'s products. In 1908 a new company was formed, Hasselblad Fotografiska AB, exclusively for photographic products.

In 1937 Victor Hasselblad (1906-1978), the grandson of Arvid Viktor Hasselblad, started his own independent shop, Victor Foto, and in 1941 he produced his first camera, the HK-7, for use by the Swedish air force. In 1943 he took ownership of the family company, including Hasselblad Fotografiska AB. By this time, he had already produced 240 HK-7 cameras, and the production continued for the duration of WWII.

In 1948 Hasselblad produced their first consumer camera, and this marked the beginning of their growth as a manufacturer of premium quality cameras. Their cameras were used to produce some iconic images. The photograph of the Beatles crossing Abbey Road was taken with a Hasselblad 500 camera, and NASA used Hasselblad cameras for many of their missions, beginning with the Mercury 8 mission in 1962, and including some of the famous photographs taken on the surface of the moon by the Apollo 11 astronauts.

Hasselblad embraced the digital era of photography, and went on to produce a range of digital cameras, as well as a *digital back* which can be attached to most of their V System cameras produced since 1957, enabling them to be used for either film or digital photography.

Hauck, W.J.

W.J. Hauck, (fl.C19[th]), was an optical, scientific, and philosophical instrument maker of IV Kettenbrückgasse 20, Vienna, Austria. He was appointed Purveyor to the Imperial and Royal Court as a mechanic during the reign of Franz Joseph I. Instruments bearing this name include: brass library telescope; barometer; laryngoscopic set; Borda circle, dated c1873, in instrument box with plaque which reads *W.J. Hauck. K K. Hof Mechanika. Wien. IV Kettenbrückgasse 20.*

Hauksbee, Francis

Francis Hauksbee, (1688-1763) [9] [18] [46] [256] [275 #553] [276 #63], was a mathematical, philosophical, and optical instrument maker, and science lecturer, of Crane Court, Fetter Lane, Near St. Dunstan's Church. London, England. In 1738 his address is recorded as 3 Crane Court, Fleet St. He served his apprenticeship with **John Marshall**, beginning in 1703, and was freed of the **Drapers' Company** by patrimony in 1714. He was the nephew of the renowned philosophical instrument maker and lecturer, Francis Hauksbee (bap.1660, d.1713) [275 #519], and set himself up in competition with his uncle, making similar instruments and offering similar lectures. From 1719 until 1726 he worked with **John**

Hadley on improvements to reflecting telescopes, and published *Proposals for Making a Large Refracting Telescope* in about 1723. Also in 1723, he was appointed Clerk-Housekeeper to the Royal Society.

Haw and Reynolds
See **Harvey, Reynolds and Co.**

Hawkins & Wale
See **George Wale**.

Hawkins, Thomas
Thomas Hawkins, (fl.1820-1840) [18] [276 #1582] [379], was an optical, mathematical, and philosophical instrument maker, and hydrometer maker of 16 Perry St., Somers Town, London, England. This address is listed in some directories as 16 Percy St., Kings Cross. This same address was also occupied by the mathematical instrument maker Frederick Hawkins from 1825 until 1865.

Hawksley & Sons
Hawksley & Sons, [53], manufacturer of laboratory equipment, of 300 Oxford Street, London, England, was founded in 1869 by Charles Hawksley. The company was the English agent for **Charles Achilles Spencer**, and sold microscopical apparatus of their own make. They took over **H.F. Angus** in 1920, (note that [320] gives this date as 1905, but the date 1920, given by [50], appears more likely). By this time their address was 83 Wigmore Street, London. During the 1930s Hawksley & Sons was acquired by the American company William Baum, and began making equipment for measuring blood pressure. The company continued, producing laboratory and clinical equipment and supplies for the medical industry.

Hay, John & James
John & James Hay, (fl.1847-1851) [263], were carvers and gilders in Aberdeen, Scotland, who also sold optical and philosophical instruments. [18] lists them as optical instrument makers. Barometers bearing their name have appeared at auction, but are believed to have been retailed by them, rather than manufactured by them. John Hay (fl.c1838) of Aberdeen was a carver and guilder who is known to have sold: sundial; microscope; compass.

Hayes Brothers
Hayes Brothers (fl.late-C19th-early-C20th) were marine opticians, compass, and lamp makers, of James St., Cardiff, Wales. They also had branches in Barry and Port Talbot. According to their trade label, in a cased two-day marine chronometer by W. Plaskett, "Hayes Brothers. Marine opticians, Compass & Lamp makers, James Street, Cardiff and at Barry. Compasses adjusted. Chronometers & Nautical instruments cleaned, repaired & accurately adjusted. Charts, nautical books, stationery, &c. &c. Time by electric telegraph direct from the Royal Observatory." Instruments bearing their name include: sextant, with scale inscribed *Hayes Brothers, Cardiff, Barry & Port Talbot*; sextant, with scale inscribed *Hayes Brothers, Barry Docks*.

H.C.R. & Son Ltd.
See **H.C. Ryland & Son Ltd.**

Hearn & Harrison
Charles Hearn, [103] [241], was originally from Exeter, England. He established his company in 1857 in Toronto, Canada, and relocated to Montreal in c1860, where he opened a shop trading as an optician and mathematical instrument maker. Charles Hearn died in 1865, upon which his wife, Susan, took over the company. She married the instrument-maker Thomas Harrison, and they renamed the company Hearn & Harrison. The company was no longer in the family by 1900, and continued business until 1936.

Instruments bearing their name include: pocket barometer; single-draw marine-style telescope; Galilean binoculars; brass surveyor's transit.

Hearn & Potter
Hearn & Potter, [417], were mathematical instrument makers, opticians, and jewellers, of Toronto, Canada. The firm was a partnership between **Charles Potter (III)** and the jeweller and watchmaker William Hearn. According to their trade advertisement: "Hearn & Potter, (from Dollond's) Mathematical instrument makers, Opticians and Jewellers, 54, King Street, Toronto. Manufacturers and importers of Theodolites, Levels, Telescopes, Microscopes, and all descriptions of Surveying, Optical, and Philosophical instruments. Hydrometers and Saccharometers, as used by the Imperial Government. Spectacles to suit all sights. Royal Admiralty charts of the St. Lawrence and the Lakes. Repairing and adjusting on the premises."

The partners went their separate ways in 1859. Charles Potter (III) started to work under his own name in Toronto, and William Hearn set up in business in Montreal, Quebec.

Hearne, George
George Hearne, (fl.1705-1750) [3] [9] [18] [276 #185a] [385], was a mathematical and optical instrument maker of Dogwell Court, White Friars, Fleet St., London, England, and the sign of *The Sphere*, Sergeant's Inn, Chancery Lane. He served his apprenticeship with Edward Barber, beginning in 1697, and was freed of the **Joiners' Company** in 1705. He was the first London instrument

maker to enter the European market, with his reflecting telescopes selling across Europe. He worked in partnership with **Joseph Jackson**, trading as **Hearne & Jackson**, and made reflecting telescopes for **John Hadley**.

Hearne & Jackson

Hearne & Jackson, [18], of London, England, was a partnership between **George Hearne** and **Joseph Jackson**. The partnership was active between 1735 and 1750.

Hearson, Charles

Charles Hearson & Co., [44] [287] [320], of Hope Works, Willow Walk, Bermondsey, London, England, were makers and suppliers of apparatus for physiology, biochemistry, agriculture, botanical, and dairy research. Their showrooms were at 27 Mortimer Street, London. Their products included microscopes, microtomes, optical accessories, and accessories for microscopists. They were the UK agents for **C. Reichert Optische Werke**.

Prior to WWII, beginning in about 1936, they manufactured **McArthur Microscope**s under license. The McArthur literature claims that Hearson was the first commercial manufacturer of the McArthur microscope. The Hearson McArthur microscope included objectives by **R.&J. Beck**.

Heath & Co.

Heath & Co., [16] [29] [44] [53] [379], were scientific, nautical, and optical instrument manufacturers, of London, England. Addresses: 28 Fenchurch St. (1883-1889); 115 & 117 Cannon St. (1891-1897); 2 Tower Royal, Cannon St. (1898-at least 1900); Observatory Works, Crayford (1888-at least 1905); New Eltham (1951); Showrooms at 79/80 High Holborn (1951); City depot at 13 Railway Approach, London Bridge (1951). They also had a depot at 52 Bothwell Street Glasgow, Scotland (1951).

The business was founded by George Heath in 1845, trading as Messrs. George Heath. George Heath's son, **George Wilson Heath**, became the proprietor of the business in 1872. In 1882 the firm was incorporated as Heath & Co. Ltd.

In 1919 the company was listed as a member of the **British Optical Instrument Manufacturers' Association**. In 1926 the business was acquired by **W.F. Stanley and Co. Ltd.**

Hezzanith was a Heath & Co. brand, which W.F. Stanley continued to use. Some examples of *Hezzanith* Galilean field binoculars bear the maker's name L. Petit, suggesting that they were made for Heath & Co. by **L. Petit** of Paris. Sextants may be seen bearing the name Heath & Co. or *Hezzanith*.

Heath, George

George Heath, (fl.1851) [18] [322], was a nautical instrument maker of Erith, Kent, England. He was the son of a grocer. He served his apprenticeship with **Henry Hughes** of the **Masons' Company**, beginning in 1840. His entry in the catalogue for the 1851 Great Exhibition [328] describes a sextant and a quadrant.

He may have been the George Heath who founded the firm Messrs. George Heath that became **Heath & Co.**

Heath, Messrs. George

See **Heath & Co.**

Heath, George Wilson

George Wilson Heath, (c1849-1937), was a scientific instrument maker of London, England. He was the son of George Heath, with whom he served his apprenticeship, commencing in 1863. He completed his apprenticeship in 1870. In 1872 he became the proprietor of his father's business, Messrs. George Heath. In 1882 the firm was incorporated as **Heath & Co. Ltd.**

G.W. Heath is credited with a number of inventions, including improvements in ship's compasses, course correctors, sextants, and thermometers. He built the firm into a highly successful enterprise. He died aged 88 in Sidcup, Kent.

Heath, Thomas (I)

Thomas Heath (I), (fl.1719, d.1773) [9] [16] [17] [18] [90] [276 #66] [361], was a scientific instrument maker, at the *Hercules & Globe*, in the Strand, London, England, from c1720. He served his apprenticeship with the mathematical instrument maker **Benjamin Scott**, and was freed of the **Grocers' Company** in 1719. His trade cards and bill-heads advertise "Mathematical instruments contrived & made in metal, ivory or wood according to the latest observations of Philosophers and Practitioners of mathematical arts by Thos. Heath at the *Hercules* next the Fountain in ye Strand, London. With books of their use." His son, **Thomas Heath (II)**, also worked in London as a scientific instrument maker.

Thomas Heath (I)'s apprentices included his son, Thomas Heath (II), as well as **George Adams (I), John Troughton (I), Tycho Wing**, and **Joseph Jackson**. Thomas Heath (I) joined Tycho Wing, who was also his son-in-law, in partnership, and they traded as **Heath and Wing**. Heath made the sea quadrant invented by **Caleb Smith**, and it appeared in the Heath and Wing catalogue of 1750.

Examples of his instruments include: telescopic sight; combined telescope and level; sea quadrants, 1735; ring dials; drawing instruments; theodolites.

Heath, Thomas (II)
Thomas Heath (II), [9] [90], was a scientific instrument maker of London, England. He began his apprenticeship with his father, **Thomas Heath (I)**, in 1746, but he did not take his freedom of the **Grocers' Company**.

Heath, William
William Heath, [53], of Plymouth, England was an optician and scientific instrument maker in the latter half of the C19th. He traded from three addresses: 46 Fore St., Devonport (1850-1852); 116 Fore St., Devonport (1857); 24 George St., Plymouth (1850). Instruments bearing his name include: telescope; microscope; Galilean binoculars; brass parallel rule; carved oak stick barometer; triple-optic Galilean binoculars, manufactured by **Lemaire** of Paris.

Heath and Wing
Heath and Wing, [18] [44], opticians, and optical, mathematical, and philosophical instrument makers of London, England. Addresses: *Hercules & Globe*, Exeter Exchange, Strand; near Exeter Exchange in the Strand (1751-67); *Hercules & Globe* next door to the Fountain Tavern, Strand (1759); near the Savoy Gate in the Strand (1771). These may have been different ways of writing the same address.

The firm was a partnership between **Thomas Heath (I)** and **Tycho Wing**. They traded from 1751 until the death of Thomas Heath (I) in 1773, at which time Tycho Wing retired. The Heath and Wing Catalogue of 1765 [242] lists an extensive, and wide range of instruments. They were succeeded in business by the mathematical and philosophical instrument maker Thomas Newman, who had been an apprentice of Tycho Wing.

Heather, William
William Heather, (fl.1763-1812 d.1812) [16] [18] [276 #604] [361], was a mathematical instrument maker, hydrographer, and engraver, and a retailer of nautical instruments, maps, and charts. He operated a Navigation Warehouse at 157 Leadenhall, London, England, under the sign of *The Little Midshipman*. He served his apprenticeship with engraver and stationer, George Michell, beginning in 1780, and was freed of the **Stationers' Company** in 1789. He was succeeded in business by **John William Norie**, who had been a teacher in his Navigation Warehouse. Instruments bearing his name include: repeating circle; Hadley's quadrant; sextant; octant, c1795, with scale marked with foul anchor symbol (see Appendix XI).

Heele, Hans
Hans Heele, (fl.1893-1923) [58] [374], was a manufacturer of precision optical and mechanical instruments, of Berlin, Germany. He worked closely with the Physiakalisch-Technische Reichsanstalt (Imperial Physical and Technical Institute), of Berlin, which was founded in 1877. He is known to have made astronomical objectives, micrometers, and spectrometers. He exhibited his work in Berlin in 1896 and in Paris in 1900. In 1923 the company was taken over by the firm of **Bamberg**, by then trading as **Askania Werke AG**. Instruments bearing his name include: spectroscopes; Nörremberg polariscope; saccharimeters; Gallenkamp colorimeter; theodolite.

Heliograph
A heliograph is:

(1) An engraving made using a process invented by **Joseph Nicéphore Niépce**. The engraving results from a chemical reaction when a photosensitive plate is exposed to light. The process is known as heliography.

(2) An instrument used to take photographs of the sun, also known as a photoheliograph.

(3) A device that uses reflected sunlight to send messages over a great distance by means of a moving mirror, or a mirror and a shutter. Messages may be sent using morse code.

Heliometer
A heliometer is a telescope equipped to measure the angular separation of heavenly bodies, or the angular size of a heavenly body.

See also double image micrometer, divided object glass micrometer, megameter, **Servington Savery**, **Pierre Bouguer**.

Helios
Helios is a UK brand of binoculars and spotting scopes. Early Helios instruments were imported from Soviet manufacturers such as **KOMZ** and **ZOMZ**. Today some of their binoculars are made in Japan and some are made by **Kunming United Optics** in China.

The Helios brand-name was also used for a range of camera lenses made by various Soviet manufacturers, including **KMZ**.

Heliostat
A heliostat, [152], is a siderostat designed for viewing the Sun. Since there is plenty of light available, a design using two or more mirrors is acceptable. The clock drive must operate at the mean solar rate for tracking of the Sun.

Generally, for a solar observatory, the telescope is mounted horizontally or vertically, with the vertically mounted solution preferred due to issues with convection currents and turbulence in the horizontal option. With the telescope orientation thus fixed, an arrangement of two or more mirrors is necessary to track the changes in the Sun's declination with the advancing seasons. The two most common solutions are the two-mirror polar heliostat and the two-mirror coelostat.

The two-mirror polar heliostat is based on the polar siderostat. Since this arrangement results in field rotation, a Dove prism may be used as a beam rotator to counter the field rotation. The Dove prism is rotated in the opposite sense to the rotation at half the rate of rotation.

The two-mirror coelostat does not require a Dove prism beam rotator, since the coelostat field does not rotate. However, the coelostat primary mirror must not be changed in declination. Two solutions are available to track changes in declination. The first solution is for primary mirror to be moved in the north/south direction, while keeping its polar orientation. The second solution, more commonly used in modern heliostats, is for the secondary coelostat mirror to be moved in the up/down direction.

Heliostate
See siderostat.

Heliotrope
A heliotrope is:

(1) A signalling instrument, invented by **Karl Friedrich Gauss**, used in long distance land surveys. It comprises a mirror, arranged with a telescope, to reflect sunlight over a great distance, and is used by surveyors in geodetic triangulation.

(2) An instrument used in astronomy to determine the time at which the sun reaches its greatest altitude above the equator.

(3) A form of heliostat used to track the sun's motion, and keep the sun's reflection correctly orientated for use in a solar camera.

Heller & Brightly
Heller & Brightly, [58], was a surveying instrument maker in Philadelphia, USA. German born Charles S. Heller (1839-1912) emigrated to the USA in the 1940s and began to work for **William J. Young**, becoming a partner in 1866. In 1870 he entered a partnership with English born Charles H. Brightly (1817-1897), who had emigrated to the USA in the 1830s. They traded as Heller & Brightly, making surveying instruments. The company was incorporated in 1926, and continued in business until 1968.

Helmholtz, Hermann von
Hermann Ludwig Ferdinand von Helmholtz, (1821-1894) [37] [155], of Potsdam, Germany, was a theoretical physicist and a physiologist. His interest in human vision led to his invention of the ophthalmoscope for inspecting the interior of the eye, and the ophthalmometer for measuring the curvature of the eye. He researched colour vision and colour blindness, and in 1867 he published *Handbuch der physiologische Optik* (Handbook of Physiological Optics). In explaining colour vision, he further developed the three-colour theory of **Thomas Young**, which became known as the *Young-Helmholtz theory*.

He was the inventor of the telestereoscope for producing an appearance of relief in the objects of a landscape at moderate distances. He also carried out research into human hearing, and published his findings in *Die Lehre von den Tönempfindungen als physiologische Grundlage für die Theorie der Musik* (On the Sensations of tone as a physiological basis for the theory of music). His work in the area of physics included thermodynamics and electrodynamics.

Hemsley & Son, Thomas
Thomas Hemsley & Son (fl.1857-1898) [18] [262] [379], were nautical instrument makers of London, England. They succeeded **Thomas Hemsley (II)** in business, and traded at 4 Tower Hill (1857-1898), and 4 King Street, Tower Hill (1887-1898). This likely began as a partnership between Thomas Hemsley (II) and his son Thomas Hemsley (III). Instruments bearing their name include: brass sextant with silvered vernier scale; brass framed sextant with wooden handle and silver scale.

Hemsley, Henry (I)
Henry Hemsley (I), (fl.1781-1794) [18] [90], was an optician and dealer in lace, of London, England. Addresses: Opposite the George on Saffron Hill (1781); Little Saffron Hill, Hatton Wall (1784-1794); 85 Fleet St. (1786). He was the son of Thomas Hemsley, a gentleman of Holborn, and may have been the brother of **Thomas Hemsley (I)**. He began his apprenticeship with **Henry Raynes Shuttleworth**, beginning in 1774, and was immediately turned over to John Graves of the Fishmongers' Company. He was freed of the **Spectacle Makers' Company** in 1786. One of his apprentices was Thomas Hemsley (I), who was turned over to him by **James Martin** in 1789 to complete his apprenticeship.

Hemsley, Henry (II)
Henry Hemsley (II), (fl.1838-1861) [18] [90] [276 #2148] [358], was an optician, and optical, mathematical, and philosophical instrument maker of London, England.

Addresses: 11 King St. Tower Hill (1823); 138 Ratcliffe Highway (1838-1843); 140 Ratcliffe Highway (1844-1845); 140 St. George St., St George's in the East (1846-1861). He was the son of **Thomas Hemsley (I)**, and the brother of **Thomas Hemsley (II)**. He was freed by patrimony of the **Grocers' Company** in 1823. Instruments bearing his name include: ebony framed octant signed *H. Hemsley 138 Ratcliffe Highway London*.

Hemsley, J.&T.

J.&T. Hemsley, (fl.1826) [18], were opticians, and optical and mathematical instrument makers of London, England. Addresses: 11 King St., Tower Hill; 11 Little Tower Hill. This was a partnership between **Thomas Hemsley (II)** and his brother Joseph Hemsley, and succeeded **Thomas Hemsley (I)** in business. A notice of dissolution of the partnership was posted in 1826 [35 #18218 p279], although [18] lists the partnership name as being in use from 1826-1828. They were succeeded in business by Thomas Hemsley (II). Instruments bearing their name include: single draw 1½-inch mahogany and brass telescope, length 38 inches extended.

Hemsley, Thomas (I)

Thomas Hemsley (I), (fl.1801-1825 d.by June 1825) [18] [90], was an optician and optical brass turner of London, England. Addresses: 1 Parson's Walk, Newington, Surrey (1801); 11 King Street, Tower Hill (1820-1825). He was the son of Thomas Hemsley, a gentleman of Holborn, and may have been the brother of **Henry Hemsley (I)**. He began his apprenticeship in 1784 with **James Martin**, and was turned over to Henry Hemsley (I) in 1789. He was freed of the **Grocers' Company** in 1801. His sons, **Henry Hemsley (II)** and **Thomas Hemsley (II)**, were both optical instrument makers. He was succeeded in business by **J.&T. Hemsley**.

Hemsley, Thomas (II)

Thomas Hemsley (II), (fl.1826-1854) [18], was an optician, and optical, mathematical, and nautical instrument maker of London, England. Addressess: 18 King St., Tower Hill; King St., Tower Hill (1825); 11 King St., Tower Hill (1840-1851). He was the son of **Thomas Hemsley (I)**, and the brother of **Henry Hemsley (II)**. He was freed by redemption of the **Spectacle Makers' Company** in 1825, and worked in partnership with his brother, trading as **J.&T. Hemsley**. By 1828 he was trading under his own name. Among his apprentices was his son, Thomas Hemsley (III), beginning in 1841.

According to an advertisement of 1828 [392], "To Merchants, owners, and captains of ships. Telescopes, sextants, quadrants, and compasses by the manufacturer, Thomas Hemsley, 11, King-Street, Tower Hill, London, near the Royal Mint. Day or night telescopes, with improved magnifying power. Charts, nautical books, and stationary. All kinds of Mathematical instruments cleaned and repaired, with correctness, dispatch, and punctuality. Spectacles to suit all sights. Captains and the trade supplied."

He was succeeded in business by **Thomas Hemsley & Son**.

Hen, Hendrik

Hendrik Hen, (1770-1819) [50] [220], of 35 Kalverstraat, Amsterdam, Netherlands, was an optical and scientific instrument maker. He served his apprenticeship with **John Cuthbertson**. He made and sold microscopes, telescopes, octants, and other nautical, optical and philosophical instruments, and also retailed microscope slides and instruments from other makers.

Henderson, Angus

Angus Henderson, (fl.1861-1884) [263] [271], of Edinburgh, Scotland, was a mathematical and optical instrument maker. He worked for both **Adie & Son** and **James Bryson** prior to setting up as an independent instrument maker in 1861. Addresses: 23 Hanover St. (1861-1868); 69 George St. (1869); 48 Frederick St. (1870-1873); 11 Teviot Place (1874-1876); 2 South St. David St. (1877); 72 Gilmore Place (1878); 51 Hanover St. (1882-1883); 5 Cathcart Place (1884).

According to his trade card: "Angus Henderson, Practical optician, Microscope, Mathematical and Philosophical instrument maker, 23 South Hanover Street, Edinburgh. For many years in the establishments of Messrs Adie & Son and Mr James Bryson."

Henry, Paul and Prosper

Paul Henry, (1848-1905), and his brother Prosper Henry, (1849-1903), [352], were astrophotographers and telescope makers. They both worked at the Paris Observatory from age 16, initially in the meteorological service, and subsequently on the astronomical staff. They had a keen interest in mapping the sky, and produced sky-charts together. At first they produced their charts by visual observation, and subsequently using astrophotography. They entered a collaboration with **Paul Gautier**, in which the Henry brothers produced the optics, and Gautier produced the mounts, for telescopes and astrographs used in a number of French and international observatories (see entry for Paul Gautier for more information). Their sky-charts, along with their astrograph design, inspired Ernest Mouchez, director of the Paris Observatory, to instigate the *Carte du Ciel* sky mapping project. The Henry brothers, together with Paul Gautier, produced many of the astrographs used in the project.

Hensoldt & Engelbert
Hensoldt & Engelbert, [47], was a partnership between **Moritz Hensoldt** and **Louis Engelbert**, making microscopes. The partnership began in 1861, and continued until Louis Engelbert's death in 1877.

Hensoldt, Moritz
Moritz Carl Hensoldt, (1821-1903) [47] [78], studied mathematics and physics, and trained as a precision mechanic. He formed a friendship with **Carl Kellner**, and in 1849 they started a company in Wetzlar called **Optisches Institut**, producing lenses and microscopes. They worked together on and off until Kellner's death in 1855. During the early 1850s he had also formed his own optical company, working under his own name. In 1861 he began to work in partnership **Louis Engelbert** producing microscopes under the name **Hensoldt & Engelbert** until Engelbert's death in 1877. Meanwhile, his own company continued to operate, and in 1896 his sons Waldemar and Karl became partners, forming M. Hensoldt & Söhne AG. By 1928, **Zeiss** was a shareholder. This Zeiss shareholding was a majority by 1954, and by 1964 Zeiss had transferred all its binocular manufacturing to the Hensoldt factory in Wetzlar. In 1968 Carl Zeiss AG completed its acquisition of Hensoldt, which became a fully owned subsidiary.

For information about binocular models by M. Hensoldt & Söhne AG, see [85].

Heron, David
David Heron, (fl.1827-1863) [18] [263] [271], was a nautical instrument maker of Greenock, Scotland. In 1827 he moved to Glasgow, where he ran the Nautical, Optical, and Stationery Warehouse at 1 William Street. This business had been started in 1797 by John Heron, who may have been his father. At this address David Heron entered a partnership with **Duncan McGregor**, trading as **Heron & McGregor**. In 1836 Duncan McGregor took over the premises, and the partnership was terminated. David Heron also conducted his business at 128 Broomielaw and, in 1836, he relocated to 212 Broomielaw, and continued his business as David Heron & Co. He traded in partnership, for a short while in about 1844, as Heron & Johnston. He was subject to sequestration proceedings in 1848, [**170** #5729 p123], then continued in business at 4 Carrick Street until at least 1863. In 1860 his daughter, Jessie, married one of his employees, **James Whyte**, who began to trade at Heron's premises in 1864.

Heron & McGregor
Heron & McGregor, [263], were nautical instrument makers of the Nautical, Optical, and Stationery Warehouse at 1 William Street Glasgow, Scotland. The company was a partnership between **David Heron** and **Duncan McGregor**, which terminated in 1836 when Duncan McGregor took over the business.

Herring, Alika
Alika Herring was a highly regarded American mirror-maker who worked for **Cave Optical Co.** until the end of the 1950s. Tom Cave required his mirror-makers to inscribe the back of the mirror with a serial number, but not the maker's name. Despite this, Herring generally inscribed his mirrors with his name, and many Cave Astrola telescopes are equipped with his primary mirrors. Upon leaving Cave Optical, he went to work for the Lunar and Planetary Laboratory to help with their lunar mapping project.

Herschel Prism
Another name for a Herschel wedge.

Herschel Wedge
A Herschel wedge, sometimes known as a Herschel prism, is an optical prism used for solar viewing or photography, and was invented by **William Herschel**. It refracts most of the sun's radiation away from the optical path, thus reducing the energy from the Sun to safe levels. It usually has a built-in heat sink in order to dissipate the excess energy. A Herschel wedge is used in place of a diagonal, usually on a small aperture refracting telescope. It is an alternative to using a solar filter mounted at the telescope objective. However, it is only used on smaller refractors since it has limited ability to dissipate the excess energy generated when observing the sun with larger aperture telescopes.

Herschel, William
Frederic William Herschel, usually known as William, (1738-1822) [24] [46] [155] [388], was a musician, composer and astronomer born in Hanover, Germany. He began his career as a military bandsman and, in 1762, upon his discharge, he came to England for what was to become a successful career as a musician. He began to study the mathematics of music, and expanded his interests in mathematics to other areas, including optics. Having read **Robert Smith**'s *Compleat System of Opticks* his interest in astronomy and the construction of telescopes blossomed. In 1772 William brought his sister Caroline to England from Hanover to train as a singer, and to collaborate in his concerts.

William's interest in astronomy soon led him to desire larger and larger aperture telescopes for increased light-gathering to see deeper into the universe. He sought to overcome the problem of limited reflection in speculum mirrors by his invention of the Herschelian telescope. Reflectors were the only practical solution, but small mirrors were not satisfying. He learned to figure his own mirrors, and became highly skilled in the construction of telescopes. By the mid-1770s he was already casting and figuring his own mirrors. He built several telescopes, including 7-foot, 10-foot, and 20-foot Newtonians. In 1781 he built a foundry for casting mirrors, and embarked on a project to build a massive mirror of 4 feet diameter. He overcame significant difficulties, eventually producing two mirrors for a 40-foot telescope, which he erected in a giant frame behind his house. He alternated the mirrors, with one in use in the telescope while the other was being polished or re-figured. This telescope was, at the time, the largest in the UK. It was only of moderate success, and Herschel preferred his 20-foot, and other telescopes, for most of his observing. Throughout his active career William constructed many fine telescopes for himself, for royalty, and for others.

William Herschel was an enthusiastic and skilful observer of the heavens, and in 1781 he discovered Uranus, making him the first to discover a planet not visible to the naked eye. In 1782 William took up a royal appointment which paid enough for him to give up his musical employment, but required him to reside near Windsor Castle. He agreed to this, thus also putting an end to Caroline's musical career. So it was that Caroline began to work with William to sweep the sky for comets. Between them, Caroline and William systematically catalogued comets, nebulae and double stars.

Herschelian Telescope

A Herschelian telescope is an off-axis reflecting telescope in which the primary mirror is tilted to focus the light at the edge of the top end of the telescope tube, where the eyepiece is located. The concept was first proposed by the French scientist, Le Maire, in 1728 [388]. **William Herschel** turned this design into a practical reality in order to avoid the need for a secondary mirror, since the speculum mirrors of his day were of limited reflectivity, and a second mirror caused considerable light loss. The compromise is that the tilt of the primary mirror introduces astigmatism into the image, which is only mitigated by a very long focal ratio. It is not a design used in modern telescopes.

Hertel & Reuss

Hertel & Reuss was founded in 1927 by Otto Hertel and Eduard Reuss, in Kassel, Germany. They produced binoculars, microscopes, opera glasses and riflescopes. In 1995 they were rescued from bankruptcy by **Gerhardt Optik GmbH**, following which their products continued to be sold under the Gerhardt brand-name Hertel & Reuss.

Hexagon AB

Hexagon AB is a Swedish multinational company providing services and products for a wide variety of industries, including surveying instruments and services. Their surveying instruments brands include **Leica Geosystems**, which they acquired in 2005, and **GeoMax**.

Heyde Gustav

Gustav Heyde, (1846-1930) [55], of Dresden, Germany, was a maker of optical and mechanical instruments. In 1872 he formed the Gustav Heyde Mathematisch-Mechanisches Institut Optische Präzisions-Werkstätten, manufacturing optical components, circular dividing engines, prisms, and photographic lenses. By the early C20th they were making astronomical instruments, optical systems, and geodesic measuring instruments. By 1945 the company was named Gustav Heyde GmbH. During WWII they marked their instruments with the **ordnance code** bwt. From 1949 the company was named Optics - Feinmess Dresden VEB, and was renamed again to Feinmess Dresden GmbH in 1990. In 1992 the company became part of the Steinmeyer Group.

Hezzanith

Hezzanith was a **Heath & Co.** brand.

H.H. & Son Ltd.

H.H. & Son Ltd., of Liverpool and London, England. Instruments bearing their name include Galilean binoculars. This is believed to refer to **Henry Hughes & Son**.

Hicks, James Joseph

James Joseph Hicks, (1837-1916) [3] [46] [143] [304], was a scientific instrument maker. He was born in Co. Cork, Ireland, and grew up in London, where he built his successful business. He served his apprenticeship with **Louis P. Casella**, and by 1860 he was foreman of Casella's plant, designing and making improvements to instruments. He set up his own business in about 1861, making clinical thermometers, syringes, hydrometers, lenses, meteorological apparatus, barometers, altimeters, and theodolites. He was most noted for his patent design of clinical thermometer, and for his aggressive marketing and protection of his intellectual property. In 1911 J.J. Hicks sold his business to **W.F. Stanley**, and remained a director until his death.

Highley, Samuel

Samuel Highley, (1826-1900) [3] [50] [113] [259], was a philosophical instrument maker and teacher of photography, of 70 Dean Street, Soho, London, England. His father was a publisher of medical books under the name Highley and Son at 32 Fleet Street, London. After his father's death, Samuel continued this business, and moved it to 70 Dean Street, Soho. He received an honourable mention in the Great Exhibition of 1851 for his specimens and models.

By 1853 Samuel was selling microscopes and other instruments under his own name. He is not known to have had any training in microscope making, and it is possible that the instruments he sold were made for him by subcontractors. His illustrated entry in the catalogue for the London International Exhibition of 1862, [258], showed: Educational and other types of microscope; Spectroscope; Polariscope; Electroscope; Educational astronomical telescope; Actinometer, and other instruments. At this exhibition he received two medals and an honourable mention. He also produced magic lanterns for educational purposes, and invented the lantern polariscope and an achromatic gas microscope lamp. He filed for bankruptcy twice; once in 1865, [35 #23028 p5007], and again in 1868, [35 #23359 p1568]. In 1869 he moved the business to 10a Great Portland Street, London, and appears to have ceased operating the business around or soon after 1870. He had a reputation for the quality of his inexpensive instruments for education and popular use, and continued teaching, writing and publishing throughout his career.

Hilger, Adam

Upon the outbreak of the Franco-Prussian war in 1870, Adam Hilger came to London from Paris with his brother Otto Hilger. Adam Hilger was employed by **John Browning (III)** and became his foreman. He established the firm of Adam Hilger, [44] [306], in 1874 in Islington, London, England, as makers of optical instruments for control of industrial processes.

Prior to WWI Hilger's main products were optical parts, sold mainly to rangefinder manufacturers **Barr and Stroud** [20] [26]. They also specialised in finished instruments, including spectrometers and spectrographs. In the years leading up to WWI they had a virtual monopoly on the fabrication of complex prisms. However, the production of optical components for Barr & Stroud suffered with quality problems. In 1898 **Frank Twyman** joined the firm as a scientific assistant and, in 1902, upon the death of Otto Hilger, Twyman succeeded him as managing director. By the end of 1915 Barr and Stroud had started to make their own optical parts, which severely impacted Adam Hilger's business. In 1916 Frank Twyman sold a controlling interest in the company to **Thomas Cooke & Sons**, not realizing that they were owned by **Vickers**, [228].

Adam Hilger Ltd. worked with Captain A.H. Marindin to produce his Marindin rangefinder, avoiding contravening the patents of Barr & Stroud and **Zeiss**. In 1919 Adam Hilger Ltd. was listed as a member of the **British Optical Instrument Manufacturers' Association**. In 1922 they were a listed exhibitor at the British Industries Fair. Their catalogue entry, [447], described them as manufacturers of: Optical and other Scientific instruments for research and control of industrial processes; Polarimeters; Saccharimeters; Spectrometers; Spectrophotometers; Refractometers; Colorimeters; Interferometers. In 1926 Frank Twyman purchased back the shares that had been sold to Thomas Cooke & Sons.

In 1946 **E.R. Watts and Son** acquired a 78% interest in the firm, and in 1948 Hilger amalgamated with E.R. Watts and Son to become **Hilger and Watts Ltd.** which was incorporated as a public company with Frank Twyman as a director.

Hilger Interferometer

See interferometer.

Hilger & Watts

Hilger and Watts, [44], were makers of optical measurement instruments in London, England. They were incorporated in 1948 as a public company to acquire three firms engaged in manufacture of theodolites, microscopes, etc. – **E.R. Watts and Son**, **Adam Hilger** and **James Swift and Son**. In 1963 Hilger & Watts took over **W. Ottway & Co. Ltd.** In 1968 James Swift and Son was purchased by the firm's ex-employee John H. Basset, and the remainder of Hilger and Watts was acquired by **Taylor, Taylor & Hobson**, part of **Rank**.

Hilkin

Hilkin was a UK company who imported Japanese binoculars and refracting telescopes during the 1960s, through to the 1980s. One example of a Hilkin 3-inch, 1000mm refractor, probably from the 1960s, bears the maker's mark of the Japanese company **Eikow**, (see Appendix IV). Other examples are made by **Carton Optical Industries, Ltd.**, or **Empire Made**.

Hilkinson

Hilkinson, of Halesworth, Suffolk, England, is an importer and retailer of binoculars, monoculars, spotting scopes, opera glasses, magnifiers, and tripods. They were acquired by **Viking Optical Ltd.**, probably in 2012. They imported binoculars from Japan and the USSR, and subsequently from Russia. More recently their binoculars have been imported from a variety of sources. **Ruper** is a Hilkinson brand-name used for their loupe magnifiers, made in Japan.

Hill & Price

Hill & Price, (fl.1855-1870) [319] [322], opticians, and chronometer, watch, and nautical instrument makers, of 1 and 2 Broad Quay, Bristol, England, was a partnership between David Brewer Hill and Charles Wetherell Price. They succeeded **J.M. Hyde** in business. Instruments bearing their name include: 5 draw wooden telescope; marine compass. The partnership was terminated in 1870 [35 #23673 p4707], after which Charles Wetherell Price continued the business on his own account. In 1884 he was declared bankrupt [35 #25352 p2114].

Hill & Son, G.

G. Hill & Son, (fl.1888-1897) [306], were camera makers of 186 Broad St., Birmingham, England. Their camera models included the *Improved Landscape Camera*, available as a half-plate, whole-plate or 10x8 field camera, and the *Improved Long Focus*, also available as a half-plate or whole-plate camera. They were succeeded at that address by P. Hill in 1897.

Hill, Nathaniel

Nathaniel Hill, (fl.1746-1764 d.1768) [18] [90] [274] [276 #363] [319] [322] [361], was a mathematical instrument and globe maker, and map engraver, of *The Sun and Globe*, Chancery Lane, London, England. The address also appears variously as Opposite Sergeant's Inn, and *The Globe and Sun*. He served his apprenticeship with James Farmer of the **Grocers' Company**, beginning in 1701, but did not take his freedom. Later, in 1730, he began an apprenticeship with the globe maker Richard Cushee, and was freed of the **Merchant Taylors' Company** in 1751. He did, however, take on an apprentice, Thomas Bateman, before his own freedom, as well as other later apprentices. In 1764 Nathaniel Hill was one of the petitioners who attempted to revoke the patent obtained by **John Dollond (I)** for the achromatic lens, and which was enforced after Dollond's death by his son, Peter (See Appendix X). His trade card, while referring to him as a globe maker and engraver, shows illustrations of a wide range of mathematical, philosophical, scientific, and optical instruments, including a quadrant, microscope, spectacles, alidade, telescope, and scioptic ball.

Hill, Peter

Peter Hill, (1778-c1845) [9] [18] [169] [263] [395], was an optician, and mathematical and optical instrument maker, of Edinburgh, Scotland. Addresses: Richmond St. (1801-1803); 7 East Richmond St. (1804-1810); 9 East Richmond St. (1811-1812); 6 Union Place (1813-1822); 7 Union Place (1820-1824); 2 Greenside Place (1825-1828); 7 South Union St. (1825).

He also retailed instruments bearing his name, but made by other makers. He was noted for making ruby and topaz jewel lenses for microscopes, and supplied jewel lenses to **Sir David Brewster**.

Hillum & Co.

Hillum & Co., (fl.1856) [18] [319], were opticians of 109 Bishopsgate Within, London, England. They succeeded the business of **Mrs. S. Hillum**.

Hillum, Joseph

Joseph Hillum (sometimes Hiltum), (c1804-1840 fl.1836-1840) [18] [276 #2149] [319], was an optician, and optical, mathematical, and philosophical instrument maker of 109 Bishopsgate Within, London, England. Instruments bearing his name include: brass telescope with 2" objective on library stand; mahogany stick barometer. He was succeeded in business by his brother, **Richard Hillum**.

Hillum, Richard

Richard Hillum (sometimes Hiltum), (1800-1880 fl.1842-1846) [18] [319], was a silversmith, jeweller, optician, and mathematical, and philosophical instrument maker of 109 Bishopsgate Within, London, England. In 1841 he occupied this address together with his mother, Sarah, then at least 80 years old, and his two sisters, Sarah and Mary, all of whom were opticians. He succeeded his brother, **Joseph Hillum**, in business, and was succeeded by **Mrs. S. Hillum**.

Hillum, Mrs. S.

Mrs. S. Hillum, (fl.1849-1851) [18] [319], was an optician, and mathematical and philosophical instrument maker of 109 Bishopsgate Within, London, England. She succeeded **Richard Hillum** in business, and was succeeded by **Hillum & Co.**

Himmler, Karl Otto

Karl Otto Himmler, (d.1899), was a German optician, renowned for making microscopes. Prior to starting his own business in Berlin, he worked for **Ernst Gundlach**, and then for **Wilhelm and Heinrich Seibert**. He started his own business in 1877, trading as Himmler and Barthning. After a few years he began to run the company alone, and by his death in 1899 the company was known as Otto Himmler Optich-Mechanische Werkstätte. The company continued in the hands of A. Himmler, producing research microscopes until the 1930s. During WWII they produced optical munitions for the German war effort, which were marked with the **ordnance code khc**.

Hinde, John
John Hinde, (fl.1802-1805) [18], was a mathematical instrument maker of London, England. Addresses: 5 Red Lion St., Clerkenwell, and 3 Plough Court, Fetter Lane (1802-1805). A brass sextant, signed by John Hinde of London c1805, to the design of **Ebenezer Hoppe**'s Improved Sextant, may be attributed to him, [16] [54].

Hinde, Roger
Roger Hinde, (fl.c1820-1834) [18] [276 #1591], was an optical, mathematical, and philosophical instrument maker of 70 Noble St., Goswell St., and 70 Noble St., Falcon Square, London, England.

Hindle, John Henry
John Henry Hindle, (1869-1942) [199] [223], was the owner of an engineering company, Hindle, Son & Co. Ltd. in Blackburn, Lancashire, England, serving the cotton, wool, paper and textile industries. As a hobby he made telescope mirrors up to 30 inches, mostly for amateur telescope makers. His largest mirror, a 30-inch Newtonian, was initially mounted at the Cambridge University Observatory, and Hindle wrote an article about it in *Scientific American* in September 1939. He devised a new test for Cassegrain and Gregorian secondary mirrors. He also designed his own *ovioid stroke* grinding and polishing machine, and instead of patenting it he had the design published in *Amateur Telescope Making* magazine. During his travels he witnessed the pouring of the disc for the **Hale** 200-inch telescope for the Palomar observatory in California.

Hindley, Henry
Henry Hindley, (1701-1771) [2] [3] [9] [18] [24] [256] [276 #186], was an English clockmaker, mathematical instrument maker and inventor, with workshops in Manchester (1722-1730), Petergate, York (1730), and Stonegate, York (1771). He invented an endless screw, a pyrometer, and first proposed the telescope mount now known as the English Equatorial in 1741. The concept was developed and brought to fruition by **Jonathan Sisson**. Hindley is believed to be the first to construct a circular dividing engine, which was designed for cutting clock-wheels, and was adapted for dividing scales. **John Stancliffe** served an apprenticeship with Hindley, and worked for him in York prior to moving to London to work for **Jesse Ramsden**. Stancliffe is alleged to have given the secrets of Hindley's dividing engine to Ramsden.

John Smith (II) was associated in his youth with Henry Hindley. According to a footnote in [284] in 1809, referring to the text of a letter dated 1748 by Henry Hindley: "The persons referred to were both bred with him. His brother, Mr. Roger Hindley, who has many years followed the ingenious profession of a watch-cap-maker in London, was so much younger as to be an apprentice to him. Mr. John Smith, now dead, had some years past the honour to work in the instrument way, under the direction of the late Dr. Demainbray, for his present Majesty."

Hinds, David
David Hinds, [199], of Tring, Hertfordshire, and Hemel Hempsted, Hertfordshire, England, was the son of **David Graham Hinds**, and was a renowned mirror maker. He co-founded, and was a director of **AstroSystems Ltd.** He produced mirrors for AstroSystems Ltd., **Broadhurst Clarkson & Co.**, **Fullerscopes**, **Astronomical Equipment**, and many others. He also provided an aluminizing service.

He stopped making and aluminizing mirrors and went on to run an astronomy equipment retail company, David Hinds Ltd. This family business traded in one form or another from 1965 until 2020.

Hinds, David Graham
David Graham Hinds, [9] [199], of Tring, Hertfordshire, England, made and aluminized fine quality mirrors during the 1940s and 1950s. He was a friend of **Henry Wildey**, and his son, **David Hinds** also became a renowned mirror maker.

Hino Kinzoku Sangyo Co.
Hino Kinzoku Sangyo Co. was a Japanese company formed in 1952 to make telescopes, binoculars, microscopes, spotting scopes, and other instruments. In 1990 they merged with **Eikow Optics** to form **Mizar Co. Ltd.** Hino Kinzoku Sangyo Co. may have been the parent company of **Hino Optical Co.**

Hino Optical Co.
Hino Optical Co. was an optical company based in Tokyo, Japan. They may have been a subsidiary of **Hino Kinzoku Sangyo Co.**, and are associated with **MIZAR**. It seems likely that Hino were the manufacturer and MIZAR was the brand-name, prior to the merger between Hino Kinzoku Sangyo Co. and **Eikow Optics** to form **Mizar Co. Ltd.**

Hinton & Co.
Hinton & Co., [61] [122] [306], of 38 Bedford Street, Strand, London, England, were pharmaceutical and photographic chemists, established in 1825. From the

1880s they manufactured cameras and photographic shutters, but prior to WWI this is believed to have ceased, and the business became focussed on chemicals and products for the photographic darkroom.

Hitch, Joseph
Joseph Hitch, (fl.1757-1773) [18], was a mathematical and optical instrument maker, who traded At the Corner of Beaufort Buildings in the Strand, London, England (1757), and at Eagle Court, St. John's Lane, Clerkenwell, London (1764-1773). He served his apprenticeship with the mathematical instrument maker John Liford. He was freed of the **Goldsmiths' Company** in 1752, and took on several apprentices of his own. In 1764 he was one of the petitioners who attempted to revoke the patent obtained by **John Dollond (I)** for the achromatic lens, and which was enforced after Dollond's death by his son, Peter (See Appendix X).

Hiyoshi Optical Company
The Hiyoshi Optical Company was a Japanese optical instrument maker. Hiyoshi instruments may bear the HOC maker's mark, (See Appendix IV). The company used maker's **JB code** JB207.

Hobcraft, William (I)
William Hobcraft (I), (fl.1817) [18] [276 #1885] [379], was an optical and mathematical instrument maker of London, England. Addresses: 28 King St., Snow Hill; 28 Cow Lane, Smthfield; 28 King St., Smithfield; 56 Whitecross St., Fore St (1817); 12 Beech St., Barbican (1822); 13 Barbican (1825); 18 Barbican (1834); 14 Barbican (1837-1852).

He served his apprenticeship with **Alexander Wellington**, beginning in 1803, and was a Freeman of the **Stationers' Company**. He was known as William Hobcraft Senior, and was the father of **William Hobcraft (II)**.

Hobcraft, William (II)
William Hobcraft (II), (fl.1845-1870) [18] [276 #1885] [379], was an optical, mathematical, and philosophical instrument maker of London, England. Addresses: 38 Princes St., Liecester Square (1845); 14 Great Turnstile (1849); 62 Dean St., Soho (1851); 419 Oxford St. (1853-1870). The latter, Oxford St. address was previously occupied by **Moritz Pillischer**.

He was also known as William Hobcraft Junior, and was the son of **William Hobcraft (I)**. He served an apprenticeship with the philosophical instrument maker and mathematical instrument seller, Thomas Staight of the **Fanmakers' Company**, beginning in 1831.

Hofmann, Dr. J.G.
Dr. J.G. Hofmann, (fl.1850-1875) [53] [113], was a German optical and philosophical instrument maker, who started his business in France, trading at rue de Bucy, 3, Paris. From 1863 the firm was known as l'Institute d'Optique Paris. Instruments bearing his name include: direct vision spectroscope, signed *Spectroscope - Hofmann Construit à l'Institute d'Optique du Dr. J.G. Hofmann à Paris*; pocket spectroscope; Porro prism telescope/monocular.

Holcomb, Amasa
Amasa Holcomb, (1787-1875) [212], was a maker of surveying instruments and astronomical and terrestrial telescopes, of Massachusetts, USA. He began his business making surveying instruments, including small transit telescopes, and expanded into astronomical telescopes in 1828, being reputedly the first American to make telescopes for sale commercially. Most of his telescopes were long focal length, 6-inch or 8-inch Herschelian reflectors, and his largest was 10 inches. He won awards for his workmanship, and his customers included Brown University, Delaware College, and Williston Academy, as well as individual astronomers. Holcomb was the only commercial telescope maker in the USA until **Alvan Clark** and **Henry Fitz** began to make refractors. He retired from telescope making in 1846.

Holland & Joyce
Holland and Joyce, (fl.1833-1834) [18] [53], was a partnership between **James Holland** and Edward Joyce. They organised an exhibition of an "improved oxy-hydrogen microscope and iriscope" at 106 New Bond St., Oxford St., London, England, in 1834. The handbills for the exhibition listed in detail the living and inanimate objects that were to be "brilliantly projected upon a disc containing 254 square feet, and are variously magnified from lowest power up to 2,500,000 times". The advertisement goes on to claim, "With the power of 2,500,000, the flea is magnified to so great an extent as to appear as large as the late Elephant 'Chunie'. The optical part is constructed upon an original plan of Mr. Holland, and under his immediate superintendence; the rest of the apparatus by eminent artists", and "Admittance 1s". Another handbill claims, "N.B.—This instrument is, in reality, what it professes to be; its powers and dimensions are not being in any respect over-stated. Its grand characteristic is distinctness.—See and Judge!!"

The partnership Holland and Joyce was terminated in 1834 [35 #19192 p1675].

Holland Circle

A Holland circle is a surveying instrument consisting of a compass attached centrally in a horizontally mounted graduated circle, to which two pairs of sights are attached. An alidade with two sights is attached in a manner concentric to the circle such that it can be rotated to take a bearing.

Holland, James

James Holland, (fl.1832-1835) [18] [264] [274] [393], was an optical, mathematical, and philosophical instrument maker of London, England. Addresses: 106 New Bond St.; 6, Manor Place, Walworth. He developed a microscopic triplet, replacing the lower lens of the **Wollaston** doublet with two plano-convex lenses placed close together. He described this innovation in a paper submitted to the Society of Arts in 1832, for which he received a Silver Medal. **Andrew Ross** commented that this enabled the errors to be reduced still further (than with the Wollaston doublet). James Holland entered a partnership with Edward Joyce to exhibit an oxy-hydrogen microscope and iriscope, trading as **Holland and Joyce** until 1834.

Holliwell Bros.

Holliwell Brothers, [3] [18], were opticians and nautical instrument makers of Liverpool, England. Addresses: 9 East Side, Salthouse Dock (1849); 64 Park Lane (1851-1855); 72 Park Lane (1857-1864); 154 Brownlow Hill (1870-1872).

The firm began in 1848 as a partnership between William Holliwell and his brother Thomas. (Note: The Willliam in this partnership may have been **William Holliwell** or a son or other relation of his by the same name). They succeeded **William Holliwell & Son** in business, and in 1851 they were joined in the partnership by another brother, Charles. In 1853, the partnership was dissolved, [35 #21490 p2943], and the business continued without Charles, once again as a partnership between William and Thomas. The company continued in business until 1872.

Holliwell, Thomas & William

Thomas and William Holliwell was another name used by the brothers, trading as **Holliwell Bros.**

Holliwell, William

William Holliwell, (fl.1795-1832) [3] [18] [276 #989] [361], was a mathematical, optical, and nautical instrument maker of Liverpool, England. Addresses: 5 Broomfield St., Salthouse Dock (1804-1816); 7 Broomfield St. (1816); West End, Old Dock (1818-1825); 1 Hurst St. (1827); 31 Stanhope St. (1827-1829); Mersey St. (1830); 34 Stanhope St. (1832).

According to his trade card, "Real manufacturer of Mathematical & Optical instruments for navigation." In 1829, he went into partnership with his son, trading as **William Holliwell & Son**.

Holliwell, William & Son

William Holliwell & Son, (fl.1829-1847) [3] [18] [276 #989], were opticians, and mathematical and nautical instrument makers of Liverpool, England. Addresses: 1 Hurst St. (1829-1832); 7 Darwen St. (1834); 7 East Side Salthouse Dock (1835-1837); 9 East Side Salthouse Dock (1839-1847).

The business was a partnership between **William Holliwell** and his son, and was succeeded by **Holliwell Bros.**

Holmes, J.K.M.

J.K.M. Holmes & Co. Ltd. (fl.c1950s-c1960s) were scientific instrument makers of Northumberland, England. Addresses: 33 Bedford Street; 4-4a Albion Rd., North Shields. Instruments bearing their name include: brass telescope with black finish, length 38¼ inches, with 3-inch objective; 40-inch single-draw telescope with 35mm objective, black finish, with brass plaque bearing the word *Moonscope*; 39-inch single draw astronomical telescope with 2-inch objective, black finish, with brass equatorial mount on wooden tripod.

Holmes-Bates Stereoscope

The Holmes-Bates stereoscope is a simple, inexpensive viewer for photographic stereo-cards. It is generally hand-held, although table-top examples exist. It consists of a hooded pair of lenses which view along an attached stem-piece, held from underneath by a handle or table-mount. On the stem-piece is mounted a sliding cross-member with attached supports for the stereo-card. This is moved back and forth to bring the images into focus when viewed through the hooded lenses.

This design, which is a variant of **Brewster**'s lenticular stereoscope, originated in about 1860 with the American physician, Dr. Oliver Wendell Holmes Sr., and was further developed by the photographer, Joseph L. Bates, of Boston, MA.

See stereoscope.

Holmes, Booth & Haydens

Holmes, Booth & Haydens, [122], of Waterbury, Connecticut, USA, was a manufacturer of photograph cases, lenses, daguerreotype silver plates and other photographic apparatus, founded in 1853. From about

1853 to 1866 they sold cameras and lenses made by **Charles F. Usener**.

Holmes Brothers

Holmes Brothers, [306], established in 1884 at 9 Pulteney Street, Barnsbury Road, Islington, London, England, were wholesale manufacturers of photographic apparatus. In 1902 they moved to Park Street, Islington. They manufactured several models of camera for other companies, including **Negretti & Zambra**, and were the makers of the **Sanderson** camera that was marketed by **George Houghton & Sons Ltd**.

In 1904 they were absorbed into George Houghton & Sons Ltd, who became Houghtons Ltd. After this merger they appear in the directories as Holmes Bros (Sanderson Camera Works).

Holmes, John

John Holmes, (fl.1817-1829 d.by 1831) [18], was an optician and spectacle maker of London, England. Addresses: 14 Redcross Square, Cripplegate; 2 Little Cloisters, Bartholomew Hospital; 14 Redcross Square, Aldersgate (1822). He was the father of **John William Holmes**, and was succeeded in business by **Mary Ann Holmes**, who may have been his wife. He is known to have sold: barometer; thermometer.

Holmes, John William

John William Holmes, (fl.1846-1859) [18] [276 #1594], was an optician and optical instrument maker of 14 Redcross Square, London, England. He was the son of **John Holmes**. He served his apprenticeship with Edward Kennard of the Innholders' Company, beginning in 1831, and was freed of the **Spectacle Makers' Company** in 1853.

Note that the Post Office directories also show John & William Holmes trading at the same address in 1845. This could have been a partnership, or it may have been a misprint.

Holmes, Julia & John

Julia & John Holmes, (fl.1839-1842) [18], were opticians of 14 Redcross Square, Aldersgate, London, England. They succeeded **Mary Ann Holmes** in business, and were succeeded by **John William Holmes**.

Holmes, Mary Ann

Mary Ann Holmes (fl.1829-1839) [18] [276 #1594], was an optician and optical instrument maker of 14 Redcross Square, Aldersgate, London, England. She succeeded **John Holmes** in business, and may have been his wife. She was succeeded by **Julia & John Holmes**. Records also show, at this same address, William Holmes, optician (1836).

Holmes, Samuel

Samuel Holmes, (b.c1832 fl.1861-1890) [319] [320], of 275 Strand, London, England, was a lamp manufacturer and optical instrument maker. The 1871 Census shows his address as 12 Brunswick Terrace, Rotherhithe, Surrey. Instruments bearing his name include: binocular microscope marked *S. Holmes's patent achromatic binocular* [53]; binocular microscope with objective cannister [270].

He was declared bankrupt in 1865 [35 #23018 p4664]. He was awarded several patents for improvements to lamps and, in 1871, a patent for "improvements in optical instruments" [35 #23714 p1342].

Holyman, John

John Holyman, (fl.1808-1840) [18] [90] [276 #1889], was an optical, nautical, mathematical, and philosophical instrument maker of London, England. Addresses: 5 Newmarket St., Wapping; 5 Old Gravel Lane, Wapping; Nightingale Lane (1808); 33 Nightingale Lane (1816).

He served his apprenticeship with **Ebenezer Rust (I)** of **Spencer, Browning and Rust**, beginning in 1784, and was freed of the **Grocers' Company** in 1808. He advertised that he made sextants and quadrants.

Holyman, William

William Holyman, (fl.1845-1859) [18] [276 #1889], was a nautical and mathematical instrument maker of 27 Sydney Place, Commercial Road, London, England. He was succeeded in business by Mrs. Holyman.

Hooke, Benjamin

Benjamin Hooke, (fl.1793-1805) [18] [276 #1158], was an optical instrument maker of 159 Fleet Street, London, England. He served his apprenticeship with **Gabriel Wright**, beginning in 1786, and was freed of the **Girdlers' Company** in 1793. In 1794 he joined the partnership **Gilbert & Wright**, and began to trade as **Gilbert, Wright & Hooke**. He continued to work in the partnership until it was terminated in 1801. He is known to have sold microscopes.

Hooke, Robert

Robert Hooke, (1635-1703) [37] [46] [155] [275 #270] [443], was an English natural philosopher, born on the Isle of Wight. He went to London to serve an apprenticeship as a painter, but he was not healthy and the paints gave him headaches. He went on to gain an education in languages and music, and in 1653 or 1654 he went to Christ Church, Oxford as a chorister. He was able to work as the assistant to the chemist Thomas Willis, and subsequently for the natural philosopher Robert Boyle. He met many learned natural philosophers, and developed an interest in astronomy. In 1662 he was appointed curator of experiments for the Royal Society of London, and soon became a fellow. Following the great fire of London in 1666 he was appointed one of the official surveyors for the reconstruction, and was the architect responsible for the design of some of the new buildings.

He was the first to describe the compound microscope, which he documented, along with his observations when using it, in his book *Micrograpia*. He also introduced the spiral spring for regulation of the balance wheel of a watch. This was part of the result of his work which led to the famous Hooke's Law, and was intended for the purpose of measuring longitude at sea. The idea was subsequently used in marine chronometers.

He was active in studying celestial mechanics, and lecturing on the subject, and was innovative in methods and devices for studying the stars. He was the first to build a Gregorian telescope, which he presented to the Royal Society in 1674. He produced what was probably the first design for a clockwork driven equatorial telescope mount, which was regulated by means of a conical pendulum. He developed a method of scale division for astronomical quadrants, and invented a circular mechanism for cutting horological gear-wheels. He also produced designs for helioscopes and novel types of telescope. **John Marshall** worked for Hooke sometime after completing his apprenticeship in 1785, and prior to starting his own business.

Hooke became embroiled in two disputes with **Sir Isaac Newton**. One arose when Newton published his work on the inverse square law of gravity, in which Hooke believed Newton had plagiarised his own work. The other occurred when Hooke published his theory of light, which resulted in a disagreement over the priority of their respective claims to the theory.

Hooker, John Daggett

John Daggett Hooker, (1838-1911), was an industrialist who made his fortune in hardware and steel pipe in Los Angeles, USA. He was a keen amateur astronomer and scientist, and a friend of **George Ellery Hale**. Hale convinced Hooker to make a major contribution to the cost of the 100-inch telescope in the Mount Wilson observatory, with additional funding from the Carnegie Institute. The telescope was named after Hooker. Later, Hooker and Hale fell out, and did not reconcile by the time Hooker died in 1911. The Hooker telescope saw first light in 1917.

Hoppe, Ebenezer

Ebenezer Hoppe, also known as Hoppé, or Hoppey, (fl.1801-1821, d.1821) [18] [276 #1159] [322] [385], was a mathematical and optical instrument maker of London, England. Addresses: Edward St., Limehouse Fields; 51 Church St., Minories (1801-1821). He served his apprenticeship with **Michael Dancer**, beginning in 1793, and was freed of the **Joiners' Company** in 1801. In about 1804, he developed the "Improved Sextant", with two scales. It was intended to enable reading of up to 265° by movement of the index glass, but probably achieved no more than 160°. This was described in the publication *An Explanation of E. Hoppe's Improved Sextant* in 1804. An example of this design of sextant was made by **John Hinde** in about 1805, [16] [54].

Horne & Thornthwaite

See **Horne, Thornthwaite and Wood**.

Horne, Thornthwaite and Wood

Horne, Thornthwaite and Wood, [18] [50] [122] [306], were optical, mathematical, and philosophical instrument makers of London, England. Addresses: 123 Newgate St. (1844-1851); 121 & 123 Newgate St. (1851); 416 Strand (1886-1893); 74 Cheapside (1888-1893).

The company was founded in 1844 when Fallon Horne, William Thornthwaite, and **Edward George Wood** took over the premises of the scientific instrument supplier Edward Palmer at 123 Newgate Street. They purchased most of Palmer's stock, and traded as Horne, Thornthwaite, & Wood, manufacturing and retailing optical instruments including cameras, microscopes and telescopes. In 1845 Thornthwaite published his book *A Guide to Photography*.

Their extensive entry in the catalogue for the 1851 Great Exhibition, [328], in London included, among other items, the following:

"Horne, Thornthwaite and Wood, 123 Newgate Street—Manufacturers. Apparatus for exhibiting dissolving views, chromatropes, etc., by the oxyhydrogen lime light, with illustrative paintings and apparatus, showing the method of producing the light, the arrangement of the lenses, and contrivance for dissolving the pictures. Oxyhydrogen microscope and apparatus, in case. Daguerreotype apparatus, consisting of an adjusting back camera, with compound achromatic lens, an improved bromine and iodine box, with contrivance for transferring the prepared plate to the frame of the camera, mercury box, plate-box, chemical-chest, buffs, plate-holders, gilding stand, tripod, etc. The parts of the apparatus are so arranged that the process may be entirely performed in the light, without the necessity of a dark room. Registered portable folding calotype camera, with achromatic lenses, for portraits and views, etc. Improved reversing frame, for producing positive pictures from calotype negatives and other photogenic processes. 'Optometer,' an instrument for ascertaining the existence of any defect in the refracting media of the eye, and for determining the range of adjustment for distances which it possesses. Transparency, exhibiting the appearance of the lunar disc when in direct opposition to the sun, as seen through Herschel's 40-feet reflecting telescope."

In 1854 Wood left the partnership to pursue his own business, and the company continued to trade as Horne & Thornthwaite until 1885 when Wood rejoined the company and they reverted to the name Horne, Thornthwaite, & Wood. They continued until 1893 when the company was taken over by William Ackland, and the brothers John and William Overstall, all of whom were already part of the company. The company once again became Horne & Thornthwaite, which it remained until it closed in 1911, when the Overstall brothers moved to Canada.

Horton & Co., George Melville

George Melville Horton & Co., (fl.1835-1843) [3] [18], were spectacle makers, and optical, mathematical, and philosophical instrument makers of London and Birmingham, England. London addresses: 17 Thaives Inn, Holborn (1835-1836); 32 Hatton Garden (1837-1839); 36 Ludgate St. (1843). The Ludgate Street address was previously occupied by **Henry Jones**. Birmingham addresses: 126 Great Hampton St. (1835, 1839-1842), and 136 Great Hampton St. (1837). In 1842 they were listed as goldsmiths, silversmiths, and jewellers.

Höschel, Christof Kaspar

Christof Kaspar Höschel, (fl.1774-1820) [256] [374], was a mathematical, philosophical, and nautical instrument maker of Augsburg, Germany. He was the son-in-law of **George Friedrich Brander**, with whom he worked from 1760. In 1774 they began to work in partnership, trading as **Brander and Höschel**. The partnership continued until Brander's death in 1783, following which Christof continued the business until 1820. He was succeeded in business by his son, C.C. Höschel.

Hoskold's Transit Theodolite

See theodolite.

Houghton & Sons, George

See **George Houghton**.

Houghton, George

George Houghton, [44] [306], was a glass merchant, maker of photographic equipment, and wholesaler of photographic materials, of 89 High Holborn, London, England, and of Glasgow, Scotland. He formed a partnership with **Antoine Claudet** in 1834, trading as **Claudet and Houghton**. In 1876 the company became George Houghton & Son, and in 1892, George Houghton & Sons. They began using the *Ensign* brand-name in 1903, while operating from Ensign House at 88/89 High Holborn. In the same year, the company was incorporated as George Houghton & Sons Ltd. In 1904, the company absorbed **Holmes Brothers** (the maker of the **Sanderson** cameras marketed by George Houghton & Sons), **A.C. Jackson**, Spratt Brothers, and **Joseph Levi & Co**, to form Houghtons Ltd., producing a variety of photographic, optical, surveying and nautical instruments.

In 1915 they were one of the companies who formed **British Photographic Industries Ltd.** In the same year Houghtons Ltd. entered a joint venture with **W. Butcher & Sons**, called Houghton-Butcher Manufacturing Co., in order to share manufacturing space. The two companies finally merged in 1926 to form Houghton-Butcher (Great Britain) Ltd. They continued to produce optical, scientific and photographic equipment, and continued to use the brand-name *Ensign*.

In 1930 Houghton-Butcher set up a sales subsidiary **Ensign Ltd.**, but in 1940 the Ensign headquarters was destroyed in enemy action. The assets were sold, but Houghton-Butcher kept the name, and used it as a brand-name for their *Ensign* cameras. In 1945 the company, together with **Elliott and Sons**, who made the *Barnet* brand of film, formed the company **Barnet Ensign**. In 1948 Barnet Ensign merged with **Ross Ltd** to form **Barnet Ensign Ross**, which was renamed **Ross-Ensign Ltd.** in 1954.

Houghton-Butcher

See **George Houghton**.

Houghtons Ltd.

See **George Houghton**.

Houston & Cameron

Houston & Cameron, (fl.1865) [263] [271] [379], were mathematical and nautical instrument makers of 19 Howard St., Glasgow, Scotland. See **Paul Cameron**.

Hovil & Sons

Hovil & Sons, (fl.1837-1842) [18], were optical, mathematical, and philosophical instrument makers of 30 Haydon St., Minories, London, England.

How, James

James How, (1821-1872) [50] [53] [319] [320], was an optical and scientific instrument maker and chemical supplier, of 2 Foster Lane, London, England. He began to work for **George Knight** in 1842, and is believed to have been Knight's manager. Following the bankruptcy and death of George Knight in 1862, James How succeeded **George Knight & Sons** in business in 1863, and traded under the name of James How & Co., Successors to George Knight & Sons.

An advertisement of 1863 reads "Microscopes, Etc. James How, Upwards of twenty years with and successor to George Knight & Sons, 2 Foster Lane, London. Now ready, free by post on receipt of four stamps, a catalogue of scientific instruments. Part I., containing Microscopes, Microscopic apparatus and objects, Telescopes, and other Optical instruments. Part II., containing Photographic apparatus and chemicals, is in preparation, and will shortly be published. A list of Entomological and Natural history apparatus may be had on application. Agent for Jones's Photographic cartoons for the Magic lantern. G. Knight & Sons, 2 Foster Lane, London."

Following his death in 1872, John How was succeeded in business by **George Smith**, who continued the firm under the name James How & Co.

Howard, Charles P.

Charles P. Howard, [185], was an amateur astronomer and lens maker, active in Hartford, Connecticut, USA in the 1880s to early 1900s. He designed his lenses to avoid reflections from the inner surfaces of the doublet or from the tube, and is known to have constructed doublet objectives flint forward.

Howe, John

John Howe, (fl.1690, d.c1747) [18] [256] [275 #466], was a spectacle maker and optical instrument maker of Threadneedle St., by the Royal Exchange, London, England. He was the son of **Joseph Howe**, and was freed of the **Spectacle Makers' Company** by patronage in 1699. He succeeded his father in business. He is known to have made telescopes, and examples of telescopes bearing the inscription *Jo. Howe Londini Fecit* are in the collection of the Science Museum, London [53]. It is unclear, however, whether *Jo* refers to John or his father, Joseph. He was succeeded by his son John Howe (II).

Howe, Joseph

Joseph Howe (sometimes How), (fl.1664-1710) [18] [275 #320], was a spectacle maker and optical instrument maker of Royal Exchange, London, England, then from 1691, Threadneedle St., Behind the Royal Exchange. He is known to have made telescopes, and examples of telescopes bearing the inscription *Jo. Howe Londini Fecit* are in the collection of the Science Museum, London [53]. It is unclear, however, whether *Jo* refers to Joseph or his son, **John Howe**. By 1664 he was a Freeman of the **Spectacle Makers' Company**, and among his apprentices in the Spectacle Makers' Company was **Thomas Gay**, beginning in 1682. In 1679 he was on the livery of the **Broderers' Company**, and among his apprentices in the Broderers' Company were **William Longland**, beginning in 1667, and **John Rowley**, beginning in 1682. He was Master of the Spectacle Makers' Company from 1672-1674 and 1689-1690. He was succeeded in business by his son, John Howe.

Howell & Green

See **Midland Camera Co. Ltd.**

Hoya

Hoya is a Japanese optical glass-maker, based in Tokyo. The brothers Shoichi and Shigeru Yamanaka established an optical glass production plant in the city of HOYA, Tokyo, in 1941, and began production of optical glass.

In 2007 Hoya acquired the **Pentax Corporation**, and announced a merger of the Pentax Corporation into Hoya, which completed in 2008. In doing so they acquired the *Pentax* brands of medical technology, cameras, and binoculars. The acquisition was made in order to gain access to its medical technology, which Hoya now markets under the *Pentax* brand-name.

In 2009 Hoya sold the *Pentax* surveying instrument business, formerly the **Fuji Surveying Instrument Company Ltd.**, to the Taiwan Instrument Co. Ltd. As a result of this acquisition the company was renamed **TI Asahi Co. Ltd.**, and continued to produce surveying instruments under the *Pentax* brand-name.

In 2011 Hoya sold Pentax Imaging Systems, the *Pentax* camera and binoculars brand, to **Ricoh**.

In 2014 Seiko Optical Products Co. Ltd. became a consolidated subsidiary of Hoya.

Hudson & Co., A.

See **Hudson & Son**.

Hudson & Son

Hudson & Son, [18] [50], were spectacle makers, and makers of mathematical, medical, nautical and optical instruments, based in Greenwich, England. Frederick T. Hudson was an optician and spectacle maker who began his business in Stockwell Street, Greenwich in 1828. Some optical instruments bear his name, but it is likely that he had them made for him. In 1851 he showed his prepared microscope slides at the Great Exhibition in London, [328]. In 1859 he was still advertising as F.T. Hudson. He began to trade in partnership with his sons, Alfred (1842-1919) and William (1845-1919), sometime between then and 1862, when they exhibited at the London International Exhibition, showing prepared microscope slides under the name Hudson & Sons, [295]. It is not clear whether they ever traded as Hudson & Sons, since most records refer to the company as Hudson & Son.

Unlike F.T. Hudson when he was trading alone, Hudson & Son were manufacturers of optical and nautical instruments. Following Frederick's retirement, the partnership of Hudson & Son continued between Alfred and William until it was dissolved in 1875, [35 #24174 p321], after which William continued to trade under the same name.

In 1885 bankruptcy proceedings were initiated for Alfred Hudson, Optician, trading as A. Hudson & Co., 34 Hatton-Garden, London, and 50 Yerbury Road, Holloway, Middlesex, [35 #25509 p4252].

An undated trade card reads: "Established half-a-century. Hudson & Son, manufacturers of Mathematical, Optical and Nautical instruments to the Lords Commissioners of the Admiralty, Royal Naval College, &c. Greenwich, London, S.E."

The business continued into the C20th.

Hudson, Frederick T.
See **Hudson & Son**.

Hudson, John
John Hudson, (fl.1817-1836) [18] [276 #1362] [319] [361], was an optician, and optical and mathematical instrument maker of London, England. Addresses: 7 Orange St., Leicester Square; 17 Ryder's Court, Leicester Square; 112 Leadenhall St.; 17 Orange St., Leicester Square (1817-1822); 9 St. Martin's Court, Leicester Square (1836). According to his trade card he was a maker of spectacles, and optical, mathematical, and philosophical instruments and apparatus. **John Thomas Hudson** worked as his assistant. He was succeeded in business by **Sarah Hudson**.

Hudson, John Thomas
John Thomas Hudson, (b.c1806 fl.1830-1845) [18] [53] [276 #1893] [319], was an optician, spectacle maker, and optical instrument maker of London, England. Addresses: 28 Henrietta St., Cavendish Square; 14 Marylebone Lane; 8 Henrietta St., Cavendish Square; 17 Ryder's Court, Leicester Square; 96 High St., Marylebone (1838-1839); 102 Waterloo Road (1845).

He worked as a journeyman optician, assistant to **John Hudson**, and for a short period he had stands where he sold opera glasses at the Drury Lane, Covent Garden, and Victoria theatres. He wrote on the subject of optics, and published at least one book on spectacles and eyesight (1833). In 1834 he was subject to insolvency proceedings [35 #19158 p978], and in 1842 he was once again insolvent, this time in debtor's prison [35 #20169 p3456]. In the 1851 Census he is recorded as living at 266 Strand, St. Clement Danes, Middlesex, pursuing a literary occupation, aged 45.

Hudson, Sarah
Sarah Hudson, (fl. 1836-1841) [18], was an optician of 9 St. Martin's Court, Leicester Square, London, England. She succeeded **John Hudson** in his business. In 1839-1840 she is recorded as trading at 17 Ryder's Court, Leicester Square.

Huet et Cie
Huet et Cie, (Huet and Co., otherwise known as Het and Co.) [9] [256], was a French optical works in Nantes, and later in Paris. The company was established in 1793 in Nantes by **Louis Huet** (or Huette). At that time, he is thought to have made navigational instruments. In 1794 he produced the first successful achromatic microscope to be made in France. In 1795-1798 he succeeded in emulating the English makers by making plane mirrors in glass, (i.e., with parallel faces), suitable for octants and other navigational instruments.

During WWII, under German occupation, Huet et Cie marked their binoculars with the **ordnance code** lww. Some binoculars bearing the name **BBT Krauss** also bear the name Huet. This was either a collaboration or some other arrangement between the two companies.

Instruments bearing their name include: telescopes; binoculars; monoculars, some dated during WW2 and into the 1950s.

Huet, Louis
Louis Huet (or Huette), (1756-1805) [256] was a French optician who travelled to Holland, London and Paris to learn his trade, before settling in Nantes in 1793. There he founded the company **Huet et Cie**, making optical instruments.

Huggins, George (I)
George Huggins (I), (1774-1852) [1] [18] [90] [319], was a mathematical, optical, and philosophical instrument maker of London, England. He was the son of William Huggins and **John Dollond (I)**'s daughter, Susan Dollond. He began his apprenticeship with **Charles Fairbone (I)**, in 1788. In 1800, upon the death of Charles Fairbone (I), he was turned over to **Charles Fairbone (II)**, and was freed of the **Grocers' Company** in 1804. Among his apprentices was his nephew, **John Huggins (II)**.

In 1805, following the death in 1804 of **John Dollond (II)**, George Huggins (I) changed his name to **George Dollond (I)**, and entered a partnership with his uncle, **Peter Dollond**, trading as **P.&G. Dollond**. In 1807 he was granted freedom by purchase of the **Spectacle Makers' Company**, and was its Master from 1811-1812.

He continued in partnership until Peter Dollond's retirement in 1819, after which he continued to operate the business under the name George Dollond. In 1819 George (I) was elected a fellow of the Royal Society. Upon his death in 1852, he was succeeded in business by his nephew, **George Huggins (II)**, who changed his name to George Dollond (II).

Huggins, George (II)
George Huggins (II), (1797-1866) [**1**] [**18**] [**319**], was a mathematical and optical instrument maker of London, England. He was the son of **John Huggins (I)**. He served his apprenticeship with his uncle, **George Huggins (I)**, who had by then changed his name to George Dollond (I). He was freed of the **Grocers' Company** in 1827. He succeeded George Dollond (I) in business in 1852 and, upon taking control of the business, he changed his name to **George Dollond (II)** [**35** #21330 p1753], as his uncle had done before him. He traded under the name George Dollond until his death, when he was succeeded in business by his son, **William Dollond**.

Huggins, John (I)
John Huggins (I), [**1**] [**18**] (d.1812), was an optician and mathematical, optical, and philosophical instrument maker of London, England. He was the son of William Huggins and **John Dollond (I)**'s daughter, Susan Dollond. He served his apprenticeship with **Peter Dollond** of the **Spectacle Makers' Company**, beginning in 1778 [**401**]. He and his brother, **George Huggins (I)** both worked for the Dollond Business. He had two sons in the optical trade – **John Huggins (II)** and **George Huggins (II)**.

Huggins, John (II)
John Huggins (II), [**18**] (fl.1812-1828), was a mathematical, optical, and philosophical instrument maker of London, England. Addresses: Surrey Street, Blackfriars Road (1812), 7 Little Knightrider St. (1821-1824); 43 Westmorland Place, City Rd. (1826-1828). He was the son of **John Huggins (I)**. He served his apprenticeship with **George Huggins (I)**, and was freed of the **Grocers' Company** in 1812. He took on **John Pallant** as an apprentice in 1826. See **William Huggins**.

Huggins, William
William Huggins, (fl.1800-1822) [**276** #1163], was a mathematical and optical instrument maker of 7 Little Knightrider St., London, England. His tenure at this address coincided with that of **John Huggins (II)**, with whom he may have been related. A beam compass by William Huggins is listed in the sale catalogue of the private estate of William Larkins of the East India Company, [**276** #620], in 1800.

Hughes, Alexander
Alexander Hughes, (1850-1924) [**18**] [**276** #1894] [**319**], was a nautical instrument maker of London, England. He was the son of **Henry Hughes**, and was freed by redemption of the **Spectacle Makers' Company** in 1890. He traded in partnership with his father as **Henry Hughes & Son**, and succeeded his father in business. He went on to trade in partnership with his own son, **Arthur Joseph Hughes**, who succeeded him on business.

Hughes, Arthur Joseph
Arthur Joseph Hughes, (1880-1961) [**276** #1894] [**319**], was a nautical instrument maker of London, England. He was the son of **Alexander Hughes**, and worked in partnership with his father, trading as **Henry Hughes & Son**. He succeeded his father in business.

Hughes, Henry
Henry Hughes, (c1816-1879) [**18**] [**276** #1894] [**319**], was an optical, nautical and mathematical instrument maker of London, England. Addresses: 3 Union Terrace, Commercial Road (1840); 120 Fenchurch St. (1845-1855); 59 Fenchurch St. (1859-1875); 36 Trinity Square, Tower Hill (1867).

He was the son of **Joseph Hughes (I)**, and the brother of **Joseph Hughes (II)**. He served his apprenticeship with his father, and was freed of the **Masons' Company** in 1840. Later, in 1867, he was freed of the **Spectacle Makers' Company** by service to his father. Among his apprentices was **George Heath**, beginning in 1840. Henry Hughes opened a Navigation Warehouse at 120 Fenchurch Street, and formed a partnership with his son **Alexander Hughes**, trading as **Henry Hughes & Son**. Henry was succeeded in business by Alexander.

Hughes and Son, Henry
Henry Hughes & Son, [**44**] [**306**], began as a partnership between the optical, nautical and mathematical instrument maker **Henry Hughes** and his son **Alexander Hughes**. They traded initially at 120 Fenchurch St., London, England. By 1859 they were trading at 59 Fenchurch St., and they continued at this address until at least 1919. After the death of Henry, Alexander and his own son, **Arthur**

Joseph Hughes, continued the business as Henry Hughes & Son.

In 1903 the firm was incorporated as Henry Hughes and Son Ltd. Their factory sites were in Forest Gate and the Husun Works in Ilford. Some instruments manufactured in the Ilford factory bore the brand-name *Husun*. In 1919 the company was listed as a member of the **British Optical Instrument Manufacturers' Association**. In 1935 **S. Smith & Sons** acquired a controlling interest in Henry Hughes & Son. Following the London office's destruction in the Blitz of 1941, they entered a collaboration with **Kelvin, Bottomley and Baird**, resulting in the establishment of **Marine Instruments**. This relationship was formalised when the companies merged to form **Kelvin and Hughes** in 1947.

Some Galilean binoculars may be seen at auction, marked with the maker's name **H.H. & Son Ltd.**, of Liverpool and London. This is believed to refer to Henry Hughes & Son.

See also Husun Marine Distance Meter.

Hughes, Joseph (I)

Joseph Hughes (I), (fl.1808, d.1845) [18] [276 #1164], was an optician, and mathematical, nautical, and philosophical instrument maker of London, England. Addresses: 16 Queen St., Ratcliff (1818-1843); 6 Cross St., Ratcliff (1829); 37 & 38 Queen St., Ratcliff (1845); 38 Queen St., Ratcliff Cross.

He served his apprenticeship with the mathematical instrument, quadrant, and compass maker, William George Cook, beginning in 1800, and was freed of the **Masons' Company** in 1808. He was a maker of sextants, quadrants, compasses, and telescopes, and is known to have sold barometers and rules. Among his apprentices were his sons, **Henry Hughes** and **Joseph Hughes (II)**, although in 1846, following his death, Joseph (II) was turned over to the mathematical instrument maker William Iliffe Carline of the **Stationers' Company** to complete his apprenticeship. Joseph (I) was succeeded in business by his son Joseph (II).

Hughes, Joseph (II)

Joseph Hughes (II), (fl.1839-1878) [18] [276 #1598], was a mathematical, philosophical, and nautical instrument maker of London, England. Addresses: 37 & 38 Queen St., Ratcliff Cross (1843-1865); 38 & 40 Queen St., Ratcliff Cross (1870-1875); Bickley Row, Rotherhithe (1870); 19 London St., Fenchurch St. (1875).

He was the son of **Joseph Hughes (I)**, and the brother of **Henry Hughes**. He served his apprenticeship with his father, although in 1846, following his father's death, he was turned over to the mathematical instrument maker William Iliffe Carline of the **Stationers' Company** to complete his apprenticeship. He succeeded his father in business.

Hughes, W.C.

William Charles Hughes, (1844-1908) [61] [109] [319] [335], was an optician, magic lantern manufacturer, and film maker of 151 Hoxton Street, London, England. He succeeded the chemist and druggist William Parberry Hughes in business. He began his career entertaining on the stage, and started to sell optical instruments in 1879. By 1884 he was trading at Brewster House, 82 Mortimer Rd., Kingsland Rd., London. In 1897 he began to sell the *Moto-Photoscope* projector and *Moto-Photograph* camera, designed by one of the Prestwich family (presumed to be **John Alfred Prestwich**), and he went on to design and produce his own cameras, projectors, and accessories. Instruments bearing his name include magic lanterns; projection microscope; monocular compound microscope; and he is known to have produced cameras and accessories.

He produced two films of Queen Victoria's diamond jubilee procession in 1897. He was awarded various patents, including "a new or improved lime light gas apparatus for dissolving views" [35 #24254 p4840], and "an improvement in a new form of lamp for burning paraffin or mineral oils, either in magic lanterns, the public streets, lighthouses, rooms, &c., to be called the *Triplexicon* lantern lamp." [35 #24509 p5517]. The company was incorporated as W.C. Hughes & Co. in 1905, and continued in business until at least 1938.

Hulbert, Richard

Richard Hulbert, (fl.1825-1832) [18], was an optical, mathematical, and philosophical instrument maker of London, England. Addresses: 2 Bowmans Buildings, Aldersgate (1825); Friar St., Blackfriars Rd. (1832). In 1825 he attended the London Mechanics' Institute.

Hulst van Keulen, Gerard

See **van Keulen**.

Hummel Optical Co. Ltd.

Hummel Optical Co. Ltd. of 94 Hatton Garden, London, England. According to the 1951 *Directory for the British Glass Industry*, [327], Hummel Optical Co. Ltd., of 94 Hatton Garden, London, had works in Rochester, Kent, and were established in 1895. They are listed as producing Lenses; Mirrors; Optical units - mounted and unmounted, for projection, photography, education, research, etc. Their trade name is given as *Leech*, and their directors, S.A. Hummel, J.H. Baker, E.S. Hayes, P.R. Hummel, and E. Baker.

See **Leech (Rochester) Ltd.**

Hunt, John
John Hunt, (fl.1837-1840) [18], was an optical, mathematical, and philosophical instrument maker of 45 Tothill St., Westminster, London, England.

Hunt & Son
Hunt & Son, [34] [304], were opticians of Cork, Ireland. Addresses: 109 George's St. (1886-1895); 6 Patrick St. (1893-1895). They succeeded the firm of **Henry Hunt** in business, and continued to trade until they ceased business in 1895.

Henry Hunt
Henry Hunt, (fl.1844-1884) [18] [34] [304] [358], was an optician and mathematical instrument maker of Cork, Ireland. Addresses: 118 Patrick St. (1844-1846); 109 Old George St. (1881); 109 George's St. (1883-1884). He was the son of **Thomas Hunt (I)**, and succeeded his brother, **Thomas Hunt (II)** in business. By 1844 Henry was trading under his own name, and the business continued under this name until 1884. By 1886 the business had started to trade as **Hunt & Son**. Instruments bearing the name of Henry Hunt include: Navigation and surveying instruments; three-draw spyglass, signed *H. Hunt, Cork*.

At the 1852 National Exhibition in Cork, he showed: achromatic telescope, 3½ feet long, with vertical and horizontal rack & pinion motion; full-mounted five-inch theodolite; fourteen-inch dumpy level; circular protractor with double verniers and silver arc; transparent dipping compass; patent metallic dipping compass; binnacles of various kinds.

Hunt, Thomas (I)
Thomas Hunt (I), (fl.1792-1812, d.c1815) [18] [34] [276 #1164a] [304], was an optician and mathematical instrument maker of Cork, Ireland. Addresses: 118 Patrick Street (c1800); Patrick Street (1805-1812). His sons were **Thomas Hunt (II)** and **Henry Hunt**. He is believed to have started his business in 1792, and was succeeded by Thomas Hunt (II). Instruments bearing his name include: A marine barometer signed *T. Hunt Cork*; A wheel barometer signed *Thos. Hunt*. (Note, these may be by Thomas (I) or Thomas (II)).

Thomas Hunt (II)
Thomas Hunt (II), also known as Thomas Hunt Jr., (fl.1820-1828 d.c1835) [18] [34] [304], was an optician and mathematical instrument maker of Cork, Ireland. Addresses: 37 Patrick St. (1820); 65 Patrick St. (1824); 20 Patrick St. (1826-1828). He was the son of **Thomas Hunt (I)**. He succeeded his father in business, and was succeeded by his brother, **Henry Hunt**.

Hunter, R.F.
R.F. Hunter, [61] [306], of Gray's Inn Road, London, England, was a photographic dealer in both wholesale and retail trade. The company sold re-branded cameras, and may have had manufacturing capability of their own. From 1936 until c1951 they marketed the cameras branded by **Purma**, and [122] suggests that they may have manufactured them.

Hunter & Sands
See **Sands & Hunter**.

Huntley, Robert
Robert Huntley, (c1780-1837) [18] [50] [276 #1365], was an optician, and optical and mathematical instrument maker, and instrument case maker, of London, England.

Addresses: 37 High Holborn; 53 Burlington Arcade; 37 Burlington Arcade; 38 Burlington Arcade; 124 Cheapside; 1 City Road, near the Turnpike; 259 Regent St.; Regent Circus, Oxford St.; 255 Regent Circus, Oxford St.; 1 Plumber's Row, City Rd. (1816); 53 High Holborn (1817-1824); 52 High Holborn (1825-1829); 118 Oxford St. (1830); 294 Regent St. (1830).

Instruments bearing his name include microscopes, telescopes, barometers and engineering tools. His wife ran a corset business, also in London. Robert Huntley left his optical business to his son, Robert Henry Huntley, but Robert Henry may have closed the business shortly afterwards, since no records have been found dating later than 1840.

Hurlimann, A.
A. Hurlimann, [360], was an optical, nautical, and mathematical instrument maker of Rue de Fontenay, à Vincennes, France. In 1880 he acquired the firm of **E. Lorieux** of 6 Rue Victor Considérant, Paris [359 1880 #44 p720]. Instruments bearing his name include: telescope (1873); rangefinder (1890); quintant (c1900); sextants. In 1900 the firm was acquired by **Ponthus & Therrode**.

Hurter & Haas
Hurter & Haas, otherwise known as Haas & Hurter, [18] [276 #843a] [362] [379], was a partnership between Johann Heinrich Hurter (1734-1799) and Jacob Bernhard Haas (d.c1840). They were mathematical, philosophical, and optical instrument makers of 53 Great Marlborough St., Oxford St., London, England.

J.B. Haas was born in Biberach, in Swabia, Germany, and worked as a philosophical instrument maker at Bull Yard, St. Ann's Court, Soho, London, England.

J.H. Hurter was a Swiss portrait painter, who moved to London in c1777. Some instruments bear his name, but are believed to have been bought wholesale by him and engraved with his name.

Although the partnership was formalised in 1792, Hurter and Haas were already working together by 1783. They are known to have sold meteorological instruments, including barometers. Instruments bearing their name include a rosewood, ebony and brass sextant. In 1792 their address was Bull Yard, St. Ann's Court, Soho. The partnership was dissolved in 1795, and their stock sold at auction. A notice, posted in 1796, advertised for anyone with demands on the joint estate of Hurter & Haas to contact the creditors [**35** #13853 p48].

Husbands and Clarke

Husbands and Clarke, (fl.1858-1870) [3] [50] [319], opticians, and optical and scientific instrument makers of 1 Denmark Street, Bristol, England. This address was at the corner of Denmark Street and St. Augustine's Parade, and was also known as 8 St. Augustine's Parade.

The firm was a partnership between **Henry Husbands** and **William Clarke (I)**. Having both been senior journeymen in the employ of **Thomas Davies King**, they took over his premises at 1 Denmark Street in 1858 to run their own business. In 1863, Henry's son **William Samuel Husbands** began to work for them as an optician's assistant, and the firm expanded into 7 St. Augustine's Parade, on the opposite corner of Denmark Street. They continued to occupy those premises until at least 1865.

Instruments bearing their name include: microscopes; binocular microscope; Sikes hydrometer; drawing instruments; prepared microscope slides.

The business continued until the retirement of William Clarke (I) in 1870, after which Henry Husbands continued the firm under his own name.

Husbands, Henry

Henry Husbands, (c1824-1900) [50] [306] [319], was an optician, and optical and scientific instrument maker, of Bristol, England. Address: 1 Denmark Street, also known as 8 St. Augustine's Parade. He was a senior journeyman for **Thomas Davies King** and, upon King's retirement in 1858, he entered a partnership with **William Clarke (I)**, another of King's senior journeymen, trading as **Husbands and Clarke**. Following Clarke's retirement in 1870 Henry Husbands continued the business, trading under his own name, and as Husbands' Optical Works.

According to a Henry Husbands letterhead, dating to the 1870s, "Optician, &c. Manufacturer of compound achromatic microscopes, and accessories; Astronomical, naval, military, and tourists' telescopes; Opera, race, and marine glasses; Metford's improved traversing theodolites; Gravatt's dumpy levels; Photographic lenses of every description; Portrait, landscape and stereoscopic cameras, stands &c; Engineers' and architects' scales and rules in ivory and boxwood; Standard barometers and thermometers; All kinds of Mathematical, Optical and Meteorological instruments. Spectacles to suit all defects of vision."

Henry's son, **William Samuel Husbands**, having worked for Husbands and Clark, continued to work for Henry until 1875, when he emigrated to Australia, and set up business in Melbourne. In 1893, Henry began to work in partnership with three of his sons, Henry James Husbands, James Wessen Husbands, and Alfred Witchell Husbands, trading as **H. Husbands and Sons**.

Instruments bearing Henry Husbands' his name include: microscopes; brass telescope on wooden tripod; wood and bellows plate camera; tortoise-shell eyeglasses; parallel rule.

Husbands and Sons, H

H. Husbands and Sons, (fl.1893-1910) [50] [306] [319], were opticians and scientific instrument makers of Bristol, England. Address: 1 Denmark Street, also known as 8 St. Augustine's-parade. The partnership was formed in 1893, when **Henry Husbands** began to work in partnership with his sons, Henry James Husbands (c1847-1916), James Wessen Husbands (c1852-1922), and Alfred Witchell Husbands (b.c1859), trading as Husbands and Sons. The partnership continued until 10 years after the death of Henry Husbands, and was terminated in 1910. [**35** #28344 p1549].

Husbands' Optical Works

See **Henry Husbands**.

Husbands, William Samuel

William Samuel Husbands, (b.1849) [319], was a manufacturing optician, and retailer of imported optical and scientific instruments of Melbourne, Australia. He was a son of **Henry Husbands**, and started his career in 1863, working as an optician's assistant in his father's business, **Husbands and Clarke**, in London. He continued to work for his father following the termination of that partnership in 1870 and, in 1875, after 12 years of learning the business, he emigrated to Australia, where he purchased the firm of John Baptiste Grimoldi, meteorological and scientific instrument maker of 81 Queen's Street, Melbourne.

William exhibited a consignment of instruments from his father's business at the Melbourne International Exhibition of 1880-1881, where he was awarded a gold medal. He operated his business at 81 Queen's Street until at least 1886, and by 1889 he had relocated to 454 Bourke St. W.

Husun
Husun was a trademark of **Henry Hughes & Son**.

Husun Marine Distance Meter
The Husun Marine Distance Meter, [4], made by **Henry Hughes & Son**, is a form of hand-held sextant rangefinder for marine use. It is used to find the range to an object up to 10 cables distant. There are two scales on the side of the instrument; the height scale reading from 30 to 150 feet, and the distance scale reading from 1 cable to 10 cables. The height of the object is set on the height scale, and the two images are brought into coincidence by means of an adjustment at the front of the instrument. The distance is then read off on the distance scale. See also rangefinder (optical).

Hutchinson, Charles C.
Charles C. Hutchinson (fl.1858-1913) [368], was an optical, nautical, and surveying instrument maker of Boston, MA, USA. He served his apprenticeship with **Frederick W. Lincoln Jr.** and, in 1858, became a partner in the business. In 1883 he purchased the company of F.W. Lincoln Jr. & Co., and continued it under the name of C.C. Hutchinson until his death in 1913. The company continued in business, using the same name, until 1940. Examples of C.C. Hutchinson's work include: refracting library telescopes in brass; marine compasses; 6½ inch diameter brass standard barometer.

Hutchinson, George
George Hutchinson, [9] [169] [263] [271], ophthalmic optician, mathematical and surveying instrument maker, began his career as an apprentice to **Adie and Wedderburn**. He continued to work for them until the company ceased business, at which time took possession of their stock. In 1888 he opened his own shop in 16 Teviot Place, Edinburgh, Scotland, where he stayed until 1907. From 1908 to 1967 he continued his business at 18 Forrest Rd., the premises that were taken over in 1968 by **Charles Frank Ltd.** From 1942 the company traded as G. Hutchinson & Sons, opticians and surveying instrument makers.

Huygens, Christian
Christian Huygens, (1629-1695) [7] [37] [41] [155], was a Dutch mathematician and physicist. He spent most of his career at the Académie des Sciences in Paris. Together with his brother Constantijn he ground and polished lenses. In 1678 the brothers met **Nicolaas Hartsoeker**, who taught them his method for making glass spheres for microscope lenses. Christian discovered Saturn's ring system in 1665-6. He also made the first working pendulum clock to a design in which he improved on that of **Galileo Galilei**. He contributed significantly to the field of optics, and developed his theory of light, in which he concluded that light slows on entering a more dense medium. During this work he also discovered the phenomenon of polarization. He was able to derive the laws of reflection and refraction, and explained the double refraction (birefringence) of calcite. His contributions to practical optics included the invention of the Huygenian eyepiece and the aerial telescope.

Hyde, J.M.
John Moor Hyde, also known as John Moore Hyde, (c1820-1883 fl.1841-1854) [3] [18] [319] [322], was an optician, and mathematical, philosophical, and nautical instrument maker of 1 Broad Quay, Bristol, England. From 1847 until 1852 he also occupied 11 Grenville Street. According to his trade card, "Manufacturer of Mathematical, Nautical, Astronomical, Philosophical, Optical, & Surveying instruments." He was awarded various patents for improvements in steam engines, as well as a "useful registered design" for a submarine communicator. Instruments bearing his name include: octant; sextant; microscope; compound microscope; telescopic level; stick barometer. He was succeeded in business by **Hill & Price**.

Hyland Optical Technologies
Hyland Optical Technologies is a US company based in California. They design, develop and manufacture optical and opto-mechanical assemblies for life science, research, and industrial applications. In 2013 they acquired **Ealing Catalog**, which has been a Hyland Optical Technologies brand since then.

Hy Score
Hy Score, [94], was a trade mark owned by S.E. Lazlo, an importer and wholesale distributor of binoculars, cameras, telescopes and other items, based in New York, USA. The Hy Score trade mark was used for imported binoculars and firearms from 1969 until 1992.

Hysom, Jim
Jim Hysom, (1933-2017) [199] [221], was an English telescope maker who specialised in mirror making. He began as an amateur, making his first mirror at the age of seventeen. He started his career with **Cox, Hargreaves and Thomson**, and left with **John V. Thomson** and J. Mortleman to form the company **Optical Surfaces**. In 1996 he left Optical Surfaces to work with **Cliff Shuttlewood** of Leicester Astronomical Centre Ltd. which, by 1997, was renamed **Astronomical Equipment**. There he worked with his brother, Robert Hysom, who was a machinist. The company continued in business until

about 1977, when it was split into AE Mechanics, run by Robert Hysom, and **AE Optics**, run by Jim Hysom. AE Optics continued in business until about 1993 when Jim Hysom went on to form the company **Hytel Optics**, which traded until sometime in the 2000s when he retired prior to his death in 2017.

Hytel Optics

Hytel Optics, [**199**], was the telescope making company run by **Jim Hysom** from around 1993, when **AE Optics** closed, until sometime in the 2000s.

I

Ibn al-Haytham, Al-Hassan

Al-Hassan Ibn al-Haytham, (965-c1040) [**41**] [**155**], sometimes known as Alhazen, was an eminent Islamic physicist whose study and conclusions in the field of optics were not only ahead of their time, but laid the foundation for much to today's understanding of basic optics. He is sometimes referred to as the father of modern optics. Using methods of experimental verification, remarkably similar to what is now known as the scientific method, he drew some important new conclusions, which he published in his influential *Book of Optics*.

Among his notable conclusions was that vision takes place when light is emitted by a luminous source and reflected from an object into the eye. This may seem obvious to the modern observer, but at the time many believed that vision was achieved by means of some kind of ray emitted by the eye. He also concluded that the phenomenon of refraction occurs because light travels at different speeds in different materials. The experimental proof of this did not come until the C19th, when **Foucault** was researching the speed of light. During the process of investigating light and vision, Ibn al-Haytham studied the reverse image formed when light propagates through a tiny hole. He concluded that light propagated in a rectilinear fashion and, in his writings, he described what is now known as the camera obscura.

ICA

International Camera Actiengeschellschaft (ICA), [**61**], of Dresden, Germany, was a manufacturer of high-quality cameras and other photographic equipment. The company was formed in 1909 by a merger of Richard Huttig & Sohn AG, Dr. R. Krugener Kamerawerk, Emil Wunsche AG and Carl Zeiss Palmos AG. In 1926 ICA merged with **Contessa-Nettel**, **Ernemann**, and **Goerz**, to form the company **Zeiss Ikon**.

ICAF

See **Antonio Bencini**.

Iceland Spar

Iceland spar is a transparent crystalline form of calcium carbonate. (Another common form of calcium carbonate is chalk). In the form of calcite, crystalline calcium carbonate, $CaCO_3$, may be transparent. Such a transparent calcite crystal exhibits the property of birefringence, and is used for optical purposes. It may be used for certain prism applications where birefringence is a required property.

IG Farben AG

Interessengemeinschaft Farbenindustrie AG, known as IG Farbenindustrie AG, or simply IG Farben, was a major German chemical and pharmaceutical conglomerate from its formation in 1925 until 1952 when it started liquidation proceedings, which concluded in 2012. IG Farben was notorious for its significant contribution to the German war effort and for the 24 key personnel who were tried for war crimes during the Nuremburg trials, 12 of whom were convicted. The protracted time taken for the company's liquidation to complete was largely due to the resulting court cases.

IG Farben was the parent company of the photographic film, lens and camera maker **AGFA** from 1925 until 1952. Bayer, the parent company of **Rietzschel**, also became part of IG Farben in 1925, at which time Rietzschel changed ownership to become a part of AGFA. During WWII all military products of IG Farben, including those of AGFA, were marked with the **ordnance code** bzz or mbv.

Ihagee Kamerawerk Steenbergen & Co.

The company was founded in Dresden, Germany, in 1912 by Dutchman, Johan Steenbergen (b.1886), as Industrie- und Handelsgesellschaft m.b.H. Kamerafabrik mit Kraftbetrieb (Industry and Trade Company Ltd. with Power Drive), producing photographic equipment and materials [**453**]. In 1913, the company name was changed to Ihagee Kamerawerk G.m.b.H., and in 1918, the name was changed again to Ihagee Kamerawerk Steenbergen & Co. The first camera with the **Exacta** brand was introduced in 1933.

John Steenbergen, whose wife was Jewish, emigrated to the USA in 1942. During WWII, Ihagee Kamerawerk Steenbergen & Co. marked their instruments with the **ordnance code** hwt, and in 1945 the factory was destroyed in a bombing raid. After WWII, production was resumed in a former **Zeiss Ikon** plant. At this time, under Soviet occupation, the company was still Dutch owned although administered by the state. Ihagee Kamerawerk AG was set up in West Germany, in competition with the original company, giving rise to a dispute over naming rights. In 1968, Ihagee became a part of **Pentacon**, and continued to use the *Exacta* brand-name. By this time, Ihagee Kamerawerk AG no longer existed separately.

Ilex Camera Works

The Ilex Camera Works, [**61**], of Stoke Newington and Hackney, England was a camera production facility set up by **A.C. Jackson**, [**306**]. This is where their *Ilex* cameras were produced. A.C. Jackson did not sell the cameras themselves, but via resellers. The company was absorbed into **George Houghton & Sons Ltd.** in 1904.

Ilex Manufacturing Co.

The Ilex Manufacturing Company was an American camera lens manufacturer, formed in 1910 by Rudolph Klein and Theodor Brueck, who had been shutter designers for **Bausch & Lomb**. Their lens brand was *Paragon*.

Ilford

In 1878 Alfred Harman started a company in London, England, making dry photographic plates. In 1891 the name was changed to the Britannia Works Company, and in 1900 the name changed again to Ilford Ltd. [44] [61] [306]. In 1912 they started to produce roll films.

In 1920 the company Selo Limited was formed incorporating Ilford, the Imperial Dry Plate Co., and the Gem Dry Plate Co., and in 1921 it became **Amalgamated Photographic Manufacturers** (initially APM 1921), including a consortium of seven more British companies. These were **Kershaw Optical Co.**, **A. Kershaw & Sons**, **Marion & Foulger**, Rotary Photographic Co., **Rajar Ltd.**, Paget Prize Plate Co., and **Marion & Co.** Ilford continued to broaden their film and plate offerings, and in 1948 they began to produce cameras.

In 1967 began a series of changes of ownership when Ilford was acquired by ICI and Ciba. Ciba also acquired **Lumière** in 1963, which in 1982 became Ilford France. Subsequent changes of ownership included a management buy-out in 2005, with the creation of Harman Technology Ltd., trading as Ilford Photo. In 2007 Harman Technology Ltd. acquired the photographic paper manufacturers Kentmere Photographic Ltd. Harman Technology Ltd. was purchased by Pemberstone Ventures Ltd. in 2015.

Ilford Imaging Europe was a branch of Ilford that did not become a part of Harman Technology. Nonetheless, they continued to operate as a separate company in Germany, using Ilford as a brand-name.

Iloca

Iloca of Hamburg, Germany, was a camera manufacturer, which began camera production c1950. The company produced various models of camera, including rangefinder cameras, and were the first to introduce an electric motor wind mechanism on an SLR. They also produced stereo cameras in various models, including the *Stereo Rapid*. They supplied this same model, but without a rangefinder or the self-timer, to the **David White Company**, of Milwaukee, USA, to be sold, branded as their *Realist 45*. Problems with suppliers led to insolvency, and Iloca was acquired by **Agfa** in 1960.

Image Plane

The image plane, sometimes referred to as the *focal plane*, is the plane in which an image is formed by a lens (which may have multiple elements) or mirror.

Examples include: the plane in which the image is formed by an objective lens, which may itself be viewed by an eyepiece, or extended further by means of a relay lens; the plane in which the image is formed by a relay lens may be either viewed by an eyepiece, or relayed further by another relay lens.

Image, Real

A real image in optics, [7], is the image formed by rays of light that converge on the image plane. A positive (converging) single lens always forms a real image, and a concave mirror generally forms a real image. A luminous image will appear on a screen, if one is placed where the real image is formed.

Image, Virtual

A virtual image in optics, [7], is the image produced by diverging rays of light from a lens or mirror. The rays appear to come from a point, but no luminous image will appear if a screen is placed at that point. A negative (diverging) single lens will always form a virtual image, and a convex mirror will generally form a virtual image.

Immersion Objective

An immersion objective is a microscope objective that is optically designed to be used with a transparent fluid filling the gap between the front element of the microscope objective and the slide's cover-slip or sample. The fluid used is specifically chosen to match the required refractive index. This enables the immersion objective to operate at its maximum numerical aperture and resolving power.

Immersion objectives are generally high-power objectives in the range x60 to x100, and are either oil immersion, or water immersion. These have a very small working distance between the front element of the objective and the slide's cover-slip. Lower powered objectives gain little from the technique and, for this reason, lower powered immersion objectives are rarely, if ever, produced.

Giovani Battista Amici is credited with the introduction of the immersion objective in microscopy in 1840. Initially the fluid used was naturally occurring cedar oil, which has a refractive index of about 1.52, closely matching that of the glass cover-slips. Soon after this he introduced water immersion as an alternative. Cedar oil has been replaced with synthetic oils for most modern applications.

Impersonal Prismatic Astrolabe
See Danjon Prismatic Astrolabe.

Imray and Son, James
J. Imray and Son, [18] [276 #2151], was a chart publishing company of 89 & 102 Minories, London, England. James Imray, (fl.1836-1842), was a chart publisher and nautical instrument maker, and sold navigation equipment. In 1836 he entered a partnership with **Robert Blachford**, trading as **Blachford & Imray** at 116 Minories, Tower Hill. Upon the termination of the partnership in 1845 he continued to trade under his own name. From 1849-1850 he additionally traded at 3 Old Fish Street Hill, and in 1851 the company moved to *Navigation Warehouse*, 102 Minories. At some point the firm's name was changed to J. Imray & Son. The company was incorporated in 1904 as Imray Laurie Norie & Wilson Ltd. at the same address, as a combination of J. Imray & Son, **Norie & Wilson**, and the map and chart sellers Laurie & Whittle, [276 #1003]. They since relocated to St. Ives in Cambridgeshire. Some telescopes may be found bearing their name.

Ingénieur Chevallier
l'Ingénieur Chevallier (Engineer Chevallier) was the name under which **Jean Gabriel Augustin Chevallier** traded.

Institut d'Optique
The Institut d'Optique of Paris, France was founded in 1919, and its first director was **Charles Fabry**. The institute began with studying optical properties such as reflection, refraction, dispersion, birefringence and homogeneity, optical component design, and aberrations. They also provided a service of checking photographic lenses, binoculars, glasses, laboratory equipment, sextants and spectacles. During WWII the institute supported the war effort, and subsequently became more focused on teaching and research. They expanded to diverse locations, and continued under various names, including Institute d'Optique, SupOptique, and Ecole Supérieure de Optique, finally settling on Institut d'Optique Graduate School. They conduct research in areas of fundamental and applied optics and related subjects.

Interference Colour Photography
Interference colour photography, [71 p668], invented by **Gabriel Lippmann** in 1891, is an early method of producing a colour photograph which is based on the interference of light waves in the photographic emulsion.

The method is as follows: A glass plate is coated with a photographic emulsion in which the silver halide crystals have been ground so fine that the individual crystals are smaller than the shortest wavelength of light that will be recorded. The coated plate is mounted in the camera with the glass side towards the objective and the emulsion side in contact with mercury, which acts as a mirror. During exposure the light passes through the emulsion, and is reflected by the mirror to pass back through the emulsion. The light, thus traversing the emulsion in both directions, forms standing waves which are recorded in the emulsion. This gives rise to a true-colour recording of the image.

The plate is developed to form the finished image. In order to view the image, it is necessary to use parallel light, and hold the plate at the correct angle. The exposures take 2-3 hours, and viewing is difficult. There is no way to reproduce the image. Using this method, Gabriel Lippmann produced the first colour photograph in 1891. For his invention of this method, he was awarded the Nobel Prize in Physics in 1908 [318].

Interferometer
An interferometer, [7], is a device that uses interference patterns to perform tasks such as making measurements or recording images. In optics and astronomy an interferometer creates interference with coherent sources of electromagnetic radiation. There are many kinds of interferometer, some of which are listed here:

Astronomical interferometer: An array of radio antennas or telescopes that are configured to be used together. The signals are combined and, in the case of radio antennas, this is achieved by means of signal processing algorithms. The resulting images have an angular resolution equivalent to that of a telescope with an aperture the size of the full sensor array. The "light-gathering", however, is limited to the sum of the apertures of the individual sensors.

Fabry-Perot interferometer: A form of interferometer, co-invented by **Charles Fabry** and **Jean-Baptiste Alfred Perot**, based on the principle of multiple reflections. See etalon for a description.

Fresnel double mirror: This wavefront-splitting interferometer consists of two plane mirrors set at a very small angle to each other. One portion of the wavefront is reflected from the first mirror, and another portion from the second mirror. Since the two optical paths differ by a fixed amount, interference will result in the region where the two resulting wavefronts are superimposed. A sensor or camera placed at this point will record the resulting interference pattern. Any movement in either of the mirrors will be detected as a change in the interference pattern.

Fresnel double prism: This wavefront-splitting interferometer consists of two thin prisms joined at their bases. The coherent light source is shone along the axis through the join, and perpendicular to the flat sides of the prisms. Each prism refracts the light slightly toward the axis, with the result that there is a region in which the two resulting wavefronts are superimposed. A sensor or camera placed at this point will record the resulting interference pattern.

Hilger interferometer: An interferometer for testing the accuracy of optical glasswork. This design was enhanced as a *microscope interferometer* for testing for aberrations in microscope lenses.

Lloyd's mirror interferometer: This wavefront-splitting interferometer consists of a single mirror placed so that the source wavefront is partially reflected from the mirror onto the screen or sensor, and partly reaches it directly. The resulting interference pattern is detected at the screen or sensor. Any movement in the mirror will be detected as a change in the interference pattern.

Lummer-Gehrcke interferometer: This interferometer uses multiple reflections within a thin plate of glass, whose faces are parallel, and accurately flat. In this sense, the operation is on a similar principle to that of the Fabry-Perot interferometer, or etalon. The light enters at a steep angle of incidence, and at each reflection a small amount of light is transmitted. A lens is used to combine these transmitted rays into an interference pattern.

Mach-Zehnder interferometer: This amplitude-splitting interferometer consists of two beam-splitters and two totally reflecting mirrors. The incoming wavefront is split by the first beam-splitter to traverse two optical paths which are combined by the second beam-splitter, and any difference in the optical paths is seen as an interference pattern. Any movement of the intermediate totally reflecting mirrors will be detected as a change in the interference pattern.

Michelson's interferometer: This amplitude-splitting interferometer is used for making very accurate measurements of small distances in terms of the wavelength of light. The incoming wavefront is passed through a beam-splitter to separate it into two optical paths. The light is returned from each optical path, by a mirror, to the beam-splitter where the two wavefronts are combined to fall on a sensor. Small changes in the length of one optical path with respect to the other result in the detection of interference fringes. The LIGO gravitational wave observatory uses laser interferometers that are a large-scale, sophisticated version of the Michelson interferometer.

Sagnac interferometer: This amplitude-splitting interferometer consists of a closed optical path, normally consisting of three or four mirrors, one of which is a beam-splitter. Light is directed by the beam-splitter to traverse in two opposite but otherwise identical paths, and are combined by the beam-splitter to coincide on a detector where interference may be detected. There are two scenarios in which interference will be detected. One is when the orientation of one of the mirrors is changed, introducing a difference in path length between the clockwise and counter-clockwise paths. The other is when the interferometer is rotated, and acts as a ring laser gyroscope.

Stellar interferometer: The stellar interferometer, invented by **Albert Michelson**, consists of four mirrors, a set of two pinholes, a positive lens, and a detector. It is used for taking high angular resolution measurements of stars and galaxies.

Young's double slit: In this wavefront-splitting interferometer a wavefront is passed through a single thin slit, and the resulting wavefront is passed through a closely spaced double slit. The two onward wavefronts are then superimposed on a screen or sensor to yield an interference pattern.

Inverted Microscope
See microscope.

Iris de Paris
Iris de Paris was a French brand of binoculars. Instruments bearing their name include: Galilean binocular models *Artillerie, Nautique, Auto, Touring, Aviation, Cycliste, Sporting, Bulldog, Regates, Hippique*, dating from c1880 to c1930. The binoculars are usually engraved with the model name, together with a graphic depiction (gun for *Artillerie*, anchor for *Nautique*, car for *Auto*, winged wheel for *Touring*, aeroplane for *Aviation*, person on a bicycle for *Cycliste*, racehorce for *Sporting*, a four-oar boat for *Regates*, just the word *Bulldog*, a horseshoe encircling a horse's head for *Hippique*), and "Marque Iris de Paris" followed by "Qualité Guarantie".

Iriscope
An iriscope is an instrument used to display the prismatic colours of light.

Irvin, Thomas & Peter
Thomas E. & Peter H. Irvin, (fl.1835-1842) [18] [379], were optical, mathematical, and philosophical instrument makers of Hatton Garden, London, England. Note that [276 #2152] lists them as Irvine. Addresses: 10 Charles St.; 19 Charles St. (1839); 11 Charles St. (1840-1842). Thomas Irvin was a mathematical instrument maker who attended the London Mechanics' Institute in 1835.

Irvine, Elizabeth
Elizabeth Irvine, (fl.1834-1846) [18] [276 #1896] [379], was a spectacle maker, optician, and optical instrument

maker of 32 Kirby St., Hatton Garden, London, England. She succeeded **John G. Irvine** in business at this address.

Irvine, John

John Irvine, (fl.1794-1821) [18] [276 #1367], was an optician, and optical, mathematical, and philosophical instrument maker of London, England. Addresses: 5 Queenhithe, Upper Thames St. (1816); 32 Kirby St., Hatton Garden (1821); 2 Newcastle St., College Hill. He was succeeded in business by **John G. Irvine**.

Irvine, John G.

John Glen Irvine, (fl.1822-1837) [18] [276 #1896] [379], was an optician, and optical, mathematical, and philosophical instrument maker of 32 Kirby St., Hatton Garden, London, England. He succeeded **John Irvine** in business at this address, and was succeeded in business by **Elizabeth Irvine**.

Irving & Sons, H.N.

Horace N. Irving and Sons of Ealing, Middlesex, England, optical instrument maker, was a highly respected maker of telescopes.

Ron N. Irving (1915-2005) was the son of Horace Irving, and the sole remaining proprietor of the instrument making firm H.N. Irving & Sons, carrying on his father's business and trade name. Ron Irving's obituary appears in the Journal of the British Astronomical Association [79].

J

Jackenkroll, A., Optische Anstalt GmbH
A. Jackenkroll Optische Anstalt GmbH, (A. Jackenkroll Optical Institution), was a manufacturer of riflescopes in Berlin, Germany. During WWII they marked their products with the **ordnance code** kxv. After WWII they used the trademark AJACK on their riflescopes.

Jackson A.C.
A.C. Jackson, [**44**] [**306**], of Hackney, London, England, was a camera manufacturer. By 1897 they were trading at 98a Amhurst Road, Hackney, making cameras, lanterns, shutters and other equipment. In the early 1900s their *Ilex* cameras were made at the **Ilex Camera Works**. A.C. Jackson was absorbed by **George Houghton & Sons Ltd.** when they became Houghtons Ltd. In 1904.

Jackson, Joseph
Joseph Jackson, (fl.1735-1760) [**9**] [**18**] [**24**] [**90**] [**276** #275], of London, England, was a mathematical instrument maker, and maker of speculum mirrors for reflecting telescopes and microscopes, and glass mirrors. He traded Opposite Exeter Exchange, Strand, and from 1736-1760 in Angel Court, Strand. He served his apprenticeship with **Thomas Heath** beginning in 1723, and was freed of the **Grocers' Company** in 1735. One of his products was a catoptric microscope and telescope, which was a Gregorian telescope that could be converted into a low power microscope. He applied the techniques originated by **John Mudge** for making speculum mirrors. He experimented with different glass compositions for glass mirrors, and he was probably the first to make and sell the **Hadley** reflecting quadrant. From 1735, he worked in partnership with **George Hearne**, trading as **Hearne & Jackson** until 1750.

Jacob, A.
A. Jacob, (fl.1830) [**18**] [**276** #1898], was an optical instrument maker of St. Anne's Place, Manchester, England.

Jaegers Optical Corporation, A.
A. Jaegers of Lynbrook, New York, USA, was the company of Al Jaegers [**185**]. The company operated during the 1950s, to the 1980s, and made large aperture telescope objectives, up to 6 inches, and eyepieces. The **Cave Optical Co.** used Jaegers lenses in their 4-inch refracting telescopes. The facility in which the company operated was destroyed by fire in the 1980s.

Jägermeister
See **Seeadler Optik**.

Jameson, James
James Jameson, (fl.1764-1784) [**18**] [**322**], was an optician and optical instrument maker of London, England. Addresses: Saffron Hill (1764); 79 Great Saffron Hill, Holborn (1784). He served his apprenticeship with **John Margas**, beginning in 1753, and was freed of the **Spectacle Makers' Company** in 1761. In 1764 he was one of the petitioners who attempted to revoke the patent obtained by **John Dollond (I)** for the achromatic lens, and which was enforced after Dollond's death by his son, Peter (See Appendix X).

Jamin Refractometer
See refractometer.

Jannsen, Zacharias
Zacharias Jannsen, [**24**] [**172**], was a C16th Dutchman who is credited by some with the invention of the refracting telescope and the compound microscope. His claim in this regard was in competition with his compatriot **Hans Lippershey** who also claimed to have invented the telescope. Historians make contradictory claims, but it seems likely that Lippershey invented the telescope, and Jannsen invented the compound microscope.

Japanese Manufacturers' Marks for Optical Products
See Appendix IV.

JB codes
See the *JB codes* section in Appendix IV.

Jecker, François-Antoine
François-Antoine Jecker, (1765-1834) [**2**] [**50**] [**230**], was an optical instrument maker, born and raised in Hirtzfelden in the Alsace region of France. Having developed an interest in science and mechanics he taught himself what he could from books. In 1784 he began an apprenticeship with a clockmaker in Besançon, then went to England in 1786, where he spent six years in the workshops of **Jesse Ramsden**. Ramsden was impressed with his abilities and enthusiasm, and Jecker learned a great deal from him, becoming his friend in the process. In 1792 Jecker returned to France where he established his own business in Paris. He made a linear dividing engine similar to Ramsden's own design, as well as a screw cutting machine. He presented them to the Bureau de Consultation des Arts, and received an award of three thousand francs. Jecker's business grew, producing astronomical, geodetic, and other optical instruments.

At the end of the French Revolution, when currency fraud was on the increase, Jecker designed and produced a coin-scale, of which he sold over eighty thousand.

Jecker's instruments were highly regarded and, at a time when the English instrument trade was dominant, he not only produced instruments to rival and equal the English, but trained many others in the trade, and began the establishment of a world-class French instrument trade. His business continued, trading as Jecker Frères from 1805, and as Jecker Fils from 1837 until ceasing business in the mid-1940s.

Jenkins, Samuel
Samuel Jenkins, (fl.1737) [18], was an optical instrument maker and mechanic of Essex Court, London, England.

Jenoptik
In 1990, following the reunification of Germany, the state-owned East German company, VEB Carl Zeiss Jena, (see **Carl Zeiss**), became Jenoptik Carl Zeiss Jena GmbH. The optical divisions of the company were re-united with the West German Zeiss company to form Carl Zeiss AG, and the state-owned company continued as Jenoptik GmbH, specialising in optoelectronics, systems technology and precision manufacturing. In 1996, the company was privatised, and became Jenoptik AG.

Jessops
Jessops is a UK retailer of photographic equipment. They sell their own-branded budget binoculars, telescopes and photographic filters, as well as other brands.

Jewel Lens
A jewel lens, [169] [395], is a lens made from a precious or semi-precious stone, such as diamond, ruby, garnet, topaz or sapphire. The idea originated with **Sir David Brewster**, who was seeking to improve the resolution of the microscope by means of an objective lens made from a material of high refractive index and low dispersion. He published the idea of using diamond and, in 1825, based on this, the physician, Dr. C.R. Goring (1792-1840), approached the London optician **Andrew Pritchard** to make a diamond lens. Although Pritchard produced such a lens, diamond proved to be too difficult to work, and suffered from unwanted polarizing and birefringence effects.

Alternatives were sought, initially using sapphire, then ruby, garnet and topaz. Ruby, garnet and sapphire lenses were made for Sir David Brewster by **Alexander Adie**, **Peter Hill**, **James Veitch** and **William Blackie**. Only a small number of jewel lenses were ever made, and their production had largely ceased by the mid- or late-1830s, following the improvements to glass microscope objectives introduced by **J.J. Lister**.

[162 Ch.5] investigates the properties of twelve known jewel lenses, and compares them with contemporary lenses of glass and fused quartz.

During the early C20th **Horace Dall** made jewel lenses for simple microscopes.

A diamond lens made by Andrew Pritchard is in the Science Museum, South Kensington, London, England.

Jewitt & Co., W.
W. Jewitt & Co., (fl.c1851-1861) [18 p187], were nautical and mathematical instrument makers and chart sellers of 39 South Castle St., Liverpool, England. They succeeded **Melling & Co.** at that address sometime after 1851. The company was a partnership between William Jewitt and Richard James, and continued in business until the partnership was dissolved in 1861 [35 #22587 p108].

According to their trade label, "Nautical & Mathematical instrument makers, Chart agents to the R^t. Honble. The Lords Commis^{rs}. Of the Admiralty. W. Jewitt & Co. Successors to Melling & Co. 39 South Castle Street, Liverpool. Mathematical, Optical, & Philosophical instruments & apparatus of every description made and repair'd on the shortest notice. Correctors of local attraction in iron vessels. Spectacles, Chronometers, &c." This trade label is identical in design, and almost identical in wording with that of Melling & Co.

Instruments bearing their name include: octants; stick barometer.

Jinghua Optics & Electronics Co. Ltd.
Jinghua Optics & Electronics Co. Ltd. was founded in 1997 and is based in Guangzhou, China. They are part-owners of **Bresser**, and provide products in China, Europe, the United States, Japan, South Korea, and internationally. The company engages in manufacturing, selling, and exporting optoelectronic instruments and optical components. It offers consumer optoelectronic products; optical lenses; lenses for digital and video cameras; digital microscopes, spotting scopes, and night vision devices; laser range-finders; lens modules of mobile phones; optical engines for micro-projectors; binoculars; telescopes.

JIS
JIS stands for Japanese Industrial Standard. The Japanese Industrial Standards are developed by the Japanese Industrial Standards Committee (JISC).

The JIS standard for the distance between the objective flange and the eyepiece flange on a compound microscope is 170mm. This is one of the two most widely used standards for this distance. The more common is the **DIN** standard of 160mm.

Jobin, Amédé
Amédé Jobin, (d.1945) [**53**] [**299**], began his career in the French army. After retiring from the army, he entered the optical instrument making business by purchasing the business of **Léon Laurent** in 1892. During his tenure at the business Jobin maintained the high standards set by his predecessors, with the firm operating at addresses in and near Paris, France. In 1911 the engineer Gustave Yvon was appointed CEO of the company, and in 1923 Jobin and Yvon formed a partnership, trading as Jobin Yvon. The company incorporated as Jobin Yvon in 1942 and, following a sequence of acquisitions, Jobin Yvon merged into the Horiba Group of Japan in 1997, becoming Horiba Jobin Yvon. The company continued as part of the Horiba Group.

Joblot, Louis
Louis Joblot, (1645-1723) [**142**] [**256**] [**274**] [**345**], was a French naturalist and microscopist. In 1718 he published *Description et Usages de Plusieurs Nouveaux Microscopes tant Simples que Composez*, (Description and Usage of Several New Microscopes, both Simple and Compound). In Part I, he describes sixteen microscopes, of varying degrees of innovation in design, with technical descriptions of the instruments as well as guidance on how they should be used. At least some of these microscopes were made for him by an engineer and mathematical instrument maker named Monsieur Le Febvre. Some of the microscopes were designed for specialised purposes while others were for more "universal" use. In Part II of the book, Joblot gives detailed descriptions of his microscopical observations. He tutored **Claude Paris** in microscope making techniques.

Johannsen, Asmus
Asmus Johannsen, [**54**], was a Danish instrument maker who started his business, Johannsen & Co., in London, England in about 1859, making sextants and chronometers. He submitted his chronometers to several of the Greenwich Time Trials [**104**], and received high marks. He made chronometers for the navies of India, Italy, Spain, Portugal, Austria and China.

Johnson, Samuel
Samuel Johnson, (1724-1772) [**18**] [**276** #371] [**322**] [**361**], was an optician and optical instrument maker of *Sir Isaac Newton & Two Pairs of Golden Spectacles*, 23 Ludgate Street, London, England. He served his apprenticeship with **James Mann (II)**, beginning in 1738, and was freed of the **Spectacle Makers' Company** in 1745. He succeeded **John Mann** in business. His trade card lists: Various kinds of spectacles; Newtonian telescopes; Gregorian telescopes; Double and single microscopes; Solar microscopes; Prisms; Camera obscuras; Optical machines for viewing perspective pictures; Convex and concave speculums; Magic lanterns; Opera glasses; Multiplying and magnifying glasses; Barometers; Thermometers; "with many other curiosities not here mentioned".

Joiners' Company
See Appendix IX: The London Guilds.

Jointed Microscope
See microscope.

Jones & Gray
Jones & Gray, [**18**], opticians and mathematical instrument makers of 26 Strand St., Liverpool, England, was a partnership between **Charles Jones** and **John Gray (I)**, formed in 1839. The partnership traded until 1841 when **Robert John Keen** joined them, and they began to trade as **Jones, Gray & Keen**. Jones & Gray are known to have sold octants.

Jones & Son, John
John Jones & Son, [**18**], was a partnership between **John Jones**, and his son **William Jones**. The partnership began in 1784, and continued until William's brother **Samuel Jones** joined the partnership in 1790, upon which they began to trade as **John Jones & Sons**. Instruments bearing their name include: microscope; theodolite; telescope. They also advertised: navigation scale; planetarium.

Jones & Sons, John
John Jones & Sons, [**18**], was a partnership between **John Jones** and his sons **William Jones** and **Samuel Jones**. The partnership began in 1790, succeeding **John Jones & Son**, and continued until John Jones presumably retired in 1791, upon which the two brothers entered a new partnership, trading as **W.&S. Jones**.

Jones, Charles
Charles Jones, (fl.1818-1842) [**18**] [**322**], was an optician and mathematical instrument maker of Liverpool, England. Addresses: 58 Stanley St. (1818); 57 Stanley St. (1821-1822); 11 East Side Dry Dock (1823-1824); 10 East Side Dry Dock (1825); 11 East Side Dry Dock (1827); Strand St. (1828); 28 Strand St. (1829); 25 Strand St. (1831-1841). He participated in two partnerships, the first being **Jones & Gray** from 1839, then **Jones, Gray & Keen** from 1841 until 1842.

A Charles Jones instrument label of 1823-1827 reads "Charles Jones, real manufacturer of sextants & quadrants, compasses, telescopes &c. No. 11 Dry Dock, Liverpool. & the above articles cleaned and repaired in the best manner. C.J., stepson & late apprentice to I. Gray." It goes on to say that the following are for sale: Sextants; Quadrants; Ships' and boats' compasses of all kinds; Day and night glasses and telescopes of all kinds; Time glasses; Parallel rules; Gunter's scales; Gauging rods; All kinds of drawing instruments; Marine and common barometers; Thermometers of all descriptions; Charts; Pilots; Navigation books.

According to [3], the reference to his apprenticeship suggests that his stepfather, with whom he served his apprenticeship, was **John Gray (I)**.

Jones, David

David Jones, (fl.c1770-1795) [274] [276 #996] [319] [361], optician, and optical, philosophical, and mathematical instrument maker, of London, England, was the father of **Thomas Jones**. Addresses: 25 Charing Cross; 35 Charing Cross (1785-1793). He is believed to have served his apprenticeship with **Benjamin Martin** of the **Goldsmiths' Company**, beginning in 1766. Records also show a David Jones being freed by redemption of the **Spectacle Makers' Company** in 1766. Instruments bearing his name include: stick barometer; 2¼" refracting telescope in a case.

Jones, Edward

Edward Jones, (fl.1780-1800) [18] [274] [276 #857] [319] [322], was an optician and microscope maker of 14 Somerset Place, New Road, Commercial Road, London, England. He made microscopes to various designs, including that of **John Cuff**. Microscopes bearing his name are in museums in Cambridge, and Munich, Germany.

Jones, Gray & Keen

Jones, Gray & Keen, [18] [322], were opticians and nautical instrument makers of 25 & 26 Strand St., Liverpool, England. This was a partnership between **Charles Jones**, **John Gray (I)** and **Robert John Keen**, formed in 1841 when Robert John Keen joined **Jones & Gray**. The partnership of Jones, Gray & Keen was dissolved in 1842, [35 #20078 p641], after which John Gray (I) and Robert John Keen continued in partnership as **Gray & Keen**.

Jones, Henry

Henry Jones, (fl.1796) [18] [276 #997], was an optical instrument maker of 36 Ludgate Street, London, England. This address was later occupied by **George Melville Horton & Co.**

Jones, John

John Jones, (1736/7-1808) [18] [319] [401], was a mathematical instrument maker of London, England. Addresses: Holborn (1776); 135 Holborn (1782-1788); 135 near Furnival's Inn, Holborn (1793); Skinner Place, Islington (1799); Wells Row, Islington (1805-1806). Note that the two Islington addresses may have been residential. He served his apprenticeship with the spectacle maker Leonard Ballett, beginning in 1751, and was freed of the **Spectacle Makers' Company** in 1758. He was also freed of the **Fishmongers' Company** in 1779. He started his business as an optician in about 1776.

His son **William Jones** served an apprenticeship with him, and by 1784 they were working in partnership as **John Jones & Son**. In 1790 John's younger son **Samuel Jones** joined the partnership, and they began to trade as **John Jones & Sons**. In 1791, William and Samuel began to trade as **W.&S. Jones**.

Jones, Samuel

Samuel Jones, (1769-1859) [18] [401], was an optician of London, England. Addresses: 30 Lower Holborn (1806); 30 Holborn (1859); Dalston House, St. Albans, Hertfordshire (1859 residence). He was the son of **John Jones**, and was freed by patrimony of the **Spectacle Makers' Company** in 1806. In 1790 he began to work in partnership with his father and his brother, **William Jones**, trading as **John Jones & Sons**. In 1791, the brothers entered a new partnership, trading as **W.&S. Jones**. Among Samuel's apprentices was **William Eden**, beginning in 1811.

Jones, Thomas

Thomas Jones, (1775-1852) [2] [18] [46] [53] [276 #998], of London, England, was an eminent maker of scientific instruments, and the son of **David Jones**. Addresses: 124 Mount St., Grosvenor Square; 13 Panton St., Haymarket; 7 Southampton St., Strand; 120 Mount Street, Berkeley Square (1806-1808), 21 Oxenden Street, Picadilly (1811-1814), 62 Charing Cross (1816-1850), and 4 Rupert Street, Coventry St. (1850-1859).

He served an apprenticeship from 1789 to 1796 with **Jesse Ramsden**, and continued to work for him for some time after. While in the employment of Ramsden he made instruments such as the dynameter and the optigraph. Jones credited Ramsden with the invention of the optigraph, and improved on the design himself. He worked under commission for **Edward Troughton**, as well as for his own customers. He produced astronomical instruments for observatories in the UK and abroad, including the Cape of Good Hope observatory, the Spanish naval observatory at San Fernando, near Cadiz, the Greenwich observatory, the Armargh observatory, and observatories at Oxford and Cambridge universities. Some of his work was criticised for its quality, notably the 6-foot diameter mural circle for the Cape of Good Hope observatory. However, he otherwise maintained a strong reputation, and continued to work until his death in 1852. His son, Thomas, worked with him, and continued his business until 1861.

Jones, W.&S.

W.&S. Jones, [18] [319] [44], were opticians, and optical, mathematical, and philosophical instrument makers of London, England. Addresses: 27 Holborn Hill; 135 Holborn, next Furnival's Inn (1792-1800); 30 Holborn (1800-1860 – sometimes listed as Holborn Bridge, Holborn Hill, or Lower Holborn); 32 Holborn Hill (1801-1805); 30 opposite Furnival's Inn, Holborn (1852); *Archimedes*, 30 Lower Holborn.

The firm was a partnership between **William Jones** and his younger brother **Samuel Jones**, both sons of **John Jones**. The partnership was formed in 1791, presumably upon the retirement of their father, and succeeded **John Jones & Sons** in business. In 1795/6, following the death of **George Adams (II)**, W.&S. Jones acquired Adams' stock, and the rights to his books.

In 1797 they published *A catalogue of optical, mathematical, and philosophical instruments, made and sold by W. and S. Jones*, including a comprehensive list of instruments, and books. The listed optical instruments included: Large range of spectacles; Opera glasses; Refracting telescopes for both terrestrial and astronomical use in various sizes; Spyglasses, 2, 3, and 4-draw; Gregorian, Newtonian, and Herschellian reflecting telescopes in various sizes; Equatorial mounts; Various models of simple and compound microscope; Solar microscopes; Lucernal microscopes; Magic lanterns and slides; Scioptic balls; Camera obscuras; Concave and convex mirrors in various sizes; Prisms; Theodolites; Levels; Hadley's quadrants; Gunter's quadrants; Transit instruments.

In 1797 William Jones also introduced the box sextant for marine navigation [175]. From 1818 until 1827, they employed **William Eden**, who had served his apprenticeship with Samuel. They exhibited at the Great Exhibition in London in 1851, [328], for which their catalogue entry listed a mountain thermometer. W.&S. Jones traded until 1859 when, upon Samuel's death, the firm ceased business.

Jones, William

William Jones, (1762-1831) [18] [401], was an optician of London, England. Addresses: Holborn (1787); 30 Lower Holborn (1803); Islington (1798 residence). He served his apprenticeship with his father, **John Jones**, beginning in 1776, and was freed by patrimony of the **Spectacle Makers' Company** in 1794. He worked in partnership with his father, beginning in 1784, trading as **John Jones & Son** until his brother, **Samuel Jones** joined the partnership in 1790, after which they traded as **John Jones & Sons** until 1791. Following this, the brothers traded in partnership as **W.&S. Jones**.

Jones-Bird Telescope

See telescope.

JTW Astronomy

JTW Astronomy design and manufacture specialised telescopes, mounts and piers as well as custom made adaptors and protective equipment. They produce custom optical designs from the smallest to telescopes over 1000mm in diameter. They also build custom mounts and tracking equipment.

K

Kahles, Karl
Kahles of Vienna, Austria, is a sport optics manufacturer, producing riflescopes, binoculars, and accessories. The company was formed in 1898 by Karl Robert Kahles, merging the companies **Plössl** and the **Karl Fritsch Opto-Mechanical Workshop**. During WWII the company used the **ordnance code** cad. The company was largely destroyed during the war, and was rebuilt afterwards by Elizabeth Kahles, the widow of the founder's son Karl, who reinvigorated the company's fortunes. The company continued in the Kahles family until it was incorporated as Kahles Ges.m.b.H.

Kalee
Kalee was a trade-name used by **A. Kershaw & Sons** for their photographic items including lanterns and projectors. The name is derived from **Kershaw, A., Lee**ds. After WWI Kershaw started a company called the **Kershaw Projector Co. Ltd.** to market these products. The company name was later changed to Kalee Ltd., and in 1948 it was acquired, along with Kershaw and other companies, by **British Optical and Precision Engineers**. Kalee Ltd. became **G.B. Kalee Ltd.** G.B. indicates **Gaumont British**, then part of **Rank**.

Kaleidoscope
A device designed to produce symmetrical patterns by means of multiple reflections. Loose fragments of glass or other material are held in a tube between inclined mirrors. The pattern is viewed through a viewing hole in one end of the tube, which is rotated to alter the pattern. The kaleidoscope was invented by **Sir David Brewster** in 1813. See also polygonoscope, symmetroscope, teleidoscope.

Kamera Werkstätten (KW)
Kamera Werkstätten Guthe & Thorsch (KW) was a camera manufacturer established in Dresden, Germany, in 1919 by Paul Guthe (d.1930) and Benno B. Thorsch. Paul Guthe had already established a small camera factory in 1915, and the new company began business at the same Dresden address. By the 1920s they were producing plate cameras and reflex cameras, and this continued until 1938 when Thorsch, by then the sole owner, being part Jewish, left Germany. He emigrated to the USA, and exchanged his business with that of Charles A. Noble of Detroit.

During the company's history the name was changed several times. In 1939, Noble changed the company name to Kamera Werkstätten Charles A. Noble. In the same year the company moved to Niedersedlitz, near Dresden, and introduced the *Prakticaflex* camera. After WWII, with the partitioning of Germany, the company was nationalised, and renamed Kamera Werkstätten VEB Niedersedlitz. In 1949 they first introduced the *Praktica*, followed in 1951 by the *Praktina*, their first 35mm SLR with interchangeable lenses and other components. Further *Praktica* and *Praktina* models followed.

In 1953 the company was renamed VEB Kamera Werke Niedersedlitz. In 1956 they absorbed the company Belca-Werk (see **Balda**). In 1959 they merged with VEB Kinowerke Dresden (see **Zeiss Ikon**) and several other companies, including VEB Kamera und Kinowerke Dresden. In 1964 the newly formed company became VEB Pentacon Dresden Kamera und Kinowerke. See **Pentacon**.

Kaspereit, Otto Karl
Otto Karl Kaspereit, (b.1882), was born in Germany and emigrated to the USA, where he settled in Philadelphia, Pennsylvania. He was awarded various patents for optical innovations during the 1930s, and is noted for his invention of the modified Erfle eyepiece design, (see Appendix II), which bears his name.

Kazan Optical and Mechanical Plant (KOMZ)
The Kazan Optical and Mechanical Plant (KOMZ) was founded in 1940 in Russia. They manufacture binoculars, monoculars, telescopes, gun sights, night vision binoculars, and magnifiers. They trade under the name **Baigish**. For Russian and Soviet makers' marks, see Appendix V.

Keary, P.J.
The Keary business, [304], were opticians and philosophical instrument makers of 13 and 17½ Wellington Quay, Dublin, Ireland. They shared both of these addresses with **Patrick Cahill**.

Patrick J. Keary (fl.1886-1890) of 17½ Wellington Quay, was an optician and philosophical instrument maker.

Also recorded at 13 Wellington Quay: P.J. Keary, (fl.1892-1909) optician; Mrs. Keary (fl.1891-1895) optician; Patrick Cahill Keary (fl.1900-1907) optician and philosophical instrument maker; and John F. Keary (fl.1908-1909) optician.

Keeler Ltd.
Keeler Ltd. of Windsor, England, is an ophthalmic instrument maker, founded in 1917 by Charles Davis Keeler who had moved to London, England from Philadelphia, USA. In 1946 the company was

incorporated as Keeler Optical Products Ltd., and in 1952 they opened a subsidiary in Broomall, Pennsylvania, USA.

Keen, Robert John

Robert John Keen, (fl.1834-1857) [18] [322], was an optician and mathematical instrument maker of London and Liverpool, England. Addresses: 8 Postern Row, Tower Hill, London (1834-1840); 4 East Smithfield, London (1840). He moved north to Liverpool where, from 1841, he traded in partnership as **Jones, Gray & Keen**. From 1842 until 1855 the partnership traded as **Gray & Keen**. Upon the termination of this partnership, he relocated to East Side Salthouse Dock, Liverpool, where he traded under his own name until 1856, when he began to work in partnership with the watch and chronometer maker Henry Frodsham.

Kellner, Carl

Carl Kellner, (1826-1855) [47], was a German optical instrument maker. He was a close friend of **Moritz Carl Hensoldt**, and together they formed a company in Wetzlar called **Optisches Institut**, in 1849, producing lenses and microscopes. Kellner is known for his eponymous eyepiece design, (see Appendix II). After Carl Kellner's death in 1855, the company was run by one of his employees, **Louis Engelbert** until 1856 when it was taken over by another employee, **Friedrich Belthle**, together with Kellner's widow. In 1865, Ernst Leitz became a partner in the company, and in 1869, upon the death of Belthle, he took control of the company and continued in business under the name **Leitz**.

Kelvin, Bottomley and Baird

Kelvin, Bottomley and Baird, [44] [263], of Glasgow, Scotland, scientific instrument makers, was formed in 1913. Since 1907, upon the death of **William Thomson**, the company **Kelvin and James White Ltd.** had had a board consisting of, among others, director **Alfred W. Baird** and chairman **James Thomson Bottomley**. In 1913 the company was reorganised and renamed Kelvin, Bottomley and Baird.

Following the change in name, the company continued to specialise in William Thomson's patents, including ship's compasses, sounding machines, general nautical instruments and electric measuring instruments. During WWI they made submarine periscopes.

In 1919 Kelvin, Bottomley and Baird was listed as a member of the **British Optical Instrument Manufacturers' Association**. James Thomson Bottomley continued as chairman of the company until his death in 1926, but the company continued to trade as Kelvin, Bottomley and Baird until 1947. In 1941 A collaboration began with **Henry Hughes and Son**, resulting in the establishment of **Marine Instruments**. In 1947 As a result of the success of this collaboration, the two companies amalgamated as **Kelvin and Hughes**.

Kelvin and Hughes

Kelvin and Hughes, [44] [263], of London, England, are nautical instrument makers. In 1941 a collaboration between **Kelvin, Bottomley and Baird** and **Henry Hughes and Son** was established as **Marine Instruments**. In 1947, the two companies amalgamated as Kelvin and Hughes, with **S. Smith and Sons (England)** as a shareholder. In 1949, the name of Marine Instruments was changed to Kelvin and Hughes (Marine). In 1953 S. Smith & Sons (England) acquired the remaining interests in the company. In 1961 the name was abbreviated to Kelvin Hughes as the marine division of S. Smith and Sons (England), known since 1967 as Smiths Industries. In 2017 Kelvin Hughes was taken over by the global technology company HENSOLDT, and was renamed HENSOLDT UK. The company specialises in navigation, surveillance and security radar systems.

Kelvin and James White Ltd.

In 1854 **William Thomson** formed an association with optical instrument maker **James White**, making instruments, trading as Kelvin & James White in Glasgow, Scotland [263]. At some point Alfred W. Baird was appointed director of Kelvin & James White. By 1900, after James White's death, and with a workforce of about 400, the company was incorporated as Kelvin and James White Ltd., and William Thomson's nephew **James Thomson Bottomley** joined the firm. In 1907, upon the death of William Thomson, James Bottomley was appointed chairman of the company. In 1913, the company organisation changed, and its name was changed to **Kelvin, Bottomley and Baird**. James Bottomley continued as chairman until his death in 1926.

Kenko Co. Ltd.; Kenko Tokina Co. Ltd.

Kenko Tokina Co. Ltd. is a manufacturer of optical equipment, including binoculars, microscopes, camera products, photographic and optical accessories, and telescopes.

Kenko Co. Ltd. was founded in 1957 in Japan, and in 1974 they founded an affiliate, Kenko Optical Co. Ltd. for manufacturing optical products. In 1989 they purchased Koei Seiki Co. Ltd. (**JB code** JB127), which began to operate as an affiliate company. Similarly, they acquired Asiaoptical Co. Ltd. in 1997, Tasco Japan Co. in 1998, and Slik Corporation in 2001. In 2006 they founded the Kenko Professional Imaging Co. Ltd. In 2008 they acquired Fujimoto Photo Industrial. In 2011 Kenko merged with

Tokina, and the company name was changed to Kenko Tokina Co. Ltd.

Their maker's mark, used by Kenko Japan on telescopes and other consumer instruments from 1959 until the 1980s is illustrated in Appendix IV. Instruments bearing this mark were sometimes branded by other companies such as **Prinz** or **Tasco**.

Keohan, Thomas
Thomas Keohan, (fl.1836-1873) [18], was an optician, and optical, philosophical, and mathematical instrument maker of London, England. Addresses: 2 Arbour Terrace, Stepney (1839); 2 Arbour Terrace, Commercial Rd. (1840-1855); 33 Upper East Smithfield (1859-1873). He served his apprenticeship with a mathematical instrument maker, William Elliott, of Ratcliff Highway, London, beginning in 1816. Instruments bearing his name include sextants.

Kepler, Johannes
Johannes Kepler, (1571-1630), was a German mathematician and astronomer, and a contemporary of **Galileo Galilei**. He earned a scholarship to the University of Tübingen to study for the Lutheran ministry, and there he learned of the heliocentric theories of Copernicus. He moved to Prague to work with the Danish astronomer Tycho Brahe and, when Brahe died in 1601, Kepler took his post as Imperial Mathematician and began to analyse Brahe's observation data. Using these data, he discovered three laws governing planetary motion, known as Kepler's Laws. Kepler also made advances in the field of optics, notably improving on the Galilean telescope with his invention of the Keplerian telescope.

Keplerian Telescope
A Keplerian telescope, named after its inventor, **Johannes Kepler**, uses a converging lens as its objective and a converging lens as its eyepiece. Both are either planoconvex or biconvex lenses. The eyepiece is a magnifying lens with short focal length. This design results in a wider field of view than the Galilean telescope, with longer eye-relief, but results in an upside-down image. Thus, for terrestrial use, it needs to have additional optical elements to invert the image.

Keratometer
A keratometer, also referred to as an ophthalmometer, is an optical instrument, used by a vision care specialist to measure the radius of curvature of the cornea of the eye. This instrument was invented by **Hermann von Helmholtz**.

Keratoscope
A keratoscope, also known as *Placido's disc*, is an optical instrument used by a vision care specialist, and consists of a disc that is marked with concentric rings. In the centre of the disc is a small lens through which the reflection of the rings in the patient's cornea may be observed, thus detecting abnormal curvature of the cornea.

Kern & Co.
Kern & Co. was a Swiss surveying instrument company, based in Aarau. The company was formed in 1819 by Jakob Kern. In 1935 **Heinrich Wild** began to produce designs for them. Wild continued to work with them until his death in 1951. In 1988 the company was acquired by **Leitz**, by then operating as the Wild Leitz AG Group. Kern & Co. ultimately became a part of **Leica Geosystems**.

Kershaw, Abraham and Sons
A. Kershaw & Sons, of Leeds, and 3 Soho Square, London, England, was an optical instrument manufacturer originally founded by Abraham Kershaw in 1888.

Abraham Kershaw, (1861-1929) [20] [29] [44] [53] [61] [133] [306], was born near Bradford, Yorkshire, son of a reed maker for the textile industry. He was known as Abram in order to distinguish him from his grandfather, also a reed maker. He undertook an apprenticeship as a philosophical instrument maker with Edwin Blakey & Co., electrical instrument makers, which then became Blakey, Emmott & Co., [44], electrical, telephone and signalling engineers, in Halifax. He completed his apprenticeship in 1882 and, soon after, went on to work in the private electrical laboratory of Louis John Crossley, where he was able to study and improve his knowledge of mechanical and electrical engineering. In 1883 he married Martha Beilby, and they had three children, Cecil (b.1884), Dorothy (b.1889), and Norman (b.1899).

In 1888 Abram and his family moved to Leeds where he started his own company as a scientific instrument maker and repairer. By 1898 he had 30 employees, and his son Cecil began to work in the company. This may have been when the name A. Kershaw and Son was first used. During the second Boer War, (1899-1902), Kershaw had government contracts to produce telegraphic communication instruments, heliographic signalling, and medical equipment.

They began to produce components for cameras for other manufacturers, and by 1900 they were producing their own cameras, projectors and accessories. They also used *Soho* as a brand-name for their cameras, most of which were sold through **Marion & Co.** They also manufactured their folding mirror mechanism for companies such as **R.&J. Beck**, **Ross Ltd.** and **J.H. Dallmeyer Ltd.** The company was incorporated as A. Kershaw and Son Ltd. in 1910, with both Abram and Cecil as directors.

In 1911 their **Kalee** cinema projectors were first announced, and by the beginning of WWI the company was selling *Kalee* projectors, marketed by the New Century Film Service Co., and their *Indomitable* projectors, marketed by the Tyler Apparatus Co.

During WWI Kershaw devoted their entire design and production effort to contracts for the Ministry of Munitions. They began by producing clinometer sights for machine guns and, in 1915, with backing from the government, they embarked on a new factory devoted to production of binoculars for the war effort. They employed three key personnel from the **Zeiss** factory in London, and initially purchased their lenses from **Taylor, Taylor & Hobson** and **Thomas Cooke & Sons**, and prisms from **Barr and Stroud**, as well as glass mouldings from **Chance Bros**. In 1917 Abram's son Norman joined the company and took a course in lens design, after which they began to produce their own lenses. The exact date when A. Kershaw and Son became A. Kershaw and Sons is unclear.

After the war they resumed peacetime production, and created two subsidiaries. The **Kershaw Optical Co.** collaborated with **Marion & Co.** to market their photographic products and binoculars. The second subsidiary, the **Kershaw Projector Co.**, marketed the *Kalee* projectors. The Kershaw Projector Co. later became **Kalee Ltd.**

In 1921 **APM** was formed from the Selo company, including a consortium of seven more British companies. These were Kershaw Optical Co., A. Kershaw & Sons, **Marion & Foulger**, Rotary Photographic Co., **Rajar Ltd.**, Paget Prize Plate Co., and Marion & Co. In 1929 APM was renamed **Soho Ltd.** as the sales division of A. Kershaw and Sons. The name *Soho* was also used as a model name for 6x24 and 8x24 binoculars by Kershaw.

During WWII Kershaw once again provided optical munitions for the war effort, producing binoculars to patterns designed for the British military. After the war, in 1946, Soho Ltd. was renamed **Kershaw-Soho Ltd.** In 1948, [**134**], the Kershaw Group was acquired by **British Optical and Precision Engineers Ltd.**, controlled by **Rank**. In 1956, British Optical and Precision Engineers Ltd. was renamed Rank Precision Industries.

Kershaw Optical Co.

The Kershaw Optical Co. was a subsidiary of **A Kershaw & Sons**, set up after WWI to market their photographic products and binoculars in collaboration with **Marion & Co.**

Kershaw Projector Co. Ltd.

The Kershaw Projector Co. Ltd. was a subsidiary of **A. Kershaw & Sons**, set up after WWI to market their *Kalee* projectors. The company name was later changed to **Kalee Ltd.**

Kershaw-Soho Ltd.

Kershaw-Soho Ltd., [**44**] [**61**], was the sales division of **A. Kershaw and Sons**. It was formed in 1946 by renaming **Soho Ltd.**, formerly **APM**. In 1948 Kershaw-Soho Ltd. was acquired by British Optical and Precision Engineers, part of **Rank**.

Keuffel & Esser

Keuffel and Esser of New York, USA was founded in 1867 by two German immigrants, William J.D. Keuffel and Hermann Esser. They sold drawing materials and instruments, and brought out their first catalogue in 1868. They set up a manufacturing facility in Hoboken, New Jersey in 1871. The firm and its products grew steadily, and they sold products made in their own factory and some bought in from Europe. In the twentieth century they also became a leading supplier of optical instruments. In 1909 they were granted a license to manufacture **Barr and Stroud** range and order indicators, and in 1910 they were given permission to begin manufacturing Barr and Stroud rangefinder stands, as the two firms worked together to recover business lost to the **Zeiss** – **Bausch and Lomb** alliance [**26** p67], known as the **Triple Alliance**. Keuffel & Esser supplied Navy periscopes in WWI and Army rangefinders in WWII. In 1982 Keuffel & Esser filed for Chapter 11 bankruptcy, and eventually was bought out by several companies including Azon Corporation and Cubic Corporation. Currently, Azon Corporation owns the Keuffel & Esser name and trademarks. Cubic Corporation acquired the optical tooling division of Keuffel & Esser, and renamed it **Cubic Precision**.

Keulen, van

Johannes van Keulen (I), (1654-1715) [**362**] [**377**] [**378**] [**380**], founded a dynasty of cartographers, map sellers, and navigation instrument makers. The firm traded under the name **de Gekroonde Lootsman** (the Crowned Pilot), in Amsterdam, Netherlands. In 1678 he was admitted to the guild of book sellers as a book seller and cross-staff maker. He is known to have sold: books and charts; quadrants; cross-staffs; compasses.

His son, Gerard van Keulen, (1678-1726), a cartographer, map seller, and navigation instrument maker, joined his father in the family business in 1704, the same year he was admitted to the guild of book sellers.

Johannes van Keulen (II), (1705-1755), scientific and navigation instrument maker, cartographer, and map seller, was the Son of Gerard van Keulen. He joined his father's firm in 1724. A year later, he was admitted to the guild of book sellers, and in 1726 he succeeded his father in running the business. He was appointed official chart-maker to the Dutch East India Company.

When Johannes (II) died, his wife, Catharina Buijs began to run the firm with their sons, Cornelis Buijs van Keulen (1736-1779) and Gerard Hulst van Keulen (1733-1801), trading as Johannes van Keulen & Zoonen. They were appointed to the Dutch Commissioner for Longitude at Sea. Upon the death of Cornelis in 1779, the firm split into two. The instrument making and chart publishing business continued, run by Gerard Hulst van Keulen, and the remainder of the business, being an anchor foundry, was run by Cornelis's widow.

Following the death of Gerard Hulst van Keulen in 1801, the instrument making part of the business ceased, and the firm continued in business as a chart publisher and retailer of navigation instruments, initially run by Gerard's widow, trading as Weduwe G. Hulst van Keulen. Following her death in 1810, her grandson continued the business until it was sold out of the family in 1823. The company continued to trade as Weduwe G. Hulst van Keulen until it ceased business in 1885.

Keyzor & Bendon

Keyzor & Bendon, (fl.1855-1873) [**53**] [**319**], were opticians and mathematical instrument makers of 50 High Holborn, London, England. The partnership was between the brothers Abraham Bandan (c1825-1873) and George Bandan (b.c1834). By the time the partnership was formed, both brothers had changed their surnames.

Abraham worked initially in Norwich with his uncle, Michael Keyzor, and changed his surname to Keyzor to work in partnership as **Michael & Abraham Keyzor**. He returned to London in 1855, and took over the shop of **William Harris & Son**, initially trading as Keyzor & Co. By 1861 his brother, who by then had changed his surname to Bendon, came to work in partnership with him as Keyzor & Bendon. The partnership was terminated three years after the death of Abraham Keyzor in 1876 [**35** #24407 p287], after which George Bendon continued the business.

Instruments bearing their name include: theodolite; measure; folding ivory rule; glass cutter; Galilean binoculars; magic lanterns; brass telescope on library stand, engraved *Keyzor & Bendon late Harris, 50 Holborn, London, Imp'*^{l} *Navy*.

Keyzor & Co.
See **Keyzor & Bendon**.

Keyzor, Michael & Abraham

Michael & Abraham Keyzor, (fl.1847-1854) [**18**] [**319**], of Castle Meadow, Norwich, England, were opticians and spectacle makers. The partnership was between Michael Keyzor and his nephew, Abraham Bandan, who changed his surname to Keyzor for the partnership. By 1854 they were trading in St. Giles St. The partnership was terminated in 1854 [**35** #21645 p4284], after which Michael Keyzor continued the business, and Abraham returned to London to form the new partnership **Keyzor & Bendon**.

Keyzor, M.&M.

M.&M. Keyzor, (fl.1870-1900) [**53**], were opticians and spectacle manufacturers, of 16 Tottenham Court Rd., London, England. According to their trade card, dated 1870-1900, they sold large compound microscopes, telescopes, and optical instruments of all descriptions. The trade card is signed "Michael Keyzor, licensed hawker". The extent to which they made these instruments, if at all, is unclear.

Kharkov Machine-Building Plant (FED)

FED is a Ukranian company, formed in 1927 by Felix E. Dzerzhinsky on the outskirts of Kharkov. Beginning in 1934 they produced portable film cameras, and later, rangefinder cameras. During WWII they produced rifle sights for the military. After WWII, while they re-established their production capability, rangefinder cameras were made for them by **Krasnogorskij Mechanicheskij Zavod (KMZ)**, using the *FED-Zorki* brand-name, (see **Zorki**). Today they produce technology for the aviation and railway industries. For Russian and Soviet makers' marks, see Appendix V.

Kiev (КИЕВ)

Kiev was a camera brand of the **Arsenal Factory**. Some cameras of this brand are marked with the name *Kiev*, and some with the Cyrillic form, Киев.

Kimbell, Isaac

Isaac Kimbell, sometimes Kimbel or Kimble, (fl.1775-1828 d.1828) [**18**] [**276** #737] [**322**], was an optician and optical instrument maker of 21 Dean Street, Fetter Lane, London, England. Note that **John Kimbell** occupied the same address. According to [**274**], Isaac served his apprenticeship with **Benjamin Martin**, and made achromatic microscope objectives during his apprenticeship.

Kimbell, John

John Kimbell (I), sometimes Kimbel or Kimble, (fl.1784-1830) [18] [276 #1372] [322], was an optician, and optical, mathematical, and philosophical instrument maker of 21 Dean Street, Fetter Lane, London, England. Note that **Issac Kimbell** occupied the same address. John was the son of husbandman Joseph Kimbell. He served his apprenticeship with **Benjamin Martin**, beginning in 1763, and was freed of the **Goldsmiths' Company** in 1771.

His son, John Kimbell (II) served an apprenticeship with him, beginning in 1799, and was freed of the Goldsmiths' Company in 1806.

Various other members of the Kimbell family worked as opticians at the same address, including James Kimbell (I) (fl.1784); James Kimbell (II) (fl.1831-1836); Josh Kimbell (fl.1835-1837).

Kinematograph

Another name for a Cinematograph.

Kinetograph

The Kinetograph was a motion picture camera invented by **Thomas Edison** and his assistant **William Dickson** in about 1888 [45] [335]. It provided stop-go movement of the film with a shutter synchronised to the motion of the film strip by a Geneva drive (otherwise known as a Maltese cross) mechanism. Kinetograph films were displayed using a Kinetoscope.

Kinetoscope

A Kinetoscope is a motion picture viewing device that utilises a film strip taken with a Kinetograph. The film strip is mounted in a continuous loop, and runs between an incandescent lamp and a shutter for viewing by a single user. The Kinetoscope was invented by **Thomas Edison** and his assistant, **William Dickson**, [335].

King

King was a Japanese brand of binoculars sold in the mid-1970s. Some King binoculars bear the Japanese **JB code** manufacturer's mark and the oval quality label. Examples of various sizes of binoculars may be found, including opera glasses and folding pocket binoculars.

King, Charles Gedney

Charles Gedney King, (1808-1858) [368], was a surveying instrument maker of Boston, MA, USA. His father, Gedney King was also a surveying instrument maker. Charles began to work for his father's company in 1837, and took it over upon his father's death in 1839. **F.W. Lincoln Jr.** served an apprenticeship with him.

King, John (I)

John King (I), (c1770-1841) [3] [18] [50] [319], was a scientific and optical instrument maker who worked as foreman to the brothers **Charles and Richard Beilby**. In 1821, upon their retirement, he took over the premises at 2 Clare Street, Bristol, England, and began his own business there. In 1822 he entered a partnership with his son **John King (II)**, trading as John King & Son. The partnership was terminated in 1831 following a disagreement [35 #18785 p526], after which John (I) continued the business as a sole trader, with John (II) as an employee.

King, John (II)

John King (II), (1797-c1871) [3] [18] [50] [319], was a scientific and optical instrument maker of 2 Clare Street, Bristol, England. He was the son of **John King (I)**, and the father of **Thomas Davis King**. He worked in partnership with his father, trading as John King & Son until 1831, after which he continued to work as his father's employee until his father's death in 1841. Soon after inheriting the business, he sold it to Thomas Davis King. John (II)'s daughter, Harriet, married **James Blake Gardiner**, who took over the premises at 2 Clare Street in about 1863.

King and Coombs

King and Coombs, [18] [50] [264] [319], optical instrument makers of 2 Clare Street, Bristol, England, was a partnership between Henry Payne Coombs (1826-1908) and **Thomas Davies King**. The partnership was terminated in 1853, [35 #21410 p385], and appears to have lasted only about two years. Following the dissolution of the partnership Henry Coombs established a cutlery and ironmongery business.

King and Son

See **John King (II)**.

King and Son, John

See **John King (I)**.

King, Thomas Davies

Thomas Davies King, (1819-1884) [3] [18] [50] [319], scientific and optical instrument maker, was the son of **John King (II)**. He purchased the family business from his father at 2 Clare Street, Bristol, England, shortly after the death of his grandfather, **John King (I)** in 1841. He received an honourable mention for his microscopes in the Great Exhibition of 1851 at Crystal Palace, London, [328], at which time he was employing five journeymen. Both **Henry Husbands** and **William Clarke (I)** worked for him as senior journeymen. He entered a partnership with

Henry Payne Coombs (1826-1908), operating as **King and Coombs**, but the partnership was terminated in 1853, and appears to have lasted only about two years. In 1853, he relocated the business to 1 Denmark St., Bristol, also known as 8 St. Augustine's Parade, since it was on the corner of Denmark Street and St. Augustine's Parade. He may have continued to occupy 2 Clare St. as a residence.

According to an advertisement of October 1853, "Thomas D. King, No. 1, Denmark Street, St. Augustine's Parade, Optical, Mathematical, Philosophical, and Nautical instrument manufacturer, (removed from No. 2, Clare Street). T.D.K. gratefully thanks his friends and the public for their liberal support during the past seven years, and respectfully informs them that he has completed his new workshops in Denmark Street, and has retained the services of all the best workmen that were previously at Clare Street, and by giving his personal superintendence to every article he manufactures, he can guarantee their accuracy. Theodolites, dumpy levels, and all surveying and mining instruments, Portable telescopes, microscopes, and opera glasses, Barometers, thermometers, and all the necessary meteorological instruments."

Known microscopes by T.D. King have serial numbers ranging from 81 (dated c1846-1850) to 223 (dated 1857).

In 1856 he filed for bankruptcy protection [**35** #21859 p1019], and in 1857 he left the business, later becoming a journalist in Canada. His business at 1 Denmark Street was taken over by **Husbands and Clarke** in 1858. The premises at 2 Clare Street were taken over by the son-in-law of John (II), **James Blake Gardiner** in about 1863.

Kinnear, C.G.H.

C.G.H. Kinnear, (fl.1857-1858), was an architect, a member of the Council of the Photographic Society of Scotland (PSS), and a colonel in the Midlothian Volunteer Artillery. In 1857 he invented a form of folding camera which he introduced to the PSS in 1858 as a "new form of portable camera for portraits and landscapes". This design went on to be much copied by other makers, and Kinnear was described as "the inventor of the modern form of camera bellows".

Kino Precision Industries

Kino Precision Industries Ltd. was a Japanese company, founded in 1959 by Tatsuo Kataoka to make lenses for 8mm cine cameras. In 1980 Kino launched its own brand of camera lenses called **Kiron**. They made lenses for brands such as **Vivitar** and **Soligor**. By 1989 the company had discontinued manufacturing of camera lenses.

Kiron

Kiron was a brand of camera lenses made by **Kino Precision Industries Ltd.**

Klein, W.

W. Klein, [**320**], of Wetzlar, Germany, was a microscope manufacturer during the early C20th.

Kleine, C.B.

C.B. Kleine, (fl.late-1870s to early-1880s) [**201**], of 274 Eighth Ave. New York, NY, USA, was an optician, and manufacturer of microscopes, microscope accessories, and other optical instruments.

Kleman & Zoon, J.M.

J.M. Kleman & Zoon, [**16**] [**256**] [**374**] [**377**], were optical, nautical, and mathematical instrument makers, of Amsterdam, Netherlands. The firm was founded by Jan Marten Kleman (1759-1845), who opened his workshop in Bergstraat 1781. He relocated to the Nieuwendijk in 1800 and, in 1808/9, he began to work in partnership with his son, Bernard (1781-1820), trading as J.M. Kleman & Zoon (J.M. Kleman & Son). In 1808 the firm was appointed Royal instrument maker, and began to advertise as "Kininklijke Mathematisch, Phijsische, Optische Instrumentmakers en Leveranciers voor den dienst ter Zee" (Royal Mathematical, Physical, Optical Instrument Makers and Suppliers for the Marine Service). They are believed to have produced sextants and octants for the Dutch navy.

The firm produced standard lengths for the government when metric weights and measures were introduced in 1812-1816. At that time, they were advertising that they made sextants to the design of **Edward Troughton**. When J.M. Klemen retired in 1830, the company was taken over by his nephew, Carl Swebilius (1808-1857), and continued in business until 1857. Instruments bearing their name include: octant; sextant; double framed quintant; quadrant; transit instrument; standard metres; drawing set; stadiometer; compass card; thunder house styled as a church.

Klingenstierna, Samuel

Samuel Klingenstierna, (1698-1765) [**1**] [**2**] [**25**], was a professor of geometry, and later physics, at Uppsala University, Sweden, and in 1754 was the first to publish a comprehensive theory of optical systems with no colour dispersion or spherical aberration. **Sir Isaac Newton** had concluded in 1672 that refraction is always accompanied by dispersion. Klingenstierna showed that these results applied only to prisms whose apex angle is small. He corresponded with **John Dollond (I)**, sharing these results, and encouraging Dollond's experimentation (See Appendix X). When Dollond published his findings in 1758 with no acknowledgement of Klingenstierna's work, their relationship faltered.

Klönne and Müller
Klönne and Müller, [50] [291], of Berlin, Germany, was a book publisher, and maker of microscopes and prepared slides. The partnership, between Julius Klönne (1844-1880) and Gustav Müller (1850-c1925), was formed in 1876 at Prinzenstrasse 56, Berlin. They moved to number 69 in 1880, then number 71 in 1885, and finally to Luisenstrasse 49 in 1888. The business continued at that address, the most recent records dating to 1921.

KMZ
See **Krasnogorskij Mechanicheskij Zavod**.

Knight, George
George Knight, (fl.1839 d.1862) [18] [50] [53] [61] [319], was an ironmonger, and optical and philosophical instrument maker of London, England. Sources are contradictory about the dates and details of his life and career. However, it appears that by 1839 George and his brother Richard were working in partnership with their father, George, trading as **George Knight & Sons**. George also worked in partnership with his brother, Richard, trading as Richard & George Knight.

Bankruptcy proceedings were initiated for George Knight in the year of his death, 1862, [35 #22639 p3341], and settled in 1863 [35 #22738 p2722]. By this time **James How**, who had worked for him since 1842, was already trading under the name James How & Co., Successors to George Knight & Sons.

Knight & Sons, George
George Knight & Sons, [18] [50] [53] [61], ironmongers, philosophical instrument makers, and manufacturers of chemical and photographic apparatus, was a partnership, formed in about 1839, between **George Knight**, his brother Richard, and their father, George. Their trading address was 41 Foster Lane, London, England. From 1843 their address was additionally 2 Foster Lane.

After the father died in 1845 the partnership continued between the brothers as George Knight & Sons. They traded as ironmongers at 40 and 41 Foster Lane, and philosophical instrument makers at 2 Foster Lane. This partnership was dissolved in 1855 [35 #21743 p2668], after which George Knight is believed to have continued to trade as George Knight & Co., although the business was also still sometimes referred to as George Knight & Sons. The company was succeeded in 1863 by **James How**.

Knott, Isaac
Isaac Knott, (c1823-1903) [319], of 18 Elliot Street, Liverpool, England, was an artist on glass, and a manufacturing optician. By the time of the 1871 census, he had moved to 18 Elliott (sic) Street, Liverpool, then, by 1881, to number 22. His name appears on hand-painted magic lantern slides. His son, James was an artist on glass, and his son Robert was an artist on glass and optician's assistant.

See **Knott & Co.**

Knott & Co.
Knott & Co., (fl.1870) [319] [225 p348], were opticians, and optical, mathematical, and philosophical instrument makers of 18 and 22 Elliot Street, Liverpool, England. They were listed as a manufacturer of microscopes and stereoscopes, as well as "Manufacturers of gold, steel, and shell spectacles, opera, field and ship binocular glasses, drawing instruments, barometers, thermometers, model engines, &c.; microscopes, magic lanterns, and slides for same of every description painted on the premises. Electric and magnetic apparatus, &c., &c."

See **Isaac Knott**.

Knox & Shain
Knox & Shain, [2] [58], of Philadelphia, Pennsylvania, USA, were manufacturers of mathematical and surveying instruments as well as telegraphic and engineering instruments. The company was formed in 1850 by Joseph Knox (1805-1877) and Charles J. Shain (1822-1891). Joseph Knox had previously been foreman to **William J. Young**. Charles Shain had served an apprenticeship with William Young, and subsequently worked for him prior to the formation of Knox & Shain.

In 1852, upon the retirement of **Nathaniel Worthington**, of London, England, Knox and Shain acquired his circular dividing engine, made by **Jesse Ramsden**. They used this until 1880 when it was acquired by Professor Henry Morton, president of the Stevens Institute of Technology, New Jersey. He, in 1890, passed it to the US National Museum, now known as the Smithsonian Museum of American History, where it remains today.

Knox & Shain continued in business until 1929.

Kolberg, Henryk
Henryk Kolberg was the founder of the **Factory of Optical and Precision Devices H. Kolberg & Co.**

Kollmorgen
Kollmorgen was founded in New York, USA, in 1916 by the German born American, Friedrich Ludwig Georg Kollmorgen, (1871-1961), who had emigrated to the USA in 1905. The company began with producing submarine periscopes for the US Navy during WWI. After the war they diversified into movie projector lenses. During WWII they produced drift-meters, bombsights, navigation

instruments, and submarine periscopes for the military. In 1960 the Kollmorgen Optical Company merged with the Inland Motor Corporation to become Kollmorgen Corporation. They continued to diversify, offering a broad range of technological products and solutions.

Kollsman Instrument Company, Inc.

The German engineer, Paul Kollsman, (1900-1982) [136], emigrated to the USA in 1923, and founded the Kollsman Instrument Company in New York in 1928. In 1939 he sold the company to **Square D**, after which it became their Kollsman Instrument division. With the advent of WWII, the Kollsman Instrument division absorbed the **Emmerich Optical** division and began manufacturing binoculars under the **Sard** trade-name. In 1950 the Kollsman Instrument division was sold, and production of binoculars ceased.

Kodak

The Eastman Kodak Company, [105] [306], is a US company specialising in imaging products. Their principal market throughout their history has been photography products. In 1879 **George Eastman** was awarded a London patent for a dry-plate coating machine for photographic plates. A US patent followed in 1880 and, in 1881, together with Henry A. Strong, he set up the Eastman Dry Plate Company. In 1883 they announced their innovative film on a roll. Their first film roll holders were made by **Yawman & Erbe**. The roll cartridge was designed to fit most models of camera in use at the time, and in 1884 the company was renamed the Eastman Dry Plate and Film Company. In 1885 George Eastman started to collaborate with **Frank Brownell**, who began to design cameras for Eastman, and manufacture parts for cameras as well as complete cameras. Eastman began selling the Kodak Camera in 1888, and in 1892 the company became the Eastman Kodak Company. Their objective was to supply the tools of photography at the lowest possible price to the greatest number of people, a philosophy that led to the development in 1900 of the *Brownie* camera, which was designed by Frank Brownell, and initially manufactured by his Brownell Manufacturing Company. In 1902 the Eastman Kodak Company took over the Brownell Manufacturing Company, retaining Frank Brownell as a design consultant until 1906.

In 1889 the Eastman Photographic Materials Company Ltd. was incorporated in London, England as an overseas distributor, and a factory was opened in Harrow, London. By 1900 they had distribution outlets around Europe, soon followed by Japan, and the Canadian Kodak Company Ltd. In 1896 they entered the motion picture market, and at the same time they entered the health imaging industry with photographic paper designed for x-ray capture.

Over the years Kodak acquired a number of other companies, some of which were as follows: In 1895 they acquired the **Boston Camera Co.**, in 1899 they acquired the **Blair Camera Co.**, and in 1912 they acquired **Wratten & Wainwright**. From 1905 until 1926 **Folmer & Schwing** was owned by Kodak. In 1927 Kodak acquired the French factory of **Charles Pathé**, where they began to manufacture cameras and film using the brand **Kodak-Pathé**. In 1928 they introduced their microfilm system for document archiving. In 1932 Eastman Kodak expanded their manufacturing into Germany with the acquisition of **Dr. Nagel-werke** in Stuttgart.

In 1965 Kodak introduced the *Super 8* film format, together with *Super 8* cameras and projectors. This 8mm film format made a significant contribution to the popularisation of home movie-making.

Kodak was slow to respond to the decline in demand for photographic film and the rise of digital photography. As a result of this, financial problems set in in the late-1990s, and in 2012 they filed for Chapter 11 bankruptcy protection. They stopped making digital cameras, pocket video cameras, and digital picture frames, and began to concentrate on the corporate digital imaging market. In 2013, having sold many of its patents, Kodak emerged from bankruptcy. Since then, Kodak has provided business printing and graphic communication services, and its offerings include print systems, enterprise inkjet systems, micro-3D printing and packaging, software and solutions, and consumer film.

Kodak-Pathé

Kodak-Pathé was a French camera and film making subsidiary of **Kodak**, formed following the acquisition of **Charles Pathé**'s French factory in 1927.

Köhler, August

August Köhler, (1866-1948), was a German scientist who made several innovations in microscopy. He is best known for his invention of Köhler Illumination in 1893. In 1900 he began to work for **Zeiss**, where he continued to work until 1945, three years before his death.

Köhler Illumination

Köhler Illumination, [142], is a form of microscope sample illumination, invented by **August Köhler**, that provides even light across the field of view, and a light cone that matches the numerical aperture of the objective lens. It is suitable for either viewing or photography. It normally consists of a collector lens near the light source, together with a field iris, and a sub-stage condenser with a condenser iris. The focus of both the collector and condenser lenses is adjustable, as is the aperture of each of the irises.

The collector lens is focused to form an image of the light source in the plane of the sub-stage condenser iris. The condenser lens is focused to form an image of the field iris in the sample plane. The two irises are adjusted to provide the required light cone. Thus, when the sample is in focus the image of the light source is completely defocused, resulting in even illumination.

KOMZ
See **Kazan Optical and Mechanical Plant**.

Konica
Konica was founded in Tokyo, Japan, in 1873 by Rokusaburo Sugiura, selling photographic and lithographic materials. They began to produce cameras in 1903, and subsequently expanded their product range to include x-ray film, photocopying machines, lenses for CD players, and films for LCD polarizers. In 2000 they entered a joint venture with **Minolta**, producing polymerised toners. In 2003 the two companies were merged to form **Konica Minolta Holdings Inc.**

Konica Minolta
Konica Minolta Holdings Inc. was formed in 2003 by the merger of **Konica** and **Minolta**. In 2007 they closed the photographic and camera part of the business, focusing on printing technology. In 2013 the company name was changed to Konica Minolta Inc.

König, Albert
Albert Adolf König, (1871-1946) [42], was a German optician. He studied mathematics and physics at the universities of Jena and Berlin. He earned his PhD under **Ernst Abbe**, after which he became a scientific employee at **Zeiss**. He worked on the development of optical systems, and was noted for his eyepiece designs, (see Appendix II).

Konus
Konus is an Italian company, founded in 1979 in Verona. They sell budget binoculars, riflescopes, spotting scopes, telescopes and microscopes.

Koristka, Franz
Francis (Franz) Koristka, (1851-1933). See **Fratelli Koristka**.

Körner & Mayer
See **Nettel Camera-werke**.

Kowa
Kowa is a Japanese maker of binoculars, spotting scopes, cameras, and camera lenses, established in 1894. They opened their optical works in 1946, and manufactured cameras from 1954 until 1978.

Koyu Company Ltd.
The Koyu Company Ltd., [136], was formed in 1954 in Tokyo, Japan, as an optical goods wholesaler. In 1970 the company name was changed to **Vixen Company Ltd.** Their maker's mark is illustrated in Appendix IV.

Kozelský, František
František Kozelský, (1913-2003) [191], was a Czech telescope maker who made Cassegrain and refracting telescopes for observatories around Eastern Europe, including the Czech Republic and Slovakia. He was well-known for his collaboration with **Vilém Gajdušek**, and asteroid 8229 is named after him. Hvězdárna Ždánice (the Ždánice observatory) was the first major project undertaken by the **Gajdušek-Kozelský** collaboration. Until its closure the observatory had several examples of their Cassegrain and refracting telescopes.

Krasnogorskij Mechanicheskij Zavod (KMZ)
Krasnogorskij Mechanicheskij Zavod (the Krasnogorsk Mechanical Plant, Красногорский Механический Завод), founded in 1941, [227], is a Russian factory in Krasnogorsk near Moscow which specialises in optical technology. KMZ manufacture the **Zenit** branded optical equipment, and **Zorki** was a brand-name they used after WWII for rangefinder cameras. Their **Mir** rangefinders were a simplified version of these. KMZ used the name **Gorizont**, (Горизонт; Horizon) for a range of panorama camera models. **LZOS** is a satellite factory of KMZ, making optical glass, and lenses for KMZ products. For Russian and Soviet makers' marks, see Appendix V.

Krauss, E.
The Krauss company was founded in Paris, France by Eugene Krauss in the 1880s. They manufactured cameras and lenses, as well as prismatic and Galilean binoculars. They were licenced by **Carl Zeiss** to produce Zeiss lens models. Eugene Krauss and **Gustav Krauss** are believed to have been brothers.

Krauss, G.A.
Gustav Adolf Krauss, maker of photographic equipment in Stuttgart in Germany from 1895. Gustav Krauss and **Eugene Krauss** are believed to have been brothers.

Kronos
Kronos is a brand of Russian made binoculars manufactured by **ZOMZ**.

Kruines

The Kruines family, [50], were scientific and optical instrument makers of Quai de l'Horloge, Paris, France. Their addresses in Quai de l'Horloge are recorded as 42 (1801, 1805), 61 (1807, 1829), and 21 (c1850 onwards). The instrument making business was founded by Mathias Kruines (c1770-1811) in 1800, and he gained a reputation for making fine microscopes. Upon his death his widow, Marie-Madeleine Kruines (c1770-c1829), took over the business. By 1829 the business was named Kruines fils, and was being run by Marie-François-Antoine Kruines (1799-1866), the son of Mathias and Marie-Madeleine. Upon his death in 1966, the firm was taken over by the proprietor of a neighbouring business, a purse maker named Eulalie Rambaud, who continued the business under the name Kruines-Rambaud successor.

Krüss, Andres

Andres Krüss, (1791-1848) [113] [298] [374] [391], married **Edmund Gabory**'s daughter, Mary Ann Gabory, in 1823, following the death of her father, and continued the Gabory optical instrument business in Hamburg, Germany, together with his brother-in-law, Edmund Nicholas Gabory. In 1844 Andres formed the Optisches Institut von A. Krüss, also in Hamburg, but died only four years later. His wife, Mary Ann, continued the business until 1851 when she passed control of it to her sons Edmund Krüss (1824-1906), and William Krüss.

The company continued, producing lenses, magic lanterns and microscopes, as well as other instruments including model steam engines and model steam boats. In 1888 control of the company passed to Edmund Krüss's son **Hugo Krüss**. During WWII they marked their instruments with the **ordnance code** hxh. The company continued as a family business, and went on to operate as A. Krüss Optronic.

Krüss, Hugo

Hugo Krüss, (1853-1925) [374], was the grandson of **Andres Krüss** and, in 1888, followed in his father's footsteps, running the optical instrument business started by his grandfather in Hamburg, Germany. He served his apprenticeship with **Steinheil** and **Dennert & Pape**, and was awarded a doctorate degree by Ludwig Maximilians University, for his work on the comparison of some objective constructions. He produced innovative new designs for various kinds of photometer, and expanded the company's products to include photometric and spectroscopic instruments. He was succeeded in managing the company by his son, Paul Krüss (1880–1976).

Kuhlmann, Franz

The company Franz Kuhlmann was established in 1873 when the clockmaker, Bernhard Friedrich Kuhlmann, took over a clock workshop in Wilhelmhaven, Germany. He began to produce navigation, locating and targeting instruments. In 1899 he was succeeded by his son Franz Wilhelm Kuhlmann, (1877-1965), and in 1906 the company was registered as Franz Kuhlmann, producing workshop machinery. During WWII they produced optical instruments for the German war effort, marking their instruments with the **ordnance code** bvu.

Kunming United Optics Corporation

Kunming United Optics Corporation (KUO) is a Chinese optical instruments company, incorporated in 2003. They are OEM manufacturers for many modern binocular brands such as **Oberwerk**, **Celestron**, and others. KUO also make binoculars, spotting scopes and telescopes that they market themselves.

Kürbi & Niggeloh

Kürbi & Niggeloh, [53], was founded in 1909 by Wilhelm Kürbi and Carl Niggeloh in Barmen-Rittershausen, Germany, producing photographic equipment. In 1911 the company moved to Radevormwald. They produced cameras from 1935 until 1975. Their cameras were sold under the brand-name **Bilora**, (*Bilora Bella*, *Bilora Boy*, *Bilora Bonita*, etc.). During WWII Kürbi & Niggeloh marked their military products with the **ordnance code** eed.

In 1994 the company split into two independent companies, Kürbi & Niggeloh BILORA GmbH, producing plastic products, and Kürbi Otto Toennes GmbH, producing photographic accessories which they continued to sell under the *Bilora* brand-name. In 2016, Kürbi & Niggeloh BILORA was renamed BILORA Kundstofftechnik GmbH, whose core business is tool making, plastic injection moulding, and finishing.

Kuribayashi Camera Works

Kuribayashi Camera Works of Tokyo, Japan, was founded in 1907, manufacturing photographic accessories. They began to produce cameras in 1919, and sold most of their camera models under the *First* brand-name. After WWII they began to use the brand-name *Petri*, and in 1962 the company name was changed to **Petri Camera Company**.

Kyocera

Kyocera began business in 1959 in Kyoto, Japan, as the Kyoto Ceramic Company Ltd. They changed their name to the Kyocera Corporation in 1982, and in 1983 they entered the camera market with the acquisition of **Yashica**. They began with a continuation of the cameras produced under the *Yashica* and **Contax** brands and, when autofocus cameras entered the market in 1985, they began an expansion of their offerings. They continued to produce cameras and lenses until 2005 when they announced that they would cease production of *Yashica*, *Contax* and *Kyocera* cameras and lenses. In 2008 the *Yashica* brand was sold to the MF Jebsen Group, a manufacturing company.

Kyowa Optical Co. Ltd.

Kyowa Optical Co. Ltd. of Hashimoto, in the Kanagawa Prefecture, near Tokyo, Japan, is a manufacturer of microscopes and optics for the semiconductor industry. Their factory is in Nanakubo, in the Nagano Prefecture. Kyowa was founded in 1940.

L

L3 Brashear
L3 Brashear is a defence, aerospace, and commercial optical component and systems company in Pittsburgh, Pennsylvania, USA. The company began as **John A. Brashear Co.**, and subsequently underwent a number of changes of ownership. It is now a subsidiary of **L3 Technologies**.

L3 SSG
L3 SSG is an aerospace optical component and systems company in Wilmington, Massachusetts, USA. The company began as **Tinsley Laboratories Inc.**, and subsequently underwent a number of changes of ownership. It is now a subsidiary of **L3 Technologies**.

L3 Technologies
L3 Technologies of New York, USA, formerly L3 Communications, develops advanced defence technologies and commercial solutions in pilot training, aviation security, night vision and EO/IR, weapons, maritime systems and space. L3 Technologies is the parent company of, among others, **L3 SSG** and **L3 Brashear**.

Laboratory Optical Co.
Laboratory Optical Co., of Plainfield, New Jersey, USA, was a manufacturer of optical instruments and components including spectrophotometers, lenses, cylindrical lenses, mirrors, and astronomical telescopes. They began advertising high-end telescopes in 1949 for amateur and professional use. An advertisement from 1995 shows a 6-inch refractor on a clock driven equatorial mount.

Lacour-Berthiot
The French optical instrument maker Lacour-Berthiot, [**192**], was formed when Eugène Lacour joined the business of his uncle **Claude Berthiot**. They traded as Lacour-Berthiot until 1906 when they were incorporated as **Établissements Lacour Berthiot S.A.** Eugène Lacour was the general director, and Charles Henri Florian was the technical director. In 1913 the company became **Société D'Optique et de Mécanique de Haute Precision (S.O.M.)**.

Ladd, William
William Ladd, (1815-1885) [**18**] [**50**] [**53**] [**264**], was a mathematical, optical, and philosophical instrument maker of London, England. Addresses: 10 Cleaver Street, Kennington (c1842-1847) 29 Penton Place (1846-1857); 31 Chancery Lane (1857-1860); 11-12 Beak Street (1860-1882); 199 Brompton Road (1872-1882).

He began trading under the name W. Ladd & Co. from about 1871. He established a strong reputation for optical instruments, including microscopes, and made his own lenses and brass-work. From the early 1860s he also made electrical apparatus.

He began to use a chain focusing control mechanism for his microscopes in 1851, and in the same year he received an Honourable Mention at the Great Exhibition. His exhibition catalogue entry, [**294**] [**328**], stated: "Ladd, W. 29 Penton Pl. Walworth, Manu. Case of instruments for pneumatic experiments. Compound microscope, with chain and spindle in lieu of rack and pinion, now in use. Registered."

He received a Medal for "excellence of construction of microscopes, induction coils, &c." at the International Exhibition of 1862, at which his catalogue entry, [**295**], stated: "Ladd, W. 12 Beak-St. W.–Focimeter for lighthouses: Induction coils and apparatus: Microscope with magnetic stage: Air pump."

He was awarded a silver medal at the Paris Exhibition of 1867 where he first exhibited the *Ladd's Dynamo-Magneto-Electric Machine*. Upon Ladd's retirement in 1882 his stock was auctioned, including microscopes, spectroscopes, electrical, optical, and other philosophical apparatus. He sold his business to W.G. Harvey and W.F. Peak, who traded as **Harvey & Peak**.

Lancaster and Son, J.
J. Lancaster and Son, [**44**] [**53**] [**306**], was an optician and camera maker of Birmingham, England. Addresses: 87 Bull Street (c1871-c1885); 37 Colmore Row (c1886-1907); Camera Buildings, 275 Broad Street (1908-c1914); 87 Parade (c1915-1934); 54 Irving Street (1935-c1955).

The company was formed in 1835 by James Lancaster, initially as an optician, and later selling telescopes, microscopes, magic lanterns, photographic enlargers, and cameras. In c1876 the company became J. Lancaster & Son. Many of the products designed and marketed by Lancaster were made for them by a number of small companies around Birmingham who worked as subcontractors. It is not clear what, if any, manufacturing capability they had of their own. In 1886 they patented their *Watch camera* which was designed to look like a hunter-type pocket watch when closed. In 1905 they became a limited company, trading initially as James Lancaster & Son Ltd., and subsequently as J. Lancaster & Son Ltd. The company ceased business in 1955.

Lane & Sons, J.L.
J.L. Lane & Sons, [**306**], of London, England, was a manufacturer of cameras and photographic equipment,

primarily for the trade. Addresses: 7 Allen Street (1878-c1887); 102 Barnesbury Road, Islington (from c1885). The company began as J.L. Lane, with the earliest references dating to the mid-1870s. The name was changed to J.L. Lane & Sons in the mid-1880s, and the company continued to appear in the directories until at least 1887.

Langley Inclinometer
Langley Inclinometer is another name for a Waymouth-Cooke Rangefinder.

Lantern
A lantern is a casing, made of transparent material such as glass, designed to contain and protect a light. See also magic lantern, Fresnel lantern.

Laser
In 1958, shortly after the invention of the *maser*, physicists Charles Townes and Arthur Schawlow defined the physical conditions necessary for *light amplification by stimulated emission of radiation*, known as *laser* [7]. In 1960 the first laser was successfully demonstrated by Theodore Maiman. A laser is a source of monochromatic, coherent, collimated light.

In a filament light-bulb, atoms in the filament are raised into an excited state by electrical energy. Each atom then drops spontaneously back to a lower state, and in doing so emits a photon with random direction and phase.

Under certain physical conditions an excited atom may be stimulated to drop to a lower state by the presence of electromagnetic radiation of appropriate frequency. In this case the emitted photon is in phase with the stimulating wave, and has the same polarization and direction. This phenomenon is used in the laser, which may be either solid-state or gas.

In a solid-state laser, the medium, e.g. a synthetic ruby rod doped with chromium ions, is pumped with energy to cause *population inversion*, meaning that more of the chromium ions are in an excited state than not. The ends of the rod are silvered as mirrors, forming an optical cavity, in which one of the mirrors is partially reflecting to allow a small percentage of light through. Initially, some atoms spontaneously drop to the ground state, each emitting a photon. The journey of a photon through the medium stimulates a cascade of new photons, all in the same direction, in phase with, and with the same polarization as the stimulating photon. Most of these cascades die down, but the optical cavity selects photon cascades that are directed along the cavity. This results in multiple reflections within the optical cavity, amplifying each time the cascades traverse the medium. The laser beam is the small percentage of light that escapes through the partially reflecting mirror.

A gas laser works on the same principle, but with a gas as the laser medium.

Latham Brothers
Grey and Otway Latham, (fl.1894-1898) [335], the sons of Major Woodville Latham, were salesmen for a drug company in New York, USA. In 1894 they began to produce films of prize fights, and opened a Kinetoscope parlour to show their films. With funding from their father, and with the help of **William Dickson**, then still employed by **Thomas Edison**, they began to develop a motion picture projector. This led to the first demonstration in 1895 of the Panoptikon projector. By 1898 the Latham brothers were no longer working in motion pictures, and by 1910 they had both died, survived by their father, Major Woodville Latham.

Laun, Hartog van
Hartog van Laun, (1732-1815) [378], was a scientific instrument maker and retailer of Amsterdam, Netherlands. He served his apprenticeship with **John Cuthbertson**, after which he set up his own business. He gave lectures, using demonstration scientific instruments he had made, which included a fine table orrery. His two sons, Abraham and Jacob, later worked with him, during which time his instruments were signed H. van Laun en Zoon (H. van Laun and son) or H. van Laun en Zonen (H. van Laun and sons). His daughter, Diena, was the mother of the instrument maker **Abraham van Emden**.

Instruments bearing his signature include: single draw brass pocket telescope with interchangeable objective lenses; horizon ring & stand; Atwood machine; double table air pump; demonstration apparatus for pulleys. Instruments made while working with his sons include: orrery; microscope; globe; frictional machine.

He was succeeded in business by his sons, Abraham and Jacob, who were in turn succeed by his grandson, Abraham van Emden.

Laurent, Léon
Léon Laurent, (1840-1909) [53] [299], began his career working for the French instrument maker Gustav Froment (1815-1865). In 1870 Laurent married **Jules Louis Duboscq**'s daughter, Marie Duboscq. Thus, by marriage, Laurent became a cousin once removed to **Henri Soleil**. In 1872, upon the retirement of Henri Soleil, Laurent succeeded him in business as an optical instrument maker. He demonstrated a great aptitude for this business, and at the French National Exposition of 1878 in Paris he was awarded two gold medals for his optical apparatus, the same year he was made a Chevalier de la Légion

d'Honneur. At the French National Exposition of 1889, he was awarded the *grand prix*. Upon his retirement in 1892 Laurent sold his thriving business to **Amédé Jobin**.

Lawley
The Lawley business, [306], of 78 Farringdon St., Ludgate Circus, London, England, was an optician and pawnbroker who sold second-hand cameras, as well as cameras and other instruments bearing Lawley's name. Records of the company go back to 1865, but it is believed to have been founded earlier. Walter Lawley, (b.1851) was a pawnbroker and optician in the 1880s, and formed a partnership with William Lawley. From about 1877 until sometime between 1881 and 1887 they traded as W. Lawley & Sons, then reverted to the name W. Lawley. William Lawley was declared bankrupt in 1881 [35 #25002 p4118]. For a period during the mid-1880s the company traded at 8 Coventry Street, London. They ceased trading in about 1897. Instruments bearing their name include: cameras; ivory protractor; microscope.

Lawrence & Mayo
Lawrence & Mayo are Indian opticians and makers of survey, scientific and meteorological instruments. The company was founded by two London, England, based families, Lawrence and Mayo, and they set up in locations all over the world including London, Cairo, Spain, Portugal, Indian subcontinent, Colombo, Rangoon and Singapore. The company became wholly Indian in 1967. Examples of sextants may be found.

Leader, Henry Francis
Henry Francis Leader, (fl.1837-1849) [18], was a telescope maker of London, England. Addresses: 49 Southampton St., Pentonville (1837-1839), and 25 Great Percy St. (1846-1849). He attended the London Mechanics' Institute from 1837-1839. The Southampton St. address was also occupied by **Robert Mills** at around that time, and Henry Leader may have been one of his employees.

Leaf
Leaf is a manufacturer of digital backs for cameras. Leaf is based in Israel, and is a subsidiary of **Phase One**. Their products include camera backs for **Mamiya** cameras, and they also market Mamiya and Phase One cameras and lenses.

Lealand, Peter H.
Peter H. Lealand, (fl.1826-) [18] [276 #1618] [279] [280 Ch.6], of 23 Phoenix Street, Somers Town, London, England, was an optical instrument maker, and brother-in-law to **Hugh Powell**. He began working as an optical instrument maker in 1826, and there are rare examples of this early work, including fine theodolites bearing his signature. During this early period, he also worked in the object-glass and optical department of Hugh Powell's workshops. In 1841 he entered a partnership with his brother-in-law to trade as **Powell & Lealand**.

Lean's Miners' Theodolite
See theodolite.

Lear, Anthony James
Anthony James Lear, (fl.1826-1831) [18] [276 #1914] [379], was an optician, and optical, mathematical, and philosophical instrument maker of London, England. Addresses: Plough Place, Fetter Lane; 34 St. John's Lane, Clerkenwell; 4 Arlington St., Clerkenwell; 36 St. John's Lane, Clerkenwell (1830). He served his apprenticeship with **Benjamin Messer** of the **Worshipful Company of Makers of Playing Cards**, beginning in 1804, but his freedom is not recorded.

Lebas, Philippe-Claude
Philippe-Claude Lebas, (fl.1669-1677 d.1677) [256], was a mathematical instrument maker of Paris, France, with a reputation for making fine lenses. He was awarded the title of optician to the King, and lived in the Louvre. He supplied telescope lenses to the Paris Observatory. Having developed his own method for polishing lenses, he was visited by **Christian Huygens**, who obtained permission to watch him at work. According to Huygens, "… but the good Lebas will not disclose his whole secret to me. He rotates the tool in which he finishes his oculars, and he has fitted up a little machine for the purpose with which his wife, instead of spinning, makes these lenses. He intends to make another which would deal with ten or twelve at one time." Following his death in 1677, Philippe-Claude Lebas was succeeded in business by his widow, **veuve Lebas**. Instruments bearing his name include: silver proportional compass dated 1669; pocket case of silver drawing instruments.

Lebas, Veuve
Veuve Lebas, (fl.1676-1689) [256], was a lens maker of Paris, France. Madame Lebas was the widow (veuve) of **Philippe-Claude Lebas**, and continued his business following his death, maintaining the secret of his methods. She established a strong reputation in her own name, and supplied lenses to the Paris Observatory. She learned from **Christian Huygens** the method for making glass spheres for microscopes and, after running the business for only two years, she had the reputation as the most skilful maker of such glass spheres. She was succeeded by her son, who continued the business until 1721.

Lee, George
George Lee, (fl.c1847) [18] [322], was a nautical instrument maker of 33 The Hard, Portsea, Portsmouth, England. He may have served an apprenticeship with the mathematical instrument maker William George Cook. He worked for **George Whitbread** before setting up his own business, then in 1847 he established the firm **George Lee & Son**. He is known to have sold: sextant.

Lee, Joseph
Joseph Lee, (fl.1839-1868) [18] [304], was an optician, watchmaker, jeweller, and silversmith of Belfast, Ireland. Addresses: 28 Gloucester St. (1835); 24 High St. (1839); 74 High St. (1840-1843); 57 High St. (1846-1849). In 1868 he was listed as Jeweller to Queen Victoria. From 1850 he worked in partnership with his son, trading as **Lee & Son** at the same address.

Lee & Son
Lee & Son, [18] [304], of 57 High St., Belfast, Ireland, was a partnership between **Joseph Lee** and his son, trading as opticians, watchmakers, jewellers, and silversmiths. The partnership began in 1850, and continued trading under this name until 1870. By 1863 they were listed as Jewellers to Queen Victoria. Instruments bearing their name include: brass single-draw marine telescope; lacquered brass surveying level; stick barometer. They are also known to have sold: microscope.

Lee & Son, George
George Lee & Son, (fl.1847-1912) [17] [18] [53] [322] [358] [361], were opticians, and nautical, mathematical, and optical instrument makers of Portsmouth, England, established in 1847 by **George Lee**. Addresses: The Hard; 3 Palmerston Rd., Southsea; Ordnance Row, The Hard, Portsea; 33 The Hard, Portsea. According to their trade card: "George Lee & Son, (from Whitbread's London) manufacturers of Mathematical, Optical, & Nautical instruments. To the Hon[ble] Corporation of the Trinity House, R.N. College & The Admiralty. Ordnance Row, the Hard, Portsea." Another trade card adds "Est[d] 1847". Instruments bearing their name include: sextants; mercury artificial horizon; Stuart's Marine Distance Meters.

Leech (Rochester) Ltd.
Leech (Rochester) Ltd. was a manufacturer of optical and scientific instruments, of 277/277a High Street, Rochester, Kent, England, and First Floor, 4 Henrietta Street, London. It was incorporated as a limited company in 1947, and was dissolved in 2002. Leech advertised as scientific optical lens manufacturers, instrument makers, and engineers. Among their products they also appear to have made medical instruments and live-steam model railway engines.

They exhibited at the 1947 British Industries Fair. Their catalogue entry [450], listed them as Leech Optical Co., of 277-277a, High Road, Rochester, Kent, with their London office at 94 Hatton Garden. It listed them as manufacturers of: Photographic enlargers; Students' optical bench; Still projection; Microscope objectives; Accessories; Projection lenses; Prisms; Condensers; Readers; Aplanatic magnifiers; Mirrors; Bloomed aluminised rhodiumised optical surfaces.

The 1951 *Directory for the British Glass Industry*, [327], contains an entry for the **Hummel Optical Co. Ltd.**, 94 Hatton Garden, London. Works: Rochester, Kent. Established 1895. Trade name: *Leech*. Their products are listed as: Lenses; Mirrors; Optical units, mounted and unmounted, for projection, photography, education, research, etc.

They also sold optical instruments produced by other manufacturers. A **Meopta** microscope, offered at auction, has a retailer's label reading, *Sales Division, Leech, Rochester Kent*.

Leech, William
William Leech, (fl.1789-1794) [18], was an optician, and optical and mathematical instrument maker of 3 St. Martin's Churchyard, Charing Cross, London, England.

Leedham & Robinson
Leedham & Robinson, [44], were opticians and optical instrument makers of Sheffield, England. They succeeded **William Ashmore** in his business in 1850 or 1851.

According to their trade card, dated 1851: "Leedham and Robinson, (late Ashmore), manufacturers of Spectacles, Telescopes, Reading glasses, and all kinds of Optical instruments. Optical works, 104 Fargate, Sheffield."

Leeuwenhoek, Antonie Philips van
Antonie Philips van Leeuwenhoek, (1632-1723) [37] [155] [274], was a Dutch microscopist who was the first to observe bacteria and protozoa. At age 16 he began an apprenticeship as a linen draper and, in c1654, he set up a business as a draper and haberdasher. He also served as chamberlain to the sheriffs in his hometown of Delft, and as a surveyor and wine-gauger. The income from these roles supported him in his interests in microscopy. He ground his own lenses, and made hundreds of simple microscopes. He invented a device for viewing aquatic specimens, commonly referred to as a Leeuwenhoek aquatic microscope. He showed that life wasn't created spontaneously, as some believed, but was the result of

breeding. He discovered human spermatozoa, and studied spermatozoa of various species.

Leeuwenhoek Aquatic Microscope
An early simple microscope with a device for viewing aquatic specimens, devised by **Antonie Philips van Leeuwenhoek**.

Legross, P.J.
P.J. Legross, also known as Le Gross, (fl.1829-1830) [18], was an optical, mathematical, and philosophical instrument maker of 1 Upper Crown St., Soho, London, England.

Leica Heerbrugg AG
In 1986, **Leitz** merged with **Wild Heerbrugg AG** to become the Wild Letiz AG Group, and in 1990 they were involved in several mergers, including one with the **Cambridge Instrument Co. Ltd.**, resulting in the formation of the Leica Heerbrugg AG group. In 1997 the Leica Heerbrugg AG group split into two independent companies, **Leica Geosystems**, and **Leica Microsystems**. In 2005 Leica Geosystems was acquired by the Swedish firm **Hexagon AB**, and in the same year Leica Microsystems was acquired by the US multinational Danaher Corporation.

Leica Camera
Leica Camera, an independent publicly owned company formed in 1986, grew from the successful **Leitz** camera brand, *Leica*. They produce digital and analogue cameras for professional and consumer applications, binoculars, spotting scopes, riflescopes and rangefinders.

Leica Geosystems
Formally a part of **Leica Heerbrugg AG**, Leica Geosystems was acquired in 2005 by the Swedish firm **Hexagon AB**, and became a part of its Geosystems division. The company's historical roots lie with **Wild Heerbrugg** and **Kern & Co.** Leica Geosystems specialise in capture, visualisation and processing of 3D spatial data for geosystems, and their products include total stations, global navigation satellite systems, laser scanners and laser tracking systems, levels, and systems for aerial use.

Leica Microsystems
Formally a part of **Leica Heerbrugg AG**, Leica Microsystems was acquired in 2005 by the US multinational Danaher Corporation. Leica Microsystems grew from the merger between the Wild Leitz AG Group with **Cambridge Instrument Co. Ltd.**, which had already acquired the **American Optical Company**, **C. Reichert**, **Optische Werke**, and the Optical Systems Division of **Bausch & Lomb**. In 2005, Leica Microsystems was acquired by the US multinational Danaher Corporation. The company specialises in microscopy, camera and software solutions for imaging and analysis of macro-, micro-, and nanostructures.

Leicester Astronomical Centre Ltd.
See **Cliff Shuttlewood**.

Leiner and Bertram
See **Ernst & Wilhelm Bertram**.

Leitz
The Leitz company was originally founded in 1849 by **Carl Kellner** and his brother-in-law **Moritz Carl Hensoldt** as **Optisches Institut** in Wetzlar, Germany. Upon Kelner's death in 1855, the company was run by **Louis Engelbert** until 1856, when, **C.F. Belthle**, together with Kellner's widow, began to run the company. In 1865 Ernst Leitz (1843-1920) became a partner, and in 1869, upon the death of C.F. Belthle, Ernst Leitz took over sole management of the firm, and continued in business as Ernst Leitz GmbH, Wetzlar, producing microscopes. In 1907 they began to produce binoculars. In 1913 they began to make cameras using the brand-name *Leica*, short for **Lei**tz **Ca**mera. In 1973 they opened their factory in Portugal. In 1986 Ernst Leitz Wetzlar GmbH merged with **Wild Heerbrugg AG** to become the Wild Leitz AG Group. Also in 1986, due to the success of the *Leica* brand, the camera division became an independent company named **Leica Camera**. In 1988 the Wild Leitz AG Group acquired **Kern & Co.** In 1990 the Wild Leitz AG Group was involved in several mergers, including one with the **Cambridge Instrument Co. Ltd.**, and became the **Leica Heerbrugg AG** group.

Lejeune & Perken
Lejeune & Perken, [306], of London, England, was a manufacturer of cameras and photographic equipment, established in 1852. Addresses: 24 Hatton Garden (1852-?); 101 Hatton Garden (?-1887); 99 Hatton Garden (from 1887); factory at 112 & 113 Great Saffron Hill.

They supplied cameras to the trade, one of their customers being **J.T. Chapman**. They registered the **Optimus** trademark in 1885. In 1887, upon their relocation, the company name changed to **Perken, Son & Rayment**.

Lekeux, Richard
Richard Lekeux, also known as Lehuix, Lecoux, or Leheux, (fl.1778-1839) [18] [53] [90] [276 #862] [322] [385], was an optical, mathematical, and philosophical instrument maker, and hardware seller of London,

England. Addresses: 103 Wapping, Near Execution Dock; 168 Wapping; 113 High St., Wapping; 137 Execution Dock, Wapping; 9 Little Aycliffe St., Goodmans Fields (1779); 36 Butcher Row, East Smithfield (1781); 138 Wapping (1785-1793); Stable, Three Tun Court, Charlotte St., Nightingale Lane, East Smithfield (1787); Chaisehouse, Stable yard, New Road, Back Lane, Ratcliffe (1787); 137 Hermitage Bridge, Wapping (1796-1817); 134 Wapping (1816); 137 High St., Wapping (1822-1839).

Richard Lekeux began his apprenticeship with **Thomas Ripley** in 1770, and was turned over to **Walton Willcox** of the **Joiners' Company** in 1775. He was freed of the **Grocers' Company** in 1777 and, a few days later, freed by patrimony of the **Fishmongers' Company**. **James Munro (I)** is believed to have started an apprenticeship with him in 1783, and to have been turned over to someone named Botten.

Instruments bearing his name include: 2-draw day-or-night telescope; sextant; octants; Hadley's quadrant. He advertised a full range of instruments.

Lemaire

Lemaire, [10], was founded in 1846 in Paris, France, rue Oberkampf 22 & 26, by Armand Lemaire (d.1885), and continued in business until 1955. When Armand died, his son in law Jean Baptiste Baille took over the company, which was by then called the Baille Lemaire group. Examples of binoculars and opera glasses may be found. Not all Lemaire made instruments are marked with their name, but they may be identified by the Lemaire logo, which is a bee, shown above.

An example of a triple-optic Galilean binocular at auction bears the name *J. Casartelli Optician, Market St. Manchester*. It also has the Lemaire bee logo and the number 3349 on the knob that turns the triple-optic, as shown here. These factors indicate that it was made by Lemaire for Casartelli. Lemaire are also known to have manufactured triple-optic binoculars, as well as conventional Galilean binoculars for a number of other opticians in Europe and further afield.

Lemardeley et Fils

F. Lemardeley et Fils, [322], was a microscope maker of 127 rue de la Glacière, Paris, France, who traded from about 1846. Microscopes bearing this name are dated to the late-C19[th] or early C20[th].

Leningrad Optical Mechanical Association (LOMO)

LOMO, the Leningrad Optical Mechanical Association (Ленинградское оптико-механическое объединение) [23], was established in 1914 as the Russian Joint Optical and Mechanical factory. The name Leningrad Optical Mechanical Association (LOMO), was assigned to the company in 1962. By the time of its inception, LOMO included other plants: KINAP (cinema equipment); **Progress** (military goods – sights, etc.); the Experimental Optical and Mechanical Plant (DIMP); the former State Optical Mechanical Plant (**GOMZ**) and others. They produce optical systems for the military and the civil markets, including high-precision optical-mechanical instruments, lenses, mirrors, prisms and diffraction gratings. Among their brand-names, *Biolam* is a LOMO brand of biological microscope, and *Polam* is a LOMO brand of polarizing microscope. For Russian and Soviet makers' marks, see Appendix V.

Lennie Opticians

Lennie Opticians, [10] [18] [263] [379], were instrument makers, retailers, jewellers and opticians of Edinburgh, Scotland. Addresses: South Bridge (1835-1840); 14 Leith St. (1840-1857); 46 Princes Street (1857-1953); 5 Castle St. (1954-1959). It is believed that many of the optical instruments bearing the Lennie name were manufactured by other makers, many of them French.

James Lennie, (1817-1854), began his career as a jeweller and optician in 1835. By 1840, he was working under his own name and, in the census of 1851, he is recorded as a master optician. Upon his death in 1854, his wife, Eliza Lennie, took over the business. In 1857 she relocated to 46 Princes Street, and began to work under her own name. An undated catalogue in her name advertises: Spectacles and eyeglasses; Achromatic telescopes; Field glasses; Opera glasses; Lenses; Cameras; Photographic material; Stereoscopes and slides; Magic lanterns; Barometers; Thermometers. During the early 1860s she advertised as "Optician and manufacturer of photographic apparatus".

The company subsequently traded under the names: J.&J. Lennie (1902-1906); Lennies (1907-1922); E.&J. Lennie (1923-1959). In 1959 the business was taken over by **Turnbull & Co.**

Instruments bearing their name include telescopes and binoculars.

Lenoir, Etienne

Etienne Lenoir, (1744-1832) [2] [10] [231], was a French instrument maker who specialised in geodetic and navigational instruments. He established a strong reputation for his navigation and observatory instruments, and made the first prototype of the Borda repeating circle. He came into possession of a circular dividing engine, made by **Jesse Ramsden**, (the first of the two built by Ramsden), which he converted to metric scale, and trained young craftsmen in its use. His son, **Paul Lenoir**, was also an instrument maker.

Lenoir, Paul Etienne Marie

Paul Etienne Marie Lenoir, (1776-1827) [**10**], was the son of **Etienne Lenoir** and, like his father, was an instrument maker.

Lens

A lens is an optical element of transparent material (frequently optical glass, but other optical materials are increasingly popular). The simple lens configurations have either spherically shaped or flat surfaces, and are *biconvex*, *plano-convex*, *positive meniscus*, *negative meniscus*, *plano-concave* and *biconcave*. A lens that is thicker at the centre is referred to as a *converging* or *positive* lens. A lens that is thinner at the centre is referred to as a *diverging* or *negative* lens.

Lenses, however, take many forms, and the surfaces of these simple lens configurations can be figured (shaped) specifically to overcome certain types of optical aberration depending on the application for which the lens is designed.

See also achromat, aplanat, apochromat, Barlow lens, Bertrand lens, Coddington lens, dioptre, focostat lens, field lens, focal reducer, Fresnel lens, jewel lens, lensball, meniscus lens, objective lens, periscopic lens, relay lens, Stanhope lens, teleconverter.

Anamorphic Lens: Anamorphosis is the effect achieved when an image is distorted so that it must be viewed in a specific manner, or through a special optical arrangement, in order to be recognizable. An anamorphic lens, [7], is a *sphero-cylindrical* (see below), or *toric* lens (see below), giving a different magnification in two principal meridians. This is used, for example, in the correction of astigmatism in vision, and in the cinematic projection of wide-screen movie pictures when the larger horizontal field of view is compacted in the film format.

Fluid lens: A lens assembly in which two lens elements are separated by a fluid of different refractive index.

Multiplying lens, multiplying glass: A lens, one side of which is plane and the other convex, in which the convex side comprises multiple plane facets angled to each-other. Thus, the image is multiplied.

Photographic lens: A lens specially designed for use in photography.

Polyoptrum: A polyoptrum, or *polyoptron*, is a lens, one side of which is plane and the other is formed of numerous concave facets. This gives a multiple view of the object, diminished in size.

Polyscope: A multiplying lens (see above).

Portrait lens: A compound photographic lens adapted to portraiture.

Rapid Rectilinear lens: A photographic lens consisting of four lens elements in two symmetrical combinations, designed to avoid distortion. Introduced by **J.H. Dallmeyer**.

Sphero-cylindrical Lens: A sphero-cylindrical lens is an anamorphic lens (see above), generally made by either forming a lens with one spherical surface and one cylindrical surface, or forming it with one spherical surface and one *toric* (see below) surface.

Stanhope lens: A lens with a focus at one of its surfaces.

Telephoto lens: A compound photographic lens, with a longer focal length than that of a simple lens of the same dimensions, giving large pictures of distant objects. The telephoto lens was patented and introduced in 1891 by Thomas Dallmeyer, the son of **J.H. Dallmeyer**, at the same time as, but independently of, its introduction by **Steinheil**.

Toric lens: A toric lens is an *anamorphic* lens, (see above), in which the lens surface is shaped as part of the surface of a torus [7]. This is used, for example, in ophthalmic lenses for the correction of astigmatism.

Wide angle lens: A photographic lens of short focal length giving a wide angle of view.

Zoom Lens: A zoom lens is one that offers continuously varying magnification, over a prescribed range, without the need to refocus.

Lensball

A lensball is a sphere made of optical glass which is used by photographers to create interesting optical effects in photography. The sphere is usually sized such that it may be held in the hand, although there are many ways to use it.

Lensman Microscope

The *Lensman* microscope was a small, hand-held disc-shaped folded optics microscope, for field work and education. The microscope was designed in 1989 by Keith Dunning and Richard Dickinson, with a novel three-dimensional folded optical design by Eddie Judd, and manufactured in the UK by **Science of Cambridge Ltd.** It was battery operated with a focus adjustment, and two magnification settings, 80x and 200x, with the higher magnification achieved by moving a Barlow lens into the optical path. It had two illumination settings, and a folding arm containing the bulb. An optional adaptor could be used to attach the microscope to a 35m camera. It was an evolution of the **McArthur** folded optics design, but with lower design specification and cost. Derivatives of the *Lensman* design include the *Micron* and the *Trekker*, both by **Enhelion**, the **Meade** *Readview* (produced for Meade by Enhelion), and the *Newton NM1* produced by **Cambridge Optronics Ltd**.

Lens Measure
See spherometer.

Lenticular Stereoscope
The lenticular stereoscope, invented by **Sir David Brewster**, is an improvement to **Sir Charles Wheatstone**'s reflecting stereoscope. Instead of mirrors it utilises lenses, and thus allows for a much smaller, hand-held instrument. *Lenticular* in this context means based on lenses. **George Lowdon** claimed to have been the first to make the lenticular stereoscope, but the design was popularised by the French instrument maker **Jules Duboscq**. Popular C20th lenticular stereoscopes included the *Tru-Vue*, the *Viewmaster*, and the *Holmes-Bates stereoscope*.

See Holmes-Bates stereoscope, mirror stereoscope, stereoscope, True-Vue, Viewmaster.

Lepetit, Albert
Albert Lepetit, (fl.1922-c1950) [322] [360], of 6 Rue Victor Considérant, Paris, France, was a nautical, mathematical and scientific instrument maker. In 1922 he succeeded **Ponthus & Therrode** in business, founding the Etablissements Albert Lepetit [**359** 1929 #9 p406]. He may have previously worked for them in the **A. Berthélemy** part of the business, and he appears to have traded under the name A. Berthélemy. In his catalogue of instruments for topography, geodesy, levelling, and mathematics, he states "A. Berthélemy. Etablissements Albert Lepetit. Anciennes Maisons A. Berthélemy, Ponthus & Therrode, P. Ponthus." In 1941 Leon Vindrier was made a partner, and the company was renamed A. Lepetit & Cie [**359** 1941 #70 p1667], and in c1950 they were succeeded in business by **Roger Poulin**.

Lerebours, Noël-Jean
Noël-Jean Lerebours, (1761-1840) [10] [24], was an instrument maker of Paris, France. He was supplied with flint glass by **Pierre Guinand**. He made refracting and reflecting telescopes, and his largest objective lenses, including a 15-inch f/8, were supplied to the Paris Observatory. He supplied achromatic telescopes with uncemented doublets to **Etienne Lenoir** for use in his geodetic and navigational instruments. He supplied Napoléon with an 11cm telescope. In 1828, on the instigation of Noël-Jean Lerebours, **Henry Guinand** founded the **Choisy-le-Roi glassworks**. Noël-Jean's son was **Noël Marie Paymal Lerebours**.

Lerebours, Noël Marie Paymal
Noël Marie Paymal Lerebours, (1807-1873) [10] [24] [70] [71], was the son of **Noël-Jean Lerebours**, and was himself an instrument maker. By 1839 he was constructing large daguerreotype cameras, producing large pictures. He established his business in Paris, France, making optical, physical, and mathematical instruments. He equipped a number of artists and writers with his daguerreotype outfits, and commissioned them to take views around the world. He also received daguerreotypes of views around the world from other daguerreotypists. He amassed a large collection of daguerreotypes, some of which he published in several volumes.

In 1840 he introduced a quarter plate camera designed especially for portraiture, fitted with single or double achromatic plano-convex lenses of very short focus. This camera could take a portrait in two minutes. In 1839, N.M.P. Lerebours advised his friend **Antoine Claudet** to seek out **Daguerre** in Paris, leading to the subsequent association between Claudet and Daguerre. In 1841 Lerebours opened a studio, and took numerous portraits, including nude studies.

He formed an association with **Marc François Louis Secrétan** in 1845, founding **Lerebours et Secretan**, which continued in business until 1955. In 1855 Lerebours retired, and Secrétan continued the business alone.

Lerebours & Secretan
Maison Lerebours & Secretan, [10] [24] [70] [71] [72] [153], was established in 1845 by **Noël Marie Paymal Lerebours** and **Marc François Louis Secrétan** in Paris, France. They made mathematical, optical and philosophical instruments, and had a strong reputation for their optical lenses. In 1855 Lerebours retired, and Secrétan continued the business alone. The eminent precision mechanical engineer **William Eichens** was a director of the company, and head of the workshop. In 1860 **Paul Gautier** came to work for Secrétan under the supervision of Eichens. In 1866 Eichens left the firm to set up his own company, and Gautier left to work with him. Marc Secrétan worked closely with **Foucault** to develop and market the reflecting telescope based on Foucault's paraboloidal silvered glass mirror [72]. After Secrétan's death his sons August and George continued the business, and by 1900 it was mostly a retail business. In 1911 the family sold the business to **Charles Épry**, who in 1913 associated with Gustave Jacquelin (1879-1939), and together they operated their business as the "sole successors" to Lerebours & Secretan. (Note that the é in Secrétan's name is unaccented on advertisements and

cards). In 1934 the firm of **Georges Prin**, the successor of Paul Gautier, who was himself the successor to William Eichens, became a part of the Lerebours & Secretan business. The Lerebours & Secretan company continued to operate until 1955.

Leupold
Leupold is a sports optics company based in Oregon, USA. They make rifle sights, spotting scopes, laser rangefinders, binoculars, thermal imagers, and accessories for the hunting market.

The company was established in 1907 by Fred Leupold and his brother-in-law Adam Volpel, trading as Leupold and Volpel, making surveying equipment. In 1942 the company name was changed to Leupold and Stevens. In 1985, in addition to their sports hunting products, they began to supply tactical optics to the US military. They sell their products under the brand-name Leupold.

Level
An instrument used to establish a horizontal line, or to adjust something with reference to a horizontal line.

See also Aldis optical level, automatic level, clinometer, Cooke's reversible level, Cowley level, Cushing's level, dumpy level, tilting level, transit level, Y Level.

Leverton, John
John Leverton, (fl.1761-1787) [3] [16] [18], was an optical, nautical, mathematical, and philosophical instrument maker of Liverpool, England. Addresses: *Hadley's Quadrant*, Water St., near the Exchange (1766); Pool Lane (1772-1773); 30 Pool Lane (1774); 31 Pool Lane (1777-1784).

He served his apprenticeship in London with the mathematical instrument maker William Parsons, beginning in 1749, and was freed of the **Goldsmiths' Company** in 1761. He is known to have sold: microscope; octant; barometer; hygrometer; thermometer. Instruments bearing his name include: octant c1777; Hadley's quadrants; Cuff-type microscope; stick barometer. He was succeeded in business by the mathematical instrument maker Susannah Leverton.

Levi & Co., Joseph
Joseph Levi & Co., [44] [61] [306], of London, England was a wholesaler of photographic equipment, and camera manufacturer. Addresses: 40 Castle Street (1858-1876); 2 Dyers Buildings, Holborn (1876-1895); 40 Furnival Street (1877-1895); 97 Hatton Garden (1890-1904). Their telegraphic address, *Leviathan*, was also shown in some of their advertisements, and was used as a model name for one of their cameras. Note, the 40 Castle St. address was also occupied in 1865 by **Joseph S. Levy & Co.**

The company was founded in 1858 by Joseph Levi. Although they traded and advertised separately from **S.J. Levi & Co.**, Joseph Levi is believed to have been the father of Samuel Joseph Levi.

In 1904 Joseph Levi & Co. merged with **George Houghton & Sons Ltd.** when they became Houghtons Ltd. Following this, Levi, Jones & Co. (Branch of Houghtons) advertised for at least three years, with the *Leviathan* telegraphic address (see S.J. Levi & Co.).

Levi & Co., S.J.
S.J. Levi & Co., [61] [306], were wholesale opticians and manufacturers of cameras, lanterns, cases, and accessories. The company was established by Samuel Joseph Levi on or before 1891, at 71 Farringdon Road, London, England, and supplied their products only to the trade. In 1892 A.J. Jones began to work for the company. In 1897 they became a public limited company, and changed the name to Levi, Jones & Co. In 1898 the company moved to 29 Hoxton Square, Old Street.

Although the company traded and advertised separately from **Joseph Levi & Co.**, Samuel Joseph Levi is believed to have been the son of Joseph Levi. The relationship between the two companies is further evidenced by the fact that for at least 3 years from 1905, following the merger of Joseph Levi & Co. with **George Houghton & Sons Ltd.**, the firm was listed as Levi, Jones & Co. (Branch of Houghtons).

Levi, Jones & Co. Ltd
See **S.J. Levi & Co.**

Levy & Co., Joseph S.
Joseph S. Levy & Co. were photographic apparatus manufacturers of London, England. The company started as Levy & Finister, trading at 56 Hounsditch (1863-1864), and went on to trade as Levy & Co., retaining the Hounsditch address. In 1865 they are also listed as trading at 40 Castle St., which was the same address as **Joseph Levi & Co.**

Levy & Finister
See **Joesph S. Levy & Co.**

Lichtenknecker Optics
Dieter Lichtenknecker, (1933-1990), was a German amateur astronomer and optical instrument maker. He learned his optical skills while working for **Askania Werke AG** in Berlin. He started his own company, Lichtenknecker Astro-Optik, in Weil der Stadt in 1959, producing telescopes. In 1973, the company moved to

Hasselt, Belguim, where together with J. Ruland he started the firm Lichtenknecker Optics. The company manufactured optical instruments for the amateur and professional astronomer. Their range of offerings grew with time, and they shifted their focus to sell mostly imported instruments, and produce specialised optical components. They sell telescopes, binoculars, spotting scopes, microscopes, stereoscopes, night vision optics, and accessories. Following the death of Dieter Lichtenknecker in 1990, the company was run by H. Ruland, a son of J. Ruland.

Liddell, John Josiah
John Josiah Liddell, (fl.1840-1858) [18] [271] [379], was a glass-worker, level maker, and optical, mathematical, and philosophical instrument maker of Edinburgh, Scotland. Addresses: 3 Hanover St. (1843-1857), and 91 South St. (1858). He exhibited at the Great Exhibition in 1851 in London [294] [328]. His catalogue entry no. 362 stated: "Liddell, John Josiah, Edinburgh—Designer and Manufacturer. Spirit levels of various kinds, used in draining, road—levelling, &c." He was succeeded in business by **James Buist & Co.**

Liddell, Robert & Co.
Robert Liddell & Co., (fl.1859-1876) [271], were level makers and optical instrument makers of Edinburgh, Scotland. Addresses: 46 Nicolson St. (1859-1860); 35A Hanover St. (1861); 35 South Bridge (1864); 1 Roxburgh Terrace (1865); Foot of Drummond St. (1866); Orchard Field, Leith Walk (1869); 10 Windmill St. (1873); 13 Leith Walk (1874); 16 Teviot Place (1876).

Liddle, William
William Liddle, (fl.1826-1833) [18] [271], was a mathematical and optical instrument maker of 3 North Bank St., Edinburgh, Scotland. He began to work in partnership with **Joseph Finnie** in 1826, trading as **Finnie & Liddle** until 1827, after which he operated the business in his own name until 1833. The dates given in [322] and [379], (1819 and 1823 respectively), suggest that he may have shared the premises of Joseph Finnie at 3 North Bank St. prior to the beginning of the partnership.

Lieberkühn, Johann N.
Johann Nathanael Lieberkühn, (1711-1756) [142], was a German physicist and physician. His research in the field of anatomy led him to the invention of a form of reflected light illumination for the microscope, known as a Lieberkühn mirror, for use when viewing opaque objects. In around 1740 he also invented the solar microscope.

Lieberkühn Mirror
The Lieberkühn mirror is a form of reflecting illumination for the microscope. It consists of a concave paraboloid mirror positioned above the stage. Historically this would have been a speculum mirror. It is used in conjunction with a low power objective, because a high-power objective would not allow sufficient room above the stage for the Lieberkühn mirror to be positioned correctly. The light used for specimen illumination is reflected up from under the stage by a conventional sub-stage mirror. The light which passes around the opaque object is reflected back onto it by the Lieberkühn mirror, providing top illumination.

The Lieberkühn mirror was invented by the German physicist and physician **Johann Lieberkühn**.

Lieberman and Gortz
Lieberman & Gortz, [61], was an English company set up in the 1940s selling binoculars, telescopes and microscopes via mail order. They sold binoculars made in France, Japan, England, and some from Hong-Kong. The first binoculars they sold were French made. In the early 1960s they sold models made in Japan and England. The English made models were made by **Ross Ltd.** of London as an economy version of Ross's own models. The Japanese models bear the quality control **JB code** denoting the maker, e.g. JB162.

Liebig, Justus von
Justus von Liebig, (1803-1873) [24 p262] [37] [45] [155], was an eminent German chemist, working in a broad range of fields, with a strong interest in chemical theory, and in organic chemistry and its applications in agriculture, pathology and physiology. He studied chemistry in Paris under Joseph Louis Gay-Lussac (1778-1850), and in 1825 he was appointed chair of chemistry at the University of Giessen, where he remained for over three decades, contributing significantly to the study of chemistry as an academic subject.

Liebig's contribution to optics came a few years after the 1851 Great Exhibition in London where Messrs. Varnish and Mellish showed their method, using silver nitrate and grape sugar, for silvering the insides of globes and vases. Liebig used a similar method to silver a mirror, with a mixture of silver nitrate, caustic potash and a small amount of ammonia. To this he added sugar as a reducing agent, and poured the solution over a suitably prepared glass substrate, resulting in the deposition of metallic silver on the glass surface, and yielding a mirror, which he then burnished. His method was first applied to astronomical mirrors in 1856 by **Carl August von Steinheil** and **Léon Foucault**.

Lietz Company, The
In 1882 the inventor, Adolph Lietz, founded The A. Lietz Company in San Francisco, California, USA, manufacturing surveying, nautical, and mining instruments. In 1916 they began to produce drafting equipment. They continued to produce instruments, building a strong reputation for quality, until 1947. They then stopped manufacturing instruments and began to operate as a distributor for other manufacturers, including Adolph Lietz patent instruments. Shortly after this the name was changed to The Lietz Company. In the 1950s they began to distribute **Sokkisha** products, and in the 1970s they became Sokkisha's exclusive US distributors. In 1984 The Lietz Company became a wholly owned subsidiary of Sokkisha.

Lilley & Gillie, John
John Lilly & Gillie are nautical instrument makers of North Shields, Tyne & Wear, England. The company originated as a merger in 1913 between **John Lilley & Son**, and **Wilson & Gillie**. The merged company was renamed John Lilley & Gillie in 1943.

Lilley & Son, Ltd., John
John Lilley & Son, Ltd. [16] [18] [44] [53] [54] [361], were English opticians, nautical and mathematical instrument makers, and compass makers. London addresses: 7 Jamaica Terrace, Limehouse (1846-1865); 9 London St., Fenchurch St. (1865-1885); 10 London St. (1890-1901). They also traded at The New Quay, North Shields, Tyne & Wear. The company started as a partnership between **John Lilley (I)** and **John Lilley (II)**.

According to their trade card dated 1865, "John Lilley & Son, 9, London Street, Fenchurch Street, London, E.C. Nautical & Mathematical instrument manufacturers to Her Majesty's Royal Navy. Inventors & Patentees of the new liquid steering compass." Another trade card states, "John Lilley & Son, Nautical and Mathematical instrument makers to Her Majesty's Royal Navy & the Hon[ble]. East India Company. Sole proprietors and manufacturers of Sir W. Snow Harris's steering compass, and Correctors of local attraction in iron ships by magnetic operations, and Tables of deviations computed. 7 Jamaica Terrace, Limehouse, London."

In 1913 the company merged with **Wilson & Gillie**.

It is not clear when the "Son" became "Sons", or whether this is a printing mistake, but in 1919 the company was listed as "John Lilley & Sons, Ltd., (Wilson & Gillie), Messrs., The New Quay, North Shields", as a member of the **British Optical Instrument Manufacturers' Association**.

In the 1929 British Industries Fair, the catalogue entry for John Lilley & Son, [448], listed them as a compass and binnacle maker, offering: Sounding machines; Nautical instruments and chronometers; Standard and steering compasses; Semaphores; Liquid compasses for trawlers and lifeboats, etc.; Morse lamps; Q.E.D. pelorus; Ships logs; Chart appliances; Clock cases; Pendulums; Sounding machines; Depth recorders. The New Quay, North Shields, and London. Telegrams: "*Gillie North Shields*". (Note, Q.E.D. refers to Q.E.D. Patents).

In 1943 the merged company was renamed **John Lilley & Gillie**.

Lilley, John (I)
John Lilley (I), (fl.1811-1845) [18] [383] [385], was an optician, and nautical and mathematical instrument maker of London, England. Addresses: Commercial Place, Commercial Rd.; Christopher St., Hatton Garden (1811); Globe St., Bethnal Green (1818-1821); 7 Jamaica Terrace, Limehouse (1823-1845); Jamaica Terrace, Commercial Rd., Near West India Docks (1826); 29 Beaumont Square, Mile End Rd. (1851) – although this last address may have been for his son, **John Lilley (II)**.

John Lilley (I) was bound as an apprentice to **Michael Dancer** in August 1801, but was re-bound as apprentice to James Archer Lilley of the **Merchant Taylors' Company** in October 1801. There are no official guild records for this as a turn-over, and it may have been for convenience. This would allow him to complete his apprenticeship under the tuition of Dancer, yet gain his freedom of the Merchant Taylors' Company, which was recorded in 1811.

In 1846 he entered a partnership with his son, John (II), and began trading as **John Lilley & Son**.

Lilley, John (II)
John Lilley (II), [18], mathematical and nautical instrument maker of London, England, was the son of **John Lilley (I)**. He was freed of the **Merchant Taylors' Company** by patrimony in 1837, and in 1846 entered a partnership with his father, trading as **John Lilley & Son**.

Limer
Limer is a Japanese brand of binoculars made by **Mirador**.

Lincoln, Charles
Charles Lincoln, (c1744-1807) [18] [274] [276 #624] [319] [322] [361], was an optician, and optical, nautical, philosophical, and mathematical instrument maker of London, England. Addresses: 11 Cornhill near the Poultry (1763); 38 Leadenhall St.; *Sir Isaac Newton's Head*,

62 Leadenhall Street (1772); 62 Leadenhall Street (1772-1805); 32 Leadenhall St. (1793).

He was the son of **Thomas Lincoln**, and was freed by patrimony of the **Spectacle Makers' Company** in 1762, and was Master of the Company from 1787 until 1789. He was also freed of the **Fletchers' Company** in 1765. Among his apprentices was **William Cox (II)**, who was turned over to him by **John Cuff** in 1771. **George Richardson** worked as foreman to Charles Lincoln prior to running his own business.

He is known to have made: telescopes; microscopes; nautical instruments; mathematical instruments; globes. Instruments bearing his name include: microscopes; telescopes; stick barometers; armillary sphere; horizontal sundial; pantograph; circumferentor.

Lincoln Jr., F.W. & Co.

F.W. Lincoln Jr. & Co., [368], were nautical and surveying instrument makers of Boston, MA, USA. The company was founded in 1839 by Frederick Walker Lincoln Jr. (1817-1898), who had served his apprenticeship with **Charles Gedney King**. In 1858, Lincoln's ex-apprentice, **Charles C. Hutchinson**, became a partner in the business, and in 1883 Hutchinson purchased the business, and continued to operate it under the name of C.C. Hutchinson.

Lincoln, Thomas

Thomas Lincoln, (fl.1720-1762) [18], was a spectacle maker of Little Britain, London, England (1737), and Parish of St. Bartholomew's the Less, London (1745-1750). He served his apprenticeship with the spectacle maker **Matthew Cooberow**, beginning in 1708, and was freed of the **Spectacle Makers' Company** in 1716. He served as Master of the Spectacle Makers' Company in 1746. He was the father of **Charles Lincoln**, and among his many apprentices were **John Cox (II)**, and **William Eastland (I)**.

Lindsay, George

George Lindsay, (fl.1728-1776 d.1776) [18] [256] [274] [276 #193] [319] [322], was a watch and clock maker, and optical instrument maker, of *the Dial*, facing Fountain Tavern, near St. Catherine Street, Strand, London, England. He was watchmaker to King George III and the Princess Dowager. He invented a portable reflecting simple microscope in 1728, which he patented in 1742 [35 #8206 p3]. The microscope folds into a wooden case, and was supplied with interchangeable objectives, one with a conical Lieberkühn mirror. He also offered a solar attachment. Instruments bearing his name include: portable microscope in silver; microscopes; telescope.

Linhof, Valentin

Valentin Linhof (1854-1929) [374], was a photographic equipment manufacturer of Munich, Germany. He started his company, Valentin Linhof OHG, in Munich in 1887, making circular leaf shutters for cameras. The firm went on to produce telescopes, goniometers, large format plate cameras, and medium format roll-film cameras, and introduced a range of large format cameras using the brand-name *Technika*. During WWII, they marked their equipment with the **ordnance code** hfo. With the rise in popularity of digital photography, they introduced a modular system whereby a camera could be used either with roll-film or for digital imaging.

Linnell, Elijah

Elijah Linnell, (fr.1725) [18], was a spectacle maker of London, England. He was freed of the **Spectacle Makers' Company** in 1725. He was father to **Joseph Linnell** and **George Linnell**.

Linnell, George

George Linnell, (fl.1769-1804) [18] [276 #625] [322], of 1 Goldsmith St., Gough Square, London, England, was an optician and optical instrument maker. He was the son of **Elijah Linnell**, and the brother of **Joseph Linnell**. He was freed of the **Spectacle Makers' Company** by patrimony in 1769. His apprentices included **Thomas Harris**, and **Charles Robert West**'s apprenticeship was turned over to him in 1790. In 1804 George Linnell is recorded as trading at 14 Crow Lane, London.

Linnell, Joseph

Joseph Linnell, (fl.1764-1775) [10] [18] [274] [276 #505] [322], was an optician and optical instrument maker of Ludgate St., London, England. He was the son of **Elijah Linnell**, and the brother of **George Linnell**. He served his apprenticeship with **Robert Linnell**, beginning in 1754. His apprenticeship was turned over to **Avis Sterrop** in 1756, then again to **Mary Sterrop** in 1760. He was freed of the **Spectacle Makers' Company** in 1763. In 1767 he succeeded **Martha Ayscough** in her business which was still trading in the name of her deceased husband, **James Ayscough**. Linnell's trading address was then *Great Golden Spectacles & Quadrant*, 33 Ludgate St., St. Paul's, London. He continued to make microscopes to the same design as Ayscough. In 1764 he was one of the petitioners who attempted to revoke the patent obtained by **John Dollond (I)** for the achromatic lens, and which was enforced after Dollond's death by his son, Peter (See Appendix X). He advertised as Joseph Linnell, Successor to the Late James Ayscough.

According to his trade card, undated, [53] [361], he made and sold: Spectacles and reading glasses;

Microscopes; Reflecting and refracting telescopes; Prisms and scioptic balls; Camera obscuras; Speculums; Magic lanterns; Opera glasses; Optical machines for viewing perspective prints; Barometers, thermometers, hygrometers, hydrometers and hydrostatical balances; Hadley's quadrants, globes, cases of drawing instruments, &c.; Sun-dials; and "all other sorts of Optical, Mathematical, or Philosophical instruments".

George Linnell was succeeded in business by **John Gilbert (III)**.

Linnell, Robert
Robert Linnell, (fl.1754-1774) [18] [322], was an optician and optical instrument maker of London, England. He began his apprenticeship in 1747, and was freed of the **Spectacle Makers' Company** in 1754. Among his apprentices was **Joseph Linnell**, beginning in 1754, and the apprenticeship was turned over to **Avis Sterrop** in 1756.

LINOS AG
LINOS AG was established in 1996, following a management buy-out of **Spindler & Hoyer**. They went on to acquire other optical instrument and component manufacturers including **Steeg & Reuter Präzisionsoptik** (1996), Franke Optik Vertriebsgesellschaft, Gsänger Optoelektronik GmbH, and **Rodenstock Präzisionsoptik** (2000). The company went public in 2000. In 2006, the **Qioptiq Group** began the process of acquiring LINOS AG, and in 2009 the acquisition was complete.

Lione & Co.
Lione & Co., (fl.1817-1822) [18], were optical, mathematical, and philosophical instrument makers of London, England. [379] gives an estimate that the business operated from 1800 until 1830. Addresses: 14 Brooke St. Holborn (1817-1822); 81 Holborn; Hatton Garden. The coincidence of address at 14 Brook St. suggests that this may have been the business of **Dominick Lione**.

Lione, Dominick
Dominick Lione, (fl.1805-1836) [18] [379], was a philosophical instrument maker of London, England. Addresses: 125 Holborn Hill (1805-1807); 14 Brooke St., Holborn (1811-1820); 16 Brooke St. (1821-1836). He worked in partnership as **Lione, Somalvico & Co.** until 1822, then continued to trade under his own name. He may have been associated with **Lione & Co.**

Lione, Somalvico & Co.
Lione, Somalvice & Co., (fl.1805-1819) [18] [379], were opticians, optical instrument makers, and barometer makers of London, England. Addresses: 125 Holborn Hill (1805-1807), 14 Brooke St., Holborn (1811-1819), and 16 Brooke St. (1820-1822). This was a partnership between **Dominick Lione** and **Joseph Somalvico**, and was sometimes known as Lione & Somalvico. They took over the business of Stampa & Co. They were succeeded by Dominick Lione at the 16 Brooke St. address.

Lippershey, Hans
Hans Lippershey, (c1570-c1619) [1] [37] [155] [388], was a Dutch spectacle maker who, in 1608, put two lenses at different distances from his eye, one concave and one convex, which gave him the unexpected effect of far-seeing. He put the words tele and scope together to make telescope. This may have been the first invention of the refracting telescope. Having made his first telescope in 1608 he began to manufacture "instruments for seeing at a distance." He petitioned the Dutch government for a patent but, due to competing claims, including one by **Zacharias Jannsen**, to the invention, his patent was denied. Hans Lippershey was also, in 1608, the first to construct a binocular telescope.

Lippmann, Gabriel
Gabriel Lippmann, (1845-1921) [37] [45] [70] [71 p668] [155], was a French physicist. He was notable for his contributions in the fields of electricity, optics and heat. He is best known for inventing interference colour photography, and for producing, in 1891, the first colour photograph. (See also photochrom). He also invented, in 1895, a form of siderostat called a coelostat.

Gabriel Lippmann was born in Luxembourg and, in 1878, after a period of study and research in Germany, he joined the Faculty of Science in Paris, France. In 1883 he was appointed professor of probability and mathematical physics at the Sorbonne. In 1886 he was appointed head of the Sorbonne's physics research group. In 1908 Lippmann was awarded the Nobel Prize for his work on interference colour photography [318].

Lister, Joseph Jackson
Joseph Jackson Lister, (1786-1869) [38] [45] [46] [82] [264], of Essex, England, was a wine merchant and microscopist. His nephew, Richard Low Beck, the father of **Richard Beck** and **Joseph Beck**, became his business partner in the wine trade. Microscopy was an amateur interest for Lister, who improved on methods of combining lenses to form a microscope objective, resulting in better image resolution due to the reduction of chromatic and spherical aberrations. He ground his own lenses, and taught his lens-grinding methods to optical instrument makers in London. In 1826 he took a novel microscope design to **Charles Tulley** to have it made. Tulley subcontracted the work to **James Smith (I)**, who

formed an association with Lister, learning his lens-making skills from him. In 1830 Lister had a paper published in the *Philosophical Transactions of the Royal Society of London*, in which he described his method of using two achromatic lens combinations, set at a given distance, to eliminate spherical aberrations in a microscope. This formed the basis of modern microscope lens design, and led to advances in the use of the microscope in scientific investigation. In 1837 he began a collaboration with **Andrew Ross** on microscope design. In 1840 they began trading in partnership as Andrew Ross and Co. Opticians. The partnership continued until 1842.

Littlewort, George
George Littlewort, (fl.1836-1843) [18] [53] [276 #2169] [379], was an optician, and optical, mathematical, and philosophical instrument maker of London, England. Addresses: 11 Ball Alley, Lombard St. (1839-1843); 14 Ball Alley, Lombard St. In 1841 he worked in partnership with **William Littlewort**, trading as **William & George Littlewort**.

Littlewort, William
William Littlewort, (fl.1821-1849) [18] [276 #1626] [379], was an optician, and optical and mathematical instrument maker of London, England. Addresses: 7 Upper East Smithfield (1821), and 7 Ball Alley, Lombard St. (1843-1849). He served his apprenticeship with the mathematical instrument maker William Glover of the **Masons' Company**, beginning in 1788. In 1841 he worked in partnership with **George Littlewort**, trading as **William & George Littlewort**. He was succeeded in business by Elizabeth Littlewort.

Littlewort, William & George
William & George Littlewort, (fl.1841) [18], were opticians, and optical, mathematical, and philosophical instrument makers of 11 Ball Alley, Lombard St., London, England. The firm was a partnership between **George Littlewort** and **William Littlewort**.

Littrow, Karl von
See **Otto von Littrow**.

Littrow, Otto von
Otto von Littrow, (1843-1864), was the son or Karl Ludwig von Littrow (1811-1877), and grandson of Joseph Johann von Littrow (1781-1840). Both his father and his grandfather held the position of director of the Vienna Observatory in Austria. Otto von Littrow was the inventor of the Littrow quartz spectrograph. He died of typhoid shortly after completing his doctorate at the age of 21. His father, Karl von Littrow devised a ghost micrometer for use in a meridional telescope [216].

Littrow Quartz Spectrograph
The Littrow quartz spectrograph was invented by **Otto von Littrow**. See Spectrograph.

Lizars, John
The optical and mathematical instrument maker John Lizars, (1810-1879) [44] [144] [263] [306], opened his first optometrists' practice in Glasgow, Scotland in 1830. Addresses: 12 Glassford St. (1830); 24 Glassford St. (1858-1874); 13 Wilson St. (1876); 16 Glassford St. (1877-1890); 260 Sauchiehall St. (1888-1890); 101-107 Buchanan St. (1891-1900).

The company went on to produce many optical instruments including telescopes, microscopes, magic lanterns and slides, cameras, stereo cameras and viewers, and binoculars. By the 1880s they were makers of chemical, mathematical, optical, and philosophical instruments, and were producing photographic apparatus and materials. In 1888 they were listed as "the only manufacturer of microscopes in Scotland". John Lizars' son-in-law, Matthew Ballantine, joined the company in 1882, and he and his descendants continued to operate the company. By 1910 the company focus was principally ophthalmic and photographic. By 1913 they had branch offices in Edinburgh, Paisley, Greenock, Aberdeen, Liverpool and Belfast. In 1999 the company merged with C. Jeffrey Black, and the resulting company, Black & Lizars continued in business as an optometrist.

Lloyd's Mirror Interferometer
See interferometer.

Loewy, Maurice
Maurice Loewy, (1833-1907), was an astronomer, and director of the Paris Observatory, France from 1896-1907. He was born in Vienna, Austria, and moved to France, subsequently naturalising as a French citizen. In 1871 he first proposed the equatorial coudé system, but it was not until 1882 that the first example was built by **Paul Gautier**, with sponsorship from the philanthropist Raphael Bischoffsheim.

Loft, Matthew
Matthew Loft, also known as Mathew, (1697-1747) [10] [18] [275 #562] [401], was an optical instrument maker of *The Golden Spectacles* behind the Royal Exchange, London, England. He served his apprenticeship with the instrument maker **Thomas Gay** beginning in 1711, and was freed of the **Spectacle Makers' Company** in 1720 or 1721. He was master of the Spectacle Makers' Company

from 1744-1745. He succeeded Gay in his business. He is known to have sold microscopes, telescopes, slide rules, and drawing instruments. His apprentices included **Edward Nairne** and **David Deane**.

Logan, James H.

James H. Logan, (fl.1869-1890), of Allegheny and Pittsburgh, Pennsylvania, USA, was a draughtsman who specialised in natural history illustrations. In 1869 he was awarded a patent for "improvements in single microscopes". According to the text of the patent [113], "In this improved microscope every part, except the lens-screw, clips, and reflecting-surface of the mirror is made of wood. The main features of the invention consist in the general construction and arrangement of the parts whereby it is possible to make them all of wood, without sacrificing strength and efficiency, together with a new and improved method of effecting the focal adjustment, and the peculiar adaptation of the microscope to the convenient and efficient use of globule-lenses"

He made microscopes to this design, one of which, according to [201], he donated to the Royal Microscopical Society. According to [264], it was presented to the RMS by James N. Logan. Two more examples are in the Billings Collection [369 p164].

LOMO

See **Leningrad Optical Mechanical Association**.

London Stereoscopic Co. Ltd.

The London Stereoscopic Company Ltd., [61] [122] [306], was founded in 1854 by George Swann Nottage (1822-1885), as the Artistic Repository & London Stereoscopic Company, located at 313 Oxford Street, London, England. The name was changed several times. By 1857 they were called London Stereoscopic Co., and had additionally taken on premises at 54 Cheapside. By 1863 they were trading as the London Stereoscopic and Photographic Co., and during the 1860s they relocated to Regent Street. They produced a large catalogue of stereo cards, for which they were renowned. The company was listed as "photographic apparatus makers". They sold cameras produced by other makers, including a *Photo-Jumelle* model by **Jules Carpentier**, and other French, German and British cameras, but it is not clear that they made optical instruments themselves. George Swann Nottage died in 1885, and in that year the company was incorporated as the London Stereoscopic and Photographic Company Ltd. In 1912 they began the process of voluntary liquidation, [35 #28640 p6480], and reformed in 1913 as the London Stereoscopic Company Ltd. They continued in business until 1922 when, due to liabilities, the company again began the process of voluntary liquidation, [35 #32753 p7083], and ceased trading. The company name has since been reused by a company selling proprietary stereo viewers, cards and books.

Long and Johnston

Long and Johnston, [18] [274] [276 #626] [322] [401], were opticians and optical, mathematical, and philosophical instrument makers, trading At the North Gate of the Royal Exchange, London, England. The firm was a partnership between **James Long** and a partner named Johnson, which began in 1785, and continued in business until 1807. Instruments bearing their name include: spyglasses. They are known to have sold telescopes.

Long, James

James Long, (c1755-1811) [18] [50] [53] [276 #626] [322] [401], was an optician, and optical and philosophical instrument maker, of London, England. Addresses: At the North Gate of the Royal Exchange; 4 Threadneedle St.; Royal Exchange (1781-1805); On the North Side of the Royal Exchange (1793); 4 Back of the Royal Exchange (1805). These are likely all different representations of the same address.

He served his apprenticeship with **Edward Nairne**, and was freed of the **Spectacle Makers' Company** in 1781, at which time he set up his own business. He was also freed by purchase of the **Vintners' Company** in 1788, and was Master of the Spectacle Makers' Company from 1805 until 1807.

In 1785 he entered a partnership, trading as **Long and Johnson**, and the partnership lasted until about 1810. He is known to have sold microscopes and telescopes. Instruments bearing his name include: shagreen spyglass. Upon his death in 1811 his premises were taken over by **Joseph Smith (II)**.

Long, Joseph (I)

Joseph Long (I), sometimes known as Josh Long, (c1797-1879) [18] [276 #1631] [319] [322] [361], was a hydrometer maker, and optical, mathematical, and philosophical instrument maker who, in 1821, began trading at 20 Little Tower St., London, England, the address previously occupied by hydrometer maker Thomas Gill, and the mathematical instrument maker, John Swan. He was freed of the **Fanmakers' Company** by purchase in 1824. He petitioned unsuccessfully for the vacant appointment to the Excise Office for the supply of hydrometers, following the closure of the firm of **R.B. Bate** [355]. In 1829 he published *Description & Use of a Slide Rule…as practised at the Port of London*. This described a slide-rule used by Customs & Excise. Instruments bearing his name include: microscope;

hydrometers; thermometers; stick barometer; slide rules; rule. See also **Joseph Long (II)**.

Joseph Long (I) died in 1879, at which time his address was still recorded as 20 Little Tower St. He was succeeded in business by his daughters, Mary Ann F. Long (b.c1835), and Amelia Long (b.c1836), both listed in 1881 as Hydrometer and Saccharometer Makers, employing 10 men and 2 boys. In 1885, the firm relocated to 43 Eastcheap, London, where it continued in business until 1936.

Long, Joseph (II)
Joseph Long (II), (fl.1827-1833) [18] [276 #1632] [319] [322] [379], was an optical, mathematical, and philosophical instrument maker of 136 Goswell St., London, England. He was appointed Maker to the Indian and Colonial Governments. Some sources suggest that he is the same person as **Joseph Long (I)**. However, directory entries have not been found to support this assertion, which appears to have originated with [276].

Longland, William
William Longland, (fl.1674-1695) [18] [54] [275 #321] [276 #87] [390] [401], was an optical instrument maker of *The Ship* in Cornhill, London, England. He served his apprenticeship with **Joseph Howe**, beginning in 1667, and was freed of the **Spectacle Makers' Company** in 1674. He served as Master of the Spectacle Makers' Company from 1686-1687 and 1694-1695. Instruments bearing his name include: six-draw parchment telescope with lignum vitae fittings; six-draw telescope made of vellum with lignum vitae fittings.

Lonsdale Bros.
Lonsdale Bros., [61] [122] [306], of Leeds, England, and later, London, were wholesale and retail manufacturers of cameras and photographic equipment, as well as components and kits for those wishing to construct their own camera. Addresses:3 Cookridge St., Leeds (1892); 40 New Briggate, Leeds (1893-1894); Grove Works, Claypit Lane, Leeds (1895); Leeds and Liverpool (1896); 22 Goswell Road, Aldersgate St., London (from 1896).

The company was established in 1891 by the brothers Adolphus and Henry Lonsdale, and used the additional trade-name of Sun Camera Co. Ltd. They published the *Cyclopedia of Photo Brasswork & Camera Furniture*. In 1896 the company relocated to London. In 1899 Adolphus left the company, and Henry continued the business alone. In 1906 it was taken over by James Christie of Sheffield.

Lorenz Scientific Ltd.
Lorenz Scientific Ltd., of Ontario, Canada, were manufacturers of "professional, astronomical and science instrumentation". A 1973 advertisement shows a modified Ritchey-Chrétien observatory telescope available from 16 inches "as large as you may require", on a fully programmable equatorial floor-mount.

Lorieux, E.
E. Lorieux père, and E. Lorieux, (fl.c1825-c1880) [322] [360], were nautical instrument makers of 6 Rue Victor Considérant, Paris, France. They were (presumably) father and son, signing their instruments respectively E. Lorieux, père, and E. Lorieux. Instruments bearing their name include: marine telescope in brass and leather; reflecting circle; sextant, signed *E. Lorieux, père à Paris*; single draw marine telescope brass and leather; cased sextant signed *E. Lorieux à Paris*.

In 1845 E. Lorieux began to work in partnership with Schwartz, trading as **Schwartz dit Lenoir**. In 1855 he began to work in a new partnership, trading as **Baudry et Lorieux** for about a year, after which he traded under his own name. In 1880 the firm of E. Lorieux was acquired by A. Hurlimann [359 1880 #44 p720].

Loriners' Company
See Appendix IX: The London Guilds.

Lorkin, Thomas
Thomas Lorkin, (fl.1789-1834) [18] [54] [90], was an optical and mathematical instrument maker, stationer, and ship chandler of London, England. Addresses: 42 Great Alie St., Goodman's Fields; Sweedland Court, Tower Hill (1789); 89 Near New Crane Stairs, Wapping (1805-1816); 89 New Crane, Wapping (1816); 89 Wapping, Hermitage Bridge (1816-1817); 89 New Gravel Lane, Wapping (1822); 89 New Crane, Shadwell (1828).

He served his apprenticeship with the mathematical instrument maker and ship chandler, John Bostock Goater, beginning in 1779, and was freed of the **Shipwrights' Company** in 1789.

Among his apprentices was the mathematical instrument maker Thomas Dench, who was turned over to him in 1789 by **Joseph Rust (I)**. He is known to have sold: octant; compass. He also advertised: telescope; quadrant; sextant; hourglass.

Loupe Magnifier
A loupe magnifier is a magnifying lens, often high powered, held close to the eye, such as those used by jewellers.

Lowdon, George
George Lowdon, (1825-1912) [18] [53] [263] [306], was a scientific and optical instrument maker, of Dundee, Scotland. Addresses: 25 Union Street (1850-1861);

Railway Buildings, South Union Street (1 Union Street) (1864-1874); 23 Nethergate (1876-1880); 65 Reform Street (1882-1900). He is not listed in any directories after 1904.

His father, George Lowdon senior, was a grocer, but had attained some skill in grinding lenses. George junior began his instrument making career in 1849 with his own workshop, employing one person. Lowdon claimed that in 1849 he was the first to make **Sir David Brewster**'s lenticular stereoscope, although the French instrument maker **Jules Duboscq** was the one to popularise it.

Lowdon also made microscopes. He made a large microscope, 4 feet tall, for the London Great Exhibition in 1851, but it was not exhibited because it was not delivered on time. He also made refracting telescopes, theodolites, cameras, sympiesometers, and other meteorological instruments, and apparatus for electricity and magnetism.

Lowe, John
John Lowe, (fl.1788-1798) [3] [18], was an optical and mathematical instrument maker, stationer, and bookseller of Birmingham, England. Addresses: 14 High St. (1788-1793); Cherry Street (1797); High Street (1798). He was also a retailer of optical and mathematical instruments, stamps, and London newspapers, and had a library of more than 3,000 books.

Lucernal Microscope
See microscope.

Lucking & Co., James
James Lucking & Co., (fl.c1870-1890) [319] [379], were English opticians, and optical and philosophical instrument makers/retailers, with branches at: 5 & 7 Corporation St., Birmingham; 37 Bond Street, Leeds (c1880); 14 Broad St., Worcester; Leicester; Walsall. They traded variously as: J. Lucking; Lucking & Co.; James Lucking & Co. (Opticians).

Their instruments are variously marked *Birmingham and Leeds*, *Birmingham and Leicester*, and *Birmingham Leicester Worcester and Walsall*. Instruments bearing their name include: barometers; black finished brass theodolite; three-draw brass telescope; brass telescope on wooden tripod; opera glass; barographs; level.

Ludolph, W
W. Ludolph, [374], was established in 1846 in Bremerhaven, Germany, developing products for the nautical and aeronautical sector. This includes sextants, compasses, and other equipment. They exhibited their work on Hannover in 1878.

Lumière
The Lumière company, [53] [70] [71] [335], was founded in 1893 by Charles Antoine Lumière (1840-1911), (known as Antoine), as a photographic plate manufacturer in Lyon, France. He had two sons, Auguste Marie Nicholas Lumière (1862-1954) and Louis Jean Lumière (1864-1948).

Having seen a Kinetoscope peepshow in 1894 Antoine proposed to his sons that they design a method of motion picture reproduction that would project the moving image onto a screen. They produced a design for a machine that performed the three combined functions of camera, projector, and printer, which they called a Cinematograph, or *Cinématographe Lumière*. The brothers carried out the development in their own workshops, with lamp-houses constructed by **Alfred Molteni**. They first demonstrated the machine to the public in 1895, and met with considerable success. They demonstrated it in London, England on the same day in 1896 that **Robert Paul** demonstrated his Theatograph projector. They then ordered 25 Cinematographs to be made by the engineer **Jules Carpentier**, and soon ordered from him a further 200.

The brothers went on to become ground-breaking film-makers, and developed the *Autochrome* colour transparency system. From the 1920s they began to produce cameras. Louis developed an interest in 3D photography and stereo films. Auguste, although he continued to help Louis with his work in films, set up a medical research laboratory, and was the first person in France to set up a working X-Ray machine.

The Lumière company also produced binoculars, and continued to produce cameras until 1961. In 1963 Lumière was taken over by Swiss company Ciba, and in 1982 it became **Ilford France**. The name Lumière was also used in the 1970s by an unrelated company.

Lummer-Brodhun Cube
A Lummer-Brodhun cube, otherwise known as a photometer cube, is a device for comparing the intensity of two light sources. It consists of two isosceles right-angled prisms, joined in optical contact by their hypotenuse faces. A small area in the centre of the face of one cube has had its surface removed so that no contact is made in that area of the join. One light source is transmitted directly through the device to an eyepiece, so that the centre of the beam is reflected out of the field of view by total internal reflection, and an annulus of light reaches the eyepiece. The second beam is transmitted into the device at right-angles to the first, so that the centre of the beam is directed into the field of view by total internal reflection. Thus, in the eyepiece, the centre of the image

(second light source) may be compared with the outside of the image (first light source).

Lummer-Brodhun cubes are used in optical instruments such as the photometer, colorimeter, and nephelometer.

Lummer-Brodhun Photometer
See photometer.

Lummer and Gehrcke Interferometer
See interferometer.

Lundin, Carl
Carl A.R. Lundin, (fl.1874-1915 d.1915) [213], was a Swedish optician who emigrated to the USA in 1873 to work for **Alvan Clark and Sons**, in Cambridgeport, Massachusetts, beginning in 1874. He rose to become a senior optician at the firm, being responsible for the construction of some of their major telescope installations. His son, Robert Lundin, also worked for Alvan Clark and Sons.

Lundin, Robert
C.A. Robert Lundin, who went by the name Robert, (1880-1962) [185] [213], was the son of **Carl Lundin** and, like his father, he worked for **Alvan Clark and Sons** in Cambridgeport, Massachusetts. Robert learned lens-making from his father when he was a teenager, and joined the firm in 1896. In 1915, upon the death of his father, Robert became head of the optical department at Alvan Clark. In 1929 the company was contracted to build a 10-inch equatorial refractor for the Elgar Weaver Observatory at Wittenberg College. Lundin had already collaborated with **Roland Sellew** previously, and did so for this instrument. Before the project was completed, Lundin left to start his own company, C.A. Robert Lundin and Associates. Together with Sellew he completed the instrument for the Elgar Weaver Observatory. He advertised making refracting and reflecting telescopes, visual and astrophotographic objectives, plane, spherical and paraboloidal mirrors, prisms, eyepieces, optical test plates, and advertised for "special optical work". He built several observatory refractors, and his company continued until a lack of orders due to the depression put an end to the business in 1933. In 1933 Lundin moved to Ohio to lead the new optical shop of **Warner and Swasey**. He retired in the mid-1940s. Lundin's work received both positive and negative criticism.

Lundy, J.F.
J.F. Lundy were makers of marine navigation instruments, based in Grimsby, England. Sextants bearing their name are dated C19[th].

Lunt
Lunt is a **Bresser** company. Andrew Lunt is CEO/Owner of Lunt Solar Systems and Lunt Engineering.

Lunt Solar Systems, located in Tucson, Arizona, designs, fabricates, assembles, and tests solar telescopes and solar filters.

Lunt Engineering, also located in Tucson, Arizona, specialises in the production of telescope optics, software and mechanics for industrial applications, professional astronomy installations, and space organizations (satellite optics, spectrographs etc.).

Lutz, Edouard
Edouard Lutz, (fl.c1870-1900) [53] [320], of 49 Rue des Noyers, Paris, France, was an optical and scientific instrument maker. Examples of his work include: microscopes; Nörremberg polariscope; Nicol prisms and other polarizing prisms; direct vision prisms; liquid prisms; Newton's rings demonstration apparatus. From about 1872 he traded at 49 Boulevard St. Germain, and from about 1880 at 82 Boulevard St. Germain.

Lyle, James
See **Gardner & Lyle**.

Lynch, James
James Lynch (I), (fl.1767-1772) [18] [34] [215] [218] [276 #743] [304], of Dublin, Ireland, was an instrument maker, but little is known of him except from an air pump, gauge and receivers in Trinity College, Dublin, dated 1767 and bearing his name. He may have worked in partnership with **Edward Spicer**, as evidenced by advertisements for a compass theodolite, and "all other instruments for surveyors, navigators, &c.", stating "as sold by Edward Spicer and James Lynch".

James (I)'s son James Lynch (II) (fl.1784-1833 d.1833), of the *Royal Spectacles*, Capel Street, and 26 Capel Street, Dublin, was an inventor and instrument maker, including astronomical and surveying instruments, and a lecturer. An undated instrument case label ascribed to James (II) [53] reads: "Made and sold by James Lynch, Mathematical, Philosophical and Optical instrument maker at the *Royal Spectacles*, Capel Street, Dublin." He is known to have traded at Capel Street from 1784 to 1807. James (II) lectured at the Royal Dublin Society for much of his career, and was a professor, lecturing in natural philosophy and astronomy at Trinity College, Dublin until about 1833. He also supplied instruments to Trinity College. James (II) worked with his son James (III) (fl.1826-1839), trading as James Lynch & Son. James (III) was an optician, and Mathematical and Philosophical Instrument Maker to the Ordnance and Trinity College. The company went on to trade as Lynch & Co., and James

Lynch & Co., and is known to have continued in business until 1846.

Lytkarino Optical Glass Factory (LZOS)

LZOS, [181] [257], is the Lytkarino Optical Glass Factory (the Lytkarinskiy Zavod Opticheskogo Stekla, Лыткаринский завод Оптического Стекла), situated north of Moscow. It is a satellite factory of **KMZ**, manufacturing optical glass and lenses for KMZ cameras. As a state-controlled company, it produced night vision devices and glass laser optics for the military market, as well as stereo microscopes, telescopes, binoculars, lenses, fibre optic devices, and other items manufactured from optical glass. For Russian and Soviet makers' marks, see Appendix V.

M

Mach-Zehnder Interferometer
See interferometer.

Mackenzie & Co.
See **Alexander Mackenzie**.

Mackenzie, Alexander
Alexander Mackenzie, (fl.1816-1858) [**18**] [**53**] [**276** #1381] [**322**] [**361**] [**383**], was an optical, mathematical, and philosophical instrument maker of 15 Cheapside, near St. Paul's Churchyard, London, England. He is also recorded at 6 Providence Place, London.

He was freed by purchase of the **Tinplate Workers' Company** in 1815. According to his trade card, "A. Mackenzie. Optical, Mathematical, and Philosophical instrument maker, No. 15 Cheapside, Near St. Pauls Churchyard, London". The engraving on the trade card shows images of many instruments, including: spectacles; telescopes; microscopes; sextant; level. During the period 1816-1818, he worked in partnership with **Joseph Swygart**, trading as **Mackenzie & Swygart**. He also published a trade card, c1817, under the name Mackenzie & Co., which may have been an alternative name for Mackenzie & Swygart.

Instruments bearing his name include: microscopes; sextant; telescope. Christies auction house (1992) listed a Cary-type botanical microscope signed Mackenzie 75 Cheapside London. He is also known to have sold: magnetic compass.

Mackenzie & Swygart
Mackenzie & Swygart, otherwise known as Mackenzie & Suggart, (fl.1816-1818) [**18**] [**276** #1381], optician and mathematical instrument maker, was a partnership between **Alexander Mackenzie** and **Joseph Swygart**, trading at 15 Cheapside, London, England. Instruments bearing their name include: sextant.

Macrae, Henry
Henry Macrae, (c1807-1889) [**18**], was an optician, and mathematical, optical, and philosophical instrument maker of London, England. Addresses: 82 Bishopsgate Within; 191 Whitechapel Road (1832); 34 Aldgate (1839-1875) 29 Royal Exchange (1865-1880); 34 Aldgate High St.

He served his apprenticeship with the mathematical instrument maker Isaac Rennoldson, and was freed of the **Merchant Taylors' Company** in 1830. By 1865 he had additionally taken on 29 Royal Exchange, the premises previously occupied by **Stephen Bithray**, and subsequently advertised this as well as the Aldgate address. He continued in business until 1884. Examples of his work include sextants and microscopes.

Magic Lantern
A magic lantern, [**109**], is an image projector, consisting of a case enclosing a light, and having suitable lenses to project an enlarged image from a transparency onto a screen. It may have two projecting lenses to allow the preparation of one view while projecting another view. The magic lantern is sometimes used in conjunction with devices that provide cross-fading and other optical effects.

Magnetograph
A magnetograph is a recording magnetometer, and is used to record, by photography or other means, the intensity or direction of a magnetic field, such as the Earth's magnetic field.

Magnusson, O.
O. Magnusson of Arvada, Colorado, USA, was a supplier of parts and accessories for telescopes and mounts. Their catalogue from the 1960s shows: Heavy duty pedestal and tripod German equatorial mount; Clock drives; Setting circles; 6-inch reflecting telescope; Focusing eyepiece holder; Mirror cell; Mirror making laps; Counterweights.

Mailhat, R.
R. Mailhat, [**197**], was a former pupil of **Paul Gautier**, and director and buyer for **Secrétan**. He was a maker of scientific, geodetic, and astronomical instruments. He operated his business in Paris, France: Addresses: 30 Rue du Faubourg Saint-Jacques, 41 Boulevard Saint-Jacques, and subsequently at 10 Rue Émile-Dubois. He is known to have built instruments for the Observatorie de Paris, La Faculte Des Sciences, the Bureau Des Longitudes, and the Observatoire De Juvisy. He was awarded a medal at l'Exposition Universelle in 1900, and a prize at the Brussels exhibition in 1910.

His catalogue, *Mécanique & d'Optique pour les Sciences et l'Industrie*, published sometime between 1910 and 1914, shows astronomical telescope objectives, mirrors, prisms, oculars, and astronomical and terrestrial telescopes. It includes alt-az and equatorially mounted refracting telescopes ranging from 61mm to 240mm objective, with or without mounts. Also reflecting telescopes up to 300mm. Mounts are portable, semi-portable, or observatory.

Maison Bardou
See **Bardou**.

Maison Chevallier

In 1842 **Pierre Queslin** took over the business of **Pierre Chevallier**, renting the premises at 1 Rue de la Bourse, Paris, France, and making cameras, microscopes, telescopes and other optical instruments. He ran the business under the name Maison Chevallier and, following his death, his daughter **Adélaïde Queslin** continued to run the business with her husband **Alfred Baserga**. In about 1898/9, A. Fontana acquired the business, and continued to use the name Maison Chevallier [50].

Maison de l'Ingénieur Chevallier

Upon the death of **Jean Gabriel Augustin Chevallier** in 1848, his daughter **Marie Chevallier** and her husband **Alexandre Ducray-Chevallier** continued to run J.G.A. Chevallier's business in Paris, France, under the new name Maison de l'Ingénieur Chevallier. Alexandre died in 1879 and, in 1883, Marie sold the business to the brothers **Charles and René Avizard**. They continued to run the business, retaining the name Maison de l'Ingénieur Chevallier, until 1900.

The brothers moved the busines to 21 Rue Royale in 1900, at which time they changed the name to Maison de l'Ingénieur Chevallier, Avizard Frères, Successeurs. Sometime between 1914 and 1921 the address changed again to 27 Avenue de l'Opéra. They continued in business until at least 1922 [50].

Maison Lerebours & Secretan

See **Lerebours & Secretan**.

Makers of Playing Cards, Worshipful Company of

See Appendix IX: The London Guilds.

Maksutov Telescope

See telescope.

Maksutov-Cassegrain Telescope

See telescope.

Mamiya

Mamiya Digital Imaging Co., of Tokyo, Japan, is a camera and lens manufacturer, founded in 1940 as Mamiya Koki Seisakusho, by Seichi Mamiya and Tsunejiro Sugawara. Their first product was the *Mamiya-6*, a medium format folding camera that used 120 format film. Following WWII they expanded their offerings, with their first 35mm camera being introduced in 1949. They continued to introduce new models until the 1980s when they refocussed their business to concentrate on medium format cameras. In 2006 they restructured the company to become Maimya Digital Imaging Co., and in 2009 **Phase One** acquired a controlling share.

Manent, Maurice

Maurice Manent was a French precision instrument maker who started his business in Rue de Parc 44, La Croix-de-Berny, Seine, France, in 1911. He produced fine telescopes and accessories for the amateur market. Several advertisements showing his work may be seen in [10]. Examples include alt-az and equatorially mounted refracting telescopes, including 75mm with focal length (f.l.) 1.05m; 81mm with f.l. 1.25m; 95mm with f.l. 1.45m; 110mm with f.l. 1.6m; 135mm with f.l. 1.9m. He also produced a *refractor reflector* in sizes from 95mm to 160mm, as well as image erecting prisms for terrestrial observing, eyepiece turrets, eyepieces and other accessories.

Mangin, Colonel Alphonse

Colonel Alphonse Mangin, (1825-1885), of the French army, was an engineer who invented the Mangin mirror, used in medial and brachymedial telescopes, as well as in lighthouse reflectors and searchlights.

Mangin Mirror

The Mangin mirror, invented in 1876 by **Col. Alphonse Mangin**, is a negative meniscus lens, with a reflective coating on the side with the larger radius (shallower curve). The incident light passes through the glass to be reflected from the mirrored surface. The lens corrects the spherical aberration normally produced by a spherical mirror. The Mangin mirror is used in medial and brachymedial telescopes, as well as in lighthouse reflectors and searchlights.

Manhattan Optical Company

The Manhattan Optical Company, [122], was founded in Creskill, New Jersey, USA, in 1892 as a manufacturer of cameras. An advertisement dated 1897 shows that by then they were a New York company, with their factory and executive offices in Creskill, New Jersey. In 1902 the company was taken over by **Gundlach Optical Co.**, and became Gundlach-Manhattan Optical Co.

Mann & Ayscough, James & James

James Mann & James Ayscough, [18], was a partnership between **James Mann (II)** and his ex-apprentice **James Ayscough**. They traded at the sign of *Sir Isaac Newton and two pair of Golden Spectacles*, near the West-End of St. Paul's, London, England from 1743 until 1747. Their trade card of 1747, [53], described improvements to the compound microscope. They are also known to have sold telescopes. After 1747 each partner continued to trade under his own name.

Mann, James (I)

James Mann (I), (c1658-c1718) [10] [18] [53] [275 #480] [401], was an optician and optical instrument maker of London, England. Addresses: *Archimedes and Two Pair of Golden Spectacles*, near St. Paul's, London; St Martins le Grand (1693); Angel St. (1695); *The Spectacles* in Butcher Hall Lane, near Christchurch (1697); Parish of Christ Church (Newgate) (1699); Fleet St. (1718).

He served his apprenticeship with the instrument maker Thomas King (fl.1667, d.c1677), beginning in 1674, and was turned over to John Hickett. He was freed of the **Spectacle Makers' Company** in 1682, and was Master of the Company from 1716-1717.

He is known to have sold spectacles, telescopes, microscopes, camera obscuras, magic lanterns, magnifying glasses, mirrors, opera glasses, prisms, stereoscopes, barometers, and thermometers. His son, **James Mann (II)**, continued his business in 1718.

Mann, James (II)

James Mann (II), (c1690-1757) [10] [18] [53] [276 #90] [319], was an optician and optical instrument maker of London, England. He succeeded his father, **James Mann (I)**, in business in 1718, initially trading at the sign of *Archimedes and Two Pair of Golden Spectacles*, near St. Paul's. He subsequently traded at the sign of *Sir Isaac Newton and Two Pair of Golden Spectacles*, near St. Paul's.

He served his apprenticeship with the spectacle-maker **Matthew Cooberow**. He was freed of the **Turners' Company** in 1706, and freed of the **Spectacle Makers' Company** in 1707, in which he served as Master from 1735-1737. Among his apprentices were **John Cuff**, beginning in 1722, **Samuel Scatliff**, beginning in 1725, and **Samuel Johnson**, beginning in 1738.

His trade card showed that he made and sold wholesale and retail: Spectacles; Refracting telescopes; Newtonian and Gregorian reflecting telescopes; Double and single microscopes; Prisms for demonstrating the "surprising theory of light and colours"; Camera obscuras; Magic lanterns; Opera glasses; Magnifying glasses; Barometers; Thermometers; and other instruments.

James Ayscough served an apprenticeship with James Mann (II) from 1732 until c1739. They worked in partnership as **James Mann & James Ayscough** from 1743 until 1747. Their trade card while working together was remarkably similar to that of James (II) when trading alone. James (II)'s son, **John Mann**, succeeded him in business.

Mann, John

John Mann, [10] [18], of London, England, was the son of **James Mann (II)**, whom he succeeded in business. He served his apprenticeship with Edward Waller, beginning in 1714, and was freed of the **Spectacle Makers' Company** in 1723 or 1724. He was succeeded in business by **Samuel Johnson**.

Manning, Charles

Charles Manning, (fl.1763) [18] [389], was an optical instrument maker of Wapping Wall, London, England. He is believed to have served his apprenticeship with the mathematical instrument maker, Nathaniel Smith, of the **Clockmakers' Company**, beginning in 1720.

Mantois, Edouard

Edouard Mantois, (1848-1900) [24], was the son-in-law of **Charles Feil**. He learned his optical glass making skills from Charles, and took over the glassworks in **Choisy-le-Roi**, near Paris, France, in 1887. By this time Numa Parra was already in a senior role in the company and, in about 1900, (the year Edouard Mantois died), the firm was renamed **Parra-Mantois**.

Marey, Étienne-Jules

Étienne-Jules Marey, (1830-1904) [45] [335], was a French physiologist. In order to facilitate his research into the movement of birds in flight he invented the chronophotographic gun which could take a sequence of 12 images per second superimposed on a single frame. The superimposed sequence of images thus produced gives a vivid sense of motion in a single image. This technique is referred to as chronophotography.

Margas, John

John Margas, sometimes Marges, Margass, Margus, (fl.1743-1767 d.1767) [18] [304] [319] [322], was an optician, and optical, mathematical, and philosophical instrument maker. He served his apprenticeship with **Nathaniel Adams**, beginning in 1735, and was turned over to **John Cox (II)** in 1741 or 1742. He was freed of the **Spectacle Makers' Company** in 1742. He began his career in London, England, trading at Charing Cross until 1745, then at *Golden Spectacles*, Long Acre, Near James St. (1745), and Rose St., near Long Acre (1753-1758). In 1758 he was declared bankrupt [35 #9780 p4], following which he moved to Dublin, Ireland, where he traded at Capel St. until at least 1767. **John Alment** worked as his foreman before setting up his own business. Among John Margas's apprentices were **James Jameson**, beginning in 1753, and **Henry Edgeworth**, beginning in 1760.

According to land surveyor John Mooney in 1759, in reference to a tellurian made by John Margas, "I have made by Mr Margas in Dublin a curious Instrument, which truly represents all the Motions of the Earth round the Sun, and by which the Longitude is illustrated to a Demonstration; and has been lately proved in that Manner by many Gentlemen of Knowledge and Learning as well in Dublin as in the County".

Instruments he is known to have sold include: microscope; magnetic compass; sundial. Instruments advertised for sale include: tellurian.

Marie, J-B
J-B Marie, (fl.1736-1760) **[256] [322] [416]**, was an optician and optical instrument maker of Paris, France. He is known to have made microscopes. He married the daughter of **Claude Paris**, and his wife succeeded him in business, trading as **veuve Marie** (widow Marie).

Marie, Veuve
Veuve Marie (widow Marie), (fl.1760) **[256] [322] [416]**, was an optician of Quai de l'Horloge, Paris, France. She was the daughter of **Claude Paris**, and married to **J-B Marie**. Following the death of her husband, she continued the business, trading as veuve Marie. She was succeeded in business at this address by **Etienne Antoine Putois**.

Marindin Rangefinder
See rangefinder.

Marine Instruments
In 1941 A collaboration between **Kelvin, Bottomley and Baird** and **Henry Hughes and Son** was established as Marine Instruments **[44]**. In 1947 the two companies amalgamated as **Kelvin and Hughes**. In 1949, the name of Marine Instruments was changed to **Kelvin and Hughes (Marine)**.

Marion & Co.
Marion & Co., **[61] [306]**, of Soho Square, London, England, was set up in about 1842 as an offshoot of the French stationery company of Auguste Marion. By the 1850s they were trading in Regent Street, and began to sell photographic items. In 1863 they moved to Soho Square. From about 1880 the company was selling photographic equipment including their own brand of cameras, although at that time they probably did not manufacture any themselves. The Marion Reflex camera, for example, was (from about 1903) made by **Kershaw & Sons Ltd**. In 1885 they opened a dry plate and photographic paper factory in Middlesex. By 1892 they were advertising cameras "made or sold by Marion & Co.", but it is not clear when they began to manufacture cameras, or which of the models they sold were made in their own factories. In 1901 the company was incorporated as Marion & Co. Ltd. In 1921 it became a part of **APM**, and subsequently **APEM**.

Marion & Foulger
Marion & Foulger, **[44]**, of Magna Works, Kathie Road, London Road, Bedford (1931), and Dents Road, Kathie Road, Bedford (1932), England, was a maker of photographic mounts, frames and mouldings, and a camera manufacturer. In 1921 they became part of **APM**, **[61]**, and they were wound up in 1932, **[35 #33796 p871]**.

Marratt & Ellis
Marratt & Ellis, **[18] [110] [319]**, of 63 King William Street, London, England, were opticians and spectacle makers. The company was a partnership between **John Symonds Marratt** and George Everest Ellis (c1841-1913). The partnership was terminated in 1879 **[35 #24850 p3287]**.

The firm continued to trade with the same name, and appears in the 1914 *Whitaker's Red Book of Commerce or Who's Who in Business*, still operating at the same address under the proprietorship of George Everest Ellis. This information was presumably submitted for listing prior to Ellis' death in March 1913. Instruments bearing their name include: opera glasses; barometers.

Marratt & Short
Marratt & Short, **[18] [319] [379]**, was a partnership between **John Symonds Marratt** and the scientific instrument maker **Thomas Watling Short**. The partnership was formed in about 1859, trading at 63 King William Street, London, England, as opticians and mathematical instrument makers, and was terminated in 1867, **[35 #23231 p1811]**.

Instruments advertised include: opera glasses; magic lanterns and slides. Instruments bearing their name include: opera glasses; marine telescope; compound microscope; binocular microscope; chondrometers; barometers.

Marratt, John Symonds
John Symonds Marratt, (c1809-1888) **[18] [44] [276 #1636] [319]**, was a spectacle-maker, and optical, mathematical, and philosophical instrument maker, of London, England. He served his apprenticeship with the London optician John Byard beginning in 1822, and was freed of the **Blacksmiths' Company** in 1829.

Addresses: 23 Meredith Street, Clerkenwell; 1 New Street, Dockheath (1829-1830); 54 Shoe Lane (1833); 15 Great Winchester Street (1839-1844); 63 King William

Street, London Bridge (1845-1851); 63 King William Street (1870-1875).

He exhibited at the Great Exhibition in London in 1851. His listing in the exhibition catalogue, [328], describes his display as: "Five-feet achromatic telescope, the vertical and horizontal motion produced by endless screws. Seven-inch transit theodolite, reading to 15" in altitude and azimuth, furnished with inverting and diagonal eye pieces, needle box, tripod staff, and locking plate, axis level, etc. It can be used with or without tripod, as may be required; and is adapted for surveying, tunnelling, magnetic, or astronomical purposes."

In about 1859 he formed a partnership with **Thomas Watling Short**, trading as **Marratt & Short**, opticians and mathematical instrument makers. The partnership was terminated in 1867.

J.S. Marratt's business, trading under his own name, appeared in the directories until 1875. He entered a partnership with George Everest Ellis, trading as **Marratt & Ellis**, opticians and spectacle makers, until 1879 when, presumably, he retired. The 1881 census listed him as retired, and he was succeeded in business by Marratt & Ellis.

Marriott, William

William Marriott, (fl.1827-1845) [18] [276 #1639], was an optician, telescope maker, mathematical and philosophical instrument maker, brass tube maker, and optical turner, of London, England. Addresses: 40 New Montague St., Brick Lane; 30 Pelham St., Spitafields; 10 King St., Bethnal Green; 38 New Montague St., Brick Lane (1839-1845). He was succeeded in business by Mrs. A. Marriott.

An optician named William Marriott is also recorded at 12 Market Street, Sheffield, in 1822 [3].

Mars

Mars is a brand-name found on French folding binoculars. They are usually marked *Brevetée France & Étranger*, or the abbreviation B^{tee} Fr & Etr, (Patented in France and overseas).

An example is a pair of folding opera glasses bearing the name **Fisher**, 188 Strand, and the markings *Jumelle "Mars" Brevetée France Étranger. Model Militaire*. They were supplied by Fisher, and are believed to have been manufactured under the Mars brand in the late-C19th.

Marsh, Rev. D.B.

Rev. Daniel Brand Marsh, (1859-1933) [309], was a Canadian minister of the Presbyterian Church. From an early age he pursued his interest in building mechanical devices. He became a member of the Royal Astronomical Society of Canada in 1901, and began to build refracting telescopes. He used objective lenses made by **Brashear** and **Watson-Conrady**, but built all of the mechanical parts himself.

Marshall, John

John Marshall, (bap.1659, d.1723) [18] [46] [53] [256] [264] [274] [275 #435] [396] [442], of London, England, was a renowned optical instrument maker. Addresses: *Two Golden Prospects*, Ludgate St.; *Prospects*, in Ludgate, the second Spectacle shop within the Gate on the left hand; *Three Keys*, Ivy Lane (1688); Gun, Ludgate St. (1688); *Archimedes & Spectacles* in Ludgate St., opposite the West End of St. Paul's (1690-1694); Ludgate St. (1695); *Archimedes and Golden Spectacles* in Ludgate St. (1701); *Archimedes & Two Golden Spectacles* in Ludgate St., near St. Paul's Churchyard (1714); *Old Archimedes & Spectacles* in Ludgate St., being the second Spectacle-shop from Ludgate.

He served his apprenticeship with **John Dunnell**, a telescope tube maker, beginning in 1673, being freed of the **Turners' Company** in 1685. Prior to starting his own business, he worked for **Robert Hooke**, and for John Flamsteed, the astronomer who founded the Greenwich Observatory and was the first Astronomer Royal of England.

Marshall was one of the two leading microscope makers in London at the time, and his business was next door to that of his rival, **John Yarwell**. According to his trade card: "John Marshall, Maker of Optic glasses to his Majesty, at *the Archimedes & the Golden Spectacles* in Ludgate Street. Sells all sorts of Double & single telescopes, Microscopes, Prospective & weather glasses, Reading and burning glasses, Spectacles etc. He was the inventor of true spectacle grinding, & the only person who had the approbation of the Royal Society."

His method of grinding lenses, which became widely adopted, involved cementing a number of pieces of glass to the surface of a large, convex spherical block, and working them simultaneously with a concave spherical tool. According to the astronomer and mathematician Edmond Halley, (1656-1742), in a letter written in 1693: "I have, by Order of the Royal Society seen and examined the method used by Mr John Marshall, for grinding glasses, and find that he performs the said work with greater ease and certainty, than hitherto has been practised, by means of an invention, which I take to be his own, and new, and whereby he is enabled to make a great number of optick-glasses, at one time, and all exactly alike, which having been reported to the Royal Society, they were pleased to approve thereof, as an Invention of great use, and highly to deserve encouragement."

Marshall invented the *double microscope*, generally referred to as the Marshall microscope. In this design the microscope tube is attached to a pillar in such a way that its height can be fine-adjusted by a screw mechanism. The pillar is attached to the weighted wooden base by means of a ball and socket joint, allowing the microscope to be tilted, or turned around to overhang the base. The Marshall microscope has interchangeable objectives.

George Willdey worked for a while for John Marshall as a journeyman. Among Marshall's apprentices were **Francis Hauksbee**, beginning in 1703, and **John Smith (I)**, beginning in 1709. John Smith (I) subsequently married Marshall's daughter Catherine (bap.1689). In 1715 Marshall was awarded a Royal Appointment to King George I. Upon Marshall's death in 1723, John Smith (I) inherited his tools and succeeded him in his business.

Martin, Adolphe Alexandre

Adolphe A. Martin, (1824-1896) [10] [24] [70] [72], was a physics teacher at the Collège Sainte Barbe, Paris, France, and inventor of the ferrotype (tintype) process and other methods of producing positive photographs. He put his inventions in the public domain, thus making them widely available, and denying himself the personal wealth his inventions could have brought.

He became a pupil of **Foucault**, and invented a silvering technique for mirrors using nitrate of ammonia. He was instructed by Foucault in the art of mirror testing, and Foucault had to dissuade him from revealing his newly acquired trade secrets to other makers – something Martin proceeded to do after Foucault's death in 1868. He is known to have figured mirrors for **Secrétan** while in the employ of Foucault. Martin collaborated with **William Eichens** in building telescopes, and figuring and supplying the mirrors.

During 1872-1873 Martin worked on telescopes, daguerreotype photography, and siderostat mirrors for the forthcoming transit of Venus expeditions.

Martin, Benjamin

Benjamin Martin, (1705-1782), [18] [46] [53] [264] [276], was a British maker of optical, philosophical and mathematical instruments. In his late-20s he established a school in Surrey, England, and wrote extensively on natural sciences, mathematics, optics and astronomy. He developed a strong interest in optics, and in 1738 he designed a portable microscope known as a drum microscope. The microscopes he made himself at this stage of his career were mainly of wood and cardboard, and did not compete with those made by professional microscope makers.

Martin continued to teach and write and, in 1756, having purchased the freedom of the **Goldsmiths' Company**, he set up a shop as an optician and instrument maker in Fleet Street, London. Most of his trade in optical instruments was retail, rather than instruments he had made himself. He acquired globe plates, and began a successful line in globe-making. Among his apprentices was **David Jones**. His son **Joshua Lover Martin** also served an apprenticeship with him beginning in 1758, and they went into partnership in 1778. **Gabriel Wright** worked for him from 1764 until 1782. In 1764 Benjamin Martin was one of the petitioners who attempted to revoke the patent obtained by **John Dollond (I)** for the achromatic lens, and which was enforced after Dollond's death by his son, Peter (See Appendix X). In 1782 he was declared bankrupt, [35 #12260 p6], and he died later that year.

Martin, James

James Martin, (fl.1784-1794 d.by 1800) [18] [90], was a mathematical instrument maker of Fetter Lane, London, England. Addresses: 25 Great New St. (1786-1787); Great New St. (1792); 23 Dean St. (1794). He was the son of Peter Martin, master of HM Ropewalks, Portsmouth. He served his apprenticeship with **John Troughton (II)**, beginning in 1768, and was freed of the **Grocers' Company** in 1784. Among his apprentices were **William Parsons (II)**, beginning in 1789, and **Thomas Hemsley (I)**, who began his apprenticeship in 1784, and was turned over to **Henry Hemsley (I)** in 1789.

Martin, Joshua Lover

Joshua Lover Martin, [2] [18] [53] [388] [396] [406] [407], was the son of **Benjamin Martin**, and served an apprenticeship with his father beginning in 1758. He was also a friend of **Jesse Ramsden**. In 1774 he was granted a patent for an improvement to **John Hadley**'s quadrant. In 1778 he entered a partnership with his father, and in 1782 he applied for a patent for a "newly-invented art of drawing tubes plated with silver or gold…for the construction of telescopes, perspectives, opera glasses…and other optical and philosophical instruments". While this patent referred to plated tubes, brass telescope tubes were almost certainly used by instrument makers earlier than this. J.L. Martin also made a telescope tube-drawing machine to his own design. Due to his father's bankruptcy in 1782, J.L. Martin had to sell the business and tools. These were acquired by **Charles Tulley**.

Martz Jr., Edwin P.

Edwin P. Martz Jr., (1916-c1966) [185], was an American optical engineer and astronomer with a keen interest in planetary astronomy and photography. He made his first telescope at an early age. He worked at the Harvard

Observatory in Jamaica, the Lowell Observatory in Flagstaff, the Steward Observatory in Tucson, the Mount Wilson Observatory and the Griffith Observatory in California, and the Dearborn Observatory of Northwestern University in Evanston, Illinois.

During WWII he worked as an optical engineer for the US Army Air Corps in Ohio, developing missile tracking systems. In 1946 he continued his work on missile tracking systems, working as an optical engineer for the Naval Ordnance Test Station in California. In 1950 he took a job as an optical physicist for the Land-Air Corporation at the White Sands Integrated Range in New Mexico, then he went on to work for the U.S. Air Force Missile Development Center, at Holloman Air Force Base, New Mexico.

In 1957 he built a telescope with a 12-inch **Alvan Clark** objective. In 1960 he was appointed Supervisor of the Space Optics Group at the Jet Propulsion Laboratory in Pasadena, California, where he worked on optical imaging systems for space missions. The two Voyager missions, launched after his death, carried video imaging equipment designed by Martz. His 12-inch telescope is now at the San Diego State University.

Mason & Co.

Mason & Co., (fl.1876) [304], of 11 Essex Bridge, Dublin, Ireland, were Opticians to the Lord Lieutenant. This was the firm of **Mason Seacombe (II)** and partner(s), and succeed **Mason & Son** in business. When the partnership was terminated, the business was continued by Mason Seacombe (II).

Mason & Son

Mason & Son, (fl.1865-1875) [304], of 11 Essex Bridge, Dublin, Ireland, were Opticians to the Lord Lieutenant. This is believed to have been a partnership between **Seacombe Mason (II)** and one or both of his sons, **Thomas Mason (II)**, and **Seacome Mason (III)**. Upon the termination of this partnership, the business continued as **Mason & Co.**

Mason & Sons Ltd., Thomas H.

Thomas H. Mason & Sons, Ltd., [304], of Dublin, Ireland, began as a partnership between **Thomas Holmes Mason** and his three sons, Barry, Alex, and **Standish Mason (II)**. The firm was incorporated as a limited company in 1935, initially run by Thomas H. Mason. Following his death Standish (II) took over, and was himself succeeded by his own son, Standish (III). The company was eventually renamed to become Mason Technology.

Mason, John

John Mason, (fl.1760-1797) [18], was a mathematical and optical instrument maker, brass turner, and brass founder of London, England. Addresses: Saffron Hill (1760); Shoe Lane, Fleet Street (1768-1797). He served his apprenticeship with Edward Warner, beginning in 1753, and was freed of the **Drapers' Company** in 1760. Among his apprentices were **William Bruce**, and **George Willson**, who was turned over to him in 1797 by James Moulding.

Mason, Jonathan

Jonathan Mason, (1784-1849) [18] [304], was an optician of Dublin, Ireland. Addresses: 8 Arran Quay (1809); 9 Ormond Quay (1810-1812); 14 Capel St. (1818-1822). He was a son of **Seacombe Mason (I),** and worked in partnership with his brother, **Thomas Mason (I)**, trading as **T.&J. Mason** from 1805 until 1808, and again from 1813 until 1817. While not working in this partnership, he traded under his own name. Sometime after 1822, having married his second wife, he moved to her home-town of Limerick, where he operated his business until his death in 1849, trading at 6 Patrick St. (1846).

Mason, R.G.

Robert George Mason, (1849-1912) [50] [109] [264] [270] [319] [320], was a manufacturing optician of London, England. Addresses: 28 Clapham Park Rd. (1881-1882); 38 Clapham Park Rd. (1882-1883); 24 Clapham Park Rd. (1883-c1890); 69 Clapham Park Rd. (c1890-c1909). He was principally known for making microscopes and prepared microscope slides. Census records list his father, also named Robert Mason, as a gun polisher. In 1871 R.G. Mason, aged 21, is listed as an optician living in Enfield. He began his career in optics working for **James Swift** until he started his own business in c1879. He advertised as "having had many years experience in the manufacture of microscopes and other scientific instruments for the trade". He devised an ingenious projecting microscope, referred to as "Mason's Patent Lantern & Table Microscopes", which sold for £4 17s. 6d., and for which he was awarded a silver medal at the Royal Cornwall Polytechnic Exhibition in 1884. He was self-employed, with no employees. He also advertised the supply of microscope parts for customers who may have wished to build their own, and produced prepared microscope slides. Instruments bearing his name include: projection microscope; compound microscope in case; combination dissecting and monocular microscope; prepared microscope slides.

Mason, Seacombe (I)

Seacombe Mason (I), sometimes known as Seacom or Seacome, (1745-1804) [18] [304], of 8 Arran Quay, Dublin, Ireland, was the founder of a dynasty of opticians, and scientific and optical instrument makers. He was a grandson of a leather worker, who moved to Dublin from Seacome, near Liverpool, England.

According to [18], Seacombe Mason (I) served his apprenticeship with **John Margas**, beginning in 1760, and was freed of the Glovers' & Skinners' guild of Dublin in 1765 – although some sources indicate that he was apprenticed to **John Alment**. He was Master of the Glovers' & Skinners' guild in 1788. Among his apprentices was **Samuel Yeates**. He set up his own company in about 1780, selling optical instruments, including spectacles, telescopes, microscopes, and opera glasses, as well as globes, electrical machines, and surveying and drawing instruments. He was succeeded in business in 1805 by his sons, **Thomas Mason (I)** and **Johatnan Mason**, trading in partnership as **T.&J. Mason**.

Mason, Seacombe (II)

Seacombe Mason (II), sometimes known as Seacome, (1808-1892) [18] [304], was an optician and mathematical instrument maker of Dublin, Ireland. Addresses: 6 Essex Bridge (1838-1844); 11 Essex Bridge (1845-1877); 11 Upper Ormond Quay (1878). He was the son of **Thomas Mason (I)**, and served as Optician to His Excellency the Lord Lieutenant. He succeeded his father in business, and worked in two partnerships, trading as **Mason & Son** from 1865 until 1875, and in 1876 as **Mason & Co.** His sons were **Seacome Mason (III)** and **Thomas Mason (II)**. During his tenure as head of the family firm, the business suffered, and was rescued when it was purchased in 1887 by Thomas Mason (II) upon his return from London.

Mason, Seacome (III)

Seacome Mason (III), (fl.1877-1879) [304], was an optician of Dublin, Ireland. Addresses: 11 Nassau St. (1877); 24 Wellington Quay (1879). He was a son of **Seacomb Mason (II)**, and worked in his father's business. He served as State Optician in 1877. He may have been a partner in **Mason & Son**, and in **Mason & Co.**

Mason, Standish (I)

Standish Mason (I), (fl.1839-1841) [18] [304], was an optician and mathematical instrument maker of 2 Upper Ormand Quay, Dublin, Ireland.

Mason, Standish (II)

Standish Mason, (d.1969) [304], was the son of **Thomas Holmes Mason**. He succeeded his father in the family business as head of **Thomas H. Mason & Sons Ltd.**, When he died, the optics and photographic sides of the business were closed, and the firm continued to produce laboratory supplies. He was succeeded by his own son, Standish Mason (III).

Mason, T.&J.

Thomas & Jonathan Mason, (fl.1805-1838) [18] [304], opticians of Dublin, Ireland, was a partnership between the brothers, **Thomas Mason (I)** and **Jonathan Mason**, sons of **Seacombe Mason (I)**. The partnership succeeded Seacombe Mason (I) in business, and operated over two periods: 1805 until 1808, trading at 8 Arran Quay, and 1813 until 1817, trading at 3 Essex Bridge. The brothers worked separately under their own names when not working in this partnership.

Mason, Thomas (I)

Thomas Mason (I), (1781-1837) [18] [304], was an optician, and mathematical and philosophical instrument maker of Dublin, Ireland. Addresses: 4 Essex Bridge (1809 and 1820); 3 Essex Bridge (1810-1812, 1818-1819, 1821-1826); 63 Essex Bridge (1827-1837). He was a son of **Seacombe Mason (I)**, and worked in partnership with his brother, **Jonathan Mason**, trading as **T.&J. Mason** from 1805 until 1808, and again from 1813 until 1817. While not working in this partnership, he traded under his own name. He served as Optician to His Excellency the Lord Lieutenant in 1837. He was succeeded in business by his son, **Seacombe Mason (II)**.

Mason, Thomas (II)

Thomas Mason (II), (1840-1913) [304], was an optician and optical instrument maker of Dublin, Ireland. Addresses: 9 Nassau St. (1866-1868); 11 Essex Bridge (1878-1883); 21 Parliament St. (1884-1895); 4 Dame St. (1896-1899); 5 Dame St. (1900-1913). He was a son of **Seacomb Mason (II)**. He is believed to have worked in partnership with his father, trading as **Mason & Son**, until he left to go to London. There, he served an apprenticeship with **Negretti & Zambra**, before returning to Dublin to purchase the struggling family business in 1887. Instruments bearing his name include a 2¾-inch brass refractor on a folding steel tripod, signed *Thos Mason Dublin*, and a 3-inch refractor on a folding steel tripod signed *Thomas Mason, 5 Dame St., Dublin* [215]. Thomas was a co-founder of the Irish Optical Society.

According to his trade card, "Estd. A.D. 1780. Spectacles made on scientific principles and adapted to every condition of sight. ... Thomas Mason, Optician, 11, Essex Bridge, Dublin, Mathematical & Philosophical Instrument Maker to His Excellency the Lord Lieutenant and the Irish Court."

Upon his death in 1913, he was succeeded in business by his son, **Thomas Holmes Mason**.

Mason, Thomas Holmes

Thomas Holmes Mason, (1877-1958) [304], was an optician, and optical, mathematical, and philosophical instrument maker of Dublin, Ireland. Addresses: 5 Dame St. (1913-1922); 6 Dame St. (1917-1922). He was the son of **Thomas Mason (II)**. He worked for his father's firm for many years, and took control of the business upon his father's death in 1913. The firm continued to expand and, by the 1920s, the optics department sold scientific instruments, surveying instruments, and laboratory equipment, as well as providing ophthalmic services. There was also a photographic department which was of particular interest to Thomas H. Mason, who was a keen photographer. He had three sons, Barry, Alex, and **Standish Mason (II)**, all of whom worked with him. He incorporated the company in 1935 as **Thomas H. Mason & Sons Ltd.**, and was succeeded in business by Standish Mason (II).

According to his trade card [53], "Established 1780. Stereoscopes, Opera glasses, Spectacles, Microscopes, Telescopes, Compasses, Ivory and boxwood scales, Rules &c. Thomas H Mason, Optician, Mathematical and Philosophical instrument maker to His Excellency the Lord Lieutenant and the Irish Court. 5 & 6 Dame Street (Near the Castle), late 21 Parliament Street. Cinematographs, Optical lanterns and slides for sale or hire. Spectacles made on scientific principles, and adapted to every condition of sight. … Photographic apparatus and materials."

Masons' Company

See Appendix IX: The London Guilds.

Mathematisch-mechanisches Institut von Reichenbach & Ertel

Mathematisch-mechanisches Institut von Reichenbach & Ertel, [374] [422], optical instrument makers of Munich, Germany, was a partnership between **Georg Friedrich von Reichenbach** and **Traugott Leberecht von Ertel**. In 1814, Reichenbach left the firm of **Utzschneider, Reichenbach & Fraunhofer in Benediktbeuern**, and a year later, in 1815, Ertel also left to join Reichenbach's new company, which was then named Mathematisch-mechanisches Institut von Reichenbach & Ertel. Among their apprentices was **Moritz Meyerstein**. In 1821 Reichenbach handed over the running of the company to Ertel who, in 1834, with his son Georg Ertel (1813-1863), re-named it **Ertel & Sohn**.

Mathematisch-mechanisches Institut von Reichenbach & Liebherr

Mathematisch-mechanisches Institut von Reichenbach & Liebherr, (fl.1802-1804) [161] [422], were mathematical and optical instrument makers of Munich, Germany. The firm, founded in 1802, was a partnership between **Georg Friedrich von Reichenbach** and Joseph Liebherr (1767-1840), a mechanic and watchmaker. In 1804, **Joseph von Utzschneider** joined the company and it was renamed **Mathematisch-mechanisches Institut von Utzschneider, Reichenbach & Liebherr**.

Mathematisch-mechanisches Institut von Utzschneider, Reichenbach & Liebherr

Mathematisch-mechanisches Institut von Utzschneider, Reichenbach & Liebherr, (fl.1804-1820) [24] [161] [422], were mathematical and optical instrument makers of Munich and, from 1806, Benediktbeuern, Germany. The firm was established in 1804 when **Joseph von Utzschneider**, joined the **Mathematisch-mechanisches Institut von Reichenbach & Liebherr**. His partners in this venture were **Georg Reichenbach**, and Joseph Liebherr (1767-1840), a mechanic and watchmaker. The objective of the Institute was to produce high quality surveying instruments for military map-making. Joseph Niggl, a skilled optician, was recruited to manufacture lenses but, to enable this, the venture required a supply of good quality optical glass.

In 1805 Utzschneider met **Pierre Guinand** and, having seen his work, and offered him a place in the Institute's new glassworks, which was set up in Benediktbeuern, about 60km south of Munich, in 1806. In 1806, the optical production was also moved from Munich to Benediktbeuern, with the warehouse and shop remaining in Munich. As a part of his terms of employment, Guinand was required to train a member of Utzschneider's staff in glassmaking. The student was **Joseph Ritter von Fraunhofer**, then only twenty years old. In 1806 **Traugott Leberecht von Ertel** began to work for the Institute. Joseph Niggl, having instructed Fraunhofer in lens grinding and polishing, left the firm in 1807.

In 1808 **Georg Merz** began to work under Fraunhofer. Then only fifteen years old, Merz helped to fire the furnaces for the glass foundry. Fraunhofer developed improved lens manufacturing and glass-making techniques, and in 1809 he was placed in charge of the glassworks, and made a partner in the Institute, which was renamed **Optisches Institut von Utzschneider, Reichenbach & Fraunhofer in Benediktbeuern**.

Mathers, Thomas
Thomas Mathers, (fl.1848) [18] [271], was a mathematical and optical instrument maker of 88 Glassford Street, Glasgow, Scotland. He succeeded **Andrew Dodd** at this address.

Matthews, Thomas B.
Thomas B. Matthews, (fl.1843-1849) [18] [319] [322], was an optician and optical instrument maker of 9 Athol Place, Pentonville, London, England. Microscopes bearing his name occasionally appear at auction.

Matthews, William (I)
William Matthews (I), (1815-1868) [53] [264] [319] [320], of 8 Portugal Street, Lincolns Inn Fields, London, England, was a surgical instrument, cutlery, and microscope maker. He was freed by patrimony of the **Goldsmiths' Company**. In his early career he traded at 11 Portugal Street, then 10 Portugal Street, and moved to number 8 in c1853. He is believed to have started making microscopes in c1855, "furnished with foreign object-glasses". An example of his work is in the History of Science Museum, Oxford [270]. Upon William's death, the firm was taken over by his sons Walter and Henry, trading as Matthews Brothers until sometime in the 1870s. They sold microscopes bearing their name, but these may have been purchased from wholesalers.

Some sources suggest that William Matthews (I) and **William Matthews (II)** are the same person. However, [50] casts doubt on this theory.

Matthews, William (II)
William Matthews (II), (fl.c1850-c1860) [319] [320], of Camden Road, London, England. Instruments bearing this name and address are of a unique design of case-mounted microscope, an example of which is in the Science Museum, London [53] and another in a private collection [113].

Some sources suggest that **William Matthews (I)** and William Matthews (II) are the same person. However, [50] casts doubt on this theory.

Mawson & Swan
Mawson & Swan, [44] [320], of 13-15 Moseley Street, Newcastle-on-Tyne, England, were suppliers of scientific apparatus, including microscopes and accessories, as well as chemical and photographic supplies.

The company was founded in about 1839 by John Mawson as a chemist and druggist. In 1846 **Joseph Swan** joined the company. He was made a partner in 1864, and the firm began to trade as Mawson & Swan. In 1867 John Mawson died in a chemical explosion, after which Joseph Swan took over management of the company. Many of Joseph Swan's innovations were produced and marketed by Mawson & Swan. These included his carbon process for photography, patented in 1864, his method of producing photographic dry plates, devised in 1877, and his gelatine bromide paper, patented in 1879. In 1880 **J.B. Payne** took over management of the Newcastle plant.

Following Joseph Swan's demonstration in 1880 of a commercially viable incandescent lamp, he collaborated with the American inventor **Thomas Edison** to commercialise it. By 1887 Mawson & Swan were advertising "Incandescence Lamps as Accessories to the Microscope".

Maxwell, James Clerk
James Clerk Maxwell, (1831-1879) [7] [37] [46] [155], of Edinburgh, Scotland, was the son of a wealthy landowner. Following his education at Edinburgh University, then at Cambridge University, he pursued his interests as a natural philosopher and, in 1856, succeeded his father as laird of the family estate in Kirkcudbrightshire.

From 1856-1860 he was professor of natural philosophy at Marshal College, Aberdeen. Following this he was professor of natural philosophy and astronomy at King's College, London, until 1865 when he returned home to manage his estate. In 1871 he took up a post as professor of experimental physics at Cambridge University, where he founded the Cavendish Laboratory for research in physics. He was renowned for developing the theory of electromagnetism, as well as advancing the kinetic theory of gases, and theory of colour.

He studied the ring system of Saturn, and concluded that it consisted of concentric rings of satellites rather than being either solid or fluid. His study of the motion of these satellites prompted his interest in the kinetic theory of gasses of which he is considered a founder.

Maxwell undertook a study of the science of electricity and magnetism. He concluded that light is an electromagnetic phenomenon, and derived a set of mathematical equations to represent the phenomenon. He was able to solve them to determine the speed of light – a result that corresponded well with the experimentally measured value. Maxwell's equations, and his electromagnetic theory of light, were major contributions to scientific theory.

His research into colour vision and optics led to his formulation of a new approach to the theory of optical instruments, also later developed independently by **Ernst Abbe**. He developed a unified theory of colorimetry, based on **Thomas Young**'s trichromatic colour theory, and projected the first trichromatic colour photograph, an image of a tartan ribbon, during a lecture at the Royal Institution in London, in May 1861.

Mayflower

Mayflower was a classic brand of Japanese telescopes, binoculars, riflescopes, and microscopes during the 1950s, '60s and '70s. The name appears to have been a trade name used by the **Colonial Optical Company** of Los Angeles, CA., USA. See their 1962 catalogue [243]. Examples seen bearing makers' marks APL and cursive HOC, and the maker's mark of **Yamamoto Seisakuyso**, (see Appendix IV).

McAll, John

John McAll, otherwise known as M'Call, (fl.1826-1850) [18] [276 #1932], was an optician, and mathematical and philosophical instrument maker of 8 Postern Row, Tower Hill, London, England. In 1826 he began to work in partnership with **James Gilkerson**, trading as **Gilkerson & McAll**, successors to **James Gilkerson & Co**. The partnership continued until 1830, after which John McAll continued in business at the same address. According to [322], he also worked in partnership with **Alfred Chislett**, c1850, trading as **Chislett & McAll**.

McAllister & Brother

McAllister & Brother, [201] [290], of Philadelphia, USA, was a partnership between microscope maker **Thomas H. McAllister**, and his brother William. The partnership was formed in 1853, and continued until 1865.

McAllister, Thomas H.

Thomas H. McAllister, (1824-1898) [201] [290], was a microscope maker of New York, NY, USA.

His grandfather, John McAllister Sr. (1753-1830) emigrated to the USA from Scotland in 1775, and set up a business making whips and canes. John Sr. entered the optical instrument making business in 1796 when he began to manufacture and retail optical instruments at his address in Chestnut Street, Philadelphia. John's son, John Jr. (1786-1877), continued the business. **James W. Queen** began to work for McAllister in about 1825. In 1835 John McAllister Jr.'s son William (1812-1896) took over the business, and soon after this, Queen became a partner. In 1852 Queen left the partnership, and in 1853 he set up his own competing business, also in Philadelphia. William McAllister formed a partnership with his brother Thomas H. in 1853, trading as McAllister & Brother. Here, Thomas began to establish a reputation as a microscope maker.

The McAllister brothers continued in partnership until Thomas moved to New York in 1865 to set up his own business making microscopes at 627 Broadway. In 1866 he moved to 49 Nassau Street, NY. Thomas sold his instruments himself, and through dealers, including James W. Queen & Co. It is believed that Thomas H. McAllister did not make the optical components for his microscopes, but it is not known who did. They were supplied with objectives by an un-named maker, with optional objectives by **J.&W. Grunow**, **William Wales**, and **Joseph Zentmayer**. Thomas' son, C.W. McAllister, joined the company sometime in the 1880s, and they continued in business until the 1890s.

McArthur Microscopes Ltd.

In the early 1930s, British physician Dr. John McArthur (1901-1996) designed a small, portable folded optics microscope using prisms to fold the optical path [264] [287]. The McArthur microscope was an inverted microscope, meaning that the slide was placed cover-slip down for viewing. Dr. McArthur formed a company which made microscopes to this design, and also licensed his design to other companies. The design evolved with time, but McArthur continued to both make the microscopes and license their manufacture to others. In 1959 his company, McArthur Microscopes Ltd., was acquired by **Cooke Troughton & Simms**, then part of **Vickers**. In 1962 the company name was changed to Vickers Instruments Ltd.

Examples of McArthur style microscopes made under license include an Open University model manufactured by **Scientific Optics Ltd.**, and models by **Charles Hearson** and **Prior Scientific**. The model *H*, and the model *HP phase contrast* microscopes, both by **Nikon** were both McArthur style microscopes. The *TWX-1* by **TaiYuan Optical Instruments** was also a derivative of the McArthur microscope.

McGregor, Duncan

Duncan McGregor, (c1803-1867) [10] [18] [44] [53] [263] [271], was a nautical instrument and chronometer maker and chart seller of Glasgow, Scotland. Addresses: 1 William St., Greenock, Renfrew (1836); 24 Clyde Place, Glasgow (1844-1854); 8 William St., Greenock, Renfrew (1853-1855); 38 Clyde Place, Glasgow (1855).

He may have served an apprenticeship with **David Heron** in Greenock. He subsequently moved to Glasgow, and they went into partnership as **Heron & McGregor** at the Nautical, Optical, and Stationary Warehouse at 1 William Street. Duncan McGregor took over the premises in 1836 to begin his own business.

McGregor advertised as an optician, nautical instrument maker, and chart seller. In about 1856, late in his career and perhaps when he retired, the company became D. McGregor & Co., and advertised as *Makers to the Admiralty*. The company continued after his death under his son, Duncan Junior, in Glasgow and Greenock, and additionally opened branches in Liverpool in 1879, and London in 1886. They continued to retail instruments under their own name. **John Morton** worked for the company as a compass adjuster prior to starting his own business. **Andrew Christie** worked for McGregor prior to starting his own business in 1890. Sometime between 1895 and 1902, at least a part of D. McGregor & Co. was taken over by **Alexander Dobbie & Son**. The McGregor & Co. business, however, continued in Glasgow, with no connections to the original family, until the 1960s.

Examples of their products include: Galilean binoculars, signed *D. McGregor & Co. Ltd., Glasgow*; Two-day marine chronometer, signed *D. McGregor & Co. Makers To The Admiralty. Glasgow & Greenock. M/7955 Auxiliary Compensation*; Eight-day marine chronometer, signed *D. McGregor & Co. Makers to the Admiralty, Glasgow, Liverpool, London. No 7357 Gold Medal Paris 1867 Prize Medals London 1862, Paris 1855.*

McIntosh, Thomas

Thomas McIntosh, also known as Macintosh (fl.1776-1784) [18] [274] [319] [322], was an optician, and optical and mathematical instrument maker of *Archimedes and Golden Spectacles*, 7 Great Queen Street, Lincoln's Inn Fields, London, England. He is also recorded as trading at 33 Near Exeter Exchange, Strand, in 1776, and 40 Great Queen Street, Lincoln's Inn Fields, in 1784. He served his apprenticeship with the spectacle-maker Thomas Hall of the **Spectacle Makers' Company**, beginning in 1756. Instruments bearing his name include microscopes.

According to his trade card, [53], "Thos. McIntosh, Optician, at the *Archimedes and Golden Spectacles*, Great Queen Street, Lincoln's Inn Fields, London. Makes & sells all sorts of Optical & Mathematical instruments wholesale and retail at the most reasonable rates. Any gentleman curious in optics may have separate glasses for any instruments at the most reasonable prices."

McIntosh, Dr. L.D.

Dr. Lyman D. McIntosh, (fl.1870s, d.1892) [113] [320] [343], of Chicago, USA, began his career as an army surgeon, after which he practiced medicine in Sheboygan, Wisconsin. He then moved to Chicago, where he founded the McIntosh Galvanic Belt and Battery Co. Among the company's products he advertised a microscope, patented in 1883, convertible to a stereopticon projection microscope. According to his advertisement, "The combination includes three distinct instruments: A solar microscope, solar stereopticon, and first-class monocular microscope. Either can be used separately or combined." The company went through two name changes to McIntosh Galvanic & Faradic Battery Co., followed by McIntosh Battery & Optical Co.

Meade Instruments

Meade Instruments, [136], is an American multinational telescope equipment manufacturer and distributor, based in Irvine, California, and is part of **Sunny Optics Inc.** Founded in 1972, Meade began business as a one-man, mail-order supplier of small refracting telescopes.

In May 2013 Meade announced a merger with JOC North America LLC, a wholly owned subsidiary of **Jinghua Optics & Electronics Co. Ltd.** However, in July the same year, they announced the termination of the JOC merger agreement, and a new merger agreement with affiliates of **Ningbo Sunny Electronics Co. Ltd.**

In 2019 Meade filed for bankruptcy, having lost an anti-trust lawsuit brought by **Orion** alleging that they colluded in price-fixing.

Meagher, P.

Patrick Meagher, (1829-1897) [61] [306], of 21 Southampton Row, High Holborn, London, England, was a camera maker with a reputation for high quality hand-crafted instruments. He learned his trade working for **Thomas Ottewill**, and in 1859 he started his own business at 1 and 1a Coppice Row, Clerkenwell. In 1865/6 he moved the business to 21 Southampton Row. An advertisement dated 1875 gives details of a binocular camera, a portable bellows camera, "improved cameras", cabinet stereoscopes, and graphoscopes. The advertisement also states that Meagher was awarded prize medals at London, Edinburgh, Berlin, and Paris. **Henry Park** worked for Meagher some time prior to starting his own company.

Medial Telescope

The Medial telescope is a kind of catadioptric telescope which uses a Mangin mirror.

Megameter

A megameter is an instrument used to determine longitude by observation of the stars, and was devised by the French Marine Lieutenant de Charniers in the mid-C18th. It was based on the principle of a heliometer, with a micrometer attached to moving wires used to make measurements.

Meiji Techno Co. Ltd.

Meiji Techno Co. Ltd. is a maker of microscopes and accessories for schools, laboratories, and industrial manufacturing facilities. They supplied instruments to be branded and sold by other companies, including **Vickers Instruments**, **Philip Harris**, and **Griffin & George**. This continued until about 1982, after which they sold instruments bearing their own brand-name [320]. During the 1990s they were based in Tokyo, Japan, and they subsequently relocated to Saitama, Japan, with subsidiaries in the USA, UK, Austria and Russia.

Meissner, A.
A. Meissner, (fl.1862-1900) [53] [374], was a surveying instrument maker of Berlin, Germany. He exhibited his work in Berlin in 1879 and 1896, and in Paris in 1900. Instruments bearing his name include: brass surveyor's level; theodolite; brass telescope, 44cm long, on wooden tripod; Galilean binoculars with compass; sundial.

Mekometer
A mekometer, is an early form of optical rangefinder.

(1) A rangefinder, used in the infantry, and operated by two users. A long wire is stretched between two angle-measuring instruments, similar to box sextants. One, known as the *right-angle instrument*, is permanently indexed at 45° and the other, known as the *reading instrument*, is used to read off the angle. The range is found using trigonometry.

(2) An optical rangefinder that analyses polarized light, reflected from a distant reflector, to determine distance.

Melling & Co.
Melling & Co, [3] [18] [383], were opticians, mathematical instrument makers, and chronometer makers of 39 South Castle St., Liverpool, England. They succeeded **Melling & Payne** in business in 1845/1846, and continued in business until sometime after 1851.

According to their trade label, "Nautical & Mathematical instrument makers, Chart agents to the R[t]. Hon[ble]. The Lords Commis[rs]. Of the Admiralty. Melling & Co. Successors to Bywater & Co. 39 South Castle Street, Liverpool. Mathematical, Optical, & Philosophical instruments & apparatus of every description made and repair'd on the shortest notice. Correctors of local attraction in iron vessels. Spectacles, Chronometers, &c." This trade label is identical in design, and almost identical in wording with that of **W. Jewitt & Co.**, who succeeded Melling & Co in business at this address.

Melling & Payne
Melling & Payne, [3] [18], were opticians, chronometer makers, and Navigation Warehouse, of 39 South Castle St., Liverpool, England. They succeeded **Dawson, Melling & Payne** at that address in 1842, by which time the firm was a partnership between Henry Melling and George Patmore Payne. Instruments bearing their name include: octant; sextant; marine barometer. In 1845/1846, they were succeeded by **Melling & Co.**

Melling, John
John Melling, also known as Mellen, Mellins, Mellin, or Malling, (fl.1672-1704) [18] [274] [275 #359] [319] [322], was an optical and philosophical instrument maker of Abchurch Lane, off Lombard Street, London, England. He was primarily known for his fine quality microscope lenses, but is also believed to have made other instruments, including barometers. He supplied a set of microscope lenses to the Royal Society, and another to the Italian biologist and physician, Malpighi, with powers up to $1/25^{th}$ of an inch.

Ménard, Guillaume (I)
Guillaume Ménard (I), sometimes known as Menard or Mesnard, (fl.1651-1667 d.c1667) [256] [426], was an optician of Paris, France. He is known to have carried out work for **Christian Huygens**, and to have made telescopes. He ground and mounted lenses used by **Chérubin** in his binocular telescopes. He had at least one son, referred to by Huygens variously as "le jeune Ménard" and "le fils Ménard". This is believed to be **Siméon Ménard**.

Ménard, Guillaume (II)
Guillaume Ménard (II), (fl.1739-1743) [256] [426], was an optician of Quai de l'Horloge, Paris, France. He may have been a son or nephew **Siméon Ménard**. A box microscope made by him bears a hand-written label, which reads, *Microscope universel fait et beaucoup augmenté par guillaume menard Md miroitier pour les faits –d'optiques: a Paris Quay de l'orloge du palais dit des morfondus a Lenseigne du Bon pasteur 1739*, which loosely translates as, "Universal microscope, made and much improved by Guillaume Menard, maker of mirrors for optical instruments; at the Quai de l'Horloge by the Boulevard du Palais, known as the Morfundus, at the sign of the Good Shepherd. 1739".

Ménard, Siméon
Siméon Ménard, sometimes known as Menard or Mesnard (fl.1667-1718) [256] [426], was an optician of Paris, France. He is believed to be the son of **Guillaume Ménard (I)**. Instruments bearing his name include: binocular telescope; telescope objective lens; binocular microscope with the inscription *SIMEON MESNARD A LA BELLE ETOILE*, and with field lens marked *Mesnard à Paris 1668*.

Meniscus Lens
A meniscus lens is one which is convex on one side and concave on the other. A negative meniscus has the concave radius smaller than the convex radius, and hence is thinner at the centre than at the edges. The negative

meniscus lens causes rays of light to diverge. Conversely, a positive meniscus has the concave radius larger than the convex radius. In this case the cross-section is crescent-shaped, being thicker at the centre than at the edges, and the lens causes rays of light to converge.

Menzies, Alexander
Alexander Menzies, (fl.1845-1847) [18] [319] [322], was an optician, cutler, and nautical instrument maker of 18 Kings Parade, Cambridge, England. Instruments bearing his name include a nautical telescope.

Meopta
Meopta is an optical engineering company in Prerov, Czech Republic. They produce medical and scientific instruments, digital cinematic projectors, aerospace technologies, military weapon systems, and consumer sports optics.

The company was formed as **Optikotechna** in 1933, [244], producing lenses and condensers. They were soon producing enlargers, composite lenses, single optics, binoculars, riflescopes, slide projectors, and the *Flexaret* range of cameras. During WWII they were required to produce military optics for the German army, including rangefinders, periscopes, binoculars and riflescopes. After the war the company was integrated with **Srb and Štys**, and renamed Meopta. During the 1950s and 1960s they became one of the biggest photographic enlarger manufacturers worldwide and the only cinema projector manufacturer in Central/Eastern Europe. During the 1970s and 1980s they largely produced cold-war military products, after which the company was privatised, and increased its focus on non-military products.

Meopta has a subsidiary in Florida, USA, where research and development is carried out on optical coatings, but no manufacturing of optical instruments takes place there.

Merchant Taylors' Company
See Appendix IX: The London Guilds.

Meredith, Nicholas
Nicholas Meredith, (fl.1789-1801) [18] [276 #1011] [319] [322], was an optician, and optical and mathematical instrument maker of 91 and 99 New Bond Street, London, England. He produced a shilling booklet entitled *A Description & Use of Mathematical and Drawing Instruments*. He was appointed Optical and Mathematical Instrument Maker to the Duke of York. Instruments bearing his name include: telescopes; microscope.

Meridian Circle
Another name for a transit circle.

Meridian Instrument
A meridian instrument is any astronomical instrument having a telescope pivoted so that its optical axis remains in the plane of a meridian. It is used to time the transit of stars across the meridian, or to measure their declination. See also transit instrument, meridian circle, transit circle.

Meridional Telescope
Another name for a transit instrument. See also meridian instrument.

Merker & Ebeling
Merker & Ebeling were microscope makers of Vienna, Austria. The firm was a partnership between **Ludwig Merker** and **Fritz Ebeling**, formed in 1886, and continued until c1892.

Merker, Ludwig
Ludwig Merker, (1856-1920) [50], was a microscope maker of Vienna, Austria. After completion of his apprenticeship in Wetzlar, Germany, he worked for **C. Reichert** in Vienna. He left this employment in 1886 with another Reichert employee, **Fritz Ebeling**, and together they traded as **Merker & Ebeling** making microscopes. This partnership continued until c1892, after which Merker continued in business alone. Following Merker's death in 1920 the company continued for some years, retaining his name.

Merklein, J.S.
J.S. Merklein, (fl.1742-1774) [322], was an optical, mathematical, and philosophical instrument maker of Dresden, Germany. Along with **J.G. Zimmer**, he served as mathematical and optical instrument maker to Count Hans Löser at Reinharz Castle in Wittenberg, where the two instrument makers worked in partnership as **Merklein und Zimmer**. Instruments bearing his name include: telescope; thermometer; artificial lodestone; a set of silver mathematical instruments marked "St. Petersburg".

Merklein und Zimmer
Merklein und Zimmer, [322], was a partnership between **J.S. Merklein** and **J.G. Zimmer**, working from 1742-1746 as mathematical and optical instrument makers to Count Hans Löser at Reinharz Castle in Wittenberg, Germany. Instruments bearing their names include: Gregorian telescope; Thermometer.

Merz & Mahler
See **Optisches Institut von Utzschneider und Fraunhofer**.

Merz & Söhne in München, G.

G. Merz & Söhne in München, (fl.1845-1858) **[161]** **[422]**, were optical instrument makers of Munich and Benediktbeuern, Germany. The firm was a partnership, between **Georg Merz**, and his son **Sigmund Merz**. The firm succeeded **Optisches Institut von Utzschneider und Fraunhofer in München** when Mahler died in 1845, and principally produced large refracting telescopes. They signed their instruments *G. Merz & Sohn in München*.

In 1847, George's other son, Ludwig, joined the partnership, and they began to trade as **Merz, Utzschneider & Fraunhofer in München**.

See Appendix VII.

Merz, G.&S.

G.&S. Merz in München, (fl.1858-1932) **[161]** **[422]**, were optical instrument makers of Munich and Benediktbeuern, Germany. The firm began as a partnership between **Georg Merz** and his son **Sigmund Merz**, formed in 1858 when they renamed their former partnership, **Merz, Utzschneider & Fraunhofer in München**, following the death of Ludwig Merz. The new partnership continued until the death of George in 1867. After this, Sigmund continued the business, trading as G.&S. Merz vormals Utzschneider & Fraunhofer in München. Also in 1867, a second workshop was set up in Munich, and the company produced large refractors, microscopes, marine binoculars, military telescopes, photographic equipment, and astronomical spectroscopes. They also produced optical components for other companies, including **Breithaupt**, **Ertel**, and **Repsold**. During the period 1858-1883, they signed their instruments *G.&S. Merz in München*, or *G.&S. Merz vorm. Utzschneider u. Fraunhofer in München*.

Sigmund Merz reserved the use of the name G.&S. Merz after transferring operation of the Munich branch of the business to his cousins, **Matthias & Jakob Merz**, in 1882. He carried on running the glassworks in Benediktbeuern, trading as **Optische Glasfabrik in Benediktbeuern, Sigmund Merz München**.

In 1903 the Merz company, including the **Optische Werkstätte von Jakob Merz** in Munich, was sold to **Paul Zschokke** who, with permission from Sigmund Merz, operated the company under the name Optisches Institut G.&S. Merz vormals Utzschneider und Fraunhofer in München. From 1908-1912 the firm traded as G.&S. Merz vormals Utzschneider und Fraunhofer in Pasing bei München, and subsequently as G.&S. Merz in Pasing. The majority of surviving instruments bearing the name G.&S. Merz were manufactured during Paul Zschokke's proprietorship of the business. Paul Zschokke continued to run G.&S. Merz until it ceased trading due to insolvency in the year of his death, 1932. The company, tools and equipment were purchased by **Georg Tremel**.

See Appendix VII.

Merz, Georg

Georg Merz, (1793-1867) **[161]** **[422]**, joined the firm **Mathematisch-mechanisches Institut von Utzschneider, Reichenbach & Liebherr** in 1808 as a fifteen-year-old boy, working under **Joseph Ritter von Fraunhofer**. He began his career working on the furnaces in the glass foundry, and underwent apprentice training in optical instrument making. After Fraunhofer's death in 1826, Joseph von Utzschneider directed the company with the assistance of Merz, trading as **Optisches Institut von Utzschneider & Fraunhofer**. In 1839, Merz purchased the company from Utzschneider, with the financial assistance of his friend Joseph Mahler (1795-1845), and continued to trade using the same name. During the 1840s **James Mackay Bryson** studied under Georg Merz. In 1845, following the death of Mahler, Georg was joined in partnership by his son, **Sigmund Merz**, and they began trading as **Merz & Söhne in München**. In 1847 they were joined by Sigmund's brother, Ludwig Merz (1817-1858), and began to trade as **Merz, Utzschneider & Fraunhofer in München**. In 1858, upon the death of Ludwig, they changed the company name to **G.&S. Merz**, and continued in business until the death of Georg in 1867. Sigmund then continued to operate the business, trading as G.&S. Merz.

See Appendix VII.

Merz, Jakob

See **Optische Werkstätte von Jakob Merz**.

Merz, Matthias & Jakob

See **Optische Werkstätte der Gebrüder Matthias und Jakob Merz**.

Merz, Sigmund

Sigmund Merz, (1824-1908) **[161]** **[422]**, optical instrument maker of Munich and Benediktbeuern, Germany, was the son of **Georg Merz** and the cousin of **Matthias & Jakob Merz**. Sigmund became a partner in his father's company in 1845. From that date they traded as **Merz & Söhne in München**. In 1847 they were joined by Sigmund's brother, Ludwig Merz (1817-1858), and began to trade as **Merz, Utzschneider & Fraunhofer in München**. In 1858, upon the death of Ludwig, they changed the company name to **G.&S. Merz**. Following the death of his father in 1867, Sigmund carried on the family business, continuing to use the name G.&S. Merz. In 1883, the Munich part of the business was taken over by Sigmund's cousins, Matthias & Jakob Merz. Sigmund, however, retained control of the name G.&S. Merz, and continued to work at the glassworks in Benediktbeuern, trading as **Optische Glasfabrik in Benediktbeuern, Sigmund Merz München**. In 1903 the Merz company, including the **Optische Werkstätte von Jakob Merz** in Munich, was sold to **Paul Zschokke**.

See Appendix VII.

Merz, Utzschneider & Fraunhofer in München

Merz, Utzschneider & Fraunhofer in München, (fl.1847-1858) [**161**] [**422**], optical instrument makers of Munich and Benediktbeuern, Germany, was a partnership, between **Georg Merz**, and his sons **Sigmund Merz** and Ludwig Merz (1817-1858). The partnership was formed in 1847 by renaming the firm **G. Merz & Söhne in München** when George's son, Ludwig, joined the partnership. They signed their instruments *Merz und Söhne in München*, or *Merz, Utzschneider und Fraunhofer in München*.

Their entry in the catalogue for the 1851 Great Exhibition [**294**], described exhibit no. 80 as: "Merz & Sons, Munich, Inv. and Manu.—Refractor, having 45″ apert., 48″ focal length, for variable latitude; equatorially mounted. Microscope, with various object-glasses and three eye-pieces, for nine magnifying powers, from 20 to 1,800 times. The instrument is provided with a screw micrometer and the necessary apparatus for holding and illuminating objects." Note that the aperture measurement of the telescope is given in French, or Paris lines (ligne), an archaic measurement equal to 2.2558mm, giving an aperture of 101.5mm.

In 1858, upon the death of Ludwig Merz, the partnership was renamed **G.&S. Merz in München**.

See Appendix VII.

Messer, Benjamin

Benjamin Messer, (fl.1794-1825) [**17**] [**18**] [**276** #1012] [**356**], was an optical and mathematical instrument maker, of London, England. Addresses: 75 Wapping (1789); 76 Wapping (1793-1797); 76 Bell Dock, Wapping (1794-1796); 155 Minories (1805-1827).

He served his apprenticeship with the mathematical instrument maker John Hussey, beginning in 1769, and was freed of the **Worshipful Company of Makers of Playing Cards** in 1801. Among his apprentices were **John Crichton**, beginning in 1820, and **Anthony James Lear**, beginning in 1804. Examples of his work include: sextant; pocket compass; double-barrelled air-pump; pocket drawing set.

Benjamin Messer's will, proved in 1825, left his worldly goods to his daughters Rebecca and Sarah. The bulk of his money and property, and his stock in trade and working tools, were bequeathed to Sarah. His stock in trade was sold at auction in 1826, and included metal sextants, quadrants, artificial horizons, day, night and lamp telescopes, universal dials, concave mirrors, microscopes, compasses, phantasmagorias with slides, binnacle and other lamps, azimuths, spectacles, sun dials, rules, barometers, and magic lanterns.

Messer, George Bracher

George Bracher Messer, (1852-1926) [**18**] [**356**], of 78 & 80 Christian St., London, England, was an optical and nautical instrument maker. He was the son of **John James Messer**, and succeeded his father in business. According to his Nautical Instruments catalogue of 1912, the business was established in 1818. The company's works was at 1 Mile End Place, Mile End Road, London. Instruments, manufactured by G.B. Messer and shown in the catalogue were various designs of sextants, including a pocket sextant, a quadrant, and an artificial horizon.

Messer, John James

John James Messer, (1812-1893) [**18**] [**356**], was an English optical and mathematical instrument maker. Addresses: 4 Harmer St., Gravesend, Kent (1845); 19 & 20 King St., Commercial Road, London (1849-1859); 78 & 80 Christian St., London (1865-1880). His signature was sometimes *JJ Messer*, and sometimes *IJ Messer*. In 1843 he was in debtor's prison as an insolvent [**35** #20257 p2946], and in 1859 he was declared bankrupt [**35** #22235 p973]. Instruments bearing his name include: mahogany mounted barometer with hygrometer, thermometer and level; octant with ebony frame and ivory scale; various mahogany and brass one, two, and three draw telescopes. He was succeeded in business by his son, **George Bracher Messer**.

Messter, Ed.

Eduard Messter, (fl.1859-c1910) [**113**] [**285**] [**320**], was an optical instrument maker of Berlin, Germany. Addresses: 94-95 Friedfichstrasse (1859-1868); 99 Friedfichstrasse (1868-c1914); 18 Schiffbauerdamm (from c1914). He founded the Optisches und Mechanisches Institut Ed. Messter in 1859. The company manufactured eyeglasses, microscopes, photographic equipment, and optical devices for magicians and showmen. Instruments bearing his name include: microscopes; opera glasses; stereo camera; barometer. His son, **Oskar Messter** worked for him, and succeeded him in business.

Messter, Oskar
Oskar Messter, (1866-1943) [285] [335], was an optical instrument maker, showman, and film producer. He was the son of **Eduard Messter**. He learned his optical manufacturing skills in his father's workshops, and while working for **Paul Waechter**, then joined his father's business in 1891. He succeeded his father in business, and grew the company considerably. He went on to become a leading figure in German film production and film equipment manufacture.

Metal
The term *metal* in the context of early optical instruments refers to a speculum metal (qv).

Metz Apertometer
See apertometer.

Meyer & Co.
Meyer & Co., [319], were opticians of 29 Market St., Manchester, England. The company was a partnership between **Abraham Franks** and Herrmann Meyer, and was dissolved in 1863 [35 #22801 p6771].

Meyer & Co., Hugo
Hugo Meyer (1863-1905) founded the Optisch-Mechanische Industrie-Anstalt Hugo Meyer & Co. in 1896 in Görlitz, eastern Germany. They produced photographic lenses under their own brand-name, and as oem lenses for other manufacturers, such as **Ihagee** for their **Exakta** model cameras. During WWII Hugo Meyer & Co. products made for the war effort were marked with the **ordnance code** ccx. In 1971 Hugo Meyer & Co. became a part of **Pentacon**, at which time the brand-name disappeared until after the reunification of Germany. The renewed brand was short lived, and closed in 1991. In 2014 the brand-name was re-registered, and used for a new company, Meyer-Optik-Görlitz, related only by name.

Meyerstein, Moritz
Moritz Meyerstein, (1808-1882) [374], was an optical, mathematical, and scientific instrument maker, and maker of microtomes, of Göttingen, Germany. He studied mathematics and physics at university, and served an apprenticeship with the **Mathematisch-mechanisches Institut von Reichenbach & Ertel**. He set up his workshop in Göttingen in 1833. There he built microscopes, surveying and astronomical instruments, and other scientific instruments, and had a close association with **Ernst Abbe**. He supplied instruments to the observatory at Göttingen. He taught **Heinrich Schröder** mechanics and optics. Among his apprentices were **Rudolf Winkel** and **Carl Diederichs**. He was succeeded in business by **August Becker**, who purchased his workshop in 1874.

Michelson, Albert Abraham
Albert Abraham Michelson, (1852-1931) [7] [155], was a Nobel Prize-winning American physicist [318]. He was born in Strelno, Poland, and came to the USA at the age of two. After graduating, he taught chemistry and physics at the US Naval Academy. Wishing to measure the speed of light, he travelled to study optics in France and Germany, before returning to the USA to become a professor of physics at the Case School in Cleveland, Ohio. He designed the echelon spectroscope and, to enable his study of the speed of light, he designed the eponymous Michelson interferometer. With it, together with Edward Morley (1828-1933), he obtained increasingly accurate measurements of the speed of light, and demonstrated that there is no luminiferous ether. Based on a suggestion by **Armand Fizeau**, he invented the stellar interferometer, which he used to make astronomical measurements, including the sizes of heavenly bodies.

Michelson Interferometer
See interferometer.

Microfiche
A microfiche is a piece of film containing multiple microphotograph images, usually being the pages of a book, journal, newspaper, or other document.

Microfilm
Microfilm is film used to capture and store photographic images in miniature size. A microfilm image is known as a microphotograph. Microphotography, [108], is believed to have first been used by **John Benjamin Dancer** in about 1839. In the late-1920s microfilm became more commercialised, and has been widely used as a document recording system since. The advent of digital storage has impacted the microfilm industry, but the technology still has its adherents.

Micrograph
A micrograph is:

(1) A photograph, or other recorded image, of a specimen as viewed through a microscope.

(2) A pantograph for executing microscopically small writing and engraving, also known as a micropantograph.

(3) The name given by **Dwight Lyman Smith** to a small, inexpensive simple microscope, produced by him in stamped tin, with a single biconvex lens, and used for viewing microphotographs, albeit reproduced as larger images than those produced by **J.B. Dancer**.

Micro Instruments (Oxford) Ltd.

Micro Instruments (Oxford) Ltd., [319], was founded in 1963 in Little Clarenden Street, Oxford, England, as a manufacturer and retailer of scientific apparatus. They remained at this address until 1989, when they relocated to Long Hanborough, Witney, Oxfordshire. Instruments bearing their name include microscopes.

Micrometer

A micrometer is:

(1) An instrument used with a telescope or microscope to measure small distances, or the apparent diameters of objects subtended by very small angles.

(2) A device for measuring small lengths in engineering or other work.

Catoptric micrometer: A form of double image micrometer, devised by **Jesse Ramsden**, [394], in which the secondary mirror of a reflecting telescope is cut into two halves across its optical centre, and the two halves are inclined to each other on an axis at right-angles to the plane that separates them. The inclination is controlled by means of a micrometer screw.

Circular micrometer: Another name for a ring micrometer.

Dioptric micrometer: A form of double image micrometer, devised by **Jesse Ramsden**, [394], similar in principle to the divided object glass micrometer, but in which the divided lens is in the eye-tube of the telescope.

Divided object glass micrometer: The divided object glass micrometer is a double image micrometer in which the double image is achieved by cutting the objective lens, and arranging the pieces to form two images of a single object. The separation is controlled and measured by means of a micrometer screw.

Double image micrometer: see separate entry.

Double refraction micrometer: A double refraction micrometer is a double image micrometer in which the two images are formed by means of double refraction in a birefringent material such as Iceland spar.

Eyepiece micrometer: A graduated scale, usually positioned at the focus of the eyepiece, or a reticule moved by a micrometer screw.

Ghost micrometer: see separate entry.

Micrometer screw: A screw with a fine thread pitch and attached scale, often a vernier scale, calibrated to indicate how far the screw has advanced when turned. Used for accurate measurements of small distances or, as in the case of a tangent micrometer, small angles.

Microscope stage micrometer: A microscope slide marked with a reticule, used for calibrating eyepiece reticules and objective lens power.

Ring micrometer: Otherwise known as a circular micrometer. An opaque ring, fixed in the focus of a telescope objective lens, used to measure the differences of declination or ascension between stars. This is achieved by measuring the times at which the stars cross the edge of the ring.

Tangent micrometer: An endless screw, engaged with a gear, with a micrometer scale on a barrel connected to the screw. The micrometer scale provides an indirect measurement of the rotation of the gear.

Microphotograph

A microfilm image is known as a microphotograph.

Micro Precision Products Ltd.

Micro Precision Products Ltd., [306], of 145 London Road, Kingston Upon Thames, Surrey, England, was a manufacturer of photographic enlargers, projectors, and cameras. The company was founded in 1941. A.J. "Jimmy" Dell was works manager, and in about 1963 he took control of the company. In 1976 they relocated to 48 Lydden Road, Wandsworth, London. They were noted for their *Micro-Technical*, *Micropress*, and *Monorail* cameras, as well as their TLR models, the *Microcord* and the *Microflex*. While they were at 48 Lydden Road, they began to trade as MPP Photographic Products Ltd., and they continued in business until 1982.

Microscope

A microscope is an optical instrument used for viewing objects that are too small to be viewed in detail by the naked eye. It consists of an arrangement of one or more lenses. See also microspectroscope.

Aquatic microscope: An aquatic microscope is a simple microscope specifically designed for viewing specimens that are immersed in water, generally in a vessel such as a watch-glass. The horizontal arm holding the objective can be slid in and out, and swivelled in an arc, thus allowing the objective to be positioned wherever desired over the specimen. See also Ellis aquatic microscope, Raspail's simple chemical microscope.

Binocular microscope: A microscope with two eye tubes, fitted with eyepieces, for viewing the object through the objective with both eyes at the same time.

Common Main Objective (CMO) stereo microscope: This uses a single, relatively large objective, the light from which is split to traverse two optical paths to the two eyepieces, giving a true stereo view. The concept derives from the design of **Prof. John Leonard Riddell** in c1853. The forerunners of modern CMO microscopes were a model introduced by **Zeiss** in 1942, and another by the **American Optical Company** in 1957. The modern CMO has a large objective which forms an image at infinity when focussed on the object. It has better light-gathering, and generally better correction of optical aberrations than the Greenough binocular microscope, albeit with higher associated cost. It also has a larger working distance.

Compass microscope: The compass microscope is a portable hand-held simple microscope in which the sample is clamped in a small pair of forceps at the end of an arm that is hinged on the instrument's handle. This hinge arrangement is similar to that of a drafting compass, hence the name. A Lieberkühn mirror mounted behind the objective provides specimen illumination. The compass microscope, invented by **Jan van Musschenbroek**, was popular during the C18th as a field microscope for viewing small opaque objects.

Compound microscope: A microscope with an objective lens to form an intermediate image, and an eyepiece lens to view the image. The objective and eyepiece may themselves comprise multiple lens elements.

Culpeper microscope: The Culpeper-type microscope, [264], is believed to have been introduced by **Edmund Culpeper** in the early 1720s. It is a compound microscope consisting of a sliding barrel in which the objective and eyepiece are mounted. The barrel is mounted upon a tripod, the feet of which are attached to the stage. The stage, in turn, is mounted on a tripod whose feet are attached to a circular base. In the centre of the circular base a mirror is mounted for illumination. The objectives may be interchangeable.

Drum microscope: A simple form of portable compound microscope invented by **Benjamin Martin**. See separate entry.

Electron microscope: A microscope that uses an accelerated beam of electrons to illuminate the object. It uses electron optical lenses, analogous to the lenses of an optical light microscope, to focus the image.

Folded optics microscope: A compact, portable microscope for field and educational purposes. See separate entry.

Greenough binocular microscope: A combination of two microscopes, each with a set of erecting Porro prisms, giving a pronounced stereoscopic effect. This is simpler in design, and less expensive to produce than the common main objective microscope.

Inverted microscope: A microscope with the light source and condenser above the stage, directed downwards, and the objective and turret below the stage directed upwards, such that the object under observation is viewed from below.

Lucernal microscope: A projecting microscope in which the illuminated object is projected onto a plate of glass that is part of the microscope, or a screen that is independent of it.

Museum microscope: A museum microscope is one which can hold multiple samples at the same time. Such a microscope is generally designed so that the samples may be swapped into view by means of a mechanism such as a carousel or a wheel. The museum microscope was first described by **Chérubin** in 1677.

Opaque microscope: An opaque, or opake microscope is one designed for viewing opaque objects. The sample is viewed in light reflected from its surface.

Oxyhydrogen microscope: A microscope in which the sample is lit by means of limelight.

Reflecting microscope: A microscope in which the image, formed by a small concave elliptical mirror or speculum rather than convex lenses, may be viewed either by the naked eye or through an eyepiece. The reflecting microscope was first successfully introduced by **Giovani Battista Amici**, and subsequently improved by **John Cuthbert**.

Screw-barrel microscope: A screw-barrel microscope is a small simple microscope, constructed so that the eyepiece barrel is threaded onto the main barrel of the microscope, and may be rotated to screw it in or out to adjust the focus. The sample slide is held in place with a spring stage. The design is generally associated with **James Wilson**, although he did not invent it, and is frequently referred to as a Wilson microscope. The concept is frequently credited to **Nicolaas Hartsoeker**, even though he never claimed to have invented it. The use of a screw-barrel for microscope focussing is believed to have originated with **Guiseppe Campani**. The use of a spring stage to hold the sample slide is believed to have originated with **Filippo Bonanni**.

Simple microscope: A microscope having a single convex lens, comparable with a magnifying glass. See also Stanhope lens.

Solar microscope: A microscope for which the illumination is provided by sunlight, and which projects an enlarged image of the object onto a screen, either in a darkened room, or in a dark box.

Wilson microscope: Another name for a screw-barrel microscope.

See also dark field microscopy, phase contrast microscopy.

Microscope Stage Micrometer
See micrometer.

Microspectroscope
A microspectroscope is an instrument that combines the functions of a microscope and a spectroscope. It us used for spectrographic investigation of microscopic objects.

The microspectroscope was invented by **Henry Clifton Sorby**.

Middlefell, Damaine
Damaine Middlefell, (fl.1739) [18], was an optical instrument maker of Newcastle upon Tyne, England. His guild records are unclear, but it is believed he may have served his apprenticeship with the London spectacle maker **Matthew Cooberow** of the **Spectacle Makers' Company**, beginning in 1705, and gained his freedom in 1719.

Midland Camera Co. Ltd.
The Midland Camera Co. Ltd., [61] [306], was established in around 1896 as Howell & Green in Birmingham, England, producing cameras. The company was renamed the Midland Camera Co. in 1899, and continued in business until 1912/13 when it was acquired by the **Thornton-Pickard Manufacturing Co. Ltd.**

Miller and Adie
Miller and Adie, [18] [263] [271], were optical and scientific instrument makers in Edinburgh, Scotland. The company was formed when **Alexander James Adie** went into partnership with his uncle, **John Miller**, in 1803. In 1822, some years after John Miller's death in 1815, the firm was renamed Alexander Adie, and from 1835 it was known as **Adie & Son**.

Miller Brothers
Miller Brothers, [201] [320], of New York, NY, USA, were manufacturers of microscopes and accessories, and retailers of microscopes by makers such as **Joseph Zentmayer**, and French imports. They also sold spectacles and other optical instruments and accessories. The partnership was formed in 1868 by the brothers E. and F. Miller. Both were former employees of **Robert Tolles**. An advertisement for Miller Brothers dated 1872 shows the addresses 69 Nassau Street, and 1223 Broadway. They later traded as F. Miller and Brother.

Miller, John
John Miller, (1746-1815) [18] [46] [263] [271] [276 #1013], was an optical and scientific instrument maker in Edinburgh, Scotland. Addresses: Nicholson St.; George IV Bridge; Back of the Fountain Well (1774); Parliament Close (1775-1794); 38 South Bridge (1795-1801); 86 South Bridge (1803-1804).

He worked for **George Adams (I)** in his London workshop prior to moving to Edinburgh in 1769, but is not known to have been Adams' apprentice. He adopted his nephew, **Alexander James Adie** who came to live with him at a young age after losing both parents, and served an apprenticeship with him. He is known to have sold: microscope; theodolite; micrometer; clinometer; globe; planetarium.

In 1803 Adie became a partner in John Miller's firm, and they traded as **Miller and Adie**.

Mills, George
George Mills, (fl.1790-1846) [18] [276 #1646] [385], was an optical, mathematical, and philosophical instrument maker of London, England. Addresses: Hermitage Yard, Wapping (1790-1830); 82 Parsons St., Ratcliff Highway, otherwise known as 82 Parsons St., East Smithfield (1826-1845); 82 St. George St. (1846).

He began his apprenticeship with **Walton Willcox** in 1782, and was turned over in 1783 to **William Spencer (I)**. He was freed of the **Joiners' Company** in 1790.

Mills, Harriett
Harriett Mills, (fl.1851-1869) [18] [406] [407], was an optician and telescope maker of 49 Southampton St., Pentonville, London, England. She succeeded **Robert Mills** in business at this address, and was succeeded in business by Alfred Mills. **Alexander Clarkson** took over the business in 1873, presumably from Alfred Mills, and with it acquired the telescope tube-drawing machine originally made by **Joshua Lover Martin**.

Mills, Robert
Robert Mills, (fl.1820-1850) [18] [276 #1647] [406] [407], was an optician and telescope maker of London, England. Addresses: 7 Southampton St., Pentonville; 31 Duke St., Lincoln's Inn Fields; 35 Duke St., Lincoln's Inn Fields (1822); 49 Southampton St., Pentonville (1839-1849). This 49 Southampton St. address was also occupied by **Henry Leader** at around that time, and Leader may have been one of Robert Mills' employees. Robert Mills acquired the business and tools of Robert & William Tulley, the sons of **Charles Tulley**, in 1782. The tools included the telescope tube-drawing machine invented and built by **Joshua Lover Martin**. Robert Mills was succeeded in business by **Harriett Mills**.

Minolta
Minolta was formed in 1928 as Nichi-Doku Shashinki Shoten to produce cameras in Japan. In 1957 they produced their first planetarium for a science exposition at Hanshin Park, in Hyōgo, Japan. They expanded their product range to include copying and other technologies. In the 1960s they provided camera technology for space flights by the American space programme. In 2000 they entered a joint venture with **Konica**, producing

polymerised toners. In 2003 the two companies were merged to form **Konica Minolta Holdings Inc.**

Minox
Minox is a brand of German manufactured sports optics, including riflescopes, binoculars, spotting scopes, night-vision scopes, trail cams, and accessories.

The company was founded as Minox GmbH in 1945 in Wetzlar, Germany, with the cigar manufacturer Rinn & Cloos as a funding partner and, beginning in 1948, was the manufacturer of the sub-miniature cameras invented by **Walter Zapp**. These sub-miniature cameras had previously been made in Riga, Latvia by the radio manufacturer Valsts Electrotechniska Fabrika, and marketed as *VEF Minox Riga*. The sub-miniature cameras formed the basis of a successful range of cameras for Minox GmbH. During the 1970s, however, business began to decline due to competition from Japanese manufacturers, and in 1988 receivers were appointed.

The company continued to operate after restructuring, and in 1996 was taken over by **Leica Camera**. Following a management buy-out in 2001, the company became once again fully independent in 2005. By this time the company had diversified into producing sports optics including riflescopes, binoculars, spotting scopes, and night-vision scopes. In 2013, together with two other companies, Minox formed the parent German Sports Optics, and they now operate as part of the Blaser Group.

Mir
Mir (Мир) was a brand-name used by **Krasnogorskij Mechanicheskij Zavod** (KMZ) for a simplified version of their *Zorki 4* rangefinder camera.

Mir was also a brand-name used by **Arsenal Factory** for a range of **Zeiss** copy camera lenses.

Mirador
Mirador is a Japanese brand of binoculars, telescopes and spotting scopes. **Limer** is a Mirador brand of binoculars.

Mirand, Antoine and Jean
Antoine Mirand, (c1810-c1888) [50], of Paris, France, was a microscope maker who started his business in 1846. Known as Mirand aîné, (the elder Mirand), he received an honourable mention for his exhibit at the Exhibition of the Industry of All Nations in New York in 1853. He also exhibited at the Exposition Universelle in Paris in 1855, and at the International Exhibition in London in 1862 [**295**].

Antoine's son, Jean Mirand (1843-c1910), also known as Mirand fils, (Mirand the son), joined his father in the business in the mid-1860s. They exhibited microscopes and lenses at the Exposition Universelle in Paris in 1867. Following his father's retirement in the 1870s, Jean won a bronze medal when he exhibited at the Exposition Universelle in Paris in 1878. He continued to make and exhibit microscopes until at least 1891.

Miranda Camera Company
The Miranda Camera Company was founded in Tokyo, Japan, in 1946, as **Orion Seiki Sangyō Y.K.**, and began to use the *Miranda* brand-name in 1955. In 1956 the company was renamed to become the Miranda Camera Company. They produced 35mm SLR cameras, and later a rangefinder camera. Their products were sold in Europe, and in the USA by **Allied Impex Corporation** (AIC). During the late 1960s Miranda became a part of AIC, and Miranda production ended with bankruptcy in 1978.

The *Miranda* brand-name was re-used in the 1980s by the British retailer, **Dixons**, for budget imported SLR cameras and binoculars manufactured by **Cosina**.

Mirror Stereoscope
The mirror stereoscope, or reflecting stereoscope, was an invention of **Sir Charles Wheatstone**, first presented to the Royal College of London in 1838. His invention consisted of a pair of mirrors set at 45° to each-other, reflecting a pair of images, set one to each side, into the user's vision. The two images represented the views, as seen one by each eye. Thus, each eye receives a reflection of one of the images, giving a stereo view with perceived depth of vision. Initially this was used with drawings, until stereo photographic images were available. A variant of Wheatstone's design utilises two mirrors set at 45° to each-other, with two further wing-style mirrors, so that the images are placed in front of the viewer rather than to the sides.

See stereoscope, lenticular stereoscope.

Mitscherlich goniometer
A Mitscherlich goniometer is a vertical circle goniometer, invented by the Prussian chemist Eilhard Mitscherlich (1794-1863). The specimen is clamped on the axis of a vertical circle. The specimen is viewed through a sighting telescope, and rotated so that the angle of rotation may be read off on the vertical circle by means of a second telescope. See goniometer.

Mittelstrass, Gebr.
Gebr. Mittelstrass, (Mittelstrass brothers) [113], were optical instrument makers of Magdeburg, Germany, founded by the brothers Carl and Otto Mittelstrass in 1867.

According to their 1903 trade advertisement (translated): "Mittelstrass Brothers. Magdeburg. Established 1867. Purveyor. Builds as a speciality: Projection apparatus of all kinds, Microscopic projection

MIZAR Optical Japan

MIZAR Optical Japan is the exclusive representative of **Mizar Tech Ltd.** of Japan, manufacturer of optical instruments including binoculars, monoculars, magnifiers, telescopes, microscopes and spotting scopes. They are a volume sales company who sell through international distributors. According to their literature, there are plentiful cheap knock-offs of their products – If the packaging bears the word Mizar with a lower-case i, it is not an original MIZAR product. Some MIZAR optical products bear the maker's mark illustrated in Appendix IV.

Mizar Tech Ltd.

Mizar Tech Ltd. was formed in 1990 as the Mizar Co. Ltd. by the merger of **Hino Kinzoku Sangyo Co.** and **Eikow Optics**. In 2013 the company changed its name to Mizar Tech Ltd. They manufacture optical instruments including binoculars, monoculars, magnifiers, telescopes, microscopes and spotting scopes, which are distributed internationally by their volume sales company **MIZAR Optical**.

M-Katera

M-Katera, [285], was a partnership between the optical engineer **Shintaro Terada**, and two others named Matsumo and Kando. The partnership was formed in 1912 to design and manufacture microscopes. In 1914 they produced the first commercial Japanese microscope. In 1934 the company was renamed **Tiyoda**.

Mogey, W.&D.

See **William Mogey & Sons Inc.**

Mogey & Sons Inc., William

According to the 1932 William Mogey & Sons Inc. catalogue, William Mogey began making corrected photographic lenses in 1882 in New York City, USA. As his business grew, he additionally began making reflecting telescopes, surveyors' transits, spectroscopes, and refracting telescopes. In 1888 he went into partnership with his brother, David Mogey, and began trading as W.&D. Mogey, specialising in refracting telescopes. In 1893 they moved to larger premises in Bayonne, New Jersey, and in 1911 they moved again to Plainfield, New Jersey, where the company remained. When David Mogey retired, William's sons joined the firm, and in 1927 they incorporated as William Mogey & Sons.

Their catalogue of 1932 shows: Achromatic doublet and astro-photographic doublet object glasses; Astronomical telescopes ranging from 3 inches to 8 inches; Equatorial and alt-az mounts, tripods, and pillars; Barlow lenses; Solar projectors; Finderscopes; Eyepieces; Diagonals; Solar wedges; Micrometers. The catalogue also shows a large photo heliograph, installed at Santa Clara College, California, with a 6-inch object glass and a focus of 37 feet.

Moginie, William

William Moginie, (1828-1881) [50] [264] [280 Ch.7] [285], was an optical instrument maker and designer, of London, England. He made microscopes of both folding tripod and bar limb designs, barometers, stereo viewers, and microphotograph slides.

Moginie designed instruments for **Charles Baker**, including the folded travelling microscope for which Moginie is famous. The travelling microscope, announced in 1867, was marketed by Baker, and was either manufactured by Baker, or made for them by Moginie. Some examples are known to have been made by Moginie. A second design followed in 1874, which was similar but larger. A custom-built example of the larger version was used by the famous microscopist Dr. M.C. Cooke. By 1868 Moginie's association with Baker had ceased, and he continued to run his own business.

Möller, Johann Diedrich

Johann Diedrich Möller, (1844-1907) [47] [50], served an apprenticeship as an artist in Hanover, Germany. During his studies he developed an interest in microscopes, and went to **Heinrich Schröder** to explore this interest. There he learned how to grind lenses and, when his art apprenticeship was over, he worked for Schröder for a short while before setting up his own business in his home town of Wedel. There he made lenses and other optical components for Schröder, and began to make microscope slides, which became a major part of his business.

Following Johann's death, his son Hugo Möller (1880-1959) took over the business. The company continued to produce microscope slides, and also made binoculars between the wars. In 1925 **Zeiss** bought a 51% interest in the company, which Hugo bought back in 1931. His sons Alfred (1910-1995) and Hans (1912-1982) ran the company after his death. During WWII they manufactured optical munitions for the war effort, which they marked with the **ordnance code** dhq, phq or qhg. In 1999 the company was acquired by **Haag-Streit**, and renamed Möller-Wedel Optical GmbH.

Molteni

The Molteni business, [109] [113] [297], of 62 and 44 Rue du Château-D'Eau, Paris, France, was founded by Joseph Antoine Balthazar Molteni, (c1746-1808), sometimes known by the gendered surname Molteno. He was an Italian, born in Switzerland, who emigrated to France, and began manufacturing optical instruments in Paris in 1782. He is believed to have made fantascopes for their inventor, Etienne Robertson. Joseph A.B. Molteni was succeeded in his business by his son Pierre François Antoine Molteni, (1786-1866), commonly known as Antoine Molteni.

Antoine Molteni continued the business, and made some of the early instruments used by **Louis Daguerre**. In 1841 the company became Molteni Et Cie, being a partnership between Antoine, his son Joseph, and the optician Ferdinand Sieger. The company made marine, mathematical, geodetic, and optical instruments, including marine spotting scopes, spectacle lenses, projection and enlargement lanterns, and others.

On the death of Joseph A.B. Molteni, Alfred Molteni joined the company. The company continued in the family, becoming Jules Molteni Et Cie in 1855, and then with Alfred, J.&A. Molteni in 1863. In 1894/5 Alfred Molteni made the lamp-houses for the early Kinetoscopes developed by the **Lumière** brothers. Jules died in 1876, after which Alfred continued the business alone until 1899 when it merged with **Radiguet & Massiot**.

Molton, Francis

Francis Molton, (fl.1822-1839) [3] [18] [276 #1936] [319] [322], was an optician, and optical and philosophical instrument maker, of Norwich, England. Addresses: 55 Lawrence Steps; Dove Lane (1822-1830); Dove Street (1830-1839). In the 1841 Census he is recorded at the Dove Street address, aged 70+. He is known to have made and sold barometers and thermometers. Instruments bearing his name include: Culpeper-type microscope c1825; stick barometer; wheel barometers.

Monolux

Monolux was a brand of Japanese manufactured telescopes, spotting scopes, binoculars, monoculars and microscopes. They were re-branded instruments made by makers such as **Astro Optical Industries Co. Ltd.**, **Apollo Labs**, and **Towa**, as well as other Japanese makers.

Moody Bell, William

William Moody Bell, [44], of Cheltenham, England, was a surgical instrument maker and optician, who established his business in Cheltenham in 1842. It is not known whether he manufactured the optical instruments that he sold, but considering that he was a surgical instrument maker, it seems likely he at least had the skill to make the metalwork, if not the optics. Instruments bearing his name include: binoculars; barometer.

Moon, John

John Moon, (fl.1826-1867) [18] [276 #1937] [322], was an optician, and optical, mathematical, and philosophical instrument maker, of London, England. Addresses: 4 Lucas Place, Commercial Road; 3 Lucas Place, Commercial Road (1830); 28 Green St., Stepney (1838-1839); 76 Minories (1845-1849); 31½ Limekiln Hill (1849-1850); 2 St. Ann's Place, Commercial Road (1850).

He succeeded his father, **William Moon**, in business. He traded in partnership as Moon & Fleming, opticians, from 1841 to 1843 at 76 Minories. Instruments bearing his name include: sextant; garden sundial; wheel barometers.

Moon, William

William Moon, (fl.1791-1817) [18] [322], was an optician and optical instrument maker, of 35 Northampton St., Clerkenwell, London, England (1811), and 49 Fleet Lane, Old Bailey (1816). He served his apprenticeship with **John Dollond (II)**, beginning in 1782, and was freed of the **Spectacle Makers' Company** in 1791. He was succeeded in business by his son, **John Moon**.

Moor, James

James Moor, (fl.1788-1801) [18] [271], was an optical instrument maker of Edinburgh, Scotland. Addresses: Opposite the Guard (1788-1790); 17 Parliament Close (1793); 15 Parliament Close (1794); Robertson's Close (1801).

Moore, John Hamilton

John Hamilton Moore, (1738-1807) [16] [18] [276 #512], of London England, was a seller of mathematical and nautical instruments, cartographer, hydrographer, and map and chart seller. He was born in Edinburgh, Scotland, and studied mathematics in Monaghan, Ireland. He served with the marines in Plymouth, after which he opened an academy in Brentford, Essex, where he taught navigation. He operated a Navigation Warehouse in London, and traded at 104 in the Minories, Tower Hill (1781-1786); King St., Little Tower Hill (1790); 2 King St., Tower Hill (1799-1805). He published *A New and Complete Collection of Voyages & Travels* in 1778, as well as numerous maps and charts. His main published work was *New Practical Navigator and Seaman's New Daily Assistant, being an Epitome of Navigation*, published in 1822. He was declared bankrupt in 1806, [35 #15943 p1021], and died the following year. A final call for creditors was made in 1811 [35 #16506 p1374]. He was succeeded in business by his son-in-law, **Robert Blachford**.

Moravian Instruments

Moravian Instruments of Zlín, Czech Republic, is a maker of specialised CMOS and CCD cameras for astrophotography.

Morgan, Francis

Francis Morgan, (fl.1764-1801, d.1803) [3] [18] [53] [276 #748], was an optical, philosophical and mathematical instrument maker of London, England and, later, St. Petersburg, Russia. Addresses: New Street Square; Carey Street (1764-1766); *Archimedes and Three Spectacles*, No. 27 Ludgate Street, Near St. Paul's (1767-1771); St. Petersburg, Russia (1772-1803).

He served his apprenticeship with his father, **John Morgan**, beginning in 1757, and was freed by patrimony of the **Joiners' Company** in 1764. In the same year he was one of the petitioners who attempted to revoke the patent obtained by **John Dollond (I)** for the achromatic lens, and which was enforced after Dollond's death by his son, Peter (See Appendix X). In 1771 he was appointed Optician and Mathematical Instrument Maker to the Empress of Russia. In 1772 he relinquished control of his business to **Samuel Whitford**, and emigrated to St. Petersburg, Russia where he served as scientific instrument maker to the Russian Admiralty until 1801. He died in St. Petersburg in 1803.

While trading at the sign of *Archimedes and Three Spectacles* his trade card gave a comprehensive list of the instruments he made and sold. These included: "Spectacles of glass or pebble; Glasses for reading, magnifying and burning; Reflecting and refracting telescopes; Double, single, solar, opake or aquatic microscopes; Camera obscuras, Sky-optic balls, Prisms to demonstrate the theory of light and colours; Concave, convex and cylindrical speculums; Magical lanthorns; Opera glasses; Air pumps and air-fountains of all kinds; Portable apparatus for electrical experiments; Barometers, diagonal, standard or portable; Thermometers; Hygrometers, hydrostatical balances, and hydrometers; Hadley's quadrant, after the most exact method, with glasses truly parallel; Davis's quadrant; Globes of all sizes; Compasses, azimuths steering, and plain; Loadstones; Nocturnal and sun dials of all sorts; Scales; Cases of drawing-instruments; Parallel rulers, proportional compasses and drawing pens; Theodolites, semicircles, circumferenters, measuring wheels, spirit levels, rules, and all sorts of the best black-lead pencils, and all other sorts of instruments of the newest and most approved invention."

Morgan, John

John Morgan, (fl.1749, d.c1758) [18] [53] [276 #385] [385], was a mathematical instrument maker of London, England. Addresses: Finch Lane, Cornhill (1755); Fleet Street (1758); Birchin Lane. He served his apprenticeship with the mathematical instrument maker Lawrence Miles, beginning in 1726, and was freed of the **Joiners' Company** in 1733. He is known to have made compasses, sextants, octants, Hadley's quadrants, and other instruments, and was the original maker of Gowin Knight's Patent Compass. **James Watt** trained under John Morgan from about 1755 to 1756. Among John Morgan's apprentices were his son, **Francis Morgan**, and **Robert Tangate**, who was turned over to **George Adams (I)** in 1758.

Morley & Cooper

See **W.H. Morley**.

Morley, W.H.

W.H. Morley, [306], of Upper Street, Islington, London, England, was a firm of camera makers, founded in 1843 by William Morley (1818-1893). Morley's son, W. Morley joined the business, working with him until 1882 when he started a music publishing business. W.H. Morley's son-in-law, H. Cooper, began to work for the business in 1886, and took control of it soon after, presumably upon the retirement of W.H. Morley. In 1890 the firm was renamed Morley & Cooper, and continued in business until at least 1909.

Morris, Thomas S.

Thomas S. Morris, (fl.1836-1842) [18], was an optical, mathematical, and philosophical instrument maker, and rule maker of London, England. Addresses: 15 Chester St., Kennington; 11 Chester St., Kennington (1839); Chester St., Lambeth (1840). He attended the London Mechanics' Institute in 1840.

Morrison, Richard

Richard Morrison, (1836-1888), was a maker of lenses for microscopes and cameras, of New York, USA, who worked for **Benjamin Pike's Sons**. He formed a partnership in about 1864 with the optical instrument maker **George Wale**, trading as Wale & Morrison. The partnership was terminated some time on or before 1871, after which he returned to work for Benjamin Pike's Sons. He went on to form his own company again but, by the mid-1880s, due to health problems, he relinquished control of his company to George Wale.

Morton & Co.
Morton & Co., (fl.1847-1851) **[18] [271] [379]**, were optical and philosophical instrument makers of 7 South College Street, Edinburgh, Scotland. They succeeded **Alexander Morton (I)** in business at this address.

Morton & Co., John
John Morton & Co., (fl.c1920) **[263]**, were nautical instrument makers and nautical publishers of Glasgow, Scotland. The company was founded by John Morton, who had been a compass adjuster, working for **D. McGregor & Co.**

Morton, Alexander (I)
Alexander Morton (I), (fl.1832-1846) **[18] [271] [379]**, was a philosophical instrument maker and model maker of Edinburgh, Scotland. Addresses: 2 Richmond Place (1822-1823); 2 Roxburgh Place (1834-1837); 9 Drummond St. (1838); 11 Hill Square (1839); 71 Adam Square (1840-1842); 7 South College St. (1843-1846). He was succeeded in business by **Morton & Co.**

Morton, Alexander (II)
Alexander Morton (II), (fl.late-1840s-after 1862) **[263]**, of Morton Place, Kilmarnock, Scotland, was a carpet machine manufacturer and amateur telescope maker. Along with his brother, Charles, he succeeded his father, **Thomas Morton**, in business. Beginning in the late 1840s, he was granted a number of patents, one of which was for a device used to determine longitude. In 1862, he published a book *The Art of Making Reflecting Telescopes made Simple and Easy*. In it, he advertised his telescopes for sale: "the author can supply any person with telescopes of any size, on very elegant stands, or any part necessary for the above in any state of progress, if they have not convenience for doing it themselves." One telescope bearing the signature of Alexander Morton (II) is known.

Morton, James
James Morton Ltd., **[44]**, were compass and nautical instrument manufacturers, of Villiers Street South, Sunderland, and South Shields, England. According to their trade label of 1951, "James Morton Ltd. Patent Compass and Nautical instrument manufacturers. Villiers Street South, Sunderland. Compasses, Binnacles, Chronometers, Sextants etc. Always in stock. Repairs promptly and carefully executed. Compass adjusters. Chronometers rated. Admiralty chart agent. Stationary." Instruments bearing their name include: large single-draw marine telescope in brass with leather cover, inscribed *James Morton Ltd. Sunderland. S.S. "Deptford"*.

Morton, Thomas
Thomas Morton, (1783-1862) **[10] [18] [262] [263]**, of Morton Place, Kilmarnock, Scotland, was a telescope maker and amateur astronomer, and an inventor and carpet machinery maker. He served his apprenticeship with turner and wheelwright, Bryce Blair, and in 1806 he started his own business. He made his name with his invention of a barrel loom for weaving carpet. He, was offered an opportunity, by the Emperor of Russia, to move to Russia and set up a carpet weaving industry. He declined the offer, and remained in Kilmarnock, where he manufactured the looms. In 1817 he began to manufacture telescopes, and began to study practical astronomy. Several telescopes bearing his name are known. He built his own observatory, which he opened to the public, and taught astronomy at the Kilmarnock Acadamy. Upon his death in 1862, the business was continued by his two sons, Charles Morton and **Alexander Morton (II)**.

Motic Group
Motic was founded in 1988, producing compound microscopes. The company has grown into a global enterprise, with subsidiary companies in the electric and energy industries, as well as digital pathology and microscopes. **Swift Optical Instruments** is a member of the Motic group.

Motion Picture Image Reproduction
Early moving image reproduction technological developments include: Aeroscope, Bioscop, Bioscope, Chronophotography, Cinematograph, Electrotachyscope, Kinetograph, Kinetoscope, Panoptikon, Phantascope, Phénakistiscope, Pleograph, Praxinoscope, Rotoscope, Theatograph, Vitascope, Zoetrope, Zoöpraxiscope.

Mountain Transit Theodolite
See theodolite.

MPP Photographic Products Ltd.
See **Micro Precision Products Ltd**.

Mudge, John
John Mudge, (bap.1721, d.1793) **[3] [10] [18] [24] [46] [276 #386]**, was a medical physician, and amateur astronomer and telescope maker, of Plymouth, England, with a strong interest in mirror-making. He was the brother of the chronometer maker Thomas Mudge, and the father of the geodetic surveyor William Mudge. He spent much of his time making telescopes. He visited the London workshop of **James Short (I)** and observed his methods of making speculum metal. In 1777, after Short's death, and the same year Mudge was elected a fellow of the Royal Society, he received the Copley medal for his article

Directions for making the best composition for the metals for reflecting telescopes, together with a description of the process for grinding, polishing, and giving the great speculum the true parabolic curve [25]. In this article he detailed his methods for making speculum metals for reflecting telescopes. He maintained that Short's methods were little different from his own.

Munford, John

John Munford, (fl.1768-1801) [18], was an optician, and optical and mathematical instrument maker and brass turner of Clerkenwell, London, England. Addresses: Parish of St. James (1768); Compton St. (1771); 15 Compton St. (1778-1799).

He served his apprenticeship with **William Bush**, beginning in 1747/8, and was freed of the **Stationers' Company** in 1755. Among his apprentices were his two sons, John, beginning in 1783, and Edward, who was freed by patrimony in 1792.

Munro, James (I)

James Munro (I), (fl.1799) [18], was an optical, mathematical, and philosophical instrument maker of 12 York Place, Lambeth, London, England. This address was later occupied by **James Munro (II)**, who went by the name James Munro junior, and may have been his son. James (I) is believed to have served his apprenticeship with **Richard Lekeux**, beginning in 1783, and to have been turned over to someone named Botten. James (I)'s grandson, **R.W. Munrow** was also an instrument maker.

Munro, James (II)

James Munro (II), (fl.1823-1867) [18] [276 #1943] [379], was an optician, and optical, mathematical, and philosophical instrument maker of Lambeth, London, England. Addresses: 72 Oakley St. (1823-1828); 27 North St. (1826-1828); 4-5 North St. (1830-1856); 4 High St. (1839-1865); 12 York Place (c1850); 5 High St., Lambeth; 4 North St., Lambeth; 14 Plough Court, Fetter Lane. He went by the name James Munro junior, and his address at 12 York Place was previously occupied by **James Munro (I)**, who may have been his father, in which case **R.W. Munro** would have been his son.

He exhibited at the Great Exhibition in 1851 in London [328]. His catalogue entry no. 282 stated: "Munro, James, jun., 4 High Street, Lambeth—Manufacturer. Model of a nine-pounder brass gun-carriage and limber, scale 1 inch to a foot; and of a twenty-four pounder, brass battering gun and carriage, scale 5/8 inch to a foot."

Munro, R.W.

R.W. Munro, [44] [53] [124] [444], were precision engineers and instrument makers of London, and later Essex, England. The company was established in Clerkenwell, London, as R.W. Munro & Co. in 1864 by Robert William Munro (1839-1912), the grandson of **James Munro (I)**. By 1870 they had expanded sufficiently to move to larger premises in Clerkenwell Green, London. Robert was an active member of the Royal Meteorological Society, and the firm quickly established themselves producing meteorological instruments, as well as producing lathes and other engineering equipment. They also produced printing machines for the Bank of England and the Post Office, as well as optical, photographic, seismographic, and hydrological instruments, and aircraft instrumentation. Their photographic products included aerial cameras and cine projectors. Their product range was diverse, including a wide variety of machines and equipment, and they sold parts to many other contemporary instrument makers.

In 1876 the firm relocated again to larger premises near Kings Cross, London. They were the sole manufacturer of the pressure tube anemometer, designed by W.H. Dines in response to the 1879 Tay River bridge rail disaster. This established a successful pattern of work in which the company designed and produced equipment in full consultation and cooperation with their clients.

In 1916 the firm was incorporated as R.W. Munro Ltd. In the 1929 British Industries Fair catalogue, [448], they advertised as an optical, scientific and photographic exhibit, and as manufacturers of: Aircraft air speed indicators and recorders; Turn indicators; Anemometers, Wind direction indicators and recorders; Gauges; Microtomes; Tube scale dividing machines.

The firm continued to expand, and they relocated to Bounds Green, London, and finally to Essex. In 2013 the company name was changed again to become Munro Instruments.

Mural Arc

A mural arc is a general term used for an instrument similar to the mural quadrant, but with an arc that may be other than a quarter-circle.

Mural Circle

A mural circle is a graduated circle, permanently mounted on a vertical wall in the plane of the meridian, usually with a telescope attached to it, pivoted at the centre of the circle. It is used in a similar manner to the mural quadrant.

Mural Quadrant

A mural quadrant is a quarter-circle graduated arc with a telescope pivoted at the circle's centre. The instrument is mounted on a wall, usually in the plane of the meridian, and is used to measure the declination of stars as they pass the meridian.

Murnaghan Industries
Murnaghan Industries of Florida, USA. See **Coulter Optics**.

Murray & Heath
Murray & Heath, [50] [53] [264] [306] [320], were optical and scientific instrument makers of London, England. The firm was a partnership, formed in 1855 between **Robert Murray** and Robert Vernon Heath (1819-1895), who usually went by the name Vernon Heath. They made spectacles, cameras, microscopes, telescopes, opera glasses, field glasses, and other optical instruments.

Heath came from a wealthy family and had a strong interest in photography. Thus, Heath was able to fund their venture, and they began business at 43 Piccadilly, London. Two years after the firm was established Robert Murray died, and Heath continued the business. By this time Murray's son, Robert Charles Murray, was working for the business. In 1862 the company was sold to Charles Heisch, (1820-1892). In 1866 they moved to 6 Jermyn Street, and in 1869 the company was re-purchased by Robert C. Murray. From the 1850s until the end of the century they held a Royal Warrant, being opticians, philosophical, chemical and photographic instrument makers by appointment to Queen Victoria.

In 1882, following bankruptcy proceedings, the firm re-emerged as Robert C. Murray, but closed in 1890, when Robert C. Murray took the position of manager of the Photographic Apparatus & Chemical Department of **John J. Griffin & Sons**. In 1892 he left Griffin to re-open his own business, trading again as Robert C. Murray. The company finally closed some time prior to Robert C. Murray's death in 1918.

Murray, Robert
Robert Murray, (1798-1857) [18], was a philosophical instrument maker of 122 Regent Street, London, England. He served an apprenticeship, beginning in 1812, with **John Newman**, a London scientific instrument maker of the **Worshipful Company of Makers of Playing Cards**. Robert stayed in Newman's service for 43 years. In 1855, two years before his death, he formed a partnership with Robert Vernon Heath, trading as **Murray & Heath**, and went on to make spectacles, cameras, microscopes, telescopes, opera glasses, field glasses, and other optical instruments. By the time Robert died, his son, Robert Charles Murray (1839-1918), was already working for Murray & Heath.

Museum Microscope
See microscope.

Musschenbroek, Jan van
Jan van Musschenbroek, (1687-1748) [16] [256] [274] [378] [425], was a mathematical, physical, and philosophical instrument maker of Leiden, Netherlands. He was the son of **Johan Joosten van Musschenbroek**, whom he succeeded in business, and was the nephew of **Samuel van Musschenbroek**. His catalogue listed a large number of instruments, many of which he almost certainly did not make himself. He may have bought in telescope tubes and other components from suppliers, but he did grind and figure his own lenses. He carried out work for the Dutch mathematician and natural philosopher, Willem Jacob 's Gravesande (1688-1742), and produced instruments to his specifications, used by 's Gravesande in his lectures and demonstrations. In 1735 he was selling octants, but it is uncertain whether he made them or imported them from England. He was in contact with **Jonathan Sisson**, from whom may have imported them. In 1738, he published a description of his air-pump. In 1748, he was selling Smith's quadrants. A simple microscope made by Jan van Musschenbroek is described in detail in [424]. He started listing telescopes for sale in 1711, and subsequent catalogues list spectacles, magnifying glasses, moongazers, reading and burning glasses, and both refracting and reflecting telescopes. He is also known to have made: air pumps; magic lantern; Leeuwenhoek style aquatic microscopes; other simple microscopes of his own design. He devised a flexible microscope sample holding arm with ball joints, and invented the compass microscope.

Musschenbroek, Johan Joosten van
Johan Joosten van Musschenbroek, (1660-1707) [256] [274] [378], was a scientific, physical, and optical instrument maker of Leiden, Netherlands. He was the brother of **Samuel van Musschenbroek**, and the father of **Jan Van Musschenbroek**, and Petrus van Musschenbroek (1692-1761), who was a highly regarded professor of natural philosophy and mathematics. He was a pupil of his brother, Samuel and, when Samuel died at a young age, Johan inherited his workshop and tools. He continued the business successfully until his death in 1707, upon which his son, Jan van Musschenbroek, took over the business. Instruments bearing his name include: air pump; microscopes; aquatic microscope; microscope compendium.

Musschenbroek, Samuel van
Samuel van Musschenbroek, (1639-1681) [256] [274] [378], was a scientific and optical instrument maker of

Leiden, Netherlands. He was the brother of **Johan Joosten van Musschenbroek**, and the uncle of **Jan van Masschenbroek**. He was the son of a brass founder, and initially took up the same profession. By c1660 he had begun to make scientific instruments, and he is known to have supplied microscopes and lenses to **Christian Huygens**, and to the Dutch naturalist and microscopist Jan Swammerdam. Instruments bearing his name also include: air pump; set of Magdeburg hemispheres. Upon his death, his business was taken over by his brother, Johan.

Muybridge, Eadweard

Eadweard Muybridge, (1830-1904) [45] [159] [335], born in Kingston Upon Thames, England, was a photographer, and an innovator in early motion picture technology. At age 20 he emigrated to America, and in San Francisco he established his reputation as a landscape photographer. Leland Stanford, former governor of California and founder of Stanford University, commissioned him to resolve the question of whether a trotting horse has, at any time in its stride, all four feet off the ground.

Muybridge's progress on the project was hampered by his trial and acquittal for the murder of his wife's lover. He resolved the question of the horse's stride by means of a sequence of cameras, each with a thread attached to its shutter. As the horse passed, it tripped each shutter in turn, giving a rapid sequence of images. He showed that a horse does indeed have all four feet off the ground for part of its stride. He devised the Zoöpraxiscope as a means to display the sequence of images as a moving image, and used it to illustrate his lectures on animal locomotion.

Muybridge returned to Kingston Upon Thames in the 1890s and, upon his death, bequeathed his equipment and prints to the Kingston Museum.

N

Nachet, Camille Sébastien
Camille Sébastien Nachet, (1799-1881) [264], optical instrument maker of Paris, France, began his career as an optician, making lenses in **Charles Chevalier**'s workshop. He set up his own business in 1840, and supplied lenses to such makers as **Lerebours**, **J.G.A. Chevallier**, **Abraham Abraham** and **Andrew Pritchard**. In 1853 he was awarded a patent for a binocular system for microscopes. Camille's son, **Jean Alfred Nachet** began to work with him in the late-1850s, and in about 1862 they went into partnership as **Nachet et Fils**.

Nachet et Fils
Nachet et Fils, [264], were optical instrument makers of Paris, France, with a reputation for producing fine microscopes. The firm was a partnership beginning in the late-1850s between **Camille Sébastien Nachet** and his son **Jean Alfred Nachet**, and later in about 1890, a partnership between Jean Alfred Nachet and his own son, Albert. In 1896 Nachet et Fils acquired the company **Bézu, Hausser, and Company**, and by 1898, they had also absorbed the company of **Georg Oberhäuser**.

Nachet, Jean Alfred
Jean Alfred Nachet, (1831-1908) [264], optical instrument maker of Paris, France, was commonly known as Alfred Nachet, and was the son of **Camille Sébastien Nachet**. He began to work with his father in the late-1850s, and in about 1862 they went into partnership as **Nachet et Fils**. When Camille died Alfred continued in business as A. Nachet. In about 1890 Alfred's son, Albert, began to work with him, and they became a partnership, again using the name Nachet et Fils.

Alfred Nachet amassed a substantial personal collection of microscopes. His collection was catalogued in 1929 by his son, Albert.

Nachet's Prism
See prism.

Nagel, Dr. August
Dr. August Nagel was a camera designer who started his company, **Drexler & Nagel**, in Stuttgart, Germany, in 1908. Soon after this the company name was changed to **Contessa Camera-werke Drexler & Nagel**. In 1919 Nagel bought the **Nettel Camera-werke** company, and renamed the combined company to **Contessa-Nettel**. In 1926 Contessa-Nettel merged with **Ernemann**, **Goerz** and **ICA** to form the company **Zeiss Ikon**. In 1928, however, Dr. Nagel formed a new company, Dr. Nagel-werke, producing folding plate cameras. This company was purchased in 1932 by the **Eastman Kodak Company**. Although Dr. Nagel continued as managing director and chief designer, his name was no longer part of the company name.

Nagler, Al
Al Nagler, (b.1948) [245], is the founder of the company **Tele Vue**. He is best known as the designer of the Nagler eyepiece (see Appendix II).

Nairne, Edward
Edward Nairne, (1726-1806) [8] [18] [44] [46] [53] [276 #388] [396] [397], was a highly regarded English optical, mathematical, and philosophical instrument maker, born in Sandwich, Kent. He served his apprenticeship with **Matthew Loft**, beginning in 1741, and was freed of the **Spectacle Makers' Company** in 1748. He succeeded Loft in his business in London.

Addresses: Lindsay Row, Chelsea; *Golden Spectacles, Reflecting Telescope & Hadley's Quadrant* in Cornhill opposite the Royal Exchange, London; *Golden Spectacles* in Cornhill opposite the Royal Exchange (1752); Corner of Bartholomew Lane, Threadneedle St. (1752); Opposite the Royal Exchange in Cornhill (1753-1774); 20 Cornhill, Opposite the Royal Exchange (1772); 20 Cornhill (1772-1796).

A trade card gave his address as *At The Golden Spectacles, Reflecting Telescope & Hadley's Quadrant* in Cornhill, opposite the Royal Exchange, London. It lists many instruments made and sold by him, including: Spectacles; Reflecting and refracting telescopes; Microscopes; Camera obscuras; Magic lanterns; Opera glasses; Pumps and electrical apparatus; Barometers; Thermometers; Hadley's quadrants; Davis's quadrants; Globes; Compasses; Sun dials; Drawing instruments; Theodolites; Levels, and "…all other sorts of Optical, Philosophical and Mathematical instruments of the newest and most approved inventions…"

He was Master of the Spectacle Makers' Company from 1769-1773, and from 1795-1796. **Thomas Blunt** served his apprenticeship with Nairne, beginning in 1760, and they formed the partnership **Nairne and Blunt** in 1774. Among Edward Nairne's apprentices were **James Long**, who was freed in 1781, and **John Field**, who was freed in 1787. Edward Nairne also employed many other workmen, and supplied instruments to eminent natural philosophers of the day including Benjamin Franklin, Henry Cavendish and Joseph Priestley, as well as to Harvard University in the USA. He supplied optical instruments to the Royal Observatory at Greenwich, and was elected a Fellow of the Royal Society in 1776.

Nairne and Blunt

Nairne and Blunt, [8] [18] [44] [69], were optical, mathematical, and philosophical instrument makers of London, England. The firm was a partnership between **Edward Nairne** and **Thomas Blunt (I)**, trading *At The Golden Spectacles, Reflecting Telescope & Hadley's Quadrant* in Cornhill, opposite the Royal Exchange, London, England. The partnership was in place between 1774 and 1793, and they had a reputation for producing fine instruments.

Nash-Kelvinator Corporation

Nash-Kelvinator was a US company formed in 1937 following the merger of Nash Motors and the Kelvinator Appliance Company. During WWII they made military binoculars to **Bausch and Lomb** patterns.

Nasmyth, James

James Nasmyth, (1808-1890) [24] [44], was a Scottish engineer, famed for the invention of the steam hammer. He had a successful career with his engineering business, from which he retired in 1856. Being the owner of a foundry and a keen amateur astronomer, he was able to cast his own specula. He experimented with this, and produced mirrors that he successfully used in his telescopes. Upon retirement he settled in Kent, England, and pursued his hobby of astronomy. He built a 20-inch modified Cassegrain telescope with which he observed the heavens. He co-wrote the book *The Moon: Considered as a Planet, a World, and a Satellite* with James Carpenter (1840-1899). Telescopes built to his modified Cassegrain design have been used in the Lick Observatory in California, the W.M. Keck Observatory in Hawaii, and the Sierra Nevada Observatory in Spain.

National Geographic

National Geographic is a **Bresser** brand of optical instruments.

National Institute of Standards and Technology (NIST)

The NIST is the US national measurement standards laboratory, founded in 1901.

National Physical Laboratory (NPL)

The NPL is the national measurement standards laboratory for the United Kingdom.

In the area of optics, NPL designs and supplies calibrated precision scales for the validation of image analysis and vision measurement systems. They provide instruments and services for measurement and calibration of optical components.

Nedinsco

Nedinsco, the **Ned**erlandse **Ins**trumenten **Co**mpagnie, was founded in 1921 by **Carl Zeiss**, in Vlissingen, Netherlands, designing and producing optical instruments, including rangefinders, periscopes, signal lamps, measurement and fire-control systems, and medical equipment. Shortly after the company was created it was moved to Venlo, Netherlands. During WWII they produced optical munitions for the German war effort, including riflescopes, and their instruments were marked with the **ordnance code** jux. In 1953 the company was acquired by the entrepreneur, G. Beusker, whose family continued to run the company. The company produces optical products for the defence, semiconductor, space, and industrial markets.

Needlemakers' Company

See Appendix IX: The London Guilds.

Negretti & Co., Henry

See **Henry Negretti**.

Negretti & Zambra

Negretti & Zambra, [18] [246] [247] [306] [379] [438], were opticians, and optical, mathematical, philosophical, nautical, and photographic instrument makers of London, England. Addresses: 11 Hatton Garden (1850-1859); 59 Cornhill (1857-1872); 68 Cornhill (1857-1859); 1 Hatton Garden (1859-1867); 107 Holborn Hill (1859-1860); 122 Regent St. (1862-1901); 153 Fleet St. (1865-1873); 103 Hatton Garden (1867-1870); 38 Holborn Viaduct (1869-after 1901); Holborn Circus, Holborn Viaduct (1870-1890); 2 Charterhouse St. (1870-1885); 45 Cornhill (1872-after 1901).

The firm was established in 1850 as a partnership between **Henry Negretti**, and **Joseph Zambra**. They initially made mostly meteorological instruments such as barometers and thermometers, and were responsible for significant improvements in the designs of these instruments. For these they won awards in the International Exhibitions in both London and Paris. Shortly after the retirement of **John Newman** in 1859, Negretti & Zambra acquired Newman's company from his son, John Frederick Newman.

Following the deaths of the founders, the families of Negretti & Zambra continued the business. They began to make optical instruments including telescopes, theodolites, levels, gunsights and others. The company produced a wide range of scientific, mathematical and optical instruments. The *Minim* binocular, patented by **John Henry Barton**, was exclusively sold by Negretti &

Zambra. In 1901 Negretti & Zambra acquired the business of **J. Hammersley**.

According to [247], following WWI they decided to discontinue manufacturing optical instruments to focus on industrial, aeronautical, and meteorological instruments. In 1919 Negretti & Zambra was listed as a member of the **British Optical Instrument Manufacturers' Association**.

The company of Negretti & Zambra prospered, moving to successively larger premises until the business was taken over in 1981 by Western Scientific.

Negretti, Henry

Enrico Angelo Ludovico Negretti, usually known as Henry, (1818-1879) [18] [46] [247] [276 #2194] [379] [438], was a glassblower and philosophical instrument maker of London, England. Addresses: Hatton Garden (1834-1835); 19 Leather Lane, Holborn (1839-1848); 20 Greville St. (1840); 9 Hatton Garden (1849).

He emigrated to England from Como, Italy in 1830. According to [247] he served an apprenticeship with **Caesar Tagliabue**. Note that [18] and [438] state that he served an apprenticeship with **F.A. Pizzala (I)**, and [46] states that he learned from both men. He attended the London Mechanics Institute from 1834 until 1835. Beginning in 1840/41, he worked in partnership with barometer and thermometer maker Jane Pizzi, trading as Pizzi & Negretti at 19 Leather Lane, London, until 1844. From 1845 until 1848 he traded as Henry Negretti & Co.

He also worked with Francis Ciceri, trading as barometer & thermometer makers at 31 Brook St., Holborn and 19 Leather Lane, Holborn, until the partnership was terminated 1846 [35 #20628 p2816]. Francis Ciceri continued in business at 31 Brook St., and Henry Negretti at 19 Leather Lane.

In 1850 he began to work in partnership with **Joseph Zambra**, trading as **Negretti & Zambra**. He continued to work in this partnership until his death in 1879. He was succeeded as partner in the business by his son Henry Paul Joseph Negretti, and the company continued to operate until 1981.

Neill Brothers

Neill Brothers, (fl.1850-1863) [18] [304], were wholesale and retail watchmakers, jewellers, silversmiths, and opticians, of 23 High St., Belfast, Ireland. They succeeded the business of **Robert Neill & Sons**. Instruments bearing their name include a 7½" sextant, a marine stick barometer, and a two-day marine chronometer. They were succeeded in business by **James Neill & Co.**, and **John R. Neill**.

It is reasonable to speculate that James Neill and John R. Neill were the sons of Robert Neill, and that they traded together as Neill Brothers until they went their separate ways, and began to trade under their own names.

Neill, James

James Neill, (1856-1880) [18] [304], was a watch and chronometer maker, jeweller, and optician, of 6 Donegall Place, Belfast, Ireland. By 1865 he was trading as James Neill & Co. at 8 Donegall Place, having succeeded **Neill Brothers** in business. From 1870 his trading address was 14 Donegall Place and, additionally in 1878, 33 Castle Lane. Instruments bearing his name include a single draw brass and leather telescope with 2" objective, 27½" length when closed.

Neill & Co, James

See **James Neill**.

Neill, John R.

John R. Neill, (fl.1865-1880) [304], was a watch and chronometer maker, jeweller, silversmith, and optician, of 23 High St., Belfast, Ireland. By 1870, he was trading at 21 High St. He succeeded **Neill Brothers** in business.

Neill, Robert

Robert Neill, (also sometimes spelled Neal, Neil, or Neale) (fl.1805-1846) [18] [304], was a watch and clockmaker, jeweller, silversmith, optical and mathematical instrument maker of 25 High St., Belfast, Ireland. From 1809 until 1818 he worked in partnership with **Henry Gardner**, trading as **Gardner & Neill**. In 1824 he was additionally listed as trading at 21 High St. In 1842 he began to trade as Robert Neill & Sons, and continued until 1846. He was succeeded by **Neill Brothers**.

Neill, Robert & Sons

See **Robert Neill**.

Neill, Sharman D.

Sharman D. Neill, (fl.1884-1921) [304], was a jeweller, silversmith, optician, and watch and clock maker of 12 Donegall Place, Belfast, Ireland, adjacent to the address previously occupied by **James Neill & Co.** By 1907 he was trading at 22 Donegall Place. Instruments bearing his name include a single draw brass telescope, 24" length when closed. He later specialised in turret clocks.

Nephelometer

A nephelometer is an instrument that measures the size and concentration of particles suspended in a liquid or gas medium. This is achieved by measurement of scattering of light from more than one angle. The design of a nephelometer may utilise a Lummer-Brodhun cube.

Nettel Camera-werke

Nettel Camera-werke was a camera maker, established as Süddeutsches Camerawerk Körner & Mayer in 1902 in Sontheim am Neckar, Germany. In 1904 or 1905 the name was changed to Camerawerk Sontheim, and in 1908 or 1909 the name was changed again to Nettel Camera-werke. In 1919 the Nettel Camera-werke was purchased by **Dr. August Nagel**, who combined the company with **Contessa Camera-werke Drexler & Nagel** to form the new company **Contessa-Nettel**.

Newbold & Bulford

Newbold & Bulford, [44], of London, England were makers of telescopes and binoculars during the late-C19th and early C20th. They began to use the trade name *Enbeeco* and, after WWII, by 1951 or earlier, the name *Enbeeco* was used extensively on their products. Research also suggests that in the early 1960s, Newbold and Bulford may have used the *Enbeeco* name as a brand for imported Japanese binoculars and telescopes.

The 1951 *Directory for the British Glass Industry* [327], lists Newbold & Bulford Ltd., of Enbeeco House, Roger Street, Grays Inn Road, London, established in 1800. It lists them as producing Ophthalmic lenses; Prisms; Spectacles; Ophthalmic instruments; Binoculars; Microscopes; Telescopes; Magnifiers. Their trade name is listed as *Enbeeco*. The managing director was B.P. Turner.

In 1960, Newbold and Bulford collaborated with **Ross Ltd.** to produce the Ross Enbeeco 13x60 Binoculars. In 1975 Newbold & Bulford bought the stock and goodwill of the company **J.H. Steward**, which was closing its business. In October 1991, Newbold and Bulford, then trading in Edenbridge, Kent, became Pyser NB (Trading) Ltd., a private limited company. This became part of what is **Pyser-SGI**, trading from the same address.

Newman & Guardia Ltd.

Newman & Guardia Ltd., [61] [306], were camera and camera shutter manufacturers of London, England. Addresses: 71 Farringdon Rd., Clerkenwell (1891-1893); 92 Shaftsbury Avenue (1893-1897); 90 & 92 Shaftsbury Avenue (1897-1909); 17-18 Rathbone Place, Oxford St. (1909-1929); 63 Newman St., Oxford St. (1929-1938); 19-23 Wells St., Oxford St. (1938-1948); 55 Berners St. (1948-1956); 60 Berners St. (1956-c1959).

The company was a partnership between the camera and shutter designer **Arthur S. Newman** and the businessman Julio Guardia. The partnership was formed following the dissolution of **Newman & Simpson** in 1891, and was trading by 1892. They produced various models of camera and, in 1897, they began to produce their own version of the Cinematograph, the *N&S Kinematograph*. Guardia died in 1906, and in 1908 Newman left the company. Newman & Guardia continued to produce a number of models of camera until receivers were appointed in 1970.

Newman & Simpson

Newman & Simpson, [61], was a partnership between **Arthur S. Newman** and Frank Lindsay-Simpson, formed in 1888, manufacturers of scientific apparatus, specialising in camera shutters and cameras. Initially the firm traded at 14 St. John's Square, Clerkenwell, London, England. From 1888 their address was 11 Albemarle Street, Clerkenwell, and from 1890 until 1891 they traded at 71 Farringdon Road. One of their customers was **Robert Abraham**. The partnership was dissolved by mutual consent in 1891, [35 #26229 p6734].

Newman & Sinclair Ltd.

Newman & Sinclair Ltd., [61] [306], of 2 Salisbury Rd., Highgate Hill, Highgate, London, England, was a partnership between **Arthur S. Newman** and James A. Sinclair, formed in 1910, producing mainly Newman-Sinclair camera shutters, and cine cameras designed by Newman. They produced cameras and shutters to be marketed by **James A. Sinclair & Co. Ltd.** In 1912 they began to manufacture **Kazimierz Prószyński**'s Aeroscope, for which Newman collaborated on the design.

Newman, Arthur S.

Arthur Samuel Newman, (1861-1943) [335], was a British inventor, camera designer, and camera shutter designer, with many patents to his name. He is believed to have worked for a London microscope maker from 1877 until 1880, at which time he left to work for **H.&E.J. Dale**. When Dale acquired **F.J. Cox**, Newman worked for their photographic department. By 1889, he had left Cox, and formed a partnership with Frank Lindsay-Simpson, trading as **Newman and Simpson**, scientific instrument makers, making camera shutters and cameras. This partnership continued until 1891, and by 1892 he had formed the partnership **Newman & Guardia Ltd.**, making camera shutters and cameras. Newman left Newman & Guardia in 1908, and in 1910 he formed a new partnership, trading as **Newman & Sinclair Ltd.**, producing cine cameras and Newman-Sinclair camera shutters. Newman worked with **Kazimierz Prószyński** on the design of Prószyński's Aeroscope.

Newman, John

John Frederick Newman, (1783-1860) [18] [46] [53] [276 #1396], was an optical, mathematical, philosophical, and nautical instrument maker of London, England. He served an apprenticeship, beginning in 1799, with the mathematical instrument maker Philip Brock of Fetter Lane, London, and was freed of the **Worshipful Company of Makers of Playing Cards** in 1807. He began to trade at 11 Windmill Road, Camberwell, and by 1809 he had moved to 7/8 Lisle Street, Leicester Square. In 1812 **Robert Murray** began an apprenticeship with Newman, and stayed in his service for 43 years prior to setting up his own business. Newman supplied laboratory equipment and chemicals to the Royal Institution, and worked there as a laboratory assistant. Meanwhile his instrument making business thrived, and he supplied the trade as well as notable scientists and explorers. His customers included Charles Wheatstone and Isambard Kingdom Brunel. In 1827 Newman moved to larger premises at 122 Regent Street, where he continued to supply scientists and explorers with a wide variety of instruments. His son, John Frederick Newman (II), joined him in a partnership which continued until John senior retired in 1859. Shortly after this, the firm was taken over by **Negretti & Zambra**.

Newton & Co.

Newton & Co., [44] [109] [403] [404] [405], were globe makers and scientific and optical instrument makers of 3 Fleet St., London, England. In their advertising literature, they claimed the company had been established by 1704, when J. Newton was head of the firm. Evidence indicates, however, that the origins of the firm were with the globe maker John Newton (1759-1844). By 1783, having worked in partnership with the globe maker William Palmer for only a few years, John Newton started his own business at 128 Chancery Lane. His son William Newton continued the business, by then at 66 Chancery Lane, and expanded it to become a patent agency, then in c1840 went into partnership with his own son, **William Edward Newton**, trading as Newton & Son. In 1851, W.E. Newton, and his first cousin **Frederick Newton**, began trading in partnership as **W.E.&F. Newton**, and in 1858 the company became Newton & Co. In c1860 W.E. Newton left the partnership, and Frederick continued the business.

The company laid claim to various royal appointments over the years, including: Queen Victoria; Albert, the Prince Consort; King Edward VII, and King George V.

A catalogue, believed to be from the 1860s, advertised "Mathematical, Optical and Philosophical instruments manufactured and sold by Newton & Co. Working Opticians and Globe makers to the Queen...", and lists their product categories as: Microscopes; Telescopes; Opera glasses, &c.; Spectacles; Surveying instruments; Nautical instruments; Drawing instruments; Barometers; Thermometers; Photographic apparatus; Air pumps, &c.; Chemical cabinets; Electric and galvanic apparatus; Orreries; Newton's globes; Magic lanterns; Dissolving apparatus; Dissolving views.

From c1902 the company took on premises at 1 Little James St., off the Grays Inn Road. Following Frederick's death in 1909, the business was continued by his son, **H.C. Newton**, who expanded into production of x-ray tubes, and formed the partnership **Newton & Wright**, which operated concurrently with Newton & Co. By 1912 the company had begun a further expansion, taking on premises at 37 King St., Covent Garden (lanterns and slides), and at 72 Wigmore St. and, from 1906, Newton Works, 471 Hornsey Road, and 193 Fairbridge Road.

In 1914 the *Whitaker's Red Book of Commerce or Who's Who in Business*, [110], listed Newton & Co. as opticians and scientific and electrical instrument makers, 72 Wigmore St., London. "In 1850 the partners were W.E. and F. Newton. Present partners: H.C. Newton and R.S. Wright. Specialities: Optical lanterns and slides; Are also ophthalmic and sight testing opticians. Hold several patents relating to optical and electrical instruments..."

The firm appeared as a listed exhibitor in the British Industries Fair in 1947, [450], advertising: Epidiascopes; diascopes for slides and filmstrip; Projection microscopes; Slides on all subjects.

The 1951 *Directory for the British Glass Industry*, [327], shows the firm at 72 Wigmore Street, London, with works at 471 Hornsey Road, producing apparatus incorporating optical glass: Diascopes; Epidiascopes; Filmstrip diascopes; Projection microscopes; Micro-attachments; Binocular-focusing magnifiers. The managing director at that time was A.E. Smith.

In 1959 the company was acquired by the Victor Corporation, based in the USA.

There is no known connection with the Liverpool firm of **H. Newton & Co.**

Newton & Co., H.

Henry Newton & Co., [306], of 16 Lord Street, and 5 South John Street, Liverpool, England, was a wholesale and export photographic apparatus manufacturer. The company was established in 1851, and succeeded in 1898 by Fred V.A. Lloyd. There is no known connection with the London firm of **Newton & Co.**

Newton & Wright

Newton & Wright, [44] [404] [405], of 72 Wigmore Street, London, England, was a partnership formed in 1911 between **Russel Wright** and **Herbert C. Newton**. They were electrical and scientific instrument makers, specialising in x-ray and electro-medical apparatus. From 1913 they traded at 417 and 496 Hornsey Road. The partnership continued until 1920 when it was dissolved, [**35** #32050 p9257], probably in order to incorporate the company, since it continued as Newton & Wright Ltd. The company went into voluntary liquidation in 1937 [**35** #34376 p1443]. This may have been for financial restructuring, because the Metropolitan Vickers Electrical Company acquired a major interest in Newton & Wright Ltd. in 1945, and in 1948 a subsequent merger formed the medical and industrial x-ray equipment company Newton Victor Ltd.

Newton NM1 Microscope

The Newton NM1 microscope, [**314**], was a folded optics microscope produced in several variants from 2007 until 2017, by **Cambridge Optronics Ltd.**, for the Millennium Health Foundation charity. The NM1 was manufactured in China, and retailed at a significantly higher price than many folded optics microscopes of the time.

Newton, E.T.

Edward T. Newton, (fl.1847-c1870) [**322**] [**358**] [**379**], was an optical, mathematical, and chemical instrument maker of Cross St., Camborne, and St. Day, both in Cornwall, England. He succeeded **William Wilton** in business, and traded in partnership with his son as E.T. Newton & Son. Instruments bearing the name of E.T. Newton include: surveyor's level with telescope, signed *E.T. Newton (late Wilton), Camborne*; Lean's miner's theodolite, signed *E.T. Newton, St. Day, Cornwall*; telescopic mining transits; altitude theodolite; sundial; surveying compass. Instruments bearing the name E.T. Newton & Son include: Theodolite.

Newton, Frederick

Frederick Newton, sometimes known as Frederic, (1824-1909) [**18**] [**109**] [**403**] [**404**] [**405**], was an optician, born in Hitchin, Hertfordshire, England. He moved to London, where he was freed by purchase of the **Spectacle Makers' Company** in 1850. He formed a partnership with his first cousin, **W.E. Newton** in 1851, trading as **W.E.&F. Newton** and, from 1858, **Newton & Co.** By c1860 W.E. Newton had left the partnership W.E.&F. Newton, and Frederick continued the business. He indicated in his advertisements that there was a family relationship with **Sir Isaac Newton**. Evidence suggests that Frederick was indeed a sixth generation descendent of Sir Isaac Newton's uncle, Robert Newton. Frederick was succeeded in the partnership Newton & Co. by his son, **H.C. Newton**.

Newton, H.C.

Herbert Charles Newton, (1859-1940) [**109**], was the son of **Frederick Newton**. In 1909, upon the death of his father, he took control of **Newton and Co.** as owner/proprietor. He worked with **Lewis Wright** on designs for optical instruments. In 1911 he formed a partnership with Lewis Wright's son, **Russel S. Wright**, trading as **Newton & Wright**.

There is no known connection with the Liverpool firm of **H. Newton & Co.**

Newton, Sir Isaac

Sir Isaac Newton, (1642-1727) [**45**] [**46**] [**176**] [**177**] [**388**], was an English Physicist and Mathematician. He was the author of, among other works, *Principia Mathematica* and *Opticks*, and is credited as the inventor of the Newtonian reflecting telescope.

Born to a farming family in Lincolnshire, Isaac Newton broke the family mould and, in 1661, embarked on a university education at Cambridge. There he began to form his interests in mathematics and natural laws and, in 1665-66, he had his *annus mirabilis*, the year in which he discovered and formed the ideas from which his later work sprung. During his career he progressed mathematical theory in many areas, most notably in what is now referred to as calculus. In mechanics he produced many fundamentally important results, including the three laws of physics known as *Newton's laws* and the *universal law of gravitation*, which have come to be collectively known as *classical mechanics*.

Newton took a keen interest in optics. He investigated refraction, and discovered that a prism would split white light into a diverging spectrum of colours, thus demonstrating that the white light from the sun is not homogeneous as was commonly thought. In 1667 he was elected a fellow of Cambridge University and, in 1669, as Lucasian professor. With few teaching duties, this enabled him to focus his attention on his research. The lectures he gave were on optics, and he went on to continue his optics research into the interference phenomenon now known as *Newton's rings*.

Based on the varying refrangibility of rays of light he concluded that chromatic aberration could be solved in a reflecting telescope more readily than in a refracting telescope, and in 1669 he constructed the first ever reflecting telescope. This design, using a concave parabolic primary mirror, and a plane secondary mirror to divert the light out of the side of the telescope to an eyepiece, was to become known as the Newtonian telescope. He presented his invention to the Royal Society and, shortly afterward, by way of explanation, submitted a condensed version of his theory of colours. The publication if his theory of light triggered a disagreeable exchange of correspondence between Newton and **Robert Hooke**, following which Newton temporarily distanced himself from the Royal Society. Following the death of Robert Hooke in 1703 Newton became once more active in the Royal Society, and later that year he was elected its president.

Newton, W.E.&F.
W.E.&F. Newton, [10] [109] [404] [405], of 3 Fleet St. London, England, was a partnership between **William Edward Newton** and **Frederick Newton**. The firm began as a globe maker, and later began to make dissolving view magic lanterns and lantern slides. The partnership lasted from 1851 to 1858 when the business name was changed to **Newton and Co.** W.E. Newton left the partnership in c1860.

Newton, William Edward
William Edward Newton, (1818-1879) [109] [403] [404] [405], began working with his father, William Newton, in c1840, in his globe making and patent agent business in Chancery Lane, where they traded as Newton & Son. In 1851, he entered a partnership with his first cousin, **Frederick Newton**, trading as **W.E.&F. Newton**, and changing the name to **Newton & Co.** in 1858. In c1860 he moved back to the family firm in Chancery Lane, where he pursued a career as a surveyor, civil engineer, and patent agent.

Newtonian Telescope
See telescope.

Nickel
Nickel was a German riflescope manufacturer, and was taken over by **Gerhardt Optik GmbH** in 1995. Nickel subsequently became a registered brand of Gerhardt Optik, used for their riflescopes.

Nicol Prism
See prism.

Nicol, William
William Nicol, (1768-1852) [155], was a Scottish geologist and physicist who lectured in natural philosophy at the University of Edinburgh. He was notable for having developed new methods of cutting thin sections of fossil woods and minerals for inspection under a microscope using transmitted light. He is best remembered, however, for his invention in 1928 of the Nicol prism for producing polarized light.

Niépce, Joseph Nicéphore
Joseph Nicéphore Niépce, (1765-1833) [37] [70], of Chalon, France, was a pioneer of photography, and a cousin of **Claude Niépce de St Victor**. He experimented with bitumen of Judea, utilising the fact that it becomes insoluble with the action of light. Thus, with a long exposure and the use of a solvent, he was able to produce the first enduring photographic image. Niépce referred to his process – that of making a picture by means of exposure to the sun – as a heliograph. In 1829 Niépce and **Louis Daguerre** formed a partnership to develop their mutual interest in capturing camera obscura images. Niépce died shortly afterwards, before the partnership could produce significant results.

Niépce de St Victor, Claude Félix Abel
Claude Niépce de St Victor, (1805-1870) [37], of Saint-Cyr, France, was a pioneer of photography, and a cousin of **J.N. Niépce**. He became interested in photographic processes after making experiments on fading dyes in military uniforms. Based on his experiments he began to use albumen, derived from egg whites, as a carrier for silver iodide. (See also photochrom). This process required long exposures, but produced sharper images than those produced using the paper negatives in **Talbot**'s calotype process. This was the first practicable glass negative process.

Niezoldi & Krämer
Niezoldi & Krämer, of Munich, Germany, was a manufacturer of hand-held 8mm movie cameras, founded in 1925 by Georg Niezoldi and Georg Krämer. In 1962 the company was taken over by **Braun AG**, the consumer products manufacturer known for its shavers, who used the brand-name *Nizo* for their movie camera products.

Night Optics
Night Optics is a US brand of night-vision riflescopes, binoculars, monoculars, and goggles. The Night Optics brand is part of the **Vista Outdoor Inc.** group.

Nihon Seiko Kenkyusho, Ltd.
Nihon Seiko Kenkyusho, Ltd., [136] [167], of Nozawa, Setagaya-ku, Tokyo, Japan, produced fine quality telescopes and other optical instruments. The company was established sometime in the early 1930s, and ceased trading in 1992. They had parts manufactured by subcontractors, and assembled the instruments, which they marketed both in Japan and abroad. They sold under the brand-names *Polarex*, *Vidhya-Bhand* and *Seiko Scope*, and were marketed under the US brand **Unitron**. They also supplied products such as eyepieces, viewfinders, and guide-scopes to *Vernonscope*, *University Optics*, and *Astro-Physics*.

Nikon
In 1917, three of Japan's leading optics manufacturers merged to form the optical company known as **Nippon**

Kōgaku Kōgyō Kabushikigaisha. In 1932 they began to use the *Nikkor* brand-name for camera lenses, and in 1946 they began to use the *Nikon* brand-name for small cameras. In 1988 Nippon Kōgaku K.K. was renamed Nikon Corporation.

In 2003 Nikon entered a joint venture with **Trimble**, which traded as Nikon-Trimble Co. Ltd. The new company was based on designing and manufacturing surveying instruments, including total stations, and represented both parent companies in the Japanese market.

Today the Nikon Corporation produces optical glass, as well as high quality camaras, rifle scopes, spotting scopes, binoculars, ophthalmic instruments, and ophthalmic lenses.

NIL
See **Nottingham Instruments Ltd.**

Nippon Kōgaku Kōgyō Kabushikigaisha
Nippon Kōgaku Kōgyō Kabushikigaisha, (Japan Optical Industry Co. Ltd.), was established in 1917 when the Optical Instruments Department of Tokyo Keiki Seisakusho merged with the Reflecting Mirror Department of Iwaki Glass Manufacturing. The newly formed company then immediately merged with Fujii Lens Manufacturing.

In 1918 Nippon Kōgaku K.K. began research and production of optical glass, and in 1927 they began mass production of optical glass. In 1932 they began to use the *Nikkor* brand-name for camera lenses, and in 1946 they began to use the *Nikon* brand-name for small cameras. The company's products included binoculars, astronomical telescopes, cameras, camera lenses, ophthalmic lenses, and microscopes.

The **JB code** used by Nippon Kōgaku Kōgyō Kabushikigaisha was JB7. In 1954 **Ehrenreich Photo Optical Industries** became the sole US importer and distributor of *Nikon* products. In 1981, after the death of Joseph Ehrenreich, Nippon Kōgaku K.K. purchased Ehrenreich Photo Optical Industries, and renamed it Nikon Inc. In 1988 Nippon Kōgaku K.K. was renamed **Nikon Corporation**.

Nishimura Co. Ltd.
Nishimura Co. Ltd. is a manufacturer of professional telescopes, domes, and instrumentation, based in Kyoto, Japan. The company began business in the last decade of the C19th, and made their first telescope in 1926. The company was incorporated in 1963. Their maker's mark is illustrated in Appendix IV.

Nitsche & Günther Optische Werke KG
Nitsche & Günther Optische Werke KG were spectacle and goggle makers of Rathenow, Germany. They advertised their meniscus lenses as the perfect eyeglass lenses. During WWII, they principally produced goggles and lenses for the German military, and marked their products with the **ordnance code** gxh. Instruments bearing their name include: spectacles; goggles; optician's case of eyesight testing lenses.

In 1945, following the end of the war, Nitsche & Günther Optische Werke KG was nationalised under Soviet occupation and, in 1946, became Rathenower Optischen Werke mbH. In 1948 Emil Busch AG Optische Industrie (see **Emil Busch**) was also nationalised, and the two companies were combined to become VEB Rathenower Optische Werke. In 1966, VEB Rathenower Optische Werke became a part of VEB Carl Zeiss Jena. In 1990, following the reunification of Germany, the company was separated from Zeiss, and privatised to become **Rathenower Optische Werke GmbH**.

Nivoc
Nivoc was a brand-name used by **F.E. Becker & Co.**, and **W.&J. George Ltd.** for laboratory apparatus and instruments during the 1930s. The brand-name continued to be used by **W.&J. George and Becker**, and subsequently by **Griffin and George Ltd.** Examples include a *Nivoc* branded spherometer.

Nobert, Friedrich Adolph
Friedrich Adolph Nobert, (1806-1881) [**280** Ch.8] [**264**], maker of microscopes and gratings, was the son of a clockmaker in Barth, Mecklenburg-Vorpommern, Germany (at that time under Swedish rule). He trained as a clockmaker, and made astronomical instruments for his own use in verifying the timekeeping of his clocks. He learned methods of scale dividing at the Technical Institute in Berlin, and made a circular dividing engine. He went on to rule spectrum plates, diffraction gratings, test objects, micrometers, and the scales on astronomical and mathematical instruments. He also made a number of fine microscopes.

In 1845 he devised a dividing engine that could be used to produce fine ruled parallel lines for use by microscopists as a micrometer. He was able to use his machine with a diamond point to rule parallel lines with spacings, ultimately, as fine as between $0.12 \mu m$ and $0.11 \mu m$. In doing so he introduced the first physical test for the resolution of an optical microscope, with line-sets of varying separations down to those beyond the resolving power of an optical microscope.

Noctovist
Noctovist is a **Ruhnke** model of binoculars.

Norie, John William
John William Norie, (1772-1843) [18] [276 #1019] [358], was and mathematics teacher at an academy attached to **William Heather**'s Navigation Warehouse at 157 Leadenhall St., London, England. He succeeded William Heather in business in 1812, and was required to settle with the late William Heather's creditors [35 #17929 p926]. He began to operate the Navigation Warehouse at 157 Leadenhall St., and traded under his own name until 1816, as well as trading in two different partnerships.

He worked in partnership with Royal Navy officer George Wilson, trading as **Norie & Wilson** at 157 Leadenhall St., and 156 Minories, from 1812 until 1820. George Wilson's son, the nautical instrument maker **Charles Wilson**, also worked in this partnership.

He also traded as **J.W. Norie & Co.**, beginning in 1816, and Charles Wilson also joined this partnership. J.W. Norie retired in 1839 and, when the partnership was dissolved in 1840 [35 #19895 p2123], it was between J.W. Norie, Charles Wilson, and James D. Welch. Following the dissolution of the partnership, Charles Wilson continued in business under his own name.

Norie & Co., J.W.
See **John William Norie**.

Norie & Wilson
See **John William Norie**.

Norman, John Thomas
John Thomas Norman, (c1807-1893) [50] [319], was a brush maker, optical and scientific instrument maker, and preparer of microscope specimen slides. He claimed that his business was established in 1846, and traded at 10 Fountain Place, City Road, London, England, the address which was later renamed to 178 City Road. According to his trade card, "Manufacturer of microscopes, telescopes, and other scientific instruments". The microscope slide preparation side of his business was carried out by himself and a number of his family members. His first son was also named John Thomas Norman – no evidence has been found that he made optical instruments, but microscope slides bearing that name are at risk of misattribution. He was succeeded in business by his sons, Alfred and Edwin. Instruments bearing his name include microscopes; microscope objectives.

Nörremberg Polariscope
A Nörremberg polariscope, (sometimes spelled Nörremberg), is one in which a polarizing reflector, analyser, eyepiece, sample mount, and horizontal mirror, are all mounted on a vertical optical axis, supported between two vertical columns.

In the Nörremberg design the light is polarized by means of a tilted glass plate, using the phenomenon of polarization by reflection (see Brewster angle). When unpolarized light is incident on a reflecting surface the horizontally polarized element is preferentially reflected, while the vertically polarized light is preferentially refracted. The tilted glass plate may be used in one of two ways. It may be used to reflect partially polarized light directly up into the optical axis. Alternatively, it can be used to reflect light down onto the vertical mirror, and back up, through the plate, into the optical axis.

The design of the Nörremberg polariscope is credited to the German physicist, Johann Gottlieb Christian Nörremberg (1787-1862), who used this arrangement in his polariscope of 1839. Some derivatives of his design are constructed with a single vertical column, and some derivatives allow for the instrument to be used at an inclined angle, rather than vertical.

Northen, Edward
Edward Northen, (fl.1830-1848) [18] [319] [322], of 50 Lowgate, Hull, England, was an optician, and optical and philosophical instrument maker, and watch maker, chronometer maker, silversmith, and jeweller. He was the son of **Richard Northen**, and succeeded **Richard Northen & Son** in business. He filed for insolvency in 1849 [35 #20980 p1704].

Northen, Richard
Richard Northen, sometimes Northern, (fl.1790-1840) [18] [276 #878] [319] [322], of Lowgate, Hull, England, was an optical, mathematical, nautical, and philosophical instrument maker, and watch maker and jeweller. Addresses: 46 Lowgate (1803-1834), and 50 Lowgate (1838-1840). He is known to have sold: hygrometer; spirit level; thermometer; drawing instruments. Instruments bearing his name include: microscope; drawing instrument set. He worked in partnership with his son, **Edward Northen**, beginning in 1830, trading as **Richard Northen and Son**, and was succeeded in business by Edward.

Northen, Richard & Son
Richard Northen & Son, [18] [319] [322], of Hull, England, was a partnership between **Richard Northen** and his son **Edward Northen**. Addresses: 46 Lowgate (1830-1834), and 50 Lowgate (1835-1841).

Northrop Grumman Corporation
The Northrop Grumman Corporation was formed in 1994 when the Northrop Corporation acquired the Grumman

Corporation. The Northrop Corporation began business in 1939 as Northrop Aircraft Incorporated in Hawthorne, California, building military aircraft. In 1959 they changed their name to Northrop Corporation. The Grumman Corporation was established in 1930 as the Grumman Aeronautical Engineering Company, building military aircraft.

In 1962, Litton Industries acquired **C. Plath**. Sperry Marine was formed in 1997 with the combination of C. Plath, Decca Marine and Sperry Gyrocompass, and C. Plath changed its name to Sperry Marine in 2000. In 2001 Sperry Marine became part of the Northrop Grumman Corporation.

The Northrop Grumman Corporation has many interests in optical instruments, carried out by Sperry Marine, and these other subsidiaries:

Cutting Edge Optronics (CEO): Laser diode arrays and laser system hardware.

AOA Xinetics (AOX): Adaptive optical systems; Deformable mirrors; Wavefront sensors; Precision actuators; Lightweight passive and active hybrid CERAFORM silicon carbide optics and precision structures.

Synoptics: Manufacturer of synthetic crystals and optical components used primarily in the solid-state laser industry.

Nottingham Instruments Ltd.

Nottingham Instruments Ltd., (NIL), [80], was a shadow company set up by the British Ministry of Supply, and incorporated in June 1941. It was operated by **Ross Ltd.** and **E.R. Watts & Son**, and the factory occupied premises in Nottingham, England, owned by the Players Cigarette Company. Here, during WWII, the two companies duplicated the manufacture of their products away from London.

Novak & Company, Kenneth

Kenneth Novak & Co. of Ladysmith, Wisconsin, USA, was a company run by Kenneth Novak (1941-2004) making and supplying parts for telescopes. His 1990 catalogue listed a large variety of parts including: Spiders and secondary mirror holders; Secondary mirrors; Newtonian and Cassegrain focusing mounts; Camera adaptors; Diagonals; Dobsonian bearing rings; Ring mounts; Cassegrain baffle sets; Newtonian and Cassegrain mirror mounts; Books.

Novosibirsk Instrument-Making Plant (NPZ)

The company was established in 1905 as the Precision Mechanics Plant in Krasnogorsk. In 1941 it was evacuated to Novosibirsk, and became the Novosibirsk Instrument-Making Plant (Новосибирский приборостроительный завод), NPZ (НПЗ). They manufacture and sell telescopes under the **Tal** brand, as well as binoculars, monoculars, spotting scopes, night-vision binoculars and sights, and other specialised optics. For Russian and Soviet makers' marks, see Appendix V.

Numerical Aperture

Numerical aperture, [7], is a term used to characterise the resolving power of an optical system, and is commonly used to describe the resolving power of a microscope objective. The medium separating the microscope slide and the objective is usually either air or, in the case of an immersion objective, immersion oil or water. If the refractive index of this medium is n_o, and the angular aperture of the objective is θ, then the numerical aperture, NA, is given by:

$$NA = n_o \sin \theta$$

The numerical aperture of an objective may be calculated based on the optical design, or it may be measured using an apertometer. See angular aperture.

Nye Optical Co.

Nye Optical Co. of Long Beach, California, USA, was a company that specialised in lenses, lens design and amateur telescope making supplies. An example of a photographic lens from them is a 150mm f/1.4 designed for imaging in ultraviolet and infrared, with a spectral range described as being from 180nm to 4500nm. This interesting lens has a Nikon bayonet mount and a very narrow depth of focus, (e.g. about 3mm depth of focus at a distance of about 1.5m). The Nye Optical Co. catalogue of 1958 shows: Mirror making kits; Materials and accessories; Mirror blanks; Spiders and mirror cells; Herschel wedges; Eyepieces and Barlows; Focusers; Telescope tubes; other parts and accessories.

O

Oakley, Joseph
Joseph Oakley, (fl.1758-1781) [3] [18], was an optician, spectacle maker, and optical instrument maker of Birmingham, England. Addresses: *near the dog* in Mount-Pleasant (1758); Upper Priory (1767-1770); Bull St. (1770); 79 Bull St. (1775-1781). In 1758 he advertised as "A spectacle maker from London" saying, "Makes temples and all other sorts of spectacles, reading glasses, telescopes, microscopes, and all other optical instruments … Any person may be supplied with any apparatus for the above instruments by enquiring for him at Mr John Hazeldine's, or at his workshop near the dog in Mount-Pleasant, Birmingham."

In 1770 he advertised, "reflecting telescopes, completely mounted in brass and neat mahogany cases, refracting telescopes, prospects and opera glasses, either covered with nourskin and brass, or black in imitation of shagreen, and some mahogany tubes mounted with brass, microscopes of different constructions, magic lanthorns and figures, diagonal mirrours (sic), concaves, glasses for short sight, convex glasses of every different size and focus &c."

Oberhäuser, Johann Georg
Georg Johann Oberhäuser, (1798-1868) [47], was German born, and emigrated to France in 1816. In 1822 he began to work as an optician in partnership with two others named Bouquet and Trecourt. He soon realised that powerful microscopes with good optics were too expensive for much of the market. To meet this need he produced large numbers of drum microscopes, based on the design of **Benjamin Martin**. In 1837 he stopped working with his partners and began to trade under his own name. In 1848 he introduced the horseshoe shaped microscope base that subsequently became popular. In 1854 he went into partnership with his nephew, **Edmund Hartnack** and, in 1864 Hartnack took over the business. By 1898 the company had become a part of **Nachet et Fils**.

Oberwerk
Oberwerk is a US brand of binoculars and spotting scopes, established in 1999 by Kevin Busarow. At least some Oberwerk branded binoculars and scopes are manufactured by the **Kunming United Optics Corporation** of China.

Objective Lens
An objective lens, sometimes referred to simply as an objective, is the lens in an optical instrument closest to the object being viewed, in the optical path between the viewer and the object. It may be a simple lens consisting of one element, or a more complex arrangement of lens elements, such as an achromat, or an apochromat. An objective with two lens elements is referred to as a doublet, and one with three lens elements is referred to as a triplet. An objective, however, may consist of more than three lens elements.

In microscopy, a *plan objective*, also known as planar, semi-plan, semi-planar, or microplan, is one that corrects for field curvature. Whereas an achromatic objective has a flat field over approximately the central 65% of the image, a plan objective achieves at least 90%, and a semi-plan 80% of the field. See also immersion objective.

Obukhov Optical Works
The Obukhov Optical Works, [27], referred to as Obhoukhoff in [26], was established in St. Petersburg, Russia, in 1905. This marked the beginning of the Russian optical industry. The company was set up to meet the needs of the navy, and employed foreign specialists and workers. In 1907 they began to assemble binoculars, and before the war they also made gun sights and other optical instruments. At this time there was no domestic production of optical glass. See **Petrograd porcelain works**.

Octant
An octant is an instrument used to make angular measurements for navigation or astronomy. It operates on a similar principle to the sextant, but has an arc of an eighth of a circle, and may be used to measure angles up to 45° or, by means of double reflection, up to 90°. An octant which can measure up to a quarter of a circle is hence also sometimes referred to as a quadrant, double-reflecting quadrant, or double-reflecting octant.

See also Davis's quadrant, Hadley's quadrant, Smith's quadrant.

OCT Scanner
An OCT scanner is used by a vision care specialist, and uses *optical coherence tomography* to scan the eye. Optical coherence tomography is a non-invasive imaging test that uses light to take a cross-section image of the retina, showing the thickness and condition of the layers of the retina.

Ocular
Ocular is another name for an eyepiece.

Ocular Instruments Inc.
Ocular Instruments Inc. of Bellevue, Washington, USA, is a manufacturer of ophthalmic, diagnostic, laser, and surgical lenses. The company was founded in the 1960s.

Oertling, Johann August Daniel

Johann August Daniel Oertling, (1803-1866) [53] [322] [374] [382], was an optical, nautical, and mathematical instrument maker of Berlin, Germany. He served his apprenticeship with **C.P.H. Pistor**, and started his own business in 1826/7. In 1840 he was the first to make an electrical dividing engine, and this was purchased by the Prussian government. He is known to have made astronomical, surveying, and physical instruments, as well as goniometers, spectrometers, and balances. A brass sextant in a mahogany case, bearing his name, is in the Science Museum, London.

Johann Oertling was the brother of the mathematical and philosophical instrument maker, Ludwig Oertling, (fl.1846-1925) [18], who traded in London during the period 1849-1888.

Off-Axis

See optical axis.

Officina Filotecnica

Officina Filotecnica, otherwise known as la Filotecnica. See **Filotecnica Salmoiraghi**.

Officine Galileo

Officine Galileo (OG), of Florence, Italy, was one of the two main Italian optical instrument companies in the 1920s [48] [374], the other being **San Giorgio** (SG) of Genoa. One of their responsibilities was to produce optical rangefinders, optical traverse measuring equipment, site and target inclination measuring equipment and fire control centres.

Optical rangefinders for the Regia Marina (Italian Navy) were supplied by **Barr and Stroud** until after WW1, after which, rangefinders were supplied by OG and SG [26, p110].

Early on, OG and SG reproduced the British and German models. In around 1928 they began to produce domestically designed and built instruments for the Regia Marina ships. They began with the coincidence type of rangefinder, and later also built stereoscopic types. In 1929 OG acquired **Fratelli Koristka**. By the end of the 1930s, the Italian rangefinders, produced by both SG and OG were considered high performance and, some of the larger ones, cutting-edge designs [32].

Officine Galileo collaborated with **Ferrania** on production of the Ferrania Condor camera range.

Ofynn, Edward

Edward Ofynn, (fl.1831-1836) [18], was an optical, philosophical, and mathematical instrument maker of 55 Upper North Place, Gray's Inn Rd., London, England.

Ogilvy & Co.

Ogilvy & Co., (fl.c1910-c1940) [319] [320], of 20 Mortimer Street, London, England, were manufacturers of microscope lamps, and agents for **E. Leitz**. A 1925 advertisement shows them at 18 Bloomsbury Square, London. A 1926 advertisement claims that they had recently installed machinery in their workshops, and were prepared to undertake the manufacture of new instruments to special designs. The advertisement listed: Microscopes; Microtomes; Dark ground condensers; Photomicrographic apparatus; Haematological appliances; Dissecting stands and lenses; Microscopical, medical, and dental illuminating apparatus; Micro-projection apparatus; Epidiascopes; Colorimiters; Prismatic field glasses; Cameras, etc.

Ohara

Ohara is an optical glass manufacturer based in Japan, with a number of global subsidiaries, including in the US, Europe, and China. The company was founded in 1935.

Oigee

Optische Anstalt Oigee GmbH, of Hauptstrasse 25, Berlin-Schöneberg, Germany, was a manufacturer of military and civilian optical instruments during the early C20th. They made telescopes, riflescopes, and binoculars, and advertised reflex sights for fixed aircraft armament, optical and mechanical sights for mobile aircraft armament, optical sights for the army and navy, optical and precision mechanical devices, and optical testing devices. During WW2 they marked their instruments with the **ordnance code** dzl.

Oil Immersion Objective

An oil immersion objective is an immersion objective that uses oil as the medium between the front element of the microscope objective and the slide's cover-slip. Originally cedar oil was used because its refractive index, about 1.52, closely matches that of the glass cover-slip. Modern applications generally use synthetic oil.

Okada Kōgaku

Okada Kōgaku, was a Japanese camera and telescope maker which began business sometime in the 1930s. In c1951 the company was renamed **Daiichi Kōgaku**.

Oleorefractometer

See refractometer.

Olivo, Angelo

Angelo Olivo, (d.1837) [75], was a contemporary of **Leonardo Semitecolo**, making telescopes in Venice, Italy in the late-C18th and early-C19th. His workshop was near the Parish of San Pietro Castello in Venice, and he had five apprentices. The firm continued to produce spectacles and telescopes until Olivo's death in 1837. His telescopes are similar in construction to those of Semitecolo, being fabricated of cardboard tube, with horn or brass mount.

Olympus

The Japanese company Olympus was founded in 1919 by Takeshi Yamashita, together with **Shintaro Terada**, who was an established microscope and thermometer maker. The company was originally called Takachiho Seisakusho, and began with making microscopes and thermometers. In 1921 they registered Olympus as a trademark. In 1923 they sold the thermometer business in order to focus on microscope making, but allowed the new owner to continue to use the Olympus brand-name for their thermometers. In 1936 they entered the camera business, making lenses and cameras. In 1942 the company was renamed Takachiho Optical Co. Ltd.

They had to rebuild their production facilities after WWII, then in 1949 the company was renamed again to become Olympus Optical Co. Ltd. The 1960s saw a rapid growth in their microscope and camera business, and also medical optics such as gastroscopes and endoscopes. In 1968 they established Olympus Corporation of America, a sales subsidiary in the USA. During the 1970s to 2000s they continued to grow their three core areas of business, microscopes, cameras and medical equipment. In 2003 the company was renamed Olympus Corporation.

The growth continued until 2011 when they were investigated for financial irregularities. The business then continued under new management, and sells medical equipment and microscopes. The camera division was sold in 2020 to the investment firm, Japan Industrial Partners, and continues to use the Olympus name as OM Digital Solutions.

Omnimeter

An omnimeter is a theodolite with a rigidly attached microscope, positioned to observe the vertical angular movement.

According to the **Negretti & Zambra** catalogue, advertising Eckhold's Patent Omnimeter: "This important Surveying Instrument measures distances and altitudes with accuracy and great economy of time, it accomplishes the work of theodolite, level and chain, and can be used as a transit theodolite."

Opaque (Opake) Microscope

See microscope.

Ophthalmoscope

An ophthalmoscope is an instrument, used by a vision care specialist, for examination of the interior of the eye. It was invented by **Hermann von Helmholtz**.

Ophthalmometer

An ophthalmometer, also referred to as a keratometer, is an optical instrument, used by a vision care specialist, to measure the radius of curvature of the cornea of the eye. This instrument was invented by **Hermann von Helmholtz**.

Oppelt, Johann Balthasar

Johann Balthasar Oppelt (sometimes I.B. Oppelt), (fl.1785-1800) [256], of Ansbach, Germany, was a maker of microscopes. Examples of his work are in various museums and collections.

Optica b/c

Optica b/c was a supplier of telescopes and related items in Oakland, California, USA, during the 1960s and 1970s. They sold imported Japanese telescopes, both reflecting and refracting, tripods and pedestal mounts, eyepieces, spectroscopes and other optical accessories, filters, astrophotography adaptors and systems, publications, orreries and planetariums, celestial globes, amateur telescope making parts, materials and accessories, and binoculars.

Their refracting telescopes included 30mm and 40mm terrestrial spotting scopes, and astronomical telescopes ranging from 2 inches 4¼ inches. Their reflecting telescopes ranged from 3¼ inches to 12 inches.

Optical Activity

A substance referred to as optically active is one that rotates the plane of polarization of a transmitted beam of plane-polarized light.

Optical Axis

The optical axis of an optical system is a straight line passing through the centre of rotational symmetry of the optical elements. Thus, for a lens or curved mirror the optical axis passes through the geometrical centre at right angles to the surface. A prism with no rotational symmetry in the optical path does not have a defined optical axis. Rays of light that are not on the optical axis are referred to as off-axis.

Optical Coherence Tomography

See OCT scanner.

Optical Craftsmen, The
The Optical Craftsmen was a telescope manufacturer in Northridge, California, USA. The company began in 1958 as **Anra Manufacturing Engineers**, and subsequently was renamed The Optical Craftsmen. The company continued to operate until at least the 1980s. They produced equatorially mounted reflecting telescopes ranging from 4¼ inches to 16 inches as well as finder scopes, a 2-inch alt-az mounted refractor, and accessories. They also produced mirrors and mirror making kits as well as eyepieces.

Optical Glass
See Appendix VI.

Optical Hardware
Optical Hardware is the UK importer and distributor of **Ostara** optical products.

Optical Surfaces
Optical Surfaces is an optical component and instrument company that specialises in large optics, beam expanders, collimators and prototypes, and custom systems and optics. The company was formed in 1961 when **John Victor Thomson**, J. Mortleman and **Jim Hysom**, left **Cox, Hargreaves & Thomson** to start the new company.

Optical Square
An optical square is an instrument used by surveyors and others for laying off lines at right angles to each other. The optical square may be constructed using two mirrors or pentaprisms.

Optical Techniques Inc.
Optical Techniques Inc., [136], of Newtown, Pennsylvania, USA, was a telescope maker founded in 1976 by John Schneck and Robert Richardson, formerly of the **Questar Corporation**. The company was set up in competition with Questar, offering portable Maksutov Cassegrain telescopes. They offered these in a choice of 4-inch f/15, 6-inch f/17, and later a 6-inch f/15. They were introducing an 8-inch model when they ceased operations in 1980.

Optical Vision Ltd. (OVL)
Optical Vision Ltd. is a UK importer and distributor of telescopes, binoculars, and accessories. They sell instruments from a number of different brands. In 2008 they re-registered the **Barr & Stroud** trade-name, and continue to use it as a brand for imported binoculars, spotting scopes and monoculars.

Opticron
Opticron is a UK firm which was established in 1970, offering sporting and nature optics. They sell spotting scopes, binoculars, range-finders, etc.

Optics - Feinmess Dresden VEB
See **Gustav Heyde**.

Optigraph
The optigraph was a small, portable optical drawing aid which preceded the **Camera Lucida** of **William Hyde Wollaston**. Its invention was credited to **Jesse Ramsden** by his employee **Thomas Jones**, who improved on the design [2]. Jones' device consisted of a vertical telescopic tube and a 45° mirror. Upon viewing through the tube, the user sees a dot, marked on clear glass. The tube is moved so that the dot is traced over the scene to be copied, causing a corresponding movement of the pencil over the paper.

Optique et Précision de Levallois (OPL)
See **Société Optique et Précision de Levallois S.A. (OPL)**.

Optikotechna
Optikotechna was the name of the company which later became **Meopta**.

Optimus
Optimus was a trademark used for cameras by **Lejeune & Perken** and, after 1887, **Perken, Son & Rayment**.

Optische Anstalt Oigee
See **Oigee**.

Optische Fabrik M. Baader
Optische Fabrik M. Baader, [374] was an optical manufacturing company founded in Munich, Germany, in 1852 by **Michael Baader**. Baader had worked for **Martin Wörle** since 1838, and had married Wörle's daughter. He continued in this employment until Wörle's death in 1851, after which he went on to start his own company. Baader's company produced refracting telescopes up to 5 inches aperture, and won several prizes in European international industrial fairs.

The corporate or family connection between Optische Fabrik M. Baader and **Baader Planetarium** is unclear [174].

Optische Glasfabrik in Benediktbeuern, Sigmund Merz München
Optische Glasfabrik in Benediktbeuern, Sigmund Merz München, (fl.1883-1889) [161] [422], succeeded

G.&S. Merz, who had been trading as G.&S. Merz vormals Utzschneider & Fraunhofer in München. The new firm was run by Sigmund Merz. Upon handing over the Munich part of the business to his cousins, who went on to trade as **Optische Werkstätte der Gebrüder Matthias und Jakob Merz**, Sigmund retained the glassworks at Benediktbeuern, where he primarily worked on the manufacture of large objective lenses. The glassworks were closed in 1889, but lenses produced in the works, presumed to be ground by Sigmund Merz, were available until at least 1901.

Optische Werkstätte der Gebrüder Matthias und Jakob Merz

Optische Werkstätte der Gebrüder Matthias und Jakob Merz, (fl. February 25th 1882 - June 21st 1883) [**161**] [**422**], optical instrument makers of Munich, Germany, was a partnership between Matthias (1826-1883) and Jakob Merz, the cousins of **Sigmund Merz**. They took over operation of the Munich based part of **G.&S. Merz**, from Sigmund in 1882, and traded as Optische Werkstätte der Gebrüder Matthias und Jakob Merz. Shortly after the partnership was formed, Matthias Merz died, and Jakob continued the business, trading as **Optische Werkstätte von Jakob Merz**.

Optische Werkstätte von Jakob Merz

Optische Werkstätte von Jakob Merz, [**161**] [**422**], were optical instrument makers of Munich, Germany. Jakob Merz, sometimes known as Jacob Merz, (1833-1906), was a cousin of **Sigmund Merz**, and brother of Matthias Merz. In 1882, Jakob and Matthias took over the Munich based part of **G.&S. Merz** from Sigmund, and began to trade under the name **Optische Werkstätte der Gebrüder Matthias und Jakob Merz**. Matthias, however, died the following year. Following the death of his brother, Jakob continued the business alone, trading as Optische Werkstätte von Jakob Merz, producing telescopes for the amateur and educational markets. The company also repaired and maintained large telescopes. He signed his optical instruments *Jakob Merz in München*, or *Jacob Merz in München*. The firm continued in business until 1903, when **Paul Zschokke** purchased the company, its machines, and tooling.

Optisches Institut

The Optisches Institut (Optical Institute) was founded in Wetzlar, Germany in 1849 by **Carl Kellner** and **Moritz Hensoldt** for the production of lenses and microscopes. Hensoldt concurrently worked on other business plans, and the Institute was operated primarily by Kellner until his death in 1855. Upon Kellner's death the company was run by one of his employees, **Louis Engelbert**, until 1856 when his wife took over the business together with another employee, **Friedrich Belthle**. In 1865, Ernst Leitz became a partner, and in 1869 he took control of the company and continued in business under the name **Leitz**.

Optisches Institut G.&S. Merz vormals Utzschneider und Fraunhofer in München

See **Paul Zschokke**.

Optisches Institut von A. Krüss

See **Andres Krüss**.

Optisches Institut von Utzschneider und Fraunhofer

Optisches Institut von Utzschneider und Fraunhofer, [**161**] [**422**], were optical instrument makers of Munich and Benediktbeuern, Germany. The firm began as a partnership between **Joseph von Utzschneider** and **Joseph Ritter von Fraunhofer**. The partnership was formed in 1814, when Reichenbach left the Institute previously known as **Optisches Institut von Utzschneider, Reichenbach & Fraunhofer in Benediktbeuern**. Upon formation of the new partnership, it was renamed Optisches Institut von Utzschneider und Fraunhofer in Benediktbeuern. By 1816 they were routinely making telescopes with apertures of 17cm and 23cm. During the period 1814-1819, they signed their instruments *Utzshneider und Fraunhofer in Benedictbeuern*.

In 1819 the optical production was moved from Benediktbeuern to Munich, while the glassmaking continued at Benediktbeuern. With this change, the firm was renamed Optisches Institut von Utzschneider und Fraunhofer in München.

Fraunhofer died in 1826, after which Utzschneider was the sole director of the Institute, with the assistance of **Georg Merz**.

In 1832 Merz was made manager of the glassworks, and in 1838 he was made director. In 1839, shortly before Utzschneider's death, Merz purchased the Institute with the financial help of his friend Joseph Mahler (1795-1845). They continued to trade as Optisches Institut von Utzschneider und Fraunhofer in München. They produced refracting telescopes and heliometers, made improvements to microscopes, and produced spectacles, magnifying lenses etc. The workshop for grinding and polishing glass was in Munich.

During the period 1819-1845, they signed their medium and large refracting telescopes *Merz & Mahler in München* and their microscopes and other telescopes *Utzschneider u. Fraunhofer in München*.

Following the death of Joseph Mahler in 1845, Georg Merz operated the business in partnership with his son, trading as **G. Merz & Söhne in München**.

Optisches Institut von Utzschneider, Reichenbach & Fraunhofer in Benediktbeuern.

Optisches Institut von Utzschneider, Reichenbach & Fraunhofer in Benediktbeuern, (fl.1809-1814) [161] [422], were optical instrument makers of Benediktbeuern and Munich, Germany. The firm was a partnership between **Joseph von Utzschneider**, **Georg Friedrich von Reichenbach**, and **Joseph von Fraunhofer**. The partnership succeeded **Mathematisch-mechanisches Institut von Utzschneider, Reichenbach & Liebherr**, and was formed when Fraunhofer was made a partner in 1809. The glassworks and optical production were in Benediktbeuern. It is unclear why Joseph Liebherr was not a partner in this firm but, in 1812, Liebherr left the firm following a dispute with Reichenbach. During the period 1809-1814, they signed their instruments *Utzschneider, Reichenbach und Fraunhofer in Benediktbeuern*.

In 1814, Guinand returned to Switzerland and, in the same year, Reichenbach left the firm to form his own company, and the Institute was renamed **Optisches Institut von Utzschneider und Fraunhofer in Benediktbeuern**.

Optisches Institut Wilhelm Steeg
See **Steeg & Reuter Präzisionsoptik**.

Optolyth
Optolyth was founded in 1856 by J.E. Sill in Nuremberg, Germany. In 1945 they relocated to Oed in Weigendorf, and began production of Simple Galilean binoculars, kaleidoscopes and telescopes for the Nuremberg toy industry. Starting in 1965 they began production of prism binoculars and scopes under the name OPTOLYTH, and sold them worldwide.

Optometer
An optometer is a vision testing instrument, used by a vision-care specialist, which determines the refractive power of the eye. It is used to identify the extent of any long- or short-sightedness.

Optron Laboratory
Optron Laboratory of Dayton, Ohio, USA, was a precision optics manufacturing company. Examples of their products shown in their catalogues include an off-axis catadioptric 4¼-inch telescope with Schmidt correcting lens, an interference viewer, astronomical solar filters, and Schmidt correcting lenses from 6 to 12 inches.

Ordnance Codes
For ordnance codes used by optical instrument manufactures in Germany and occupied territories during WWII, see Appendix III.

Orion
Orion is an American company based in California, USA, producing telescopes and binoculars. Most Orion branded optical equipment is manufactured by **Synta** of Taiwan.

Orion Camera Co.
See **Orion Seiki Sangyō Y.K.**

Orion Optics UK
Orion Optics UK was formed in 1984 by Barry Pemberton, and in 1987 they began production of their first 150mm Schmidt-Newtonian telescope. The business, based in Newcastle Under Lyme, Staffordshire, England, has grown since then, and they went on to produce a range of high-quality telescopes, optical components and accessories. Barry's son John Pemberton succeeded his father running the business.

Orion Seiki Sangyō Y.K.
Orion Seiki Sangyō Y.K., (Orion Precision Products Industries Co., Ltd.), was founded in 1946 in Japan, making lenses and other camera components. They began production of their first camera in 1954, a 35mm SLR that was to be named *Phoenix*, but by the time of production was called the *Miranda T*. Their early cameras had the name *Orion Camera Co.* on the front, as well as the brand-name *Miranda*. In 1956 the company was renamed to become **Miranda Camera Company**.

Orthochromatic
Orthochromatic photographic emulsion or film is one with a near uniform sensitivity to light across the whole visible range.

More generally, the word orthochromatic is used to describe to a process, or optical system or component, that represents the intensity of colours across the entire visible range of light in their correct proportions. See also panchromatic.

Orthoscope
An orthoscope is an instrument for viewing the interior of the eye. Distortion is minimised by viewing through a layer of water.

Orthoscopic
Relating to the production of a true and undistorted image. See orthoscopic eyepiece in Appendix II.

Osborne, Thomas P.G.

Thomas Paine Gerald Osborne, (fl.1825-1854) [18] [276 #1949, #1767], was an optician. From 1825 to 1837 he traded in partnership with **William Ashmore** under the name **Ashmore & Osborne** at 42 Burgess Street, Sheffield, England. Following this he traded under his own name, with the directories showing him at 57 Scotland Street, in 1841, and 97 Scotland Street in 1854. He advertised spectacles, microscopes and telescopes.

Evidence suggests that, as with Ashmore & Osborne, Thomas P.G. Osborne had a branch in London, England. Instruments bearing his name include: three-draw brass telescope with mahogany barrel is engraved *Osborne Day or Night London*; 2½-inch brass day or night telescope with mahogany barrel, 16½ inches closed, 37 inches fully extended, by Osborne, London.

Ostara

Ostara products are imported, perhaps from China, and branded by the UK company **Optical Hardware**, who are the importers and distributors. Their range includes binoculars, spotting scopes, telescopes, and eyepieces.

Otoscope

An otoscope is a medical instrument used for inspection of the ear. It provides a view of the external ear canal and tympanic membrane, or eardrum. Also known as an auriscope.

Ottewill & Morgan

Ottwill & Morgan was a partnership between the camera maker **Thomas Ottewill** and William Morgan. The partnership began in 1854, and was terminated in 1855, [35 #21701 p1641].

Ottewill, Collis & Co.

Ottewill, Collis & Co. was a partnership between the camera maker **Thomas Ottewill** and an ex-employee of **Ross**, Mr. Collis. The partnership began in 1863, and continued until about 1868.

Ottewill, Thomas

Thomas Ottewill, (fl.c1851-c1868) [61] [122] [306], of 23 & 24 Charlotte Terrace, Barnsbury Road, Islington, London, England, was a highly regarded camera maker whose early cameras had a reputation for high quality and innovation. He began to produce cameras at 24 Charlotte Terrace in about 1851, trading as Thomas Ottewill. He traded in partnership as **Ottewill & Morgan** from 1854 until 1855. In 1855, the company also expanded next-door into 23 Charlotte Terrace, and Ottewill went on to trade in partnership with Frederick Aaron Gush as Thomas Ottewill & Co. This partnership was dissolved in 1858, [35 #22142 p2538], following which Ottewill continued the company alone. He was declared bankrupt in 1861, and discharged in 1862, [35 #22609 p1527]. In 1863 he entered a new partnership, trading as **Ottewill, Collis & Co.** However, following a period in debtor's prison, [35 #22908 p5244], Ottewill was declared bankrupt again in 1864, [35 #22917 p6429]. The firm of Ottewill, Collis & Co. continued in business until about 1868. The camera makers **John Garland**, **George Hare**, and **Patrick Meagher** all learned their trade in the workshops of Thomas Ottewill.

Ottway & Co. Ltd., W.

W. Ottway & Co. Ltd., [20] [44] [53] [269], were optical instrument makers of Orion Works, Ealing, London, England. Reference [11] states that the company was established in the C19[th], but the company's advertisements showed 1640 as the date they were established. They made telescope mountings for mirrors by the eminent mirror maker, **Calver** (c1870). During the period 1905-1907, they supplied a large number of sighting telescopes to the British Admiralty, in five models: fixed magnification 3x and 6x, and variable magnification 3-9x, 5-15x, and 7-21x. They continued to supply sighting telescopes through WWI, until at least the 1920s. In 1919 the company was listed as a member of the **British Optical Instrument Manufacturers' Association**.

During WWII they manufactured parts for the De Havilland Mosquito. The trade-name **Britex** was used for some W. Ottway & Co. Ltd. products.

They exhibited at the 1947 British Industries Fair. Their catalogue entry, [450], indicated that they supplied prisms, and showed them as manufacturers of: Surveying, mining and geophysical instruments; Observation, reconnaissance and gun sighting telescopes; Astronomical telescopes and allied equipment; Optical alignment and measuring instruments; Students' and workshop microscopes; Optical components; General scientific equipment.

The 1951 Users section of the *Directory for the British Glass Industry* [327], lists Ottway (W.) & Co. Ltd., of Orion Works, Ealing, London, established in 1640. Their products are listed as: Optical lenses and prisms; Magnifiers; Graticules; Spectroscopes; Microscopes; Telescopes. Their managing director is listed as W.T. Ottway.

In 1953 the managing director, Walter Thomas Ottway, died. The company continued under the direction of Geoffrey Ottway, who was the son of Charles Edward Ottway. In 1963 W. Ottway & Co. Ltd. was taken over by **Hilger & Watts**. They continued to trade under the name

W. Ottway & Co. Ltd. until the company was finally closed in 1968.

Owen, Charles

Charles Owen, (fl.c1860-1876) [319] [320], was an optician, retailer, and maker of optical instruments, of 283 Strand, London, England. He advertised microscopes, object glasses, and microscopic accessories, some of which he claimed to have manufactured. He filed for bankruptcy in 1876 [35 #24356 p4705].

Oxyhydrogen Microscope

See microscope.

P

Pacific Instruments
Pacific Instruments of Pacoima, California, USA, was a supplier of parts for amateur telescope making. Their 1970 catalogue shows telescope mounts, parts for making telescope mounts, and mirror cells for amateur telescope making.

Palar
Palar is a Korean brand of binoculars, microscopes, and camera lenses.

Paillard Bolex
E. Paillard & Co. was established in Switzerland in 1814 by Möise Paillard. The company became known for their music-box mechanisms, and their *Hermes* typewriters. In 1930 they acquired the **Bolex** company, manufacturers of 16mm cine cameras and projectors. This became the Paillard Bolex division of the company, [346], who retained **Jacques Bogopolsky** on their staff for about five years. They went on to produce a number of models of 8mm and 16mm projectors and cameras, as well as a small number of 9.5mm models. In 1969 the Paillard Bolex division was acquired by the Austrian company **Eumig**.

Pallant, John
John Pallant, (fl.1839-1869) [18] [322], was a philosophical, optical, and mathematical instrument maker of London, England. Addresses: 4 Little Russell St., Drury Lane (1839); 35 Tavistock St., Covent Garden (1839); Trafalgar St., Walworth Rd. (1840); 9 Great May's Buildings (1849-1852); 44 Museum St. (1863-1869), and 51 Strand. He served his apprenticeship with **John Huggins (II)** of the **Grocers' Company**, beginning in 1826. According to his trade label in a telescope box: "Philosophical, Optical, and Mathematical instrument maker to the local and government boards. Binocular field glasses for India. Spectacles and eye glasses adapted for all sights. Pallant optician 51 Strand, London WC".

Instruments bearing his name include: brass Gregorian telescope on library stand; 3 draw mahogany and brass telescope; brass telescope on brass and wood tripod, 73mm aperture with focal length 1m; meridian telescope; aneroid barometer; portable transit instrument with 2-inch aperture [53]. He is also known to have sold: level.

John Pallant worked in partnership with W. Muxton, trading as Muxton & Pallant, at 44 Museum St., prior to 1843. Instruments bearing this name include: bubble level.

Paludan, D.A.
Paludan Optik, opticians, have traded in Århus, Denmark, since 1899. Instruments bearing this name include: Galilean binoculars inscribed *D.A. Paludan, Århus*.

Panchromatic
Panchromatic photographic emulsion or film is one that is sensitive to the whole visible range of light, although not necessarily equally across the range.

More generally, the word *panchromatic* is used to describe to a process, or optical system or component, that utilises or relates to the entire visible range of light. See also *orthochromatic*.

Pancratic
Having variable magnification. The term is used to describe an instrument, such as a microscope or telescope, of adjustable power.

Panoptic
All-seeing. Allowing the viewing of all parts or elements within one view. All encompassing.

Panoptikon
A panoptikon, [335], is an early form of motion picture projector, first used commercially in 1895, in which a two-inch film is in continuous motion. The panoptikon was developed by the **Latham Brothers** with funding from their father, Major Woodville Latham, and with design help from **William Dickson**.

Parallactic Ladder Mount
A parallactic ladder mount is a simple form of equatorial mount for portable telescopes, and is designed to be cheaper and easier to construct than a fixed mount. It is a simplified version of the English equatorial mount.

The parallactic ladder mount resembles a tall tripod in which one leg is formed of two limbs with ladder-steps between, positioned to form the right ascension axis, and may be rotated on its axis. The other two legs are so positioned to ensure that the ladder-leg is aligned parallel to the Earth's axis. The telescope is cradled upon one of the ladder steps, and can be turned by hand. Tilting the telescope up and down alters its declination, and rotating the ladder-leg of the mount alters its right ascension.

Such a mount [1 p100] was constructed by the British optician and instrument maker **George Dollond (I)**, and presented to the British astronomer William Henry Smyth (1788-1865), and was a modified version of a mount that Dollond had seen in operation.

Parallactic Telescope

Parallactic telescope is an historical term for an equatorially mounted telescope used for measuring stellar parallax.

Paraxial

Paraxial, [7], refers to rays of light in an optical system that are close to the optical axis and make a small angle θ to the optical axis, such that θ satisfies the small angle approximation - i.e. that sine θ ≈ θ.

Paris, Claude

Claude Paris (1703-1763) [256] [416], was an optical instrument maker of Paris, France. He was tutored in microscope making techniques by **Louis Joblot**. By the age of seventeen he had made several mathematical instruments, and began to specialise in optics, and by 1733 he was skilled at making telescopes. He married **Jean-Baptiste Gonichon**'s sister, Marie-Anne Gonichon (**Marie-Anne Paris**), and his sister, Marie Michelle Paris (**Marie Michelle Gonichon**), married Jean-Baptiste Gonichon. In c1733 he began to work in partnership with Jean-Baptiste Gonichon, although no instrument bears both signatures. They traded at rue Bordet, près la Fontaine sainte Geneviève, à l'Hôtel S. Paul. Together they designed and built a machine for polishing mirrors. In 1757 they presented a 6-foot Gregorian telescope to the Royal Academy of Sciences. The business may have moved to Place de l'Estrapade prior to Claude's death. He was succeeded in business by his wife, Marie-Anne, trading in partnership with their son, **M. Paris**.

Paris, M.

M. Paris, (fl.1763-1774) [416], was a telescope maker of Paris, France. He was the son of **Claude Paris** and **Marie-Anne Paris**, and began his career working for his father. Upon his father's death, he carried on the business in partnership with his mother, trading at Place de l'Estrapade until he was declared bankrupt in 1774.

Paris, Marie-Anne

Marie-Anne Paris, (fl.1763-1774 d.1784) [416], was an optician of Paris, France. She was born Marie-Anne Gonichon, the sister of **John-Baptiste Gonichon**, and married **Claude Paris**. Upon the death of her husband, she continued his business in partnership with their son, **M. Paris**, trading at Place de l'Estrapade until M. Paris was declared bankrupt in 1774.

Park, Henry

Henry Park, (1848-1907) [61] [306], of Acton Street, Kingsland Road, London, England, advertised as a photographic apparatus manufacturer. He established his business in 1877, after having worked for five years with **P. Meagher**, and for eight years with **G. Hare**. He manufactured primarily for the trade, and advertised various models of camera bearing his own name for retail, with many being sold by **J. Fallowfield**. His business remained in the directories until his death in 1907.

Parker & Co., A.G.

A.G. Parker & Co., [44], was founded in 1880 in Birmingham, England. They made guns and accessories, including riflescopes. Arthur T.C. Hale joined the company, and was manager by 1910. A.G. Parker's son, Alfred J. Parker, also worked for the firm. By 1925 Arthur Hale was a director and, by 1928, A.J. Parker had set up his own company, **Alfred J. Parker & Co.** Around 1930, Arthur Hale's involvement in A.G. Parker & Co. had reached such a significant level that the name of the company was changed to **Parker-Hale**.

Instruments bearing the name A.G. Parker & Co. include: brass and leather four-draw telescope marked *"The Century" A.G. Parker & Co. Ltd., Birmingham*; brass and leather three-draw hand-held telescope, with end-caps, tripod, and extra lens, marked *100 Yards Spotter, A.G. Parker & Co. Ltd., Birmingham*.

Parker & Co., Alfred J.

Alfred J. Parker & Co., [44], was a maker of guns and accessories, including riflescopes, of Birmingham, England. The company was formed by 1928, by Alfred J. Parker when he left the firm of his father, **A.G. Parker & Co.** His *Twin Zero* aperture sight was a popular model of gun sight. The company continued in business until 2007.

Parker Hale

Parker Hale, [44], of Birmingham, England, is a manufacturer of shooting accessories, including riflescopes. The company was formed in 1880 as **A.G. Parker and Co.**, and re-named Parker Hale in around 1930. They continued in business until the 1990s, when Parker-Hale was sold to the Midlands engineering group, Modular Industries. In November 2000, John Rothery (Wholesale) Co. Ltd. purchased the brand-name Parker-Hale and use it for their gun care items and accessories.

Parkes and Son, James

James Parkes and Son, [3] [18] [44] [50] [110], of Birmingham, England, manufacturers of optical and mathematical instruments, was established 1815. The company's specialities included various scientific apparatus among which were: astronomical telescopes; microscopes; magnetic compasses and sundials;

mathematical instruments of all grades including school drawing compasses; drawing pins and scales.

James Parkes, (1786-1877), started his business under his own name in 1815, variously described as a gilt toy manufacturer and a stamper and piercer. In 1846 his only son, Samuel Hickling Parkes, (1817-1896), became a partner, and they became James Parkes and Son. An 1848 catalogue referred to them as "Manufacturers of improved measuring tapes, land chains, mathematical instruments, miners' and mariners' compasses, watch keys, seals, &c." In an 1849 Birmingham city directory they were referred to as "Manfs. of mathematical instruments and gilt toys." By the 1850s they were producing microscopes.

Their entry in the 1851 Great Exhibition catalogue [**328**] includes: mathematical drawing instruments; tapes and surveyors' measures; compasses and sundials; *botanist's companion*, consisting of a double lens microscope, with tweezers, dissecting needle, fine scissors, etc., in neat morocco pocket-book.

In the 1862 International Exhibition in London, [**258**], their catalogue entry includes: microscopes; astronomical telescopes; mathematical, philosophical, and surveying instruments.

In 1864 they were awarded patents for "impts. in opera glasses, telescopes, microscopes, spectacles, and other optical instruments."

The 1871 census showed that the company had 38 employees. After James Parkes' death in 1877, Samuel continued the business under the same name. Upon Samuel's death, his son, Samuel Thomas Hickling Parkes, (1856-1939) and his nephew James Ebenezer Moulton, (1844-1924), continued the business until 1908 when James Moulton left the partnership. After this, Samuel continued the business alone.

In the 1914 *Whitaker's Red Book of Commerce or Who's Who in Business*, [**110**], their entry includes the following information: Parkes (James) & Son. Manufacturers of Optical and Mathematical instruments, 16 and 18 Vesey Street, St. Mary's, Birmingham. Established in 1815 by James Parkes in St. Mary's Row. Continued by James Parkes, Samuel Hickling Parkes. Present Principal, S.T.H. Parkes. Specialities: Various scientific apparatus, including Astronomical telescopes, Microscopes, Magnetic compasses and sundials, Mathematical instruments of all grades, including school drawing compasses, Drawing pins, and scales. Patents: Various improvements in mathematical instruments, microscopes, binocular glasses, and reading lamps.

Parkes and Sons, John

John Parkes and Sons of 11 St George's Crescent, Liverpool, was a partnership between John Parkes senior, John Parkes junior and Albert Job Parkes.

According to the London Gazette, July 15th 1910, [**35** #28395 p5071]: "Notice is hereby given, that the partnership heretofore subsisting between John Parkes, senior, John Parkes, junior, and Albert Job Parkes, carrying on business as Chronometer Makers and Opticians, at 11 St. George's-crescent, in the city of Liverpool, under the style or firm of John Parkes and Sons, has been dissolved as from the 1st day of January, 1910, so far as concerns the said John Parkes, senior, who retires from the said firm.—Dated this 7th day of July, 1910." The notice is signed, John Parkes, Senr. John Parkes, Junr. Albert Job Parkes.

According to a 1951 advertisement, [**44**]: "Established over a century. John Parkes and Sons. Compass adjusters and Nautical instrument makers. Chronometers and sextants. Manufacturers and repairers of Thermometers, Barometers, Hydrometers, Gauges, Engine counters, etc., Compasses, Azimuth mirrors, Sounding machines, Clocks, Watches, Telescopes, Binoculars, Ships logs and accessories supplied and repaired. 74 South Castle St., Liverpool. Works: 6, 8, 8a Windsor St., Liverpool."

Since the 1951 advertisement post-dates the London Gazette notice, this suggests that John Parkes Jr., continued the business under the name John Parkes and Sons, possibly in partnership with Albert Job Parkes.

Instruments bearing their name include: ship's chronometer; sextant; Galilean binoculars.

Reference [**3**] lists the following as opticians in Birmingham:

John Parkes, Sand Pits, (1815-1835), succeeded by John Parkes and Son, Sand Pits, (1837), succeeded by John Parkes and Sons, Sand Pits, (1839), Succeeded by William Parkes, 4 Court, Digbeth, (1845-1846). What, if any, relationship these businesses have with the Liverpool family, or with the Birmingham firm of James Parkes & Son, is unclear.

Parks Optical

Parks Optical, of Simi Valley, California, USA, was a brand of refracting, reflecting, and Newtonian Cassegrain telescopes, binoculars, spotting scopes, microscopes, accessories, and telescope making supplies. The company was established in 1954, and operated until at least 2003.

Parnell & Son, William

William Parnell & Son, [**18**], were opticians, and mathematical and philosophical instrument makers of London, England. The firm was a partnership between

William Parnell and his son, **Thomas Parnell (II)**. The partnership began in 1839, and continued until 1840, trading at 2 Lower East Smithfield.

Parnell, Thomas (I)

Thomas Parnell (I), (fl.1784-1811) [18] [53] [276 #1023] [322] [361], was a mathematical and nautical instrument maker, and ship chandler of London, England. Addresses: *Mariner & Quadrant*, 94 Near the Hermitage Bridge, Lower East Smithfield; *Mariner & Quadrant*, 2 Lower East Smithfield; 93 Lower East Smithfield; 94 near the Hermitage Bridge, Lower East Smithfield; 11 Cannon St., Ratcliff Highway (1784); 25 East Smithfield (1793).

He served his apprenticeship with the mathematical instrument maker, John Blake, beginning in 1768, and was freed of the **Joiners' Company** in 1776. He was the father of **William Parnell**.

According to his trade label, "Thos Parnell. Mathematical instrument maker & Ship chandler, at the Mariner & Quadrant, No 94, Near the Hermitage Bridge, Lower East Smithfield, London. Makes & sells every article in the above branch on the lowest terms. Viz^t. Sextants & quadrants; Telescopes; Steering compasses; Cases of instruments; Sliding Gunter scales; Rules of all sorts. Likewise all kinds of Stationary, Sea charts, Navigation books, Pilots for all parts of the world, & Coasting directions of all sorts. Wholesale & retail. NB. Quadrants clean'd, & compasses touch'd in the best manner." He also advertised octants for sale.

Parnell, Thomas (II)

Thomas Parnell (II), (fl.1839-1869 d.c1869) [18] [322], was an optician and mathematical instrument maker of 2 Lower East Smithfield, London, England. In 1839 he began working in partnership with his father, **William Parnell**, trading as **William Parnell & Son**. In 1840 he began to work under his own name, and continued to do so until 1869.

Parnell, William

William Parnell, (fl.1811-1840) [18] [54] [276 #1023] [322], was a mathematical and nautical instrument maker, and ship chandler of London, England. Addresses: *Mariner & Quadrant*, 91a Near the Hermitage Bridge, Lower East Smithfield; 2 Lower East Smithfield; 94 Lower East Smithfield, Butcher Row (1816-1839); 94 Near the Hermitage Bridge, Lower East Smithfield.

He was the son of **Thomas Parnell (I)**, and was freed by patrimony of the **Joiners' Company** in 1811. The design and wording of his trade label was exactly the same as that of Thomas Parnell (I), but with William Parnell's name in place of his father's, and the address 91a instead of 94. From 1839 until 1840 he worked in partnership with his son, **Thomas Parnell (II)**, trading as **William Parnell & Son**. William Parnell's name appears on an octant bearing the scale divider's mark for **Spencer, Browning and Rust**, (see Appendix XI).

Parra, Numa

Numa Parra – See **Parra Mantois et Cie**.

Parra Mantois et Cie

Parra Mantois et Cie was a French optical glass manufacturer. **Edouard Mantois** was the son-in-law of **Charles Feil**. He learned his optical glass making skills from Charles, and took over the glassworks in **Choisy-le-Roi** near Paris, France in 1887. By this time Numa Parra was already in a senior role in the company and, in about 1900 (the year Edouard Mantois died), the firm was renamed Parra Mantios et Cie [24 p252] [28 p345].

Parra Mantois manufactured special optical glasses for photo and film cameras, binoculars, range-finders, microscopes, surveying instruments, and precision instruments that demanded high specifications. Parra Mantois products were considered to be the best scientific glass in France, and were used in French Military submarine periscopes, radar and telescopes.

Barr and Stroud, [26 p73], were prohibited by Royal Proclamation to export instruments of war during WWI. France, however, was exempted from this ban in return for a guarantee that supplies of optical glass would continue to be sent to Britain by Parra Mantois.

Parsons Optical Glass Company

The Parsons Optical Glass Company, [44], was an optical glass manufacturer of Little Chester, Derby, England. In 1921 **Sir Charles Algernon Parsons** acquired a controlling interest in the optical firm of **Ross Ltd.**, of Clapham. In the same year, he also purchased the **Derby Crown Glass Company** to improve the basic glass. With these two acquisitions, he formed the Parsons Optical Glass Company, which produced many different types of glass for optical purposes.

Parsons, James

James Parsons, (fl.1837-1890) [18] [276 #2199], was an optical, philosophical, and mathematical instrument maker, and divider of mathematical instruments, of London, England. Addresses: 22 Bull & Mouth St., Aldersgate St. (1837-1839); 50 Red Lion St., Clerkenwell (1840-1846); 1 Rodney St., Pentonville (1849-1851); 43 Stamford St., Blackfriars Rd. (1851); 11 Upper Stamford St., Blackfriars Rd. (1855); 111 Upper Stanford St. (1861-1865); 73 Stamford St. (1869-1880).

He served his apprenticeship with his father, **William Parsons (II)**, beginning in 1812, and was freed of the

Grocers' Company in 1840. He attended the London Mechanics' Institute from 1826 until 1828. From 1827 until 1836 he worked in partnership with his brother, George Parsons, trading as **James & George Parsons**, after which he worked under his own name. His son, James William Parsons, served an apprenticeship with him, beginning in 1844.

Parsons, James & George

James & George Parsons, (fl.1827-1836) [18], were optical, mathematical, and philosophical instrument makers, and dividers of mathematical instruments, of 22 Bull & Mouth St., Aldersgate St., London, England. The firm was a partnership between two of **William Parsons (II)**'s sons, **James Parsons**, and George Parsons, who was an engraver of mathematical instruments.

Parsons, Sir Charles Algernon

Sir Charles Algernon Parsons, (1854-1931) [37] [44], youngest son of **William Parsons (I)**, third Earl of Rosse, was a British engineer, best known for his invention of the multi-stage (compound) steam turbine – the Parsons Steam Turbine. He studied pure and applied mathematics at Cambridge, and went on to an apprenticeship with the engineering company Sir W.G. Armstrong & Co., in Newcastle upon Tyne.

In about 1921, after acquiring the firm **Ross Ltd.**, he formed the **Parsons Optical Glass Company**. In 1925 he purchased the firm of **Sir Howard Grubb and Sons**, makers of large astronomical telescopes. The re-named company, Grubb Parsons, had works at Walkergate, adjacent to the turbine works, where many large telescopes were produced. In 1938 Grubb Parsons took over the telescope making part of the **Cooke, Troughton and Simms** business.

Parsons, William (I)

William Parsons (I), (1800-1867) [46], of York, England was a politician and astronomer, and he succeeded his father in the title Lord Oxmontown in 1801, becoming the third Earl of Rosse. His youngest son was the renowned **Charles Algernon Parsons**. William Parsons (I) endeavoured to improve on the techniques of making large mirrors for reflecting telescopes, and of the mounting for the telescopes. Unlike his contemporaries, such as **William Herschel**, he made his knowledge and methods public rather than keep them proprietary. He set up his own team and equipped them with ovens for casting mirror blanks. He made large mirrors, including composite mirrors with longitudinal adjustments to reduce spherical aberration. He stopped producing composite mirrors once he was able to produce the desired size in a single casting. He also designed a steam-driven mirror polishing machine.

In 1841 Parsons began construction of a massive six-foot aperture Newtonian telescope, for which the mirror weighed 4 tons. For more than sixty years it was the largest telescope in the world, and became known as *the monster telescope*, or *Leviathan of Parsonstown*, with a 15m long tube suspended between masonry walls. Parsons produced two six-foot mirrors for this telescope. They tarnished quickly, and polishing them was time-consuming, so he used one for observation while the other was being polished. The tube could be adjusted by only 15°, and its movement was controlled by winches, chains and counterweights.

In 1848 Parsons became president of the Royal Society, and he won the society's Royal Medal in 1851. He continued to juggle his political career with his interest in astronomy, and continued in his endeavours to improve on mirror-making and the making of telescopes and mounts, as well as his observing activities. He was succeeded by his oldest son, Laurence Parsons, in his title, and in management of the ongoing work with the telescopes.

Parsons, William (II)

William Parsons (II), (fl.1801-1832 d.by 1840) [18] [90], was an optician and mathematical instrument maker of London, England. Addresses: 22 Bull & Mouth St., St. Martin's le Grand; 32 Bull & Mouth St.; New St., Fetter Lane (1801); Salisbury Court, Fleet St. (1808-1809); Paul's Alley, St. Paul's Church Yard (1812); 8 Paul's Alley, St. Paul's Church Yard (1822); Bull & Mouth St. (1822); 9 Kirby St., Hatton Garden (1832).

He served his apprenticeship with **James Martin**, beginning in 1789, and was freed of the **Grocers' Company** in 1799. Among his apprentices were his sons, **William Parsons (III)**, **James Parsons**, and George Parsons, who was an engraver of mathematical instruments. William (II) was succeeded in business by William (III).

Parsons, William (III)

William Parsons (III), (fl.1834-1840) [18] [276 #1660], was an optical, philosophical, and mathematical instrument maker of Salisbury Court, Fleet St., London, England (1824), and 9 Kirby St., Hatton Garden, (1834-1840). He served his apprenticeship with his father, **William Parsons (II)**, beginning in 1809, and was freed of the **Grocers' Company** in 1824. He succeeded his father in business.

Passemant, Claude-Siméon

Claude-Siméon Passemant, (1702-1769) [11] [256] [276 #296a] [382], was an optical instrument maker and precision mechanic whose address was *A la Pomme d'Or*, Rue de la Monnaie, Paris. He was the son of a tailor and, after failing to qualify in law, and subsequently working for a draper, he started a haberdashery. When he was thirty-one, he handed over the business to his wife, and focussed his efforts on optics and mechanics.

He began to produce optical instruments and, in 1738, he published *Construction d'un Télescope de Réflexion* (the Construction of a Reflecting Telescope). In 1746 he presented a reflecting telescope to the Académie des Sciences, and in 1749 he presented an astronomical clock to Louis XIV. For this latter effort he was awarded a pension, given accommodation at the Louvre in Paris, and appointed Ingenieur du Roy (Engineer to the King). He produced both reflecting and refracting telescopes, astronomical clocks, microscopes, celestial and terrestrial globes, and devices to create optical illusions. He made an equatorial telescope with a clock drive for the King. He is also known to have made: brass protractor; silver transit instrument; sundials; barometers.

He was succeeded in business at the Louvre by his brother-in-law, Ollivier, and one of his workers, Nicollet.

Passey, D.

D. Passey, (fl.c1790) [18] [319] [322], was an optical instrument maker of Bristol, England. Instruments bearing his name include a solar microscope.

Pastorelli & Co., Francis

Pastorelli & Co., [18] [379], were optical, mathematical, and philosophical instrument makers of London, England. Addresses: 4 Cross Street, Hatton Garden (1851-1869); 208 Piccadilly (1857-1878); Great Warner St. (1870-1871); 10 New Bond St. (1879-1880). This was one of several partnerships formed by **Francis John Pastorelli**.

Instruments bearing their name include: 3-draw brass telescope with leather cover and case; Coddington lens; thermometers; pedometer.

Pastorelli & Rapkin

Pastorelli and Rapkin, [11] [53] [379], of 46 Hatton Garden, London, England, began as a partnership between **Francis John Pastorelli**, and Alfred Thomas Rapkin, manufacturing barometers and thermometers. Their partnership was dissolved in 1876 [35 #24361 p4935]. The business continued under the same name, however, as wholesalers of optical and mathematical instruments, and was incorporated in 1901. They were listed in the 1951 *Directory for the British Glass Industry*. The company remained in business until 1988.

Instruments bearing their name include: 3-draw military telescope; 3-draw brass telescope with sun-shield and sliding eye-lens cover; brass microscope in wooden case with accessories; barometers; thermograph, hydragraph.

Pastorelli, Francis John

Francis John Pastorelli, [18] [379] (fl.1851-1880, d.1897), was an optical, surveying, mathematical, philosophical, and meteorological instrument maker and a wholesale optician of London, England. Addresses: 4 Cross Street, Hatton Garden (1849-1869); Hatton Garden (1869); 208 Piccadilly (1856-1878); 10 New Bond St. (1879-1880).

He worked in several partnerships. He traded as **Francis Pastorelli & Co.** from 1851 until at least 1871. He had a long-standing partnership with Alfred Thomas Rapkin, trading as **Pastorelli and Rapkin** until 1876. He also formed a partnership with Edward Cetti, trading as Pastorelli & Cetti, barometer and thermometer makers and opticians, at 11 Brook Street, Holborn. The partnership was terminated in 1852 [35 #21398 p14]. He formed a partnership with John Gibb, trading as opticians and philosophical instrument makers at 208 Piccadilly, and 4 Cross Street, Hatton Garden. This partnership was terminated in 1859 [35 #22332 p4616].

In 1863 he was granted a patent for "improvements in the construction of surveyors' levels, and other surveying instruments". The patent was declared void in 1866 due to non-payment of his fee [35 #23182 p5910].

Instruments bearing his name include: Brewster stereoscope; brass astronomical telescope on tripod, length 110cm.

Several other members of the Pastorelli family of Hatton Garden, and other London addresses, were philosophical instrument makers – mainly barometer and thermometer makers.

Pastorelli, John

John Pastorelli, [3] [18] [126] [276 #2201] [379] was an optician, barometer maker, and mathematical instrument maker of Liverpool, England. Some directories list him as an optical instrument maker. Addresses: 28 Cable Street (1834-1837); 55 Cable Street (1839-1847); 61 Cable Street (1848-1853); 49 Pitt Street (1855); 10 South Castle Street (1857-1860).

John Pastorelli was preceded in Liverpool by Joseph Pastorelli & Co., of 43 Artherton St., who terminated their business in 1800, [60]. According to their notice on 8th May 1800: "Joseph Pastorelli and Co. respectfully inform their friends and the public, that as they mean to decline business, they have now selling at their shop, No. 43 Atherton-Street, Liverpool, a large quantity of picture frames, and engravings, prints, drawing books, weather-glasses, barometers, thermometers, looking-glasses, &c. which they intend to dispose of at various reduced prices."

Pathé, Charles

Charles Pathé, (1863-1957) [**335**], was a French film producer and manufacturer of motion picture equipment. His interest began in 1893 when he bought an **Edison** phonograph and, realising that selling equipment would be profitable, bought three more and sold them. In 1895, because Edison's patent did not apply in Britain, he had some Kinetoscopes made for him by the English inventor and film producer **Robert Paul**. He sold these on to fairgrounds and, in order to widen the scope of available films, he had a camera made for him by the French inventor Henri Joli.

With the advent of **Lumière**'s Cinematograph Charles Pathé and his brother Émile began producing films, having formed the company Pathé Frères, trading at 98, rue de Richelieu in Paris. In 1897 the company became Compagnie Générale de Cinématographes, Phonographes et Pellicules, with the two sides of the business, phonograph and Cinematograh, run by Émile and Charles respectively. From 1902 the company began to expand, opening branches in various other countries, some of which were formed as independently operating film production companies. In 1912 they introduced a 28mm film format using the brand-name **Pathéscope**, and the Pathéscope subsidiary company was formed in London, England.

In 1918 the operation of the French company was split into two. Pathé-Cinema, whose business was films and newsreels, as well as the manufacture of movie cameras and projectors, was run by Charles. Pathé Records was run by Émile. In 1921 the Pathé-Cinema subsidiary in the USA was sold, ultimately ending up in the ownership of RKO Radio Pictures who ceased their film production activities. Subsequently the international structure and ownership of the various Pathé companies became more complex due to a number of changes of ownership. In 1927, the year Charles Pathé retired, **Kodak** took control of the Pathé factory in France, re-branding it **Kodak-Pathé**. Also in 1927, the Pathé-Cinema subsidiary in England was sold, ultimately becoming British Pathé [337].

Pathéscope

Pathéscope Ltd, [**44**] [**111**], was formed in 1912, with offices in Piccadilly Circus, London, England, for selling Pathétone gramophones and the 28mm K.O.K. camera and projector manufactured by the French parent company of **Charles Pathé**. In 1922 the 9.5mm film format was introduced for home use. The 9.5mm format was ultimately overtaken by the 8mm format introduced by **Kodak**. In about 1958 the firm was taken over by a UK businessman, and became Pathéscope (Great Britain). In 1961 they went into receivership. They were taken over by Great Universal Stores, and renamed Pathéscope (London). By 1964 the sales of 9.5mm items was discontinued, but they continued to sell many different items under their new name. Although primarily known for their moving film cameras and projectors, Pathéscope did also sell binoculars.

Patrick, George

George Patrick, (fl.1823-1827) [**18**] [**276** #1665], was an optician and optical instrument maker, of 16 Finch Lane, London, England. He served his apprenticeship with **George Richardson**, beginning in 1814, and was freed of the **Spectacle Makers' Company** in 1823.

Patroni, Pietro

Pietro Patroni, (c1676-1744) [**48**] [**256**] [**372**], was a highly regarded optical instrument maker of Milan, Italy. He was particularly noted for his binocular microscopes and telescopes, based on those of **Chérubin**. His instruments bear the signature "Petrus Patronus". They are finely decorated, and some show innovative construction. Several telescopes and microscopes bearing his name, dating from 1711 to 1736, and exist in museums and private collections. Both **James Mann (II)** and **Samuel Johnson** advertised, comparing their telescopes with "those of the celebrated Pietro Patrone at Milan".

Patten, Richard

Richard Patten, (1792-1865) [**16**] [**58**] [**368**], was a surveying and nautical instrument maker, who started his business in New York, USA, in 1813. He advertised that his instruments were "… made to order & warranted being divided on an Engine after the Plan of Ramsdens." During the 1830s his premises were taken over by two of his ex-employees, Edmund Brown and Harvey W. Blunt. In c1842 Richard Patten moved to Washington DC, and began to work in partnership with his son, George Patten, trading as Richard Patten & Son. In c1853, the firm relocated again to Baltimore, where George died a few years later. Richard continued the business alone, returning to New York in c1862, where he traded until his death in 1865. He is known to have made: sextant; quadrant; theodolite; level; surveyor's compasses.

Instruments bearing his name include, presumably exported to England: octant with scale dividers' mark showing a foul anchor with initials JA (see Appendix XI).

Pattenmakers' Company

See Appendix IX: The London Guilds.

Paul, Robert William

Robert William Paul, (1869-1943) [335] [454], was a scientific instrument maker, and pioneering film producer, of London, England. He learned his instrument making skills working for **Elliott Brothers**, and subsequently working for the Bell Telephone Company in Antwerp, Belguim. He started his own company in 1891, trading as Robert W. Paul Instrument Co. at 44 Hatton Garden, London.

Because **Thomas Eddison**'s patent for the Kinetoscope did not apply in Britain, Robert Paul was able to make Kinetoscopes for his customers. By 1894, he was making them for the Greek entrepreneurs Georgiades and Tragides, and in 1895 for **Charles Pathé**. He then wanted to make a moving picture camera that could make films for the Kinetoscope so, working with the photographer Birt Acres, he developed the *Paul-Acres Camera*. This was the first English made moving picture camera, and was used to make the first English moving pictures. Robert Paul then went on to produce a movie projector, which he called the Theatograph, and in 1896 he demonstrated it on the same day **Lumière** demonstrated their Cinematograph in London. Robert Paul was innovative, not only in his subsequent design improvements to his Theatograph, but also in his film productions.

He withdrew from the film industry in 1910 to focus on his scientific instrument making business. In 1919/20 his company was taken over by the **Cambridge Scientific Instrument Company**, and became the Cambridge & Paul Instrument Company, which was renamed the **Cambridge Instrument Company Ltd.** in 1924.

Payne & Chapman

Payne & Chapman, [61] [122] [306], chemist and druggist, and retail photographic dealers of 63 Piccadilly, Manchester, England, was a partnership between **J.B. Payne** and **J.T. Chapman**. The partnership, formed in 1871, succeeded **Robert Hampson** in business, and was dissolved in 1874 [35 #24083 p1994]. J.B. Payne continued the business under his own name, and J.T. Chapman started his own business.

Payne, J.B

John Buxton Payne, (fl.late 1860s-1880) [306], was a chemist and druggist, and photographic equipment maker of 63 Piccadilly, Manchester, England. In the late 1860s he began to work for **Robert Hampson** and, upon Hampson's retirement in 1871, Payne, along with another of Hampson's employees, **J.T. Chapman**, took over the business. They continued the business in partnership, trading as **Payne & Chapman** until 1874, after which Payne continued the business alone. In 1880 he took over management of the Newcastle plant of **Mawson & Swan**.

Peacock, James

James Peacock, (fl.1835-1836) [18], was an optical, mathematical, and philosophical instrument maker of 32 Fore St., Limehouse, London, England.

Pearson, Richard

Richard Pearson, (fl.1827) [18] [276 #1666] [322], was an optical, mathematical, and nautical instrument maker of North Side, Old Dock, Hull, England. **Robert Pearson**, later occupied the same address.

Pearson, Robert

Robert Pearson, (fl.1834-1841) [18], was an optician, ship chandler, and nautical and mathematical instrument maker, of North Side, Old Dock, Hull, England. **Richard Pearson** had previously occupied the same address.

Pellin, Philibért François

Philibért François Pellin, (1847-1923) [53] [299], was a scientific and optical instrument maker of Paris, France. He began his career in instrument making when he started a collaboration with **Jules Louis Duboscq** in 1883. In 1885 they formed a partnership, trading as **Duboscq & Pellin**. In 1886 Duboscq died, and Pellin continued the business himself, building further international success. He expanded the company's product line into areas such as meteorological instruments and wireless.

Pellin's son, Félix Marie Pellin, (1877-1940), joined the company in 1900. By 1903 he was general director, and in 1912 they went into partnership as Philibért and Félix Pellin. In 1921 Félix was made Chevalier de la Légion d'Honneur, and in 1927 he became the owner of the firm. Soon after his death the firm merged with **Société Industrielle d'Instruments de Précision**, which was formerly the firm of **Louis Joseph Deleuil**. The company continued to trade until the 1950s.

Pelorus

A compass-like device, usually used on a ship or an aircraft, for taking a bearing relative to the direction of the craft. The device is mounted in a fixed orientation on the craft, and the operator uses it to sight a target and take a relative bearing.

Pentacon

VEB Pentacon Dresden was a German photographic equipment manufacturer, founded in Dresden as VEB Kamera und Kinowerke Dresden. In 1959 they merged with VEB Kinowerke Dresden (see **Zeiss Ikon**) and several other companies, including VEB Kamera-

Werke Niedersedlitz (see **Kamera Werkstätten**) and VEB Welta Kamera Werke (see **Welta**), and in 1964 the newly formed company, together with **Ihagee**, became VEB Pentacon. In 1971 **Hugo Meyer & Co.** became a part of VEB Pentacon. In 1976 VEB Kamerafabrik Freital (see **Beier**) became part of VEB Pentacon which, in 1985, became a part of VEB Carl Zeiss Jena (See **Carl Zeiss**). They operated until 1991 when, after liquidation, their assets were taken over, and the company re-emerged with a new owner. The camera and lens brands of the new Pentacon were **Praktica**, **Exakta**, and **Schneider Dresden**.

Pentaprism
See prism.

Pentax (Pentax Corporation)
Pentax began as a **Zeiss Ikon** brand-name, derived from the words *pentaprism* and *Contax* (a Zeiss Ikon camera model). The Pentax brand-name was sold to the **Asahi Optical Company** in 1957. In 2002 the Asahi Optical Company name was changed to Pentax Corporation.

In 2004 Seiko acquired the Pentax Vision subsidiary from Pentax Corporation, together with the use of the *Pentax* brand-name for its ophthalmic lenses.

In 2007 **Hoya** acquired the Pentax Corporation, and announced a merger of the Pentax Corporation into Hoya, which completed in 2008. In doing so they acquired the *Pentax* brands of medical technology, surveying instruments, cameras, and binoculars. The acquisition was made in order to gain access to its medical technology, which Hoya went on to market under the *Pentax* brand-name.

In 2009 Hoya sold the *Pentax* surveying instrument business, formerly the **Fuji Surveying Instrument Company Ltd.**, to the Taiwan Instrument Co. Ltd. As a result of this acquisition, the company was renamed **TI Asahi Co. Ltd.**, and continued to produce surveying instruments under the *Pentax* brand-name.

In 2011 **Ricoh** purchased Pentax Imaging Systems, the *Pentax* camera and binocular brand, from Hoya, [**112**], and they continued to use the *Pentax* brand-name for their products.

Periscope
A periscope is an instrument that enables a user to view an object that is not in their direct line of vision. It generally consists of an extension tube, at each and of which is an arrangement of mirrors or prisms to alter the optical path by 90°. A periscope may make use of one or more relay lenses to extend its reach. See also altiscope.

Periscopic Lens
A periscopic lens has a wide field of view.

(1) A meniscus lens. When used in spectacles, this allows the wearer to see clearly when turning their eyes from side to side, rather than needing to turn their head.

(2) A form of non-achromatic photographic lens consisting of two meniscus lenses arranged symmetrically, separated by an air gap in which may be placed a shutter. This was first introduced by **Steinheil** in 1865.

Periscopic Prism Company (PPCo)
The Periscopic Prism Company, [**20**], of Camden Town, London, England, chiefly made lenses and prisms for the optical instruments trade. They also made snipers' riflescopes for the War Office during WWI. By the end of the war the company had been taken over by the state, and was producing artillery sighting telescopes.

Perken, Son and Rayment
Perken, Son and Rayment, [**44**] [**306**], of Hatton Garden, London, England, was a maker of cameras, lenses, magic lanterns, and photographic apparatus, as well as other optical equipment. The company began in 1852 as **Lejeune & Perken**. They registered the **Optimus** trademark in 1885. In 1887 the company name changed to Perken, Son & Rayment, and from 1900 they traded as Perken, Son & Co. Ltd. Instruments bearing their name include a pocket sextant.

Perkin Elmer
Perkin Elmer [**193**], is an American corporation based in Massachusetts. Their products and services include diagnostics and imaging for medical and life sciences applications. In 1965 they acquired **Boller & Chivens**, and continued to produce large telescopes and mounts until the late-1980s. Perkin Elmer famously manufactured the optical components for the Hubble Space Telescope, for which they constructed the main mirror between 1979 and 1981. *The Hubble Space Telescope Optical Systems Failure Report*, [**194**], blamed Perkin Elmer for the flaw in the optics that resulted in the inability to focus the telescope, and necessitated a repair mission, carried out by Space Shuttle Endeavour in 1993.

Perot, Jean-Baptiste Alfred
Jean-Baptiste Alfred Perot, (1863-1925), was a French physicist renowned for co-inventing, in 1896, the Fabry-Perot interferometer with **Charles Fabry**.

Perry, Charles
Charles Perry, [**113**] [**280** ch.6], was the foreman for **Powell & Lealand**. In 1904 he left to work for **C. Baker**,

where he made microscopes for Thomas Powell until the Powell & Lealand business ceased in about 1914. Following this he started his own company at 41 Northolme Road, Highbury, London, England, where he traded under his own name.

Perspective; Perspective Glass

The term *perspective*, or *perspective glass*, was originally used to refer to any single lens or mirror used to view an object. The use of the term widened to mean an optical instrument used to assist the sight, such as a telescope or binoculars, and was commonly used to refer to a Galilean telescope, [24]. The term appears to have been interchangeable with the terms *prospect* or *prospect glass*, and is now obsolete.

Peter & Zoon, G.W.

G.W. Peter & Zoon (G.W. Peter & Son), [378], was a retailer of navigation instruments in Rotterdam, Netherlands. The company was a partnership between Gerrit Willem Peter (1798-1857) and his son, Hendrik Johannes Peter (1827-1896), who continued the business after his father's death. Instruments bearing their name include: brass marine telescope; azimuth compass; sextant.

Petitdidier, Octave Leon

Octave Leon Petitdidier, (1853-1918) [185], of Villemagne, France, emigrated to the USA in about 1873, and there began his career as a civil engineer. After working in Cincinnati, St. Louis, and Indianapolis, he began building telescopes. He issued a telescope catalogue in 1894 in Mount Carmel, Illinois, before settling in Chicago to produce optical instruments. He produced telescopes, optical parts for telescopes, and precision echelon gratings for **Albert Michelson**'s design of echelon spectroscope. He provided optics and other components for telescopes made by **William Gaertner**. Upon his death in 1918 his business was taken over by Gaertner.

Petit, L.

L. Petit was an optical instrument maker based in Paris, France. The firm's products generally bore the mark *L Petit Fab^T Paris* (L. Petit, made in Paris). An example of Galilean field binoculars by L. Petit bears the trade name **Hezzanith**, suggesting that they were made by L. Petit for **Heath & Co.** Instruments bearing their name include Galilean binoculars and opera glasses dating around the late-C19th and early-C20th. Some were military issue.

Petri Camera Company

Petri Camera Company was founded in 1907 as **Kuribayashi Camera Works** in Tokyo, Japan. They began to use the *Petri* brand-name in 1962 and, in the same year, the company name was changed to Petri Camera Works. They produced a variety of models of camera under the *Petri* brand, and continued to trade until bankruptcy in 1977.

Petrograd Porcelain Works

With the onset of war in 1914, the Russian Company for Optical and Mechanical Production was formed for the production of optical equipment and detonators for shells, and was mainly owned by **Schneider-Creusot**. The outbreak of war, however, resulted in the loss of their main suppliers of optical glass. They tried unsuccessfully to produce their own optical glass, and approached the **Chance** company in Birmingham, UK for help. Chance sold them the technology for 600,000 gold rubles, and the resulting production of optical glass took place at the Petrograd Porcelain Works [27].

Petzval, Josef Max

Josef Max Petzval, (1807-1891) [37] [70], of Hungary, was a professor of higher mathematics at the University of Budapest. Photographic camera lenses at the time, such as **Chevalier**'s achromatic meniscus lens, had small apertures, and thus required long exposure times. Petzval's design used a well-corrected telescope lens for the front component, with a mathematically designed air-spaced doublet behind it, which gave sharp definition and corrected for spherical aberration. He took his design to **Voigtländer** who manufactured the lens. This Petzval lens, unlike others at the time, had a large aperture, and improved exposure times by a factor of 20, making portrait photography a more realistic endeavour. This much imitated lens became a standard design, and enjoyed enduring success.

Pfister, Hermann

Hermann Pfister, (1847-1911), was a mathematical and surveying instrument maker of Cincinnati, USA. He was born in Switzerland, and emigrated to the USA in the late 1860s. By 1870 he had moved to Cincinnati, and started his business making mathematical instruments. In 1878 he purchased the business of **R. Whitcomb**, and continued making mathematical and surveying instruments until his death in 1911. His son, William Henry Pfister became workshop foreman in 1905, and succeeded his father in the business.

Phantascope; Bio-Phantascope

The phantascope, or bio-phantascope, was a magic lantern with adaptation to project a series of seven images from a circular gallery, which was moved around the body of the magic lantern in a stop-go manner, with a shutter that opened while the frame was still. This produced an illusion of motion. The phantascope was invented by **John Rudge**.

Phase Contrast Microscopy

In conventional light microscopy it is the difference in amplitude of light across the image that creates the contrast in the image. Some specimens, however, appear transparent. One method of viewing such specimens is to first kill, fix, and stain them. Because of this, they cannot be viewed, and their behaviour studied, as living specimens. However, transparent specimens such as this generally cause diffraction, and a phase shift in the light, which varies with the detail of the specimen. These phase differences are not visible in a conventional light microscope, but may be viewed using phase contrast microscopy [7], invented in 1934 by the Dutch physicist Fritz Zernike.

Phase contrast microscopy translates phase differences into amplitude differences in viewing an image of a specimen, hence providing a high-contrast image without loss of resolution. The substage illumination uses partially coherent light, and a specially designed plate, the *phase annulus*, located at the back focal plane of the condenser. This obscures part of the light to yield an annulus of illumination. A phase plate, located at the back focal plane of the objective lens, retards the un-diffracted light by ¼ wavelength. This goes on to form an image at the image plane by means of constructive or destructive interference with the diffracted light that has passed through the specimen.

The phase annulus and the phase plate must be properly aligned with respect to each other. This is normally achieved using fine adjustments to the position of the phase annulus plate. The adjustment is made visually using either a Bertrand lens or a phase telescope.

See also dark field microscopy.

Phase One

Phase One is a digital imaging company, founded in 1993 in Copenhagen, Denmark. Their subsidiaries include **Leaf** and, from 2009, **Mamiya**. Their products include digital imaging software, and medium format digital camera products. This includes Mamiya cameras, Leaf digital backs, and camera lenses by Mamiya, some of which are designed by **Schneider Kreuznach**.

Phase Telescope

A phase telescope is a specialised microscope eyepiece, with an additional lens element that brings the back focal plane of the microscope objective into focus at the image plane viewed by the eyepiece. This allows the objective back focal plane to be observed through the microscope eyepiece, and is used in a phase contrast microscope to facilitate the accurate alignment of the condenser annulus with the phase plate in the objective. Some phase contrast microscopes are equipped with a Bertrand lens built into the eyepiece tube, in which case it is used for the same purpose instead of a phase telescope.

Phelps & Gurley

See **W.&L.E. Gurley**. An advertisement for Phelps & Gurley of Troy, New York, USA reads:

"Phelps & Gurley. Manufacturers and dealers in Mathematical & Philosophical instruments, No. 319 River Street, Troy, N.Y. Surveyors compasses, Levels and theodolites, Transit instruments, Telescopes, Microscopes, Drafting instruments, Barometers and thermometers, Globes, Chemical apparatus, Electrical machines, Air pumps, Mechanical powers, Magnets, Electro magnets, Electro magnetic apparatus for medical purposes, Galvanic batteries &c., &c., Builders levels, Spirit level tubes, Brass, lead and pewter castings, Brass finishing, Models of machinery, and repairing to order."

Phénakistiscope

The Phénakistiscope is a device that allows a single user to view an animation based on a sequence of images around one face of a cardboard disk. The disk has slots cut at regular intervals in its edge. It is set spinning with its back to the user, and the user views through the slots to a mirror. The slots act as a shutter so that the user sees the illustrated face of the disk at intervals depending on how quickly the disk is spinning. Each view through a slot is a near-still view of the illustrated face, progressed around the axis from the previous view. This results in the illusion of a moving image.

Eadweard Muybridge devised a projection version of the Phénakistiscope, called a Zoöpraxiscope, which he used to illustrate his lectures on animal locomotion.

Phillips, Solomon

Solomon Phillips, (fl.1839-1846) [18] [276 #2206] [322], was an optician, and optical, mathematical, and philosophical instrument maker, of 43 Rathbone Place, London, England. From 1840 he traded at 231 Tottenham Court Road. At this latter address he entered a partnership, trading as Phillips & Jacobs from 1844 until 1846. According to his trade card, "Several years lecturer in optics and astronomy to the Canterbury Literary and Philosophical Society" [53].

Phoneidoscope

A phoneidoscope is an instrument used to observe the variations of colour and vibration in thin liquid films, such as soap, when acted on by sound waves.

Photochrom, Photochrome, Photochromy

A photochrom, or photochrome, [70] [71 p664], is a method of producing a colour photograph. The term generally refers to the methods of early innovators such as **Claude Félix Niépce de St Victor** and **Gabriel Lippmann**. The term is also used to refer to a print produced using such a method. Photochromy is the process of producing such images.

Photograph

A photograph is a recorded image produced by an optical process, usually in a camera. Throughout history photographs have been produced by many different processes and technologies [70].

See also ambrotype, autochrome, calotype, collodion process, daguerreotype, heliograph (3), interference colour photography, photochrom, photographic film, ferrotype (tintype), trichromatic photography. Modern photographs are usually digital, or produced using photographic film.

Photographic Film

Photographic film consists of a thin, transparent plastic sheet coated on one side with an emulsion containing silver halide crystals. It is used in a camera to record an image, and subsequently developed to produce an image. The image produced is either a positive or a negative image, depending on the film type and process. The positive image may be used in film slides and still or moving image projectors. The negative may be used to produce multiple positive prints.

Photographic Zenith Tube (PZT)

The photographic zenith tube, [53] [236] [399], is a form of zenith telescope. It consists of a vertically mounted telescope which can view stars within about 15 minutes of the zenith. The image of the star is reflected from a standing pool of mercury, and back up to a photographic plate. The telescope is accurately aligned to the vertical in a permanent mount, and the pool of mercury ensures a horizontal mirror surface. Hence the instrument is capable of highly accurate measurements. The PZT was an observatory instrument used for accurate time measurement as well as measurement of the variation of latitude. A PZT was built in 1955 for the Royal Greenwich Observatory in Herstmonceaux, England, by **Grubb Parsons**. It is now in the Science Museum, London. Another PZT was made by Grubb Parsons for the Neuchâtel Observatory in Switzerland.

Photographometer

A photographometer is an instrument used to determine the sensitivity of photographic plates to light.

Photoheliograph

A photoheliograph is an instrument used to take photographs of the sun. It consists of a camera with a telescope that is adapted for the purpose. See heliograph, siderostat.

Photometer

A photometer is an instrument for measuring or comparing the intensity of two sources of light. There are several kinds of photometer.

Lummer-Brodhun photometer: A photometer in which the light from each of the two sources is reflected from a disc into a Lummer-Brodhun cube. Thus, a comparison may be made, via an eyepiece, between one light source in the centre of the view, and the other as an annulus around it.

Bunsen's Photometer: A photometer disc, consisting of a paper disc with a grease-spot, is mounted on an optical bench. It is illuminated by one light source on each side. The disc is moved along the optical bench until the two light sources are the same brightness on the disc. At this position, the intensities of the two light sources are proportional to the squares of their distances from the disc. If one light source is of known intensity, the other may be calculated.

Bunsen's Mirror Photometer: This is a variant of Bunsen's photometer in which the grease-spot is viewed by means of angled mirrors so that both sides may be seen simultaneously in an eyepiece.

Dibdin's Hand Photometer: A hand-held photometer, designed by the English scientist, William Joseph Dibdin (1850-1925), used to measure light sources in-situ, such as street-lights, railway, and school lights.

Fire-Damp Photometer: An instrument used to detect the presence of flammable hydrocarbons in the air. It uses an electric current to ignite the fire-damp, and a photometer to observe the intensity of the resulting flame.

Flicker photometer: A photometer that enables the comparison of two light sources by rapidly alternating between the two. If the view flickers, the intensities are different. If the view does not flicker, the two light sources are of equal intensity, irrespective of any difference in colour between them.

Prism Photometer: A photometer in which the two light sources to be compared are brought together by means of total internal reflection in two prisms.

Richie's Photometer: A photometer, invented by William Richie, based on the principle that absorbed light will heat a chamber, and thus increase the pressure within the chamber. Two chambers, each illuminated by one of the two light sources, are connected by a u-tube containing a small amount of fluid that moves, depending on the difference in pressure of the two chambers, and hence the difference in intensity of the two light sources.

Simmance-Abady Photometer: Another name for a Flicker Photometer.

Rumford's Photometer: A simple photometer in which a rod casts two shadows on a white screen, one from each light source. When the two shadows are of equal darkness, the intensities of the light sources are proportional to the squares of their distances from the rod.

Wheatstone's Photometer: Invented by **Sir Charles Wheatstone**, uses the phenomenon of persistence of vision. The two sources of light are reflected in a small sphere, rapidly moving in a given pattern. The effect of persistence of vision is to smear out the reflections, and reduce their intensities. This allows for a sensitive comparison of the intensities of the sources. It consists of one or more silvered beads or convex mirrors attached to a disc. The disc spins as it rotates around the edge of a toothed wheel, resulting in a complex orbital motion, driven manually by a handle.

Photophone

A photophone is an instrument used to communicate sounds by means of a beam of light.

Photoscope

A photoscope is:

(1) An optical instrument used to exhibit photographs.

(2) An instrument consisting of a photosensitive device, such as a selenium cell, which is used to detect varying intensities of light.

(3) An instrument used to examine some internal parts of the human body, using sunlight for illumination.

Photo-Survey Camera

A photo-survey camera is one that produces survey photographs with an integral photo-record of compass bearing and horizontal level.

Photo-Theodolite

A photo-theodolite is a theodolite with an integral camera. The photographs, rather than direct observation through the telescope, are used to determine angles.

Pickart, Alexandre Auguste

Alexandre Auguste Pickart, (1834-c1910) [50] [287], was an optical and scientific instrument maker of 20 Rue Mayet, Paris, France. He began his own business around 1875, and made instruments, including mineralogical instruments, microscopes, petrological microscopes, polarizers, goniometers, spectroscopes, heliostats and other physics instruments. He developed a particular interest in mineralogical and chemical apparatus. The business was listed as Pickart (A) et Fils for a period in the late-1890s, but by 1900 was again listed as Pickart (A). The business continued until at least 1907.

Pickering, Thomas

Thomas Pickering, (fl.1788-1847) [18] [385], was an optical, mathematical, and philosophical instrument maker of London, England. Addresses: 1 Regent St., Lambeth, and 36 Regent St, Lambeth (1839-1847). He served his apprenticeship with **Michael Dancer** of the **Joiners' Company**, beginning in 1781, but there is no record of his freedom.

Piggott, William Peter

William Peter Piggott, (fl.1838-1859) [18] [276 #2208], was an optical, mathematical, and philosophical instrument maker of London, England. Addresses: 13 Arnold Place, Walworth (1838); 20 Wardrobe Place, Doctor's Commons (1840); 11 Wardrobe Place, Doctor's Commons (1845-1846); Great Carter Lane, Doctor's Commons (1848); 523 Oxford St. (1851).

He attended the London Mechanics' Institute from 1826 until 1832. He served his apprenticeship with his father, the mathematical instrument maker Peter William Piggott, beginning in 1829, and was freed of the **Merchant Taylors' Company** in 1838. Among his apprentices was **Thomas Boddy**, beginning in 1840.

He worked in partnership with Thomas Boddy, trading as Pigott & Boddy, opticians and mathematical instrument makers of Wardrobe Place, Doctor's Commons, and New Oxford Street. The partnership was terminated in March 1849 [35 #20593 p769]. He also worked in partnership with Robert Weare and Thomas Weare, trading as watchmakers, opticians, and mathematical instrument makers. They traded as Weare & Piggott in Birkenhead in the county of Cheshire, and as Piggott & Weare in New Oxford St., London. This partnership was terminated in October 1849 [35 #21039 p3507], after which William Peter Piggott continued the business at 523 Oxford St., trading as Piggott & Co. until 1859.

W.P. Piggott was awarded various patents, including for dials and nautical instruments. According to the London Gazette patent notices in 1855 [35 #21823 p4592], "William Peter Piggott, of 523, Oxford Street, in the county of Middlesex, Medical Galvanist, has given the like notice in respect of the invention of 'improvements, in galvanic, electric, and electro-magnetic apparatus, and in the mode of applying the same as a curative and remedial agent.'"

Pignons SA
Pignons SA was a Swiss company, founded in 1918, which created the **Alpa** brand of cameras. They also manufactured the first cameras to be designed and sold by the **Bolsey** company of the USA.

Pike, Benjamin
Benjamin Pike, (1777-1863) [50] [201], was an Englishman who emigrated to the USA in 1804. He established his business as an optician in New York in 1806. Although his catalogue listed a large number of instruments, many of them were imported. It is likely that they manufactured at least one model of microscope.

Benjamin Pike operated the company with his three sons, Benjamin Jr. (1809-1864), Daniel (1815-1893), and Gardiner (1824-1893). Over the following decades the company operated under a variety of names including Benjamin Pike, Benjamin Pike & Son, Benjamin Pike & Sons, Benjamin Pike, Jr., Benjamin Pike's Son & Company, and Benjamin Pike's Sons.

Walter H. Bulloch served his apprenticeship with Benjamin Pike & Son. Both **Richard Morrison** and **George Wale** worked for Benjamin Pike's Sons producing microscope and camera lenses.

Pilkington Brothers
In 1827 William Pilkington (1800-1872) and his brother-in-law Peter Greenall became the owners of the St. Helens Crown Glass Company in St. Helens, Lancashire, England, and changed the name to Greenall and Pilkingtons. In 1849 the company, by then fully controlled by the Pilkington family, was renamed Pilkington Brothers [44] [46].

In 1938 The Pilkington glass company acquired an interest in the **Chance Brothers** glass-fibre facility in Firhill, Glasgow. Chance had been making glass-fibres at Firhill since the late-1920s, and had been involved in the manufacture of optical glass in Birmingham since 1838. Prior to WWII, Pilkington erected a shadow optical factory for Chance at St. Helens in case of war damage to the Chance Birmingham plant. By 1945 Pilkington had acquired a 50 percent shareholding in Chance, and by 1951 Chance became a wholly owned subsidiary. In 1957, the optical side of both companies was combined in the new **Chance-Pilkington Optical Works** at St. Asaph, North Wales.

In 1997 **Barr and Stroud** merged with Pilkington, a relationship which lasted 3 years until 2000, when Barr and Stroud and Pilkington became subsidiaries of the Thales Group, **Thales Optronics**. In 2005 Thales Optronics was sold, and became **Qioptiq**. In 2006 Pilkington was acquired by the NSG Group, and became an NSG brand.

The NSG Group manufactures and sells glass in the UK & Ireland under the *Pilkington* brand-name. They manufacture glass for various applications, but not optical glass.

Pilkington, George
George Pilkington, (fl.1835-1849) [18] [276 #2209] [379], was an optician, and optical, mathematical, and philosophical instrument maker of London, England. Addresses: 7 Vine St., Laystall St.; 48 St. James's St., Walworth; 4 St. James St., Clerkenwell (1839); 14 Clarence Place, Pentonville (1839-1849). He was succeeded in business by **John Pilkington**.

Pilkington, John
John Pilkington, (fl.1850-1851) [18] [379], was an optician, telescope maker, and mathematical and optical instrument maker, of 14 Clarence Place, Pentonville, London, England. He succeeded **George Pilkington** in business, and was succeeded by **Mary A. Pilkington**.

Pilkington, Mary A.
Mary A. Pilkington, (fl.1851-1859) [18] [379], was an optician, spectacle maker, and mathematical and philosophical instrument maker of London, England. Addresses: 14 Clarence Place, Pentonville (1851-1855), and 205 Pentonville Rd (1859). She succeeded **John Pilkington** in business.

Pillischer, Moritz
Moritz Pillischer (sometimes spelled Morrice), (c1819-1893 fl.1851-1887) [11] [18] [44] [113] [264] [319] [320], was an optical, mathematical, and philosophical instrument maker of London, England, with a reputation for making fine microscopes. Addresses: 419 Oxford St. (prior to 1851); 398 Oxford St. (1851-1853); 88 New Bond St. (1854-1887). The premises at 419 Oxford St. were subsequently occupied by **William Hobcraft (II)**. Moritz was born in Hungary, and came to England in 1845, reputedly after gaining experience in "various continental centres".

Moritz exhibited at the 1851 Great Exhibition in London. His catalogue entry, no. 269, [328], listed him at 398 Oxford Street, as a designer and manufacturer. It listed his exhibit as: Large and small achromatic microscopes, with the stage movements simplified; Students' microscope, capable of forming a portable dissecting, as well as clinical microscope, with all the necessary apparatus; Double achromatic opera glass; Opera glasses, mounted in tortoiseshell and gilt, and mounted in ivory; Newly invented compasses, for describing ellipses of any size; Six's thermometer in ivory, for registering maximum and minimum temperatures.

Moritz's nephew, **Jacob Pillischer**, came from Hungary to work with him in 1857.

A catalogue of 1873 lists Moritz as a manufacturer of achromatic microscopes, opera, race, and field glasses, and other optical, philosophical, mathematical, surveying, and standard meteorological instruments.

An advertisement of 1883 lists Moritz as optician and scientific instrument maker to the Queen, the Prince and Princess of Wales, the Royal Family, Her Majesty's Government, &c.

Jacob succeeded him in business in 1887.

Instruments bearing his name include: microscopes; microscope lamps; telescopes; stereoscope; barometers; sunshine recorder; barograph. Recorded serial numbers on microscopes are in the range 33 to 5261.

Pillischer, Jacob

Jacob Pillischer, (1838-1930) [**11**] [**18**] [**44**] [**113**] [**264**] [**319**], was a scientific and optical instrument maker of 88 New Bond St., London, England, with a reputation for making fine microscopes. He came to England from Budapest, Hungary, in 1857 to work with his uncle, **Moritz Pillischer**. He succeeded his uncle in business in 1887, and subsequently traded under the names Jacob Pillischer & Sons, and J. Pillischer Ltd. He was naturalized as a British citizen in 1911.

In 1929, in the year before his death, Jacob exhibited at the British Industries Fair [**448**]. His advertisement named the *International* Microscope as one of their products, and listed the company as manufacturers of microscopes and accessories, as well as optical and other scientific instruments.

Upon his death in 1930, Jacob left the business to his three children, Edward (1873-1966), Leopold (b.c1876), and Bertha (c1878-1952). The company, J. Pillischer Ltd. continued in business until it was voluntarily wound up in 1947 [**35** #37938 p1788], at which time Bertha was chairman of the company.

Instruments bearing his name include: microscopes; Galilean binoculars.

Pilot, Etienne

Etienne Pilot, (fl.1900) [**341**], of 60 Rue Pernéty, Paris, France, was a scientific and optical instrument maker. He is known to have made microscopes, as well as scientific instruments, such as gyroscopes, electrical instruments, and other instruments to demonstrate physical phenomena.

Pinhole Camera

See camera obscura.

Pistor, Carl Philipp Heinrich

Carl Philipp Heinrich Pistor, (1778-1845) [**47**] [**374**], began his career in the Prussian postal service. He attended lectures in astronomy, chemistry and physics, and worked for a while in Berlin in the mechanical workshop of Nathan Mendelssohn. He opened his own workshop in Berlin in 1813, and established his reputation for making large astronomical and geodetic instruments. Among his apprentices was **Johann August Oertling**. In 1824 he took on **Friedrich Wilhelm Schieck** as a journeyman who went on to gain a reputation for his fine microscopes. Schieck rose to become a partner in the company, from which time they traded as Pistor and Schieck. In 1830 Pistor supplied most of the telescopes and signalling equipment for the optical telegraph between Berlin and Koblenz. In 1836 Schieck left to start his own workshops, and in 1841 Pistor formed a partnership with his son-in-law Carl Otto Albrecht Martins (1816-1871), with whom he traded as Pistor and Martins. After Carl Pistor's death in 1845 the company was run by his son Gottfried together with Martins. The firm continued after the death of Martins in 1871, but ceased trading in 1873.

Pistor & Martins

A partnership between **Carl Philipp Heinrich Pistor** and his son-in-law Carl Otto Albrecht Martins.

Pistor & Schieck

A partnership between **Carl Philipp Heinrich Pistor** and **Friedrich Wilhelm Schieck**. Microscope makers.

Pixii Père et Fils

See **Nicolas Constant Pixii**.

Pixii, Nicolas Constant

Nicolas Constant Pixii (1776-1861), [**17**] [**256**] [**322**] [**423**], was a manufacturer of philosophical instruments, and physical apparatus for electrodynamics, electromagnetism, optics, and physical acoustics, of 2 rue du Jardinet, Paris, France. He was the son-in-law of one of the **frères Dumotiez**, whom he succeeded in business. His son, Antoine-Hippolyte Pixii (1808-1835), worked with him in partnership, trading as Pixii père et fils, and was famous for his invention of the magneto-electrical generator, named after him. In 1838, shortly after the death of his son, Nicolas moved the business to 18, rue de Grenelle St. Germain. There he continued to expand on the catalogue and reputation of his predecessor. Upon Pixii's retirement in 1855, the firm was acquired by Fabre et Kunemann and, in 1858, renamed Fabre de Lagrange.

Pizzala, Augustus
Augustus Pizzala, (fl.1837-1853) [18] [322] [379], was an optician, and mathematical and philosophical instrument maker of London, England. Addresses: 22 Leather Lane, Holborn (1837); 7 Charles St., Hatton Garden, Holborn (1839-1846); 19 Hatton Garden (1847-1853). He succeeded **Francis Augustus Pizzala (I)** in business, and was succeeded by **Francis Augustus Pizzala (II)**.

Pizzala, Francis Augustus (I)
Francis Augustus Pizzala (I), (fl.1837-1839) [18] [276 #2211] [322] [379], was a philosophical instrument maker of London, England. Addresses: 7 Charles St., Hatton Garden, Holborn (1837-1839); 22 Leather Lane, Holborn (1839). Reference [18] states that **Henry Negretti** served an apprenticeship with him starting in 1838. He was succeeded in business by **Augustus Pizzala**.

Pizzala, Francis Augustus (II)
Francis Augustus Pizzala (II), (fl.1851-1865) [18] [322] [379], was an optician, and a mathematical and philosophical instrument maker of London, England. Addresses: 19 Hatton Garden, Holborn (1851-1860); 25A Hatton Garden (1865). He succeeded **Augustus Pizzala** in business. He worked in partnership with Matthew Charles Greene, trading as Pizzala & Greene, philosophical instrument makers and looking-glass manufacturers at 19 Hatton Garden, until the partnership was declared bankrupt in 1860 [170 #7017 p693].

Instruments bearing his name include a brass and mahogany three-draw telescope with sliding sunshield, signed *Pizzala, 19 Hatton Garden Holborn London*.

Pizzala, Joseph
Joseph Pizzala, (fl.1809-1830) [18] [322] [379], was a looking-glass maker and philosophical instrument maker of 84 Leather Lane, Holborn, London. He is known to have sold barometers.

Placido's Disc
Another name for a keratoscope.

Plan objective
See objective lens.

Plath, Carl
Carl Christian Plath, (1825-1910) [44] [114], of Hamburg, Germany, was a manufacturer of nautical instruments. He served an apprenticeship with a local instrument maker in Hamburg, and in 1857 he started his own business manufacturing surveying instruments. In 1862 Carl Plath sold his business to one of his employees, **J.C. Dennert**, and purchased another Hamburg business, started by David Filby in 1837, that traded in nautical literature, charts and sextants imported from England. Having established himself in his new business he concentrated on the manufacture of sextants, magnetic compasses, binnacles and barometers.

In 1889 Carl Plath's son, Theodor Christian Plath (1868-1960), became co-owner of the company. Carl Plath retired in 1905, and in 1908 he handed over his share in the business to Theodor, making him the sole owner. Theodor's only son, Johann Christian, died at the age of only 27 in 1929 and, wishing to keep the firm in the family, Theodor handed the business over to his son-in-law Johannes Boysen in 1937, but required first that Boysen, at the age of 32, served an apprenticeship as an instrument maker.

At the beginning of 1953 Johannes Boysen, together with the American navigator P.V.H. Weems, founded in Washington the firm of **Weems & Plath Inc.**, whose main purpose was to step up the sale of C. Plath sextants in the USA.

Theodor Plath retired in 1950 at the age of 82, and gave up his limited partnership. In 1962, Litton Industries acquired C. Plath. Sperry Marine was formed in 1997 with the combination of C. Plath, Decca Marine and Sperry Gyrocompass, and C. Plath changed its name to Sperry Marine in 2000. In 2001 Sperry Marine became part of the **Northrop Grumman Corporation**.

Plaubel
Plaubel & Co. was founded in Frankfurt, Germany, in 1902. Initially the company produced lenses, and in 1909 they began to produce cameras. In 1911 they began to use the name *Makina* for a series of camera models. In 1975 the company was sold into Japanese ownership, and many of their later products were manufactured in Japan. Production of *Makina* cameras ceased in 1986.

Pleograph
The Pleograph was an early kind of motion picture camera devised by the Polish inventor **Kazimierz Prószyński** in 1894. By 1898 Prószyński had also developed a motion picture projector, the Bio-pleograph.

Plössl, Gustáv Simon
Gustáv Simon Plössl, (1794-1868) [39], was born in Vienna, Austria. He served his apprenticeship with **Voigtländer** in 1812, and in 1823 he established his own company in Vienna. He made microscope objectives of his own design, and opera glasses. In 1839 he is reported to have made a daguerreotype camera using a modified

Chevalier landscape lens. This was a modification of the Chevalier achromatic doublet which preceded those of **Steinheil** and **Dallmeyer** by more than two decades. In 1860 he developed an improved version of the Ramsden eyepiece resulting in a class of orthoscopic, achromatic, wide-field eyepieces of three types; the Symmetrical, the Dial-Sight, and the Plössl, (see Appendix II). In 1898 the Plössl company was merged with the **Karl Fritsch Opto-Mechanical Workshop** to form **Karl Kahles**.

Polam

Polam is a brand of polarizing microscope produced by the **Leningrad Optical Mechanical Association** (LOMO).

Polar Clock

The polar clock, designed by **Sir Charles Wheatstone**, is an optical instrument used to ascertain the time of day by detecting the plane of polarization of the sunlight. It uses the fact that the plane of maximum polarization of sunlight is always at right angles to the direction of the sun's position in the sky. The instrument is a conical tube with thin films of selenite at the objective end, and a rotatable Nicol prism at the eyepiece. The polar clock is aligned to the polar axis of the Earth, and the eyepiece rotated until the plane of polarization of the light is found. The degree of rotation of the eyepiece thus indicates the time of day. It does not give as accurate a reading as a sundial, but has the advantage that it can be used under cloud cover, and when the sun is below the horizon.

Polarex

Polarex was a trade mark of the company **Nihon Seiko Kenkyusho Ltd.**

Polarimeter

A polarimeter is:

    (1) An instrument that measures how much the plane of polarization in plane-polarized light is rotated when passing it through a liquid or solution, or other optically active substance.

    (2) An instrument that measures the degree to which light is polarized.

Polariscope

The polariscope is an optical inspection device used to detect internal stresses in glass and other transparent materials such as plastics, synthetic resins, etc. A polariscope is composed chiefly of a light source and two crossed polarized lenses. Material to be examined is placed between the two polariscope lenses and viewed through the lens opposite the light source lens. It is commonly used in detecting the optical properties of gemstones. A microscope with cross-polarizing filters serves as a polariscope. See also Nörremberg polariscope.

Polarization

In its simplest representation, light may be treated as a transverse electromagnetic wave, in which the electric field varies sinusoidally in a plane of vibration (direction and amplitude given by the electric field vector), and propagates in a direction along that plane (direction and speed given by the propagation vector). In this case the light is referred to as *linearly* or *plane-polarized* light [7].

In general, natural light consists of large numbers of waves, emitted by natural atomic processes. A plane-polarized wave is emitted for about a hundred-millionth of a second, and the light we see is the combination of a large number of these waves, each with an unpredictable polarization angle. This is referred to as *unpolarized*, although it is in fact a combination of a large number of randomly polarized waves. Some physical processes, such as reflection or atmospheric scattering, result in the polarization of light. In most practical situations, light is neither completely polarized nor completely unpolarized, and is referred to as *partially polarized*.

Two plane-polarized waves, propagating in the same direction, may be superimposed, although the direction of the plane of vibration of one may be at an angle to the direction of the plane of vibration of the other. If the two waves are in phase (the wave crests occur at the same time), this combination results in a wave that is also plane polarized. If the direction of the plane of vibration of the two waves is the same, the resulting wave is in the same plane. If the direction of the plane of vibration of the two waves is different, the direction of the resulting wave plane is given by the sum of the two electric field vectors.

If the two waves are out of phase by 90° (the crests of one wave occur at the same time as the troughs of the other), the resultant wave is in a plane that rotates about the direction of propagation. This is referred to as *circular polarization* – right circular or left circular depending on the direction of rotation. If they are not in phase, but the phase difference is other than 90°, the resultant wave rotates in an elliptical manner, and is referred to as *elliptical polarization*.

In optical instruments, the polarization of light is utilised or manipulated in: Glan-Foucault Prism, Glan-Thompson Prism, Nicol prism, Nörremberg polariscope, polar clock, polarimeter, polariscope, Rutherford Prism, Thompson prism, Wollaston Prism.

See also birefringence, **Sir David Brewster**, Brewster angle, Iceland spar.

Polarization by Reflection

See Brewster angle.

Polaroid

The company was formed in 1932 by the American, Edwin Land, (1909-1991), as the Land-Wheelwright Laboratories, and in 1937 it was incorporated as Polaroid. Initially they marketed their innovative plastic-sheet polarizer in sunglasses and other optical applications. After WWII they introduced their first self-developing film camera, the Polaroid Land Camera, and the company continued to produce so-called instant cameras and films, introducing their instant motion pictures in 1977.

In 2001 began a series of bankruptcies and changes of ownership which saw the Polaroid camera discontinued in 2007, and Polaroid film discontinued in 2009. In 2017 the company was acquired by the Impossible Project, a former Polaroid film factory in the Netherlands, and Polaroid film was re-launched under the *Impossible* brand. The film was re-branded *Polaroid Originals* shortly afterwards. The company offers instant and digital still cameras, and high-definition and mountable sports action video cameras.

Polar Scope

A polar scope is an instrument used for polar alignment of an equatorial telescope mount. It is a specialised telescope, normally integrated into the equatorial mount, used to align the mount's right ascension axis with the celestial pole.

Polar Siderostat

See siderostat.

Polemoscope

A polemoscope is a device, similar to a monocular opera glass, but which has an internal mirror, mounted at 45°, allowing the user to view through a hole in the side of the instrument. This enables the user to view objects that are not in their direct line of sight. Otherwise known as a *jealousy glass*, this device was often used for observing people without their knowledge.

Polish Optical Industries (PZO)

Polish Optical Industries, or Polskie Zakłady Optyczne (PZO), [106] [115], was formed in 1930 when shares in **H. Kolberg & Co.** were sold by **Henryk Kolberg** to **Optique et Précision de Levallois (OPL)**, **Krauss**, and **Barbier Bénard et Turenne (BBT)**. This, together with other investors, resulted in H. Kolberg & Co. being re-formed as PZO. Production continued uninterrupted, except for the re-branding, and during WWII their main customer was the Polish army. Their military production included 6x30 and 8x30 binoculars, trench periscopes, sights for heavy machine guns, anti-tank cannons, anti-aircraft cannons, and aircraft sights for bomb dropping.

At the end of the Nazi occupation the Germans withdrew, taking with them what machines and equipment they could, and destroying the rest along with the factory buildings. Despite the ongoing war, the PZO production facilities were subsequently re-built, with government help, by a group of former PZO workers and others. This effort was assisted partly by the availability of machines and materials from the now closed H. Kolberg factory, which had not been destroyed.

After the war, PZO opened a series of optical retail shops where they sold their own brand of instruments along with those from other manufacturers. In the years following the war, PZO made microscope accessories, microscopes, 6x30 and 8x30 binoculars, dumpy levels, loupe magnifiers, projection lenses, and more. In 1948 PZO was nationalized and made a part of the Union of Precision and Optical Industries. By the early 1950s, production levels were high, and included exports, mostly of microscopes and accessories, to China, Hungary and Romania. In 1953 they introduced coatings on their optical lenses, and by 1956, with a growing catalogue of products, they were exporting additionally to Great Britain, Canada, Bulgaria, Italy, Spain, and by 1965, to 46 countries globally. In about 1968 **Warsaw Photo-Optical Works (WZFO)** was merged to become a part of PZO.

Pollard, Joseph

Joseph Pollard, (fl.1820-1840) [18] [276 #1675], was an optician, hydrometer maker, and optical and philosophical instrument maker, of London, England. Addresses: 42 Princes St., Leicester Square; 59 Millbank St., Westminster (1822); 62 Millbank St., Westminster (1822); 42 Brewer St., Golden Square (1839).

Pöller, Franz

Franz Pöller, sometimes Pœller or Poeller, (fl.late-C19th) [113], was a retailer of optical, physical, and philosophical instruments, and the proprietor of a physical-optical institute in Munich, Germany. Instruments he sold were supplied to him wholesale, and branded with his name. According to his trade advertisement of 1888, he sold: Marine telescopes; Binoculars for theatre, hunting, travel, and regatta; Magnifying glasses for inspection of coins and banknotes; Barometers; Compasses; Hygrometers, etc. He advertised that he had connections with the European courts and armies, and that between 1874 and 1888 he had received over 20,000 orders from state offices all over Europe.

Polygonoscope

A polygonoscope is a form of kaleidoscope, utilising an arrangement of hinged mirrors.

Polyoptrum; Polyoptron
See lens.

Ponder and Best
See **Vivitar**.

Ponthus & Therrode
Ponthus & Therrode, [360], were surveying and nautical instrument makers of 6 Rue Victor Considérant, Paris, France. The firm was established in 1895, and in the same year they acquired **A. Barthélemy**. In 1900 they acquired **A. Hurlimann**, and in the same year they exhibited at the Exposition Universelle in Paris. Instruments bearing their name include: rangefinders; theodolites; sextants; drawing instruments. They were succeeded in 1922 by **A. Lepetit**.

Poole, Thomas
Thomas Poole, (fl.1813-1818) [18] [276 #1408] [379], was an optician, and optical, mathematical, and philosophical instrument maker of Upper North Place, Gray's Inn Lane, London, England.

Porro, Ignazio
Ignazio Porro, (1801-1875) [148] [162 p194-197] [374], was an Italian inventor in the field of optical devices. During his military service he made improvements to the surveying instruments he was using and, after he had retired from the military, he set up his own workshop in Turin, Italy, and subsequently in Paris, France. Here he developed improved lens systems for cameras. The invention for which he is most noted is the Porro prism image erecting system.

In 1865, at the age of 64, Porro set up **La Filotecnica** in Milan, a combined training school and production laboratory where students and their mentors produced and sold optical instruments, principally for topographical and geodetic use. Angelo Salmoiraghi (1848-1939) graduated from the Politecnico di Milano, the technical university in Milan, in 1866. Porro, who already knew Salmoiraghi, engaged him in the Filotecnica, and soon promoted him. Salmoiraghi's influence in the Filotecnica increased until he took over ownership, and changed its name to Filotecnica Ing. A. Salmoiraghi.

Porro Prism
The Porro prism, [43] [148], invented by **Ignazio Porro**, consists of a right-angled prism in which the light enters near one end of the hypotenuse and, after two internal reflections, exits near the other end of the hypotenuse. The resulting image is reversed from top to bottom, but not from left to right. The incident ray of light is normal to the surface of the prism, as is the exiting ray. The two internal reflections are at such an angle that total internal reflection takes place.

Two such prisms may be brought together orthogonally, resulting in an outgoing ray that is parallel to the incident ray, but shifted laterally and inverted. This arrangement is used in binoculars and stereo microscopes. The two prisms may be cemented together, (see *Bonding Optical Glasses* in Appendix VI), thus reducing the number of air-glass interfaces.

The *Porro-II*, or *Porro-Abbe* prism is a modification to the Porro prism configuration. It can be constructed with either two or three prisms, depending on their shape. In this configuration the resulting ray is parallel to the incident ray, but shifted laterally, and inverted.

Porst
Porst was a camera retailer of Nuremberg, Germany, founded in 1919. They re-branded cameras supplied to them by manufacturers of the time. From the 1930s to the late 1950s they used the brand-name *Hapo* for cameras manufactured by makers such as **ADOX**, **AGFA**, and **Balda**. Later they used the brand-names *Porst* and *Carena* for cameras manufactured by makers such as **Cosina**, Balda, **Fujifilm**, **Mamiya**, and **Yashica**.

Porter & Hunt
Porter & Hunt, [358], was a partnership between **James Porter** and an optician named John Hunt. The partnership endured during the 1830s, and was dissolved in 1838 [35 #19686 p2899]. According to their trade card, "Porter & Hunt, Polytechnic Institution, No 309, Regent Street. Spectacle makers. Practical opticians."

Porter, Henry
Henry Porter, (c1832-1902) [18 p51] [50], was an optical instrument maker of London, England. He served his apprenticeship with **William Cary**. After **Henry Gould**'s death in 1856, his wife, **Charlotte Gould** continued their business as Gould, late Cary. In about 1863, Henry Porter acquired an interest in the business, after which it operated as Gould and Porter. The business continued as Gould and Porter after Charlotte's death in 1865, until sometime between 1874 and 1876 when Henry Porter is believed to have become the sole owner (presumably having bought out Charlotte's heirs). Henry Porter continued to use the name Cary, and is known to have been trading as Cary & Co., at 7 Pall Mall, London, until at least 1898.

On Henry's death in 1902, his sons Sydney and Clement inherited the business of Cary Porter, Ltd., and continued to maintain the reputation of their predecessors. By 1931, however, the business was no longer in operation.

Porter, James

James Porter, (fl.1831-1843) [18] [319], was an optical, mathematical, and philosophical instrument maker of London, England. Addresses: 14 York Street, York Road, Lambeth; 282 Strand; 4 West Smithfield; 309 Regent Street (1840); 126 Great Portland Street (1840); 8 Brownlow Street, Drury Lane (1843).

The 309 Regent Street address was used for scientific lectures and demonstrations, advertised under the name Polytechnic Gallery Incorporated, during the 1830s. This was initially operated as the partnership **Porter & Hunt**.

Potter, Charles (I)

Charles Potter (I), (1799-1864) [18], was a mathematical and optical instrument maker of London, England. Addresses: 61 Paternoster Row (1824-1825); 133 Albany Rd., Camberwell (1847); 108 Albany Rd., Camberwell (1864). He may have been the same person as **Charles Potter (III)**, in which case some of the birth/death dates are in error, and this last address may have been for his son, Charles (II).

He was the brother of **John Dennett Potter**. He served his apprenticeship with George Dollond (I) (born **George Huggins**), beginning in 1813, and was freed of the **Grocers' Company** in 1847. He attended the London Mechanics' Institute from 1824-1825. His son, **Charles Potter (II)** served an apprenticeship with him, beginning in 1847.

Potter, Charles (II)

Charles Potter (II) (fl.1847) [18], was a mathematical instrument maker and optician of London, England. He was the son of **Charles Potter (I)**, and served an apprenticeship with his father, in the **Grocers' Company**, beginning in 1847. He may have been located at 108 Albany Rd., Camberwell in 1864.

Potter, Charles (III)

Charles Potter (III), (1831-1899), was a mathematical and optical instrument maker of Toronto, Canada. According to [417] and [418], he was born in London, England, and trained with **George Dollond (I)**. He moved to Toronto in Canada in 1853, and entered a partnership with the jeweller and watchmaker William Hearn, trading as **Hearn & Potter**. The partnership lasted until 1859, after which Charles Potter (III) began to trade under his own name at 84 King Street West, Toronto. The business underwent several address changes, and by 1864 he was trading at the sign of *The Spectacles*, King Street East, Toronto. There he was listed as manufacturing "surveying instruments, philosophical apparatus, globes, mathematical instruments, and school apparatus", and supplying the Educational Department of Ontario. He later traded at 31 King Street East. The business continued to expand, diversify, and enjoy success until Charles Potter (III) died in 1899, after which the business was acquired by Charles Petry, who ran it under the name of Charles Potter until it was sold to W. Harry Landon in 1951.

The similarity between his early history and that of **Charles Potter (I)** suggests that either they may have been the same person, with a possible error in some of the listed birth/death dates, or that the references contain erroneous information conflating the histories of two different instrument makers.

Potter, John D.

John Dennett Potter, (1810-1882) [18] [50] [53] [54] [276 #1966] [319] [361], of 31 Poultry, London, England, was a mathematical, optical and philosophical instrument maker. He was the brother of **Charles Potter (I)**. In 1850 he succeeded **Robert Brettell Bate** in his role as chart agent to the Admiralty. He had previously been Bates' shopman.

According to his trade card: "Hydrometer and Mathematical instrument maker. Chart agent to the R[t] Hon[ble] the Lords Commiss[rs] of the Admiralty. J.D. Potter. (Successor to R.B. Bate) 31 Poultry, London. Mathematical, Optical & Philosophical instruments & apparatus of every description upon the most simple & accurate construction."

In 1854 he additionally took on premises at 11 King Street, and he continued his business at both addresses. In 1856 he was listed as an optical, mathematical, philosophical and drawing instrument maker, sole agent for the sale of admiralty charts, publisher of nautical works, and manufacturer of Sykes' hydrometer & saccharometer. Upon his death in 1882, he was succeeded in business by his son, Septimus C. Potter (b.c1853).

By 1900 the firm of J.D. Potter, nautical publishers and chart agents of 31 Poultry and 11 King St., was a partnership between Edward Octavius Potter (b.c1855), David Alexander Potter (b.c1859), and Bruce Hersee Potter (b.c1851). In that year, David retired from the firm, and it continued as a partnership between Edward and Bruce [35 #27242 p6649].

Potts, Thomas

Thomas Potts, (fl.1805-1814) [18] [276 #1198] [379], was an optician and optical instrument maker of London, England. Addresses: 371 Strand (1805), and 18 St. Martin's Court, St. Martin's Lane (1807-1814). The premises at 18 St. Martin's Court were also occupied by the optician William Potts in 1809.

Poulin, Roger
Établissements Roger Poulin were surveying instrument makers of Paris, France. They succeeded **Albert Lepetit** in business in c1950. Instruments bearing their name include a theodolite in a box with label reading "Établissements Roger Poulin. Instruments de précision. Marque Lepetit-Poulin. 27. Rue de Verdun, Bagneux (Seine)."

Pouilly, J.
J. Pouilly, (fl.late-C17th) [256], was a scientific and optical instrument maker. He established a workshop in 1683 at the sign *Au Compas Marin* in rue Dauphine, Paris, France. Examples of his work include an undated microscope, and a graphometer dated 1686.

Pouzet, Gustav
Gustav Pouzet, (fl.late-C19th), was an optician of Geneva, Switzerland. According to his trade advertisement in 1886: "G^{ve} Pouzet, Opticien, 8 Rue du Mont-Blanc, 8, Genève. Assortiment le plus complet pour tout ce qui concerne l'optique et les sciences." (The most complete assortment of everything related to optics and science).

Powell & Lealand
Powell and Lealand, [53] [142] [172] [264] [280 Ch.6], of 24 Clarendon St., Somers Town, London, was a major English manufacturer of microscopes during the mid-to-late C19th. It was a partnership formed in 1841 between **Hugh Powell** and his brother-in-law **Peter H. Lealand**. From 1846-1857 the business was located at 4 Seymour Place, Euston Square and, due to renaming of streets, this same address became known as 170 Euston Road, where they continued until 1905. Powell & Lealand earned a reputation as makers of the finest microscopes of the era. **Charles Perry** was the foreman of Powell & Lealand until he left in 1904 to work for **C. Baker**. Upon Hugh Powell's death in 1883 the firm continued under the management of his son, **Thomas Hugh Powell**, and remained in business until about 1914.

According to the microscopist Henri van Heurck in 1891 [172], "Messrs. Powell and Lealand occupy quite a unique position in the microscopic world. Their workshops are small, the number of instruments which they produce are few, but every piece of apparatus, marked with their name, is an artistic production, perfect in all its details. Moreover, both instruments and objectives of these makers are in the greatest request, and are used in England by all serious microscopists."

Powell, Hugh
Hugh Powell, (1799-1883) [18] [142] [264] [280 Ch.6], of 24 Clarendon St., Somers Town, London, England, was an optical and philosophical instrument maker. He was making microscopes from the 1830s, and he supplied microscopes to, among others, retailers such as **Andrew Pritchard**. In 1840 he began to sign microscopes with his own name. In 1841 he formed a partnership with his brother-in-law **Peter H. Lealand**, trading as **Powell & Lealand**. Hugh Powell was a pioneer of the production of high-power microscope objectives, having made a 1/16 in about 1840, and in subsequent years powers up to 1/25 and 1/50. In 1840 he was elected a member of the Microscopial Society of London. Upon Hugh's death the firm continued under the management of his son, **Thomas Hugh Powell**, and remained in business until about 1914.

Powell, Samuel
Samuel Powell, (fl.1832-1836) [18], was an optical, mathematical, and philosophical instrument maker of 33 Judd St., Brunswick Square, London, England.

Powell, Thomas Hugh
Thomas Hugh Powell, (1834-1925) [18] [53] [142] [264] [280 Ch.6], the son of **Hugh Powell**, continued to run **Powell and Lealand** following his father's death in 1883. In 1880 he was elected a Fellow of the Microscopical Society of London. The business of Powell and Lealand continued until about 1914.

Praktika
Praktika was a brand-name used by **Kamera Werkstätten** (KW), and subsequently **Pentacon**, for cameras and binoculars.

Praktina
Praktina was a brand-name used by **Kamera Werkstätten** (KW) for cameras.

Praxinoscope
A Praxinoscope, [45], is an improved version of the Zoetrope. Instead of having slots cut in the side of the drum, the images on the inner surface of the drum are viewed by means of mirrors mounted close to the spindle at an angle convenient for viewing. The Praxinoscope was invented by **Charles-Émile Reynaud** in 1877. He also patented a version of the Praxinoscope that, by means of a lamp and a magic lantern type projection lens, projected the image onto a screen.

Precision Optical Co. Ltd.
Precision Optical Co. Ltd., [29], of 5 Wigmore Street, London, England, was a part of **Theodore Hamblin Ltd.**

Prentice, James
James Prentice, (1812-1888) [58] [201], was a mathematical, philosophical and optical instrument

maker. He served his apprenticeship with **John Beale** in London, England, probably from the late-1820s to the early 1830s. In 1842 he emigrated to the USA where he settled in New York, NY, and conducted his business as an instrument maker. In 1887 he advertised "Transits, levels, microscopes, drawing instruments and all kinds of mathematical instruments, made to order and warranted." He worked in partnership with his son, Charles F. Prentice (1854-1946), from 1883, trading as James Prentice & Son. Charles continued to trade under the same name after his father's death until 1897, following which he traded as James Prentice & Son Company until 1927.

Preston, Grant
Grant Preston, (fl.1813-1851) [18] [54] [262], was an optical, philosophical, mathematical, and nautical instrument maker, compass maker, and nautical brazier, of London, England. Addresses: 4 Ebenezer Place, Commercial Rd; Burr St., Wapping (1813-1826); 108 Minories (1827-1840); Union Row, Tower Hill (1841-1845); 2 Union Row, Minories (1846-1851). He was freed by purchase of the **Armourers and Braziers' Company** in 1813. He was awarded a patent for a compass in 1813, and had a royal appointment to Queen Victoria.

Prestwich, John Alfred
John Alfred Prestwich, (1874-1952) [44] [335], of Kensington, London, England, was a scientific instrument maker, engineer, and inventor. In 1895 he founded the Prestwich Manufacturing Company in London, manufacturing scientific instruments and experimental cinematographic equipment. In 1896 he began a collaboration with **William Friese Greene** to produce a projector, designed to overcome the problem of flicker by means of two vertically arranged projecting lenses, with a mechanism to show the film through one while the next frame was readied for the other.

The *Moto-Photoscope* projector and *Moto-Photograph* camera were designed by one of the Prestwich family, presumed to be John Alfred. They were sold by **W.C. Hughes**, and won a silver medal at the Glasgow International Photographic Exhibition. The firm continued to grow and to produce a variety of cinematographic equipment. By 1903 Prestwich had designed and produced his first internal combustion engine, after which production of motorcycle engines, including the *JAP* cycle motor, increasingly dominated the business. In 1918 the company was incorporated as J.A. Prestwich & Co. In 1957 the company merged with Villiers Engineering, and in 1964 the Prestwich name ceased to be used.

Price, William
William Price, (fl.1786-1844) [18] [276 #1969], was an optical, mathematical, and philosophical instrument maker of Fetter Lane, London, England. Addresses: 115 Fetter Lane (1786-1844); 44 Fetter Lane, Fleet St. (1793). He served his apprenticeship with **William Archer**, beginning in 1771, and was freed of the **Stationers' Company** in 1778. He had several apprentices, including his sons, William Archer Price, beginning in 1801, and Charles Price, beginning in 1809. In 1786 William Archer turned over his last apprentice, **John Johnson Evans**, to William Price.

Prin, Georges
Georges Prin, (1885-1959) [11] [72], was a French instrument maker. In 1910 he succeeded **Paul Gautier** in his instrument making business and, in 1934, the Prin firm became part of the **Lerebours & Secretan** business.

Prince, Abraham
Abraham Prince, (fl.c1839) [34] [304], of Waterford, Ireland, was an optician and mathematical instrument maker. An example of his work is a brass three-draw spyglass in a leather case signed *A. Prince, Waterford*.

Prinz
Prinz was a brand of imported Japanese cameras, projectors, telescopes and binoculars, ranging from budget quality upwards, marketed by the UK retailer, **Dixons**. Some higher quality instruments are considered collectible, notably those with **Towa** or **Kenko** optics. Some Prinz telescopes were branded *Astral*, and some Prinz products were branded *Prinzlux*. Prinz cameras were re-branded products by manufacturers such as **Cosina**, **Chinon**, **Halina**, **Zenit**, and **Mamiya**.

Prior Scientific
Prior Scientific began as W.R. Prior & Co. Ltd., [53]. The company was formed in 1919 in London, England, by Walter Robert Prior and Andrew Physicks, making microscopes and accessories. In 1941, after their premises were destroyed in the blitz, they moved to Bishop's Stortford, Hertfordshire. The company was acquired in 1978 by Douglas Fielding, and its name changed to Prior Scientific. Fielding was an investor in **McArthur Microscopes**, who were, at that time, already licensing Prior to make the McArthur microscope. In 1981 the company merged with **James Swift & Son**, and in 1988 the company relocated to Cambridge, England. In 1991 they opened a manufacturing facility in Rockland, Massachusetts, USA. Their key products are: microscope automation equipment including high precision motorized microscope stages, automated slide and well plate loaders,

Piezo nano-positioning systems, motorized filter wheels, high-speed shutters, laser autofocus systems, custom and OEM electro-mechanical and optical systems.

Prior & Co. Ltd., W.R.
See **Prior Scientific**.

Prism
A prism is a piece of glass, or other transparent optical material, in the form of a geometrical prism, used for deflecting rays of light by either reflection or refraction [7] [22] [30].

Corner Cube Prism: Otherwise known as a tetragonal prism. A prism having three triangular surfaces at right angles to each other and one surface at equal inclination to the other three. It has the property of reflecting light incident at any angle on the large surface back along the line of incidence.

Diatom Prism: A triangular prism used in microscopy to provide oblique lighting for small objects.

Direct Vision Prism: A combination of two or more prisms that provides dispersal of light into its spectral components without deviation of the central wavelength. Hence the spectrum may be viewed in the same optical axis as the incoming light.

Double Image Prism: A prism of birefringent material, such as calcite or Iceland spar, giving a double image of an object.

Erecting Prism: A prism which, by means of internal reflection of the beam of light, provides an erect image when it would otherwise be inverted.

Fresnel Prism: A thin prism having the optical properties of a thicker prism. It consists of a sheet of optically transparent material, flat on one side, and on the other side are regular, parallel angular grooves, creating a profile of a series of small prisms. Fresnel prisms may be used by optometrists to correct double vision in patients resulting from trauma or stroke.

Fresnel Biprism: A thin prism with an obtuse apex angle slightly less than 180°. This may be used to produce interference in light.

Glan-Foucault Prism: A polarizing beam splitter, using birefringence to split a polarized beam. It is similar to a Glan-Thompson prism, but the two right-angled calcite prisms are separated by a thin air-gap instead of cement.

Glan-Thompson Prism: A polarizing beam splitter, using birefringence to split a polarized beam. It consists of two right-angled calcite prisms cemented together along their long sides. As with a Nicol prism, one of the two polarized beams of light produced by birefringence is thrown out of the field by total internal reflection from the internal cemented surface, and the other is transmitted.

Nachet's Prism: A form of prism used in microscopy to provide oblique lighting for small objects.

Nicol Prism: A Nicol prism is a rhomb of Iceland spar for transmitting polarized light. It is bisected obliquely at a certain angle, and the two parts again joined with transparent cement, originally Canada balsam. Thus, one of the two polarized beams of light produced by birefringence is thrown out of the field by total internal reflection from the internal cemented surface, and the other is transmitted. The Nicol prism was invented by the Scottish physicist and geologist **William Nicol**. An example of its application is the polar clock invented by **Sir Charles Wheatstone**. Nicol prisms were also used in polarized light microscopy.

Pentaprism: A pentaprism is a five-sided prism that reflects light through an angle of 90°.

Rutherford Prism: A double-image prism used in polarizing experiments.

Tetragonal Prism: The same as a *corner cube prism*, (see above).

Thompson Prism: A polarizing prism, invented by **Prof. Silvanus Thompson**, passing a very wide angle of light.

Wollaston Prism: A Wollaston Prism is a polarizing beam-splitter, invented by **William Hyde Wollaston**. Usually made from calcite or quartz, it comprises two prisms cemented together, and uses birefringence to split a beam into two orthogonally polarized beams. The Wollaston Prism is used in microscopy, and as a component used in CD players.

See also Dove prism, Lummer-Brodhun cube, Porro prism, Roof prism.

Prismatic Astrolabe
The prismatic astrolabe, [277], bears no resemblance to the traditional astrolabe. It is an instrument that measures the precise time a star passes a vertical circle. If the exact position of the star is known, the instrument may be used to determine the exact time – conversely, if the exact time is known, the instrument may be used to determine the exact location of the star. It was invented in 1899 by the French astronomer Auguste Claude then, in 1910, he improved the design with the collaboration of Ludovic Driencourt, to make it more useful to geodetic engineers.

The instrument consists of a horizontal telescope with a 60° prism mounted in front of the objective. Beneath the prism a pool of mercury forms a horizontal mirror. The light, both directly from the star, and from its reflection in the mirror, passes through the prism into the telescope, forming two images at the eyepiece. The angle of the prism determines that when the two images of the star coincide, the star is at an altitude of 30° from the zenith.

The design of the prismatic astrolabe makes it vulnerable to *personal errors* introduced by the observer. These are overcome by the improvements made by André-Louis Danjon, resulting in the Danjon Prismatic Astrolabe.

Prismatic Compass

A prismatic compass is a compass with a sighting prism so arranged that the user can simultaneously observe the bearing reading on the compass, and the view through a sighting vane to a distant object. The bearing reading is given by a graduated circle attached to, and rotating freely with the compass needle so that, at any time, the angle relative to magnetic north is directly under the sighting prism. It is used to find a bearing to a distant object, such as in surveying tasks that do not require the accuracy of a theodolite. It was invented by **Charles Augustus Schmalcalder**, and patented in 1812.

Prism Photometer

See photometer.

Pritchard, Andrew

Andrew Pritchard, (1804-1882) [18] [50] [169] [264], was an optician, spectacle maker, draftsman, and patent agent, of London, England. Addresses: 32 Upper Thornhaugh St. (1824-1826); 18 Pickett St., Strand (1827-1829 and 1831-1835); 312 Strand (1829-1831); 263 Strand (1835-1838); 162 Fleet St. (1838-1854); 150 Fleet St. (1839).

He served his apprenticeship with his uncle, **Cornelius Varley**, and was freed by purchase of the **Spectacle Makers' Company** in 1839. He employed **Alfred Frederick Eden**, who studied under him, and may have been his apprentice. He was renowned for making microscopes and microscope slides, and retailed microscopes by other makers, such as **Hugh Powell**. In 1824 he was the first to attempt making a diamond lens, but at the time, given the difficulties of grinding and polishing the diamond, he shifted his efforts to sapphire. He collaborated with Cornelius Varley in making these jewel lenses, and by about 1827 he was making and selling sapphire jewel lenses in London. A diamond lens made by Andrew Pritchard is in the Science Museum, South Kensington. Some of his lens production is known to have been carried out by Hugh Powell and **Camille Nachet**.

He exhibited an achromatic microscope at the Great Exhibition in London in 1851, [294] [328].

Pritchard, Charles

Charles Pritchard, [11] [24 p296], was a C19th astronomer at the University Observatory, Oxford. In about 1881 he devised the extinction wedge, which could be used to measure the brightness of a star.

Pritchard, Edward

Edward Pritchard, (fl.c1800) [2] [16] [18], was a divider of mathematical instruments of London, England. He was an apprentice of **Jesse Ramsden**, and continued to work for Ramsden until Ramsden's death in 1800. Ramsden bequeathed his small circular dividing engine to Edward Pritchard, who continued to use it. The scales on marine octants and sextants, engraved by Pritchard using Ramsden's small dividing engine were marked with a foul anchor symbol, similar to that of Ramsden, but with the letters E P, as illustrated here. See Appendix XI.

Pritchard, James

James Pritchard, (fl.1829-1830) [18], was an optical, mathematical, and philosophical instrument maker of 9 Great Newport St., Long Acre, London, England. These same premises were also occupied by George Pritchard, optician, and mathematical and philosophical instrument maker in 1826-1838, and John Pritchard, optician, in 1833-1836.

Proctor & Beilby

Proctor & Beilby, [3] [18], were scientific instrument makers of Sheffield and Birmingham. The company was founded by the two brothers Charles and Luke Proctor, initially making surgical instruments. Luke left the firm, but Charles continued the business and had two sons, George and William, who also worked in the family business. The company first appears in the Sheffield directories in 1781, and in Birmingham in 1788. By 1800 the Sheffield firm was called George & William Proctor. By 1788, **Thomas Beilby** had joined the company, and the Birmingham company was listed as Proctor & Beilby. In 1800 the Sheffield company was called Proctor & Beilby; in 1809 Proctor Beilby & Co.; in 1815 George & William Proctor; in 1818 George Proctor. By 1821 the firm was no longer operating. The changes of company name, however, appear to have affected the Sheffield and Birmingham branches of the business at different dates. William Proctor restarted the business in 1825, but only continued to operate until 1834.

J.P. Cutts is reputed to have served his apprenticeship with Proctor & Beilby. Some instruments by Proctor & Beilby are marked *London*. Instruments bearing their name include: early C19th brass universal equinoctial ring dial; single-draw telescope signed *Proctor, Beilby & Co, London. Day or Night* with a wooden barrel; three-draw telescope with a mahogany barrel.

A fascinating contemporary account of the Proctor & Beilby works during the early C19[th] can be seen in [325] and [326].

Progress Optics
Progress Optics, (Прогресс), began as a microscope manufacturer, set up by **Carl Zeiss** under contract with the Soviet government. They were part of **LOMO**, and established a reputation for fine quality optics. For Russian and Soviet makers' marks, see Appendix V.

Projector
A projector is a device that projects an image onto a surface that acts as a screen. The image may be still, as with a slide projector, or moving, as with a cine projector.

See anamorphic lens, Bio-pleograph, Bioscop, Bioscope, cine, Cinematograph, epidiascope, fantascope, Fresnel lantern, magic lantern, Panoptikon, stereopticon, solar microscope, Theatograph, Vitascope, Zoöpraxiscope.

Prokesch, Wenzel
Wenzel Prokesch, (fl.c1820-c1873) [219], was an optician and optical instrument maker of Vienna, Austria. Instruments bearing his name include: telescopes; microscopes; camera obscura; linen tester. One of his employees was **Karl Fritsch**, who succeeded him in business.

Prontor
Prontor was a camera shutter brand-name used by **AGC**. Subsequently the company was renamed Prontor AG, and in 2014 the company was taken over by Hitech, to become Hitech Prontor GmbH.

Prospect; Prospective Glass; Prospect Glass
The terms *prospect*, *prospective glass*, and *prospect glass*, now obsolete, appear to have been interchangeable with the term *perspective glass*.

Prószyński, Kazimierz
Kazimierz Prószyński, (1875-1945) [335], was a Polish inventor and cinematographer best known for his invention of the Pleograph, the Bio-pleograph, and the Aeroscope the latter being designed in collaboration with **Arthur S. Newman**. He completed his engineering degree in 1908, and worked on the problem of reducing flicker in motion pictures. He promoted the three-bladed shutter system that was adopted by companies such as **Gaumont**. Prószyński continued with his innovations in motion picture technology until WWII. He died in a German concentration camp in 1945.

Pseudoscope; Pseudoscopic
A pseudoscope is an optical instrument which shows objects with the proper depth of relief reversed. Thus the points on the object that are nearer to the observer appear farther away, and vice versa. The effect is usually achieved in one of two ways by means of prisms or mirrors. Either the left-eye view is presented to the right eye, and the right-eye view to the left eye, or the views of both eyes are rotated by 180°.

A pseudoscopic view or image is that produced by such an instrument.

Pulfrich Refractometer
See refractometer.

Pullin and Co., R.B.
R.B. Pullin and Co., [44], were electrical engineers and scientific instrument makers, of Phoenix Works, Great West Road, Brentford, Middlesex, England. The company was founded in 1932 in West Ealing. They supplied instruments to the aviation industry. In 1939 they established the **Pullin Optical Company**, also in Brentford, Middlesex. In 1957 they acquired **Aldis Brothers**. In 1959 they acquired **Neville Brown and Co.** and Milbro Photographic. In 1964, **Rank** made a bid for R.B. Pullin and Co., and soon after that, Stanley Cox, a subsidiary of Pullin, became a part of Rank Medical Equipment.

Pullin Optical Company
Pullin Optical Company, [44], were photographic apparatus manufacturers of Phoenix Works, Great West Road, Brentford, Middlesex, England. The firm was a subsidiary of **R.B. Pullin & Co.**, formed in 1939.

They exhibited at the 1947 British Industries Fair, where their catalogue entry, [450], listed them as manufacturers of: Photographic lenses; Film strip and slide projectors; X-ray film projectors; Projection lenses; Photographic enlargers; Photographic rangefinders; Optical prisms; Magnifiers; Focimeters for spectacle lens measurement; Aneroid barometers.

Their entry in the 1951 *Directory for the British Glass Industry*, [327], additionally listed: Optical lenses; Object glasses; Exposure meters; Cine tripods, and listed their trade names as *Pulnar* and *Pulkino*.

Purma Cameras Ltd.
Purma Cameras Ltd., [61] [306], of 7 Queen Street, Mayfair, London, England, was a camera brand, with the *Purma* trade-mark registered in 1935. The company was founded by Tom Purvis and Alfred C. Mayo, and their *Purma* branded cameras were exclusively marketed by **R.F. Hunter** from 1936 until c1951. They were

manufactured in the UK, and [122] suggests that they may have been manufactured by R.F. Hunter. Beginning in the mid-1950s Purma Cameras Ltd. marketed the cameras themselves. There were three models of camera, the *Purma Speed*, *Purma Special*, and the *Purma Plus*.

Purser and Brother, H.F.
H.F. Purser & Brother, [29] [44], were optical instrument makers of 35 Charles Street, Hatton Garden, London, England. On 6 June 1918 they received a government order for 600 No 3 Mk1 military binoculars (apparently amended to 1120). By 10 February 1919, 349 had been delivered. The contract was considered completed and compensation was paid.

They were a Listed Exhibitor at the 1922 British Industry Fair. Their catalogue entry, [447], listed them as manufacturers of Cinematograph projection lenses, binoculars, and field glasses.

Purus
See **Astro-Mechanik**.

Putois, Etienne Antoine
Etienne Antoine Putois, (1753-1828) [11] [50] [416], was an optical instrument maker at Quai de l'Horloge 81, Paris, France, the premises also known as *au Griffon*. He succeeded **Veuve Marie**, the widow of **J-B Marie**, in business at this address. He was married to the optical instrument maker **Marie Putois**. He made the objective lens for the Lenoir transit instrument at rue de Paradis, Paris (c1789). The instrument is now in the Paris Museum.

Putois, Marie
Marie Angélique Catherine Putois, (1746-1831) [50] [416], was an optical instrument maker at Quai de l'Horloge 81, Paris, France, the premises also known as *au Griffon*. She was the daughter of **Jean-Baptiste Gonichon** and **Marie Michelle Gonichon**, and married to the optical instrument maker **Etienne Antoine Putois**. She outlived her husband and, from 1807-1817, she worked in partnership with **Rochette Jeune**, who succeeded her in business, and continued until 1845.

Pye & Co., W.G.
William George Pye, (1869-1949) [44] [46], was the son of William T. Pye (c1847-1921), who was foreman of the **Cambridge Scientific Instrument Co. Ltd.** In 1892 W.G. Pye became an employee of the Cavendish Laboratory, and in 1896 he started his own company in Cambridge, England, initially part time. In 1898, together with his father, he formed the W.G. Pye Instrument Co., which then became W.G. Pye and Co. He left the Cavendish Laboratory in 1899 to focus on his company.

In 1909 W.T. Pye left the partnership, [35 #28217 p606], and W.G. Pye continued to run the company as W.G. Pye and Co.

During WWI the company produced optical equipment including precision gunsights, and Aldis signalling lamps. The Pye company entered the businesses of radio and television, and their subsidiary Pye Instruments, itself had subsidiaries including W.G. Pye & Co., and others including, in 1957, **W. Watson & Sons Ltd.** In 1919 W.G. Pye & Co. was listed as a member of the **British Optical Instrument Manufacturers' Association**. In 1936, upon the retirement of William G. Pye, the company continued under the direction of his son, Harold John Pye (1901-1986). During WWII the company supplied equipment for the war effort. In 1947 H.J. Pye sold the business to Pye Ltd., and the family association with the company ceased.

By 1949 Pye had acquired **Unicam Instruments** as a subsidiary. In 1968 Pye merged with Unicam Instruments to form Pye Unicam.

Pyefinch, Henry
Henry Pyefinch, (fl.1763 d.1790) [18] [276 #642] [319] [322], was an optician, and optical, philosophical, and mathematical instrument maker, of London, England. Addresses: *Golden Quadrant, Sun & Spectacles*, No. 67 between Bishopsgate St. and the Royal Exchange in Cornhill; 64 Cornhill (1765); *Golden Sun, Quadrant & Spectacles*, 67 Cornhill (1768); 67 Cornhill (1772-1782); 45 Cornhill (1782-1791).

He served his apprenticeship with **Francis Watkins (I)**, beginning in 1753, and was freed of the **Spectacle Makers' Company** in 1763.

In 1764 he was one of the petitioners who attempted to revoke the patent obtained by **John Dollond (I)** for the achromatic lens, and which was enforced after Dollond's death by his son, Peter (See Appendix X). Despite the failure of the petition, he continued to make achromatic telescopes, and in 1768 he was successfully sued by **Peter Dollond** for infringement of the patent.

His trade card, [53] [361], lists spectacles, and a wide range of optical, mathematical, and philosophical instruments that he produced. The optical instruments included: Refracting and reflecting telescopes; Compound microscope; Wilson's microscope; Solar microscope; Opaque microscope; Prisms; Magic lantern; Various types of mirror; Pocket camera; Camera obscura in the form of a book; Diagonal mirror or optical machine.

According to [276] Elizabeth Pyefinch is recorded at 45 Cornhill with an umbrella and mathematical instrument maker's business in 1786. She also had a China Warehouse at a different address.

Pyne, J.J.

Joseph J. Pyne, (fl.1856-mid-1860s) [**306**], was a photographic dealer of 63 Piccadilly, Manchester, England. He succeeded George Dawson in business in 1856. Cameras bearing his name are believed to date from c1856 to sometime between 1865 and 1867. He was succeeded in business by **Robert Hampson**.

Pyrex

Pyrex is a brand-name, introduced in 1915 by the glass maker **Corning** for a kind of borosilicate glass with low thermal expansion properties. Since 1998 the brand has been licensed by a spinoff company of Corning's throughout the world. Because of its low thermal expansion properties, Pyrex is used as a substrate for astronomical mirrors.

Pyser-Britex Group

The Pyser-Britex Group, of Fircroft Way, Edenbridge, Kent, England, was a sales company. It was part of the Pyser Group, (see **Pyser-SGI**), and had at least two subsidiaries:

Pyser-Britex (Swift) Ltd., marketed binoculars of the US **Swift Optical Instruments** brand. According to [**320**], during the 1960s, they also marketed microscopes, manufactured in Japan, and branded by Swift Instruments Inc., of San Jose, California, USA (later known as Swift Optical Instruments).

Pyser-Britex (Sales) Ltd., marketed optical instruments such as microscopes, manufactured by Britex (Scientific Instruments). See **Britex**. Some of these Britex microscope models were marketed for use by children. According to an advertisement of 1963, "Specialists in projection microscopy, and suppliers of microscopes by Britex and all other leading manufacturers, including equipment by Swift," together with the logo for Swift Optical Instruments.

Pyser-SGI (Pyser Optics Ltd.)

Pyser-SGI of Edenbridge, Kent, England, is a privately owned British Company dating back to 1848. In 2018 they changed their name to Pyser Optics Ltd.

They manufacture and supply the following: Thermal weapon sights; Hand held thermal imaging long range surveillance systems; Thermal imaging monoculars; Hand held/helmet/head mounted image intensified night vision monoculars and image intensified goggles; Image intensified monoculars for photographic/video evidence gathering; Image intensified red dot weapon sights and clip-on in-line universal image intensified long range weapon sights; Gyro-stabilised day and night vision binoculars and conventional military binoculars; Small arms collimators; Prismatic compasses; Survival compasses; Portable rapid-deployment observation and monitoring systems for long distance day and night surveillance; Day and night observation systems; Day/night vision surveillance camera systems and covert cameras; Mini thermal imaging cameras. They also produce graticules and optical gratings, and broadcast and CCTV lenses.

The **Pyser-Britex Group** was part of the Pyser Group. **Newbold and Bulford** became part of the Pyser Group in October 1991.

PZO

See **Polish Optical Industries**.

Q

QHYCCD
QHYCCD of Beijing, China, is a manufacturer of specialised CMOS and CCD cameras and accessories for astrophotography.

Qioptiq
Qioptiq is an optical and photonics company that was formed in 2005 by the purchase of **Thales Optronics Ltd.**, bringing on board **Pilkington, Barr & Stroud** (albeit without the trade-name), and **Avimo**. In 2006 Pilkington was sold to the NSG group. In the same year Qioptiq began the process, completed in 2009, of acquiring the company **LINOS AG**, bringing on board **Spindler & Hoyer**, **Steeg & Reuter Präzisionsoptik**, Franke Optik Vertriebsgesellschaft, Gsänger Optoelektronik GmbH, and **Rodenstock Präzisionsoptik**. Other aquistitions include Optem and Point Source. In 2013 Qioptic was acquired by the US company Excelitas Technologies Corp.

Quadrant
A quadrant is an instrument used to make angular measurements, with applications in astronomy, surveying, gunnery, etc. A number of different forms of quadrant exist:

In its earliest form, a quadrant consisted of a quarter-circular piece of sheet-brass or wood, with the curved edge graduated in degrees. A hand-held quadrant of this kind would have a plumb-line attached close to the right-angled corner, and two sight vanes on one of the straight edges. The instrument could be used to measure, for example, the angle of altitude of an astronomical body.

An octant that uses double reflection to measure angles up to 90° is also sometimes referred to as a quadrant.

Other forms of quadrant include: Davis's quadrant (see also backstaff), gunner's quadrant, Gunter's quadrant, Hadley's quadrant, mural quadrant, Smith's quadrant.

Quartz Spectrograph
See spectrograph.

Queen & Company, James W.
James W. Queen (1813-1890) began to work for **John McAllister Jr.** in Philadelphia, USA, at the age of 14, rising to become a partner 11 years later. He left the partnership in 1852 to set up his own company. In 1853 he established the scientific instrument supply company, James W. Queen and Company, [201], in Philadelphia. They manufactured some instruments themselves, and acted as agents for companies such as **R.&J. Beck, Acme, Joseph Zentmayer, Edmund Hartnack, Nachet et Fils, Carl Reichart, Thomas H. McAllister, John T. Norman**, and others.

Queslin, Adélaïde Amanda
Adélaïde Amanda Queslin, (1838-1898) [50], was the daughter of **Pierre Queslin**. In 1860 she married **Alfred Baserga** and, following the death of her father, they continued his business in Paris, France in the name **Maison Chevallier**.

Queslin, Pierre Louis Amédée
Pierre Louis Amédée Queslin, (1819-1883) [50], took over the optical business of **Pierre Chevallier** in 1842, renting the premises at 1 Rue de la Bourse, Paris, France, and making cameras, microscopes, telescopes and other optical instruments. He ran the business under the name **Maison Chevallier** and, following his death, the business was continued by his daughter **Adélaïde Queslin** and her husband **Alfred Baserga**.

Questar
The Questar Corporation of Pennsylvania, USA, was established in 1950 by Lawrence Braymer. He designed a compact telescope that was a derivative of the Maksutov-Cassegrain design, and production began in 1954. This design evolved into a range of products including telescopes for astronomical and terrestrial use, microscopes, and other optical instruments, and they maintain a reputation for high quality.

In 1976 two employees, John Schneck and Robert Richardson, left Questar to form the company **Optical Techniques Inc.** in competition with Questar.

Quintant
A quintant is an instrument, similar in construction and use to the sextant, enabling the measurement of larger angles. Whereas a sextant has an arc of 60°, the quintant has an arc of 72°. Thus, the quintant can measure an angular separation of up to 72° or, by means of double reflection, 144°. The quintant was historically used for measuring angles in astronomy and navigation. For an example of a quintant by **Carl Plath**, see [54].

R

Radford, John (I)
John Radford (I), sometimes Redford, (fl.1659-1703 d.by 1711) [18] [274] [401], was a spectacle maker of London, England. Addresses: *The Golden Spectacles*, without Temple Bar (1668); St. Clement Danes (1670, 1695-1703); Fleet Street (1693). He served his apprenticeship with Thomas Copeland, beginning in 1633/4, and was a Freeman of the **Spectacle Makers' Company**. He was Master of the Spectacle Makers' Company from 1669-1670, and in 1687. His son, **John Radford (II)**, was also a spectacle maker.

Radford, John (II)
John Radford (II), (fl.1718-1722) [18], was a spectacle maker of *The Great Golden Spectacle* against St. Clement's Church in the Strand, London, England. He was the son of **John Radford (I)**, and served his apprenticeship with Thomas Berry of the **Pattenmakers' Company** beginning in 1711. From 1718 until 1722 he worked in partnership with Robert Cocks and **William Radford**, trading as Cocks, Radford & Radford, spectacle makers.

Radford, William
William Radford, (fl.1718-1740) [18] [401], was a spectacle maker of *The Great Golden Spectacle* against St. Clement's Church in the Strand, London, England. He served his apprenticeship with **Matthew Cooberow**, and was freed of the **Spectacle Makers' Company** in 1720. Among his employees was journeyman **Oliver Coombs**. From 1718 until 1722 he worked in partnership with Robert Cocks and **John Radford (II)**, trading as Cocks, Radford & Radford, spectacle makers.

Radiguet & Massiot
Radiguet & Massiot, [109] [296] [297], of 13 and 15 Boulevard des Filles-du-Calvaire, Paris, France, was a company formed in 1899 by the scientific instrument maker Arthur Radiguet (1850-1905), and his son-in-law Georges Jules Massiot (1875-1962). Within a few days of forming the company they merged with **A. Molteni**, to trade as Radiguet & Massiot, keeping on the Moltini workshops at 44 Rue du Château-D'Eau. They continued to produce projectors, and also produced medical equipment, specialising in radiology. The company continued in business until 1960 when Phillips acquired an interest. The name was changed to Massiot-Phillips, and subsequently it was absorbed into the Phillips brand.

Rainbow
Rainbow was a brand of Japanese-made binoculars in the mid-1950s to late-1960s [94]. Examples, 16x50, 8x30, 7x50.

Rajar
Rajar, [44] [61] [306], was a trade-name of the Brooks-Watson Daylight Camera Company Ltd., of Liverpool and London, England. In 1904 the company changed its name to Rajar Ltd. Their specialisation was a daylight changing system for plate cameras. They also produced various formats of film and, for a time, cameras and photographic accessories. In 1921 they became a part of **APM**.

Ramage & Co.
Ramage & Co., (fl.1837-1838) [18] [271], were optical instrument makers of Aberdeen, Scotland. Addresses: 41 St. Nicholas St. (1837), and 6 St. Nicholas Lane (1838). They succeeded **John Ramage (II)** in business, and were succeeded by **Smith & Ramage**.

Ramage, John (I)
John Ramage (I), (1783-1835) [11] [18] [250] [271] [276 #1202], was an optical instrument maker of Aberdeen, Scotland. Addresses: 85 Broad St. (1824-1829) and 39 Union St. (1831-1835). He had a particular interest in the production of large telescopes. In c1820 he produced a telescope with a 15-inch speculum mirror and a focal length of 25 feet. It was exhibited at the Royal Astronomical Society in 1825, by which time he had made two more telescopes of this size. At the time, the only larger telescope in the UK was the 40-foot reflector built by **William Herschel**. He made a 21-inch speculum mirror with focal length 54 feet, but never mounted it in a telescope. John Ramage (I) was succeeded in business by **John Ramage (II)**.

Ramage, John (II)
John Ramage (II), also known as John junior, [18] [271], was an optical instrument maker of Aberdeen, Scotland. Addresses: 39 Union St. (1835), and 104 Union St. (1836). He succeeded **John Ramage (I)** in business. He was succeeded in business by **Ramage & Co.**

Ramsden Disc
Ramsden disc is another name for exit pupil, and is named after **Jesse Ramsden**.

Ramsden, Jesse
Jesse Ramsden, (1735-1800) [1] [2] [18] [24] [129] [46] [397] [443], was a renowned English maker of scientific instruments. He served an apprenticeship as a clothworker, and went to work in London as a clerk in a

cloth warehouse. At the age of twenty-one he bound himself to a four-year apprenticeship with the mathematical instrument maker **Mark Burton** in the Strand. He became highly skilled at engraving mathematical scales and, having set up his own business, he learned about optical instruments by spending his spare time with the Dollond family. In 1766 he married **John Dollond (I)**'s youngest daughter, Sarah, and some sources say that as part of her dowry he received a share in the Dollond patent for the achromatic lens (see Appendix X).

Having set up shop in Haymarket, under the sign of the *Golden Spectacles*, he constructed a dividing engine which could produce more accurately engraved circular scales. This improved the accuracy of instruments used for astronomical measurements and navigation. **John Stancliffe**, who had worked for **Henry Hindley** in York prior to moving to London, worked for Ramsden at this time, and is alleged to have revealed to him the secrets of Hindley's dividing engine.

Marine sextants and octants divided on Ramsden's dividing engine were marked with a **foul anchor** symbol with the letters I R, as illustrated here.

In about 1775 Ramsden devised the dynameter. Most dynameters made in Ramsden's workshops were probably made by his former apprentice, **Thomas Jones**. Jones also credited Ramsden with the invention of the optigraph, a design which he himself improved. Ramsden's apprentices also included **William Cary**, and **Simon Spicer**. For a full discussion of the instrument makers who served apprenticeships with Ramsden, and/or worked for him as journeymen, see [**2** pp.60-65].

In 1780 Ramsden expanded his business, taking larger premises in Piccadilly to accommodate his workforce of about fifty people, and providing a home for his foreman and leading craftsman, **Matthew Berge**. Here, Ramsden and his workforce produced large instruments such as quadrants and circles measuring up to eight feet, as well as smaller instruments including refracting and reflecting telescopes, zenith sectors, sextants, portable quadrants, precision balances, levels, pyrometers, and barometers. Ramsden's eponymous eyepiece design, (see Appendix II), aimed at producing a flat, achromatic field, is described in his paper *A Description of a new Construction of Eye-glasses for such Telescopes as may be applied to Mathematical Instruments* [**130**].

Ramsden's dividing engine, could only accommodate smaller instruments. His manual techniques for engraving a full circle or a large quadrant for use in major astronomical installations led to his reputation for perfectionism, equalled only by his reputation for slowness. Sometimes it took years to complete an instrument, and some were never delivered. This proved frustrating for his customers, especially since his engraved circles and quadrants were considered to be the best available at the time. The work involved in engraving such large instruments could only be carried out in conditions of good light and even temperature. Any lamp used for illumination would cause uneven warming and expansion of the metal.

Ramsden moved to Brighton to benefit his ill-health, aggravated by overwork and the poor air-quality in London. He left Matthew Berge running his business and, when he died in 1800, Berge took control of the business.

Rangefinder (Optical)

An optical rangefinder is an instrument designed to find the distance from the observer to a remote object. In the past, rangefinders generally used trigonometry to estimate the measurement, and this meant that the longer the instrument's baseline, the higher the accuracy that may be achieved. The development of rangefinder technology has been largely driven by the needs of surveying, and the military in the areas of artillery, airborne, and marine targeting.

The first optical rangefinders in common use were based on the principle of the Watkin Mekometer, invented by **Major H.S. Watkin** of the British Royal Artillery. This was introduced in the UK in the 1870s and was a two-operator device. The operators stood at each end of a cord of known length (usually 25 or 30 yards), holding it taut. One observer viewed the target at right angles to the cord. The other observer used a device similar to a box-sextant to measure the angle to the target, and thus calculate the distance. This device was adopted by the British Infantry in 1891, but it was cumbersome, slow to use, and susceptible to damage when used in the field, resulting in reduced accuracy. Watkin Mekometers were made by companies such as **T. Cooke and Sons**. A rangefinder of this design was also made by **J.H. Steward**, and was called the Steward Telemeter.

In 1908 the mekometer began to be replaced with the Marindin rangefinder as an infantry rangefinder for the British army. This was the invention of Captain A.H. Marindin of the Royal Highlanders, (the Black Watch). For the invention he worked in conjunction with **Adam Hilger & Co. Ltd.** to avoid contravening the patents of **Barr & Stroud** and **Zeiss** (see below). The design is a coincidence rangefinder in which the prism on the right is moveable, and this movement is a function of the range.

The first single-observer rangefinder was invented by **Patrick Adie**. It consisted of a metal tube with mirrors set 3 feet 6 inches apart. The mirrors reflected the two images to a central eyepiece. One mirror was set at right-angle, and a micrometer screw adjustment altered the angle of the second mirror to bring the two images into coincidence. The angle between the mirrors was scaled and calibrated to provide a reading of the range to the target. The instrument was delicate, and not suited to field use for infantry or artillery.

Another invention of Major H.S. Watkin was the Depression Rangefinder. This instrument consists of a telescope, similar to a theodolite telescope, mounted on a level base and adjustable to tilt in a horizontal axis perpendicular to the telescope. The Depression Rangefinder is mounted at a height above sea-level that is accurately known. This height acts as the baseline for the range measurement. A sighting is taken to the target, perhaps a ship, and the angle downwards to which the telescope is tilted to sight the target is measured on a graduated arc. For a given height above sea-level, this reading can be calibrated for range rather than angle. The Depression Rangefinder was designed for coastal artillery, and adopted by the British Army in 1881.

In 1888 the British War Office placed an advertisement in *Engineering* magazine for an infantry rangefinder that could withstand rough usage, would meet stringent requirements for accuracy, and with which a man of average intelligence could take four readings a minute. **Dr. Archibald Barr** and **Dr. William Stroud** responded by designing an innovative rangefinder to meet these needs. After experiencing problems in trials of their early designs, they finally received an order to provide the Admiralty with five rangefinders. They ordered their prisms, lenses and reflectors from **Adam Hilger & Sons** of London, and assembled and tested the rangefinders themselves [26]. This signalled the beginning of the company **Barr and Stroud** (initially called Barr and Stroud Patents).

Barr and Stroud's rangefinder business boomed, and they provided a large number to both the British military, and to foreign powers. By the time of WWI, they were supplying rangefinders to Japan, France, Italy, the US, Germany and other foreign powers. After 1918, some of these countries began to produce their own rangefinders: In the US by **Bausch and Lomb**; in France by **Société Optique et Précision de Levallois** and **Société d'Optique et de Mécanique de Haute Précision**; in Italy by **San Giorgio** and **Officine Galileo**; and in Germany by **Zeiss**.

There were principally two different rangefinder designs that competed at this time – the **Zeiss** stereoscopic type and the **Barr and Stroud** coincidence type. While the Zeiss design provided serious competition for the Barr and Stroud design, the latter was found in trials to be more successful. This was largely due to the fact that a user must have good eyesight, and be unstressed and in good health to use the stereoscopic type of rangefinder successfully. The physical and optical principles of the coincidence rangefinder are explained in [6 p253]. The method of operation of the Barr & Stroud small-base coincidence rangefinder are explained in [4 p301].

In 1935 the Scottish scientist R. Watson Watt successfully demonstrated that radar could be used to find and measure the range of a target at great distances. This signalled the beginning of the end of the optical rangefinder and, had radar been more reliable during WWII, the Admiralty would have had no need for them. However, radar was susceptible to enemy damage, and the instrumentation was unreliable, so optical rangefinders were still carried, and maintained in readiness. This proved to be very successful as a back-up range-finding method, but inevitably, after the war, radar technology became more reliable, and optical rangefinders became obsolete.

Other companies who have manufactured optical rangefinders include **Emil Busch**, **Thomas Cooke and Sons**, **A.&R. Hahn**, **Keuffel and Esser**, **C.P. Goerz**, **Rathenower Optische Werke**.

Rangefinder cameras traditionally used a form of coincidence rangefinder, with the obvious limitation that they have a very short baseline. The simplest form of rangefinding, found in instruments such as binoculars, theodolites, and early rangefinding periscopes, uses a graticule in one of the optical tubes. If the size of the target object is known the range may be calculated from the measured size against the graticule. Modern rangefinder binoculars generally have an integral laser rangefinder.

See also Husun Marine Distance Meter, Langley Inclinometer, sextant rangefinder, Stewart's Distance-Finder, Waymouth-Cooke Rangefinder, Waymouth-Ross Sextant Rangefinder, Reeve's distance finder.

Rank, J. Arthur

Joseph Arthur Rank, (1888-1972) [44] [46], was born in Kingston upon Hull, England, son of a prominent Methodist and wealthy owner of flour mills. Considered unpromising at school he was brought into the family business. Himself an enthusiastic Methodist, he founded the Religious Films Society in 1933 with the objective of promoting films with a religious and moral message. In 1934 he formed the film production company British National, and shortly afterwards used his wealth to acquire an interest in Universal Pictures and a controlling interest in Charles Woolf's General Film Distributors (whose symbol was a man striking a gong).

He also entered the cinema business by purchasing the Leicester Square Theatre and, in association with Lady Yule, he formed Pinewood Studios. "I am in films," he said, "because of the Holy Spirit." He intended to use films as a force for good, and wished to rid the British film industry of Hollywood values. He began to produce news documentaries, and continued to do so during WWII, with these and his films becoming an important part of wartime propaganda. In 1941 he acquired control of the **Gaumont British Picture Corporation** and, in 1942, Odeon Theatres. In 1944 the Ealing Studios became a satellite of Rank, whose trademark by this time was Bombardier Billy Wells beating a gong. By 1946 Rank owned 5 studios, 2 newsreels, and 650 cinemas, and had a staff of 31,000.

In 1947 the new name J. Arthur Rank Organisation was used to encompass the Gaumont British Picture Corporation and Odeon Theatres. In the same year, the firm **British Optical and Precision Engineers** was formed as a subsidiary of the J. Arthur Rank Organisation, including the acquisition of **Taylor, Taylor & Hobson Ltd.** [134]. In 1948 it was converted to a public company to acquire **A. Kershaw and Sons Ltd.** [134], along with **British Acoustic Films Ltd.**, **G.B. Kalee Ltd.**, and Gaumont-Kalee Seating Ltd. In 1956 the name British Optical and Precision Engineers was changed to Rank Precision Industries, [44], which included **G.B. Equipments Ltd.**, G.B. Kalee Ltd., Taylor, Taylor & Hobson Ltd., and A. Kershaw and Sons Ltd.

Also in 1956 Rank diversified into copying equipment (Rank Xerox Ltd.), and flour-milling and baking (Rank Hovis McDougal). In 1962 J. Arthur Rank resigned as chairman of the Rank Organisation.

Rapid Rentilinear Lens
See lens.

Rapson
Rapson, (fl.1834-1842) [18], was an optical, mathematical, and philosophical instrument maker of Lambeth, London, England. Addresses: 30 Stangate (1839); 20 Stangate (1849-1851); 4 Bridge Rd. (1852-1859); 229 Westminster Bridge Rd. (1865-1885).

RAS Thread
The RAS thread was a standard thread, established by the Royal Astronomical Society in Britain, for telescope eyepieces. It has a diameter of 1¼ inches with 16 threads per inch. This standard of eyepiece fitting is found on some very old telescopes. See the *Telescope Eyepieces* section in Appendix II.

Raspail's Simple Chemical Microscope
Raspail's simple chemical microscope is a form of aquatic microscope which is an improvement on the Ellis aquatic microscope, providing for more precise movement of the objective. It was invented in about 1830 by the "father of histochemistry", François-Vincent Raspail (1794-1878), and first manufactured by **Louis Joseph Deleuil** of Paris.

Rathenower Optische Werke GmbH
Rathenower Optische Werke GmbH, of Rathenow, Germany, is a manufacturer of ophthalmic frames, lenses, and sunglass lenses.

Following WWII, Emil Busch AG Optische Industrie (see **Emil Busch**), whose majority shareholder was **Zeiss**, was nationalised under Soviet occupation to become VEB Rathenower Optische Werke. The new company also incorporated Rathenower Optischen Werke mbH, formally known as **Nitsche & Günther Optische Werke KG**. In 1966, VEB Rathenower Optische Werke became a part of VEB Carl Zeiss Jena.

In 1990, following the reunification of Germany, the company was privatised, and split again. Rathenower Optische Werke GmbH was formed as a separate company, and VEB Carl Zeiss Jena was privatised and purchased by its West German counterpart, Carl Zeiss Oberkochen. In 1991, the microscope technology division of Rathenower Optische Werke GmbH was split off to form **ASKANIA Mikroskop Technik Rathenow GmbH**.

Raxter, John
John Raxter, also known as Raetor, Roxter, or Raxtor, (fl.1801-1829) [3] [18], was an optical, surveying, mathematical, and philosophical instrument maker, of Birmingham, England. Addresses: 49 Cheapside, Deritend (1801); 6 John St. (1815-1822); 6 Old John St. (1816-1817); 32 John St. (1823-1825); 20 Court, Lichfield St. (1829). He was succeeded in business by the mathematical instrument maker Robert Raxter, who traded in Livery St. (fl.1829-1839).

Ray & Co., J.W.
J.W. Ray & Co. (Liverpool) Ltd., (fl.1937) [44], were scientific instrument makers of Liverpool and 76 Mark Lane, London, England. Instruments bearing their name include: sextants; ships' telegraphs; ships' clocks.

Rayleigh Criterion
See angular resolution.

Reed International Group
Reed International was originally formed in 1895 by Albert E. Reed as a newsprint manufacturer in Tovil, near

Maidstone, Kent, England. The company underwent name changes, and expanded into many other areas of business. In 1983 Reed International Group took over the **British Optical Lens Co.**, and renamed it British Optical Lens Company (1983) Ltd. In 1986 they formed Euro-Optics Asia Ltd. in Hong Kong through a buy-out of the manufacturing division of **Hanimex**. The complex company structure also included, at various stages, company names Caradon British Optical Ltd. (1986); Caradon Lenses (1992); Optar UK Ltd. (1990); British Optical Ltd. (1991); Bronzechain Ltd. (1991); International Optical Group Ltd. (1995). The company continues in business with diverse interests as RELX Plc., with headquarters in London.

Reeve's Distance Finder

A Reeve's distance finder is a kind of alidade that may be used for range-finding without the use of a surveyor's rod placed at the distant point to which the range is to be found. It is equipped with two small sighting telescopes that can be attached, one at each end of the rule. It is used to construct, on the surveyor's plane table, a scale triangle from which the range is calculated using trigonometry. In their catalogue of 1911, [238], **C.F. Casella** give details for two methods of use, and claim to be the sole maker of the Reeve's distance finder.

Reeves, John

John Reeves (sometimes Reeve), (fl.1657-1689) [18] [275 #412], was an optical instrument maker of London, England, and the son of **Richard Reeves**. He is known to have made telescopes. **Christopher Cock**, who began his apprenticeship in 1657 with John Stonehall, was turned over to John Reeves to complete his apprenticeship.

Reeves, Richard

Richard Reeves (sometimes Rives, or Reeve), (fl.1641-1679) [18] [256] [275 #199], was a glass-grinder and optical instrument maker of London, England. His first recorded address was *Over Against the Foot & Leg*, Long Acre, then, from 1632 at The East End of Henrietta St., near the Piazza, and finally, from 1663 in Long Acre. He was the father of **John Reeves** and Richard Reeves (II). He is known to have made long focal length refracting telescopes, microscopes, and magic lanterns. He introduced a field lens into his compound microscopes, and may have been the first to do so. His instruments had a good reputation, and he sold a microscope to Samuel Pepys. Later he showed Pepys the moons of Jupiter through a 12-foot telescope, and supplied him with a magic lantern. In 1663 he worked with **Christopher Cock** to make a speculum for **James Gregory**, but the result was unsatisfactory. Richard Reeves' birth and death dates are not known, but it is believed that he may have died in 1679.

Reflectance

Reflectance is a measure of the reflected or scattered light from a surface as a proportion of the incident light.

Reflecting Circle

A reflecting circle is an instrument, invented in the mid-C18th, used for measuring altitudes and angular distances. It works on a principle similar to the sextant, but its graduated arc is a full circle, rather than a sixth of a circle as in the sextant. An improved version of this instrument was the repeating circle, invented by Jean-Charles de Borda, a mathematician in the French navy. The first prototype Borda repeating circle was made by **Etienne Lenoir**.

Reflecting Microscope

See microscope.

Reflecting Telescope; Reflector

See telescope.

Reflection Grating

A reflection grating is a form of diffraction grating that produces interference in reflected light rather than transmitted light.

Reflectometer

A reflectometer is an instrument used to measure the reflectance of a surface by measuring how much light is either reflected or scattered by the surface. See also total reflectometer.

Reflex Mirror

A reflex mirror is one that is positioned behind a lens to redirect the image to a viewfinder. It is most commonly used in twin-lens reflex (TLR) and single-lens reflex (SLR) cameras. In the TLR this is usually a fixed mirror. However, in the SLR it is in the path between the objective and the film or sensor. In this case, either the mirror is lifted as the shutter is fired, or some other method is used, such as a partially reflecting mirror.

Refracting Telescope; Refractor

See telescope.

Refraction

Refraction is the phenomenon in which a ray of light changes direction when transmitted across the optical boundary between materials of different refractive index. This occurs because the speed of light in a material depends on its refractive index. The relationship between the incident angle and the refracted angle of the light ray is given by Snell's law (qv). See also the section *properties of optical glass* in Appendix VI.

Refractive Index
See the *Properties of Optical Glass* section in Appendix VI.

Refractometer
A refractometer is an instrument for measuring the refractive index of transparent substances. Various kinds of refractometer exist for specialised purposes. When an instrument is referred to simply as a *refractometer*, this generally describes an instrument for measuring the refractive index of optical glass, or other materials used in place of glass in optical instruments.

Abbe Refractometer: Measures the refractive index of liquids. The sample to be measured is held as a thin layer between two prisms, and illuminated through one of the prisms. The prisms are hinged together so that one may be swung back to introduce the sample or clean the prisms.

Butyro Refractometer: A form of immersion refractometer that measures the refractive index of butter, lard and oils.

Dipping Refractometer: Another name for an immersion refractometer.

Féry Refractometer: Measures the refractive index for sodium light of oils, solutions and other liquids.

Immersion Refractometer: For use by immersion in the liquid whose refractive index is to be measured.

Jamin Refractometer: Measures the refractive index of gasses or liquids, by means of the interference patterns formed by two beams of light, one of which passes through the sample.

Oleorefractometer: Measures the refractive index of butter, lard and oils.

Pulfrich's Refractometer: For solids and liquids. Solid material to be measured is placed on the top prism surface, over a few drips of liquid of high refractive index. A liquid sample to be measured is placed in a glass cell.

Thornoe's Refractometer: Measures the refractive index of beer.

Regent
Regent was a post-war brand of classic Porro prism binoculars, either made in Japan, or **Empire Made**. Binoculars marked *Regent Chance Pilkington* are Regent binoculars made using optical glass from the **Chance-Pilkington Optical Works**.

Reichenbach & Ertel
See **Mathematisch-mechanisches Institut von Reichenbach & Ertel**.

Reichenbach & Liebherr
See **Mathematisch-mechanisches Institut von Reichenbach & Liebherr**.

Reichenbach, Georg Friedrich von
Georg Friedrich von Reichenbach, (1772-1826) [37] [161] [422], was a German engineer, and optical and mathematical instrument maker. He designed mathematical instruments while at Military School in Mannheim and, sponsored by the Bavarian government, he studied mechanical engineering for two years in Britain. He manufactured arms in Munich during the Napoleonic Wars. In his spare time, he built precision instruments, and designed a circular dividing engine. **Ulrich Schenk** worked for him prior to setting up his own business in Switzerland.

In 1802, together with Joseph Liebherr (1767-1840), a mechanic and watchmaker, he formed the company **Mathematisch-mechanisches Institut von Reichenbach & Liebherr** in Munich, Germany. In 1804, **Joseph von Utzschneider** joined the partnership, and it was renamed **Mathematisch-mechanisches Institut von Utzschneider, Reichenbach & Liebherr**. In 1809, Joseph von Fraunhofer was made a partner, and the firm was renamed **Optisches Institut von Utzschneider, Reichenbach & Fraunhofer**. There, Reichenbach made high quality optical instruments until he left in 1814 to start his own company. **Traugott Leberecht von Ertel** also left the Institute a year later, and joined Reichenbach in his new enterprise, which was named **Mathematisch-mechanisches Institut von Reichenbach & Ertel**. In 1821 Reichenbach, with many other commitments for his time, handed over the running of the company to Ertel who, in 1834, with his son Georg, re-named it **Ertel & Sohn**.

In 1807 Reichenbach was given technical responsibility for the salt-works in Reichenhall. He applied his engineering skills to introduce innovations in the brine transport line. He was generously rewarded for this work by the Bavarian king when the line opened in 1817, and in 1820 he became head of the department for roads and waterways.

See Appendix VII.

Reichenbach, Utzschneider & Liebherr
See **Mathematisch-mechanisches Institut von Utzschneider, Reichenbach & Liebherr**.

Reichert, C., Optische Werke
C. Reichert Optische Werke, [53] [172], of 26 Benogasse, Vienna, Austria, was founded in 1876 by Carl Reichert. The company specialised in manufacturing microscopes and microtomes. **Ludwig Merker** and **Fritz Ebeling**

worked for Reichert before leaving to begin their own partnership, **Merker & Ebeling**, in 1886.

C. Reichert Optische Werke also produced cameras and projectors. The company was incorporated in about 1909 as C. Reichert Optische Werke AG. During WWII they marked their military products with the **ordnance codes** bvf, pvf or pwf. In 1963 the **American Optical Company**, of Massachusetts, USA, acquired an interest in the company, and in 1972 they took full ownership.

Reid and Sigrist

Reid and Sigrist, [44] [306], of New Malden, Surrey, England, was a precision engineering company manufacturing, among other things, aircraft instrumentation. The company was founded in 1924 by Major G.H. Reid of the Royal Flying Corps, and Frederick Sigrist, an aircraft industry designer and production manager. After WWII they began manufacturing cameras to the **Leica** designs taken from the **Leitz** factory in Germany. They expanded their optical product range, and in 1954 they were taken over by the Decca Record Company, and continued production of cameras until 1964.

Reinfelder & Hertel

Reinfelder & Hertel, [219] [374] [422], were optical and astronomical instrument makers of Munich, Germany. The company was a partnership between Gottlieb Reinfelder (1836-1898) and Wilhelm Hertel (1837-1893), both of whom had served their apprenticeships with **Steinheil & Söhn**. The firm was established in 1865 by Gottlieb Reinfelder and, in 1867, Wilhelm Hertel became his partner. In 1869 Reinfelder's son, Karl Reinfelder (1869-1915) became a partner, as did **Paul Zschokke** in the same year. In 1903 Paul Zschokke left, and purchased the **Merz** company.

Reinfelder & Hertel are known to have produced: telescopes; geodesic instruments; spectroscopes; prisms; microscopes. Many of their instruments were made using French optical glass.

Reiss, R.

R. Reiss, of Bad Liebenwerda, Germany, was founded in 1882 by Robert Reiss, initially as a mail order retailer of office supplies and measuring equipment. In 1886 he began to manufacture equipment for sale. He produced drafting equipment and surveying equipment. The company continued to expand successfully until Robert's death in 1911, upon which his son, Paul Reiss, took over as sole shareholder. In 1949 the company stopped making surveying instruments, and concentrated on producing drafting equipment and office furniture. In 1990 they re-focussed again, this time producing only office furniture.

Relay Lens

A relay lens is a lens, or a group of lenses, that is placed in the optical path to invert the image, to extend the optical path, or both. It is positioned on the eyepiece side of the objective image plane, and forms an image, more distant from the objective, that is either viewed by the eyepiece, or further extended by another relay lens. This is used in such applications as early field glasses, hand-held refracting telescopes and spyglasses, endoscopes, and periscopes.

Repeating Circle

The repeating circle, also known as a Borda circle, or a Borda repeating circle, was invented by Jean-Charles de Borda, a mathematician in the French navy. It is an improvement on the reflecting circle, and works on the principle of a sextant, but its graduated arc is a full circle, rather than a sixth of a circle as in the sextant. The first prototype Borda repeating circle was made by **Etienne Lenoir**.

Repsold, A.&G.

See **A. Repsold & Söhne**.

Repsold & Söhne, A.

Johann Georg Repsold, (1770-1830) [374], was a firefighter and an amateur astronomer. In 1799 he set up a small company in Hamburg, Germany, to make astronomical and geodetic instruments. After his death in a fire in 1830 the company was run by his sons Adolf and Georg, calling it A.&G. Repsold. During the 1840s **James Mackay Bryson** studied under A.&G. Repsold, and worked for them. Adolf's sons Johan Adolf and Oscar took over the company in 1871, calling it A. Repsold & Söhne, and the company continued in business until 1919.

Research Enterprises Ltd. (REL)

REL was a Canadian company, based in Toronto, that built electronics and optical instruments during WWII. The company was established in 1940, and stayed in business until 1946. When the factory closed, some of its assets formed part of the **Corning** glass plant. REL made gunsights, telescopes, and binoculars, including 7x50 and 6x30 binoculars.

Revue; Revueflex

The retailer Foto-Quelle, of Nuremberg, Germany, used the brand-names *Revue* and *Revueflex* for a number of models of camera, re-branded by them, and manufactured by camera makers such as **Chinon**, **Cosina**, and **KMZ**.

Rew, Robert
Robert Rew, (fl.1755-1764) [18] [256] [276 #528], was an optical instrument maker of Coldbath Fields, London, England. Early on he was an employee in the workshop of **John Dollond (I)**, and some sources say that he was the person who made John Dollond (I) aware of Chester Moor Hall's work on the achromatic lens. However, by 1764 he had his own business and, in that year, he was one of the petitioners who attempted to revoke the patent obtained by John Dollond (I) for the achromatic lens, and which was enforced after Dollond's death by his son, Peter (See Appendix X).

Reynaud, Charles-Émile
Charles-Émile Reynaud, (1844-1918) [45] [335], was a French precision engineer, inventor, and showman who invented the Praxinoscope. He served an apprenticeship as a precision engineer in Paris, and went on to study under a sculptor who was also a photographer. He started to produce photographic and hand-drawn lantern slides, which led him to gain experience in projection. He patented the Praxinoscope in 1877, and began to produce them commercially. He exhibited at the Paris Exposition of 1878, where he received an honourable mention.

He made various improvements to his design, leading to a patent in 1888 for the Théâtre Optique, which was a projection Praxinoscope for public audiences. He went on to produce a number of reels for the Théâtre Optique, giving public shows until finally in 1918 he gave up in the face of competition from the Cinématographe of the **Lumière** brothers.

Reynolds and Branson
In 1886 (some sources say 1883) the firm **Harvey, Reynolds and Co.** became Reynolds & Branson, [53] [61] [306], trading at 14 Commercial Street, Leeds, Yorkshire, England. They were manufacturing chemists and opticians, and they dealt in and produced cameras, photographic equipment, and scientific apparatus. They used the trade-name *Phœnix* on some items, and they used the trade-name *Rystos* for dark room and other equipment.

In about 1901 they incorporated as Reynolds & Branson Ltd., and they continued in business until the 1970s.

Reynolds & Wiggins
Reynolds & Wiggins, [304], were the successors to **Thomas Bennett** in his instrument making business at 124 Patrick Street, Cork, Ireland. By 1870 they had taken over the premises from Bennett, and continued in business as Reynolds & Wiggins (Late Bennett) until 1872. The premises were taken over by the company Purcell, who published the *Commercial Cork Almanac*.

Ribright, George
George Ribright, (fl.1765-1778 d.c1783) [18] [274] [276 #529] [322] [401], was an optician and optical and mathematical instrument maker of 40 The Poultry, London, England. He served an apprenticeship beginning in 1744 (guild unknown), and in 1751 he was freed of the **Spectacle Makers' Company** as a Foreign Brother. He succeeded his father, **Thomas Ribright (I)**, in business. In 1764 he was one of the petitioners who attempted to revoke the patent obtained by **John Dollond (I)** for the achromatic lens, and which was enforced after Dollond's death by his son, Peter (See Appendix X). He took on his son, **Thomas Ribright (II)** as an apprentice in 1768, and in 1778 he began to work in partnership with Thomas (II), trading as **Ribright & Son** until his death in 1782 or 1783.

Ribright, Thomas (I)
Thomas Ribright (I), (1712-1781 fl.1735-1772) [18] [53] [274] [276 #299] [319] [322], was an optician, and mathematical and optical instrument maker, of *Golden Spectacles*, The Poultry, London, England. He served his apprenticeship with **Thomas Sterrop (II)**, beginning in 1726/7, and was freed of the **Spectacle Makers' Company** in 1734. He served as Master of the Spectacle Makers' Company in 1758-1759. He had a royal appointment as optician to George, Prince of Wales, who was crowned King George III in 1760. He also had an appointment to the Office of Ordnance. His business card has illustrations of spectacles, microscope, scioptic ball, and telescopes. He is known to have sold: telescope; gunner's level. Instruments bearing his name include: Gregorian telescope; spyglass; Culpeper-type microscope; case of instruments with perspective glass (for which he obtained a patent in 1749). He was succeeded in business by his son, **George Ribright**.

Ribright, Thomas (II)
Thomas Ribright (II), (fl.1783-1806 d.1811) [18] [53] [274] [319] [322], was an optician, and optical, mathematical, and philosophical instrument maker, of 40 The Poultry, London, England. He served his apprenticeship with his father, **George Ribright**, beginning in 1768, and was freed of the **Spectacle Makers' Company** in 1775. He worked in partnership with his father, trading as **Ribright & Son** from 1778 until his father's death in 1782 or 1783. He was awarded a patent for an artificial horizon in 1790. He worked with Thomas Whitney in a partnership which was dissolved in 1792 [35 #13461 p750], after which he continued in business under his own name. Instruments bearing his name include: wheel barometer; spyglass.

Ribright & Son
Ribright & Son, (fl.1778-1783) [18] [322], was a partnership between **George Ribright** and his son **Thomas Ribright (II)**, trading at 40 The Poultry, London, England until the death of George Ribright in 1782 or 1783.

Richard, Félix-Max
Félix-Max Richard was the brother of **Jules Richard** who, after working in partnership with his brother from 1882 to 1891, set up his own photographic and optical equipment company, **Comptoir Général de Photographie**. Since he was competing with his brother, a legal battle followed, resulting in the sale of his company to **Léon Gaumont** in 1895.

Richard, Jules
Jules Richard, (1848-1930) [53] [270] [324], was the son of Félix Richard, who started an instrument making workshop in Lyons, France in 1845. Upon his father's death in 1876, Jules took over the company. From 1882 to 1891 he ran the company as Richard Frères in partnership with his brother **Félix-Max Richard**.

Félix-Max went on to start his own company, the **Comptoir Général de Photographie**, but set himself in competition with Jules. A legal battle between the brothers resulted in Félix-Max selling his company to **Léon Gaumont** in 1895.

After his brother left in 1891 Jules ran the company until his death in 1930. The company specialised in stereo photography, and their *Vérascope* camera was a stereoscopic hand camera, invented by Jules Richard, which became their core product.

In 1925 Jules Richard was awarded the honour of Commandeur de la Légion d'Honneur. In 1930, just a few months before his death, he founded l'école d'Apprentis Mécaniciens Précisionnistes, which continues today as Lycée Technologique Jules Richard. See Vérascope.

Richardson, Adie & Co. Ltd.
Richardson, Adie & Co. Ltd., [263], in Edinburgh, Scotland, were cutlers and opticians, silversmiths, watchmakers, photographic dealers, and meterological, surveying, and mathematical instrument makers. The company was established in 1913, incorporating **Adie & Wedderburn** and R.S. Richardson & Co., cutlers, (established 1828). Richardson, Adie & Co. traded at 99 Princes St., and 15 South St. Andrew St., Edinburgh [173]. They ceased trading as Richardson, Adie & Co. Ltd. in November 1949 [170 #16695 p510].

Richardson, George
George Richardson, (fl.1807-1830) [18] [276 #1205], was an optician, and optical and mathematical instrument maker of London, England. Addresses: 48 Leadenhall Street (1807); 7 St. Catherine's Street, near the Tower (1808-1817); 12 Upper East Smithfield (1820-1826); 38 Windham Street, New Road, Marylebone (1826).

He was the son of optician Winstanley Richardson of the **Spectacle Makers' Company**, and worked as foreman for **Charles Lincoln** prior to running his own company. He was freed by patrimony of the Spectacle Makers' Company in 1807, and succeeded his father in business. Among his apprentices was **George Patrick**, beginning in 1814.

According to his trade card of c1807 [200]: "G. Richardson, Optical & Mathematical instrument maker, N°48 Leadenhall Street, London. (Late foreman to Mr Chas Lincoln). Real manufacturer of the improved telescopes for day or night. Sextants. Quadrants and compasses of the latest improvements, also spectacles & reading glasses curiously adapted to suit all sights either on glass or Brazil pebbles. Merchants and captains supplied on the most moderate terms"

Richardson, Matthew
Matthew Richardson, (1716-1740) [8] [18] [276 #400], was an optician and optical instrument maker, of *Sir Isaac Newton's Head and Golden Spectacles*, opposite York Building, Strand, London, England. He served his apprenticeship with **Edward Scarlett (I)**, and was freed of the **Spectacle Makers' Company** in 1737. He advertised spectacles, camera obscuras, microscopes, telescopes, mirrors, barometers, and thermometers.

Richie's Photometer
See photometer.

Ricoh
Ricoh Company Ltd., based in Tokyo, Japan, was founded in 1936. They develop imaging devices, industrial products and network systems solutions. The company started its camera business in 1937, and in 1950 it began Japan's first mass production of cameras. In 2011 Ricoh purchased Pentax Imaging Systems, the **Pentax** camera and binocular brand, from **Hoya**, [112]. *Pentax* became a brand-name used by Ricoh for cameras and binoculars.

Riddell, Prof. John Leonard
Prof. John Leonard Riddell, (1807-1867) [285], of New Orleans, USA, was a physician, biologist, inventor, and science-fiction writer who pursued a wide range of interests. In about 1853 he designed a binocular microscope using dual illuminating mirrors directing light

up through a single objective to two prisms, one for each of two eye-tubes. At the top of each eye-tube was another prism producing a horizontally viewed, correct orientation view of the object. The first microscope of this design was made by **J.&W. Grunow**. Riddel's design was the predecessor to the modern Common Main Objective microscope. A variant on Riddell's design was invented independently in 1870 by the English microscopist **John Ware Stephenson**.

Rider, Job
Job Rider, (fl.1791, d.1833) [34] [304], of Belfast, Ireland, was a clockmaker and optical instrument maker. In 1791 he began his business at the sign of *The Reflecting Telescope*, Shambles Street. From 1805 until 1807 he worked in partnership with **Henry Gardner**.

Rienks, S.J.
Sieds (or Syds) Johannesz Rienks, (1770-1845) [17] [220] [256] [264], was a Dutch optical instrument maker. He is known to have made reflecting telescopes, including Gregorian telescopes, as well as microscopes. He traded in Hallum from 1798, and in Lieden from 1826. In 1822, (sources differ on this date – it may have been 1826), he began production of catoptric (reflecting) microscopes following its introduction in 1813 by **Giovani Battista Amici**. Examples of microscopes by Rienks also include tripod-mounted compound microscopes.

Rietzschel, A.H.
Alexander Heinrich Rietzschel learned his craft as an optician with **Rodenstock**, **Zeiss** and **Steinheil** before founding Optische Anstalt A. Hch. Rietzschel (Optical Institute of Alexander Heinrich Rietzschel) in Munich in 1896. He began by producing lenses, and in 1900 he started producing cameras. In 1921 the company was acquired by Bayer. In 1925, when Bayer became a part of **IG Farben**, Optische Anstalt A. Hch. Rietzschel changed ownership to become a part of **AGFA**, also a part of IG Farben.

Riflescope
A riflescope is generally a small refracting telescope, designed to be mounted on a rifle, that provides an upright image with corrected left/right image orientation and long eye-relief. A riflescope will generally also have crosshairs.

Riggs and Brother
William H.C. Riggs, a Philadelphia clock and watch maker, founded Riggs & Brother, [116] [117], in 1818 as a chronometer and nautical instrument maker. Subsequent family members continued the firm, keeping its name over the years. Riggs & Brother published a nautical almanac, and retailed jewellery, silverware, clocks and watches, nautical chronometers and other nautical instruments.

Ringfoto
Ringfoto is a German photographic products retailer. In 1997 they acquired the **Voigtländer** brand-name and, in 1999, began to license the name to **Cosina** who make *Voigtländer* branded products exclusively for them.

Ring Laser Gyroscope
A ring laser gyroscope is a Sagnac interferometer, [7 p323], containing a laser in one arm of the closed optical path, and is used to measure rates of rotation. The laser light traverses the optical path in both directions. The interferometer is rotated about an axis perpendicular to the plane of the optical path, and passing through the centre of the interferometer. The rotation effectively shortens the time taken for the light to traverse one optical path with respect to the other, resulting in an interference fringe shift that is proportional to the rate of rotation.

This effect is the result of a small shift in the lasing frequency of each of the two light beams such that in both cases the length of the optical path is an integral number of wavelengths. If the rotation rate is below a threshold the two beams suffer from a coupling effect known as *lock-in*. This results in the frequencies of the two beams locking to become the same, with the result that no fringe shift is detected. This problem may be overcome by *dithering* which is a forced oscillatory rotation about its axis – backwards and forwards – so that the rotation rate is mostly above the lock-in threshold.

The Sagnac interferometer was first demonstrated by the French physicist Georges Sagnac in 1913. The laser was invented in 1958, and first demonstrated in 1960. The ring laser gyroscope began to replace mechanical gyroscopes in inertial navigation systems in the early 1980s.

Ring Micrometer
See micrometer.

Ripley & Son, Thomas
Thomas Ripley & Son, [16] [18] [276 #760], mathematical, optical, and nautical instrument makers of London, England. Addresses: 335 Hermitage; 364 Hermitage; Hermitage Bridge (1795); 364 Hermitage Bridge, Wapping (1800-1805); 335 High St., Wapping (1805).

The firm was a partnership between **Thomas Ripley** and his son **James Ripley**. Instruments bearing their name include: quintant; Hadley's quadrant; octant; sextant. The partnership was terminated in 1805, after which James Ripley continued to trade under his own name.

Ripley, James
James Ripley, (fl.1790-1844) [**18**] [**90**] [**276** #1208], was an optical, mathematical, and philosophical instrument maker of London, England. Addresses: 13 Warkworth Terrace, Commercial Road; Mill Place, Commercial Rd.; Parish of St. John, Wapping (1805); 335 High St., Wapping (1807-1822); 15 Warkworth Terrace, Commercial Rd. (1839-1843).

He served his apprenticeship with his father, **Thomas Ripley**, beginning in 1782, and was freed of the **Grocers' Company** in 1796. He traded in partnership with his father as **Thomas Ripley & Son**, from 1790 until 1805, after which he began to trade under his own name. Among his apprentices was his son, Thomas Brand Ripley, beginning in 1805. In 1822 he was declared bankrupt, [**35** #17819 p860], and a final call for creditors was made in 1840, [**35** #19836 p681].

According to his trade card, [**361**]: "Ripley, No. 335 Wapping, London. Mathematical, Optical & Philosophical instruments for sea & land, Navigation books and sea-charts." On this example of his trade card, the address is crossed out, and replaced in handwriting with "Mill Place, Commercial Road, Limehouse."

Ripley, Thomas
Thomas Ripley, (fl.1765-1795) [**16**] [**18**] [**90**] [**276** #760], was a mathematical and optical instrument maker of London, England. Addresses: 364 Hermitage Bridge Rd., Wapping; the *Globe, Quadrant & Spectacles* near Hermitage Bridge below the Tower; Bakers Buildings, New Broad St. (1763); Near the Hermitage, Wapping (1765-1770); Bakers Buildings (1770); Near the Hermitage (1773); 364 Hermitage (1774-1793).

He served his apprenticeship with **John Gilbert (II)**, beginning in 1755, and was freed of the **Grocers' Company** in 1763. Among his apprentices were his sons, Thomas Ripley (II), beginning in 1778, **James Ripley**, beginning in 1782, and **Richard Lekeux**, who began his apprenticeship in 1770 and was turned over to **Walton Willcox** of the **Joiners' Company** in 1775. Instruments bearing his name include: Hadley's quadrant; octant; four-draw telescope. In 1790 he formed a partnership with James Ripley, trading as **Thomas Ripley & Son**.

Ritchey-Chrétien Telescope
See telescope.

Ritchey, George Willis
George Willis Ritchey, (1864-1945) [**165**], was an American astronomer, inventor, designer and engineer. He joined the staff at the Yerkes Observatory in Wisconsin on the invitation of the astronomer **George Ellery Hale**. There he produced a 24-inch reflector, and a coelostat with a 24-inch horizontal telescope. He went on to work with Hale at the Mount Wilson Observatory in Pasadena, where he produced both a 60-inch telescope and a 100-inch telescope.

Ritchey came up with many original ideas, some of which were successful, some less so, and some of which never came to fruition. His most notable innovation was the joint invention with the French astronomer **Henri Chrétien** of a variant of the Cassegrain telescope known as the Ritchey-Chrétien telescope.

Riva, Ferdinando
Ferdinando Riva, also known as Ferdinand, (fl.1830-1834) [**3**] [**18**] [**276** #1979] [**379**], was an optical instrument maker and barometer maker of Sheffield, England. Addresses: 7 Watson Walk (1830-1834), and Court 3, High St. (1833).

RMS Thread
The RMS thread is the Royal Microscopical Society's standard screw-thread for microscope objective lens attachment. This provided for interchangeability of objectives by British makers and any others who chose to conform. The standard, adopted in c1860, is a Whitworth thread, outer diameter 0.7965 inches x 36TPI.

Robbins, J.&C.
J.&C. Robbins, [**264**], opticians and microscope makers, of 146 Aldersgate Street, London, England, was a partnership between John Robbins and Charles Robbins. The partnership was dissolved in 1876, [**35** #24377 p5810], after which John Robbins continued the business alone.

Roberts & Co., Austin
Austin Roberts & Co., [**199**], was a company in Birkenhead, Cheshire, England, specialising in mirror making kits for amateur telescope makers. The company was established in 1967 by a group of amateur telescope makers, and in 1973 they advertised reflecting telescopes of 115mm and larger.

Robertson, William
William Robertson, (fl.c1730-c1760) [**18**] [**264**] [**271**], was an optical instrument maker of Edinburgh, Scotland. An example of his *Catadioptric Microscope* survives in the collection of the Royal Microscopical Society. Catadioptric in this case refers to the use of reflected light from a substage mirror for illumination.

Robinson & Barrow

In 1842, following the death of **Thomas Charles Robinson** in 1841, **Henry Barrow** took over his business at 28 Devonshire Street, Portland Place, London, England. He began to trade as Robinson & Barrow, opticians, [18] [53]. From 1843, his business was located at 26 Oxenden Street, Haymarket, London. He traded as Robinson & Barrow until at least 1845 and, by 1849, he was trading as **Henry Barrow & Co.**

Robinson & Sons, James

James Robinson & Sons, [11] [18] [34] [215] [304] [306], were instrument makers and photographic suppliers of 65 Grafton Street, Dublin, Ireland. The firm was a partnership between James Robinson, (fl.1845-1884), and his sons. The partnership traded under this name from 1885-1903, and then as J. Robinson and Sons Ltd., from 1904 until 1910. They are known to have retailed cameras and telescopes.

Robinson, John

John Robinson, (fl.1783-1795) [11] [18] [304], was a mathematical instrument maker of Dublin, Ireland. He is known to have traded at 4 Drogheda Street from 1791-1792, and at 21 Hawkin's Street from 1794-1795, and to have sold telescopes. He may have been a predecessor to **James Robinson & Sons**.

Robinson, Thomas Charles

Thomas Charles Robinson, (1792-1841) [18] [53] [276 #1687], was an optician, and mathematical and philosophical instrument maker of 38 Devonshire Street, Portland Place, London, England. He traded at this address from 1825 until his death in 1841. He produced innovative designs, including an improved quadrant with a spirit level, an improved balance, and an optical instrument for taking the lines of soundings, by means of which the user could simultaneously view the near and forward marks. He sold balances, barometers, dip circles, and transit instruments, and was appointed compass maker to the Admiralty. Following his death his business was taken over by **Henry Barrow**, who initially traded as **Robinson & Barrow**.

Rochette, Gaspard

See **Rochette Père**.

Rochette Jeune

Rochette Jeune (young Rochette), [11] [50], optical and scientific instrument maker, was the son of **Rochette Père**. From about 1807 to 1817 he worked in partnership with **Marie Putois** at Quai de l'Horloge 81, Paris, the premises also known as *au Griffon*. By 1817 he was the sole owner of the business, which he continued until 1845. Rochette Jeune is known to have made graphometers, telescopes, drawing sets, mining pocket compasses, optical squares and Davy lamps. He had a strong reputation for his telescope objective lenses, and made the objective for the equatorial refractor at l'Ecole Militaire, Paris.

Rochette Père

Rochette Père (Rochette the father), [50], was an optical and scientific instrument maker of 75 Quai de l'Horloge, Paris. He was the father of **Rochette Jeune**, and is believed to be Gaspard Rochette (fl.1794-1822), whose signature appears on instruments of that time.

Rodenstock

Rodenstock, [47] [374], is a German company, producing ophthalmic lenses. The company was established by Josef Rodenstock (1846-1932) in Würzburg, Germany, in 1877, producing barometers, precision scales, measuring instruments and spectacle lenses and frames. He named the company after his father, G. Rodenstock. In 1878, his brother, Michael, became a partner in the business. In 1882, he opened a branch in Munich, and in 1884 the firm relocated to Munich. By the early 1890s, the company had over 100 employees, and were producing ophthalmic lenses, photographic lenses, binoculars, telescopes, microscopes, and cameras. **A.H. Rietzschel** worked for Rodenstock prior to setting up his own company in 1896.

In 1905, Michael left the business, and Josef's son, Alexander, joined the company. In 1906 the company opened their own glass-works, producing optical glass for lens production. In 1919, Josef Rodenstock retired, and Alexander took over running the business. During WWII, operating with the name G. Rodenstock Optische Werke, the company marked its military products with the **ordnance code** eso.

In 1996 the precision optics production was split from the ophthalmic lens production to form a separate company, trading as Rodenstock Präzisionsoptik GmbH, leaving the core Rodenstock company to specialise in ophthalmic lenses.

In 2000, Rodenstock Präzisionsoptik was acquired by **LINOS AG**. In 2006, the **Qioptiq Group** began the process of acquiring LINOS AG. The acquisition was completed in 2009 and, since then, Rodenstock Präzisionsoptik has operated as their Rodenstock Photographic Optics division.

Roessler, Paul

Paul Roessler, (1829-1904) [50], was a German born American mathematical, optical and philosophical instrument maker and dealer. He began his business in

New Haven, Connecticut, in 1855, shortly after emigrating to the USA. In 1859 he won a silver medal at the State Fair in New Haven for his optical and mathematical instruments, microscopes, opera glasses, steel spectacles and eye glasses. Upon the formation of a partnership with W.F.&R.P. Sternberg in 1890 they began trading as Paul Roessler & Company. Upon Paul's retirement in 1893 he was succeeded by his son Fritz Roessler, at which time the company became Roessler & Company. The partnership continued until 1894 after which Fritz continued the business as Paul Roessler's Son. In 1900 he entered a new partnership as Fritz & Hawley, and the company remained in business as Fritz & Hawley Guild Opticians.

Rollei

In 1962 **Franke and Heidecke**, of Brunswick, Germany, was renamed to Rollei-Werke Franke & Heidecke. The company went on to produce new and existing models of camera and projector using the Rollei name, with frequent releases of new models. In 1979 they became Rollei-Werke Franke & Heidecke GmbH & Co. KG. In 1982 the company filed for bankruptcy, and re-emerged as Rollei Fototechnic GmbH, and was renamed again in 1987 to become Rollei Fototechnic GmbH & Co. KG. In 1991 they began to produce digital cameras, and continued to produce new models. By 2009 the company was in financial trouble again and, between then and 2015, following two insolvencies, the company sold its inventory and ceased trading. Rollei GmbH & Co. KG. was formed as the owner of the brand rights.

Ronchetti

The Ronchetti family, [3] [18] [53] [126] [127] [384], originally from Tavernerio, Italy, were scientific and optical instrument makers in Manchester, England. The family is known to have intermarried with the **Casartelli** family, also from Tavernerio, in at least two generations.

Giovanni Battista Ronchetti, known as Baptist Ronchetti came to Manchester from Tavernerio at about the time of the birth of his son, Charles Joshua Ronchetti (1790-1850). There he started a barometer making business, and later sent to Italy for Charles Joshua, and his nephew Louis Casartelli. He retired to Italy in c1810, at which time Louis took over his business.

Charles Joshua set up business in Liverpool in c1814, but returned to Manchester within a about a year, upon exchanging businesses with Louis. He initially worked for the carver and guilder, Vittore Zanetti, but as a skilled glass-blower he presently set up his own business selling barometers, thermometers and hydrometers. By the late 1820s he was also selling optical and mathematical instruments, and had expanded his business in philosophical instruments. In about 1842 he opened a chemical plant, also in Manchester, and at this time his two sons, Joshua and John Baptist (1812-1880), took over the business making scientific and optical instruments. His daughter, Jane Harriet Ronchetti, married Joseph Louis Casartelli, who took over the business in c1852 when Charles Joshua retired to Italy.

Roof Pentaprism

See roof prism.

Roof Prism

A roof prism, otherwise known as a dach prism, is an optical prism in which one side comprises two optical surfaces at 90° to each other forming a shape reminiscent of a roof. This surface acts to reverse the image along the axis where the surfaces meet. Roof prisms come in many forms, the most common being *Amici roof prism*, *roof pentaprism*, *Schmidt-Pechan prism*, and *Abbe-König prism*. Inherent in the roof prism concept is the need for high precision in prism shaping and production - much more so than for Porro prisms to achieve the same optical quality. Roof prisms introduce a phase shift between the two halves of the resulting image. Because of this they need phase correction coatings to be used in instruments such as binoculars.

An *Amici roof prism*, named after the Italian astronomer **Giovani Amici**, is sometimes known as a *right-angle roof prism*. It reflects an image through 90° and inverts it. Applications such as spotting scopes commonly use this kind of prism to produce an erect image.

A *roof pentaprism* flips an image laterally, but does not invert it. This kind of prism is used in SLR cameras to provide a true view through the viewfinder.

A *Schmidt-Pechan prism* is used to provide an erect image in binoculars. With this design the exit beam is coaxial with the entry beam. It is a combination of two prisms, resulting in five internal reflections and four glass/air interfaces. The internal reflections and glass/air interfaces introduce optical losses. Because of this, and the need for high precision, quality comes at a higher price than with the Porro prism design.

An *Abbe-König prism* is also a combination of two prisms, used to provide an erect image in a pair of binoculars. As with the Schmidt-Pecan prism the exit beam is coaxial with the entry beam. It is less compact than the Schmidt-Pecan prism, but has the advantage of only three internal reflections.

Ross, Andrew

Andrew Ross, (1798-1859) [44] [18] [82] [264] [320], was the son of John Ross, a corset-maker of Fleet Street, London, England. At the age of 15 he began an apprenticeship with John Corless, a mathematical instrument maker of 19 Newcastle Street, Strand, London. Upon completion of his apprenticeship, he worked as a mechanical engineer for three years. In about 1823, the year in which his son **Thomas Ross** was born, he began to work for **W.&T. Gilbert**, who was a manufacturer of levels, theodolites and astronomical instruments, with whom Ross rose to become works manager. Here he demonstrated his skills as an instrument maker, constructing and dividing an astronomical circle that was sent for use in the Cape of Good Hope.

In 1830 Andrew Ross set up his own company in London where he soon began to produce microscopes. By 1832 he was trading at 15 St John's Square, Clerkenwell where, in 1833, he appears in the directories as a mathematical instrument maker. During the 1830s **Henry Anderson** came to learn his trade as a microscope maker from Ross, working for him until the 1860s or 1870s when he left to begin his own company.

In 1837 Ross began a collaboration with **J.J. Lister** on microscope design. He continued to make mathematical and optical instruments, and in 1839 he was one of the founders of the Microscopical Society. At this time, he is known to have corresponded with **William Henry Fox Talbert** about matters relating to silvered surfaces, microscopes and camera obscuras. In 1840 he began trading as Andrew Ross and Co. Opticians, in partnership with J.J. Lister until 1842, after which he began to trade as Andrew Ross Optician. It was at this time that he began to put a serial number on his microscopes, and he moved premises to 21 Featherstone Buildings, Holborn. His catalogue showed microscopes, terrestrial and astronomical telescopes, a binocular night telescope, spectacles, magic lanterns and hygrometers. By 1848 his trading address was 2 Featherstone Buildings, Holborn.

In 1851 **J.H. Dallmeyer** started to work for Ross, but he soon left, unsatisfied with his position. At this time Ross was making telescopes, microscopes and camera lenses. In 1852 J.H. Dallmeyer returned to work for Ross as Scientific Adviser. In 1854 he married Andrew Ross' daughter Hannah, and took over management of the manufacturing of telescopes. Ross' son **Thomas Ross**, heading the photographic side of the business, had a difficult relationship with Dallmeyer.

James Swift also worked for Andrew Ross. Records differ on when he left to start his own business, but it was sometime between 1853 and 1857.

In 1859 Andrew Ross died, leaving two thirds of his estate to his son Thomas, and one third, being the telescope making part of the business, to J.H. Dallmeyer.

Ross, Thomas; Ross & Co.

Thomas Ross, (1818-1870) [44] [82] [172] [264] [320], was the son of **Andrew Ross**, and in 1854 he took over management of the photographic part of his father's business. When his father died in 1859, he continued to run the family business, absent the telescope making side of the business, which was bequeathed by his father to **J.H. Dallmeyer**.

After his father's death Thomas continued the business as Thomas Ross Optician, producing binoculars, telescopes, microscopes and photographic lenses. In 1869 he moved the business to 7 Wigmore Street, Cavendish Square, London, England. Upon his death in 1870 he left the business to his wife Mary Anne Ross. In that same year John Stuart, a landscape photographer, joined the company to manage the photographic section. In 1872 John Stuart married Mary Anne Ross, and they continued to run the company as Thomas Ross and Co. By 1874 the company was trading as Ross and Co., opticians, manufacturers of microscopes, telescopes, photographic lenses, cameras and apparatus, spectacles, race, field and opera glasses, and philosophical instruments. In the early 1880s **Heinrich Schröder** moved to England from Germany to become the technical director. The company continued in business, expanding their offerings, until 1897 when the company **Ross Ltd.** was registered at 111 New Stock Exchange, Bond Street, London West.

Ross Ltd., London

Ross optical works, [44] [82] [320], has a history going back to 1830 with the business started by **Andrew Ross** in London, England, and continued by his son, **Thomas Ross**. In 1897 the company Ross Ltd. was incorporated, and in 1906 they stopped manufacturing microscopes. In 1917, when **Carl Zeiss (London)** was closed down under the Trading with the Enemy Act, the factory at Mill Hill was sold, and Ross Ltd. acquired all their assets. In 1919 Ross Ltd. was listed as a member of the **British Optical Instrument Manufacturers' Association**. In 1921 **Charles Parsons** obtained a controlling interest in the company

In 1922 Ross Ltd. was listed as an exhibitor at the British Industries Fair. Their catalogue entry, [447], described them as manufacturers of: Cinematograph projectors; Photographic lenses; Lenses for aeronautical cameras; Photographic cameras; Prism field glasses; Telescopes, sporting, military and naval.

In 1929 the Ross Ltd. advertisement in the British Industries Fair Catalogue, [448], described their stand as an optical, scientific and photographic exhibit. It listed their products as: Photographic lenses; Cameras; Prism binoculars; Field glasses; Opera glasses; Terrestrial and astronomical telescopes; Cinematograph projectors; Search-light arc lamps; Equipment; Optical lanterns; Aeronautical, astronomical and nautical instruments; Lenses; Prisms of all kinds.

In 1941 the Ministry of Supply set up a shadow company in Nottingham to be operated by **E.R. Watts & Son** and Ross Ltd. The factory was called **Nottingham Instruments Ltd.**, and it was there that Ross Ltd. and E.R. Watts & Son duplicated their manufacturing in London. In 1948 Ross Ltd. and **Barnet Ensign Ltd.** were merged to form **Barnet Ensign Ross Ltd.**, however the company Ross Ltd. continued under the same name. In 1954 Barnet Ensign Ross Ltd. was Renamed **Ross-Ensign Ltd.** In 1961 they ceased production of cameras, but continued to sell optical instruments such as binoculars and enlarging lenses. In 1969 the company name appears to have changed to Ross Optical Ltd. 1978 is the last year they are recorded as holding a Royal Warrant as suppliers of binoculars, [35 #47732 p60], and in 1982 proceedings were started to liquidate the company, [35 #48888 p1941].

For information on Ross camera models see [306]. For information on Ross binocular models and serial numbers, see [80] [92] [93].

Ross-Ensign Ltd.
See **Barnet Ensign**, **Ross Ltd.**

Rothwell
Rothwell, [18] [319] [322], was an optical, mathematical, and philosophical instrument maker, and retailer of Manchester, England. Little is known about this maker, and dates of work are uncertain, but were some time c1780-c1850. Instruments bearing this name include: microscope; Culpeper-type microscope; stick barometer; barometer; stellar goniometer. The extent to which these instruments were made by Rothwell is unclear.

Rotoscope
The Rotoscope is a device used by animators to trace live action, from a motion picture film, frame by frame. The Rotoscope was invented in 1915 by the Polish born American animator Max Fleischer, (1883-1772) [158]. He used it for animation of classics such as *Out of the Inkwell*, *Koko the Clown*, *Betty Boop*, and *Popye the Sailor*. His patent expired in the early 1930s, and in 1937 Walt Disney used a Rotoscope to animate *Snow White and the Seven Dwarves*.

Rouch & Co., W.W.
W.W. Rouch & Co., [61] [122] [306], makers of cameras and photographic equipment, of 180 Strand, London, England, was first established as Burfield & Rouch in 1854 when William White Rouch (d.1899) entered a partnership with Henry Burfield. Burfield had, since the 1830s, an established business as a chemist at that address. In 1863 the company was renamed W.W. Rouch & Co. **John Garland** later advertised that he had worked as foreman for Rouch.

In 1873 the company expanded to additionally occupy the next-door property at 43 Norfolk Street, which they used as their factory (their showrooms at 180 Strand being on the corner of Norfolk Street). In 1894 the company moved to 161 Strand. Following the death of W.W. Rouch his son, W.A. Rouch, continued the business, but the camera production began to decline, and the photography and retail side of the business began to dominate their trade.

Roux, Ange-Joseph Antoine
Ange-Joseph Antoine Roux, also known as Antoine, (1765-1835) [16] [54], was a hydrographer and publisher and seller of charts, as well as a manufacturer and retailer of navigation instruments. He was also a highly regarded marine watercolour artist. He was the third generation of his family to run the business. According to his trade label (translated), "Ant^{ne} Roux, of the port of Marseilles. Makes and sells all kinds of charts and navigation instruments, also canvas flags of any colour."

He took over the chart-publishing and navigation instrument making business from his father, Joseph Roux (1725-1793), who was also a hydrographer, and a marine artist who painted in oils. Joseph was appointed Hydrographer to the King.

Rowland Concave Diffraction Grating
See concave diffraction grating.

Rowland, Henry Augustus
Henry Augustus Rowland, (1848-1901) [45], was an American physicist known for his invention of the concave diffraction grating.

Rowley, John
John Rowley, (c1668-1728) [18] [90] [234] [275 #507] [396] [442], was a mathematical instrument maker of London, England. Addresses: Behind the Exchange, Threadneedle Street (1691); *The Globe Under St. Dunstan's Church*, Fleet Street (1702-1715); *Under the Dial of St. Dunstan's Church*, Fleet Street (1714). This may have been the premises previously occupied by the mathematical instrument maker **John Worley**.

He served his apprenticeship with **Joseph Howe**, beginning in 1682, and was freed of the **Broderers' Company** in 1690

John Rowley was appointed Master of Mechanicks to King George I. He was highly regarded, and is known to have made many kinds of instrument, including: Drawing instruments; Pocket dials; Proportional compasses; Parallel rulers; Protractors; Artillery scales; Dialling spheres; Gunter's quadrants; Globes; Planetariums (Ptolemaic and Copernican); Levels; Sextants; Orreries, and other instruments. Although the orrery was invented by the clock-maker George Graham (1673-1751), Rowley made a fine orrery for the Earl of Orrery, as well as others, including one for the East India Company, and one for King George I, and he is credited with bringing the word *Orrery* into the English language. One of Rowley's apprentices was **Benjamin Scott**, who began his apprenticeship in 1702 with James Anderton of the **Grocers' Company**, but was turned over to Rowley in 1706. **Thomas Wright (I)** was also one of his apprentices. Wright succeeded Rowley in his business in 1718, trading at the *Orrery and Globe*, Fleet Street, London. Note, it is likely that *The Globe*, the address occupied by John Rowley's business, was renamed *Orrery & Globe* following Rowley's involvement in the development of orreries.

Royal

Royal was a brand-name used by **Astro Optical Industries Co. Ltd.**

Rubergall, Thomas

Thomas Rubergall, (1775-1854) [**18**] [**264**], was a mathematical and philosophical instrument maker of London, England. Addresses: Princes St., Soho; 10 Crown Court, Pulteney St. (1800); 27 Coventry St., Haymarket (1805-1822); 21 Coventry St. (1839); 24 Coventry St. (1840-1851). His trade card, [**145**], advertised: "Thos. Rubergall, Optician, Mathematical & Philosophical Instrument Maker, To his Royal Highness the Duke of Clarence. 27 Coventry Street, Haymarket, London."

Upon his death in 1854 two of his employees, **William Bithray** and Thomas Steane, took over the business and continued to operate in partnership, trading as **Bithray & Steane**.

He is known to have sold: chronometers; microscopes. Examples of his work include: barometers; thermometers; telescopes.

Rudge, John Arthur Roebuck

John Arthur Roebuck Rudge, (1837–1903) [**71**] [**335**], was a philosophical instrument maker and inventor of Bath, England. He was also an entertainer, giving magic lantern shows. He designed and demonstrated a mechanism for serial projection of lantern images, giving the illusion of motion. The device, which he called the *Phantascope*, or *Bio-phantascope*, used a circular gallery containing seven frames, which was moved in a stop-go manner with a shutter that opened while the frame was still. He collaborated with **William Friese-Greene** on producing the images and demonstrating the machine, work which stimulated Friese-Greene's continued interest in moving picture technology. Rudge also produced other moving picture demonstrations, including a mechanism with four converging lenses and a rotating shutter, which projected the four lantern slides in quick succession, giving the illusion of motion. He was the subject of bankruptcy proceedings in 1882 [**35** #25086 p1317].

Ruhnke

Ruhnke was a company founded by Optikermeister Carl Ruhnke, (1874-1922), in April 1896, when he was 22 years old. He opened a shop for "comfortable vision aids" in Leipziger Strasse, near the Spittelmarkt in Berlin, Germany. Soon after, he opened a factory in Rathenow, and by 1920 he had 200 employees. He then acquired new premises in Zehlendorf. In 1922 he died in a motor accident on the journey to Rathenow, and his widow, Martha, took over the business, followed by their son, Fritz in 1933. Fritz Ruhnke moved their production to Hamburg, and his son Jörg took over in 1975. Jörg died in a plane crash in 1986, at which time Christa Ruhnke's husband Peter Barg ran the company until it was taken over by the Danish company Synoptik in 2002. At that time the shop had 29 branches in Berlin, but nobody in the family wished to continue the business [**118**]. Synoptik is a high-street optician, and the brand-name Ruhnke did not survive the take-over.

Ruhnke was a high-street optician, and they also manufactured their own brand of binoculars. *Noctovist* is a Ruhnke model of binoculars. Examples of barometers may also be found.

Rumball, Samuel

Samuel Rumball, (fl.1834-1839) [**18**] [**276** #2226] [**379**], was an optical, mathematical, and philosophical instrument maker of London, England. Addresses: 5 Crane Court, St. Peter's Hill (1834); Crane Court, Doctor's Commons (1837); 21 St. Peter's Hill, City (1839). He attended the London Mechanics' Institute in 1834.

Rumford's Photometer

See photometer.

Ruper

Ruper is a **Hilkinson** brand-name used for their loupe magnifiers, made in Japan.

Russian and Soviet Makers' Marks
See Appendix V.

Rust, Ebenezer (I)
Ebenezer Rust (I), (fl.1777-at least 1800) [18] [90], was a notable English mathematical instrument maker. Addresses: Amsterdam, Netherlands (1779-1782); Green Churchyard, St. Catherine's, London (1782); Eaton Ford, St. Neots, Huntingdonshire (1784-1787); Eaton Socon, St. Neots, Huntingdonshire (1792); Wapping, London (1795); St. George's in the East, London (1798); 20 Sampson's Gardens, Wapping, London (1799). This last address may have been a residence.

He was the son of Joseph Rust, a yeoman of Great Staughton, Huntingdonshire, and served his apprenticeship with **Richard Rust**, who was probably his uncle. He was freed of the **Grocers' Company** in 1777. He worked in Amsterdam, Netherlands, from about 1779-1782. Also apprenticed to Richard Rust were **William Spencer (I)** and **Samuel Browning (I)**, whose partnership with Ebenezer (I) was formed in 1784 to trade as **Spencer, Browning and Rust**.

His two sons, **Ebenezer Rust (II)** and **Joseph Rust (II)** served their apprenticeships with him, and both were freed of the Grocers' Company in 1809. Also among his apprentices was **John Holyman**, beginning in 1784. He bequeathed his share of Spencer, Browning & Rust to his son Ebenezer Rust (II).

See Appendix VII.

Rust, Ebenezer (II)
Ebenezer Rust (II), (fl.1809, d.1838) [18] [90], was an optician and mathematical instrument maker of London, England. Addresses: 66 Wapping (1809); Burr St. (1809).

He was the son of **Ebenezer Rust (I)** and the brother of **Joseph Rust (II)**. He served his apprenticeship with his father, being freed of the **Grocers' Company** in 1809. He inherited his father's share of the business, **Spencer, Browning and Rust**, and continued working in the company, along with **Richard Browning** and **William Browning**, until his death in 1838.

See Appendix VII.

Rust, Joseph (I)
Joseph Rust (I), (fl.1786-at least 1793) [18] [90], was a mathematical and optical instrument maker of London, England. Addresses: Corner of Catherine Stairs (1786-1793); 15 Free School St., Southwark (1793).

He was the son of **Richard Rust**. He served an apprenticeship with his father, beginning in 1777, and was freed of the **Grocers' Company** in 1786, at which time he began to trade at the Corner of Catherine Stairs. (See **Richard Rust & Son**). Among his apprentices was Thomas Dench, whom he turned over to **Thomas Lorkin** in 1789. He was declared bankrupt in 1789, [35 #13094 p360].

See Appendix VII.

Rust, Joseph (II)
Joseph Rust (II), (fl.1809) [18] [90], was a mathematical instrument maker of Burr St., London, England. He was the son of **Ebenezer Rust (I)** and the brother of **Ebenezer Rust (II)**. He served his apprenticeship with his father, beginning in 1798, and was freed of the **Grocers' Company** in 1809.

Sources [18] and [276 #1424] suggest that he may have moved to Liverpool, and be the same person as Joseph Rust (fl.1811-1816), who kept a Navigation Warehouse at 28, 29, and 31 Pool Lane, and a Mathematical Warehouse in Pool Lane (1816).

See Appendix VII.

Rust, Richard
Richard Rust, (d.1785) [18] [90], was a renowned mathematical instrument maker of London, England. Addresses: At Mr. Thomas Rust's, *Anchor & Belles*, Minories (1752); The Minories (1753-1778); 125 Minories (1776); St. Catherine's High St. (1780); Corner of St. Catherine's Stairs, near the Tower (1781).

He was the son of John Rust, a farmer of Kimbolton, Huntingdonshire. He served his apprenticeship with the mathematical instrument maker John Parminter, being freed of the **Grocers' Company** in 1752. He was famously apprentice master to **William Spencer (I)**, **Samuel Browning**, and **Ebenezer Rust (I)**, the latter of whom was probably his nephew.

Also among his apprentices were also **John Browning (I)**, beginning in 1768, and Richard's son, **Joseph Rust (I)**, who began his apprenticeship in 1777, and was freed of the Grocers' Company in 1786. Among his customers was **James Watt**, who bought rules from him for resale. He was succeeded in business by **Richard Rust & Son**.

See Appendix VII.

Rust, Richard & Son
Richard Rust & Son, (fl.1785-1794), were mathematical instrument makers of Catherine Stairs, London, England. The firm succeeded **Richard Rust** in business. This was presumably a partnership between Richard and his son **Joseph Rust (I)**, and the name continued to be used after Richard's death in 1785.

Rutherford, Lewis Morris
Lewis Morris Rutherford, (1816-1892) [24] [184] [187], was an American lawyer noted for his pioneering work in

astrophotography and solar and stellar spectroscopy. He was a keen astronomer, who built his own observatory, and made his own diffraction gratings for spectroscopy. He was supplied with several telescopes by **Henry Fitz** with whom he worked to design and build a telescope for astrophotography. The telescope was finished after the death of Henry Fitz, whose son **Henry Giles Fitz** completed the work, and helped develop another, larger instrument.

Rutherford's Prism
See prism.

Ryland & Son Ltd., H.C.
H.C. Ryland & Son Ltd., of London, England, were Makers of military telescopes and gun sights. Instruments bearing their name have been seen from WW1 and WW2, including examples marked *H.C. Ryland & Son Ltd.*, and *H.C.R. & Son Ltd.*

S

Saegmuller, George N.
George N. Saegmuller, (1847-1934) [12] [189] [207] [208], was an inventor and maker of scientific instruments. He was born in Bavaria, educated in Nürnberg, Germany, and apprenticed to **Repsold** in Hamburg. In 1865 he travelled to England and began to work for **Thomas Cooke & Sons**. In 1870, together with his brother-in-law **Camill Fauth**, he went to Washington DC, USA, having been recruited by the instrument maker **William Würdemann**, to work in his workshop. He worked for Würdemann until 1874 when George Saegmuller, Camill Fauth, and another brother-in-law, Henry Lockwood, formed their own company, **Fauth & Co.**

Saegmuller & Co., George N.
Upon the retirement in 1887 of **Camill Fauth**, **George Saegmuller** became owner and sole proprietor of **Fauth & Co.** Saegmuller continued to use the name Fauth & Co. on his instruments until about 1892, when he changed the name to Geo. N. Saegmuller & Co., and moved to a new location. In 1905 his company merged with **Bausch & Lomb** in Rochester, New York, and became Bausch, Lomb, Saegmuller Co. In 1907 the name was changed back to Bausch & Lomb. In 1908 **Zeiss** formed a commercial agreement with Bausch & Lomb and, shortly after this, the Bausch & Lomb, Zeiss, and Saegmuller subsidiaries merged to form the Bausch & Lomb Optical Company, which became known as the **Triple Alliance**, and lasted until WWI.

Sagnac Interferometer
See interferometer.

Salanson & Co., H.
H. Salanson & Co., [119], of Cardiff, Wales, and Bristol and Gloucester, England, was a photographic and audio-visual equipment dealer and retailer. The company was established in 1887 by Alfred Salanson, selling electrical accessories, and in due course diversified into selling microscopes, spectacles, cameras and photographic equipment. A 1938 advertisement shows cameras, cine cameras and projectors, and other photographic equipment of several well-known makes as well as: Barometers; Field glasses; Microscopes; Telescopes; Opera glasses; Compasses; Magnifying glasses; Thermometers. Some optical instruments sold by them bore their name, but may have been re-branded instruments from other makers.

Instruments bearing their name include: Galilean and Porro prism binoculars; brass and leather 3-draw telescope.

Saldarini, Joseph
Joseph Saldarini, sometimes Salderini, (b.1803 fl.1830-1841) [18] [276 #1996] [319] [322], was an Italian born optician and gilder, and optical and philosophical instrument maker, of Long Causeway, Peterborough, England. Instruments bearing his name include: wheel barometer.

Salmoiraghi, Angelo
Angelo Salmoiraghi, (1848-1939), was an Italian optical instrument maker. See **Filotecnica Salmoiraghi**.

Salmoiraghi and Viganò
See **Filotecnica Salmoiraghi**.

Salmon, William John
William John Salmon, (fl.1838-1881) [18] [320], was an optician, and mathematical, philosophical and optical instrument maker of London, England. Addresses: 105 Fenchurch Street (1838-1845); 254 Whitechapel Rd. (1846-1853); 100 Fenchurch Street (1854-1862); 48 Lombard Street (1858-1861); 85 Fenchurch St. (1865-1877); 2 Aldgate High Street (1878-1881).

He is known to have sold telescopes and microscopes. At the 1851 Great Exhibition in London, he showed day or night telescopes for ships' use, [294] [328]. Two microscopes by W.J. Salmon are in the collection of the Royal Microscopical Society, [264].

Salom & Co.
Benjamin Salom, (fl.1840-1842) [18], was an optician of 31 St. Andrew's Square, Edinburgh, Scotland.

He was succeeded in business by Salom & Co., [53] [322] [379], optical, mathematical, and philosophical instrument makers. Addresses: 98 Princes Street, Edinburgh (1920s and early 1930s); 52 Regent Street, London, England; 137 Regent Street, London (1867-1882).

Instruments bearing their name include: 3-draw brass *Reconnoit'rer* telescope with mahogany barrel; 3-draw brass and leather *Reconnoit'rer* telescope. 5-draw brass and leather telescope; student microscope.

Salyut
Salyut was a camera brand of the **Arsenal Factory**, manufactured from the 1950s until 1980.

Sanders & Co.
See **Sanders & Crowhurst**.

Sanders & Crowhurst
Sanders & Crowhurst, [306], of 71 Shaftsbury Avenue, London, and 55 Western Road, Hove, Sussex, England,

was a camera manufacturer established in 1900. It was a partnership between Harold Armytage Sanders and Harry Arthur Crowhurst. The partnership was dissolved by mutual consent in 1908, [35 #28189 p7791], following which Crowhurst continued the Hove business alone, and Sanders continued the London business as Sanders & Co. In 1910 Sanders & Co. was acquired by **James A. Sinclair & Co. Ltd.**

Sanderson, F.H.
Frederic Herbert Sanderson, (1856-1929) [44] [61] [306], of Cambridge, England, was a cabinet maker, a wood and stone carver, and an architectural photographer. In 1895 he was granted a patent for a bellows camera with movable lens-board arrangement. This Sanderson camera was originally manufactured by **Holmes Brothers**, and was marketed by **George Houghton & Sons Ltd.**

Sands & Hunter
The company Sands & Hunter, [50] [61] [306], began in the mid-1870s as a partnership between John Hunter (1831-1880), his son John J. Hunter (b.1858), and his brother-in-law Charles Sands (1835-1908). The Hunters had already been producing microscope slides. They traded at 20 Cranbourne Street, London, England, as Hunter & Sands, selling optical and photographic equipment. They made some instruments themselves, including microscopes and cameras, and retailed instruments by other makers. During the 1880s they had a factory at 146 Holborn, London. In 1883 they changed the company name to Sands & Hunter, then in 1890 it was changed again to Sands, Hunter & Co. In the same year the partnership was dissolved and the company was sold to J.J. Foster, under whom it continued to trade with the same name. In 1905 the company moved to 37 Bedford Street, Strand, London. In 1915 they formed the limited company Sands, Hunter & Co. Ltd. They appear to have narrowed their business focus mainly to cameras and other photographic equipment.

San Giorgio
San Giorgio (SG) of Genoa, Italy, was one of the two main Italian optical instrument companies in the 1920s, the other being **Officine Galileo** (OG) of Florence [32]. One of their responsibilities was to produce optical rangefinders, optical traverse measuring equipment, site and target inclination measuring equipment and fire control centres.

Optical rangefinders for the Regia Marina (Italian Navy) were supplied by **Barr and Stroud** until after WW1, after which, rangefinders were supplied by OG and SG [26, p110].

Early on OG and SG reproduced the British and German models. In around 1928 they began to produce domestically designed and built instruments for the Regia Marina ships. They began with the coincidence type of rangefinder, and later also built stereoscopic types. By the end of the 1930s, the Italian rangefinders, produced by both SG and OG were considered high performance and, some of the larger ones, cutting-edge designs.

Sankei Koki Seisakujo Inc.
Japanese optical equipment manufacturer. Example is **Viper** 10x50 binoculars with **JB code** JB16. These binoculars were produced sometime during the 1950s to the early 1970s.

Sard
Sard, [136], was a trade-name used by the **Kollsman Instrument** division of the **Square D** company of the USA, for a range of binoculars. With the advent of WWII, the Kollsman Instrument division absorbed the **Emmerich Optical** division and began manufacturing binoculars under the Sard trade-name. After the war, they continued to make binoculars for the commercial market. In 1950 the Square D Kollsman Instrument division was sold, and production of binoculars ceased. Notable Sard binocular models include 6x42 and 7x50 models.

Sartorius, Florenz
Florenz Sartorius, (1846-1925) [374], served an apprenticeship as a precision mechanic, and started his own business, Feinmechanische Werkstatt F. Sartorius, at the age of 24, manufacturing precision balances in Göttingen, Germany. In 1906, he acquired the firms of **August Becker** and **Ludwig Tesdorph**, expanding the firm's products to include astronomical, surveying, and physical instruments. By the end of WWI, the company was focussing mainly on production of balances, incubators, and microtomes. The company has since expanded and diversified.

Sartory Instruments
Sartory Instruments Ltd., [320] [344], of 7 Steele Road, Chiswick, London, England, was a maker of microscopes and accessories. The company was founded in 1949 when Peter Karel Sartory (1908-1982) acquired the firm of R.S. Alldridge Ltd., formerly **Chapman & Alldridge**. They ceased manufacturing in 1963, and the company was dissolved in 1968.

Savery, Servington
Servington Savery, (c1670-c1744) [18] [24] [46] [276 #302], was a natural philosopher and inventor, of Shilston, near Modbury, Devon, England. He was born to

a wealthy family, and lived on the family estates for his whole career. He was related to the inventor Thomas Savery and, in 1691, he married Elizabeth Hale, a close relative of Lord Chief Justice Hale. He is primarily remembered for his work on magnetism. He devised a method of magnetising steel bars, which were marketed in London and Exeter by William Lovelace and other members of the Lovelace family.

Servington Savery invented a heliometer, in the form of a double image micrometer. He wrote a paper on his invention, and communicated it to the Royal Society in 1743. In 1753 **James Short (I)** arranged for Savery's paper on the subject to be published in the *Philosophical Transactions of the Royal Society of London*, accompanied by a letter from himself.

Sawyers

Sawyers, [338], of Portland, Oregon, USA, was founded in 1914 as a photo-finishing service. In 1926 Howard Graves, an ex-army photographer, bought a stake in the company. In 1938 he formed a partnership with William Gruber, a photographer and organ maker, and they began to produce the **Viewmaster**, a hand-held lenticular stereoscope. In c1952 Sawyers took over their competitor, **Tru-Vue**. In 1952 they also began to produce stereocameras and, soon after that, projectors. The company then began an expansion into Belgium, Australia, Japan, and France. In 1966 Sawyers was acquired by the General Aniline & Film Corporation, (GAF). From the early 1980s a series of acquisitions culminated with the company being sold to the Ideal Toy Company in 1984, then to Tyco Toys in 1989, and finally becoming part of the Mattel owned *Fisher-Price* brand.

Scarlett, Edward (I)

Edward Scarlett (I), (c1688-1743) [12] [18] [275 #473] [276 #115], was an optical instrument maker and spectacle maker of London, England. Addresses: *Archimedes and Globe*, Dean St., Near St. Anne's Church, Soho; *Archimedes and Globe*, Near St. Anne's Church, Soho (1705-1743); *Archimedes and Globe*, King St., Soho (1722); *Archimedes and Globe*, in Market St., nr. St. Anne's Church, Soho (1724).

He served his apprenticeship with **Christopher Cock** of Long Acre, London, and was freed of the **Spectacle Makers' Company** in 1705. He became Master of the Spectacle Makers' Company from 1720-1721. He is known to have made reflecting telescopes, and had a strong reputation for his mirrors. He also sold microscopes, camera obscuras, magic lanterns, and barometers. **Nathaniel Adams** and **Matthew Richardson** served their apprenticeships with him, as did his son, **Edward Scarlett (II)** who continued his business.

Scarlett, Edward (II)

Edward Scarlett (II), (fl.1724-1779) [12] [18] [276 #209], was an optician of London, England. Addresses: *The Spectacles*, 2nd House from Essex St., nr. Temple Bar; Macclesfield (or Maxfield) Street (1749); Near St. Anne's Church, Soho (1763).

He was the son of **Edward Scarlett (I)**, and succeeded his father in business. He began his apprenticeship with his father in 1716/17, and was freed of the **Spectacle Makers' Company** in 1724. Instruments dating from between 1724 and his father's death in 1743 are hard to attribute, since the markings are similar to those of his father. He became Master of the Spectacle Makers' Company in 1745. At some point **Francis Watkins (I)** worked for him. **John Davis (III)** served an apprenticeship with him, and was freed of the Spectacle Makers' Company by purchase in 1777.

Scatliff, Daniel (I)

Daniel Scatliff (I), (fl.1760-1767) [18], was a mathematical instrument maker of Wapping, London, England. In 1760 he was jointly awarded a patent for a quadrant with **John Dollond (I)**. He was declared bankrupt in 1777 [35 #11782 p4].

Scatliff, Daniel (II)

Daniel Scatliff (II), (fl.1831-1835) [18], was a mathematical instrument maker of 4 Wapping Wall, Shadwell, London, England. He served his apprenticeship with the mathematical instrument maker, David Bale, of the **Cordwainers' Company**, beginning in 1804. He succeeded his father, the mathematical instrument maker, ship chandler, and compass maker, Simon Scatliff (I), in business.

Scatliff, Daniel (III)

Daniel Scatliff (III), (fl.1796-1830) [276 #1998] [322] [379], was a mathematical, nautical, and philosophical instrument maker of 6 Wapping Wall, Shadwell, London, England. He made a Hadley's quadrant used by the explorer of Canada, Samuel Hearne. He may be the same person as **Daniel Scatliff (II)**, having moved from 6 Wapping Wall to number 4 in 1830/1831.

Scatliff, John

John Scatliff, (fl.1765) [18] [274] [322], was an optical instrument maker of London, England, and a son of **Samuel Scatliff**. He was freed of the **Spectacle Makers' Company** in 1765 by patrimony, and succeeded his father in business.

Scatliff, Samuel

Samuel Scatliff, (fl.1737-1764) [18] [274] [276 #303] [319] [322], of *Friar Bacon's Head*, Corner of St. Michael's Alley, Cornhill, London, England, was an optician, and optical and mathematical instrument maker, and 'optik-glass maker'. He served his apprenticeship with **James Mann (II)**, beginning in 1725, and was freed of the **Spectacle Makers' Company** in 1734. He was the son of mathematical instrument maker and ship chandler, Simon Scatliff (II), and he succeeded his uncle, the mathematical instrument maker James Scatliff (I), in business. He served as Master of the Spectacle Makers' Company in 1751. In 1764, then trading at St. Paul's Churchyard, he was one of the petitioners who attempted to revoke the patent obtained by **John Dollond (I)** for the achromatic lens, and which was enforced after Dollond's death by his son, Peter (See Appendix X). His son, James Scatliff (II), served an apprenticeship with him, beginning in 1749. Another son, **John Scatliff**, succeeded Samuel in business.

According to his trade card [361]: "Makes and sells wholesale and retail ye finest chrystal spectacles & reading glasses ground to the utmost perfection on brass tools in ye method approv'd by the Royal Society. He likewise makes telescopes and perspective glasses for sea or land of all lengths & compos'd with the nicest accuracy & exactness. As also all sorts of microscopes, single or double, which amazingly magnifie ye minutest objects, discovering the circulation of the blood in animals, & many other surprising phenomena."

Scheffel, Mark

Mark Scheffel, [94], was a UK brand of Japanese made binoculars. The company is believed to have operated from c1968 to c1980 in Essex, England, and may have changed their name to Scientific & Technical Ltd. in 1978/9. Examples of binoculars include: 10x50, 20x50, 35x50, and 10-30x50 zoom.

Schenk, Ulrich

Ulrich Schenk, (1786-1845) [283], was a Swiss mathematical, philosophical and optical instrument maker. Early in his career he worked with **Georg Friedrich von Reichenbach** in Munich, Germany, and subsequently set up his own workshop in Berne, Switzerland, trading as Ulrich Schenk et Cie, méchaniciens. He developed and built his own circular dividing engine. His catalogue of 1815 included Borda repeating circles, various models of repeating theodolite, glasses, compasses, barometers, and other instruments. Beginning in 1817 he established a pump making factory in Berne, which largely, if not completely, diverted his attention from instrument making.

Schieck, Friedrich Wilhelm

Friedrich Wilhelm Schieck (sometimes spelled Schiek), (1790-1870) [320], of 14 Halle'sche Strasse, Berlin, Germany, was a microscope maker who began work as a journeyman for **Carl Philipp Heinrich Pistor** in 1824. He rose to become a partner in 1830, trading as Pistor and Schieck until 1836 when Schieck left to start his own workshops. He was succeeded in business by his son, F.W.H. Schiek (1843-1916), who continued to operate the company until at least 1900.

Schiefspiegler

The Schiefspiegler is a kind of tilted component telescope. It is based on a Cassegrain, but with a tilted primary and secondary mirror to avoid the need for a hole in the primary mirror.

Schlegel, L.

L. Schlegel, (fl.1875) [374], was a German maker of microscopes and spectroscopes. He exhibited his work in Dresden in 1875.

Schleiffelder & Cie, Otto

Otto Schleiffelder & Cie Optiker, of Stadt am Graben 22, Wien (Vienna), Austria, were retailers and distributors of optical, physical, mathematical, and meteorological instruments, established in 1889. They retailed instruments by manufacturers of the time. They also sold instruments under their own name, which would have been supplied to them wholesale by contemporary makers. By 1901 they were offering facilities for recording speech and music, and even had a piano player available on request. A 1906 advertisement offered the "best and brightest" crystal glasses with a free professional eye examination. Instruments bearing their name include: opera glasses; field glasses; telescopes; opticians' lens set.

Schleussner Fotowerke GmbH, Dr. C.

Dr. C. Schleussner (1830-1899), was a chemist and photographic film and camera manufacturer of Frankfurt, Germany. In 1860, he established a photochemical company in Frankfurt, named Dr. C. Schleussner Fotowerke GmbH. He soon began to produce photographic plates, and began to use *ADOX* as a brand-name. The firm was an early adopter of x-ray technology, and went on to become a major producer of x-ray plates. Dr. Schleussner was succeeded in management of the company by his sons Carl Moritz Schleussner (1858-1943) and Carl Friedrich Ludwig Schleussner (1864-1928), and his son-in-law Eduard Ritsert (1859-1946).

In 1938 Dr. C. Schleussner Fotowerke GmbH assumed control of the **Wirgin** factory in Wiesbaden. Based on the acquired knowledge and technology, they began camera production in 1939 at the former Wirgin factory, naming the firm **ADOX Kamerawerk GmbH**, and using the *ADOX* brand-name. During WWII, Dr. C. Schleussner Fotowerke GmbH marked their products with the **ordnance code** mca. In 1948, the Wirgin factory was sold back to Henry Wirgin, who reinstated his business. With this sale, the ADOX camera works were relocated to another factory in Wiesbaden. Cameras marketed in the USA were sold under the *Schleussner* brand-name. The firm continued in production until the factory was sold in 1962.

Schmalcalder, Charles Augustus

Charles Augustus Schmalcalder, (c1786-1843 fl.1806-1840) [16] [18] [53] [276 #1215] [319] [322], was an optical, mathematical, and philosophical instrument maker of 82 Strand, London, England and, from 1827, 399 Strand. He was born in Stuttgart, Germany, and by 1804 he had moved to England. He was the inventor of the prismatic compass, which he patented in 1812. He marked his instruments "London" or "82 Strand, London".

Instruments bearing his name include: 3-draw brass telescope with mahogany barrel; boxed microscope after the style of **W.&S. Jones**; Gould type microscope; pocket box sextant; patent portable theodolite; reflecting circle; double-frame brass sextant; box sextant; prismatic compasses; surveyor's compass; patent prismatic clinometer; stick barometers; wheel barometer; case of silver mathematical drawing instruments; plotting protractor; pantograph; miniature terrestrial globe.

He suffered paralysis following an accident, and died in 1843 in a workhouse in St. Martin in the Fields. He was buried in Camden Town. His sons, **John Thomas Schmalcalder** and George Schmalcalder served their apprenticeships with **Thomas Gilbert**. Charles was succeeded in business by his son, John Thomas Schmalcalder.

Schmalcalder, John Thomas

John Thomas Schmalcalder, (fl.1841-1845) [18] [379], was an optician, and an optical, mathematical, and philosophical instrument maker of 2 Fairfax's Court, and 400 Strand, London, England. He served his apprenticeship with **Thomas Gilbert** of the **Grocers' Company**, beginning in 1829. He succeeded his father, **Charles Augustus Schmalcalder**, in business.

Schmidt & Haensch, Franz

Franz Schmidt & Haensch, [53] [113] [374], was founded in Berlin, Germany in 1864 as a manufacturer of optical, physical, and chemical instruments. The founders were the mechanic, Franz Schmidt, and the mechanic and optician, Hermann Haensch, both of whom had trained under Wilhelm Langhoff. By 1871 the company had 32 employees. In 1881 they were commissioned by **Albert Michaelson** to make the interferometer he famously used to study the speed of light.

Instruments bearing their name include: heliostat; microscope; direct vision spectroscope; spectrometer; interferometer; quartz wedge polarimeter; circle polarimeter; refractometer; projector; colour mixing apparatus; drafting tables.

During WWII they marked their instruments with the **ordnance code** crh. In 1969 part of their production was moved to Cologne. Today the company produces refractometers, polarimeters, density meters, and colour measuring devices, both as off-the-shelf instrumentation, or built to custom requirements.

Schmidt Camera

The Schmidt camera, designed by **Bernhardt Schmidt**, is a telescope that comprises a spherical primary mirror and a *Schmidt corrector lens* to correct for spherical aberration. The Schmidt corrector lens is flat on one side with a small aspheric curve on the inside. The arrangement suffers from field curvature, and to correct this Schmidt proposed that a thin plano-convex lens be placed in front of the photographic plate (or sensor), with its convex surface facing the mirror. Alternatively, and more commonly in modern instruments, the secondary mirror is figured to act as a field flattener. Schmidt's first camera was an f/1.7 system with a corrector lens of 36cm aperture, and a mirror of 44cm diameter [24]. This would be referred to as f/1.7 36cm x 44cm. The Schmidt camera design is a variant of the *Cassegrain* telescope, and is generally referred to as a *Schmidt-Cassegrain telescope*, or SCT.

Schmidt, Bernhardt

Bernhardt Schmidt, (1879-1935) [24] [150], was of Finnish descent, and born in Naissar, Estonia, at that time part of the Russian Empire. As a child he lost his right hand while experimenting with gunpowder. He developed an interest in amateur astronomy, and subsequently attended the Technical Institute in Mittweida, Germany to study mathematics, drafting, design, and working with precision tools. During his studies he also took on jobs in astronomical observatories where he learned to figure the optical surfaces of lenses and mirrors.

He went on to offer his services to various observatories, offering to make paraboloidal mirrors at least as good as the same sized refractor. He quickly established a strong reputation in this field, receiving testimonials from astrophysicists H.C. Vogel and Karl Schwarzchild.

During WWI he was interned in Germany as an enemy alien, due to his Russian nationality. This began a downturn in his optical business, and in 1923 he closed his workshop. His reputation as an optical technician, however, was not forgotten, and he was offered a job by Richard Schorr, director of Hamburg Observatory. There Schmidt had a workshop in which he did work for the observatory and private work for his own clients.

In 1929 Schmidt devised his eponymous Schmidt camera design for which he is famous.

Schmidt-Cassegrain Telescope
See Schmidt camera.

Schmidt-Pechan Prism
See roof prism.

Schneider Dresden
Schneider Dresden is a **Pentacon** brand of photographic equipment.

Schneider Kreuznach
Schneider Kreuznach is a German developer and manufacturer of high-performance lenses, industrial optical equipment, and precision engineering. They produce lenses and filters for both the home photographic market and the industrial market. They also produce cine and projection equipment for commercial and home users. The company is based in Bad Kreuznach in Rhineland-Palatinate, Germany.

Schneider Creusot
Schneider Creusot, [27], was a French iron and steel mill which became a major arms manufacturer. The firm was founded by Eugène Schneider in the town of Le Creusot in the Bourgogne region in eastern France. After WWII, Schneider-Creusot evolved into Schneider Electric.

In 1914 in Petrograd the Russian Company for Optical and Mechanical Production was founded, with majority ownership in the hands of Schneider-Creusot. Their production of optical glass was organised at the **Petrograd porcelain works**.

Schott AG; Schott & Associates
Schott AG [21] is a glass technology company. The Schott & Associates Glass Technology Laboratory in Jena, Germany was founded by **Otto Schott**, **Ernst Abbe** and Carl and Roderich **Zeiss** in 1884. In 1930 they acquired an 80% interest in **Deutsche Spiegelglas AG**, and in 2002 they took 100% ownership. The company became Schott AG in 2004. The Carl-Zeiss-Stiftung (Carl Zeiss Foundation), located in Heidenheim an der Brenz and Jena, Germany, is the sole shareholder of the two companies **Carl Zeiss AG** and Schott AG. The Carl-Zeiss-Stiftung was founded by **Ernst Abbe** and named after his business partner Carl Zeiss.

Schott, Otto
Otto Schott, (1851-1935), was the son of a glassmaker in Witten, Westphalia, Germany. He studied chemistry, mineralogy and physics and, in 1879, he began his own research into the melting, glass forming, and crystallization behaviour of different chemical compounds. He collaborated with both **Ernst Abbe** and **Carl Zeiss**, and to further this relationship he moved to Jena in 1882. In 1884 the three of them together formed the Schott & Associates Glass Technology Laboratory, the company that ultimately became **Schott AG**. Schott used his scientific knowledge to design glasses with defined properties for specific specialised purposes.

Schrauer, Leopold
Leopold Schrauer, (fl.late-1850s to late-1880s) [113] [201], was a microscope maker who began his business in the late-1850s in Boston, Massachusetts, USA. By 1877 his advertisements gave his address as 50 Chatham Street, New York, NY, and by 1879 his address was 42 Nassau Street. In 1879 he advertised his *New Universal Microscope*, with objectives by **William Wales**. By 1881 Schrauer's business address was 228 East 34th Street.

Schröder, Heinrich Ludwig Hugo
Heinrich Ludwig Hugo Schröder, (1834-1902) [12] [50], was an optical instrument maker born in Hamburg, Germany. He learned mechanics and optics from the instrument maker **Moritz Meyerstein** and the mathematician J.B. Listing. He began his own business in Hamburg, making primarily microscopes and telescopes. **Rudolf Fuess** worked for him as a journeyman, making astronomical and scientific instruments, before leaving to start his own business in Hamburg. **Johann Diedrich Möller** came to Schröder during his apprenticeship as an artist, and Schröder taught him the art of lens-making, and employed him for a short while after his apprenticeship prior to Möller starting his own business in Wedel.

After a failed attempt to raise funds for a public observatory for the city of Hamburg, Schröder moved his business to Oberursel, near Frankfurt am Main, but it closed in 1882. He moved to England to become technical director of **Ross & Co.**, and Schröder and his family naturalised as British citizens in 1888.

Schupmann, Ludwig

Ludwig Schupmann, (1851-1920), was a German architect and optical designer renowned for his Medial and Brachymedial telescopes. The asteroid *5779 Schupmann* is named after him.

Schwaiger, A.

A. Schwaiger, (fl.1854) [374], was a spectacle maker and telescope maker of Augsberg, Germany. He exhibited his work in Berlin in 1854.

Schwartz dit Lenoir

Schwartz dit Lenoir, [322] [360], nautical and mathematical instrument makers of Paris, France, was a partnership between **E. Lorieux** and Schwartz, formed in 1845. The name, which means "Schwartz known as Lenoir", appears itself to be a translation, the words schwartz and noir both meaning black in German and French respectively. Instruments bearing their name include reflecting circles; a quintant; octants; sextants. The company was succeeded by **Baudry et Lorieux** in 1855.

Science of Cambridge Ltd.

In December 1977 Sinclair Instruments Ltd., the company of entrepreneur Clive Sinclair, was renamed Science of Cambridge Ltd., [315]. The Sinclair company underwent further name changes to become Sinclair Research Ltd., but Science of Cambridge Ltd. continued as a company – although it is not clear what interest Clive Sinclair retained in it.

In 1989 Richard Dickinson, who had previously been Head of Design for Sinclair, worked together with Keith Dunning to produce the *Lensman*, a folded optics microscope with a novel three-dimensional folded optics design by Eddie Judd. The *Lensman* was manufactured in the UK by Science of Cambridge Ltd. In 2007 the company name was changed to **Cambridge Optronics Ltd.**

The *Readview*, *Micron*, and *Trekker* microscopes, derivatives of the *Lensman*, were produced by **Enhelion**, another of Dickinson and Dunning's companies.

Scientific Instrument Manufacturers' Association of Great Britain (SIMA)

The Scientific Instrument Manufacturers' Association of Great Britain, (SIMA), of Buckingham House, Buckingham Street, Adelphi, London, and River Plate House, 12-13 South Place, London, was the representative body of the manufacturers of scientific instruments and apparatus in Great Britain. They exhibited at the 1947 British Industries Fair, [450].

The historical relationship between SIMA and the **British Optical Instrument Manufacturers' Association** (BOIMA) is unclear. According to [331], SIMA was formed as the successor to BOIMA in 1953. However, their entry in the British Industries Fair predates this, as does the SIMA Bulletin of September 1948, [330], which lists 95 SIMA member companies, including most of those listed as BOIMA members in 1921. It also lists the past presidents of SIMA dating back to 1915/1916. These include **Conrad Beck** (1915-1922), **F. Twyman** (1925-1926), and **W.E. Watson-Baker** (1930-1932).

An unconfirmed source, [332], states that SIMA was an alternative name for the BOIMA. The SIMA was listed in the Associations and Societies section of the 1951 *Directory for the British Glass Industry*, [327].

According to [331], SIMA merged in 1981 with British Industrial Measuring & Control Apparatus Manufacturers (BIMCAM), and the Control and Automation Manufacturers' Association (CAMA), to form GAMBICA.

Scientific Optics Ltd.

Scientific Optics Ltd., of Ponswood Industrial Estate, Drury Lane, Hastings, East Sussex, England, were optical instrument makers who manufactured the McArthur microscope under license for the Open University during the 1970s. The OU McArthur microscope is a low-cost instrument made in abs plastic. Three versions of the OU microscope were produced: the basic model with a x10 eyepiece and two objectives of power x8 and x20; a model with three objectives yielding magnifications of x80, x200 and x400; and a polarizing version with a revolving stage. See McArthur microscope.

Scioptic Ball

A scioptic ball, or sky-optic ball, is a wooden sphere or globe with a hole through it containing a lens. The ball is mounted in a socket so that the lens may be turned to face in any direction. This arrangement was sometimes used in magic lanterns, aerial telescopes, and other instruments.

Scott, Benjamin

Benjamin Scott, (fr.1712, d.1751) [3] [18] [90] [276 #116], was a mathematical instrument maker of London, England. He began his apprenticeship in 1702 with James Anderton of the **Grocers' Company**, but was turned over to **John Rowley** of the **Broderers' Company** in 1706. He was freed of the Grocers' Company in 1712, and traded at the *Mariner & Globe*, Exeter Exchange, Strand. One of his apprentices was **Thomas Heath (I)**. In 1747 he emigrated to St. Petersburg in Russia where he continued to make instruments, and trained Russian craftsmen at the St. Petersburg Academy of Sciences.

Screw-Barrel Microscope
See microscope.

Seagull
See **Shanghai Camera Factory**.

Secrétan, Marc François Louis
Marc François Louis Secrétan, (d.1868) [12] [24] [70] [71], of Paris, France, was a professor of optics at Lausanne University, and an instrument maker. He had a reputation for constructing excellent mathematical, optical and philosophical instruments. In 1845 he formed an association with **Noël Marie Paymal Lerebours**, forming the company **Lerebours et Secretan**. His sons, August and George continued the business after his death. (Note that the é in Secrétan's name is unaccented on advertisements and cards).

Search, James
James Search, (fl.1771-1781) [18] [274] [276 #888] [319] [322], was a mathematical and optical instrument maker of Crown Court, Westminster, London, England. The address was also variously known as Crown Court, Pultney St., Golden Square, London (usually spelled Pulteney St.), and Crown Court, Soho, London.

He served his apprenticeship with **John Bennett (I)** and, upon Bennett's death in 1770, was turned over to the tin-plate worker Peter Balchin to complete the apprenticeship. He was freed of the **Stationers' Company** in 1771, and succeeded John Bennett (I) in business. He is known to have sold: circumferentor; drawing instruments; telescope. Instruments bearing his name include: circumferentor; cased simple theodolite; pair of proportional compasses; stick barometer; magnetic compass; parallel rules. Among his apprentices was **Alexander Wellington**, who succeeded him in business.

Sears
Sears is a US retailer with a history of selling through stores and mail-order. The company was founded in 1886, and moved to Chicago in 1887 where they were headquartered in the famous Sears Tower. They sold cameras, telescopes, and binoculars under their *Sears* brand-name as well as under their *Tower* brand-name. They sold a large number of models, which were re-branded, having been manufactured by a large number of manufacturers globally. They no longer use these brand-names for cameras, telescopes or binoculars.

Seeadler Optik
Seeadler Optik is a German company, founded in 1928, selling binoculars, riflescopes and spotting scopes. The German word Seeadler translates as Sea Eagle. Their brands include *Eagle Optics* and *Jägermeister* (which translates as master hunter).

Seibert & Krafft
See **Wilhelm and Heinrich Siebert**.

Seibert, Wilhelm and Heinrich
Wilhelm and Heinrich Seibert were brothers who worked for **Kellner** in the 1850s, along with **Ernst Gundlach**. In 1859 Gundlach left to start his own company making microscopes, and the Seibert brothers came with him as employees. In 1872, with the help of businessman Georg Krafft, the Seiberts bought Gundlach's company, following which Gundlach emigrated to the USA. The Seibert brothers relocated to Wetzlar, and the company was initially called Seibert and Krafft. **Otto Himmler**, who had also worked for Gundlach, worked for Wilhelm and Heinrich Seibert until leaving to start his own company in 1877. In about 1884 the company was renamed W.&H. Seibert, and continued in business until about 1925.

Seiko Scope
Seiko Scope was a trade mark of **Nihon Seiko Kenkyusho, Ltd.**

Seiwa Optical Company Ltd.
Seiwa Optical Company Ltd., of Tokyo, Japan, is a manufacturer of microscopes, cameras, and lenses. The company was established after WWII. From the late-1950s until the 1980s their products were marked with the **JB code** JB191.

Self-Levelling Level
Self-levelling level is another name for an automatic level.

Sellew, Roland W.
Roland W. Sellew, [185] [213], was a consulting engineer in Connecticut, USA, who designed telescopes, observatory domes, telescope mounts and other observatory equipment. He did not build equipment to his own designs, instead having them built for him, and acknowledging the builder. He entered a collaboration with **Robert Lundin**, a lens and mirror maker initially with **Alvan Clark and Sons**. Their first works together were produced during Lundin's tenure at Alvan Clark, but Lundin left Alvan Clark to form his own company. Sellew and Lundin continued to work together until Lundin's company failed in 1933 due to the depression. Sellew's work received both positive and negative criticism.

Selligue
Alexandre François Gilles, (1784-1845) [50], was generally known by the name Selligue (derived from his

surname spelled backwards) for his professional work. He was a French engineer, famed for being the first to extract oil from shale, and generating gas for use in lighting. He was innovative in the area of optics for microscopes and is credited with making instruments himself as well as having his inventions constructed by others. He devised a method of combining a series of low powered achromatic lenses to create a high-powered objective for a microscope. The microscope was made in 1824 by **Vincent and Charles Chevalier** and, after adjustment by the Chevaliers for spherical aberration, proved successful up to magnifications of 200x. The results of this work were presented to the French Academy of Sciences without any credit to the Chevaliers. This resulted in the termination of the working relationship between Selligue and the Chevaliers, who went on to market their own version of it. Selligue later worked with **J.G.A. Chevallier** who went on to produce microscopes to Selligue's design.

Selo Ltd.
See **Ilford**.

Selsi
Selsi was a brand of telescopes and binoculars from the 1920s to the end of the 1980s. From the 1920s until the 1940s Selsi imported and retailed French telescopes in the USA, made by **E. Vion**. Some Selsi telescopes dating from the late-1950s to the 1980s are Japanese made, and bear the **Towa** maker's mark (see Appendix IV).

Selva, Domenico (I)
Domenico Selva, (d.1758) [74], moved to Venice, Italy, from Maniago in 1696, and began to trade there as an optician. He had two sons, **Lorenzo Selva** and Giovanni Selva, both of whom continued the business. The Selva family were considered to be the finest optical instrument makers of C18th Venice.

Selva, Domenico (II)
Domenico Selva (II), [74], was an instrument maker of Venice, Italy. Domenico (II) was the son of **Lorenzo Selva**, and continued his father's business.

Selva, Guiseppe
Guiseppe Selva, [74], was an instrument maker of Venice, Italy. Guiseppe was the son of **Lorenzo Selva**, and continued his father's business.

Selva, Lorenzo
Lorenzo Selva, (1716-1800) [74], was an instrument maker of Venice, Italy. He was the son of **Dominico Selva (I)**, and continued his father's business. He signed all his instruments with his father's name, *Domenico*, making it impossible to distinguish his work from his father's prior to his father's death in 1758. The Selva family were considered to be the finest optical instrument makers of C18th Venice, and Lorenzo is reputed to have sold his instruments in the Orient, Spain and Portugal. Lorenzo claimed that he mostly manufactured spectacles and refracting telescopes, but is also known to have made Newtonian and Gregorian reflecting telescopes, microscopes, camera obscuras, magic lanterns, optical toys, and other optical instruments. He had seven children, at least two of whom, **Giuseppe Selva** and **Domenico Selva (II)** continued his business.

Semitecolo, Leonardo
Leonardo Semitecolo, (d.1869) [12] [75] [388], owned an optical firm near the Parish of San Zaccaria, Fondamenta dell'Osmarin n. 4100 in Venice, Italy. His son is known to have continued his business, and may also have been named Leonardo. Instruments attributed to Leonardo Semitecolo include low-cost spectacles with horn or white metal frames, and low-cost telescopes fabricated with a cardboard tube and horn or brass mount. Such telescopes, appearing in collections or on auction sites, appear to date from the late-C18th and C19th. The signatures most frequently seen on telescope tubes are *SEMITECOLO, LEONARDO SEMITECOLO, DA SEMITECOLO*, and sometimes followed by *VENEZIA*. The lacquered cardboard tubes of Semitecolo telescopes were often in bright colours such as red, yellow or orange, with tooled ornamentation.

Several other Venetian telescope makers who were contemporaries of Semitecolo, or near-contemporaries, made telescopes that bore a resemblance to Semiteloco's work in that they were fabricated with a cardboard tube with horn, bone or brass mount. These makers include: **Angelo Deregini, Angelo Olivo, Fratelli Dolci**.

Sewill, Joseph
Joseph Sewill, (b.1800 fl.1837-1900) [18] [71] [322], was an optician, and optical, mathematical, and nautical instrument maker of Liverpool, England. Addresses: 31 South Castle St. (1837); 8 Duncan St. (1839); 61 South Castle St. (1841-1895); 54 Canning Place and 5 Parr St. (1857); 15 Canning Place (1872-1937); 14 & 16 Canning Place (1887-1895); 15 & 16 Canning Place (1898-1905).

He is referred to in [3] as an itinerant, or travelling optician; one of many firms which used numerous temporary addresses to sell their wares and reach a wider clientele. To this end, additional branches of his business were run by his sons: Frank, who traded at 126 Broomiclaw and 2 York St., Glasgow, and John, who traded in c1875 and later at 30 Cornhill, Royal Exchange, London.

Joseph Sewill was appointed "Maker to the Admiralty". Instruments bearing his name include: sextant; octant; single-draw brass and leather day and night telescope; brass and leather telescope on wooden tripod; barometers; stick barometer; aneroid barometers; chronometers.

Sextant

A sextant is an instrument used to measure angular distances between objects. It consists of a graduated arc of one sixth of a circle, and may be used to measure angles up to 60° or, by means of double reflection, up to 120°. Various kinds of sextant exist for use in astronomy, navigation, surveying, etc.

Marine sextant: A hand-held double reflecting sextant designed for use in navigation at sea. As stated in [**5** chXVII p141]: "The principle on which the sextant is built is that if a ray of light is reflected twice in the same plane by two plane mirrors, the angle between the first and last directions of the ray is twice the angle between the mirrors. For this reason the arc of the sextant is graduated up to 120°." The double reflection is achieved using a part-silvered plane mirror *horizon glass*, fixed to the sextant frame, and a silvered plane mirror *index glass*, attached to the *index arm*, which is pivoted at the geometrical centre of the graduated arc, and used to make the angular measurement. Shades are used to enable viewing into the sun, and to reduce unwanted reflections from external sources. Using the sighting scope, the horizon is viewed through the un-silvered part of the horizon glass. The index arm is moved until the celestial object is reflected by the index glass, onto the silvered part of the horizon glass, and into the sighting scope. Once the index arm is adjusted to bring the celestial object into exact correspondence with the horizon, the position of the index arm on the graduated arc is read to yield the angular reading. The sextant can be used in a similar manner to measure the angle between any two objects separated by up to 120°. For a description of how to use a marine sextant, see [**5** chXVII]. The physical and optical principles of the marine sextant are explained in [**6** p247]. The first marine sextant was introduced by **John Bird** and Captain John Campbell in 1757.

Box Sextant: A compact pocket sextant enclosed in a flat tubular box. The box sextant was first introduced in 1797 by **William Jones** [**175**].

Double Sextant: A sextant used by hydrographers for survey work, which enables the simultaneous measurement of two angles between three objects. It has two sets each of index glasses, scales, verniers, and index arms. The double sextant was first introduced in c1784.

Stellar Sextant: A sextant designed for observation of stars. The sighting telescope has a large objective diameter for additional light-gathering.

Bubble Sextant: A sextant designed for aerial navigation that uses a bubble to establish the horizontal level. The bubble sextant was first introduced in 1939.

For a concise history of the sextant see [**236** pp57-60]. For a detailed history, with photographs of examples, see [**16**].

Sextant Rangefinder

A sextant rangefinder is an optical rangefinder which uses double reflection, with an *index glass* (a plane mirror), and a *horizon glass* (also a plane mirror), in the same manner as a marine sextant. It is used to find the range to an object of known height, such as, for example, the next ship in a convoy. The user views the object through a sighting scope. This gives a split view, and an adjustment allows the direct view of the object, and the double-reflected view, to be brought into coincidence, the bottom of one with the top of the other. The angular size of the object is twice the angle between the two mirrors. Thus, given the known height of the object, the distance may be calculated or read from a scale.

See **Husun Marine Distance Meter**, **Langley Inclinometer**, **Waymouth-Cooke Rangefinder**, **Waymouth-Ross Sextant Rangefinder**.

Shadbolt, G.

G. Shadbolt, (fl.1851) [**319**] [**328**], was an optical instrument maker of 2 Lime Street Square, London, England. At the 1851 Great Exhibition in London, [**328**], he exhibited a "Sphaero-annular condenser, for condensing light in a peculiar manner, on transparent objects while under examination by the microscope".

Shanghai Camera Factory

The Shanghai Camera Factory was founded in 1958 in Shanghai, China, with sister factories being opened as the business grew. They produced *Shanghai* branded cameras until 1967, when the brand-name was changed to *Seagull*. In 1978 three of the factories merged into a single factory in the Songjiang district of Shanghai. The company continued to grow, and increase its product range, and went on to operate as Shanghai Seagull Digital Camera Co.

Shanghai Seagull Digital Camera Co.

See **Shanghai Camera Factory**.

Sharp, Abraham

Abraham Sharp, (c1653-1742) [**46**] [**275** #361] [**276** #118] [**322**] [**358**] [**363**] [**396**], was a mathematician and astronomer, and an optical and mathematical instrument maker, of Little Horton, near Bradford, Yorkshire, England. From 1684 to 1685, and 1688 to 1690 he was assistant to John Flamsteed, the first Astronomer Royal, at the Royal Observatory in Greenwich. He returned to Little Horton in 1694, where he pursued his interests in mathematics, astronomy, and instrument making. He is reputed to have been prolific in his instrument-making, producing telescopes, quadrants, sextants, sundials, armillary spheres, micrometers, lathes, and watch-making tools. He ground and figured telescope lenses for his own instruments and for others. He also produced books of mathematical tables, and other works on mathematics, and contributed astronomical calculations and data for publication in Flamsteed's work, both before Flamsteed's death, and subsequently in his name. Surviving instruments by Sharp are in the Bolling Hall Museum in Bradford, the National Maritime Museum, Greenwich, and the Science Museum, London.

Sharp, Benson

Benson Sharp, (fl.1901-1929) [53], of 39 Levens Grove, Blackpool, England, is the name on the maker's label of a camera of wooden construction. By 1929 the maker's address was 231 Church Street, Blackpool. The maker's plate reads "The Essanbee automatic ferrotype camera. Pat. Nº 164,422. Benson Sharp 39, Levens Grove Blackpool, Eng." It also has a diamond shaped logo with the inscription "S&B". The lens patent number 164,422 was awarded in 1901. A camera of this description, dated 1901-1911, is in the Science Museum, London, and one, dated 1901, was sold by Flints auction house in 2018.

There is a gravestone in the Blackpool Jewish Cemetery for Benson Sharp, b.1870, d.1934. This may be the maker of the Essanbee cameras.

Sharp, Samuel

Samuel Sharp, (fl.1851) [3] [18] [319], was an optical, mathematical, and philosophical instrument maker of New George St., Sheffield, England. He exhibited "Set of ten lenses for a single microscope, from 1-10th to 1-100th of an inch focal length." at the Great Exhibition, London in 1851 [328].

Sharp, Samuel Charles

Samuel Charles Sharp, (fl.1825-1848) [3] [18] [276 #1699] [322], was an optical, mathematical, and philosophical instrument maker of 29 Brownlow St., Long Acre, London, England (1827), and 16 Church Row, Pancras Rd., Somers Town (1839-1846).

Shephard, Felix

Felix Shephard, (fl.1829-1834) [18], was a spectacle maker, and optical, mathematical, and philosophical instrument maker of London, England. Addresses: 1 Great Sutton St., Clerkenwell (1829), and 40 Wellington St., Goswell St. (1834). He may have served his apprenticeship with **John Ustonson**. He was succeeded in business by the spectacle maker Francis Shephard, (fl.1835-1837).

Sherwood and Co.

Sherwood & Co., [20] [29], were optical instrument makers of Verulam Street, Grays Inn Road, London, England. The company was set up in 1915, and produced about 50 pairs of military binoculars a week during WWI. After WWI, the company was renamed to become the **Endacott Scientific Instrument Co.**

Shew & Co., J.F.

J.F. Shew, [61] [122] [306] [351], was a retailer and manufacturer of cameras, of London, England, Addresses: 32 Rathbone Place (1851-1857); 30 Oxford St. (1857-1863); 89 Newman St., Oxford St. (1863-1873, 1877-1882); 132 Wardour St. (1881-1885); 88 Newman St., Oxford St. (1881-1919); 87 Newman St., Oxford St. (1890-1899); 21 Bartlett's Buildings., Holborn Circus (1919-c1922).

The firm was established in 1849 by James Fludger Shew (1810-1873) as a retailer of photographic supplies and equipment. He was declared bankrupt in 1869, [35 #23522 p4351].

Following the death of James Shew in 1873, his son, James John Shew (1840-1922), started the firm of J.J. Shew. Addresses: 89 Newman St., Oxford St. (1873-1877); 28 Wardour St. (c1878); 132 Wardour St. (1878-1882); 132 Camberwell Rd. (1882). J.J. Shew were listed as photographic apparatus manufacturers, and are known to have marketed some models of camera. However, J.J. Shew were later listed as picture frame makers.

By 1877, J.J. Shew had moved to a different address, and the newly incorporated firm of J.F. Shew & Co. began to trade again at the 89 Newman Street address. By 1879 they were also manufacturers of cameras. They marketed a number of models of camera bearing their name, some of which they manufactured themselves. In 1915 they were taken over by **A.E. Staley & Co.**, and the firm continued as **Staley, Shew & Co.** until 1919, after which the name was once again changed to J.F. Shew & Co. In 1920, the firm incorporated as J.F. Shew & Co. Ltd., and they remained in business until at least 1922, the year J.J. Shew died.

Shew, J.J.
See **J.F. Shew & Co.**

Shimadzu
In 1875 Genzo Shimadzu Sr. began to manufacture educational physical and chemistry instruments in Kiyamachi-Nijo, Kyoto, Japan. In 1915 the company began to manufacture optical measuring systems, and they went on to offer diverse products ranging from optical imaging, medical, x-ray, optical and laser systems to measuring systems and vacuum, hydraulic and aviation technology.

Shipwrights' Company
See Appendix IX: The London Guilds.

Short & Mason
Short & Mason, [44] [53] [319] [379], were aeronautical, meteorological, surveying, and scientific instrument makers of London, England. Addresses: 62 Hatton Garden (1873-1875); 40 Hatton Garden (1876-at least 1900); Aneroid Works, Macdonald Road, Walthamstow (1910-1929); The Instrument House, Walthamstow (1937-1945); Aneroid Works, 280 Wood St., Walthamstow (1957).

The firm began as a partnership between **Thomas Watling Short** and the engineer, William James Mason, as aneroid barometer makers, and makers of other scientific instruments. The partnership was formed in 1864, and dissolved in 1900 [35 #27328 p4351]. At about this time, they began to work in partnership with Taylor Instrument Companies of Rochester, New York, USA, whose trademark *Tycos* they began to use, and the company continued to trade as Short & Mason.

They exhibited in the 1929 British Industries Fair, for which their catalogue entry, [448], listed their entry as an optical, scientific, and photographic exhibit. They were listed as specialists in temperature regulators, and manufacturers of: Aeronautical, medical, meteorological, and surveying instruments; Barographs; Thermographs; Rain gauges; Pocket and military compasses; Prismatic and reflecting levels; Gas pressure gauges; *Tycos* sphygmomanometers.

In 1969 they merged with Taylor Instrument Companies (Europe) Ltd., of Leighton-Buzzard, Bedfordshire, England, and the name Short & Mason ceased to be used.

Short, James (I)
James Short, (1710-1768) [3] [12] [18] [46] [263] [388] [396] [440], was an optical instrument maker of Edinburgh, Scotland, and later London, England. He was inspired and guided by his mathematics lecturer, Colin MacLaurin, at the University of Edinburgh, to an interest in mathematical optics. He was making telescopes by 1732, and he improved on the metallic specula devised by **Newton** for the reflecting telescope. He mastered the art of grinding and polishing paraboloidal mirrors, and by 1734 he was making sophisticated telescopes, principally Gregorian in design. Based on his skill as a telescope maker he was elected to the Royal Society in 1737, and he was a founding member of the Philosophical Society of Edinburgh. He produced about 180 telescopes in Edinburgh, then moved to Surrey Street, Strand, London, England in 1738, where he built on his already substantial reputation, becoming an eminent telescope maker with international clientele. In all he made about 1370 reflecting telescopes.

Following James Short's death in 1768, his brother **Thomas Short**, himself an instrument maker, continued his business until 1776.

Short, James (II)
James Short (II), (b.1752, d.c1774-6) [12], nephew of **James Short (I)**, lived at Short's Surrey Street address in London until 1773. He may have assembled and retailed James (I)'s telescopes. He returned to Edinburgh at around the time of James (I)'s death and founded the Calton Hill Observatory.

Short, Thomas
Thomas Short, (1711-1788) [12] [18], was an optical instrument maker, and brother of **James Short (I)**. He made telescopes at his shop in Broad Wynd, Leith, Edinburgh, Scotland from 1748, and later in Surrey Street, Strand, London, England where he took over his brother James' business upon his brother's death in 1768. He operated the business until 1776, and was succeeded in business by his grandson, **James Douglas**.

Short, Thomas Watling
Thomas Watling Short, (1833-1906) [44] [319] [379], was an optician and scientific instrument maker of 62 Hatton Garden, London, England. Census data shows him as an optician in 1861, and a scientific instrument maker from 1871. In about 1859 he began to work in partnership with **John Symonds Marratt**, trading as **Marratt & Short** until 1867. In 1864 he began to work in partnership with the engineer, William James Mason, trading as **Short & Mason** until 1900 when, presumably, he retired.

Showa Koki Seizou Co. Ltd.
Showa Koki Seizou Co. Ltd. was a Japanese company, established in 1954 to manufacture lenses for cameras, binoculars, and precision optical measuring instruments. In 1994 the company changed its name to **Showa Optronics Co. Ltd.** From the late-1950s until the 1980s

their binoculars were marked with the **JB code** JB172. Subsequent to the company's name change their binoculars bore the maker's mark SOC, (see Appendix IV).

Showa Optronics Company Ltd. (SOC)
Showa Optronics Company Ltd. (SOC), which was founded as **Showa Koki Seizou** in 1954, manufactured lenses for cameras, binoculars, and precision optical measuring instruments. Their consumer optics such as binoculars and telescopes bore the maker's mark SOC (see Appendix IV).

Shuron Optical
Shuron Optical of Geneva, New York, USA, was formed as a merger between **Standard Optical Company** and **Geneva Optical Company**, both of Geneva, New York. The company continued in business until 1963.

Shuttlewood, Cliff
Cliff Shuttlewood, [199], was a member of the Leicester Astronomical Society, England, in the early 1960s. He began building 6-inch f/8 reflectors with another member, Mike Goddard. They turned this into a business, Leicester Astronomical Centre Ltd., producing 6-inch f/8 and 8½-inch f/6 reflecting telescopes designed by Cliff Shuttlewood. **Jim Hysom** of **Optical Surfaces** made the mirrors, and in 1996 he was hired by Leicester Astronomical Centre Ltd. By 1997 the company had been renamed **Astronomical Equipment**.

Shuttleworth, Henry Raynes
Henry Raynes Shuttleworth, (c1732-1798) [18] [50] [276 #537] [401], was an optician and optical instrument maker of London, England. Addresses: *Sir Isaac Newton & Two Pairs of Golden Spectacles*, the Old Mathematical Shop, Near the West End of St. Paul's; Ludgate St. (1774-1788); 23 Ludgate St. (1780-1796).

He served his apprenticeship with **John Cuff**, beginning in 1746 or 1747. He was freed of the **Spectacle Makers' Company** in 1756, and served as Master from 1782-1786. **William Eastland (II)** was turned over to him in 1762. **John Bleuler** was one of his apprentices, and worked with him for some years after completing his apprenticeship. **Henry Hemsley (I)** began an apprenticeship with him in 1774, and was immediately turned over to John Graves of the Fishmongers' Company. His son, Henry Shuttleworth (II), also served an apprenticeship with him, and succeeded him in business.

Siderostat
A siderostat, [152], is an astronomical instrument, consisting of a plane mirror and a drive mechanism that rotates it in order to keep the reflection of an astronomical body orientated in a constant direction. This can be used to illuminate any optical instrument, such as a telescope, microscope, spectroscope, etc.

Certain circumstances favour mounting a telescope in a manner such that its optical axis is fixed in relation to the ground. For example, a telescope with a very long focal length that requires support along its length to avoid flexure of the tube. Such a mounting does not track the apparent movement of the stars. This can be solved using a siderostat.

A *polar siderostat* is one in which the telescope is mounted in a fixed position with its optical axis parallel to the axis of rotation of the Earth. In this arrangement the movement of the mirror is a simple rotation about the polar axis.

When the telescope is not mounted with its optical axis parallel to the axis of rotation of the Earth, a complex biaxial rotation of the mirror is required. This would be the case, for example, if the telescope is mounted parallel to the ground (except at the equator, where this orientation is polar). This concept was first put into practice by the Dutch mathematician and natural philosopher, Willem Jacob 's Gravesande (1688-1742), in 1742. He referred to his device as a *heliostate* and, while it was suitable for illumination of a microscope, its optical performance was insufficient for astronomy. The *Foucault siderostat*, invented by **Jean Bernard Foucault**, solved this problem, and siderostats of this design have been used in many applications. A Foucault siderostat was used in the 49-inch telescope built by **Paul Gautier** for the 1900 Great Paris Exhibition.

A *uranostat* is a type of siderostat. It is a plane mirror so mounted that it has two perpendicular axes of rotation. A common implementation of this is that in which one axis of rotation is the polar axis. This can be used to reflect any point in the sky in any direction and, given appropriate rotation on the mirror's axes, the reflected light can be maintained in the same direction. A uranostat in combination with a horizontally mounted long-focal-length telescope is frequently used as a photo heliograph.

Every kind of siderostat suffers from field rotation in the field of view, with the exception of the coelostat (qv). This is an inescapable consequence of the fixed telescope mounting and the rotation of the mirror. The mirror also causes some degree of loss of light and, for this reason, the siderostat is infrequently used for observations other than the Sun. See also heliostat.

Sidle & Company, John W. (Acme Optical Works)

John W. Sidle & Company, [201], of Pennsylvania, USA, was a microscope manufacturer. They began as a manufacturing company, Sidle & Poalk, which started to manufacture microscopes following a meeting between John W. Sidle of Sidle and Poalk, and the vice-president of the National Microscopical Congress in 1878. Their aim was to produce an improved microscope at an affordable price. The company's first microscopes were produced in 1880. The company name became John W. Sidle & Co., and they began to refer to their microscope production facility as Acme Optical Works. In 1881 John W. Sidle & Co. appointed **James W. Queen & Company** as their business agents, and began to sell all their products through them.

Sidle & Poalk
See **John W. Sidle & Company**.

Sigma Corporation
Sigma Corporation, (Kabushiki-gaisha Shiguma), is a Japanese manufacturer of cameras, lenses and accessories. The company was formed in 1961 as Sigma Research Centre in Setagaya, Tokyo, Japan. In 1970 the company name was changed to Sigma Corporation, and since then they have grown into a multi-national company.

Sigma Research
Sigma Research of George Washington Way, Richland, Washington, USA were makers of observatory class Cassegrain telescopes. A 1976 advertisement in *Sky and Telescope* magazine shows a large fork-mounted Cassegrain, and another advertisement shows a 32-inch fork-mounted Cassegrain. A similar telescope by Sigma is in the Three College Observatory in North Carolina.

Silberrad, Charles
Charles Silberrad, (fl.1799-1834) [18] [53] [276 #1218] [319] [322], was an optician, and optical, mathematical, and philosophical instrument maker of 34 Aldgate Street, London, England. He served his apprenticeship with **Charles Lincoln**, beginning in 1791, and was freed of the **Fletchers' Company** in 1798. He worked at the Aldgate address until his death in 1834. Instruments bearing his name include: Culpeper style microscopes; 3 draw brass and mahogany telescope; double barometer; stick barometers; wheel barometer; terrestrial pocket globes; universal ring sundial.

In 1829 he advertised: "Polyscopic Spectacles–This entirely new and elegant invention, adapted to all sights, comprises the properties of the common spectacles, of a microscope for surveying minerals, medals, flowers, insects or any small object, and with adjustment of an aeriel or French opera glass for the theatre or for viewing paintings; the whole contained within the small compass of a pair of spectacles of the usual size. This useful and excellent discovery, which has met with the highest patronage, and is universally esteemed and admired, may be purchased only at the Inventer's, C. Silberrad, Optician, 34, Aldgate, opposite the pump."

Simmance-Abady Photometer
See photometer.

Simmons Optics
Simmons is a US brand of riflescopes, binoculars, spotting scopes, rangefinders and trail cameras. The Simmons Optics brand is part of the **Vista Outdoor Inc.** group. An example of the Simmons Optics model 1200 spotting scope is marked as made in Taiwan.

Simms Wilson, James
James Simms Wilson, (1893-1976) [228] [234], was the son of **James Simms (IV)**'s sister Eleanor and her husband James Wilson. He became a director of **Troughton & Simms** in 1922. Along with James (IV)'s son, Arthur Davison Simms (1891-1976), he was the last of the Simms family to work in the firm. By the time James Simms Wilson retired in 1956, he was joint managing director of **Cooke, Troughton & Simms**, with **Wilfred Taylor**. In the same year, Wilfred Taylor and Arthur Davison Simms also retired, marking the end of an era for the company. See Appendix VII.

Simms, George
George Simms (1799-1886), [18] [228] [234], was a nautical instrument maker and compass maker of London, England. Addresses: 4 Broadway, Blackfriars (1822); 9 Greville St., Hatton Garden (1840-1855). He was a son of **William Simms (I)**, and brother of **James Simms (II)** and **William Simms (II)**. He served his apprenticeship with James (II) in the **Clothworkers' Company**. He succeeded his father in his business, working in partnership with James (II) as **James & George Simms** until 1855, after which he operated the business in his own name. See Appendix VII.

Simms, James & George
James & George Simms, (fl.1820-1855) [18], were opticians, mathematical and nautical instrument makers, compass makers, and sundial makers of London, England. Addresses: 4 Broadway, Blackfriars (1822); 9 Greville St., Hatton Garden (1840-1855). The firm was a partnership between **George Simms** and **James**

Simms (II), and succeeded **William Simms (I)** in business. Upon termination of the partnership, George continued the business under his own name. See Appendix VII.

Simms, James (I)

James Simms (I), (1710-1795) [**18**] [**234**] [**276** #309], was a drawing compass maker in Birmingham, England (1710-1760), and later in the Parish of St. Giles, Cripplegate, London. He was the father of **William Simms (I)**. See Appendix VII.

Simms, James (II)

James Simms (II) (b.1792), [**18**] [**228**] [**234**], was an optical, nautical and mathematical instrument maker of London, England. Addresses: 4 Broadway, Blackfriars (1822); 9 Greville St., Hatton Garden (1840-1855). He was a son of **William Simms (I)**, and brother of **George Simms**, and **William Simms (II)**. James (II) served his apprenticeship with Edward Barker, a shagreen case maker, beginning in 1806, and was freed of the **Clothworkers' Company** in 1813. Among his apprentices were his son, **William Simms (III)**, and his brother, George. He succeeded his father in his business, working in partnership with George, trading as **James & George Simms** until 1855. See Appendix VII.

Simms, James (III)

James Simms (III), (1828-1915) [**18**] [**228**] [**234**], was the son of **William Simms (II)**. He served his apprenticeship with his father in the **Goldsmiths' Company**, beginning in 1843. Upon the death of **William Simms (II)** in 1860, James Simms (III) and **William Simms (III)** ran the **Troughton & Simms** business successfully until business declined after WWI. They were succeeded in 1915 by James (III)'s sons, **William Simms (IV)** and **James Simms (IV)**. James Simms (III)'s daughter Eleanor was married to James Wilson, and was the mother of **James Simms Wilson**. See Appendix VII.

Simms, James (IV)

James Simms (IV), (1862-1939) [**228**] [**234**], was a son of **James Simms (III)**, and brother of **William Simms (IV)**. In 1915, William (IV) and James (IV) took control of the business **Troughton & Simms**. In 1922 the business merged with **T. Cooke and Sons**, then part of **Vickers**, to form **Cooke, Troughton & Simms**. The last of the Simms family to work in the business were James (IV)'s son, Arthur Davison Simms (1897-1976), who was sales manager, and **James Simms Wilson**, both of whom retired in 1956. See Appendix VII.

Simms, William (I)

William Simms (I), (1763-1828) [**18**] [**228**] [**234**] [**276** #889], was an English mathematical instrument maker, gold and silversmith, dial, and compass maker. Addresses: 44 Colehill Street, Birmingham (1780-1781); Birmingham (1793); London (1794); Bowman's Buildings, Aldersgate Street, London (1808); 4 Broadway, Blackfriars, London (1818-1822). He was the son of **James Simms (I)**. He served his apprenticeship with Charles Simms, and was freed of the **Butchers' Company** in 1802. In all, William (I) had eight children, of whom **James Simms (II)** and **George Simms** continued his business, and the mathematical instrument makers, Alfred Septimus Simms and Henry Simms, both also worked in the family business. Among William (II)'s apprentices was another son, **William Simms (II)**, who was turned over to him in 1809 by Thomas Penstone of the **Goldsmiths' Company**, and went on to form the partnership **Troughton & Simms**. See Appendix VII.

Simms, William (II)

William Simms (II), (1793-1860) [**18**] [**44**] [**228**] [**234**] [**276** #1430] [**443**], was an optician, mathematical instrument maker, and mariner's compass maker of London, England. Addresses: 1 Bowman's Buildings, Aldersgate St. (1821-1826); 136 Fleet St. (1826-1843); 138 Fleet St. (1843-1846); 2 & 4 Peterborough Court, Fleet St. adjoining the rear of 138 (1843).

He began his apprenticeship in 1808 with the London goldsmith Thomas Penstone. He was turned over to his father, **William Simms (I)**, the following year, and freed of the **Goldsmiths' Company** in 1815. (Note, Ref. [**234**] claims he was apprenticed to "Mr. Bennett, a mathematical instrument maker of considerable ability and reputation". This may refer to **Leonard Bennett**). Early in his career William (II) made a number of theodolites for the Ordnance Survey. William (II)'s brothers included **James Simms (II)**, and **George Simms**. Among his apprentices were his son **James Simms (III)**, his brother Alfred Septimus Simms, and **Joseph Beck** beginning in 1846.

In 1826 he entered a partnership with **Edward Troughton** to trade as **Troughton & Simms**. In 1831 he was elected a fellow of the Royal Astronomical Society. In 1833, upon the retirement of Edward Troughton, William (II) continued as sole proprietor of the business, and continued to produce fine instruments, including equipment for observatories around the world. William (II) took advantage of the significant improvements in toolmaking and precision engineering since the days of **Jesse Ramsden**, and built a dividing engine, completed in 1843, which was driven by steam. It could function unattended, and was 100 times more accurate than Ramsden's engine. In 1852 he was elected a fellow of the Royal Society. Following the death of William (II) in 1860, the business of Troughton and Simms was continued by his son, James Simms (III), and his nephew, **William Simms (III)**.

See Appendix VII.

Simms, William (III)

William Simms (III), (1817-1907) [18] [228] [234], was an optician and mathematical instrument maker of London, England. Addresses: Bowman's Building, Aldersgate St. (1831); 138 Fleet St. (1860-1871). He served his apprenticeship in the **Clothworkers' Company** with his father, **James Simms (II)**, beginning in 1831. In 1851 he was elected a Fellow of the Royal Astronomical Society. Upon the death of **William Simms (II)** in 1860, **James Simms (III)** and William Simms (III) ran the **Troughton & Simms** business successfully until business declined after WWI. They were succeeded in 1915 by the brothers, **William Simms (IV)** and **James Simms (IV)**. See Appendix VII.

Simms, William (IV)

William Simms (IV) (1860-1938) [228] [234], was a son of **James Simms (III)**, and brother of **James Simms (IV)**. In 1915, William (IV) and James (IV) took control of the business **Troughton & Simms**. In 1922 the business merged with **T. Cooke & Sons**, part of **Vickers**, to form **Cooke, Troughton & Simms**. William (IV) and James (IV) were succeeded in the business by **James Simms Wilson**, and James (IV)'s son, Arthur Davison Simms who was sales manager. See Appendix VII.

Simons, James

James Simons, also known as Simonds, Symons, (fl.1771-1794) [18] [274] [276 #1042a] [319] [322] [361], was a mathematical, philosophical, and optical instrument maker of *Sir Isaac Newton's Head*, the corner of Marylebone St., opposite Glasshouse St., London, England. The address was also variously listed as Marylebone St., Golden Square, and *Isaac Newton's Head*, 17 Marylebone St. He served his apprenticeship with **John Bennett (I)**, beginning in 1757, and was freed of the **Stationers' Company** in 1771. A microscope, made by James Simons, is described in [264] as "massive, and poorly designed; the foot is not adequate for the weight and height of the body." Instruments bearing his name include: compound microscope; theodolites; sundials; pantograph; stick barometer. He is also known to have sold: theodolite.

Sinclair & Co. Ltd., James A.

James A. Sinclair & Co. Ltd. [61] [306], of 54 Haymarket, London, England, were photographic and scientific instrument makers, founded in 1903 by James A. Sinclair (1864-1940). From 1910 James A. Sinclair also worked in partnership with **Arthur S. Newman**, trading as **Newman & Sinclair**, making cameras and shutters that were marketed by Sinclair. In 1910 Sinclair also acquired **Sanders & Co.** In 1926 the company moved to 3 Whitehall, London, although initially this address was known as 9 & 10 Charing Cross Road.

Sine Condition

For a lens to be free of spherical aberration and coma it must satisfy the sine condition. Such a lens is referred to as an aplanat. The sine condition [7], in geometrical optics is credited to **Ernst Abbe**, and is stated as follows:

$$(\sin \theta_o)/(\sin \theta_i) \approx (\theta_{op})/(\theta_{ip}) = \text{constant}$$

where θ_o is the angle of the ray in the object space and θ_i is its angle in the image space. Similarly, θ_{op} is the angle of the equivalent paraxial ray in the object space, and θ_{ip} is the angle of the equivalent paraxial ray in the image space.

Singlet

Singlet is the term usually used in optics to refer to a lens arrangement consisting of a single lens element.

Sisson, Jeremiah

Jeremiah Sisson, (1720-1783) [18] [50] [256] [276 #411] [443], was the son of **Jonathan Sisson**. He was a mathematical instrument maker, producing astronomical, surveying and navigation instruments as well as a small number of microscopes, and succeeded his father at *The Sphere*, Corner of Beaufort Buildings, Strand, London, England. He was declared bankrupt in 1751 [35 #9082 p4]. Upon the death of **George Adams (I)** in 1772, Jeremiah Sisson took over his ordnance trade, and supplied instruments by appointment to the Board of Ordnance until 1775, when he was again declared bankrupt [35 #11620 p8].

Sisson, Jonathan

Jonathan Sisson, (1690-1749 fl.1722-1737) [3] [18] [24] [46] [256] [276 #120] [396], originally from Lincolnshire, England, was a mathematical instrument and astronomical mirror maker of *The Sphere*, Corner of Beaufort Buildings, Strand, London. Early in his career he worked as an assistant to **George Graham**. In 1729 he was awarded a Royal appointment to the Prince of Wales. In 1740 **John Bird** began to work for him, dividing standard measures. By 1745, Bird had left to start his own company.

Jonathan Sisson became known for his large mural quadrants, the accuracy of his arcs, and for his improved azimuth theodolites which included the addition of a telescopic sight and a spirit level. He also developed the English Equatorial telescope mount, first proposed by **Henry Hindley** of York in 1741. He produced large astronomical instruments for European observatories. His son, **Jeremiah Sisson**, succeeded him as an instrument maker at the same address.

Skybolt
Skybolt was a brand of Japanese telescopes and binoculars. Examples include small refractors and hand-held telescopes dating from the 1970s.

Sky-optic ball
Another name for a scioptic ball.

Skywatcher
Skywatcher is a **Synta** brand of telescopes and other related equipment. The UK importer for Skywatcher is **Optical Vision Ltd** (OVL).

Slade, W.
W. Slade, [199], of Bristol, England, was a C20th maker of astronomical telescopes, mirrors and flats, who also re-silvered mirrors.

Slater, Thomas
Thomas Slater, (1817-1889) [13] [24] [312], was an optical technician and telescope maker of 4 Somers Place, West Euston Square, London, England. He famously figured the 24-inch f/38 lens for the Craig Telescope, a 75-foot behemoth mounted on a brick tower on Wandsworth Common. The construction of the telescope was superintended by **William Gravatt** on behalf of Rev. John Craig. Once commissioned, the lens was discovered to suffer from severe spherical aberration. It was, however, never corrected, and the telescope was dismantled a few years later.

Instruments bearing his name also include: 3-inch lacquered brass refracting astronomical telescope, length 36½ inches; 2½-inch brass refracting telescope, length 35 inches; 15-inch object glass and equatorial mount for the Downside School observatory at Downside Abbey in Stratton-on-the-Fosse. This was finished in 1859, and destroyed by fire in 1867.

Slit Lamp
A slit lamp is an instrument, used by a vision care specialist, for examination of the cornea and the interior of the eye. It usually consists of a bright source of light focused to shine through a slit of variable width into the eye, and a binocular microscope for inspection of the interior of the eye. A smaller, hand-held variant may be binocular or monocular.

SLR Camera
See camera.

Slugg, Josiah Thomas
Josiah Thomas Slugg, (1814-1888) [44] [53] [286], of Lancashire, England, was a chemist and druggist who served his apprenticeship in Manchester where, by 1840, he had his own business. Addresses: 214 Stretford Road (1860-1862), 242 Stretford Road (1872). His nephew, the camera maker, **J.T. Chapman**, served an apprenticeship with Slugg as a chemist and druggist, and learned his optical skills from him [61].

Slugg developed a keen interest in astronomy and, by the late 1850s, he was producing and advertising for sale "cheap telescopes" for viewing the night sky. He wrote several books and leaflets on the subject of astronomy, including details for the construction and use of a telescope, and a book on observational astronomy.

According to a catalogue, published in 1861, [432], he supplied: Achromatic telescopes; Eyepieces; Achromatic object glasses mounted in lens cells; Tourists' telescopes, varying from 1 to 9 draws; Binocular telescopes, marine, and race glasses; Opera glasses; Monoculars; Magnifying glasses; Stereoscopes. He is also known to have produced parallactic stands for telescopes and microscopes.

In 1866, he was elected a Fellow of the Royal Astronomical Society. Alongside his business he wrote two historical books relating to Manchester. When he died his business was continued by his assistant, Harry Kemp, until 1896.

Smethwick
Smethwick, (fl.1674-1675) [275 #379], was a turner, telescope tube maker, and lens maker of London, England. He was referred to by **Robert Hooke** as "Smethwick, turner" in order not confuse him with **Francis Smethwick**. As well as making telescope tubes, he was a glass-grinder, and made lenses for telescopes and other optical instruments. He is believed to have worked in partnership with **John Dunnell**, and may have worked with **John Marshall**.

Smethwick, Francis
Francis Smethwick, (fl.1667-1685) [256] [275 #323], was an amateur optical instrument maker of London, England. He was a pupil of William Oughtred (1575-1660), mathematician, and inventor of the slide-rule. Francis Smethwick invented a method of grinding aspherical lenses, and made: microscope; telescope; burning glasses. He showed his work to the Royal Society, and was elected

a Fellow in 1667. **Robert Hooke** was highly critical of his work.

Smith & Beck
Smith & Beck, [18] [264], were opticians and microscope makers of London, England. Addresses: 6 Coleman St., City (1848-1857); Peartree Cottage, Holloway Rd. (1855-1857). The firm was a partnership between **James Smith (I)** and **Richard Beck**, formed in 1847. **William Wales** worked for Smith & Beck prior to moving to the USA, where he set up his own business in 1862. They advertised for sale: spectacles; microscope; binoculars; magic lantern; stereoscope. In 1857, **Joseph Beck** joined the partnership, and they began to trade as **Smith, Beck & Beck**.

Smith & Ramage
Smith & Ramage, (fl.1841-1861) [18] [271], were nautical and optical instrument makers of 45 Regent's Quay, Aberdeen, Scotland. The firm was a partnership between Charles Ramage, and Charles Smith, who was a watch maker, clock maker, and nautical instrument maker. They succeeded **Ramage & Co.** in business, and were succeeded by **John Grant**.

Smith & Sons, S. (England)
S. Smith & Sons (England), [44], of Cricklewood and Cheltenham, founded by Samuel Smith in 1852, began as a clock and watch business, and in 1904 the company's diversification began with making motor accessories. By 1944 it was an engineering company specialising in motor accessories, industrial instruments, aircraft accessories, and watches and clocks.

In 1935 S. Smith & Sons acquired a controlling interest in **Henry Hughes & Son**. In 1947, upon the formation of the company **Kelvin & Hughes**, S. Smith and Sons were shareholders in the business. By 1951 the subsidiaries of S. Smith & Sons included Kelvin & Hughes, **Kelvin Bottomley & Baird**, and Henry Hughes & Son. In 1953 they acquired the remaining interests in Kelvin & Hughes.

In 1966 the name of the company was changed from S. Smith & Sons (England) to Smiths Industries, and in 1967 they took over **K.G. Corfield**.

Smith, Addison
Addison Smith, (fl.1750-1789) [18] [276 #538], was an optical, mathematical, and philosophical instrument maker, of London, England. Addresses: Opposite Northumberland St., Strand; St. Martin's Lane, near Charing Cross (1764); Near Charing Cross, Strand (1774); 481 Strand (1779-1783); 79 Charlotte St., Rathbone Place (1783).

He served his apprenticeship with **Francis Watkins (I)**, and was freed of the **Spectacle Makers' Company** in 1763. That year, he began to work in partnership with Francis Watkins (I), trading as **Watkins & Smith** until 1774.

In 1764 Addison Smith was one of the petitioners who attempted to revoke the patent obtained by **John Dollond (I)** for the achromatic lens, and which was enforced after Dollond's death by his son, **Peter Dollond** (See Appendix X). Despite the failure of the petition, Watkins and Smith continued to make achromatic telescopes, and in 1766 they were successfully sued, as joint defendants, by Peter Dollond for infringement of the patent.

After the partnership was terminated, Smith continued to work under his own name. He is known to have sold: balances; barometers; hygrometers; thermometers; optical and surveying instruments.

Smith, Beck & Beck
Smith, Beck & Beck, [264], was a partnership formed in 1857 between **James Smith (I)**, his former apprentice **Richard Beck**, and Richard's younger brother **Joseph Beck**. They were microscope makers, and traded at 6 Coleman St., London, England. **Henry Crouch** served an apprenticeship with them, and worked for them until he left in 1862 to set up his own business. The partnership of Smith, Beck and Beck continued until 1866 when James Smith (I) left to work in partnership with his eldest son, James John Smith. The Beck brothers then reorganised, and continued to trade together as **R.&J. Beck**.

Soon after **Charles Coppock** graduated in 1856, he began to work for Smith, Beck & Beck, and he stayed with the firm when Smith left, becoming a partner in R.&J. Beck shortly after Richard Beck's death in 1866.

Smith, Caleb
Caleb Smith, (fl.1734) [13] [256] [276 #311], was an insurance broker and inventor of London, England. He was a contemporary of **John Hadley** and **Thomas Godfrey**. He designed a sea quadrant, similar to Hadley's quadrant, but using glass prisms instead of mirrors, which he referred to as an astroscope, and petitioned the Board of Longitude for an award for his invention. He had his sea quadrant made by **Thomas Heath (I)**. It was, however, the better performing reflecting quadrants of Hadley and Godfrey that came into more general use.

Smith, Dwight Lyman
Dwight Lyman Smith, (1839-1905) [50], of Waterbury, Connecticut, USA, worked for a belt buckle company, and was the inventor of a device he called a Micrograph. It was a small, inexpensive simple microscope, produced by him

Smith, E.&W.
Egerton & William Smith, (fl.1803-1807) [18] [276 #1220a] [319] [322] were opticians, mathematical instrument makers, printers, and stationers of Navigation Shop, 18 Pool Lane Liverpool, England and, from 1805, 19 Pool Lane. They succeeded their mother, Ann Smith, the wife of and successor to **Egerton Smith (I)**, in business. The business was run by Egerton Smith (II) and his brother William. They retailed optical and navigation instruments bearing their name. They are known to have retailed microscopes.

Smith, Egerton
Egerton Smith (I), (fl.1766 d.1788) [3] [18] [276 #769] [319] [322], was a mathematical and nautical instrument maker of Liverpool, England. Addresses: Church St. (1766-1772); Cable St. (1766); *Newton's Head*, 17 Pool Lane, (1774-1776); Navigation Warehouse, Pool Lane (1774-1783); 18 Pool Lane (1780-1787).

He began his career as a schoolmaster, and gave annual lectures on geography, astronomy, and navigation. By 1773 he was running a 'Mathematical, Philosophical, and Optical Shop' at the Pool Lane address. Instruments bearing his name include: two ebony framed octants with brass fittings [54].

Upon the death of Egerton (I) the business was run by his wife Ann Smith [18], who is listed as a mathematical and optical instrument seller, bookbinder, and publisher. In 1800 she entered a partnership with her son, the printer and stationer Egerton Smith (II) (1774-1841), who ran the Navigation Warehouse with his brother, William, as **E.&W. Smith**.

Smith, George
George Smith, (1821-1872) [50] [319], succeeded **James How** in business following How's death in 1872. He traded as James How & Co., chemical, optical, and general scientific instrument manufacturers, of 2 Foster Lane, London, England. He is known to have sold microscopes and accessories. In 1876 the business was relocated to 5 St. Bride Street, London, and from 1879, at 73 Farringdon Street. The last records of the company are dated 1891.

Smith, James (I)
James Smith (I), (1800-1873) [18] [50] [264], of 50 Ironmonger Row, Old St., London, England, was a notable microscope maker. (Note, [18] states that he served an apprenticeship with **Daniel Weeden** beginning in 1806. However, this appears to be in error, since he would have been 6 years old at the time). He began by producing the bodies of microscopes, without optics, for other sellers. In 1826 **Joseph Jackson Lister** asked **Charles Tulley** to produce a microscope body to a design he had originated. Tulley subcontracted this to James Smith (I) who proceeded to learn from Lister the art of lens-making. Smith began to make microscopes and retail them using his own name in 1839. His association with Lister brought him in contact with Lister's nephew and business partner, Richard Low Beck, whose son **Richard Beck** served an apprenticeship with Smith. In 1847, around the time Richard Beck would have completed his apprenticeship, he formed a partnership with James Smith (I), trading as **Smith & Beck**. They traded at 6 Coleman St., London, England, until 1857 when **Joseph Beck** joined the partnership, and they began to trade as **Smith, Beck & Beck**. This continued until 1866, at which time James Smith (I) left to work in partnership with his eldest son, James John Smith (1833-1877).

Smith, James (II)
James Smith (II), (fl.1821-1838) [18] [53] [361], was an optical instrument maker of 17 Bath Place (later, 15 Palace Row), New Road, Fitzroy Square, London, England. At some point in his career, he worked for **William Fraser**. According to [276 #1707] he formed a partnership with **Joseph Smith (I)**, trading at the same address.

According to his trade card: "Established 1817. J. Smith, practical optician. 17 Bath Place, New Road, Tottenham Court Road. From Fraser's, Bond Street, Opticians to the Royal Family. Inventor and sole proprietor of the Compound Engraver's Glasses…"

Smith, James (III)
James Smith (III), (fl.1820-1825) [276 #1706] [322], was an optical, nautical, and mathematical instrument maker of 126 Wapping, London, England. He succeeded **John Smith (III)** at this address. Instruments bearing his name include a telescope. He is also known to have made octants. The premises at 126 Wapping were the same as those occupied by **John Smith (III)** and **John Soulby & Co.**

Smith, Jeppe
Jeppe Smith, (1759-1821) [3] [16], was an instrument maker of Copenhagen, Denmark. In 1802, he took over the workshops and assets of **Jesper Bidstrup**. Although he did not have the first-hand knowledge that Bidstrup had acquired in his visit to London, he had a thriving business, producing a wide range of instruments for the Danish market until his death in 1821. He is known to have made

octants and sextants. He was succeeded in his business by his nephew.

Smith, John (I)

John Smith (I), (fl.1716-1730) [18] [46], was an optical instrument maker of *the Archimedes*, Ludgate Street, London, England. Also of *the Archimedes & Three Golden Prospects*, and optician to King George I. He served his apprenticeship with the instrument maker **John Marshall**, beginning in 1709, and was freed of the **Turners' Company** in 1716. He married Marshall's daughter, Catherine (bap.1689). Upon John Marshall's death, John Smith inherited his tools and succeeded him in his business.

According to his trade card: "John Smith. Servant to his late Majesty at *the Archimedes* in Ludgate Street London. Makes and sells all sorts of Telescopes, Microscopes, Spectacles, Thermometers, Barometers, Prospective, Optical and reading glasses, and also all manner of Optical instruments according to ye best and latest improvements."

Smith, John (II)

John Smith (II), (fl.1740-1780) [18] [276 #413], was an English mathematical instrument maker. Addresses: York; King's Private Observatory, Kew, Richmond, Surrey; Royal Exchange, London. He was associated in his youth with **Henry Hindley**. According to a footnote in [284] in 1809, referring to the text of a letter dated 1748 by Henry Hindley: "The persons referred to were both bred with him. His brother, Mr. Roger Hindley, who has many years followed the ingenious profession of a watch-cap-maker in London, was so much younger as to be an apprentice to him. Mr. John Smith, now dead, had some years past the honour to work in the instrument way, under the direction of the late Dr. Demainbray, for his present Majesty."

Dr. Demainbray was the superintendent of King George III's private observatory at Kew, Richmond, Surrey. As referred to in the footnote, John Smith (II) became assistant to Demainbray at the observatory.

Smith, John (III)

John Smith (III), (fl.1817-1839) [18] [53] [276 #1432], was an optical and mathematical instrument maker of 126 Wapping, London, England (1817-1839), and 35 Leicester Square (1836). Instruments bearing his name include an ebony and brass quadrant with inlaid ivory scale. The premises at 126 Wapping were the same as those occupied by **James Smith (III)** and **John Soulby & Co.**

Smith, Joseph (I)

Joseph Smith (I), (fl.1817-1846) [18], was an optician, and mathematical and philosophical instrument maker of 42 Threadneedle Street, London, England. Some confusion exists between him and **Joseph Smith (II)** in the literature, and this is clarified to some extent in [50].

[276 #1433] states that Joseph Smith (I) entered a partnership with **James Smith (II)**. The partnership traded at 17 Bath Place (later 15 Palace Row), New Road, Fitzroy Square.

Smith, Joseph (II)

Joseph Smith (II), (1766-1825) [18], was an optical instrument maker of London, England. He is likely to have been the Joseph Smith, son of Richard Smith a linen draper, who served an apprenticeship with his father. He was freed of the **Spectacle Makers' Company** in 1811, perhaps some years after completing his apprenticeship, to enable him to trade under his own name in London. Some confusion exists between him and **Joseph Smith (I)** in the literature, and this is clarified to some extent in [50].

In 1811, upon the death of **James Long**, Joseph Smith (II) took over Long's premises at 4 Back of the Royal Exchange. It is not clear what employment he had prior to this. However, it is possible that he was a journeyman optical instrument maker, working for another maker. Having started his own business, he is known to have made microscopes and telescopes. One of his employees was **Stephen Bithray**, who is believed to have started working for him soon after Smith opened his business. Joseph Smith (II) continued in business at the same address until his death in 1825, following which Stephen Bithray succeeded him in business at the same premises.

Smith, Robert

Robert Smith, (bap.1689, d.1768) [3] [46] [274] [276 #121], was professor of astronomy and experimental philosophy at Cambridge University. He was author of the *Compleat System of Opticks* (1738) [248], which was considered a standard natural philosophy textbook of its day.

Smith, Thomas (I)

Thomas Smith (I), (fl.1790-1823) [18] [276 #1434] [358], was an optical and mathematical instrument maker, quadrant maker, and compass maker, of 53 Old Gravel Lane, Ratcliff, London, England. Instruments bearing his name include an ebony octant with brass index arm and ivory scale, radius 13 inches.

Smith, Thomas (II)
Thomas Smith (II), (fl.1823-1835) [18], was an optician, and optical, mathematical, and philosophical instrument maker of London, England. Addresses: 6 Worcester St., Ratcliff (1823-1829); 1 Hope St., Hackney Rd. (1829-1830); 61 Greenfield St., Commercial Rd. (1829-1835).

Smith's Quadrant
The Smith's quadrant is a form of quadrant invented by **Caleb Smith**. It is similar to Hadley's quadrant, but uses glass prisms instead of mirrors. It was invented by **Caleb Smith**, who referred to it as an astroscope.

Smiths Industries
See **S. Smith & Sons (England)**.

Snart, John
John Snart, (fl.1799-1832) [18] [274] [276 #1221] [319] [322] [361], was an optical, mathematical, and philosophical instrument maker of 122 Tooley St., London, England (1799), and 215 Tooley St. (1805-1827). He served an apprenticeship with **Edward Chester** of the **Merchant Taylors' Company**, beginning in 1783. His trade label, in a microscope case dated c1820, stated "Telescope maker to the trade". Instruments bearing his name include: microscope; wheel barometers.

According to his trade card "Optical, Mathematical, and Philosophical instruments made and sold by John Snart", followed by an extensive list of instruments, including: Sextants and quadrants; Refracting and reflecting telescopes for day and night; Theodolites; Simple and compound microscopes; Spectacles; Concave and convex reading and burning glasses; Plain mirrors; Magic lanterns; Prisms for elucidating the laws of refraction and refrangibility; Improved camera obscura.

He was succeeded in business by Miss Neriah Snart, (sometimes Neariah or Nehomiah), optician, and mathematical and philosophical instrument maker.

Snell's Law
Snell's law of refraction was discovered experimentally in 1621 by the Dutch mathematician and physicist, Willebrord van Roijen Snell (1580-1626). The discovery remained unpublished until René du Perron Descartes (1596-1650), who had determined the law independently, published it in *La Dioptrique* in 1637.

The law defines the relationship between the incident angle and the refracted angle of a light ray when it is transmitted across the optical boundary between materials of different refractive index. Snell's law states:

$$n_1 \sin \theta_i = n_2 \sin \theta_r$$

where n_1 is the refractive index of material 1 in which the incident ray reaches the optical boundary; n_2 is the refractive index of material 2 in which the refracted ray propagates away from the optical boundary; θ_i is the angle of incidence, measured from the normal to the optical boundary in material 1; θ_r is the angle of the refracted ray, measured from the normal to the optical boundary in material 2. See refraction. See also the section *properties of optical glass* in Appendix VI.

Società Scientifica Radio Brevetti Ducati (SSRD)
Società Scientifica Radio Brevetti Ducati (SSRD), in Bologna, Italy, [55], was formed by the Ducati family in 1926, producing radio equipment based on the patents of Adriano Ducati. The company was highly successful, and expanded quickly. In 1939 they created an optical department, making optical munitions for the Italian Army, with optical components supplied by **Officine Galileo**. During WWII, under German occupation, the company made optical instruments for the Axis war effort. They marked these instruments with the **ordnance code** mlr. After the war they continued to produce some optical instruments until the optical department closed in 1953. It was in the post-war period that they began to produce the motorcycles for which they are famous today.

Société des Etablissements Gaumont
In 1906 **Léon Gaumont** raised new financing, reorganised, and renamed L. Gaumont et Cie to become Société des Etablissements Gaumont [53]. The company owned a chain of cinemas, and was a major player in French film-making, and a manufacturer of still and motion picture equipment. Léon Gaumont retired in 1930 with the advent of talking movies and, after a merger with Franco-Film-Aubert to form Gaumont-Franco-Film-Aubert, the company went into liquidation in 1934. The company re-emerged in 1938 as Société Nouvelle des Etablissements Gaumont.

Société des Etablissements Krauss (SEK)
Société des Etablissements Krauss was a French optical instrument company. They worked closely with the Polish company **PZO**, and sold them technical documentation for the production of binoculars. Later they cooperated with PZO over military photographic equipment. In 1935 **Barbier, Benard, et Turenne** merged with SEK to form **BBT Krauss**.

Société Genevoise pour la Construction des Instruments de Physique
Société Genevoise pour la Construction des Instruments de Physique, (SIP), [53] [58] [113] [320], of Geneva, Switzerland, was founded in 1862 by Auguste de la Rive

and Marc Thury. Initially they produced instruments for physics and optics. In 1870 they began to produce electrical equipment and precision measuring devices. Their products also included rotary tables, jig-boring machines, air and refrigeration compressors, and other machinery, as well as telescope mounts, scientific and surveying instruments, and microscopes.

Société Industrielle d'Instruments de Précision

In 1911 the company Maison Deleuil Velter & Cie, successors to **Louis Joseph Deleuil**, was renamed Société Industrielle d'Instruments de Précision. Following the death of Félix Pellin in 1940, the company Philibért and Félix Pellin, successors to **Philibért François Pellin**, merged with Société Industrielle d'Instruments de Précision. The company continued to trade until the 1950s.

Société Optique et Précision de Levallois S.A. (OPL)

Société Optique et Précision de Levallois S.A. (OPL), [26 p110] [444], was established in 1919 by Armand de Gramont, duc de Guiche. The founding objective of the company was the production of optical instruments for the French military, to reduce their dependence on imported instruments. In the years immediately following WWI, they mainly produced military rangefinders, with the navy being a major customer. They began to produce cameras in 1945, and **Foca** was a brand-name they used for their rangefinder cameras and other cameras. In 1964 they ceased production of cameras and merged with **Société d'Optique et de Mécanique de Haute Précision (SOM)** to form Société d'Optique, Précision, Electronique et Mécanique (SOPEM, then Sopelem).

Société d'Optique et de Mécanique (SOM)

Société d'Optique et de Mécanique de Haute Précision (SOM), [26] [31] [192] [444], a French optical instruments firm, was founded in 1857 by **Claude Berthiot** making photographic lenses. In 1884 he was joined in his business by his nephew Eugène Lacour, and the name was changed to **Lacour-Berthiot**. They traded as Lacour-Berthiot until 1906 when they were incorporated as **Établissements Lacour Berthiot S.A.**, at which time Eugène Lacour was the general director, and Charles Henri Florian was the technical director. In 1913 the company became Société D'Optique et de Mécanique de Haute Precision (SOM). At this time, they began to mark their products with the brand-name **SOM-Berthiot**. After WWI, SOM began to compete with **Société Optique et Précision de Levallois (OPL)**, producing rangefinders for the French military, with the air force being a major customer. They continued to use the SOM-Berthiot brand-name until 1964 when the company merged with OPL to form Société d'Optique, Précision, Electronique et Mécanique (SOPEM, then Sopelem).

Société d'Optique, Précision, Electronique et Mécanique (SOPEM, Sopelem)

See **Société d'Optique et de Mécanique (SOM)**.

Soho Ltd.

Soho Ltd., [44] [61] [306], of 3 Soho Square, London, England, was the sales division of **A. Kershaw and Sons**. It was formed in 1929 by renaming **APM** and, in 1946, shortly before becoming a part of **Rank**, it was renamed again to become **Kershaw-Soho Ltd.**

Sokkia

Sokkia was a manufacturer of surveying and measuring instruments. The company was founded in 1920 in Tokyo, Japan, as **Sokkisha**, and changed its name to Sokkia in 1992. In 2008 Sokkia was acquired by **Topcon**, and became their *Sokkia* brand. Their surveying products include electronic total stations and field books, electronic distance meters, digital micrometer theodolites, transits, and transit levels. Their construction products include laser level planers, laser plummets, laser distance meters, digital levels, and automatic levels.

Sokkisha

Sokkisha was a manufacturer of surveying and measuring instruments, founded in 1920 in Tokyo, Japan, initially making a 12-inch Y level. The company grew, and expanded its range of products. In 1934 it became a limited liability company, and in 1943 it became a corporation. In the 1970s **The Lietz Company** became Sokkisha's exclusive US distributor. In 1982 Sokkisha established its European headquarters in the Netherlands. In 1984 they acquired The Lietz Company as a wholly owned subsidiary. In 1992 Sokkisha was renamed **Sokkia**.

Solar Camera

A solar camera was a solar-illuminated instrument used to make photographic enlargements from a negative, and was a predecessor to the more modern darkroom enlarger. See also heliotrope (3).

Solar Microscope

See microscope.

Soleil, François

François Soleil, (d.1846) [53] [299] [300], was an optical instrument maker of Paris, France. Earliest records for his business date to 1799. He made telescopic rangefinders, microscopes, achromatic telescopes, and other optical instruments. He is remembered best for his collaboration with **Augustin Jean Fresnel** in the development of Fresnel lenses for use in lighthouses. His son, **Jean Baptiste François Soleil**, began his career in the continuation of François Soleil's work.

Soleil, Henri
Henri Soleil, (d.1879) [53] [299], was an optical instrument maker of Paris, France, and the son of **Jean Baptiste François Soleil**. In 1850 he purchased his father's retail shop, and the workshop where he had made optical components. He retailed instruments including microscopes, binoculars, barometers, thermometers, and hydrometers, and continued to make optical glasses, lenses, prisms and crystal plates. He won medals at the French National Expositions of 1855 and 1867 for his optical components. Upon his retirement in 1872, Henri sold the business to **Léon Laurent**.

Soleil, Jean Baptiste François
Jean Baptiste François Soleil, (1798-1878) [13] [53] [299], was an optical instrument maker of Paris, France. In his early career he continued the work of his father, **François Soleil**, constructing and improving lighthouse optics. He broadened his interests, however, and established a strong reputation for optical instruments designed for laboratory experiments. He collaborated with scientists in the design and creation of instruments for the investigation of phenomena such as diffraction, interference, and polarization. He also made polarizing microscopes, an improved saccharimeter, an improved polar clock, and other innovations. When the invention of the daguerreotype was announced he began to produce daguerreotypes. His instruments won medals at the French National Expositions of 1839, 1844, and 1849, the latter being a gold medal, awarded shortly before he was made Chevalier de la Légion d'Honneur.

Upon his retirement in 1850, Jean Baptiste Soleil divided his business into two parts. His son, **Henri Soleil**, purchased the retail shop and the workshop for making optical glasses and crystal. Jean Baptiste's son-in-law, **Jules Louis Duboscq**, purchased the workshop for making scientific instruments.

Soligor GmbH
Soligor GmbH was a German company, founded in 1968 as A.I.C. Phototechnik GmbH, and changed its name to Soligor GmbH in 1993. The company produces camera lenses and accessories, and has had some lenses made by **Kino Precision Industries Ltd.** A.I.C. Phototechnik GmbH was originally set up as a subsidiary of the American distributor Allied Impex Corporation, who used the brand-name *Soligor* for cameras and lenses imported from Japan. Allied Impex Corporation also imported **Miranda** cameras, and took control of the Miranda company at some time in the 1960s.

Solomons, Elias
Elias Solomons, (fl.1838-1864) [18] [276 #2007] [319] [322] [361], was an optician, and optical instrument maker of 37 Old Bond Street, London, England and, 36 Old Bond St. (1839-1846), and 27b Old Bond St. (1849-1859). He advertised spectacles, and various kinds of telescope, as well as other instruments.

An advertisement in the Times newspaper in 1852 reads "Telescopes–Mr. E. Solomons, patentee of spectacles, 27, Old Bond-street, Piccadilly, has made a new and most important improvement in telescopes for the waistcoat pocket, shooting, deer stalking, military and sea purposes, possessing such extraordinary powers that a pin's head may be clearly discerned from 10 to 15 miles' distance; and some 3 inches long, with an extra eye-piece, will show distinctly Jupiter's satellites, Saturn's ring, &c. Also a most powerful small waistcoat pocket-glass, price 15s., to discern minute objects at a distance of from four to five miles. Only to be had at 27, Old Bond-street, London, and 19, Nassau-street, Dublin." The dubious claims in this advertisement were repeated in several other advertisements during the 1850s.

The address of 19 Nassau St., Dublin appears in several of his advertisements during the period 1851-1856. This address was occupied from 1856 until 1905 by the optician M.E. Solomons [304].

Solomons, S.&B.
Samuel and Benjamin Solomons, [13] [18] [53] [276 #2252] [319], were mathematical, philosophical and optical instrument makers and opticians of London, England. The firm was a partnership between Samuel Solomons and Bemjamin Solomons (b.c1792-1793). Addresses: 5 New Road, St. George's East (1838); 39 Albermarle Street, Piccadilly (1840-1875); 76 King Street, City (1843). They are known to have sold microscopes, telescopes, and magic lanterns. Samuel Solomons was declared bankrupt in 1866 [35 #7703 p1529]. Benjamin Solomons was awarded various patents: "Improvements in telescopes and other glasses in their application to the measurement of distance" (1856); "Improvements in transparent slides for magic lanterns, and other similar purposes" (1865); "Improvements in telescopes" (1868); "Improvements in the construction and arrangement of meteorological indicators" (1868); "An improvement in prismatic compasses" (1871). S.&B. Solomons continued in business until at least 1879.

Somalvico & Co.

Somalvico & Co. [16] was a philosophical and nautical instrument maker of London, England. They may have been the same company as **Joseph Somalvico & Co.** Instruments bearing this name include: marine barometer; octant with scale showing a scale divider's mark formed of the initials FW (See Appendix (XI).

Somalvico & Co., Joseph

Joseph Somalvico & Co., (fl.1839-1913) [18] [276 #2252a] [322] [438], were opticians, and philosophical and mathematical instrument makers of London, England. Addresses: 2 Hatton Garden (1839-1865); 16 Charles St., Hatton Garden (1869); 18 Charles St., Hatton Garden (1901); 81 Holborn (1901). See also **Somalvico & Co.** The company is one of a whole dynasty of C19th barometer makers named Somalvico, in Hatton Garden and Holborn [379].

According to their trade card, [361]: "Patronized by the Royal Family and most of the nobility and gentry. Much approved by gentlemen travelling and frequenting watering-places, &c. Newly invented portable walking-stick telescope with achromatic object glasses of exceeding power, by which views of nature's scenery and animate objects, warranted to be seen clearly, at a distance of six miles. Made in handsome foreign woods and sold wholesale, at extremely moderate prices, by the manufacturers, J. Somalvico & Co., opticians, Hatton Garden, London."

They exhibited in the Great Exhibition in London in 1851 [328], for which their catalogue entry no. 681a listed them as manufacturers, with items on display including: Various kinds of barometer; Engineers' guide gauges; Vacuum steam-pressure gauge; Steam engine indicator; engines; "Walking-stick telescopes, with compass and hygrometer, and with double eyeglasses to spring out of the stick"; Improved sextant; Solid limb sextant; Model representing the circulation of the blood; Improved self-generating coffee-pot; Case of mathematical instruments, &c.

Joseph Somalvico also traded as Joseph Somalvico & Son, [18] [276 #2229], at 41 Kirby St., Hatton Garden (1820) and 37 Charles St., Hatton Garden (1833-1839), and subsequently under his own name at 37 Charles St., Hatton Garden (1840-1841). He may have been the Joseph Somalvico in the partnership **Lione, Somalvico & Co.**

Somalvico & Son, Joseph

See **Joseph Somalvico & Co.**

SOM-Berthiot

SOM-Berthiot, [192], was a brand-name used by **Société D'Optique et de Mécanique de Haute Precision (SOM)**. They began to use the brand-name in 1913, when SOM was incorporated, and the name stayed in use until 1964 when the company merged with the **Société Optique et Précision de Levallois** to form Société d'Optique, Précision, Electronique et Mécanique (SOPEM, then Sopelem).

Sony

Sony Corporation is a Japanese multinational with many facets to its business.

The company was founded in 1946 as Tokyo Tsushin Kogyo K.K. (Tokyo Telecommunications Engineering Corporation). In 1958 the company name was changed to Sony Corporation. In the early 1980s they produced the world's first CCD video camera, and they went on to produce both video and still digital cameras, as well as lenses and photographic accessories.

Sorby, Henry Clifton

Henry Clifton Sorby, (1826-1908) [46] [264] [320], of Yorkshire, England, was a geologist, and a keen microscopist. He was prolific in his research, and was considered by some to be "the father of microscopical petrology". He made a large number of thin sections of rocks for microscopic inspection, and pioneered the use of the microscope in petrography. In the 1860s he invented the microspectroscope. He established the Sorby Fellowship of the Royal Society, and he helped to found Firth College, which became the University of Sheffield.

Soulby, John

John Soulby, (fl.1830-1856) [18] [322], was an optical, mathematical, and nautical instrument maker, and ship chandler, trading at the London, England address variously listed as 126 Wapping; 126 Wapping New Stairs; 126 High Street, Wapping. These were the same premises as occupied by **James Smith (III)** and **John Smith (III)**.

He exhibited in the 1851 Great Exhibition in London. His catalogue entry [294] stated: "Soulby, J. 126, High St. Wapping, Manu.–Hely's salvage boat. Life girdle. Safty windlass. Captain Cook's quadrant and compass, the identical instrument used by that celebrated mariner in his voyage round the world."

From 1842 he traded as J. Soulby & Co. According to his trade card, "J. Soulby & Co., successors to J. Smith. Mathematical & Optical instrument makers & Ship chandlers. 126 Wapping New Stairs, London. Stationary, Charts, Maps, Navigation books &c. Compasses, Quadrants, Sextants, &c. repaired, and the modern improvements applied to them …" (the last word being unreadable).

Soulby & Co., John
See **John Soulby**.

Southern Precision Instrument Company (SPI)
The Southern Precision Instrument Company of San Antonio, Texas, USA, was an importer and seller of Japanese microscopes, telescopes and accessories beginning in the 1950s or before. The Japanese manufacturers included **Daiichi Kogaku**, **Astro Optical Industries Co. Ltd.** and **Towa**, as well as others.

Spear, Richard
Richard Spear, (fl.c1791-1837) [18] [34] [276 #1046] [304], was an optical and mathematical instrument maker of Dublin, Ireland. Addresses: 29 Capel St. (1791-1792); 23 Capel St. (1793-1809); College Green (1809); 35 College Green (1810-1811); 27 College Green (1812-1814).

He worked in partnership with **Edward Clarke** from 1815 until 1817, trading as Spear & Clarke. Richard Spear is known to have retailed instruments from the London makers.

Examples of instruments bearing Richard Spear's name include: banjo barometer in walnut with mother of pearl inlay signed *R. Spear, College Green, Dublin*; single-draw ½-inch spyglass with tapered mahogany barrel, extending from 25 inches to 31 inches, signed *R. Spear, Instrument Maker to His Majesty's Crown of Customs in Ireland*; three draw spyglass with mahogany barrel, signed *R. Spear, Dublin*.

Spectacle Makers' Company
See Appendix IX: The London Guilds.

Spectra Geospatial
The laser technology company Spectra Physics, of Santa Clara, California, USA, was founded in 1961. Its optical surveying and laser business was Spectra Precision, whose subsidiary, Spectra Geospatial, makes products for surveying, geographic information systems, and construction. In 1999 Spectra Precision entered a joint venture with **Carl Zeiss AG** to manage the Zeiss surveying business in Jena, Germany. In 2000 Spectra Precision was acquired by **Trimble Inc.** Spectra Geospatial continues today as a Trimble brand.

Spectra Pricision
See **Spectra Geospatial**.

Spectrograph
A spectrograph is a recording spectroscope. The spectra under investigation are recorded by photography or other means.

Concave Grating Spectrograph: A spectrograph in which the spectrum is produced by a Rowland concave grating instead of a prism.

Féry Spectrograph: A spectrograph which uses a prism with curved surfaces.

Quartz Spectrograph: A spectrograph with optical components made of quartz to enable use with ultra-violet light.

Littrow Quartz Spectrograph: A compact quartz spectrograph designed to provide a widely dispersed spectrum. The Littrow quartz spectrograph was invented by **Otto von Littrow**.

Ultra-Violet Glass Spectrograph: A spectrograph with optical components made of glass, designed for use with ultra-violet light.

Spectroheliograph
A spectroheliograph is an instrument, invented by **George Ellery Hale**, for taking a monochromatic image of the sun's disc at a chosen wavelength.

Spectrometer
A spectrometer is a spectroscope that is equipped to make measurements of the spectral properties under observation.

Spectrophotometer
A spectrophotometer is an instrument for measuring the intensity of light with respect to wavelength.

Spectroscope
A spectroscope is an instrument for the production and examination of spectra. It normally consists of a collimator to produce a parallel beam of light, a prism or diffraction grating to produce the spectrum, and a viewing telescope.

Thorp's diffraction spectroscope, manufactured by **R.&J. Beck**, is a spectroscope in which a diffraction grating is used to produce the spectrum.

A *direct vision spectroscope* is one in which a direct vision prism is used to give an un-deviated spectrum.

The *echelon spectroscope*, invented by **Albert Michelson** in 1878, uses an echelon grating instead of a standard diffraction grating, yielding considerably higher resolution.

The *Evershed spectroscope* is one in which the spectrum is produced by two or more prisms, and is used for the examination of the spectra of solar prominences. It

Speculum; Speculum Metal

A speculum is:

(1) A mirror or looking glass; especially a metal mirror.

(2) A reflector of polished metal used in astronomical telescopes.

(3) A medical instrument used to dilate certain bodily passages.

Mirrors for instruments such as reflecting telescopes and microscopes were originally made of polished metal, generally consisting of a mixture of copper and tin. This was referred to as a speculum, speculum metal, or metal. The exact mix of metals was subject to considerable research and experimentation in the early days of reflecting telescopes, with the aim of attaining the highest possible reflectivity as well as the best possible thermal properties. In the mid-nineteenth century silvered glass mirrors took over from metal mirrors in astronomical telescope-making [24].

Spencer & Browning

Spencer & Browning, [18] [90], were optical and mathematical instrument makers of 327 Wapping, London, England. The partnership between **William Spencer (I)** and his brother-in-law, **Samuel Browning (I)**, was formed in 1781. In 1784 they were joined by **Ebenezer Rust (I)**, and the partnership became **Spencer, Browning & Rust**.

See Appendix VII.

Spencer, Browning & Co.

Spencer Browning & Co., [18] [90], were wholesale manufacturers of mathematical, philosophical, meteorological, optical, nautical, and surveying instruments of London, England. Addresses: 66 Wapping (1840); 111 Minories (1840-1870); 6 Vine St. (1848-1870); 6 America Square (1852); 6 Vine Street, America Square (1856).

They succeeded the business of **Spencer Browning and Rust** in 1840. It is likely that neither the family of **William Spencer (I)**, nor the family of **John Spencer (I)**, had any involvement in the partnership. In 1855 the partnership, then consisting of **William Browning** and two of his sons, **Samuel John Browning** and **John Browning (III)**, was dissolved [35 #21831 p4843]. From this date Samuel John traded at 52 High Street, Portsmouth, while William and John (III) each worked "on their own account" at 111 Minories. The name of Spencer, Browning & Co., however, continued to be used until 1870, after which John (III) traded as **John Browning & Co.**

An 1856 advertisement, [44], reads "Established 90 years. Spencer, Browning & Co. (Late Spencer, Browning and Rust). Manufacturers of Sextants, Quadrants, Telescopes, Compasses, Barometers, Binnacles, Brass side & deck lights, Lamps & lanterns, Time glasses, Glass deck lights, &c., &c., and all sorts of Mathematical, Optical and Surveying instruments; wholesale Ship Chandlers, Bunting factors and Flag makers, 111 Minories, and 6 Vine Street, America Square, London...." The advertisement also contains a warning about counterfeit instruments bearing their name, or part of their name.

See Appendix VII.

Spencer Browning & Rust

Spencer Browning & Rust, (fl.1784-1840) [18] [90], was a manufacturer of marine and navigational instruments of London, England. Addresses: 66 & 327 Wapping; 66 & 67 Wapping; 327 Wapping (1784-1796); 66 High Street, Wapping (1797-1839); 66 Wapping (1816).

The partnership began as **Spencer & Browning**, formed in 1781 by **William Spencer (I)** and **Samuel Browning (I)**, and continued until 1784 when they were joined by **Ebenezer Rust (I)**, after which they traded as Spencer Browning and Rust.

They inscribed their instruments SBR. They also used a cursive SBR signature on scales divided on their dividing engine, many of which were for other instrument makers (see Appendix XI). They traded under the same name until 1840. All of the original partners of Spencer, Browning & Rust had died by 1819. By the time of his death, William Spencer (I) had already ceased his involvement with the firm, and begun to withdraw his capital in the firm in annual instalments, leading to the end of his family's involvement in the firm. After this it is not clear what, if any, participation the family of William Spencer (I), or the family of **John Spencer (I)**, subsequently had in the company. However, Ebenezer Rust (II), Richard Browning, and Richard's brother **William Browning**, continued the firm, retaining Spencer's name. In 1838 Ebenezer Rust (II) died, and in 1840 the name was changed to **Spencer, Browning & Co.**

See Appendix VII.

Spencer, Charles Achilles

Charles Achilles Spencer, (1813-1881) [185] [201] [264] [320], was an optical instrument maker, initially working in Canastota, New York, USA. He was known for his refracting telescopes as well as his Gregorian and Newtonian reflecting telescopes and microscopes, and particularly for his high-quality microscope objectives. Spencer entered a partnership with Professor Asahel Eaton, trading as Spencer and Eaton. They built a 13½-inch refractor with focal length 16 feet for Hamilton College. In order to make the setting circles for this telescope Spencer built a dividing engine.

In 1843 Spencer took on **Robert B. Tolles** as an apprentice. Eaton convinced Spencer to teach Tolles his techniques and formulae for making microscope objectives and in 1858, armed with this knowledge, Tolles began his own business in Canastota. In 1865 Spencer formed a partnership with his son, **Herbert Spencer**. Spencer and Eaton continued to make microscopes and telescopes until the Canastota shop burned down in 1873. In 1875 Spencer went to Geneva, New York, to work for the **Geneva Optical Company**, with whom he worked for about two years, during which time he continued to make microscope objectives under his own name.

Spencer and Eaton
See **Charles Achilles Spencer**.

Spencer, Herbert
Herbert Spencer, (1849-1900) [185], was a son of **Charles Achilles Spencer**. He worked for his father, and they entered a partnership in 1865. In 1880 he formed his own company, Herbert Spencer & Co. in Geneva, New York, USA, making microscopes, telescopes, field glasses and opera glasses. He moved his company to Cleveland, New York, in 1889, and continued in business until 1891 when he moved to Buffalo, New York, and formed the company Spencer & Smith Optical Co. This was renamed to become the Spencer Lens Company in 1895, and was taken over by the **American Optical Company** in 1935.

Spencer, John (I)
John Spencer (I), [90], was the father of three members of the Spencer family who were mathematical instrument makers: Samuel, John (II), and **William (II)**.

Samuel and John (II) both served their apprenticeships with **William Spencer (I)**, Samuel beginning in 1778 and John (II) beginning in 1793. Neither, however, took their freedom of the **Grocers' Company**.

See Appendix VII.

Spencer, John (II)
See **John Spencer (I)**.

Spencer, John (III)
See **William Spencer (II)**.

Spencer, John (IV)
John Spencer (IV), (fl.1838-1863) [18] [304], was an optical, mathematical, and philosophical instrument maker of Dublin, Ireland. Addresses: 128 Summerhill (1838); 3 Aungier St. (1845-1851); 13 Aungier St. (1852-1863). In 1864, he began to work in partnership with his son, trading as **John Spencer & Son**.

Spencer & Son, John
John Spencer & Son, [304], were optical, philosophical, and engineering instrument makers, and mechanical engineers, of Dublin, Ireland. Addresses: 13 Aungier St. (1864-1868); 19 Grafton St. (1866-1883); 23 Nassau St. (1884-1886). The firm was a partnership, formed in 1864 between **John Spencer (IV)** and his son. By 1873 they were Opticians and Engineering Instrument Manufacturers to the Board of Works. They are known to have made: microscopes; surveying levels; optical saccharometer; Jellett's saccharimeter; cathetometer, railway transit theodolite; air pumps, Cruise endoscope; equatorial stand; heliostat. They also retailed microscopes from other makers. John Spencer and Son were the first to make the Stoney Heliostat, and they built a binocular spectroscope for **Charles Burton**, with parts by **Sir Howard Grubb**.

Spencer Lens Company
See **Herbert Spencer**.

Spencer & Smith Optical Co.
See **Herbert Spencer**.

Spencer, William (I)
William Spencer (I), (c1751-1816) [18] [90], was a mathematical instrument maker of London, England. Addresses: 26 Wapping St. (1777); 327 Wapping St. (1778-1795); Near Union Stairs in Wapping (1786).

He served his apprenticeship with **Richard Rust** beginning in 1766, and was freed of the **Grocers' Company** in 1773. In 1778 he took on Samuel Spencer, son of **John Spencer (I)**, as an apprentice. William (I) married the sister of **Samuel Browning (I)**, and in 1781 he joined in partnership with Samuel Browning (I), trading with his brother-in-law as **Spencer & Browning**. In 1784 **Ebenezer Rust (I)**, joined the partnership, and they began to trade as **Spencer Browning & Rust**. In 1793 William (I) took on John Spencer (II), son of John Spencer (I), as an apprentice. **George Mills** was turned over to him by **Walton Willcox** in 1783 to complete his apprenticeship.

Prior to his death in 1816, William Spencer (I) ceased participating in Spencer, Browning & Rust, and began to withdraw his capital from the firm in annual instalments, thus ending his family's interest in the firm. During the time taken to withdraw all of his capital it is not clear what, if any, involvement his family, or that of **John Spencer (I)**, had in Spencer, Browning & Rust and its successors.

See Appendix VII.

Spencer, William (II)

William Spencer (II), (fl.1811-at least 1839) [**18**], mathematical instrument maker of London, England, was a son of **John Spencer (I)**. He served his apprenticeship with **Samuel Browning (I)**, and was freed of the **Grocers' Company** in 1811. William (II) took two of his sons as apprentices: John (III) and William (III), both beginning in 1839.

See Appendix VII.

Spherical Aberration

Spherical aberration is so named because it arises with the use of a spherically figured mirror or lens. It is, however, present to varying degrees in mirrors or lenses of other figures. Spherical aberration occurs because rays of light parallel to the optical axis have different focus depending on their distance from the optical axis. When using a lens or mirror with uncorrected spherical aberration there is no point in the image at which perfect focus can be achieved. This is because at every point on the image the focus points for different incoming rays of light are at different distances.

Sphero-cylindrical Lens

See lens.

Spherometer

A spherometer is an instrument for measuring the diameter of curvature of spherical surfaces, such as lenses. Several kinds of spherometer exist.

The simplest type of spherometer, known as a *lens measure* consists of a linear arrangement of three equally spaced pins. The outer two are fixed, defining a straight line, and the centre pin is sprung to move in and out, and attached to a dial showing how far from the straight line it is. A lens measure is usually calibrated to read in dioptres for glass of refractive index 1.523, as is usually used for measuring the lenses in spectacles.

The most common type of spherometer consists of three legs with pointed ends, positioned at the points of an equilateral triangle, thus defining a plane. A fourth pin is mounted centrally on a micrometer, and may be adjusted up and down until it touches the surface upon which the three legs are placed. The micrometer reads the distance from the plane of the legs, perpendicular to it, to the point at the centre. The instrument is placed on a flat surface to calibrate the zero reading, and when placed on a spherical surface it then measures the radius of curvature. The radius of curvature, R, is calculated using the spherometer formula:

$$R = (L^2)/(6h) + h/2$$

where L is the length of side of the equilateral triangle formed by the spherometer's three legs, and h is the height reading indicated on the scale on the spherometer. The spherometer formula is derived using basic trigonometry.

An *Aldis spherometer*, invented by Mr. A.C. Aldis of **Aldis Brothers** improves on the more standard design by using three small spheres instead of three pointed legs. In this design the lens is usually placed onto the three spheres, and a micrometer measure is adjusted up from underneath to touch the centre and thus give a reading.

An *Abbe spherometer*, invented by **Ernst Abbe**, makes contact with the spherical surface with, instead of three pointed feet or spheres, a single circular edge. A micrometer attached to a central measuring pin works in the same way as in other models.

Spicer, Edward

Edward Spicer, (fl.1760-1772 d.c1775) [**18**] [**276** #655] [**304**] [**322**], was a mathematical, surveying, and nautical instrument maker of Plunket Street, Dublin, Ireland. His surveying instruments were recommended by the Irish surveyor, Benjamin Noble, for professional use. He may have worked in partnership with **James Lynch (I)**, as evidenced by advertisements for a compass theodolite, and "all other instruments for surveyors, navigators, &c. ... as sold by Edward Spicer and James Lynch".

Spicer, Simon

Simon Spicer, (fl.1777-1784) [**2**] [**18**], was an optician and mathematical instrument maker of London, England. Addresses: Three Falcon Court, Fleet Street (1777); 36 Little Britain (1780); Charing Cross (1784); At the latter address, his utensils, stock and goods in trust were insured for £450. He served a seven-year apprenticeship with **Jesse Ramsden**, beginning in 1768.

Spider; Spider Vane

A spider, or spider vane, is an arrangement of thin material struts that hold the secondary mirror in position on a reflecting telescope. They are made as thin as possible in order to reduce the optical effect of their presence in the optical path for the light that enters the telescope to fall on the main mirror.

Spina, Alessandro

Alessandro Spina, [**1**] [**121**], was an Italian friar from the Dominican convent of St Catherine in Pisa. He is credited (as are others) with the invention of spectacles in the 12th century.

Spindler & Hoyer

Spindler & Hoyer of Göttingen, Germany, was established in 1898 by August Spindler (1870-1927) and Adolf Hoyer (1874-1943), and succeeded the company of **Carl Diederichs**. They produced a range of optical and scientific instruments, and were a major supplier of binoculars to the German army during WWI and WWII. During WWII they marked their military optics with the **ordnance code** fvs. In 1996, following a management buy-out, the company became **LINOS AG**. They went on to acquire other optical instrument and component manufacturers including **Steeg & Reuter Präzisionsoptik**, Franke Optik Vertriebsgesellschaft, Gsänger Optoelektronik GmbH, and **Rodenstock Präzisionsoptik**. The company went public in 2000, and in 2009 became part of the **Qioptiq Group**.

Spinoza, Benedict de

Benedict de Spinoza, (1632-1677) [45], [289], was a Dutch Jewish philosopher, variously known by his Hebrew name Baruch de Spinoza, his Latin name Benedictus de Spinoza, and his Portuguese name Bento de Espinosa. He was a major contributor to the philosophy of Rationalism, and a key part of the Enlightenment movement. His profession as an optical instrument maker is less widely known than are his contributions to philosophy. He specialised in grinding lenses, and made telescopes and microscopes, although it is not known whether he made the complete instruments, or made the optical components for instruments constructed by others. He gained a good reputation for his lenses, and received acclaim from natural philosophers including Gottfried Wilhelm Leibniz (1646-1716), and **Christian Huygens**.

Spinthariscope

The spinthariscope is an educational device or toy that enables the user to view individual scintillations as alpha particles from a radioactive sample strike a fluorescent screen. The sample and fluorescent screen are generally mounted in a short hand-held tube, at the end of which is a viewing lens. The device was invented in 1903 by the British chemist and physicist Sir William Crookes (1832-1919). Early spinthariscopes were made using radium compounds as the radioactive source. Modern instruments are made to higher safety standards and are more likely to use thorium.

Split Image Telescope

A split image telescope is one that uses a double image micrometer to form two images of an object. It is designed to measure angular distances such as the separation between stars, or the diameter of the Sun. A split image telescope was used by the eminent German astronomer and mathematician Friedrich Wilhelm Bessel, (1784-1846), in his measurements of the angular separation of stars as small as less than 0.00002°.

Spotting Scope

A spotting scope, or spotting telescope, is a kind of refracting telescope. Spotting scopes are generally for terrestrial use, and usually make use of erecting prisms to produce an upright image with corrected left/right image orientation. Spotting scopes may have an integral eyepiece or interchangeable eyepieces, and often make use of zoom eyepieces.

Sprenger, Eduard

Eduard Sprenger, (fl.1876-1896) [374], was a mathematical and optical instrument maker of Berlin, Germany. He is known to have made telescopes, surveying instruments, and drawing instruments. He exhibited his work in London in 1876, and in Berlin in 1879 and 1896. He was a member of the founding committee of the Deutsche Gesellschaft für Mechanik und Optik (German Society for Mechanics and Optics) in Berlin in 1881. During WWII his company marked their instruments with the **ordnance code** cln.

Square D

The Square D company of the USA is a major manufacturer of electrical equipment, founded in 1902. In 1939 they acquired the firm **Kollsman Instrument Company, Inc.**, which became their Kollsman Instrument Division. In 1940 they acquired **John H. Emmerich Optical Company**, which was subsequently merged into the Kollsman Instrument Division. They began manufacturing binoculars under the **Sard** trade-name, and continued until 1950, when the Kollsman Instrument Division was sold and production of binoculars ceased.

Springer, Joshua

Joshua Springer, (fl.c1760) [3] [18], was a scientific instrument maker, trading at *Hadley's Quadrant*, in St. Stephen's Lane, Bristol, England. He succeeded **John Wright** at this address, which is almost certainly the same address as the *Sphere and Hadley's Quadrant*, Near St. Stephen's Church, Bristol. By 1774 he was trading at 2 Clare Street, Bristol. He was succeeded at this address in 1808 by **Richard and Charles Beilby**, and in 1809 he is recorded at Kings down, Bristol.

According to his trade card: "Joshua Springer, Mathematical, Philosophical, and Optical instrument-maker, At Hadley's Quadrant...(Lately the Shop of Mr. John Wright,) Makes to the utmost degree of accuracy, and sells wholesale and retail, every individual instrument, that has been invented (or improv'd) in the above branches of science. Particularly, Hadley's quadrants, mounted either in silver, brass or wood, whole glasses are truly plane, parallel, and rightly adjusted; Also Smith's, Davis's, and other quadrants; Azimuth, amplitude, and other compasses, either for cabin, steerage, or pocket; Artificial magnets; Measuring wheels; Coach and chaise way-wisers; Theodolites; Circumferenters, plain tables and spirit levels; Curious cases of instruments; Magazine cases with sectors, scales, elliptical, proportional, triangular compasses, &c. Horizontal sundials for any latitude; Gunner's quadrants, and all other instruments for fortification; Gunter's scales; Gauging rods; Parallel and all other kinds of rulers; Orreries, spheres and globes of any size; Electrical machines; Barometers and thermometers of all sorts; Air pumps either double or single barrel; Hydrostatic balances; Reflecting telescopes either Newtonian or Gregorian with great improvements; Refracting telescopes either for sea or land; Various sorts of microscopes, single or compound; Diagonal mirrors, and camera obscuras; Convex and concave mirrors; Spectacles; white crown, or Venetian green glass, &c. &c. Merchants, captains of ships, and others, may be supplied with any of the above goods on the shortest notice, and as cheap as in London."

Springer, William
William Springer, (fl.1775-1811) [3] [18], was an optician, and optical, mathematical, and philosophical instrument maker of Bristol, England. Addresses: 24 Charles St. (1775); Quay (1783-1784); Charles St. (1785-1811); John St. (1787); St. John's Bridge (1793-1795).

Spyglass
A spyglass is a hand-held refracting telescope with an integral eyepiece.

Srb and Štys
Srb and Štys was founded in 1919 by Jaroslav Srb (1893-1967) and Josef Štys (1889-1950) in Prague, Czechoslovakia. They repaired optical instruments, including microscopes, binoculars and projectors, and constructed instruments from parts supplied by other companies. They expanded their business to manufacture geodetic and medical instruments, epidiascopes, microscopes, telescopes and gunsights. Under German occupation during WWII, they were required to produce instruments for the German military. These were marked with the **ordnance code** bmk. In 1945, after occupation, the company was nationalised and became part of **Meopta**.

Stackpole & Brother
Stackpole & Brother, [58], was a partnership between the Irish brothers William Stackpole (1819-1895) and Robert Stackpole (1823-1873), who emigrated to the USA in 1833. The company produced mathematical, astronomical, navigation and surveying instruments until they ceased trading in 1910.

Stadimeter
A stadimeter is an instrument used to estimate the distance to an object of known size by measuring its angular size. The distance is calculated by means of trigonometry.

Staley & Co., A.E.
A.E. Staley & Co., [306], was established by Alfred Edward Staley in about 1895, as dealers and importers of photographic apparatus, trading at 35 Aldermanbury, London, England. In 1904 they began to advertise as wholesale manufacturers, and soon began to use the *Royal* brand-name. They marked their cameras with the initials *AES & Co*. In 1905 the company moved to Thavies Inn, Holborn Circus, London, trading initially at number 19, and from 1914 at number 24. In 1915 they took over the company **J.F. Shew & Co.**, and began to trade as **Staley, Shew & Co.**, with **Walter Docktree** in charge of production.

Staley, Shew & Co.
Staley, Shew & Co., [306], camera manufacturer of 88 Newman Street, London, England, was formed when **A.E. Staley & Co.** acquired **J.F. Shew & Co.** in 1915, with **Walter Docktree** in charge of production. The company traded as Staley, Shew & Co. until 1919.

Stancliffe, John
John Stancliffe, (fl.1770-1807) [2] [3] [18] [276 #896], was a mathematical instrument maker of York, England, and later, London. Addresses: 26 Little Marylebone Street, Cavendish Square, London (1779); Little Marylebone Street (or Marybone Street), London (1793-1812). He served his apprenticeship with **Henry Hindley** of York, England. Hindley is believed to have been the first to construct a circular dividing engine, which was designed for cutting clock-wheels, and was adapted for dividing scales on astronomical instruments. Stancliffe moved to London to work for **Jesse Ramsden**, and is reputed to have become his foreman. It is not certain that Stancliffe transferred the dividing engine technology from Hindley to Ramsden, but historians have speculated that it is likely. Stancliffe is said to have finished building his own dividing engine in 1788. Some sources say that this was while he was still an apprentice, but this seems unlikely, and if it were true, the date must be wrong. He

went on to set up in business making sextants, and one of his apprentices was his nephew, Benjamin Stancliffe, beginning in 1796. John Stancliffe worked with **John Bird** making instruments for the Radcliffe Observatory at the University of Oxford. Benjamin Stancliffe succeeded him in business, trading at 13 Bennet St., Blackfriars Rd. (1816-1822).

Standard Optical Co.
Standard Optical Co., of Geneva, New York, USA, was a sales company for the lenses produced by the **Geneva Optical Company**. Geneva Optical and Standard Optical later merged to become **Shuron Optical**, which continued in business until 1963.

Stanhope Lens
A Stanhope lens, [30] [287], is handheld magnifier that acts as a simple microscope. It consists of an objective attached to a handle. The objective consists of a cylindrical lens, at each end of which is a convex surface. The focus of the lens is at the surface of the least convex lens. This necessitates bringing the lens into contact, or near-contact with the specimen being viewed.

The Stanhope lens was invented by Charles Stanhope, Lord Mahon, (1753-1816) [276, #744], who was an amateur scientist and inventor. Examples were made by, among others, **Francis West**.

Stanley, William Ford
William Ford Stanley, (1829-1909) [50] [44] [46] [53] [306], was a British engineer and inventor, born in Islington, Middlesex, England, (now a London borough). His family's finances did not allow him a good education, but at age 18 he began to work in a pattern shop at an engineering company in Whitechapel. He spent his spare time studying, and designing various contraptions. He invented the steel-wired spider wheel, which was soon to be in common usage, but he was discouraged from patenting the idea by his father who believed it would not be strong enough for use on the roads. After three years he went into partnership with his uncle, from whom he learned the value of business-sense.

In 1854 he opened a shop at 3 Great Turnstile, Holborn, London, and began to design and make drawing instruments. His cousin, Henry Robinson, became his partner, bringing much needed funding, and Stanley supplemented his business by making an improved version of the *panoptic* stereoscope, a popular toy. At the 1862 London Exhibition Stanley was awarded a gold medal for a straight-line dividing engine that he designed and built. This boosted his sales both in the UK and abroad, and began his rise to success. Stanley moved to Norwood in Surrey in 1865, and at his factory in South Norwood he manufactured a variety of instruments including microscopes and improved designs of theodolite and other surveying instruments. He was one of the first instrument makers to work in aluminium.

In 1900 the firm was incorporated as a limited company, W.F. Stanley & Co. William Stanley was a member of the Physical, Geological, Royal Meteorological, and Royal Astronomical societies, and of the British Astronomical Association. In 1919 W.F. Stanley & Co. was listed as a member of the **British Optical Instrument Manufacturers' Association**. By 1937 W.F. Stanley & Co. had acquired **Heath & Co.**

In 1947 Stanley was listed as an exhibitor at the British Industries Fair. The catalogue entry, [450], listed them as manufacturers of: Theodolites; Levels; Surveying equipment; Planimeters; Pantographs; Integrators; Drafting machines; Drawing instruments; Drawing office equipment; Binnacles; Compasses; Binoculars; Sextants; Sounding machines; Navigational equipment; Thermometers; Barometers; Meteorological equipment; Optical components; Light engineering.

W.F. Stanley & Co. continued to produce instruments until the company was liquidated in 1999. There are many fake Stanley products available through auctions and other web sites. There is an American company currently operating as "Stanley London" selling, among other things, budget reproductions of genuine W.F. Stanley & Co. instruments.

Stanton Brothers
Stanton Brothers, [18], telescope makers, tube makers, and metal warehouse, was a partnership between **John Stanton**'s three sons, John Alfred, Robert, and George William. The business operated from c1845 at 73 Shoe Lane, Holborn, London, England. All three brothers were optical turners. John Alfred (d.1880) [35 #24999 p3711] served his apprenticeship with his father, beginning in 1829, and was freed of the **Drapers' Company** in 1837. Robert and George William were both freed of the Drapers' Company by patrimony, in 1841 and 1843 respectively. George William stayed in the partnership until at least 1854, but by the time of John Alfred's death in 1880 Stanton Brothers was a partnership between himself and Robert. Stanton Brothers was still operating as a metal warehouse at the same address until at least the 1960s [44].

Stanton, John
John Stanton, (fl.1813-1843) [18] [276 #1226] [319] [322], was a telescope maker, brass tube maker, and optical turner of London, England. Addresses: 111 Shoe Lane, Holborn; 82 Shoe Lane, Holborn (1813); 73 Shoe Lane (1816-1843). He served his apprenticeship with the

brass founder and brass turner Philip Hewes of London, beginning in 1804, and was freed of the **Drapers' Company** in 1813. Among his apprentices were **Robert Gogerty**, and his son John Alfred. He retired in 1843, and was succeeded in business by his three sons, John Alfred, Robert, and George William, trading as **Stanton Brothers**.

Starlight Xpress
Starlight Xpress of Binfield, Berkshire, England, is a supplier of cooled CCD cameras and accessories for astrophotography and industrial imaging. The company was founded in 1991, and the cameras are designed by the amateur astronomer Terry Platt.

Star-Liner Company
Star-Liner Company, of 1106 South Columbus Blvd., Tucson, Arizona, USA, made reflector telescopes and mounts, including a pedestal mounted German equatorial mount. The business began in the mid-C20th, and was still going in the 1980s. A 1965 advertisement in *Sky and Telescope* magazine shows a 20-inch equatorially mounted observatory f/16 Cassegrain, and lists 12.5-inch to 24-inch Cassegrains. A similar, undated advertisement shows the 24-inch observatory model, and lists a Cassegrain/Newtonian, and sizes 10, 12.5, 14, 16, 20 and 24-inch models. They also sold smaller models aimed at the amateur market.

Statham, William Edward
William Edward Statham, (1816-1899) [50], of London, England, was a chemist who sold educational chemistry sets and other related products. He also sold microscopes and other optical instruments bearing his name, although it is unlikely he made them himself. His business was continued by his son, Francis.

Stationers' Company
See Appendix IX: The London Guilds.

Stebbing & Son, George
George Stebbing & Son, [18], was a partnership between **George Stebbing** and his son, **Horatio Nelson Stebbing**, trading at 66 High Street, Portsmouth, England, until the partnership was terminated in 1845 [35 #20521 p3245]. George Stebbing & Son continued to trade, however, until the year of George's death in 1847.

Stebbing & Wood
Stebbing & Wood, [18] [322] [358], optical, mathematical, and nautical instrument makers of 47 High St., Southampton, England, was a partnership between **Joseph Rankin Stebbing** and Albert Wood, trading from 1851 until 1853, and succeeded the business of **J.R.&H. Stebbing**. They were awarded a royal appointment to Queen Victoria.

Stebbing, George
George Stebbing, (1775-1847) [13] [18] [276 #1435] [322], of Portsmouth, England, was an optical, nautical, mathematical, and philosophical instrument maker. He was born in Holborn, London, and may have learned his trade in London, prior to moving to 29 Broad St., Portsmouth where, by 1804 he was advertising as a working optician, and manufacturer of optical and mathematical instruments. He continued to trade at this address until 1845, and also had premises at 60 High Street (1810); 99 High St. (1816); 66 High St. (1830); and 5 Common Hard, Portsea. In 1810 he patented an improved sea and land compass. He was freed of the **Vintners' Company** in 1816, and among his apprentices were his sons Frederick George, Richard William, and **George James Stebbing**. He is known to have made: telescopes; microscopes; sextants; quadrants; octants; clinometers; compasses; marine barometers; thermometers; globes. He was awarded a royal appointment to the Duchess of Kent. In 1832 he acted as an agent for **Robert B. Bate**. He worked in partnership with another son, **Horatio Nelson Stebbing**, trading as **George Stebbing & Son**. The partnership was terminated in 1845, [35 #20521 p3245], but George continued to trade under this name until his death. His son **Joseph Rankin Stebbing** was also a successful instrument maker.

Stebbing, George James
George James Stebbing, (1803-1860) [18], of Portsmouth, England, was an optical instrument maker, and a son of **George Stebbing**. He served his apprenticeship with his father, beginning in 1817, and was freed of the **Vintners' Company** in 1825. He accompanied Charles Darwin, as Librarian and Instrument Maker, on his trip around the world in HMS Beagle from 1831 until 1836. He went on to set up business in Portsmouth in direct competition with his father.

Stebbing, Horatio Nelson
Horatio Nelson Stebbing, (1812-1883), English optical and mathematical instrument maker, was a son of **George Stebbing**. He worked in partnership with his brother **Joseph Rankin Stebbing**, trading in Southampton as **J.R.&H. Stebbing** until 1836. He later worked in partnership with his father in Portsmouth, trading as **George Stebbing & Son** until 1845.

Stebbing, J.R.&H.

J.R.&H. Stebbing, [18] [319] [322] [358], optical and mathematical instrument makers of 47 and 63 High St., Southampton, England, was a partnership between **Joseph Rankin Stebbing** and his brother **Horatio Nelson Stebbing** during the 1830s. They were awarded a royal appointment to Queen Victoria and the Duchess of Kent. The partnership was terminated in 1836 [35 #19385 p935]. The business was succeeded by **Stebbing & Wood**.

Stebbing, Joseph Rankin

Joseph Rankin Stebbing, (1810-1874) [18] [53] [322], was an English optical, mathematical, and nautical instrument maker, and a son of **George Stebbing**. He moved from his home town of Portsmouth to start his business at 47 High St., Southampton. He is known to have produced sextants, marine and stick barometers, and telescopes. He worked in partnership with his brother **Horatio Nelson Stebbing**, trading at 47 and 63 High St., as **J.R.&H. Stebbing** until 1836. From 1851 until 1853 he worked in partnership with Albert Wood, trading as **Stebbing & Wood**. He was awarded a royal appointment to Queen Victoria. Joseph went on to become mayor of Southampton in 1867.

Stedman, Christopher

Christopher Stedman, (fl.1747-1774 d.c1774) [18] [276 #541] [322], was a mathematical, optical, surveying, and navigational instrument maker of London, England. Addresses: *The Globe* on London Bridge (1747-1750); Leadenhall St. (1758-1763); 24 Leadenhall St. (until 1774). He served his apprenticeship with T. Devonish, beginning in 1737, then Samuel Austin from 1738, and was freed of the **Stationers' Company** in 1745. In all he trained about 18 apprentices, and by 1759 he was employing at least 15 people in addition to apprentices. Among his apprentices was **James Chapman (I)**.

According to his trade card [361], "Mathematical instrument maker at The Globe on London Bridge. Makes & sells all sorts of Mathematical instruments for sea or land gauging, surveying, measuring, geometry, navigation, arithmetik, &c. Wholesale or retail." Instruments bearing his name include: small telescope; backstaff; slide-rule; magnetic compass.

In 1765 he was successfully sued by **Peter Dollond** for infringement of the Dollond achromatic lens patent (see Appendix X). He was succeeded in business at 24 Leadenhall St. by his wife, the mathematical instrument maker Elizabeth Stedman in 1774, and she, in turn, was succeeded by their son, the mathematical instrument maker also named Christopher Stedman, in 1784.

Steeg & Reuter Präzisionsoptik

Steeg & Reuter Präzisionsoptik, [374], were optical instrument makers of Bad Hamburg, Germany. The company was established in 1855 by Wilhelm Steeg, trading as Optisches Institut Wilhelm Steeg, and specialising in polarising optics. In 1866, Peter Reuter began to work for the company, and in 1877 he joined Steeg in partnership, at which time they renamed the company Dr. Steeg & Reuter. The firm was later renamed Steeg & Reuter Präzisionsoptik. In 1996 the firm was acquired by **LINOS AG**, formally **Spindler & Hoyer** and, in 2006, LINOS AG became a part of the **Qioptic Group**.

Instruments bearing their name include: saccharimeter; Nörremberg polariscope; polarizing microscope; polarimeter; 2¼-inch refracting library telescope; contact goniometer; mounted birefringent specimens; crystal sections; crystal oscillator; microphone.

Steinheil & Söhne

Carl August von Steinheil (1801-1870), [24] [37] [70] [374], was a German physicist, and an innovator in the fields of photography and lens design. When **W.H.F. Talbot** announced his negative-positive photographic process in 1839, Steinheil began his own experiments, working with Franz Von Kobel, on the photographic process. He designed and built his own cameras, and originated some minor improvements to the daguerreotype process. In 1855 C.A. Steinheil started an optical workshop with his son Hugo Adolph Steinheil, and this was the birth of the company C.A. Steinheil & Söhne in Munich, Germany.

Among those who served their apprenticeship with Steinheil were: **Paul Zschokke**, who rose to become a director of the company; Eugen Hartmann, who went on to found the company **Hartmann & Braun**; Hugo Krüss, the grandson and successor to **Andres Krüss**; Gottlieb Reinfelder and Wilhelm Hertel, who later worked together in the partnership **Reinfelder & Hertel**; and Otto Brugger and B. Breitsamer, who later worked together in the partnership **Brugger & Breitsamer**.

Initially the company produced mostly astronomical telescopes and spectroscopic optics, mainly for big observatories. In 1865 they patented their periscopic wide-angle photographic lens, and in 1866 Adolph Steinheil designed and introduced an improved aplanat, at the same time as **Dallmeyer** introduced a similar lens, the *Rapid Rectilinear*. These lenses became a widely used standard for photographic camera objectives. In 1891 he introduced the telephoto lens, at the same time as, but independently of, Thomas Dallmeyer. After C.A. Steinheil's death in 1870, his son Adolph Steinheil (1832-1893) continued to design and market high quality lenses until his death. Adolph's son, Rudolph Steinheil (1865-1930) took ownership of the company in 1892. The firm remained a family business until 1962, and continued in business until c1995.

Stellar Interferometer
See interferometer.

Stendicke, August
August Stendicke, (fl.late-1870s to early-1880s) [201], was an optical instrument maker of 329 East 23rd Street, New York, NY, USA. He advertised student's spectroscopes and microspectroscopes, and may have made microscopes.

Stephenson, John Ware
John Ware Stephenson, (fl.1870) [264] [285], was an English microscopist, a council member of the Royal Microscopical Society, and a contributor to Encyclopaedia Britannica. He designed a binocular microscope, which was similar to that of **Prof. John Riddell**, but independently conceived, and published his design in his paper *On an Erecting Binocular Microscope* in the *Monthly Microscopical Journal* in 1870. His concept differed from Riddell's in the prism design, being smaller and in closer proximity to the rear element of the objective lens. Microscopes to this design, with various modifications, were manufactured by makers such as **Ross & Co.**, **John Browning (III)**, **Charles Baker**, and **James Swift & Son**.

Stereographoscope
See graphostereoscope.

Stereopticon
A stereopticon is a projector or magic lantern, for use with transparent slides, having two projection lenses to enable dissolving views.

Stereoscope
A stereoscope is an instrument designed to produce a stereoscopic view by means of two pictures of an object taken from slightly different points of view - corresponding to the separation of the eyes. The result is a three-dimensional view as seen by the user.

The instrument was invented by **Sir Charles Wheatstone** in the form of a mirror stereoscope, or reflecting stereoscope, in which the images are viewed by means of mirrors. The more modern form, invented by **Sir David Brewster**, is known as a refracting or lenticular stereoscope, and utilises lenses instead of mirrors.

See lenticular stereoscope, mirror stereoscope, graphostereoscope, Holmes-Bates stereoscope, Tru-Vue, Viewmaster.

Sterrop & Yarwell
Sterrop & Yarwell, [18] [322], were optical instrument makers of the *Archimedes and Three Pair of Golden Spectacles* in Ludgate St., the second Shop from Ludgate, London, England. The firm was a partnership between **Ralph Sterrop** and **John Yarwell**, formed in 1697, which continued until Yarwell's retirement in 1708, after which Ralph Sterrop continued the business under his own name.

According to their trade card, [145], (note the two different spellings of Sterrop's name): "Sterrop's True Spectacles, Made and Sold by Ralph Stirrop, and John Yarwell, at the *Archimedes and Three Pair of Golden Spectacles* in Ludgate St., the second Shop from Ludgate, London." There follows a description of the construction, quality and benefits of their spectacles, after which: "Also telescopes of all lengths for day or night; Perspective great and small; A new double microscope invented by the said Sterrop, fitted up for all uses, particularly that admirable curiosity of seeing the circulation of the blood in small fishes, and other animals, and thousands of living creatures in a drop of pepper water; Magnifying, multiplying and weather glasses; Speaking trumpets, Reading glasses of all sizes, with all other sorts of glasses, both concave and convex, of the newest and most useful invention, all made by the above-named Ralph Sterrop."

Sterrop, Avis
Avis Sterrop, (fl.1756, d.1760), of St. Paul's Churchyard, London, England, [18] [322] [401], was an optician and optical instrument maker. She was freed of the **Spectacle Makers' Company** in 1756 or before. She succeeded her husband, **George Sterrop**, in his business following his death in 1756. The apprenticeship of **Joseph Linnell** was turned over to her in 1756 from **Robert Linnell**, and subsequently turned over again to her mother-in-law, **Mary Sterrop**, when Avis died in 1760.

Sterrop, George
George Sterrop (sometimes Stirrup or Stirrop), (fl.1716, d.1756) [18] [54] [256] [264] [274] [276 #122] [322] [401], was an optician and optical instrument maker of St. Paul's Churchyard, London, England. He was the husband of **Avis Sterrop**, and the nephew of **Ralph Sterrop** whom he succeeded in business. He was the son of **Thomas Sterrop (II)**, and served an apprenticeship with his mother, **Mary Sterrop**. He was freed of the **Spectacle Makers' Company** in 1737. From 1737 until

1747 he worked in partnership with his mother as **Mary Sterrop & Son**, after which he continued to trade under his own name. He was Master of the Spectacle Makers' Company in 1750. Examples of his work include: microscopes; octant; sextant; quadrant; telescopes.

Sterrop, Jane

Jane Sterrop (sometimes Sturrop), (fl.1708-1726) [18] [322] [401], was an optical instrument maker and retailer, of Little Britain, London, England. She was freed of the **Spectacle Makers' Company** in 1709 or before. Upon the death of her husband, **Thomas Sterrop (I)** in 1708 she succeeded him in business, which she continued to run until 1726. Among her apprentices was **John Day**.

Sterrop, Mary

Mary Sterrop, (fl.1730-1763) [18] [322] [401], was an optical instrument maker and retailer of St. Paul's Churchyard, London, England. She was freed of the **Spectacle Makers' Company** in 1730 or before. She succeeded her husband, **Thomas Sterrop (II)** in his business following his death in 1728. Her son, **George Sterrop** served an apprenticeship with her, being freed in 1737. From 1737 until 1747 she worked in partnership with George Sterrop as **Mary Sterrop & Son**. In 1760 the apprenticeship of **Joseph Linnell** was turned over to her following the death of her daughter-in-law, **Avis Sterrop**.

Sterrop, Mary & Son

Mary Sterrop & Son, [18] [322], was a partnership between **Mary Sterrop** and her son, **George Sterrop**, trading between 1737 and 1747 at *Archimedes & One Pair of Golden Spectacles*, the North Side of St. Paul's Churchyard, London, England.

Sterrop, Ralph

Ralph Sterrop (sometimes Stirrup or Stirrop), (1686-1736) [18] [264] [274] [275 #513] [276 #123] [322] [401], was an optical instrument maker of London, England. Addresses: *The Archimedes* in Ludgate St.; St. Paul's Churchyard (1693-1695). He was the uncle of **George Sterrop**. He was freed of the **Spectacle Makers' Company** in 1685. In 1697 he entered a partnership with **John Yarwell**, trading as **Sterrop & Yarwell** until Yarwell's retirement in 1708, following which Ralph Sterrop continued the business under his own name. He was Master of the Spectacle Makers' Company from 1702-1708. Among his apprentices were **Timothy Brandreth**, beginning in 1693 and being freed in 1701, and **George Bass**, beginning in 1706 and being freed in 1716/17. Ralph was succeeded in his business by his nephew George Sterrop.

Sterrop, Thomas (I)

Thomas Sterrop (I) (sometimes Stirrop), (fl.1683-1708) [18] [274] [322] [401], was an optical instrument maker of Little Britain, London, England. He was the son of Ralph Sterrop, a clothier of Worcestershire. He served his apprenticeship with the mathematical instrument maker Thomas King (fl.1667, d.c1677), beginning in 1672, and was freed of the **Spectacle Makers' Company** in 1679, after which he took on at least six apprentices. He was Master of the Spectacle Makers' Company in 1701. He was the father of **Thomas Sterrop (II)**. Upon his death in 1708 he was succeeded in business by his wife, **Jane Sterrop**.

Sterrop, Thomas (II)

Thomas Sterrop (II), (fl.1711, d.c1728) [18] [274] [322] [401], was an optical instrument maker of London, England. Addresses: *Archimedes* in St. Paul's Churchyard (1714); *Archimedes and One Pair Golden Spectacles*, St. Paul's Churchyard (1724). He was the son of **Thomas Sterrop (I)** and **Jane Sterrop**, and the father of **George Sterrop**. He served his apprenticeship with his father beginning in 1701, and was freed of the **Spectacle Makers' Company** in 1708, subsequently taking on at least eight apprentices. Among his apprentices was **Thomas Ribright (I)**, beginning in 1726/7. Upon his death in c1728 he was succeeded in business by his wife, **Mary Sterrop**.

Steward, J.H.

James Henry Steward, (1817-1896) [13] [44] [98], was a telescope-maker, watch and barometer-maker, and jeweller, who established his business in London, England in 1852.

London addresses: 406 The Strand (1856-1971); 54 Cornhill (1866-c1893); 63 St. Paul's Churchyard (1867-c1930); 66 The Strand (1869-c1930); 457 West Strand (1886-1928); 7 Gracechurch St. (1893-c1905); Catherine St. (1971-1973). Sussex address: 154 Church Road, Hove. (1973-1975).

The company was incorporated as J.H. Steward Ltd., in 1913. They were both a retailer and maker of optical, scientific and mathematical instruments. The instruments made by them included binoculars, telescopes, rifle sights, surveying and scientific instruments, and calculators. After 1900 their strategy began to change, and by 1930 they were a purely retail company. Even before this, some of the items they sold marked with the name J.H. Steward were probably made for them by other companies, and branded by J.H. Steward.

Throughout the history of the company, they maintained strong ties with the National Rifle Association, and in 1866 they began to claim that they were Official Optician to the National Rifle Association, and later also Optician to British and Foreign Governments. These were, however, marketing claims by the company rather than official recognition. An advertisement dated 1882 shows: The *Luke* bi-unial lantern for lecturers; the *Bridgeman* triple lantern; the *Lord Bury* telescope; the *Duke* binocular. The advertisement says: "To the British and Foreign Governments, the National Rifle Associations of England, Ireland, Canada and America, and the National Artillery Association, by Appointment." In 1900 J.H. Steward brought out a pocket calculator designed by Professor R.H. Smith.

James Henry Steward had four sons, all of whom continued the business. They were James Henry Charles, Henry William Lake, John James, and William Jesse. Each of these brothers had a shop of their own, and hence the company had several addresses in London. William Jesse had two sons who continued the business: James Henry and William Malcolm. William Malcolm Steward's son David Michael continued the business with his wife Theodora Daphne Lloyd Hood, being the last of the family to run the business until its closure in 1975. Upon closing the business David Michael Steward sold the company and goodwill to **Newbold and Bulford**.

Steward Telemeter
A Steward telemeter was an early form of optical rangefinder made by **J.H. Steward**.

Stewart's Distance-Finder
A Stewart's distance-finder, also known as a *Stuart's distance meter*, is a hand-held optical rangefinder, used in marine applications.

It consists of a wedge-shaped lens mounted to a frame. A long, curved lens, fixed to the edge of a distance scale on a slider, can be moved up and down the frame, adjacent to the wedge-shaped lens. A graduated height bar is mounted on the fixed frame across the scale. A sighting telescope views through both lenses, and the slider is moved to bring the view of the top of the distant target into coincidence with the view of the waterline immediately beneath it.

According to [4, p311], a Stuart's distance meter provides a means of measuring the range of an object of known height at ranges between one quarter of a cable and about 15 cables. Readings at a greater distance cannot be considered reliable.

Stiassnie, Maurice
Maurice Stiassnie, (1851-1930) [50] [264] [285], was an optical instrument maker of Paris, France. He worked for **Constant Verick** prior to taking over the business in the early 1880s. Verick and Stiassnie may have operated the business together in partnership for a few years prior to Stiassnie continuing alone. By 1882 Stiassnie was operating the business at 43 Rue des Écoles, and the Rue De La Parcheminerie 2 address was relinquished in 1885. In 1905 Stiassnie moved his business to 204 Boulevard Raspail. In 1922 he entered a partnership with his brother, trading as Stiassnie Frères. Following the death of Maurice Stiassnie, sometime in the 1930s the firm moved to 67 Boulevard Blanqui. The company closed due to bankruptcy after WWII. In 1970 the name was resuscitated as Societe Nouvelle de Microscopie Stiassnie, but the venture was not a success.

Stoney, George Johnstone
George Johnstone Stoney, (1826-1911) [53], of Leinster, Ireland, was a physicist and university administrator. He designed a small, portable, low cost, clock-driven heliostat. This instrument was first made by **John Spencer & Son** of Dublin.

Stoney Heliostat
See **George Johnstone Stoney**.

Storer, William
William Storer, (fl.1778-1789) [18] [53] [319] [322], was an optician, and optical and mathematical instrument maker of Lisle St., Leicester Fields, London, England, and Great Marlborough St. (1784). He was awarded various patents, including in 1778 for the *Accurate Delineator*, an arrangement of lenses, speculae and mirrors used for drawing the human face; and in 1783 for a method of preparing a special type of glass for optical instruments including telescopes, microscopes, and opera glasses. Instruments bearing his name include: camera obscura; *Artificial Eye* camera obscura; *Royal Delineator* camera obscura; refracting telescope; compound microscope. He was declared bankrupt in 1784 [35 #12594 p7].

Street, Samuel
Samuel Street (I), (fl.1771-1802) [18], was an optician, and optical and mathematical instrument maker of Lomes Court, Coldbath Fields, London, England. He served his apprenticeship with **William Bush**, beginning in 1752, and was freed of the **Stationers' Company** in 1759. Among his apprentices was his son, Samuel Street (II), beginning in 1786. By 1786 he was trading at 26 Plough Court, Fetter Lane, where he continued in business until 1802.

Street, Thomas
Thomas Street, (fl.1829-1880) [18] [276 #2015] [322], was an optician, and optical, mathematical, and

philosophical instrument maker of London, England. Addresses: 30 Commercial Rd., Lambeth; 4 Charles St., Blackfriars Rd; 39 Commercial Rd, Lambeth (1839-1880). One of his employees was the mathematical instrument maker B. Senior, who advertised that he was "from Mr. Street, Principal Worker to **Troughton & Simms**, London".

Instruments bearing his name include: telescopic level; theodolite; pocket compasses; dumpy level.

Stroud, William

William Stroud, (1860-1938) [26] [44], was a co-founder of **Barr and Stroud**. He was born in Bristol, England, and graduated with a master's degree from the schools of mathematics and natural science, Balliol College, Oxford. He continued his scientific studies at the Universities of Heidelberg and Würzburg, Germany, after which he obtained his doctorate in Physics at the University of London.

In June, 1885, Dr. Stroud was appointed to the position of Professor of Physics at the Yorkshire College of Science, Leeds, and there in 1885 he met **Dr. Barr**. In 1887 Dr. Stroud began his first collaboration with Dr. Barr when they devised a lantern-slide camera, which they patented in 1889.

In 1888 the War Office advertised for an infantry rangefinder. This was another opportunity for collaboration between Dr. Barr and Dr. Stroud, and they produced a design which they submitted to the War Office. By 1895 they had opened a workshop producing the rangefinders. This success gave birth to what was to become **Barr & Stroud**, in Anniesland, Glasgow, and a long history of producing optical instruments, largely for the military.

In 1909 Dr. Stroud resigned his chair at University of Leeds and moved to Glasgow to work for the company full-time. He continued in this role, protecting, expanding and diversifying the company's interests, until his death in 1938.

Stuart, George

George Stuart, (fl.1834-1847) [18] [276 #2257], was an optical and mathematical instrument maker of Newcastle Upon Tyne, England. Addresses: 82 Westgate St. (1834-1838) and 17 Clayton St. (1847).

Stuart's Distance Meter

Stuart's distance meter is another name for a Stewart's Distance-Finder.

Süddeutsches Camerawerk Körner & Mayer

See **Nettel Camera-werke**.

Sun Camera Co. Ltd.

Sun Camera Co. Ltd. was a trade-name used by **Lonsdale Bros**.

Sunny Group Co. Ltd.

The Sunny Group is a large, complex Chinese multinational. Some of its component companies are:

Sunny Group Co. Ltd. is an optical equipment manufacturer. It specialises in aspherical lens applications, auto-focus, zoom, and multi-layer coating, as well as other core optical technologies. The group is mainly engaged in optical product development, manufacturing and sales of glass/plastic lenses, mirrors, prisms, lens modules, microscopes, metrology, analytical instruments, cell phone camera modules etc.

Sunny Optics, Inc., designs, manufactures, imports, and distributes telescopes, telescope accessories, binoculars, and spotting scopes. The company was incorporated in 2013 and is based in the United States. Sunny Optics, Inc. operates as a subsidiary of Ningbo Sunny Electronics Co. Ltd. In 2013 **Meade Instruments** announced that it had completed a merger agreement with Sunny Optics, Inc.

Ningbo Sunny Electronics Co. Ltd. develops, manufactures and sells sport and outdoor optical products such as binoculars, telescopes, spotting scopes, riflescopes and diverse optical components and accessories.

Sunny Optical Technology (Group) Company Limited is an integrated optical device manufacturer and optical imaging system solution provider. The Company went public in 2007.

Ningbo Sunny Instruments Co. Ltd., established in 1991, is engaged in product development, manufacturing and sales service. Their main products include digital microscopes, stereo microscopes, biological microscopes, polarizing microscopes, metallurgical microscopes, inverted microscopes, fluorescence microscopes, etc.

Super 8

Kodak introduced the *Super 8* motion film standard in 1965, together with *Super 8* cameras and projectors. This highly influential 8mm film format was subsequently adopted by many other makers of film, cameras, and projectors, and contributed greatly to popularising home movie-making. In 1973 the addition of a magnetic strip made sound recording possible, along with the pictures, on an appropriately equipped camera.

Surgett, Charles

Charles Surgett, (fl.1830-1838) [276 #2016], was an optical instrument maker of 12 Beauchamp Street, London, England.

Sutcliffe Self-Recording Keratometer

A Sutcliffe self-recording keratometer is one which gives a reading in dioptres for the astigmatic error in the human eye.

Swan, Sir Joseph Wilson

Sir Joseph Wilson Swan, (1828-1914) [37] [46] [70], was an English chemist and inventor, best known for his invention of the incandescent lamp. At age 14 he began an apprenticeship with a firm of druggists in Sunderland. He was released from this after three years, and in 1846 he joined the company of John Mawson. In 1864 he became a partner, and the company began to trade as **Mawson & Swan**. Following John Mawson's death in 1867 Joseph Swan took over the management of the company.

Joseph Swan was the first to introduce a commercially viable carbon process for photography, which he patented in 1864. This was a process whereby a sensitised carbon tissue was exposed under a photographic negative, providing the ability to produce multiple prints. In 1877 he devised a method of producing photographic dry plates, with gelatine emulsion, which had an exposure time that compared with that of the wet collodion process. In 1879 he was awarded a patent for gelatine bromide paper, which had been first produced in 1873 by Peter Mawdsley of the Liverpool Dry Plate & Photographic Printing Company.

In 1880 he first demonstrated a commercially viable incandescent lamp, and went on to commercialise it together with the American inventor **Thomas Edison**. In 1881 he patented the cellular lead plate for rechargeable batteries. In 1894 he was elected a Fellow of the Royal Society, and in 1904 he was knighted.

Swann, Thomas

Thomas Swann (sometimes Swan), (fl.1790-1837) [18] [53] [276 #1718], was a mathematical and optical instrument maker, of Liverpool, Lancashire, England.

Addresses: 3 Mann's Island; 14 Strand St. (1790); 5 Murray Square, Atherton St. (1796); 10 Gibralter Row, Canal (1800-1803); 8 Freemason's Row (1807-1810); 31 Freemason's Row (1810); 37 Freemason's Row (1813-1815); 43 Banastre St. (1816-1820); 44 Banastre St. (1821); 54 Banastre St. (1821-1827); 64 Banastre St. (1823); 2 Bridgewater Place (1831-1834); 3 Bridgewater Place (1837).

According to his trade card, "Makes and repairs all sorts of Quadrants, Sextants, Telescopes, Ship compasses, and Time glasses on the most reasonable terms."

Swarovski Optik

Swarovski Optik is an Austrian company, founded in 1949 by Wilhelm Swarovski, the son of Daniel Swarovski, who founded the Swarovski jewellery business in 1895. Today, Swarovski Optik manufactures optical glass, and high-end binoculars and spotting scopes.

Sweeney, Edward

The Sweeney family, [18] [34] [304], were instrument makers of Cork, Ireland. Edward Sweeney (fl.1763), instrument maker, was the son of Nathaniel Sweeney, (fl.1702, d.1763), who was an instrument maker of Paul Street, Cork. Edward succeeded his father in his business, and married Mary Green in 1772. The instrument maker Widow Sweeney (fl.1795) may have been Mary Green. Edward Nathaniel Sweeney (fl.1798), was a mathematical and optical instrument maker of 27 Patrick Street.

According to an advertisement in *The Harp of Erin* paper of Cork on 10th March 1798, Edward Nathaniel Sweeney made and sold: Barometers; Thermometers; Theodolites; Circumferentors; Electrifying machines; Concave and convex glasses mounted in gold, silver, tortoiseshell and horn; Spectacles; Hydrometers; Sextants; Quadrants; Compasses; Cases of drawing instruments; Sun dials; Telescopes; Opera glasses; Linen provers.

A mahogany framed octant, c1780, with brass index arm and fittings, inscribed *Sweeny Cork Fecit For Capt Iohn Leworthy*, is in the National Maritime Museum [16].

Swift Optical Instruments

Swift Optical Instruments, of Texas, USA, started as Swift Instruments Inc., founded in San Jose, CA, USA, in 1926, and incorporated in Massachusetts. During the 1960s their microscopes and binoculars were marketed in the UK by the **Pyser-Britex Group**. The company operates today as Swift Optical Instruments, a member of the **Motic Group**, selling microscopes and accessories. Binoculars, spotting scopes, and riflescopes are sold under their **Swift Sport Optics** brand.

Swift and Son, James

James Swift and Son, [44] [82] [113] [172] [264] [319] [320], were English microscope makers. James Powell Swift (1828-1906) learned his microscope-making skills while working for **Andrew Ross**. In 1854 he left Ross to trade under his own name, James Powell Swift. Addresses: 15 Kingsland Rd., London (1854-1870); City Rd., London (1870-1872); 43 University St., London (1872-1877).

In 1877 James Swift began to work in partnership with his son, Mansell James Swift (1854-1942), trading as Swift & Son. Addresses: 43 University St., Tottenham Court Rd., London (1877-1881); *University Optical Works* 81 Tottenham Court Rd., London (1881-1912).

The business was incorporated in 1912 as James Swift & Son Ltd. Addresses: 81 Tottenham Court Rd., London (1912-1952); 113 Camberwell Rd., London (1952-1968);

43 Pentlands Close, Mitcham, Surrey (1968-1973); 113-115a Camberwell Rd., London (1954); Joule Rd., Basingstoke (1973-1980).

Robert G. Mason worked for Swift prior to setting up his own business in around 1879. During the 1880s and 1890s, Swift produced camera lenses, and some models of camera [306]. In 1891, James Swift & Son were the first to manufacture the petrological microscope designed by **A.B. Dick**. In 1901 the company supplied Captain Scott with their *Discovery* model microscopes for use on his 1901-1904 Discovery expedition. In 1903 James' grandson, Mansell Powell John Swift (1885-1942), joined the company, just three years before James died.

In 1909 Mansell James Swift began to also work in partnership with **Herbert F. Angus**, trading as H.F. Angus & Co. That partnership was dissolved in 1913, [35 #28735 p4910]. In 1919, James Swift & Son Ltd. was listed as a member of the **British Optical Instrument Manufacturers' Association**.

Mansell James Swift and his son, Mansell Powell John Swift, both died in 1942 and shortly after, in 1946, the company was taken over by **E.R. Watts & Son**. In 1947 they exhibited at the British Industries Fair where their catalogue entry, [450], listed them as manufacturers of microscopes for biology, mineralogy, metallurgy, and accessory apparatus. In 1949 John H. Bassett joined the company. He left the company in 1959, but came back and purchased James Swift & Son Ltd. in 1968. In 1981 the company merged with **Prior Scientific**.

Swift, James
See **James Swift & Son**.

Swift, Mansell James
See **James Swift & Son**.

Swift Sport Optics
Swift Sport Optics is a **Swift Optical Instruments** brand of binoculars, spotting scopes, pistol, rifle and "tactical" scopes. Some early Swift telescopes bore the AvA maker's mark, (see Appendix IV), indicating that they were manufactured by **Takahashi**. Swift binoculars were marketed in the UK by Pyser-Britex (Swift) Ltd. (see **Pyser-Britex Group**).

Swygart, Joseph
Joseph Swygart, sometimes known as Swyggett, Suygart, or Suggart, (fl.1805-1849) [18] [276 #2018], was an optician, and optical, mathematical, and philosophical instrument maker of London, England. Addresses: 10 Ely Court, Holborn, (1805); 43 Edmund St., Battle Bridge (1839-1849). During the period 1816-1818, he worked in partnership with **Alexander Mackenzie**, trading as **Mackenzie & Swygart**.

Syeds, Agnes
Agnes Syeds, (fl.1811-1853) [18] [276 #1047a], was a nautical, mathematical, and philosophical instrument maker and compass maker of 379 Rotherhithe St., Near King Stairs, London, England. Upon the death of her husband, **John Syeds**, in 1811, she succeeded him in business, and operated as **A. Syeds & Co.** until 1817 after which she worked in partnership, trading as **Syeds & Davis** until 1833. Following the termination of this partnership, she traded under her own name until 1853.

Syeds, A. & Co.
Agnes Syeds & Co, (fl.1810-1817) [18] [53] [322], optical, nautical, and mathematical instrument maker of 379 Rotherhithe St., Near King Stairs, London, was the company of **Agnes Syeds**. Her trade card stated "Mathematical instrument makers to His Majesty's Navy" and "inventor of the *Boats Betticle*," as well as "Sextants, Quadrants, Telescopes, Compasses, etc., on the most improved principle, made and repaired with care. Manufacturers of brass fittings of every description."

Syeds & Davis
Syeds & Davis, (fl.1817-1835) [18] [276 #1047] [322], were mathematical and nautical instrument makers, compass makers, and ship chandlers of London, England. Addresses: Fountain Stairs, Rotherhithe; Bermondsey Wall; 379 Rotherhithe St., Near King Stairs (1822-1833). The partnership was between **Agnes Syeds** and an instrument maker named Davis. Instruments bearing the name Syeds & Davis include: octants; Hadley's quadrant; compasses; 3-draw brass and mahogany marine telescope; single draw mahogany and brass marine telescope.

Syeds, John
John Syeds, (fl.1788-1811) [18] [53] [276 #1047] [322], was a mariner, compass maker, and nautical instrument maker of London, England. Addresses: 25 Parker's Row, New Rd., Dock Head, Southwark; 367 Rotherhithe Wall; 17 Bermondsey Wall; Mill St, Bermondsey (1791); 17 Rotherhithe Wall (1802); Fountain Stairs, Rotherhithe Wall (1805).

He was the husband of **Agnes Syeds**, who succeeded him in business. He patented a quadrant for obtaining the altitude when the horizon is obscured by fog, etc. According to his trade card he "Makes repairs, and cleans Quadrants, with or without his Artificial horizon; and Steering compasses…" and "an improvement on Azimuth compasses." Instruments bearing his name include:

compass dated 1805; quadrant dated 1791; four-draw brass and mahogany telescope.

Sym & Co., James

James Sym & Co., (fl.1817-1825) [18] [263] [271], were optical and mathematical instrument makers of Glasgow, Scotland. Addresses: 236 High St. (1817-1824) and 82-85 High St. (1825). They succeeded **James Sym (I)** in business, and were succeeded in business by **James Sym (II)**.

Sym, James (I)

James Sym (I), (fl.1792-1816) [18] [263] [271], was an optical and mathematical instrument maker of Glasgow, Scotland. Addresses: Ayton Court, Old Venal (1792); Bell St. (1799-1801); 2 Bell St. (1802-1805, 1808); 16 Bell St. (1806); 236 High St. (1807, 1809-1811, 1815-1816); 266 High St. (1812-1814).

He served his apprenticeship with **John Gardner (I)**. In 1792 he became a Burgess, enabling him to trade in the city, and set up his own company. He was succeeded in business by **James Sym & Co.**

The register of *Burgesses & Guild Brethren of Glasgow, 1751-1846,* [398], contains two entries for him. The first, dated 26 July 1792, reads: "Sym, James, mathematical and optical instrument maker, B. and G.B., as serving appr. with John Gardner, maltman and mathematical instrument maker, B. and G.B." The second, dated 19 September 1804, reads: "Sym, James, previously (26 July 1792) admitted as a mathematical and optical instrument maker, is readmitted B. and G.B. as hammerman, by purchase."

Another James Sym, who may have been his father, worked with John Gardner (I) as a journeyman for **James Watt**.

Sym, James (II)

James Sym (II), (fl.1826-1846) [18] [263] [271] [276 #2260], was an optical and mathematical instrument maker of 167 High St., Glasgow, Scotland. He succeeded **James Sym & Co.** in business, and may have been the same person as **James Sym (I)**.

Symmetroscope

A symmetroscope is a kind of toy kaleidoscope for which the American optical instrument maker **George Wale** applied for a patent in 1899. It was initially marketed by the Wale-Irving Company, then after Wale's death by Frank P. Irving, and by the American Symmetroscope Company.

Synta Technology Corporation of Taiwan

The Synta Technology Corporation is a Taiwan based manufacturer and distributor of telescope equipment. They manufacture telescopes for their own brands, **Celestron** and **Skywatcher**. They also distribute under the **Acuter** name, and manufacture products for **Orion**.

T

Tacheometer
A tacheometer is a transit theodolite with features that enable it to act as a telemeter. It has an anallatic telescope, and stadia lines at the focus of the telescope eyepiece to enable measurements to be made against a distant surveyor's rod. Based on the measurements taken, the range is calculated using trigonometry.

Tagliabue & Casella
Tagliabue & Casella, [18] [379] [438], were optical and philosophical instrument makers of 23 Hatton Garden, London England. The firm was a partnership between **Caesar Tagliabue** and **Louis Pascal Casella**, established in 1838. Following Tagliabue's death in 1844, Casella bought out the remaining interest in the business from his wife, Maria's sister, and in 1848 the firm became **Louis Casella & Co.**

Tagliabue & Co., Caesar
Caesar Tagliabue & Co., (fl.1806-1814) [18] [276 #1229] [379] [438], were opticians, and optical and philosophical instrument makers of 26 High Holborn, London, England. The firm was a partnership involving **Caesar Tagliabue**, and succeeded **Tagliabue & Torre**. Following the termination of this partnership, Caesar Tagliabue continued the business, trading under his own name.

Tagliabue & Torre
Tagliabue & Torre, (fl.c1800-1806) [18] [379] [438], were opticians, and optical and philosophical instrument makers of 294 High Holborn, London, England. They may have also traded under the name Tagliabue, Torre & Co. who are recorded at the same address. The firm was a partnership between **Caesar Tagliabue** and **Anthony de la Torre**, and was succeeded by **Caesar Tagliabue & Co.**

Tagliabue & Zambra
Tagliabue & Zambra, [18] [379] [438], were philosophical instrument makers of 11 Brook St., Holborn, London, England. The firm was a partnership, established in 1847, between **Joseph Zambra** and **John Tagliabue**. The partnership was dissolved in 1850, [35 #21084 p1005], after which Joseph Zambra began to work in partnership with **Henry Negretti**, trading as **Negretti & Zambra**.

Tagliabue, Caesar
Caesar Tagliabue, (1767-1844) [18] [276 #1720] [379] [438], was an optician, and optical and philosophical instrument maker. He was born near Como in Italy, and settled in London, England in 1799. Addresses: 294 Holborn (1799-1800); 26 Holborn (1897-1809); 11 Brook St., Holborn (1816); 28 Cross St., Hatton Garden (1822-1829); 23 Hatton Garden (1829-1839).

In c1800 he began to work in partnership with one of the Torre family (believed to be Anthony), trading as **Tagliabue & Torre** (who may have also traded under the name **Tagliabue, Torre & Co.** at the same address) until 1806, following which he traded as **Caesar Tagliabue & Co.** until 1814. He employed **Louis Pascal Casella** who, in 1838, married his daughter, Maria Louisa Tagliabue. In the same year, Caesar Tagliabue and Louis Pascal Casella began to trade in partnership as **Tagliabue & Casella**. This partnership continued until the death of Caesar Tagliabue in 1844. He was succeeded by Louis Pascal Casella.

Instruments bearing his name include: 3-draw brass telescope with mahogany barrel; barometers; thermometers.

John Tagliabue, is believed to have been Caesar's son.

Tagliabue, J.&J.
J.&J. Tagliabue, (fl.1817-1819) [18] [379] [438], were opticians, and optical and philosophical instrument makers of 11 Brook St., Holborn, London, England. The firm was a partnership between **John Tagliabue** and Joseph Tagliabue. The partnership was dissolved in 1818, [35 #17324 p150], with the business being continued by John Tagliabue. The name of Joseph, however, appeared on an insurance document for the firm in 1823.

Tagliabue, John
John Tagliabue, (fl.1817-1852) [18] [276 #1440] [379] [438], was an optician and philosophical instrument maker of 11 Brook St., Holborn, London, England. He is believed to have been a son of **Caesar Tagliabue**. He worked in partnership with Joseph Tagliabue from 1817 until 1819, trading as **J.&J. Tagliabue**, after which he continued the business under his own name. He later worked in partnership with **Joseph Zambra**, from 1847 until 1850, trading as **Tagliabue & Zambra**.

Tagliabue, Torre & Co.
Tagliabue, Torre & Co., (fl.1800-1807) [18] [379] [438], were opticians and optical instrument makers of 294 High Holborn, London, England. This may have been another name used by **Tagliabue & Torre**, who traded at the same address. Instruments bearing their name include: 3-draw brass telescope with mahogany barrel.

Taisei Optical Equipment Manufacturing
Taisei Optical Equipment Manufacturing was established in 1950 in Urawa city in Japan, manufacturing cameras

and binocular lenses and, from 1957, interchangeable camera lenses. From 1959 their exported optical instruments bore the maker's **JB code** JB45. In 1958 they registered the *Tamron Brand* trade-mark. In 1959 they moved to Hasunuma, Omiya city. They continued to expand their product lines, and in 1970 the company name was changed to **Tamron Co. Ltd.**

TaiYuan Optical Instruments
TaiYuan Optical Instruments, of Taiyuan, Shanxi Province, China, was the company that produced the *TWX-1* folded optics microscope [316]. The design was derived from the **McArthur microscope**, but differed in several ways. Most notably the optics were folded in a different path, so that the microscope slide is viewed from above rather than below as with the McArthur design and its other derivatives. The *TWX-1* was robust for field use, and developed to a very high standard of quality.

The *TWX-1*, manufactured during the 1970s, was built exclusively for the Chinese army (PLA) front-line field hospital use, although it may also have been supplied to the armies of other countries considered to be suitable recipients by the Chinese.

Takachiho
For Takachiho Seisakusho and Takachiho Optical Co. Ltd., see **Olympus**.

Takahashi
Takahashi is a Japanese maker of high-quality telescopes, mounts and accessories. Some early Takahashi telescopes were sold under the **Swift** brand bearing the AvA maker's mark, (see Appendix IV).

Takahashi also used to sell binoculars. See their 1982 binoculars catalogue [249].

Tal
Tal is an **NPZ** brand of telescopes and accessories made in Russia.

Talbot & Eamer
Talbot & Eamer, [61] [122] [306], of Blackburn, Lancashire, England, was a camera manufacturer established in 1884 by Henry Percy Tattersall. He registered the *Talmer* and *Miral* trade-marks during the 1890s. He sold the company in 1901 to G. Jones (I), whose son G. Jones (II) ran the company, having relocated it to Liverpool. The company changed hands again in 1906. It was renamed Talbot & Eamer Mirals Ltd. in 1909, and ceased trading in 1923.

Talbot, William Henry Fox
William Henry Fox Talbot, (1800-1877) [37] [70] [71], of Melbury, England, was a scientist and inventor. He determined that he could expose a piece of paper coated with silver salts, and capture an image with the resulting chemical reaction. This produced an image with light and dark reversed, and in 1835 he realised that he could produce a positive image from this negative by placing the negative over another sheet of coated paper, and exposing it to sunlight. In 1840 Talbot further discovered that by using a mixture including a solution of gallic acid, he could produce a latent image that could be made visible by development. He called this improved process the calotype, and patented it in 1841. Thus, Talbot had invented the technique of producing an unlimited number of positives from a single negative, which continued to be the basic principle of popular photography until the advent of digital photography.

Louis Daguerre's photographic process was announced earlier than Talbot's, and was the first practicable photographic process invented. However, it was Talbot's technique that led into the future of popular photography.

Tamaya Technics Inc.
Tamaya Technics Inc. is a Japanese company, based in Tokyo, who manufacture surveying and drawing instruments, weather equipment, and measuring, navigation, and optical instruments. The company began as a retail shop, which was renamed Tamaya Shoten in 1901, importing and distributing surveying instruments as well as manufacturing transits, levels, theodolites and sextants. The company was renamed Tamaya Technics Inc. in 1983.

Tamron Co. Ltd.
In 1970 the Japanese company **Taisei Optical Equipment Manufacturing** changed its name to Tamron Co. Ltd. From 1959, for about three decades, their exported optical instruments bore the maker's **JB code** JB45. The company has since grown to become a multinational, producing lenses for a wide variety of applications, from consumer cameras to automotive applications.

Tangate, Robert
Robert Tangate, (fl.1766 d.1808) [18] [53] [385], was an optician, and optical and mathematical instrument maker of London, England. Addresses: 9 West Square; Shoe Lane, Fleet St. (1770-1773); Bride Lane, Fleet St. (1776); 4 Bride Lane (1778); Bride Lane (1789-1794); 42 Elliott's Buildings, St. George's Fields (1800). He began his apprenticeship with **John Morgan**, and was turned over to **George Adams (I)** in 1758. He was freed of the

Joiners' Company in 1761. Among his apprentices were his son, Robert Tangate (II), beginning in 1781, and **Michael Dancer**, beginning in 1766. Instruments bearing his name include: brass theodolite.

Tanzutsu
Tanzutsu was a Japanese maker of telescopes. Their maker's mark is illustrated in Appendix IV.

Tasco
Tasco is a US based manufacturer and distributor of budget telescopes and binoculars, and is one of the **Vista Outdoor Inc.** brands. The company is based in Utah, USA. Prior to the 1980s, Tasco sold rebranded telescopes manufactured by makers such as **Kenko**, **Towa** and **Astro Optical**.

In 1998 Tasco acquired **Celestron**, and in 2005 Celestron was sold on to **Synta**.

Taunton, Frederick Adolphus
Frederick Adolphus Taunton, (fl.1845-1849) [18], was an optician and telescope maker of London, England. Addresses: 4 Norfolk St., Islington (1845); 22 Norfolk St., Islington (1846); Brown's Cottage, Canonbury Square (1846-1849).

Tavistock Theodolite
The Tavistock theodolite is a design of theodolite that resulted from a specification drawn up in 1926 during a meeting of British instrument makers and surveyors in Tavistock, Devon, England. The concept was the brainchild of **Heinrich Wild** of **Zeiss**. The objective was to design a theodolite so that the user could read the micrometers on both sides of the instrument without leaving their position at the telescope. The user would need to observe two opposite portions of the circles through a single eyepiece, and read them by means of a single screw micrometer.

The companies **Cooke, Troughton & Simms Ltd.**, **E.R. Watts & Son Ltd.**, and **C.F. Casella & Co. Ltd.** were invited to submit designs. The design by Watts incorporated Zeiss patents. The design by Cooke, Troughton & Simms, largely conceived by their optical designer **Wilfred Taylor**, was an original design that became known as the Tavistock Theodolite. Casella did not submit a design.

Taylor, Harold Dennis
Harold Dennis Taylor, (1862-1943) [20] [46] [190] [228], was an optical designer, born in Lockwood, Yorkshire, England. He began to train as an architect but, rather than finish this training, he instead took up a job with **Thomas Cooke & Sons** in York. He was self-taught in optical design, and rose to the position of optical manager. His son, **Wilfred Taylor**, later served an apprenticeship with the company.

In 1892 Taylor designed the *Cooke Photo Visual* telescope objective. This was a triplet apochromat that used innovative glass from **Schott & Associates** glassworks in Jena. This was a highly influential design, being probably the first triplet apochromat telescope objective. In 1893 he designed the *Cooke triplet* photographic lens which achieved a remarkably flat field and aberration free image. Thomas Cooke & Sons did not manufacture the lens themselves, since they were not in the photographic business. Instead, Taylor licensed the production to **Taylor, Taylor & Hobson** (with whom he had no previous connection).

In 1894 Taylor published the results of his research in improving lens transmission by means of a layer of haze or tarnish on the lens surface. In 1904 he patented a process of leaching lens surfaces using acids and other chemicals. This was the forerunner of modern lens coatings.

In 1901 his book *On the Adjustment and Testing of Telescopic Objectives*, [251], was published by T. Cooke & Sons, giving details on star-testing and the adjustment of telescope objectives. This publication proved valuable to professionals and amateurs alike. In 1906 his book *A System of Applied Optics* was published. Both of these books became widely read classics.

In all, H. Dennis Taylor was awarded nearly fifty patents for his designs throughout his career, and his numerous contributions to the field of optics did not end with his retirement in 1915. He continued to act as a consultant to Taylor, Taylor & Hobson during his retirement, and was recognised with industry awards, including those from the Royal Photographic Society and the Physical Society.

Taylor, Janet (Mrs)
Mrs. Janet Taylor, (1804-1870) [16] [18] [46] [54] [276 #2023] [322] [358], of London, England, was a nautical and mathematical instrument maker, as well as being an astronomer, mathematician, navigator, and teacher. She was born Jane Anne Ionn, in County Durham, England, the daughter of Revd Peter Ionn, a school teacher who taught her mathematics and navigation. In 1830, then living in London, she married George Taylor Jane, and they adopted the surname Taylor.

In 1831 they were located at 6 East St., Red Lion St. She used her inheritance to fund publication of several books on navigation and astronomical tables. In 1834 she patented her "mariner's calculator". By 1835 she was already teaching navigation, and she opened a navigation warehouse and nautical academy at 103 Minories, with the company, Janet Taylor & Co., registered in her husband's name, but run by herself.

In 1845 she briefly traded at both 103 and 104 Minories, then relocated the business to 104 Minories, where she employed a staff of mathematical and nautical instrument makers. She exhibited in the Great Exhibition in London in 1851, for which her two catalogue entries, [**328**], stated, "105 Taylor, Janet, 104 Minories—Manufacturer. A bronze binnacle, with compass, designed from the water lily.", and "350 Taylor, Janet, 104 Minories—Manufacturer. Sextant for measuring angular distances between the heavenly bodies."

According to her trade card, "Mrs Janet Taylor. Mathematical instrument manufacturer, Chart, Map & Bookseller. Nautical Acadamy. By appointment. Navigation Warehouse. 104 Minories, opposite the rail road. London. Agent for Mr. Dent's chronometers. Nautical instruments made and repaired on the premises. Under the patronage of the Admiralty, East India Company, and Trinity House. Agent for the sale of Admiralty charts. Chronometers rated & repaired. The deviation of Compasses in iron vessels found & *corrected*." (Note, the final word is barely legible).

In 1860 she was awarded a civil list pension of £50 per year, and by 1868 she was retired. Instruments bearing her name, or that of her company, include: octants; sextant; compass; barometers; telescope; quintant, with fine ornamentation, which belonged to the Prince of Wales, later King Edward VII [**16**].

Taylor, Janet & Co.
See **Mrs. Janet Taylor**.

Taylor, Wilfred
Edward Wilfred Taylor, (1891-1980) [**46**] [**190**], of York, England, generally known as Wilfred, was the son of **H. Dennis Taylor**, and followed in his father's footsteps as an optical designer. In 1908 he began an apprenticeship at **Thomas Cook and Sons** for whom he rose to the position of optical manager, then technical director, and finally joint managing director of **Cooke, Troughton & Simms** with **James Simms Wilson**. He played a major part in the development of the Tavistock theodolite, and in the expansion of **Cooke, Troughton and Simms** during WWII, developing optical munitions for the military.

Taylor, Taylor & Hobson
Taylor, Taylor & Hobson, [**44**] [**306**], were optical instrument makers. The company was formed in 1886 as T.S.&W. Taylor (1865-1937) when brothers William [**46**] and **Thomas Smithies Taylor**, started a company to make lenses. In 1887 W.S.H. Hobson joined as head of sales, and in 1888 the company name was changed to Taylor, Taylor & Hobson.

In 1893 **H. Dennis Taylor**, who had no previous connection with the company, patented the *Cooke Triplet* lens. Since **T. Cooke & Sons**, for whom he worked, had no experience in manufacturing photographic lenses, he took his design to Taylor, Taylor & Hobson, and had them manufacture and market the lens. Taylor, Taylor & Hobson went on to produce a series of photographic lenses, over a number of years, marketed as *Taylor-Hobson Cooke Lenses*, including cinematic camera and projector lenses.

The partnership of Taylor, Taylor & Hobson was dissolved in 1894, and in 1901 the company was incorporated as Taylor, Taylor & Hobson Ltd. During WWI, as a part of their wartime production, the company supplied lenses for binocular production by **Kershaw**. In 1919 Taylor, Taylor & Hobson Ltd. was listed as a member of the **British Optical Instrument Manufacturers' Association**. By 1925 their advertisements bore the name Taylor Hobson. In 1947, [**135**], the company was acquired by **British Optical and Precision Engineers**, part of **Rank**, and renamed Rank Taylor Hobson. In 1997 Rank Taylor Hobson was sold to Schroders Ventures under the name Taylor Hobson, and in 1998 the optics division of Taylor Hobson was sold to an American distributor, Les Zellan, who began to operate the company as **Cooke Optics Ltd.** [**446**].

Taylor, Thomas Smithies
Thomas Smithies Taylor served his apprenticeship with **Joseph Beck**, beginning in 1879. In 1886, together with his brother, William, he formed a company to make optical lenses. In 1887 W.S.H. Hobson joined as head of sales, and in 1888 the company name was changed to **Taylor, Taylor & Hobson**.

Tecnomasio Italiano
Tecnomasio Italiano, [**374**], was a company founded by **Carlo dell'Acqua**, along with engineer and mathematician Luigi Longoni, and photographer Alessandro Duroni, in 1864 in Milan, Italy. They quickly established a strong reputation for manufacturing scientific instruments and, by 1865, their catalogue listed over 1,000 items. They also boldly claimed, "…over and above the instruments indicated, the firm will accept to construct any machine or instrument described in the various scientific or artistic works, as well as any only imagined by lovers of mechanics, physics or chemistry and that has as yet never been executed."

By 1867, when they exhibited at the Exposition Universelle in Paris, the company already employed more than 50 people. In 1879, the engineer Bartolomeo Cabella took over running the company, and by the end of the century they employed over 500 people. In 1903 the firm was acquired by Brown Bovery, at which time it changed its name to Tecnomasio Italiano Brown Bovery (TIBB), and ceased production of scientific instruments to concentrate on industrial electrical products.

Tectron Telescopes

Tectron Telescopes [214], of Sarasota, Florida, USA, was a company formed in 1983 by Tom Clark, building large telescopes for the amateur market. He built telescopes ranging from 15 inches to 42 inches, but most were 20 inches or larger. He established the Tectron Machine Corporation to carry out the manufacturing. An undated advertisement shows "lightweight" 16-inch and 20-inch Newtonians on clock-driven fork mounts. He also produced parts and components for amateur telescope makers. After retirement in 1997 he began to produce portable travel Newtonians in wooden Dobsonian mounts. He founded *Amateur Astronomy* magazine in 1994, and published his book *The Modern Dobsonian* in 1992 for amateur telescope makers. The book includes details for making portable Dobsonians.

Teinoscope

A teinoscope is an optical instrument in which an arrangement of prisms is so designed that the apparent linear dimensions of a viewed object are increased or decreased, while chromatic aberration is corrected.

Teleconverter

A teleconverter is a diverging lens, or arrangement of lens elements, that increases the effective focal length of a camera lens. The teleconverter is mounted between the camera body and the lens. It is a form of Barlow lens, and serves the same optical purpose.

Teledyne-Gurley

See **W&L.E. Gurley**.

Teleiconograph

A teleiconograph is a telescope combined with a camera lucida, such that distant objects may be sketched or measured.

Teleidoscope

A teleidoscope is a variant of the kaleidoscope. Instead of viewing arrangements of fragments of broken glass, etc., the view is open, seen through a strong lens. Any scene can be viewed through the teleidoscope, and the view is subject to kaleidoscopic mirror reflections, forming interesting patterns. The teleidoscope gives more even illumination than a kaleidoscope, and may be attached to a camera or a projector. A patent was granted in 1972 to American inventors John Burnside and Henry Hay.

Telemeter

Telemeter is another word for rangefinder.

Telephoto Lens

See lens.

Telescope

A telescope is an optical instrument used to view distant objects. It generally consists of a tube or frame, with an arrangement of mirrors and/or lenses used to focus incident light, and magnify the resulting image.

There are many kinds of telescope for the consumer market, ranging from small hand-held telescopes to large aperture mounted telescopes. Most telescopes are either *refracting*, *reflecting*, or a combination of the two, known as *catadioptric*.

Refracting Telescope

A refracting telescope has an objective and an eyepiece, each of which may be a multiple-element lens design, and may have other intermediate optical elements such as lenses or erecting prisms. The most common objective designs are achromatic or apochromatic, using carefully selected optical glass. A refracting telescope generally suffers from field curvature, which may be corrected with the use of a field flattener.

Hand-held telescopes, or *spyglasses*, frequently have an integral eyepiece. They make use of an intermediate relay lens to produce an upright image with corrected left/right image orientation. Some hand-held telescopes are pancratic, with one of the sliding tubes altering the magnification.

See also Galilean telescope, Keplerian telescope, split image telescope, anallatic telescope, aerial telescope, spotting scope, spyglass, riflescope.

Reflecting Telescope

A *Newtonian* telescope uses a concave primary mirror and a flat secondary mirror mounted diagonally to bring the image out to an eyepiece at the side of the telescope. The concave primary mirror is usually paraboloidal, since a spherical mirror suffers from spherical aberration. A Newtonian telescope suffers from coma, an optical aberration which may be corrected with the use of a coma corrector.

The *Cassegrain* telescope uses a concave primary mirror, and a convex secondary mirror. This design was first published in 1672. Both mirrors share the same optical axis, and the image is directed back through a hole in the centre of the primary mirror to an eyepiece. Generally, the primary mirror is paraboloidal, and the secondary is hyperboloidal. A variant of this is the

Ritchey-Chrétien telescope in which both the primary and secondary mirrors are hyperboloidal. See also Dall-Kirkham Cassegrain telescope.

The Newtonian, Cassegrain and Ritchey-Chrétien designs are the most common for reflecting telescopes, but many other variants exist. When a reflecting telescope has refracting optical elements included, it is referred to as a *Catadioptric* telescope.

See also Brachyte, Gregorian telescope, Dobsonian telescope (mount), Herschelian telescope, Schiefspiegler, three mirror telescope.

Catadioptric Telescope
A catadioptric telescope is one which combines both reflecting components and refracting components to reduce aberrations. The word catadioptric is derived from *catoptric* (of or related to reflection, or forming an image using a mirror), and *dioptric* (of or related to refraction or refracted light - related to the word dioptre).

A *Schmidt-Cassegrain* telescope, or *Schmidt camera*, is a Cassegrain telescope with a spherical primary mirror, and a Schmidt corrector lens that corrects the spherical aberration. The secondary mirror is figured to act as a field flattener. The Schmidt corrector lens is flat on one side with a small aspheric curve on the inside.

A *Maksutov* telescope has a spherical mirror and a corrector plate consisting of a negative meniscus lens. The most common variant used is a *Maksutov-Cassegrain* which is a Cassegrain telescope incorporating those two elements. Generally, in this design the secondary mirror is spherical, and is often integral to the corrector plate, being a mirrored area at its centre.

The key difference between a Maksutov-Cassegrain and a Schmidt-Cassegrain is in the corrector plate. In the Schmidt-Cassegrain the plate is flat with a small aspheric curve. The Maksutov-Cassegrain corrector plate, in contrast, is much more curved, and generally considerably thicker. A consequence is that the Maksutov-Cassegrain will require more time to equalise temperature prior to use.

The Medial and Brachymedial telescope designs are catadioptric telescopes that use a Mangin mirror.

A *Bird-Jones* (or Jones-Bird) telescope is a variant of the Newtonian, but with a spherical primary mirror and a correcting lens generally mounted either inside the focuser tube or in front of the secondary mirror. While the Bird-Jones design is not inherently flawed, this kind of telescope has generally been aimed at the low-end of the market.

Telespectroscope
A telespectroscope is a combined telescope and spectroscope, used to observe heavenly bodies, and perform a spectroscopic analysis on the observed light. It may be used, for example, to analyse the composition of stars' atmospheres.

Telestereoscope
The telestereoscope, invented by **Hermann von Helmholtz**, is a binocular instrument which yields an enhanced stereo relief in objects viewed at a moderate distance, by means of objectives placed wider apart than the interocular distance.

Tele Vue
The company Tele Vue was founded by **Al Nagler** in 1977. The company makes eyepieces to the eponymous **Nagler** design as well as others. They also make telescopes and mounts.

Telstar
Telstar, (not to be confused with the **Meade** *Telstar* telescopes), was a brand of budget telescopes and binoculars. Examples: Telescopes - 500x114 reflector; 900x114 reflector. Binoculars - 8x30 bearing Japanese **JB code**; 13x60 of French manufacture.

Tento
Tento was a Soviet era Russian brand of binoculars, manufactured by **ZOMZ**. The ZOMZ binoculars now sell under the brand **Kronos**.

Terada, Shintaro
Shintaro Terada, (fl.1912), was a Japanese optical designer, and thermometer and microscope maker. In 1912 he began to work in partnership with Matsumo and Kando, trading as **M-Katera**, producing microscopes. In 1914 he won a bronze prize for one of his microscopes at the Taisho Exhibition in Tokyo. In 1919 he went on to work in partnership with Takeshi Yamashita, with whom he co-founded the company Takachiho Seisakusho, which was the forerunner of **Olympus**, becoming their chief engineer.

Terrascan
Terrascan is a brand of reflecting binocular telescopes by the astronomer and optical designer **Peter Drew**. Peter Drew has made many large binocular telescopes up to 200mm aperture with refracting optics and up to 300mm aperture with mirrors.

Tesdorph, Ludwig
Ludwig Tesdorph, (1856-1905) [**374**], was a surveying and astronomical instrument maker of Stuttgart, Germany. He established his company in 1861 and, in 1906, the firm was purchased by **Florenz Sartorius**.

Tetragonal Prism
See prism.

Texereau, Jean
Jean Texereau, (1919-2014) [**164**], was a renowned French optical engineer and astronomer. He was skilled at making, figuring and testing the optics for large aperture telescopes, and well known for promoting amateur telescope making through his role in the Société Astronomique de France. He wrote the book *La Construction du Télescope d'Amateur*, published in 1951, for which the English language edition, *How to Make a Telescope* was published in 1957.

Texereau designed a variant of the **Plössl** eyepiece, and took the design to **Clavé** in Paris, who produced them in various focal lengths from 1954.

He began his career in optics making amateur telescopes, and in about 1946 was offered a position at the Paris Observatory. He went on to make, figure, or re-figure the optics for telescopes at the Paris Observatory, the Haute-Provence Observatory, Pic du Midi, and the McDonald Observatory in Texas.

Thales Optronics Ltd.
Thales Optronics Ltd. was a subsidiary of the French multinational Thales Group. Their headquarters and main production facility was in Glasgow, Scotland. In 2000 **Barr and Stroud**, by then already merged with **Pilkington**, became a subsidiary of the Thales Group. In 2001 Thales merged with **Avimo**, and were re-branded as Thales Optronics Ltd. In 2005 Thales Optronics was sold, and became **Qioptiq**. The Thales Group, however, continued to develop specialised optronics products within the subsidiary parts of their Defence group.

Thate, Paul
Paul Thate, (fl.1836) [**374**], was an optical instrument maker of Berlin, Germany. He specialised in making microscopes, microtomes, and micro-photographic apparatus, and exhibited his work in Berlin in 1836.

Theatograph
The Theatograph, invented by **Robert Paul** in 1896, was the first 35mm moving film projector to be produced commercially in Britain, [**335**].

Theodolite
A theodolite is a surveying instrument used to accurately measure horizontal and vertical angles by means of a telescope and graduated circles. Normally the telescope can move through an arc of about 45° upwards or downwards from the horizontal plane. The theodolite is usually mounted on a strong tripod and is provided with compass, level, and verniers for reading the circular scales. There is no universally agreed difference between a theodolite and a transit level. However, in general, a transit uses exposed vernier scales, and a theodolite uses an enclosed vernier scale.

See also dioptra, Dunbar-Scott auxiliary top and side telescope, gyro-theodolite, photo-theodolite, Tavistock theodolite.

Lean's Miners' Theodolite: A specialist theodolite used in mining applications.

Transit Theodolite: A theodolite in which the telescope can rotate through 360° vertically.

Hoskold's Transit Theodolite: A specialist theodolite or tacheometer used in applications such as mining and underground railway work.

Mountain Transit Theodolite: A theodolite with all the features of a full-sized transit theodolite, but smaller and lighter for ease in carrying on mountainous terrain.

Everest Theodolite: Invented by Sir George Everest (1790-1866), together with **William Simms (II)** of **Troughton & Simms**. Everest was working on the Great Trigonometrical Survey of India in the 1820s, and wanted a rugged theodolite for surveys carried out by the East India Company. The telescope of the Everest theodolite cannot transit, since its pivot is low in the body of the theodolite. The vertical circle is formed of two arcs, rather than a full circle.

In modern surveying the theodolite is being superseded by the total station.

Thompson, John (I)
John Thompson (I), (fl.1835-1844) [**18**] [**322**], was an optical, mathematical, philosophical, and surgical instrument maker, lath maker, and tool maker, of 26 Glasshouse St., Nottingham, England, and from 1839-1844, Pelham St., Nottingham. Instruments bearing his name include: three-draw telescope.

Thompson, John (II)
John Thompson (II), (fl.c1830-1860) [**18**] [**319**] [**322**], was an optician, and optical, mathematical, and philosophical instrument maker of Manchester St., Liverpool, England, and 85 Lord St., Liverpool. Instruments bearing his name include: theodolite; aneroid barometer.

Thompson, Joseph
Joseph Thompson, (fl.1827-1838) [**18**] [**276** #2026] [**322**], was an optical, mathematical, and philosophical instrument maker of 36 High St., Wapping, London, England. Instruments bearing his name include: Hadley's quadrant.

He succeeded the optician, mathematical instrument maker, and compass maker, Johathan Thompson in business, and was succeeded by the mathematical and philosophical instrument maker Joseph Berry Thompson.

Thompson Prism
See prism.

Thompson, Silvanus P.
Professor Silvanus P. Thompson, (1851-1916) [44], was a physicist and a teacher. He was the inventor of the Thompson prism, which is a polarizing prism passing a very wide angle of light.

Thomson, Elihu
Elihu Thomson, (1853-1937) [45] [185], was a British born American electrical engineer and inventor who moved to the USA during his childhood. He was prolific in his inventions, and established the American Electric Company, which became the Thomson-Houston Electric Company, and later merged with the Edison General Electric Company to become General Electric.

He developed an interest in telescope making at an early age, and learned the skills of figuring mirrors and lenses. He maintained this interest throughout his life, investing a considerable amount of time and money into increasingly large telescopes. He was innovative in his use and processing of substrate materials for the fabrication of mirrors. He is known to have corresponded with many eminent astronomers and optical instrument makers, including **George Ellery Hale**, **George Willis Ritchey**, **John Brashear**, and **Sir Charles Parsons**.

Thomson, John Victor
John Victor Thomson, [9] [24] [199] [221], was a British optical engineer. He made a 36-inch mirror for the spectrograph on the **Hale** 200-inch telescope at the Palomar observatory in California. In 1947 he was a co-founder of the optical and engineering company **Cox, Hargreaves & Thomson**, along with **H.W. Cox** and **Frederick J. Hargreaves**. In 1961 John Thomson, together with two other members of the company, J. Mortleman and **Jim Hysom**, left Cox, Hargreaves and Thomson to start the company **Optical Surfaces**.

Thomson, William
William Thomson, Lord Kelvin, (1824-1907) [37] [44] [46] [263], was a mathematician, physicist, and prolific inventor. He was born in Belfast, Ireland, and raised in Glasgow, Scotland after his father was appointed to the Glasgow College Chair of Mathematics. He studied at Glasgow and Cambridge universities and in 1846, at the age of 22, was appointed Professor of Natural Philosophy at Glasgow University, a position he held until the age of 75 in 1899. William's brother James was also a brilliant academic, and was professor of engineering at Glasgow. From about 1876 to 1884 James was assisted by **Archibald Barr**, who also attended William Thomson's lectures, and collaborated with him in research.

William Thomson was an enthusiastic sportsman and musician, and his core academic interests were mathematics and natural philosophy. He was prolific in publishing papers throughout his academic career, and worked tirelessly on his interests in thermodynamics and electric theories and practice, and navigation. He developed an absolute scale of temperature which, named after him, was to be called the Kelvin scale. In order to facilitate his work, and to help in teaching, he also developed a keen interest in the design and manufacturing of precision instruments. In 1854 he formed an association with the scientific and optical instrument maker **James White**, who was largely responsible for equipping his laboratory, and together they manufactured a variety of instruments.

Thomson applied his ingenuity to the problem of submarine cables which were experiencing problematic retardation effects. By 1856 he was director of the Atlantic Telegraph Company, and he was knighted for his contribution to the cable-laying effort. During this period, he secured several patents relating to telegraphy, leading to considerable personal wealth.

He went on to develop and patent a new sounding machine and compass for marine use, and began to produce them, selling to mail-liner companies and the British Admiralty. In 1892 he was elevated to the peerage as Baron Kelvin of Largs, with the title Lord Kelvin. After James White's death, Thomson became a partner in White's company, and in 1900 it was incorporated as **Kelvin and James White Ltd.** By then the range of instruments produced was much wider. In 1914, after William Thomson's death, the company name was changed to **Kelvin, Bottomley and Baird**.

Thornoe's Refractometer
See refractometer.

Thornton Ltd., A.G.
A.G. Thornton Ltd, [44] [138] [141], were drawing and surveying instrument specialists based in Manchester, England. The firm began in 1878 when Alexander George Thornton entered a partnership with **Joseph Halden**, and set up business in 8 Albert Square, Manchester. The partnership only lasted about two years, after which A.G. Thornton continued in business supplying drawing and surveying instruments and materials. In the early C20[th] the company expanded considerably, increasing their

range of drawing and surveying instruments, accessories and supplies. In 1919 the company was listed as a member of the **British Optical Instrument Manufacturers' Association**. In 1967 they changed their name to British Thornton. At about the same time they increased their focus on drawing office furniture, and began to sell electronic calculators, reducing their sales of slide-rules. British Thornton merged with the company Education & Science Furniture in 1992, and went on to specialise in the manufacture of furniture for the education sector.

Thornton-Pickard Manufacturing Co. Ltd.

The Thornton-Pickard Manufacturing Co. Ltd., [44] [61] [306], were camera makers of Manchester, England. The company was formed as J.E. Thornton in 1886 by John Edward Thornton (1865-1940). Their early cameras were manufactured by the **Billcliff Camera Works**. In 1888 Edgar Pickard joined the company, and the name was changed to the Thornton-Pickard Manufacturing Co. In 1897 it became a limited company. In 1898 Thornton was voted off the board, and removed from running the company, which remained in the hands of the Pickard family. In 1912/13 they acquired the **Midland Camera Co. Ltd.** Prior to WWI, the company began to go into decline. During WWI they produced aerial cameras for the war effort, and after the war the company's decline continued. By 1921 **APM** were the principal shareholders, and Thornton-Pickard Manufacturing Co. Ltd. continued to trade under its own name until 1940.

Thorp's Diffraction Grating

A ruled diffraction grating on a celluloid substrate.

Thorp's Diffraction Spectroscope

See spectroscope.

Three Mirror Telescope

The Three Mirror Telescope (3MT), [224], designed by **Dr. Roderick Willstrop** of the University of Cambridge's Institute of Astronomy, uses three mirror surfaces to obtain a wide field of view in small, sharp photographic images. Its design is an improvement on an earlier design by the French optician Maurice Paul in 1935, also invented independently in 1945 by the American James Baker.

The 3MT, like the Schmidt camera, can produce a wide field of view. However, since the 3MT has all reflecting optics, it may be constructed with a larger aperture than a Schmidt camera, in which the aperture is limited by the need for a correcting lens. A working model of the 3MT was first built by the Institute of Astronomy in 1985 with an aperture of 4 inches. A prototype was then built in 1989 with an aperture of 20 inches. The primary mirror for the prototype 20-inch 3MT was made spherical by **Jim Hysom**, and figured by Dr. Willstrop.

TI Asahi Co. Ltd.

TI Asahi Co. Ltd. began as the **Fuji Surveying Instrument Company Ltd.** In 1967 the company was acquired by the **Asahi Optical Company**, and became an Asahi brand. In 1976 Asahi changed the *Fuji* brand-name to **Pentax**. In 2009 the Pentax surveying instrument business was acquired by the Taiwan Instrument Co. Ltd. and became TI Asahi Co. Ltd., producing surveying instruments under the *Pentax* brand-name. Their products include total stations, global navigation satellite systems, scanning systems, electronic theodolites, optical levels, and unmanned aircraft systems.

Tilted Component Telescope

A tilted component telescope (TCT) is one in which one or more of the optical components is tilted to redirect the optical axis. The earliest TCT was the Herschelian Telescope. Later designs include the Brachyte and the Schiefspiegler.

Tilting Level

The tilting level is a form of surveyor's level, similar to a dumpy level. The main difference is that the base plate need only be levelled approximately because the telescope can be tilted up to a few degrees with respect to the base plate. The tilt is adjusted using a spirit level, and must be re-adjusted for every reading. This is more time-consuming for multiple readings than using a dumpy level, but has the advantage that any error in level applies only to a single reading.

Tinplate Workers' Company

See Appendix IX: The London Guilds.

Tinsley, Clayton R.

Clayton R. Tinsley, [185], formed his company, Tinsley Laboratories, in 1926 in Berkeley, California, USA. They specialised in telescopes, mirrors and telescope making kits and components. The company's offerings increased in scope over the years. Their catalogue *Telescopes for School and Home* published in the 1930s shows them using a *Saturn* trademark. The catalogue includes pier mounted Cassegrains from 10 to 20 inches, and Newtonians from 6 to 12 inches; tripod mounted smaller Newtonians, and telescope mounts, parts, materials and tools, including a spherometer. The catalogue also shows eyepieces and other accessories and hand-held spyglass refractors.

In 1937 Clayton Tinsley sold the company to Donald A. Jenkins. During WWII they made binoculars and optical components for the war effort. In 1957, Donald Jenkins retired and sold the company to a group of employees and investors. In 1961 they incorporated as Tinsley Laboratories Inc. A catalogue from the 1960s shows a 5.65-inch Maksutov-Cassegrain, an 8-inch Cassegrain, and a 5-inch f/15 refractor.

In 2001 Tinsley Laboratories Inc. was acquired by Massachusetts based SSG Precision Optronics Inc., who specialised in space-based and other high-performance optical subsystems. In 2006 SSG-Tinsley was taken over by L3 Communications (now **L3 Technologies**), to become their subsidiary **L3 SSG**.

Tinsley Laboratories Inc.
See **Clayton R. Tinsley**.

Tintometer
See colorimeter.

Tintype
Tintype, [70], is another name for a ferrotype.

Tisley & Spiller
Tisley & Spiller, (fl.1870s-1880s) [53] [270] [319] [322], of 172 Brompton Road, South Kensington, London, England, was a manufacturer of optical and philosophical instruments, aimed at the science education market. The company was a partnership between Samuel Charles Tisley (b.c1830) and George Spiller. The partnership was dissolved in 1877 [35 #24499 p5067], after which Samuel Charles Tisley continued the business alone. The company was renamed S.C. Tisley & Co., and continued to trade at the same address. Instruments bearing their name include: electrical instruments; polarimeter; oscillating prism; Nicol prism; apparatus for observing polarized light through glass.

Tisley, S.C. & Co.
See **Tisley & Spiller**.

Tiyoda
Tiyoda, [285], was a microscope manufacturer, formerly known as **M-Katera**, which was renamed in 1934. Tiyoda began to supply field-microscopes to the Japanese military for use in their field-hospitals. During WWII, as a result of the Japanese cooperation with Germany, Tiyoda were able to use some aspects of **Zeiss** microscope designs in their own instruments. Following the end of WWII, Tiyoda continued production of microscopes, but did not expand its production in the way its competitors did. In 1949 they manufactured the first Japanese made phase contrast microscope. In 1976 Tiyoda's parent company, the medical technology company Sakura Finetek, shut down Tiyoda's microscope production and absorbed its manufacturing facilities.

TLR Camera
See camera.

Tohyoh
Tohyoh was a Japanese brand of binoculars and monoculars based in Tokyo. Examples may be found in various sizes including zoom binoculars and large sizes such as 30x70. Examples date from around the 1950s through to about the 1980s, including some bearing a Japanese **JB code** and oval quality sticker.

Tokina Co. Ltd.
Tokina Co. Ltd. is a manufacturer of camera lenses and camera security products. The company was founded in 1950 as Tokyo Optical Equipment Manufacturing in Shinjuku, Tokyo. In 1971 the company name was changed to Tokina Optical Co. Ltd., and in 1995 it was changed again to Tokina Co. Ltd. In 1999 they were commissioned by **Kenko Co. Ltd.** to operate its overseas interchangeable lens sales division. In 2011 Tokina Co. Ltd. was merged with Kenko Co. Ltd. to become **Kenko-Tokina Co. Ltd.** The head office was transferred to Tokyo.

Tokyo Kogaku Kikai KK
Tokyo Kogaku Kikai KK, (Tokyo Optical Company Ltd.), was established in 1932 in Tokyo, Japan, to manufacture surveying instruments, binoculars, and cameras for the Japanese army. The company was temporarily closed after WWII, following which it re-opened to manufacture surveying instruments, cameras and binoculars for civilian use. In 1947 they began to produce ophthalmic and medical instruments. In 1953 they began to use the *Topcon* brand-name, and in 1957 they began to produce their first *Topcon* branded SLR camera. From 1959, for about three decades, their exported optical instruments bore the maker's **JB code** JB18. In 1960 the company became an affiliate of Tokyo Shibaura Electric Co. Ltd., (currently known as Toshiba Corporation). In 1970 they established the companies Topcon Europe NV in the Netherlands, and Topcon Instrument Corporation in the USA. Between 1975 and 1986 they established Topcon subsidiaries in Japan, Singapore, and Hong Kong. In 1989 the company changed its name to **Topcon Corporation**.

Tokyo Optical Co. Ltd.
See **Tokyo Kogaku Kikai KK**.

Tolles, Robert Bruce
Robert B. Tolles, (c1822-1883) [50] [184] [185] [186] [201], was an American optical instrument maker, born in

Connecticut, USA. In about 1843 he served his apprenticeship with **Charles Spencer**. Spencer's partner, Professor Asahel Eaton, convinced Spencer to teach Tolles his techniques and formulae for making microscope objectives and in 1858, armed with this knowledge, Tolles began his own business in Canastota. By 1859 Tolles was making the optical components of microscopes, with the mechanical parts made by Charles E. Grunow, (who may have been a relation of **J.&W. Grunow**). Tolles was innovative in microscope optics and design, and had a number of patents to his name. He also made binoculars and telescopes. **E. Miller** worked for Tolles prior to setting up his own business in New York in 1868.

In 1867, at the instigation of Charles Stodder, who became his business manager and agent, Tolles moved to Boston and became a partner in the newly formed Boston Optical Works. In 1871 Tolles became the sole manager of the business, employing Charles Spencer's son, Clarence Spencer, and son-in-law, O.T. May, **John Green**, and a small number of others. Charles Stodder, marketed the products from Tolles' workshop. When Robert Tolles died in 1883 his employee **Charles X. Dalton** took over the company and continued to run it as Charles X. Dalton, successor to the late R.B. Tolles, Boston Optical Works, located at 30 and 48 Hanover Street, Boston, until at least 1895.

Tomlinson, James

James Tomlinson, (fl.1741-1773 d.1777) [18] [401], was an optical instrument maker of London, England. He served his apprenticeship with **Matthew Loft**, beginning in 1725, and was freed of the **Spectacle Makers' Company** in 1739. He was Master of the Spectacle Makers' Company from 1764 to 1768, and had to apologise to them for employing people who were not free of a London guild. **James Watt** bought optical instruments from him for resale. In 1756 he employed his nephew, also named James Tomlinson.

Topcon Corporation

Topcon Corporation was established in Tokyo, Japan, in 1932 as **Tokyo Kogaku Kikai KK**, and began to use the *Topcon* brand-name in 1953. In 1960 the company became an affiliate of Tokyo Shibaura Electric Co. Ltd., (now known as Toshiba Corporation). In 1989 the company name was changed to Topcon Corporation. In 2008 Topcon acquired the company **Sokkia**. In 2015 Toshiba Corporation sold all their shares in Topcon. Topcon's optical products today include surveying and construction instruments sold under both their *Sokkia* brand and their own *Topcon* brand, as well as ophthalmic instruments under their *Eye Care* brand.

Toric Lens

See lens.

Torre & Co.

Torre & Co., (fl.1805) [18] [145] [379] [438], were opticians, and optical and philosophical instrument makers of 12 Leigh St., Red Lion Square, London, England. Instruments bearing their name include: stick barometers; wheel barometers.

The company is estimated to have operated at this address from c1805-c1823. Sources differ on who the participants in this partnership may have been, but one seems likely to have been **Anthony de la Torre**.

Torre & Co., in a different guise, or as a branch of the above company, also traded at 12 Holborn, London and is estimated to have operated at this address from c1820 until at least 1840.

Torre, Anthony de la

Anthony de la Torre, also known as della Torre, (fl.1805-1823) [18] [379] [438], was an optician, looking glass maker, and optical and philosophical instrument maker of London, England. Addresses: 12 Leigh St., Red Lion Square (1805-1811); 4 Leigh St., Red Lion Square (1815-1823). Instruments bearing his name include: 2-draw brass telescope with mahogany barrel.

From c1800 until 1806, Anthony de la Torre worked in partnership with **Caesar Tagliabue**, trading as **Tagliabue & Torre**. He may also have been a partner in **Torre & Co**.

By 1826, the company had been succeeded by della Torre & Barelli, looking glass makers, barometer and thermometer makers, and print makers of 9 Lamb's Conduit St., London. They, in turn, were succeeded by Joseph della Torre & Co., merchants, trading at the same address from 1834 until 1851.

Torrelli, Christopher

Christopher Torrelli, (fl.1830) [276 #2020] [379], was an optical instrument maker of Wood Hill, Northampton, England.

Tosco Co. Ltd.

Tosco Co. Ltd. was an optical equipment manufacturer of Tokyo, Japan. During the 1950s to the early 1970s, they marked their instruments with **JB Code** JB111. Example is **Viper** 20x80 centre focus binoculars.

Total Internal Reflection; Total Reflection

Total internal reflection (sometimes referred to as total reflection), [7], is a phenomenon that can occur when an incident ray of light strikes an optical surface at the boundary between two materials of refractive indices n_1

(e.g., glass), and n_2 (e.g., air), at an angle θ_i greater than the *critical angle*.

The incident angle θ_i is measured from the normal to the surface in the medium whose refractive index is n_1, (i.e., a ray of light at a higher incident angle is angled closer to the optical surface). The transmitted ray of light is refracted at angle θ_r similarly measured from the normal to the surface in the medium whose refractive index is n_2. The ray of light will be bent at the surface by an amount governed by the ratio of the two refractive indices and the incident angle (see Snell's law):

$$\sin \theta_r = (n_1/n_2) \sin \theta_i$$

As the incident angle θ_i increases, so the angle of the refracted ray θ_r increases, bringing it closer to the air-glass surface. As θ_i increases, the refraction will at some point be sufficient to bring the refracted ray so that it is angled along the air-glass surface. This happens when the incident angle reaches the *critical angle* θ_c.

θ_c is the value of the angle θ_i for which $\theta_r = 90°$.

If the incident angle is further increased, instead of the ray going through the glass-air surface and being refracted, the entire ray is reflected back into the glass. This is known as total internal reflection (or total reflection), and can only happen if the refractive index n_1 is greater than n_2, as in the example, in which n_1 is for glass and n_2 is for air.

Total Reflectometer

A total reflectometer is an instrument used to measure the refractive index of a solid or liquid by measuring the *critical angle* at which *total reflection* occurs at its surface. See also total internal reflection, reflectometer.

Total Station

A total station is a surveying instrument that performs all the functions of a theodolite, together with electronic distance measuring, recording and computation features. The electronic distance measuring feature uses either microwave, infrared, or visible light, directed to a target. It is generally accurate to within 5mm or 10mm per km over distances of up to about 4km. The total station is replacing the theodolite in modern surveying.

Towa

Towa was a Japanese telescope maker, using maker's mark illustrated in Appendix IV, although the use by Towa of this mark was not exclusive. Telescopes manufactured by Towa were often branded **Prinz**, **Tasco**, or **Meade**, as well as other brands.

Tower

Tower was brand-name used by **Sears** for cameras, telescopes, and binoculars, manufactured by a large number of manufacturers, and re-branded for sale by Sears.

Transit

A transit is:
 (1) Another name for a transit level.
 (2) Another name for a transit instrument.

Transit Circle

A transit circle, sometimes called a meridian circle, is a transit instrument equipped with a full graduated circle to measure the declination of a star as it transits the meridian. It may be used in conjunction with a clock to time the transit.

Transit Instrument

A transit instrument, or transit telescope, is a telescope mounted on a horizontal axis so that its view is always in the plane of the meridian. It is used in conjunction with a clock to time the transit of a star across the meridian. If equipped with a graduated arc, it can be used to measure the declination of the star during transit. If equipped with a full graduated circle, it is referred to as a transit circle or meridian circle.

Transit Level

A transit level, sometimes referred to as a transit, is an optical instrument used by surveyors to establish a level line or plane, and to provide accurate readings of angles both horizontally and vertically. There is no universally agreed difference between a transit level and a theodolite. However, in general, a transit uses exposed vernier scales, and a theodolite uses an enclosed vernier scale. The transit level should not be confused with the dumpy level.

Transit Telescope

Another name for a transit instrument.

Transit Theodolite

See theodolite.

Tremel, Georg

Georg Tremel, (d.1979) [219], was an optical instrument maker of Munich, Germany. He began to work for **G.&S. Merz** in 1918 while it was under the proprietorship of **Paul Zschokke**. Paul Zschokke died in 1932, at which time the company was insolvent. In 1933, together with another Merz employee, August Lösch, Georg Tremel took over the company, machines, and tooling of G.&S. Merz. He started his own company, Astro-Optische Werkstaetten Georg Tremel, offering small school refracting telescopes, 54mm f/12 and 61mm f/12, and accessories, as well as larger telescopes. He produced doublet objectives up to 200mm, and paraboloidal mirrors up to 350mm. He produced various kinds of oculars, prisms, and flats. He produced Cassegrain optics, and a *Neo-Brachyt* model similar to a modern anastigmatic Schiefspiegler.

Georg Tremel continued to run the business until it ceased trading in 1976.

Trichromatic Photography

Trichromatic photography is a process, invented by **J.C. Maxwell**, which was used to make the first colour photograph – an image of a Scottish tartan ribbon. The process is based on the *trichromatic colour theory* of vision, originated by **Thomas Young**.

Maxwell's trichromatic process involves taking three monochrome images, one each through red, green, and blue filters. These photographs are made into transparent prints. Each print is projected using the same colour filter it was taken with, and the three projected images are superimposed on a screen. The resulting image on the screen is a trichromatic colour reproduction of the original photographed scene.

Maxwell projected the first trichromatic image during a lecture at the Royal Institution, London, in May 1861.

The concept of trichromatic photography is the basis of most modern digital photography, which uses a matrix of red, green, and blue sensors to record an image.

Trimble Inc.

Trimble Inc. is a multinational corporation, based in Sunnyvale, California, USA, which was founded in 1978 by Charlie Trimble along with two others who were from Hewlett Packard. Initially they focussed on marine navigation products. They quickly ventured into GPS which was, at the time, a new technology, and they expanded into positioning and geodetic surveying instruments, followed by many other product markets for GPS technology. In 2000 Trimble acquired **Spectra Precision**, with its subsidiary **Spectra Geospatial**. This began a period of rapid expansion and acquisitions.

In 1999 Spectra Precision had entered a joint venture with **Carl Zeiss AG** to manage the Zeiss surveying business in Jena, Germany. In 2001 Trimble acquired the remaining interest in the Zeiss surveying business, which continued as Trimble Jena GmbH, although retaining its location at the Zeiss plant in Jena. In 2003 Trimble entered a joint venture with **Nikon**, which traded as Nikon-Trimble Co. Ltd. The objective of the new company was the design and manufacturing of surveying instruments, including total stations, and represented both parent companies in the Japanese market. Trimble Inc. continued to produce surveying instruments, with a large number of subsidiaries providing products and services in many areas of business, including agriculture, construction & operations, geospatial, natural resources, transportation, utilities, and government.

Triple Alliance

The Triple Alliance was the name used for a commercial arrangement between **George N. Saegmuller Co.**, **Bausch and Lomb** and **Zeiss**. The agreement began around 1908 and lasted until WWI.

This is not to be confused with: (a) The Triple Alliance which was a political alliance between Germany, Austria-Hungary and Italy, formed in 1882, and terminated early in WWI; or (b) The War of the Triple Alliance, fought between Paraguay and an alliance of Argentina, Brazil and Uruguay, between 1864 and 1870.

Triplet

Triplet is the term usually used in optics to refer to a lens arrangement consisting of three lens elements.

Trochon, François

François Trochon was a renowned French optician, trading at 1 Quai de l'Horloge, Paris, France. He established his business in 1740. His grandson, **Jean Gabriel Augustin Chevallier**, inherited his premises at 1 Quai de l'Horloge, Paris [50].

Troughton, Edward

Edward Troughton, (1753-1835) [3] [18] [24] [46] [90] [178] [228] [234] [397] [443], of *The Orrery*, 136 Fleet Street, London, England, was a highly respected optician, and optical, mathematical, and scientific instrument maker. He learned his trade as an apprentice to his brother, **John Troughton (II)**, beginning in 1773, and was freed of the **Grocers' Company** in 1784. One of his apprentices was **Joseph J. Dallaway**. Like his brother, Edward reputedly suffered from colour blindness, and thus did not specialise in working with lenses, a task he delegated to contemporary opticians such as **George Dollond (I)**. In 1788, upon the death of **John Troughton (I)**, Edward and John (II) entered a partnership, trading as **J.&E. Troughton**. This partnership continued until the retirement of John (II) in 1804, after which Edward began a period of working alone. Edward made a dividing engine, simpler in construction than that of his brother, which he completed in 1793, and used only to mark scales on instruments made by himself. **John Cail** worked for Edward Troughton for a period which is believed to have been in the early 1820s. **James Fayrer (I)**, and his son James (II) also worked for Edward Troughton.

In 1826 Edward took **William Simms (II)** into partnership, and began trading as **Troughton and Simms**.

See Appendix VII.

Troughton, John (I)

John Troughton (I), (c1716-1788) [18] [46] [90] [178] [228], was a mathematical instrument maker of London, England. Addresses: Standgate Lane, Lambeth (1752); Surrey St., Strand (1755-1777); Strand Lane near Surrey St. (1764-1774).

He was the son of William Troughton, a yeoman farmer from Corney, Cumberland. He served his apprenticeship with **Thomas Heath** beginning in 1734, and was freed of the **Grocers' Company** in 1756. In 1764 he was one of the petitioners who attempted to revoke the patent obtained by **John Dollond (I)** for the achromatic lens, and which was enforced after Dollond's death by his son, Peter (See Appendix X). John Troughton (I) took on as apprentices three sons of his brother Francis, the brothers, **John Troughton (II)**, beginning in 1757, Francis Troughton (bap. 1748), beginning in 1763, and Joseph Troughton (d.1770), beginning in 1765. He retired in 1778.

See Appendix VII.

Troughton, John (II)

John Troughton (II), (c1739-1807) [18] [24] [46] [90] [178] [228] [397] [443], was a mathematical instrument maker of London, England. Addresses: Surrey St., Strand (1764); Crown Court, Fleet St. (1768-1771); 17 Dean St., Fetter Lane (1771-1778); 1 Queen's Square, Bartholomew Close (1778-1782); 136 Fleet St. (1782-1788).

He served his apprenticeship with his uncle **John Troughton (I)**, and was freed of the **Grocers' Company** in 1764. He made a circular dividing engine to **Ramsden**'s design. Among his apprentices were his brother **Edward Troughton**, Joseph Troughton, son of Joseph, and **James Martin**. In 1782 he purchased the business of **Benjamin Cole (II)**, and began to trade from the same address, known as *The Orrery*, 136 Fleet Street. Upon the death of his uncle in 1788, John (II) went into partnership with his brother Edward, and they traded as **J.&E. Troughton**. Like his brother, John (II) reputedly suffered from colour blindness, and thus did not specialise in working with lenses, a task he delegated to contemporary opticians such as **George Dollond (I)**.

See Appendix VII.

Troughton, J.&E.

J.&E. Troughton, [18] [46] [90] [178] [228], were mathematical instrument makers of 136 Fleet Street, London, England. The firm was a partnership between **John Troughton (II)** and his brother **Edward Troughton**, which lasted from 1788 until the retirement of John (II) in 1804.

See Appendix VII.

Troughton and Simms

Troughton & Simms, [3] [18] [24] [44] [90] [228], were opticians, and optical, mathematical, and philosophical instrument makers, of London, England. Addresses: 136 Fleet St. (1839-1844); 138 Fleet St. (1844-1915); factory at 340 Woolwich Rd., Charlton (1866-at least 1919).

In 1826 **Edward Troughton** took **William Simms (II)** into partnership, and they began trading as Troughton and Simms. They manufactured mathematical and optical instruments, including transit circles and theodolites, and attracted customers from around the world. During the 1820s William Simms (II) worked with George Everest of the Great Trigonometrical Survey of India to design a rugged theodolite for use in surveys by the East India Company, and the resulting design was called the Everest Theodolite. Troughton & Simms supplied two ten-foot Ordnance Standards to the Ordnance Survey Department for use in the Principal Triangulation of Great Britain and Ireland.

Edward Troughton retired in 1833 and died in 1835, after which William Simms (II) continued in sole control of the business, still trading as Troughton & Simms. Not only were orders for surveying instruments still strong, but also orders from observatories, including the Royal Observatory at Greenwich where Professor Airy, newly appointed Astronomer Royal, began a programme of re-equipment. William Simms devised a much-acclaimed self-acting circular dividing engine which substantially reduced the time needed to engrave a graduated arc. Among his major astronomical projects were several large mural circles, and in 1832 Simms built the Northumberland telescope for Cambridge University, using optics produced by **Robert Cauchoix**. The company also produced telescopes for the amateur astronomy market. In 1838 Troughton & Simms took over from **William Gilbert (II)** as supplier of theodolites to the East India Company for the Survey of India, following a comparison of the accuracy of their instruments. In 1845, when the duty on glass was abolished, Simms began making his own optical components using French manufactured optical glass.

Following the death of William Simms (II) in 1860 the business of Troughton and Simms was continued by his son and his nephew, **William Simms (III)** and **James Simms (III)**. William (III) and James (III) ran the Troughton and Simms business successfully until William retired in 1871, then James alone until 1915, from which time **William Simms (IV)** and **James Simms (IV)**, the sons of William (III), continued the business. During WWI they produced optical munitions, including rangefinders. In 1916 The firm incorporated to become

Troughton and Simms, Ltd., but experienced a downturn in trade during the following years. In 1919 they were listed as a member of the **British Optical Instrument Manufacturers' Association**. In 1922 the company merged with **T. Cooke and Sons**, part of **Vickers**, to form **Cooke, Troughton and Simms** which became a wholly owned subsidiary of Vickers. In 1938 **Grubb Parsons** took over the telescope making part of the business. In 1963 Cooke, Troughton and Simms became part of the new company, **Vickers Instruments**, which continued in business until 1988.

See Appendix VII.

Tru-Vue

Tru-Vue Inc., [338] [339], of Rock Island, New York, USA, established in 1931, was a subsidiary of the Rock Island Bridge and Iron Works. They produced lenticular stereoscopes under the brand-name *Tru-Vue*. Their device was hand-held, and used a 35mm film, containing 14 stereo images, which passed through the viewer horizontally, and was progressed by a lever. After 1950 the film-reels contained 10 stereo images, and in the same year they introduced colour film-reels, which contained 9 stereo images. The design of the viewer evolved over the years, but the most radical change came after the company was taken over in c1952 by **Sawyers**. After this the design was changed to use a vertical strip of card containing seven stereo views.

TRY ME

The firm of **I.P. Cutts, Sutton & Son** used a trade mark in which the words *Trade Mark* appear over a **foul anchor**, [225 p348]. Under the anchor are the words *TRY ME*. The company used this trade mark for all products other than ships' logs and sounding machines, for which they used a different trade mark.

Tulley & Sons, C.

See **Charles Tulley**.

Tulley, Charles

Charles Tulley, (1761-c1831) [18] [24] [264] [276 #903] [406] [407], was an optician and optical instrument maker of London, England. Addresses: Goswell Street Road; 11 Pierpoint Row, Islington (1805); 4 Terrett's Court, Upper Street, Islington (1826); 7 Church Row, Islington (1845-1846).

He acquired the business and tools of **Joshua Lover Martin** in 1782, including the telescope tube-drawing machine invented and built by Martin. Charles Tulley was skilled in grinding and figuring lenses and mirrors. He produced fine quality object lenses, some from **Guinand** glass blanks. He also made Newtonian, Gregorian and Cassegrain telescopes. In 1826 he was asked by **Joseph Jackson Lister** to produce a microscope body which Lister had designed, a task which he subcontracted to **James Smith (I)**.

Charles Tulley had three sons, **Henry Tulley**, William Tulley (1789-1935), and Thomas Aston Tulley (1791-1846). From 1826 Charles worked in partnership with William and Thomas, trading as Charles Tulley & Sons at 4 Terrett's Court, Upper Street, Islington. William was the first optician to market achromatic microscope lenses. After the death of Charles, the brothers continued to trade as William & Thomas Tulley. By 1845 they were trading at 7 Church Row, Islington, and the business finally closed upon the death of Thomas in 1846. The brothers were succeeded in business by **Robert Mills** who, in turn, acquired the telescope tube-drawing machine.

Tulley, Henry

Henry Tulley, (fl.1822-1833) [18], was an optician and philosophical instrument maker of Bath, Somerset, England. Addresses: *Near the Pump Room*; 5 Kingston Buildings (1822-1826); 3 Pulteney Bridge (1826-1833). He was the son of **Charles Tulley**, and the brother of William and Thomas. He is known to have sold barometers and pantographs.

Tulley, William & Thomas

See **Charles Tulley**.

Turnbull, J.M; Turnbull & Co.

John Miller Turnbull, (fl.1865-1904) [263] [320] [379], was an optician, philosophical instrument maker, and chemical dealer of Edinburgh, Scotland. Addresses: 14 Nicolson Square (1865); 19 South St. David Street (1881-1884); 6 Rose Street (from 1885); 60 Princes Street (from 1894).

In 1886 he devised an improved sliding nosepiece and adaptor for the microscope. He also retailed instruments by other makers.

In 1894 a second branch of the business was opened at 60 Princes Street, succeeding **James Bryson** at that address. This business traded as Turnbull & Co. J.M. Turnbull's two sons, William and Frederick, both worked as opticians also. In 1959, Turnbull & Co. took over the optical business of **Lennie** in Edinburgh, trading at that time as E.&J. Lennie.

Turners' Company

See Appendix IX: The London Guilds.

Tuther, John Penn

John Penn Tuther, (1774-1827) [18] [50] [53] [274], was an optical, philosophical, and mathematical instrument maker of London, England. Addresses: 209 High Holborn; 64 King St., Bloomsbury (1816-1817); 221 High Holborn (1822).

He produced a large variety of instruments, including microscopes and custom scientific and philosophical instruments. His business was one of the larger of the mathematical and optical instrument makers in London at the time.

Twyman, Frank

Frank Twyman, (1876-1959) [20] [44] [46], of Canterbury, Kent, England, began his career testing telephone cables. Only a year later he took a post as scientific assistant at **Adam Hilger** in London. There he learned his trade as an optical designer from Adam's brother Otto Hilger. In 1902, when Otto Hilger died, Twyman became managing director of the firm. He became a recognised expert on the design and manufacturing of optical instruments, and industrial spectrochemical analysis. He was President of the **Scientific Instrument Manufacturers' Association of Great Britain** from 1925 to 1926. In 1946 he became chairman of Adam Hilger. He also took on a role as technical advisor to **E.R. Watts & Son**, and in 1948 when Adam Hilger Ltd. merged with E.R. Watts & Son to become **Hilger & Watts**, he became a director. He stayed in this role until his retirement in 1952 after which he continued to act as an advisor to the company until his death.

U

Ulrich
Ulrich were French opticians whose business was in Paris, Vichy and Nice, France. Instruments bearing their name include opera glasses variously marked *L. Ulrich, Paris, Vichy, Nice*, and *C. Ulrich, Paris, Vichy, Nice*.

Ultra-Violet Glass Spectrograph
See spectrograph.

Underwood, E.&T.
E.&T. Underwood, [306], was a camera maker of Brunswick Works, Granville Street, Birmingham, England. The firm was established in c1885 by the brothers E. Underwood and T. Underwood. They produced a number of different models of camera, some of which were premium quality. T. Underwood died in 1894, and the company continued in business until at least 1905.

Unicam Instruments
Unicam Instruments, [44] [53] [320], were scientific instrument manufacturers of Cambridge, England. The firm was established in 1934 by Sidney W.J. Stubbens, formerly the foreman of **Cambridge Instrument Co. Ltd.** By 1949 Unicam Instruments had become a subsidiary of **W.G. Pye & Co.**

Their entry in the 1951 Directory for the British Glass Industry, [327], listed them as Unicam Instruments (Cambridge) Ltd., of Arbury Works, Cambridge, established 1934. Their products were listed as: Ultra-violet, visible and infra-red spectrophotometers; Colorimeters; Crystallographic instruments. Trade Name: *Unicam*. Managing Director: S.W.J. Stubbens.

In 1968 they merged with W.G. Pye & Co. to become Pye Unicam.

Union Optical Co. Ltd.
Union Optical Co. Ltd., [320], of Tokyo, Japan, is a manufacturer of microscopes, objective lenses, and measuring systems. The company was established in 1948.

Uniscope
Uniscope was a Japanese maker of binoculars, exporting to the UK, US, Europe, Australia and New Zealand in the 1960s [94].

Uni-Scope Optical Systems Ltd.
Uni-Scope Optical Systems Ltd. is an optical instrument maker based in Karmiel, Israel. They make periscopes, rifle sights, binoculars and telescopes, as well as accessories.

United Kingdom Optical Company Ltd.
The United Kingdom Optical Co. Ltd., [44], of White Hart Lane, Tottenham, London, was established in 1919. The company's aim was to strengthen the British ophthalmic lens making industry based on the experiences of WWI. In 1919 the United Kingdom Optical Co. was listed as a member of the **British Optical Instrument Manufacturers' Association**.

In 1922, at the British Industries Fair, their catalogue entry' [447], listed them as manufacturers of all kinds of ophthalmic lenses, photographic lenses, and complete optical systems for scientific instruments.

In the 1929 British Industries Fair Catalogue, [448], their listing described their exhibit as an optical, scientific and photographic exhibit. It listed them as manufacturers of: All kinds of ophthalmic lenses (*Univis*, *Akro-Univis*, *Twofo* and *Akro* fused bifocals); *Pervextor* toric lenses; Photographic lenses; Readers; Condensers; Lenses and prisms for binoculars; Instruments of precision; Laboratory lenses and mirrors; Galvanometer mirrors.

In 1936, Frederick W.W. Baker of **W. Watson & Sons Ltd.** was listed as Chairman of the United Kingdom Optical Co. Ltd. During WWII the company produced optical components for military purposes. Towards the end of the fifties, they signed a technical agreement with **Bausch & Lomb**. In 1976 the company employed about 3,500 staff in the United Kingdom and about 500 overseas. By the 1990s the company was dormant, and in 2007 it was dissolved.

United Trading Co.
United Trading Co., [167], See **Unitron**.

Unitron
Unitron, [136] [167], was founded in Boston, Massachusetts., USA, as the United Trading Co. in the early 1930s by Lawrence Fine. High quality telescopes were imported from **Nihon Seiko Kenkyusho, Ltd.** in Japan and sold under the Unitron brand.

In 1975 the company was bought by **Ehrenreich Photo Optical Industries**, and moved to New York. Within two years the company name was changed to Unitron Instruments, and they continued to expand their astronomy equipment offerings.

In 1981 Ehrenreich Photo Optical Industries was acquired by **Nippon Kōgaku Kōgyō Kabushikigaisha**, the makers of *Nikon* cameras, and Unitron USA was a part of the Nikon Instruments group until a management buyout in 1986, at which time it became a private company. In addition to fine quality telescopes, they produced binoculars, theatre glasses, microscopes, and other optical instruments.

In 1992 Nihon Seiko Kenkyusho, Ltd. notified Unitron USA that they planned to discontinue production. Unitron products were advertised for sale until at least 1993, but sales ceased at around that time.

Universal Camera Corporation
Universal Camera Corporation was an optical instrument manufacturer based in New York, USA. The company was established in 1932, and continued in business until 1964. They manufactured still and cine cameras, binoculars, and projectors. During WWII they made binoculars to **Bausch and Lomb** patterns for military use.

University Optics Inc.
University Optics Inc. of Ann Arbor, Michigan, USA, was a maker and supplier of telescope accessories, binoculars, and telescope making supplies. The company was founded in about 1959/60, and continued in business until about 2017.

UQG Optics
UQG Optics is an optical glass manufacturer based in Cambridge, England. They produce optical and technical glasses, flat plano windows and substrates, optical colour glass filters and dichroic filters for heat and light control, plano flat and concave glass aluminised mirrors, plano-convex lenses in standard focal lengths for UV and visible light, and right-angle prisms for UV and visible light.

Uranostat
A uranostat is a kind of siderostat.

Urban, Charles; Urban Bioscope
Charles Urban, (1867-1942) [46] [335], was born in Ohio, USA. He was an innovative motion picture film producer who co-invented the Bioscope projector with the engineer Walter Isaacs in 1896. In 1897 he moved to the UK to become managing director of the Warwick Trading Company, where his Bioscope was marketed as the *Warwick Bioscope*. He left the Warwick Trading Company in 1903 to form his own company, and continued to market his invention as the *Urban Bioscope*. He worked with the film-maker George Albert Smith, (1854-1959), to develop the Kinemacolour two-colour film and projection system, which he first demonstrated in 1908. This system was highly successful until **William Friese-Greene** challenged his patent. Urban and Smith lost the challenge, bringing an end to the use of the Kinemacolour system. Charles Urban continued his career as an innovative film-maker in both the UK and USA.

Ure, John
John Ure, (fl.1821-1833) [18] [271], was an optical and mathematical instrument maker of Glasgow, Scotland. Addresses: 40 Stockwell (1821) and 136 High St. (1831-1833).

The premises at 40 Stockwell were also occupied by the mathematical instrument maker William Ure from 1819-1820, and by the mathematical instrument makers William Ure & Son from 1822-1823.

Urings, John
John Urings, (fl.1738-1772, d.1773) [18] [274] [276 #315] [319] [322], was a ship chandler, and mathematical, optical, and nautical instrument maker of London, England. Addresses: Minories; Hermitage St., St. Catherine's; St. Catherine's, Tower Hill (1759-1768); 174 Fenchurch St. (until 1773). He was freed by patrimony of the **Joiners' Company** in 1738. Among his apprentices was **Walton Willcox**, beginning in 1765. Instruments bearing his name include: Gregorian telescope; Culpeper type microscope; backstaff; Hadley's quadrants; octant; sundial.

His stock in trade was auctioned following his death in 1773, and included a large quantity of finished instruments as well as materials, tools, and fittings.

Usener, Charles F.
Charles F. Usener, (1823-1900) [122] [187], of Württemberg, Germany, emigrated to the USA in the early 1850s where he settled in New York. There his profession is listed as "Daguerreian and Optics Manufacturer", and he began to make lenses and cameras for **Holmes, Booth & Haydens**. In 1860 he naturalised as a US citizen and in 1866 his business was acquired by the photographic wholesalers **J.W. Willard & Co.** Charles Usener continued to produce his optical products exclusively for J.W. Willard & Co. Usener is reported to have additionally made telescope objectives.

U.S. Naval Gun Factory Optical Shop Annex
In 1917, when the US entered WWI, the **Crown Optical Company** could not keep up with domestic demand for military optics, and were nationalised. From that time, they ceased using the name Crown Optical Company, and became known as the U.S Naval Gun Factory Optical Shop Annex [136]. After WWI, and through WWII, they continued repairing and servicing optical instruments as well as designing and manufacturing lenses, prisms, telescopes and optical rangefinders.

Ustonson, John

John Ustonson, (fl.1804-1830) [18] [385], was a spectacle maker, optician, silversmith, and optical instrument maker, of London, England. Addresses: 21 George Yard, Old St. (1804-1811); Fleet St., near St. Dunstan's Church (1817); Cross St. (c1818); 15 Whiskin St., Spa Fields (1830). He served his apprenticeship with the renowned fishing tackle maker Onesimus Ustonson, beginning in 1787, and was freed of the **Turners' Company** in 1794. He took over the business of the optician William Ustonson (fl.1805). Among his apprentices was Felix Samuel Sheppard (this may have been **Felix Shephard**).

Utzschneider & Fraunhofer

See **Optisches Institut von Utzschneider und Fraunhofer**.

Utzschneider & Merz

See **Optisches Institut von Utzschneider und Fraunhofer**.

Utzschneider, Joseph von

Joseph von Utzschneider, (1763-1840) [24] [374], was an entrepreneur and financier. In 1804, he joined the **Mathematisch-mechanisches Institut von Reichenbach & Liebherr**, which became **Mathematisch-mechanisches Institut von Utzschneider, Reichenbach & Liebherr**. In 1809, the Institute was renamed **Optisches Institut von Utzschneider, Reichenbach & Fraunhofer in Benediktbeuern**, and in 1814, the it was again renamed **Optisches Institut von Utzschneider und Fraunhofer**. Following the death of **Joseph Ritter von Fraunhofer**, Utzschneider continued to run the company, and in 1838 he entered a partnership with **Georg Merz**, which continued until 1839, when he sold the company to **Merz & Mahler**, who continued to trade under the name Optisches Institut von Utzschneider und Fraunhofer in München.

See Appendix VII.

Utzschneider, Reichenbach & Fraunhofer in Benediktbeuern

See **Optisches Institut von Utzschneider, Reichenbach & Fraunhofer in Benediktbeuern**.

Utzschneider, Reichenbach & Liebherr

See **Mathematisch-mechanisches Institut von Utzschneider, Reichenbach & Liebherr**.

V

Valley Microscope
See **Carton Optical Canada Inc.**

Van Cort Instruments
Van Cort Instruments of Massachusetts, USA, were scientific and optical instrument makers, producing reproductions of old instruments, and newly designed instruments inspired by them. The company was founded in 1979, and operated for around 20 years.

Vanguard
Vanguard is a Chinese manufacturer of binoculars, rifle scopes and spotting scopes, based in Guangdong, China, and founded in 1986.

Varley, Cornelius
Cornelius Varley, (1781-1873) [18] [46] [53] [264] [319] [320] [388], was a landscape painter, optical instrument maker, and inventor, of London, England. Addresses: 22 Charlotte St., Fitzroy Square; 228 Tottenham Court Road; Junction Place, Paddington (1811); 1 Charles St., Clarendon Square (1811-1856); 42 Newman St. (1815); 51 Upper Thornhough St., Tottenham Court Rd. (1825); 7 York Place, Kentish Town (1857-1863); 337 Kentish Town Rd. (1864-1873).

His father died when he was ten years old, after which he was cared for by his uncle, Samuel Varley, who was a watchmaker, jeweller, and natural philosopher. During this time Cornelius started to make lenses and microscopes, and assisted his uncle in his lectures and experiments. In 1800, when his uncle ceased his business, Cornelius began to teach himself sketching and watercolour painting. Although his art was appreciated less during his lifetime than later, he had a moderately successful career as an artist and teacher. To assist himself in his art he invented, in 1809, the graphic telescope, and patented it in 1811.

From 1814 C. Varley's main occupation was manufacturing optical instruments. In 1831 and 1833 he was awarded silver medals by the Society of Arts for his optical instruments. His nephew, **Andrew Pritchard**, served an apprenticeship with him, after which Varley and Pritchard collaborated on making jewel lenses. In 1839 he was one of the co-founders of the Royal Microscopical Society. In 1841 he patented his microscope lever-stage movement for following aquatic organisms, and in the same year he was awarded an Isis gold medal for his improvements to the microscope. By 1848 he was working in partnership with his son, trading as Varley & Son. In 1851 Varley & Son exhibited at the Great Exhibition in London, [328]. Their exhibit included graphic telescopes, a reversing camera by means of which pictures or objects may be traced in reverse, a microscope with lever-stage movement, reflecting telescopes, air pumps, and electrical apparatus. The partnership Varley & Son continued until 1854.

Varley, Samuel Alfred
Samuel Alfred Varley, (fl.1850-1862) [18] [319], was an optical instrument maker of London, England. Addresses: 1 Charles St., Clarendon Square (1850-1851), and 7 York Place, Kentish Town (1862). In the 1851 Census, he is listed as 19 years old, living at 1 Charles St., the same address as **Cornelius Varley**, who may have been his father.

He attended the London Mechanics' Institute from 1850-1851, and was Engineering Superintendent of the Metropolitan District of the Electric & International Telegraph Company until 1862.

Vasseur & Cie, A. le
A. le Vasseur & Cie, [320] [334], of 33 rue de Fleurus, Paris, France, was a seller of microscopes and accessories. Their catalogue, dated c1900, shows microscopes, objectives, and accessories, and states that all instruments shown are made by **M. Dumaige**. The microscopes in the catalogue are shown with a price to purchase outright, or by monthly payments.

Vaughan, C.B.
C.B. Vaughan were vendors of guns, rifles, binoculars, etc., and specialists in gun fitting, of 39 Strand, London, England. The company was established in 1782, and was a dealer in second-hand guns, and own-branded guns and binoculars. An advertisement from 1933, [44], shows that they sold **Zeiss** binoculars. An example pair of 8x24 binoculars bears the inscription *C.B. Vaughan, 39 Strand, London* and *Telacht*. Telacht is a binocular model name used by Zeiss/**Hensoldt**.

Vavilov State Optical Institute (GOI)
The Vavilov State Optical Institute, [23], in St. Petersburg, Russia, otherwise known as Gosudarstvennyy Opticheskiy Institut (GOI), was established in 1918 to set the standards for all the optics institutes, plants, and enterprises in the former Soviet Union. Roughly 60 percent of their research and design efforts have been historically dedicated to military and space projects.

They manufacture a wide range of both civil and military optical products and components, and publish the journals *Optical Journal* and *Works of the Vavilov Institute*.

Veitch, James

James Veitch, (1771-1838) [18] [24] [169] [263] [395], of Inchbonny, Scotland, was a professional plough-maker, a weights and measures inspector, a highly skilled amateur telescope maker, and a friend of **Sir David Brewster**. He was self-taught in the sciences, with a particular interest in mathematics and astronomy. He mentored Brewster in telescope making. He was skilled at making mirrors and lenses, and supplied lenses, including jewel lenses in garnet, to Brewster. He sold his telescopes to many eminent people of the time, including the Scottish author, Sir Walter Scott, who owned a clock and a telescope made by him. In addition to telescopes, which were the principal instruments he made, he is known to have made microscopes, barometers, and other philosophical instruments.

Vérascope

The *vérascope* was a brand of stereoscopic hand cameras invented and manufactured by **Jules Richard**, introduced in 1893. It used 45x107mm photographic plates in a magazine that could hold multiple plates. The design was compact, enabling it to be used hand-held, and played a large part in popularising stereo photography. The *vérascope* camera models continued in production until the 1950s. Jules Richard also went on to use the *vérascope* brand-name for stereoscopes and accessories.

Verick, M. Constant

Marie Constant Verick, usually known as Constant Verick, (1829-1892) [50] [53] [264], was an optical instrument maker of Rue De La Parcheminerie 2, Paris, France. He worked for **Edmund Hartnack** prior to starting his own business in 1866, and claimed to have been Hartnack's "special student". He received a medal of merit at the Vienna international exposition of 1873, and a gold medal at the Paris international exposition of 1878. Verick's catalogue of 1885 shows 9 different microscope models, as well as microtomes and other accessories. During the early 1880s one of his employees, **Maurice Stiassnie**, who was also his nephew-in-law, took over his business. Verick and Stiassnie may have operated the business together in partnership for a few years prior to Stiassnie continuing alone.

Vernier Reading Theodolite

A vernier reading theodolite is one in which the graduated scales are equipped with a vernier scale for increased accuracy.

Vernonscope & Co.

Vernonscope & Company was formed in 1958 in Candor, New York, USA, by Donald Yeier. A 1989 advertisement shows a 94mm f/7 refractor on an equatorial mount with a wooden tripod. In 2013 the company was acquired by Tony and Liz Mansfield, and renamed Vernonscope LLC, selling spotting scopes, eyepieces, Barlow lenses, diagonals, binoviewers, filters, and other accessories.

Vertical Illuminator

A vertical illuminator is an apparatus used in conjunction with a microscope to illuminate opaque sample surfaces for examination. The light is directed downward, so that the direction of illumination coincides with the viewing axis, thus avoiding shadows.

Vial, Jules

Jules Vial, [72] [197], succeeded Albert Denis Bardou in the company **Bardou & Son**, at 55 rue de Chabrol, Paris, France, in 1896. The company made terrestrial and astronomical telescopes, microscopes, and binoculars for field, marine and theatre. He continued to use the Bardou name for his business. Vial produced a catalogue showing up to 6-inch pier mounted telescopes. By 1899 the company had moved to 59 rue Caulaincourt, Paris. A 1911 catalogue for Bardou and Son, [237], shows refracting telescopes ranging from 2¼ inches to 4¼ inches aperture, as well as accessories.

Vickers

Vickers, [44], was a British engineering conglomerate that merged many of its engineering and armaments assets with those of Armstrong Whitworth to form Vickers-Armstrongs in 1927. Areas of activity included shipbuilding, weapons, flight, and scientific instruments. **Sir Howard Grubb & Company** began to build and supply periscopes for Vickers submarines in 1901 [20]. At that time Vickers had a virtual monopoly on supplying submarines for the Royal Navy.

In 1915 Vickers acquired a 70% share in **Thomas Cooke & Sons** [228] [234]. In 1916 they purchased a controlling interest in **Adam Hilger & Co.**, but in 1926 Hilgers purchased the shares back. In 1922 Thomas Cooke & Sons merged with **Troughton & Simms** to form **Cooke, Troughton & Simms**. In 1924 Vickers liquidated Cooke, Troughton & Simms, and re-floated it under their direct control. In 1959 they acquired **McArthur Microscopes Ltd.**, the **C. Baker Ltd** microscope factory, and Casella (Electronics) Ltd. In 1962 the expanded company was renamed Vickers Instruments. This continued as a profitable business for many years, mainly selling microscopes, surveying instruments and micro measurement apparatus. Among their products were binoculars imported under the name **Alderblick**, and branded as *Vickers Alderblick*. By 1988, in a changing market, Vickers Instruments were focused on high-precision measuring apparatus and laser rangefinders. At that time the Vickers Group sold the ordnance instruments part of the company to British Aerospace, and the rest of Vickers Instruments Ltd. to the US company Bio-Rad Laboratories Inc.

Video Camera
See camera.

Vidhya-Bhand
Vidhya-Bhand was a trade mark of **Nihon Seiko Kenkyusho, Ltd.**

Viewfinder Camera
See camera.

Viewmaster
The *Viewmaster*, [**338**], was a popular hand-held lenticular stereoscope first produced by **Sawyers** in 1938. The stereo photos were transparencies mounted in a circular disc containing seven stereo images. The *Viewmaster* was developed as an educational tool, and used by the US military for training in fields such as aircraft identification. It soon became popular, however, for children's entertainment. Several models were produced over the years, some marketed directly for children.

Viganò, Angelo
Angelo Viganò founded the Viganò chain of opticians in Milan, Italy in 1880. See **Filotecnica Salmoiraghi**.

Vignette
n: A photograph with edges that are shaded off.
v: To finish a photograph with fading borders in the form of a vignette.

Vignetting
Vignetting, sometimes known as "light fall-off", is a reduction in the brightness of an image with increasing distance from the centre. Vignetting is either caused by the optical arrangement in use or introduced by a vignetter. Vignetting can result from the design and figure of the lens(es) in use, or from the use of filters or physical restrictions in the optical path.

Vignetter
A vignetter is:

(1) An optical instrument used as part of, or in conjunction with a photographic enlarger, to produce photographic prints with a vignetting effect.

(2) Any tool, physical or digital, that is used to introduce a vignetting effect into a photograph.

Viking Optical Ltd.
Viking Optical is an importer and distributor of optical equipment, including binoculars, monoculars, spotting scopes and telescopes. They are based in Suffolk, England. They acquired the **Hilkinson** brand, probably in 2012. They sell many brands of optics, including their own *Hilkinson* and *Viking* brands.

Vintners' Company
See Appendix IX: The London Guilds.

Vion
The business of Vion, [**197**], in Paris, France, was established in 1832 by Th. Vion. He had two sons who continued the business as Vion Frères, and showed their work at international exhibitions in 1878, 1889, 1893 and 1900. Their exhibition listings included: The *longview* telescope for terrestrial and marine use; Terrestrial and astronomical telescopes; Equatorial mounts; Surveying instruments; Microscopes; Spherical and paraboloidal mirrors; Prisms; Doublet objectives; Prismatic astrolabe.

By 1922 the company was named Etablissement Vion, abbreviated to Etablt. Vion or, more commonly, E. Vion. They continued to produce instruments including telescopes, spyglasses, compasses, depth gauges, and student microscopes. Their products were retailed in the USA by **Selsi** from the 1920s until the 1940s, and were available via European retailers until at least 1960.

Viper
Viper was a Japanese brand of binoculars. A 10x50 example bearing the maker's **JB code** JB16, was made by **Sankei Koki Seisakujo Inc.** A 20x80 (centre focus) example bearing the maker's JB code JB111 was made by **Tosco Co. Ltd.**, of Tokyo. Other examples include 8x30, 9x40, 12x50, 20x50, and 15x70. Viper binoculars were made from around the 1950s to the early 1970s, and are not to be confused with *Viper* binoculars, produced by Vortix Optics of Wisconsin, USA.

Visiometer
A visiometer is:

(1) An instrument used to determine visibility, used in weather stations. It comprises a light source, and an off-axis detector to measure scattered light.

(2) An instrument used to determine the power of spectacle lenses.

Vision Engineering Ltd.
Vision Engineering Ltd., [**53**] [**320**], of Send, Woking, Surrey, England, is a manufacturer of optical and digital stereo microscopes and non-contact measuring systems. The company was founded in 1958 by ex-Jaguar Racing

employee Rob Freeman. Initially the company was an optical contractor to companies such as Rolls Royce, Vickers, Ferranti, and GEC. In 1972 they introduced a patented design of stereo microscope which gave the viewer more freedom for positioning the eyes. In 1994 they introduced the *Mantis* "eyepiece-less" stereo microscope. In 2019 they introduced their *Deep Reality Viewer* 3D visualisation technology.

Vista Outdoor Inc.

Vista Outdoor is a US company which manufactures and markets outdoor and recreation products. They operate in two groups, shooting sports and outdoor products. They own a large number of brands. Those brands that sell optical products are: **Bushnell**, **Tasco**, **Simmons**, and **Night Optics**.

Vitascope

The Vitascope was an early type of motion picture projector invented by the American, Thomas Armat, in 1895. It used sprocketed film with a stop-go mechanism, and projected the moving image onto a screen. The Vitascope was marketed by **Thomas Edison** as the Edison Vitascope. The Vitascope used a single blade shutter, resulting in flicker in the projected image. It required connection to a direct current supply, and could only show a maximum of fifty feet of film before changing the film roll. These limitations led **Charles Urban**, together with the engineer Walter Isaacs, to make improvements to it, resulting in the Bioscope.

Vivascope

A vivascope, [30], is a telescope for viewing objects at close ranges. Such instruments were made by **J.H. Dallmeyer**, and **Heath & Co.**

Vivitar

Vivitar is a US based company, established in 1938 by Max Ponder and John Best as **Ponder and Best**. They imported photographic equipment from Germany, Taiwan and Japan. In 1960 they created the *Vivitar* brand, and through it they developed and marketed lenses which were manufactured for them by Japanese makers such as **Kino Precision Industries**. In 1979 they changed their name to Vivitar Corp. and became multinational. Subsequently their optical products have included budget photographic equipment, binoculars, microscopes, and telescopes.

Vixen Co. Ltd.

The Vixen Company Ltd., [136], was formed in Tokyo, Japan in 1970. They manufacture and distribute telescopes, binoculars, microscopes, and related accessories. The company's predecessor was **Koyu Company Ltd.** which was renamed to become Vixen.

Voigt & Hochgesang

Voigt & Hochgesang, (fl.1869-c1915) [113] [320], of Göttingen, Germany, were manufacturers of specialist optical equipment for crystallography and petrography. The company was co-founded by **F.G. Voigt** in 1869. In 1886 the company was taken over by Richard Brunnée, who had previously worked for **F. Sartorius** and **Leitz**.

Examples of instruments by Voigt & Hochgesang include: Mitscherlich goniometer, petrological microscope, polarizing microscope, and prepared microscope slides of thin sections of rocks, minerals, and fossils.

Voigt, F.G.

F. Gustav Voigt, (fl.1858-1886) [374], was a mathematical and optical instrument maker of Göttingen, Germany. He served an apprenticeship with **Breithaupt**, and was the first apprentice of **Rudolf Winkel**, beginning in 1858. He co-founded **Voigt & Hochgesang** in 1869.

Voigtländer

Johann Christoph Voigtländer, (1732-1797) [36] [37] [70] [71], was a carpenter, who founded the Voigtländer company in Vienna in 1756. He established a strong reputation making mathematical and mechanical instruments and, upon his death in 1797, his three sons, Christian Wilhelm, Johann Siegmouth and Johann Friedrich Voigtländer (1779-1859), continued the business. During the period 1801-1806 Johann Friedrich travelled in Europe, training as an optician, visiting Stuttgart, Frankfurt and London, and it is believed that he worked for some of that time in the workshops of **Dollond** in London. In 1806, upon the death of his mother, Johann Friedrich returned to Vienna and began to operate his business independently of his brothers. Initially he made mathematical and mechanical instruments as well as optical instruments, but his optical instrument production became an increasingly important part of his business.

In 1837, Johann Freidrich retired, and his son, Peter Wilhelm Friedrich Voigtländer (1812-1878), took over the business. At this time the technology of photography, and the daguerreotype had reached the stage that there was a demand for objective lenses, and subsequently complete cameras, a demand which Voigtländer met with high quality products. In 1849, he opened a branch in his wife's home-town, Braunschweig (Brunswick), Germany. Following the revolution in Vienna in 1848/9, Voigtländer relocated his business to Braunschweig. As their photographic business grew, so did their network of international business partners, and hence their trade. In

1843 the Langenheim brothers opened a daguerreotype studio in Philadelphia, USA, and acted as agents of the company Voigtländer & Sohn. In 1868 the Vienna branch of the business was closed. During the latter years of his tenure at Voigtländer Braunschweig, Peter Wilhelm Friedrich resisted technical innovation, and as a result, the company's fortunes suffered.

Peter Wilhelm Friedrich was succeeded by his son Friedrich V. Voigtländer, who was head of the company from 1876 to 1898. Friedrich steadily rebuilt the company's fortunes, embracing technical advance, and adopting **Zeiss** lenses in the late-1880s. In 1898 Voigtländer became a public company, Voigtländer & Sohn AG, and in 1923 the majority of shares were acquired by Schering AG. The company continued as a market leader until Schering AG sold its share of the company to the **Carl Zeiss Foundation** in 1956, and in 1965 they integrated with **Zeiss Ikon**. In 1971 they ended production of cameras at the Voigtländer factory, consolidating to become Optische Werke Voigtländer (Voigtländer Optical Works). The company failed in 1982.

Plusfoto took over the name when the company failed in 1982, and sold it to **Ringfoto** in 1997. In 1999 Ringfoto licensed the name to the Japanese optical glass maker **Cosina**, who went on to use the Voigtländer brand-name to market their photographic products.

Volanterio, Joshua

Joshua Volanterio, otherwise known as Volanteno, (fl.1818-1841) [**18**] [**276** #1730] [**379**], was an optician, optical instrument maker, looking glass maker, and barometer maker, of Doncaster, England. Addresses: Fishergate (1818); Frenchgate (1822); High St. (1834); Baxter Gate (1841). He particularly advertised his barometers.

Volk Optical Ltd.

Volk Optical Ltd. of Mentor, Ohio, USA, is a manufacturer of ophthalmic lenses, including specialist lenses for ophthalmic instruments and imaging. The company was established in 1974 by Dr. David Volk, and has since opened additional operations in China and India.

W

Wade, John Creswell
John Creswell Wade, (fl.1809-1836) [18], was an optical, mathematical, and philosophical instrument maker, and measure maker, of London, England. Addresses: Pump Court, Union St., Southwark (1829-1836); The Grove, Great Guildford St. (1809).

Waechter, Paul
Paul Carl Friedrich Waechter, (1847-1893) [50] [113] [287] [374], of Berlin, Germany, was an optical instrument maker who primarily made microscopes and photographic objectives. He is best known for his microscopes designed for detecting parasitic roundworms in meat. Waechter trained as an optician and mechanic with **Carl Zeiss** in Jena, and left in 1872 to set up his own business in Berlin. **Oskar Messter** worked for Waechter before joining his father's business in 1891.

Berlin addresses: 14 Wartenburg Strasse (1872-1874); 19 Grüner Weg (1874-1880); 16 Grüner Weg (1880-1882); 115 Köpnicker Strasse, alternatively known as Cöpnicker Strasse (1882-1889); 112 Köpnicker Strasse (1889-1890); 21 Albe Strasse, Friedenau (1890-c1940).

Following Waechter's death in 1893, his wife continued the business with one of his employees, August Pulcher, as manager. In 1900 August Pulcher purchased the business together with Paul Prasser. They operated the Paul Waechter business in partnership until 1907. The business moved to Potsdam sometime around WWII, and finally to Wetzlar sometime in the 1950s.

Wake, Charles
Charles Wake, (b.1806 fl.1833-1875) [18] [276 #2269], was an optical, mathematical, and philosophical instrument maker, and drawing instrument maker, of 11 Silver St., Golden Square, London, England. He was the brother of the mathematical instrument maker Francis Henry Wake (1817-1883 fl.1851-1855), also of London.

Wale, George
George Wale, (1849-c1903) [50] [201], of New York, USA, was an optical instrument maker, renowned for his microscopes, photographic lenses, and for an attachment for a magic lantern, called a *College Lantern*, that produced an image from a horizontally held specimen, and thus could be used with liquid specimens. He was a cousin of **William Wales**, although they spelled their surnames differently. In his early career he worked with **Benjamin Pike's Sons**, with whom he learned to make lenses. He then formed a partnership in about 1864 with **Richard Morrison**, who had also worked for Benjamin Pike's Sons, trading as Wale & Morrison. This partnership was dissolved after only a few years, and by 1871 he had entered a new partnership, this time with an instrument maker named Hawkins, trading as Hawkins & Wale in Hoboken, New Jersey, producing microscopes, spectroscopes, and other instruments. This partnership ceased in 1874, after which Wale began to trade as George Wale & Company. He continued to produce microscopes and related instruments, and came up with several innovations.

Upon the decline of Richard Morrison's health Wale took over Morrison's business, and continued Morrison's production of photographic lenses. Sometime between 1895 and 1899 he moved to New York, and in 1899 he applied for a patent on a toy kaleidoscope which he referred to as the Symmetroscope. This was marketed by Wale in partnership with Frank P. Irving, trading as the Wale-Irving Company. By 1900 Wale was manufacturing optical instruments in Boston, Massachusetts.

Wale & Morrison
See **George Wale**.

Wales, William
William Wales, (fl.1862-) [201], was the cousin of **George Wale** (Note, however, the different spelling of the surname). He was born in England, and learned the optics trade while working for **Smith & Beck**. In 1862 he moved to the USA, and settled in New York City, where he established a strong reputation for making microscope objectives. He supplied these objectives to American microscope makers such as **Walter H. Bulloch**, **Leopold Schrauer**, and **Thomas H. McAllister**. He entered a partnership with Walter Bulloch in about 1864, in which Bulloch made the microscope stands and Wales made the optics. This partnership continued until 1867. After this Wales continued his own business, making objectives.

Walker & Son, M.
M. Walker & Son, [379], were chronometer and nautical instrument makers of Glasgow, Scotland. Addresses: 44 & 45 Clyde Place (1885-1891). Branches at: Cathcart St., Greenock (1886-1891); Bath St., Liverpool, England (1887-1891). During 1887-1888, the Liverpool branch was the company **Thomas Bassnett & Co.**, but may have also traded under the name Walker & Son, since a 13-inch radius vernier octant, with ebony T-frame and brass index arms and fittings, sold at auction, bears the inscription *Walker & Son, Liverpool*.

Walker, Francis
Francis Walker, (fl.1810-1859) [18] [276 #1453] [379], was an optician, and optical, mathematical, and

philosophical instrument maker of London, England. Addresses: 35 Wapping Wall (1810-1845); 17 Wapping Wall (1846-1850); 77 Broad St., Ratcliff Cross (1851-1859). He served his apprenticeship with the mathematical instrument maker William George Cook of the **Masons' Company**, beginning in 1818, and was freed by purchase of the **Spectacle Makers' Company** in 1845.

Walker, J.&A.
John & Alexander Walker, (fl.1823-1859) [**3**] [**276** #1731, 1731a] [**322**] [**379**], were optical, mathematical, and nautical instrument makers, Navigation and Stationary Warehouse, and chart sellers, of Liverpool and London, England. Addresses: 33 Pool Lane, Liverpool (1827); 34 Castle St., Liverpool (1847); 47 Bernard St., London. Many of their instruments were supplied to them wholesale and sold retail.

Wall, John
John Wall, (1932-2018) [**179**], was an engineer and amateur telescope builder. He was born in Crayford, Kent, England, and served an engineering apprenticeship with **Vickers** in Crayford. In 1969 he invented what is now known as the Crayford focuser. Throughout his career he indulged his interest in amateur telescope building, experimenting with dialyte lenses. Later in his career he worked for the National Maritime Museum at Greenwich as an educator, and as an optical designer for the Old Royal Observatory. In his retirement he began to pursue his interest in retrofocally corrected dialyte telescopes, a purely refracting design, [**180**], that improves on the Medial telescope designs of **Ludwig Schupmann**.

Warner & Swasey Co.
Warner & Swasey Co., [**196**], was established in Chicago, Illinois, USA in 1880 as a partnership between Worcester Reed Warner (1846-1929) and Ambrose Swasey (1846-1937). The company then moved to Cleveland, Ohio, in 1881. The company's primary business was producing machine tools but, because of W.R. Warner's interest in astronomy, they also produced telescopes. They soon became renowned for their telescopes and mounts, and built the 36-inch telescope at the Lick Observatory in California, which saw first light in 1888, at which time it was the world's largest refracting telescope. In 1895 **Gottlieb Fecker** began to work for the company. He became their chief designer and manager of their instrument department, and was responsible for the designs of many of their fine instruments until his death in 1921. Warner & Swasey continued to produce telescopes until 1970.

Warsaw Cine-technical Works
See **WZFO**.

Warsaw Photo-Optical Works (WZFO)
WZFO, of Warsaw, Poland, started after WWII as Warsaw Cine-technical Works, with management supervision by the Central Cinematography Committee, producing film projectors. In 1952/3, management was passed to the Heavy Industry Ministry, and the name was changed to Warsaw Photo-Optical Works (WZFO) [**137**]. Their plan was to produce cameras and photographic equipment, and they began production of their *Start* cameras in 1953/4, with an objective lens produced by **Polish Optical Industries** (PZO). They went on to produce enlargers and other photographic equipment, and the camera models *Druh*, *Fenix* and *Ami* (launched 1954-57), and *Alfa* (1962). In 1963 they produced an innovative enlarger lens with integral colour correcting filters called the *Janpol Color*. The company also produced slide projectors, flash units, microscopes and loupe magnifiers. In about 1968 WZFO was merged into PZO, after which the name Warsaw Photo-optical Works was no longer used.

Warwick Bioscope
See **Charles Urban**.

Warwick Trading Company
See **Charles Urban**.

Washbourne
Washbourne, (fl.c1800) [**18**], of London, England, is the name on the label of a lucernal microscope which, since it can be dismantled, is thought to have been made for a travelling lecturer. The instrument is described in [**274**].

Wasserlein, Rudolph
Rudolph Wasserlein, (1830-1905), of Berlin, Germany, was a microscope maker. In about 1850 he entered a partnership with **Charles Louis Bénèche**, trading as **Bénèche & Wasserlein**. The partnership continued until about 1860, after which Wasserlein continued his business alone.

Water Immersion Objective
A water immersion objective is an immersion objective that uses water as the medium between the front element of the microscope objective and the slide's cover-slip.

Watkin Clinometer
See clinometer.

Watkin Depression Rangefinder
A depression rangefinder is an early form of optical rangefinder, invented by **Major H.S. Watkin** of the British Royal Artillery. See rangefinder (optical).

Watkin Mekometer
A Watkin mekometer is an early form of optical rangefinder, invented by **Major H.S. Watkin** of the British Royal Artillery. See rangefinder (optical).

Watkin Mirror Clinometer
See clinometer.

Watkin, Major H.S.
Major H.S. Watkin, of the British Royal Artillery, was the inventor of the Watkin depression rangefinder, and the Watkin mekometer.

Watkins & Hill
Watkins and Hill, [14] [18] [44] [53] [264] [435], were opticians, and optical, mathematical, and philosophical instrument makers of 5 Charing Cross, London, England. The firm succeeded **Jerimiah & Walter Watkins** in business in 1818, and was a partnership between **Francis Watkins (II)**, the son of **Jeremiah Watkins**, and the company's manager, William Hill. **Edward Marmaduke Clarke** worked for the company briefly during the 1830s prior to starting his own instrument making company. With the deaths in 1847 of both Francis Watkins (II) and William Hill, the company continued under the management of Abraham Day.

In the Great Exhibition of 1851, [294] [328], they exhibited in three classes – Class V: Sectional models of steam engines. Class VIII: Rifle fitted with sighting telescope. Class X: Electrical machine; Galvanometer; Electro-magnetic engine; Microscopes; Aerometric balance; Sextants; Rain-gauge; Polariscope; Theodolite; Levels; &c. This was considered by some to be a "poor showing".

In 1856, following a period of decline, Watkins and Hill were taken over by **Elliott Brothers**.

Watkins & Smith
Watkins and Smith, [18], of Charing Cross, London, England, was a partnership between **Francis Watkins (I)** and his former apprentice, **Addison Smith**. The partnership traded from 1763 until 1774, and they are known to have sold: barometer; callipers; hygrometer; perpetual calendar; telescope; theodolite.

Watkins, Francis (I)
Francis Watkins (I), (bap.1723, d.1791) [14] [18] [46] [53] [264] [276 #423] [319] [400], was an optician and mathematical instrument maker of London, England. He served his apprenticeship with the instrument maker **Nathaniel Adams** at the *Golden Spectacles*, Charing Cross, London, and was freed of the **Spectacle Makers' Company** in 1746. At some time, he worked for **Edward Scarlett (II)**. In 1747 Francis Watkins (I) was trading at *Sir Isaac Newton's Head*, 4/5 Charing Cross, London, and he continued to trade at 5 Charing Cross until at least 1784. Among his apprentices were **Henry Pyefinch** and **Addison Smith**.

Francis Watkins (I) was a contemporary of **John Dollond (I)**, and entered a financial partnership with him to market refracting telescopes based on Dollond's patent for the achromatic lens. After John Dollond (I)'s death, **Peter Dollond** bought out Watkins' part of the agreement. Watkins, however, continued to make telescopes in violation of the Dollond patent. This led to a dispute, and Peter Dollond sued Watkins successfully (see Appendix X). From 1763 until 1774, Francis Watkins (I) traded in partnership with his former apprentice, Addison Smith. The partnership traded as **Watkins and Smith**. In 1766, Peter Dollond once again took successful legal action to enforce his patent, this time against Francis Watkins (I) and Addison Smith as joint defendants.

By 1784 two of his nephews were trading as **Jeremiah and Walter Watkins** at 5 Charing Cross. When Francis Watkins (I) died in 1791 they inherited his business.

Watkins, Francis (II)
Francis Watkins (II), (1800-1847) [18] [319] [435], was an optician of 5 Charing Cross, London, England, and was the son of **Jeremiah Watkins**. He served his apprenticeship with Richard Frisby, beginning in 1810, and was freed of the **Vintners' Company** in 1818. He succeeded his father in the business of **Jeremiah and Walter Watkins**, following which he traded, together with an employee, William Hill, as **Watkins and Hill** until his death in 1847.

Watkins, Jeremiah
Jeremiah Watkins, (c1758-1810) [14] [18] [44] [53] [319] [435], was an optical, mathematical, and philosophical instrument maker of 5 Charing Cross, London, England. He was a nephew of **Francis Watkins (I)**, and the father of **Francis Watkins (II)**. In 1784 he began to work in partnership with **Walter Watkins**, trading as **Jeremiah & Walter Watkins** until Walter's death in 1798. Jeremiah continued the business alone, trading under the same name until his death in 1810.

Watkins, Jeremiah & Walter
Jerimiah & Walter Watkins, otherwise known as J.&W. Watkins or I.&W. Watkins, [14] [18] [53] [319]

[361], were optical, mathematical and philosophical instrument makers of 5 Charing Cross, London, England. **Jeremiah Watkins** and **Walter Watkins**, were the nephews of **Francis Watkins (I)**. They traded together in this partnership from 1784 to 1798. According to an instrument repair label, "Repaired by Jeremiah and Walter Watkins, Optical, Mathematical and Philosophical instrument makers to his Royal Highness the Duke of Clarence, No. 5, Charing Cross, London."

They took over the business of Francis Watkins (I) when he died in 1791. In 1798, Walter Watkins died, and Jeremiah continued the business alone. After Jeremiah's death in 1810 the company was run on behalf of his widow by an employee, William Hill, until her son, **Francis Watkins (II)**, became a partner in 1818. The company then traded as **Watkins and Hill** until 1856 when they were taken over by **Elliott Brothers**.

Watkins, Walter
Walter Watkins, (d.1798) [18] [319] [435], was an optician of 5 Charing Cross, London, England, and was the nephew of **Francis Watkins (I)**. He served his apprenticeship with the printer, William Owen, beginning in 1774, and was freed of the **Stationers' Company** in 1781. In 1784 he began to work in partnership with **Jeremiah Watkins**, trading as **Jeremiah & Walter Watkins** until his death in 1798.

Watkins, William (I)
William Watkins (I), (fl.1784-1809) [18] [53] [276 #1053a] [322], was an optical, mathematical, surveying, and philosophical instrument maker of London, England. Addresses: 21 St. James Street; 22 St. James Street (1784-1799); 70 St. James (1800-1809). Instruments bearing his name include: microscope; telescope; waywiser; hodometer; magnetic compass; barometers; sundial.

Watkins, William (II)
William Watkins (II), (fl.1803-1832) [3] [18] [53] [276 #1734] [322] [358], was an optical, mathematical, nautical, and philosophical instrument maker of Bristol, England. Addresses: 12 St. Michael's Hill (1803-1811); 13 Clare Street. (1812); 21 Clare Street. (1813); 16 St. Augustine's Back, at the Lord Nelson (1814-1832); 16 St. Augustine's Parade (1830).

Instruments bearing his name include: telescope; horizontal dial; wheel barometer; sundials. He is also known to have sold: magnetic compass.

Watson-Baker, Wilfred Ernest
Wilfred Ernest Watson Baker, (1891-1951) [319], was the son of Frederick William Watson Baker, and a great-grandson of **William Watson (I)**. He was a director of **W. Watson & Sons Ltd.** from 1922-1930, managing director from 1930-1934, and was made chairman in 1934. In 1935 his book, *The World Beneath the Microscope* was published. In 1940 he changed his surname from Baker to Watson-Baker. By 1941 he had formed the company **Watson-Baker Co. Ltd.** In 1949, when he retired, he sold his interest in W. Watson & Sons Ltd. to a mariner and financier named Captain James Cook.

He served as honorary curator of the Royal Microscopical Society, and in 1914 he was elected Fellow. He was President of the Queckett Microscopical Club from 1938-1944, and again from 1950 until his death in 1951.

He was a tireless promoter of the British optical industry. He was President of the **Scientific Instrument Manufacturers' Association of Great Britain** from 1930-1932. In 1941 he formed Scientific Exports (G.B.) Ltd., known as SCIEX, promoting the British scientific export trade.

Watson-Baker Co. Ltd.

Watson-Baker Co. Ltd., [319], of 41 Boardwater Rd., Welwyn Garden City, Hertfordshire, England, produced military binoculars during WWII. The company first appeared in the directories in 1941, and was formed by **W.E. Watson-Baker** of **W. Watson & Sons Ltd.** The company was formally dissolved in 1968.

Watson Barnet
In 1906 **W. Watson & Sons** of London, England, expanded their manufacturing capacity with a new facility in Barnet, Hertfordshire. Subsequently, many instruments manufactured there were labelled *Watson Barnet*.

Watson Brothers
Watson Brothers, [319], was a partnership formed in 1884 between **Thomas W. Watson** and his brother Arthur Henry Watson (1857-c1936). They traded at 4 Pall Mall, London, England, as opticians, gun-makers, and optical and scientific instrument makers. By 1890 they were also advertising as electricians and electrical engineers. From 1894-1898 the brothers were listed at Observatory House, 31 Cockspur Street, Charing Cross. From 1895 Watson Brothers were listed at 29 Old Bond Street. Following this move they appear to have ceased making optical, mathematical and scientific instruments. In 1898 Watson Brothers auctioned their stock of optical, mathematical and scientific instruments. In 1935 the firm Watson Brothers was sold to the gun-makers Stephen Grant & Joseph Lang Ltd., who continued in business until 1959.

Watson-Conrady

Watson-Conrady is a marking found on some optical instruments manufactured by **W. Watson & Sons Ltd.** with one or more optical component designed by **A.E. Conrady**. Conrady's association with Watson lasted from 1902 until 1917. Watson-Conrady optical components were also used by other makers, such as **Rev. D.B. Marsh**.

Watson, Thomas W.

Thomas William Watson, (1848-1933) [**319**], was a gun-maker, optician, and mathematical instrument maker of 4 Pall Mall, London, England. His father was Thomas Watson, a coach-builder who was the brother of **William Watson (I)**. Thomas W. Watson is believed to have begun making microscopes in 1875. In 1884 he entered a partnership with his brother, Arthur Henry Watson, and they began to trade as **Watson Brothers**. By 1911 he had retired from the business.

Watson, William

William Watson (I), (1814-1881) [**319**], of Brentford, Middlesex, England, was the son of William and Lucy Watson. His father was a staymaker in Middlesex. In 1837 William (I) founded the London firm of W. Watson, which later became **W. Watson & Sons**.

William (I) had 13 children from two marriages. His daughter, Mary Ann (b.1841), married William Baker, a bank clerk. Their son, Frederick William Watson Baker (1866-1952), became a director of the Watson business in 1883. F.W.W. Baker's son, **Wilfred Ernest Watson-Baker** became a director of the Watson business in 1922, and founded **Watson-Baker Co. Ltd.** F.W.W. Baker's brothers, Henry Herbert Baker and Frank Leopold Baker, started the firm **W. Watson & Sons, Melbourne** in 1888.

William (I)'s son William (II) (1844-1869) became a partner with his father in 1866 when they began trading as W. Watson & Son. When, in 1878, the company became W. Watson & Sons, William (I)'s sons, Thomas Parsons Watson (1855-1897) and George Frederick Watson (1857-1883), were already directors. His fourth son, Charles Henry Watson (1866-1938) became a director of the company in 1883.

William (I)'s brother, Thomas Watson, was a coach-builder. Thomas's sons Thomas William Watson and Arthur Henry Watson started **Watson Brothers**.

Watson & Sons Ltd, W.

W. Watson & Sons Ltd., [**29**] [**44**] [**113**] [**122**] [**306**] [**319**] [**334**], was founded 1837 as W. Watson by the gun-maker and optician **William Watson (I)**. The firm was established at 12 City Road, London, England, at which time Watson traded as a general salesman and miscellaneous dealer. In 1859 City Road was renumbered, and number 12 became number 74. Watson began to advertise the business address as trading at 74 City Road.

In 1861, census records show William (I)'s eldest son, William Watson (II), at 313 High Holborn, and by 1863 the W. Watson business was trading there, dealing in guns. By 1865 W. Watson was advertising optical instruments by makers such as **Dollond**, **Smith & Beck**, **Ross**, and others, as well as mathematical and surveying instruments.

In 1866 William (I) entered a partnership with, William (II), and the firm began to trade as W. Watson & Son. (Note: This date is given as 1867 in the 1937 Watson Centenary publication [**285**]). By this time, they were advertising mathematical, surveying and optical instruments, musical instruments, guns, and second-hand stock. In 1876 they began to manufacture microscopes. This may have been their first venture into manufacturing optical instruments. In 1878 the firm began to advertise as W. Watson & Sons. By this time William (I)'s sons Thomas Parsons Watson and George Frederick Watson were already directors. Charles Henry Watson, William (I)'s youngest son, became a director in 1883, and in the same year, William (I)'s grandson, Frederick William Watson Baker, also became a director.

In 1888 they set up steam factories at 9, 10, 11, 16 and 17 Fulwoods Rents, Holborn. In the same year they displayed their products at the Centennial Exhibition in Melbourne, Australia, following which F.W.W. Baker's brothers, Henry Baker and Frank Leopold Baker started the subsidiary **W. Watson & Sons, Melbourne**. By 1894 W. Watson & Sons were advertising as manufacturers of field glasses, binoculars, microscopes, telescopes, and photographic cameras and accessories. In 1900 they acquired **John Browning & Co.** By 1900 the company was producing medical equipment, including x-ray equipment. In 1902 **A.E. Conrady** began his association with W. Watson & Sons as scientific advisor and lens designer, a position he held until 1917. Watson instruments with one or more optical component designed by Conrady were sometimes marked **Watson-Conrady**. In 1906 Watson built an additional manufacturing facility in Barnet, Hertfordshire. Subsequently, many instruments manufactured there were marked *Watson Barnet*. Charles Henry Watson became Chairman of the company in 1908, and in the same year the company was incorporated as W. Watson & Sons Ltd.

During WWI they were contracted to manufacture a large number of binoculars for military use, with lenses supplied by **Guaranteed Lens Co.** They continued to make binoculars after the war. In 1919 W. Watson & Sons Ltd. was listed as a member of the **British Optical Instrument Manufacturers' Association**. **Wilfred**

Ernest Watson-Baker became a director of the Watson business in 1922.

In 1929 they exhibited at the British Industries Fair. The catalogue, [448], listed their entry as an optical, scientific and photographic exhibit, and listed them as manufacturers of: Microscopes for medical, industrial, educational purposes and for the amateur; Prism binoculars; Astronomical and portable telescopes; Photographic lenses and cameras; Surveying and measuring instruments; Photometers; Scientific apparatus of every description.

In 1930 Wilfred E. Watson-Baker became the Managing Director. From 1930-1932 he was also President of the British Optical Instrument Manufacturers' Association, and in 1934 he became Chairman of W. Watson & Sons. In 1936 Frederick W.W. Baker was listed as the chairman of the **United Kingdom Optical Company Ltd.** By 1941 Wilfred E. Watson Baker had formed the company **Watson-Baker Co. Ltd.**

In 1947 W. Watson & Sons exhibited at the British Industries Fair. The catalogue, [450], listed them as manufacturers of: Microscopes for all purposes; Auxiliary optical and mechanical accessories; Photometers; Telescopes; Prism binoculars; Photographic lenses (all types); Optical elements in every form.

In 1949, upon his retirement, Wilfred E. Watson-Baker sold his interest in W. Watson & Sons to a mariner and financier named Captain James Cook. This was the end of the Watson family's involvement in the business. In 1957 the company was sold to **Pye**, and in 1967 they were taken over by Philips. The last microscope manufactured by W. Watson & Sons was made in 1970. In 1971 the company became the Optical Division of one of Philips' subsidiaries, and the Watson name ceased to be associated with the firm.

For information on dating Watson microscopes, refer to [320]. See the 1912 W. Watson & Sons Ltd. catalogue of microscopes and accessories [253].

Watson & Sons, W., Melbourne

W. Watson & Sons of Melbourne, Australia, [51] [319] was a branch of **W. Watson & Sons** of London, England.

Henry Herbert Baker (1867-1940) was the grandson of **William Watson (I)**, and brother of Frederick W.W. Baker and Frank Leopold Baker (1870-c1953). In 1888 Henry Baker travelled to Melbourne, Australia to promote W. Watson & Sons at the Centennial Exhibition. He saw the Australian market as an opportunity, and stayed on to set up a branch of the company in Melbourne. His brother Frank Leopold Baker joined him six months later. They were opticians, and sold optical, scientific, x-ray and medical equipment and accessories imported from their English parent company. They opened a workshop for adjustments and repair to the equipment they sold, and began to manufacture custom and specialised instruments as well as x-ray and other medical equipment. They were appointed as contractors to the Australian government for surveying instruments and other goods.

During WWI, supplies of binoculars from England were unavailable, since they all went to the British War Office. W. Watson & Sons in Melbourne advertised to buy second-hand prism binoculars for re-sale to military personnel. They also increased their second-hand offerings of other instruments. After WWI they increased their manufacturing capacity and opened branches in Brisbane, Adelaide and Perth, as well as Wellington and Auckland in New Zealand. By the mid-1940s they had formed the company Watson Victor Ltd., and in 1969 W. Watson & Sons of Melbourne changed their name to Watson Victor Holdings Ltd. By the year 2000 Watson Victor, then already a part of the Medic Corporation of New Zealand, was taken over by the healthcare company EBOS.

Watt, James

James Watt, (1736-1819) [3] [37] [46] [155] [263] [271], was a mathematical instrument maker, engineer, and scientist, best remembered for his improvements to the steam engine. He was born in Greenock, Renfrewshire, Scotland, and did not excel at school until he began to study mathematics. He began his training with a Glasgow instrument maker, but was advised to go to London to improve on his training. He spent a short period training under a watchmaker called Mr. Neale then, following an introduction to **James Short (I)**, he underwent a period of training, from about 1755 to 1756, with **John Morgan**. He returned to Glasgow at the end of this training and set up his business as an instrument maker and retailer. He made instruments, including Hadley's quadrants, and is known to have been selling compasses, burning glasses and microscopes. He also made musical instruments. He had an interest in surveying, and made an improved surveying level and a dividing engine. By 1764 he had sixteen employees in his workshop, and by 1769 his senior journeyman was **John Gardner**.

He went on to develop an interest in steam engines, leading to improvements including the separate condenser and the air-pump, for which he was awarded a patent. He also pursued a career in surveying, and was involved in surveying for many of Britain's canals, presumably leaving the instrument making workshop under the charge of John Gardner. Following the death of his wife in 1773, Watt relocated to Birmingham for his partnership with Matthew Boulton for the production and further development of engines based on Watts' patent. He continued in this partnership until his retirement in 1800.

Watts & Son, E.R.
E.R. Watts & Son, [29] [44], were makers of theodolites and other surveying instruments, based in London, England. The company was established in 1856 by Edwin Watts. In 1919 they were listed as a member of the **British Optical Instrument Manufacturers' Association**. In 1941 the Ministry of Supply set up a shadow company in Nottingham to be operated by E.R. Watts & Son and **Ross Ltd**. The factory was called **Nottingham Instruments Ltd.**, and it was here that Ross Ltd. and E.R. Watts & Son duplicated their manufacturing in London. In 1946 **Frank Twyman** became a technical advisor to the firm, and in the same year E.R. Watts & Son took over **James Swift & Son Ltd**. In 1948 E.R. Watts & Son amalgamated with **Adam Hilger** to form **Hilger & Watts**, with Twyman as a director.

Watts, James
James Watts, (fl. late-C19th) [14] [34], of 29 Eden Quay, Dublin, Ireland, was an instrument maker and retailer. Examples of his work include: A refracting telescope signed *J. Watts, 29 Eden Quay, Dublin*; A barometer signed *James Watts, Dublin*.

Waugh, James
James Waugh, (fl. late-C18th) [14] [34] [304], was from a family of clockmakers in Armagh, Ireland. In c1793 he made a 2-inch transit instrument for the Armagh Observatory. He also made a 35-inch radius astronomical quadrant for the astronomer Robert Hogg.

Waymouth-Cooke Rangefinder
The Waymouth-Cooke Rangefinder, introduced in British naval service in 1914, was a form of sextant rangefinder for marine use. The instrument could be used to find the range of an object of known height, or could be used as a Langley Inclinometer.

In use as a rangefinder the known height of the object to be viewed is set on the instrument's height scale, and the object is sighted through the sighting scope. This gives a split view, and an adjustment allows the direct view of the object, and the double-reflected view, to be brought into coincidence, the bottom of one with the top of the other. The range is then read off on the range scale.

In use as an inclinometer an object, such as a ship, of known range and length is viewed. The range and length are first set on the instrument's scales. The views of both ends of the ship, or two masts, are brought horizontally into coincidence, and this gives a reading of the observed length of the ship. Thus, due to the amount of foreshortening, any angle between broadside and bow-on or stern-on may be read off on the inclination scale.

By the 1950s the Waymouth-Cooke Rangefinder was superseded in British naval use by the Waymouth-Ross Sextant Rangefinder. See also rangefinder (optical).

Waymouth-Ross Sextant Rangefinder
The Waymouth-Ross Sextant Rangefinder, [4], is a form of sextant rangefinder for marine use which, by the mid-1950s, had replaced the Waymouth-Cooke Rangefinder in British naval service. It was used for measuring ranges between 1,000 and 18,000 yards.

The instrument has a sighting telescope, giving a horizontally split view, allowing the viewer to observe the object directly in one half, and via the double-reflection in the other. At the side of the instrument are one fixed, and three independently rotatable concentric discs, similar to a circular slide-rule. The fixed, outer disc, is marked with the height scale, the indices, and the infinity mark. The three rotatable scales are the height ring, marked with a scale from 14 to 200 feet; the range ring, marked with a scale from 1,000 to 18,000 yards; and the inclination ring, marked with a scale from 5° to 90°.

To measure the range to an object of known height the height is dialled into the height scale, and the instrument is adjusted to bring the top of one image into coincidence with the bottom of the other. The range is then read off on the range scale. Conversely, the instrument may be used to measure the height or length of an object at a known range; the range of an object using the Horizon Method; or when turned horizontally, the inclination (any angle between broadside and bow-on or stern-on) of a ship of known length and range. Its use as an inclinometer is similar to that of the Waymouth-Cooke Rangefinder.

See also rangefinder (optical).

Webb, Joseph Benjamin
Joseph Benjamin Webb, (fl.1825-1840) [18] [276 #1737], was an optical, philosophical, and mathematical instrument maker, and brass turner of London, England. Addresses: 18 St. James' St., Clerkenwell; 42 York St., St. Luke's; 15 St. James' St. (1824-1825); 13 Charles St., City Road (1839). He attended the London Mechanics' Institute from 1824 until 1825.

Webb, John
John Webb, (fl.1760-1847) [18] [53] [276 #665] [322] [361], was an optician, and optical and mathematical instrument maker of London, England. Addresses: 327 Oxford St. (1800); 403 Oxford St. (1805); 408 Oxford St. (1808); 192 Tottenham Court Road (1809-1822), 28 Francis Street, Tottenham Court Road (1834-1847).

One of his trade cards reads, "John Webb, Optician, removed from the old established shop, Oxford Street to Tottenham Court Road, opposite the Chapel, London. All sorts of Optical, Mathematical, & Philosophical instruments made & repaired in the most accurate manner, according to the latest improvements." His trade cards include illustrations of a microscope, an octant, and a telescope. Instruments bearing his name include: theodolite; stick barometer; wheel barometer.

Wedderburn, Thomas
Thomas Wedderburn, (d.1886) [8] [169] [263], was workshop foreman for **Richard Adie** in Edinburgh, Scotland. Following the death of Adie, he took over the firm, and changed its name to **Adie and Wedderburn**, a name which was used until 1913. Thomas Wedderburn was a "master optician". He was elected a member of the Scottish Meteorological Society in 1882, and as a member of the Society of Arts in 1884.

Weeden, Charles Cartwright
Charles Cartwright Weeden, also known as Weedon, (1843-1904) [319] [409], was a microscope maker of London, England. Addresses: 14 Brook St., Holborn (1871-1881), and 15 Merlins Place, Clerkenwell (1891-1901). He shared both addresses with his father, **Michael Weeden**, and step-mother, **Dinah Weeden**. He served his apprenticeship with his father in the **Merchant Taylors' Company**, beginning in 1859.

Weeden, Charles Fox (I)
Charles Fox Weeden (I), also known as Weedon, (b.1782 fl.1811) [18] [319] [409], was an optician and microscope maker of London, England. He served his apprenticeship with his father, **Daniel Weeden** of the **Turners' Company**, beginning in 1799.

Weeden, Charles Fox (II)
Charles Fox Weeden (II), also known as Weedon, (1812-after 1880) [409], was a microscope maker of London, England. He was the son of **William John Weeden**.

Weeden, Daniel
Daniel Weeden, also known as Weedon, (1748-1809) [18] [409], was an optical turner of Clerkenwell, London, England, and a Freeman of the **Turners' Company**. His apprentices included his son, **Charles Fox Weeden (I)**, beginning in 1799. Another son was the optician and optical instrument maker **William John Weeden**.

Weeden, Dinah
Dinah Weeden, also known as Weedon, (1837-1892) [319] [409], was a microscope maker of London, England. Addresses: 14 Brooks St., St Andrew (1881); 15 Merlins Place, Clerkenwell (1891). Born Dinah Reilly, she was married to **Michael Weedon** in 1874, after which they both worked as microscope makers, and shared the same addresses. By the time of the 1891 census, Michael was described as a pensioner, and Dinah was listed without an occupation.

Weeden, Edward John
Edward John Weeden, also known as Weedon, (1842-1932) [319] [320] [409], was a microscope manufacturer and wholesaler, of London, England. Addresses: 51 Penton Street, Pentonville, Clerkenwell (1861); 9 Williams Street, Islington (1871); 22 St. James Walk, Clerkenwell (1881); 68 Myddelton Square, Clerkenwell (1891-1895); 18 Myddleton Square, Clerkenwell; 96 Petherton Road, Islington (1901); 104 Petherton Road, Highbury (1911).

He was a son of **William Weeden**. In 1877 he married Matilda Sarah Ada Cooper, the daughter of the optician, **Michael Cooper**. He published *An Illustrated Catalogue of Microscopes, Objectives, and Accessory Apparatus*, in which he stated that he sold wholesale only.

Weeden, John
John Weeden, also known as Weedon, (fl.1819-1828) [18] [276 #1738], was an optician, optical, mathematical, and philosophical instrument maker, optical turner, and brass tube maker, of 18 Baldwin St., City Rd., London, England.

Weeden, Michael
Michael Weeden, also known as Weedon, (1818-1892) [18] [319] [409], was an optician and optical instrument maker of London, England. Addresses: Lower Rosomon St., Clerkenwell (1847); 3 Camden Passage, Clerkenwell (1852); 27 Camden St., Islington (1855-1865); 14 Brook St., Holborn (1871-1881); 15 Merlins Place, Clerkenwell (1891).

He is known to have made microscopes. He was the son of **William John Weeden**, and was freed by patrimony of the **Merchant Taylors' Company** in 1847. Among his apprentices was his son, **Charles Cartwright Weeden**. In 1874, by then a widower, he married Dinah Reilly (**Dinah Weeden**), after which they both worked as microscope makers, and shared the same addresses. By the time of the 1891 census, Michael was described as a pensioner, and Dinah was listed without an occupation.

Weeden, Walter Henry
Walter Henry Weeden, also known as Weedon, (1845-1884) [319] [409], was an optician and microscope maker of London, England. He was the son of **William Weeden**.

Addresses: Bethnal Green Road (1870); 23 Shepperton St., Islington (1871); 24 Georges Road (1877-1881).

Weeden, William
William Weeden, also known as Weedon, (b.1814) [319] [409], was a microscope maker of London, England, and the son of **William John Weeden**. Addresses: 36 Barton St., Clerkenwell (1851); 51 Penton St., Pentonville, Clerkenwell (1861); 6 Camden Passage, Islington (1871); 33 Camden Passage, St. Mary (1881). By 1851 he was employing 1 man, and by 1881 he was employing two men. His sons were **William Edwin Weeden**, **Edward John Weeden**, and **Walter Henry Weeden**. His daughter, Maryann Elizabeth (b.1848) married the mathematical instrument maker John William Webber (b.1841). By 1880, William Weeden was listed as a pauper in a workhouse.

Weeden, William Edwin
William Edwin Weeden, also known as Weedon, (1839-1905) [409], was a microscope maker, turner, and brass finisher of London, England. He was the son of **William Weeden**.

Weeden, William John
William John Weeden, also known as Weedon, (1775-1847) [18] [276 #1739] [409], was an optician and optical instrument maker of London, England. Addresses: 18 Redcross St., Cripplegate; 14 St. James' St., Clerkenwell; 7 Norman St., St. Luke's (1817-1822); 1 Wenlock St., St. Luke's (1825); 68 Chapel St., Pentonville (1844).

He was a son of **Daniel Weeden**. He served his apprenticeship with John David Hutchins, beginning in 1790, and was freed of the **Merchant Taylors' Company** in 1797. William John Weeden's sons, **Michael Weeden**, **Charles Fox Weeden (II)**, and **William Weeden**, were also opticians and optical instrument makers, as was his grandson, **Edward John Weeden**.

Weems & Plath
Weems & Plath, "Manufacturers of fine nautical and weather instruments", was founded in Washington, USA, in 1953 by Johannes Boysen, the husband of **Carl Plath**'s granddaughter, and the American navigator P.V.H. Weems. They formed the company initially to sell Carl Plath sextants in the USA. The company, since then relocated to Maryland, sells nautical instruments and accessories including binoculars and sextants. Their sextants are manufactured in Japan by **Tamaya**. They also sell binoculars under the Weems & Plath brand.

Welch Allyn
Welch Allyn is a US based manufacturer of medical diagnostic equipment and patient monitoring systems. The company was founded in 1915 when Dr. Francis Welch and William Noah Allyn developed a handheld, direct illuminating ophthalmoscope. They were taken over in 2015 by the Chicago based company Hillrom.

Wellington, Alexander
Alexander Wellington, (fl.1784-1805 d.1812) [16] [18] [53] [276 #1056] [319] [322], was an optician, and optical, mathematical, and philosophical instrument maker of 20 Crown Court, St. Ann's, Soho, London, England, the address also variously known as 20 Crown Court, Wardour St.; 20 Crown Court, Princes St., Soho; 20 Crown Court, Golden Square. He is also recorded as trading at Sherbourne Lane, Lombard St. in 1784. He served his apprenticeship with **James Search**, beginning in 1774, and was freed of the **Stationers' Company** in 1781. Among his apprentices were **William Hobcraft (I)**, and **Samuel Wells**. He succeeded James Search in business.

According to his trade card, "Alex^R Wellington, Mathematical instrument maker to their Royal Highnesses the Dukes of Gloucester & Cumberland. Successor to the late MR James Search, at *The Globe*, Crown Court, S^T Ann's, Soho, London. Sells all sorts of Spectacles, Reading glasses, Telescopes, Opera glasses, Drawing instruments, Rules for measuring and gauging & all sorts of Surveying instruments, with all other Optical, Mathematical & Philosophical instruments on the most reasonable terms. Books of the use of the above instruments."

Instruments bearing his name include: octant; telescope; microscope; pantograph; cased sundial; drawing instrument sets; stick barometer.

Wells, Samuel
Samuel Wells, (fl.1817-1839) [18] [276 #1455], was an optical, mathematical, and philosophical instrument maker, and ironmonger, of London, England. Addresses: 137 Old St., St. Luke's, and 3 Clerkenwell Green, Aylesbury St. (8117). He served his apprenticeship with **Alexander Wellington** of the **Stationers' Company**, but there is no record of him taking his freedom.

Welta
Welta was a camera manufacturer, founded as the Weeka Camera Werke in 1914, near Dresden, Germany, by Walter Waurich and Theodor Weber. They used *Welta* as a trademark, and in 1919 they changed the name of the company to Welta Camera Werk. At the end of WWII, the company was nationalised, and became VEB Welta

Wenckebach, Eduard

Eduard Wenckebach, (1813-1874), [17] [374] [375] [377] [378], was an optical, scientific, and mathematical instrument maker of Nieuwebrugsteeg Amsterdam, Netherlands. He learned his craft as an instrument maker in the leading workshops of Germany, France, Austria, and England and, in 1839, he set up his business in Amsterdam. He produced astronomical instruments, including various pieces of equipment for the Leiden Observatory, [376], for whom he also converted a theodolite and levelling device into a portable transit telescope for astronomical use. The director of the Leiden Observatory, Professor Frederik Kaiser, developed a reflecting circle based on prisms, for which at least four prototypes were made by Eduard Wenckebach.

Wenckebach founded the Nederlandse Telegraafmaatschappij (Dutch Telegraph Company) in 1850, and was its first director. He went on to design and implement an extensive Dutch telegraph network and, by 1857, all the key national and international connections were complete.

Instruments bearing his name include: sextant; finder telescopes; standard 1kg weight; He is also known to have made: meteorological instruments; reflecting circles.

Wenham, Francis Herbert

Francis Herbert Wenham, (1824-1908) [44] [53] [264] [285], was an English marine engineer and designer with a keen interest in aviation and nautical research. He studied aerofoil sections, seeking to establish the most efficient aircraft wing configuration. In 1866 he delivered a paper *Aerial Locomotion* to the Aeronautical Society of Great Britain, which became a seminal work. To test his ideas, he designed and built the first wind tunnel, together with **John Browning (III)**, in 1871 in Greenwich.

Wenham was a skilled microscopist, with an interest in binocular microscopy. He invented a binocular viewing mechanism for a microscope with a single objective. The circle of light from the objective's rear element is viewed in two halves. One half is viewed, unobstructed, through a straight eyepiece tube. The other half is viewed through a prism which directs the light into an angled eyepiece tube. Since a compound microscope produces an image optically rotated by 180°, the view from the left of the objective is directed to the right eye, and from the right of the objective to the left eye, to avoid producing a pseudoscopic image. In Wenham's design, the prism can be slid out of the optical path, with the result that the two eye-tubes view a simple monocular image. In 1870, following the death of **Thomas Ross**, Wenham began to work as a consultant for the Ross company.

Wenham's design was used by most major microscope manufacturers in the UK and USA during the remainder of the C19[th], and many of these microscopes used prisms made by **Carston Diederich Ahrens**.

Werkstätten für Präzisions Mechanik und Optik

See **Askania Werke AG**.

West, Charles (I)

Charles West (I), (fl.1814-1825) [18] [276 #1456], was an optician, and optical and mathematical instrument maker of London, England. Addresses: 5 Cursitor St. (1817-1822), and 83 St. James St. (1817). He supplied lenses to **George Dollond (I)**, and **Francis Watkins (II)**. He shared the premises at 5 Cursitor St. with **Charles Robert West**.

West, Charles (II)

Charles West (II), (fl.1833-1867) [3] [18] [50], was an optician and mathematical instrument maker of Liverpool, England. Addresses: 20 Lord St. (1839-1851) and 3 Paradise St. (1857-1867). He began to work for **Abraham Abraham** in about 1833 as manager of Abraham's business in Lord Street. In about 1849, Abraham entered a partnership with Charles West (II) and **George Smart Wood**, trading as Abraham & Co. The partnership was terminated in 1855 [35 #21840 p216], after which Charles West (II) set up in business, working under his own name.

West, Charles Robert

Charles Robert West, (fl.1801, d.1824) [18] [276 #1242] [329], was an optician and optical instrument maker of London, England. Addresses: 23 Plough Court, Fetter Lane (1801-1805); 1 St. James St. (1805-1808); Cursitor St. (1813); 5 Cursitor St., Chancery Lane (1817-1822); Serle's Passage, Serle St. (1817); 83 St. James St. (1822); End of St. James St., Pall Mall (1824); Gateway of Lincoln's Inn.

He began his apprenticeship with the optician Joshua Davies, but was turned over to **George Linnell** in 1790. He was freed of the **Spectacle Makers' Company** in 1801. He is known to have produced telescopes, and was awarded a patent in 1806 for a method of using a bearing to keep the telescope tubes parallel. He collaborated with **William Bruce**, and together they were awarded a patent for more easily portable telescopes. Charles' brother, **Francis West**, served an apprenticeship with him. He shared the premises at 5 Cursitor St. with **Charles West (I)**.

West, Francis

Francis West, (fl.1822-1852) [18] [53], was an optical, mathematical, and philosophical instrument maker of London, England. He served his apprenticeship with his brother, **Charles Robert West**, beginning in 1806, and was freed of the **Spectacle Makers' Company** in 1828. Addresses: Serle's Passage, Lincoln's Inn; Cursitor St., Chancery Lane; 17 Russel Court, Drury Lane (1822-1828); 83 Fleet St., near St. Bride's Church (1829-1845); 41 Strand (1845-1849); 92 & 93 Fleet St. (1849-1852).

He advertised as *Successor to Mr. Adams*, but this claim may not have a sound basis. His trade card bearing the address 83 Fleet Street states "Optical, Mathematical, and Philosophical instruments of every description, wholesale and for exportation, of the best workmanship, and on the most reasonable terms."

In 1836 he published, [311], *A Descriptive Account of a Variety of Intellectual Toys, Made for the Instruction and Amusement of Youth, of Both Sexes, as Recommended by Dr. Beddoes, Miss Edgeworth. Illustrated by 30 Engravings of the Various Instruments*. The book describes instruments that he made, and how they should be used. Among the instruments shown are: Orrery; Achromatic telescope; Optical diagonal machine for viewing pictures; Cosmorama, or show-glass; Single, double, and triple magnifiers; Pocket microscope; Prism; Camera obscura; Magic lantern; Claude-Lorraine glasses; Multiplying glasses; Small globes; Electrical machine; Voltaic pile; Galvanic battery; Air pump; Condensing syringe; Artificial fountain; Chemical chests; Small stills; Spirit lamp; Magnetic toys; Philosophical crystal, or new perpetual rotary motion; Small chests of minerals; Newly invented sun-dial; Mariner's compass; Porter's magnetic sun-dial. The book's title page describes him as "Optical, Mathematical and Philosophical instrument maker to His Majesty", referring to King William IV.

Francis West had three sons, all of whom served apprenticeships with him – the optical instrument maker John George West (1819-1873), beginning in 1833, the optical, philosophical and mathematical instrument maker **Francis Linsell West**, beginning in 1835, and the microphotograph slide maker, Henry West (1825-1899), beginning in 1840 [50].

Examples of Francis West's work include: heliostat; stereoscopic spectacles for use with stereoscopic illustrations; camera lucida; barometer; marine barometer; microscope; Stanhope lens (microscope) [287].

West, Francis L.

Francis Linsell West, (fl.1849-1885) [18] [53] [113], was an optician and optical, mathematical and philosophical instrument maker of 39 Southampton Street, Strand, London, England. He traded at this address from 1848 to 1885, and at 31 Cockspur Street, Charing Cross, London, from 1859 to 1885. He served his apprenticeship in the **Spectacle Makers' Company**, with his father, **Francis West**, beginning in 1835. He is known to have sold microscopes, telescopes, sextants, and drawing instruments. Instruments bearing his name include: microscopes; telescopes with the address 39 Southampton Street, Strand, London, or 31 Cockspur Street, Charing Cross; telescopes; sundials; brass cased compass; barometer.

Westinghouse Electric and Manufacturing Co.

The Westinghouse Electric and Manufacturing Co., [44], is a US company that was formed in 1886 by George Westinghouse, making turbines, generators, motors, and switch gear for generation, transmission, and use of electricity. During WWII they manufactured **Bausch and Lomb** pattern binoculars for military use.

Wheatstone, Sir Charles

Sir Charles Wheatstone, (1802-1875) [37] [155], was an English physicist. He began his career as a musical instrument maker. He patented and developed the concertina, and carried out scientific research in acoustics and optics. He invented the *kaleidophone* to illustrate harmonic motions of different periods. He also invented the Wheatstone's photometer, the polar clock, and the mirror stereoscope, or reflecting stereoscope. In 1834 he was appointed Professor of Experimental Philosophy at Kings College, London. He is most famous for producing, together with Sir William Cooke (1806-1879), the first working electric telegraph system which he demonstrated in 1837 in conjunction with the London and Birmingham Railway Company. He made many further contributions to telegraphy, and was knighted in 1868.

Wheatstone's Photometer

See photometer.

Wheeler, Edmund

Edmund Wheeler (I), (1808-1884) [50] [319], was born in Kent, England. He spent his early career there, and subsequently in Basingstoke, as an ironmonger. In 1848 he moved to London and began a career as a lecturer. Beginning around 1855, as well as lecturing, he opened an optical business. In 1862 he moved his optical business to

48 Tollington Road, London. Among his staff, his son, Edmund (II) (1836-1930), who had previously worked for **Smith & Beck**, worked for him producing microscopes, lenses, and telescopes.

Edmund (I)'s advertisement of 1863 shows him as a manufacturer of achromatic objectives, microscopes and telescopes, of 48 Tollington Road, Holloway, London. He lists: Achromatic microscopes, including binocular microscopes; Microscopic objects; Astronomical telescopes; Telescopes for tourists, seaside, and general use; First class binocular, opera, field and marine glasses; Standard barometers and thermometers; E. Wheeler's pocket mountain aneroid barometer. Also, microscopes, optical instruments &c by **Smith, Beck & Beck**.

He sold prepared microscope slides made by himself and other makers. In 1864 he began to make some of his optical instruments in aluminium. Edmund (II) moved to Brighton in 1870, and set up a photography studio. Edmund (I) had recently retired due to ill-health when he died in 1884.

Wheelwrights' Company
See Appendix IX: The London Guilds.

Whitbread, George
George Whitbread, (fl.1828-1877) [18] [276 #1741] [322] [358], was an optician, and mathematical, nautical, philosophical, and surveying instrument maker of 11 Exmouth St., Commercial Road, London, England. He served his apprenticeship with the mathematical instrument maker, William George Cook, beginning in 1812, and was freed of the **Spectacle Makers' Company** by purchase in 1828. **George Lee** was one of his employees. Among his apprentices was his son, George Thomas Whitbread, who succeeded him in business. Instruments bearing his name include: quadrants; octants; quintants; sextants; artificial horizons. He is also known to have sold: theodolite.

Whitcomb, R.
Rasselas Prince Whitcomb, (1808-1884), was a mathematical and surveying instrument maker of New York, USA. Initially he worked as a civil engineer, and surveyed part of Wisconsin for the US Government. By the mid-1850s he had set up his own company in Cincinnati, where he continued in business until 1878, at which time he sold his business, by then named R. Whitcomb & Co., to **Hermann Pfister**.

White, James
James White, (1824-1884) [18] [44] [46] [53] [263], was a scientific and optical instrument maker of Glasgow, Scotland. Addresses: 24 Renfield St. (1850-1852); 14 Renfield St. (1853-1856); 1 Renfield St. (1860-1863); 60 Gordon St. (1860-1863); 95 Buchanan St. (1864-1868); 78 Union St. (1869-1875); 241 Sauchiehall St. (1876-1883); 209 Sauchiehall St. (1884-1890); 16, 18 & 20 Cambridge St. (1884-1900); 18 Cambridge St.

In 1839 he began his apprenticeship with the mathematical, philosophical and optical instrument makers **Gardner and Co.** of Glasgow. In 1850 he set up his own firm, trading as James White, and from 1854 he supplied instruments to **William Thomson** for his teaching laboratory at Glasgow University, leading to his appointment as optician and philosophical instrument maker to the university. From 1857 to 1859 James White worked in partnership with John Haddin Barr, and during this time they traded as White and Barr. Their partnership was terminated when John Barr emigrated to New Zealand, and the company continued as James White.

James White was subject to sequestration proceedings, and declared bankrupt in 1861 due to "various unprofitable contracts" [170 #7167 p1304], at which time his business was both retail and instrument making. Later that year, after satisfying his creditors, the company resumed business. They continued to produce William Thomson's patent compass and sounding machines as well as other marine instruments, and continued to equip Thomson's teaching laboratory, as well as producing electrical apparatus of Thomson's invention. In 1883 two of White's assistants, Mathew Edwards and David Reid, became partners in the business, and they continued the business after his death. Following this, William Thomson also became a partner, and in 1900 the company name was changed to **Kelvin and James White Ltd.** Upon incorporation of the new company, **Alfred W. Baird** was a director, and **James Thomson Bottomley**, nephew of William Thomson, also joined the firm.

Whitehouse, John T.
John Thomas Whitehouse, (fl.1829-1844) [18] [276 #2042], was a spectacle maker, optician, and optical instrument maker, of 13 St. John's Lane, West Smithfield, London, England.

Whitehouse, M.
M. Whitehouse, (fl.1829-1830) [18], was an optical, mathematical, and philosophical instrument maker of 3 Silver St., Clerkenwell, London, England. He succeeded the optician David Whitehouse, (fl.1826), in business at this address.

Whitehouse, Nathaniel
Nathaniel Whitehouse, (fl.1825-1882) [276 #1742] [379], was an optical and philosophical instrument maker of London, England. Addresses: 3 Cross St., Hatton Garden

(1825-1837); 1 Castle St., Leicester Square (1838-1846); 2 Cranbourn St., Leicester Square (1847-1882); 1 Castle St., Great Newport St.

He exhibited at the Great Exhibition in 1851 in London [294]. His catalogue entry no. 280 lists a solid silver opera-glass, artificial eye, artificial silver nose, and several kinds of spectacles.

[18] lists Nathaniel Whitehouse at the 3 Cross St. address, as well as at 1 Cross St., Hatton Garden, and there is a separate entry for Nathaniel W. Whitehouse, at the Castle St. and Cranborn St. addresses, as well as 27 Grafton St., Soho (1836).

Instruments bearing his name include: telescope; wheel barometer. He is also known to have sold: opera glasses; spectacles.

Whitford, Samuel

Samuel Whitford, (fl.1762, d.1789) [3] [18] [53] [276 #549], was an optical, philosophical, and mathematical instrument maker of *Archimedes & Three Spectacles*, 27 Ludgate Street, near St. Paul's Churchyard, London, England, and was the son of **Thomas Whitford**. He served his apprenticeship with London scale-maker John Lind, and was freed of the **Blacksmiths' Company** in 1762. His occupation is not documented between 1762 and 1771 when he went to St. Petersburg, Russia. Upon his return in 1772 he took over the business of **Francis Morgan** at 27 Ludgate Street. It is not clear when his father, Thomas Whitford, began working at the same address, but Thomas outlived him, and was succeeded in business at 27 Ludgate Street in 1790 by **John Bleuler**.

According to Samuel Whitford's trade card, he made and sold, among other items: Spectacles; Reflecting and refracting telescopes; Double, single, solar, opake, and aquatic microscopes; Camera obscuras; Sky-optic balls; Prisms; Magic lanterns; Opera glasses; Air pumps and air fountains; Glass pumps; Electrical apparatus; Thermometers, hydrostatical balances, and hydrometers; Hadley's quadrants; Davis's quadrants; Globes; Compasses, and azimuths; Nocturnals and sun-dials; Scales; Drawing instruments; Theodolites; Circumferenters; Measuring wheels; Spirit levels.

Whitford, Thomas

Thomas Whitford, (fl.1743, d.1792) [18], of St. Martin's le Grand, London, England, was an optician, and was the father of **Samuel Whitford**. In 1743 he was freed of the **Spectacle Makers' Company** as a Foreign Brother. In 1782 he was working at St. Ann's Parish, Aldersgate, London, and by 1790 he was working at 27 Ludgate St., the address of his recently deceased son. He was succeeded at this address by **John Bleuler**.

Whyte, James (I)

James Whyte (I), (b.c1836) [263] [271], was a nautical instrument maker of Glasgow, Scotland. He began his career working for **David Heron** and, in 1860, married Heron's daughter Jessie. In 1864 he began to trade at 4 Carrick Street, the address previously occupied by David Heron's business. At this time James Thomson was working for the firm. Addresses: 4 Carrick St. (1864-1870); 144 Broomielaw (1864-1873); 3 James Watt St. (1871-1873). In about 1875 the company started trading as Whyte & Co. At this time Whyte's son **James Whyte (II)** was working for the firm. In 1889 James Whyte (II) formed a partnership with James Thomson, and began trading as **Whyte, Thomson & Co.**

Whyte, James (II)

James Whyte (II), [263] [271], was the son of **James Whyte (I)**. In 1889 he formed a partnership with James Thomson, trading as **Whyte, Thomson & Co.**

Whyte, Thomson & Co.

Whyte, Thomson & Co., [44] [263] [271], were chronometer, watch, and nautical, optical, and philosophical instrument makers, admiralty chart sellers, and "adjusters of compasses in iron vessels", of 144 Broomielaw, Glasgow, Scotland. The company was formed in 1889 by **James Whyte (II)** and James Thomson. Among the specialists working for the company was James Wilson, who later formed the company **Christie & Wilson** with **Andrew Christie**.

Wiesel, Johann

Johann Wiesel (sometimes Wiessel or Wiselius), (fl.C17th) [14] [256] [274], of Augsburg, Germany, was an optician, known for making spectacles, telescopes and microscopes, which were sold throughout Europe. He was the father-in-law of **Guiseppe Campani**.

Wijk, Cornelius van

Cornelius van Wijk, (fl.1783-1794) [256] [322] [374] [378], was a scientific and optical instrument maker of Utrecht, Netherlands. A microscope dated 1783, and a circle dated 1785 bearing his signature were made in Utrecht. From 1786 until 1790 he trained in the workshops of **frères Dumotiez** in Paris, France. An air pump dated 1787 bearing his signature was made during this period. From 1792 until 1794 he worked for the Teyler's Foundation in Haarlem, Netherlands, [362], as a scientific instrument maker.

Wijk, Jacobus van (I)
Jacobus van Wijk (I) (1706-1766) [378], was a watch maker of Amsterdam, Netherlands. He was the father of **Jacobus van Wijk (II)** and **Jan van Wijk**.

Wijk, Jacobus van (II)
Jacobus van Wijk (II) (1734-1791) [322] [378], was a navigational and mathematical instrument maker of Amsterdam, Netherlands where, from 1757 his address was in the Kalverstraat. He was the son of **Jacobus van Wijk (I)**, and the brother of **Jan van Wijk**. He shared an address with **John Cuthburtson** from 1773 until 1777. Instruments bearing his name include: octant.

Wijk, Jan (Johannes) van
Johannes van Wijk, commonly known as Jan, (1732-1795) [322] [378], was a nautical instrument maker of Amsterdam, Netherlands where, from 1767 his address was in the Kalverstraat. He was the son of **Jacobus van Wijk (I)**, and the brother of **Jacobus van Wijk (II)**. He began work in about 1760, and is known to have made quadrants. Instruments bearing his name also include circumferenters; Holland circle.

Wild Heerbrugg
Wild Heerbrugg, [254], was a company formed in Heerbrugg, Switzerland, by **Heinrich Wild**. Wild was a successful and influential Swiss optical designer who had been working for **Zeiss** in Jena, Germany. In 1921 he returned to Switzerland with his family and started his own company, specialising in surveying instruments. Wild left the company in 1932. In 1986 Wild Heerbrugg AG merged with **Leitz** to become part of the Wild Leitz AG Group, and ultimately **Leica Geosystems**.

Wild, Heinrich
Heinrich Wild, (1877-1951), was a Swiss surveyor and optical instrument designer. He designed a theodolite with which the user could read the micrometers on both sides of the instrument without leaving their position at the telescope. With this design the user was able to observe two opposite portions of the circles through a single eyepiece, and read them by means of a single screw micrometer. Wild took his design to **Zeiss** and was offered a job with them in Jena, where he brought his designs to fruition. The design became the stimulus for the design of the competing Tavistock theodolite. Following WWI, Wild returned to Switzerland and started his own company, **Wild Heerbrugg**, continuing his work designing surveying instruments. In 1932, during the depression, he quit the company and, in 1935, began to produce designs for **Kern & Co.**. His work with Kern continued until his death in 1951.

Wild Leitz AG Group
See **Leitz**.

Wildey, Henry
Henry Wildey, (1913-2003) [14] [83] [123], of Hampstead, London, England, was a highly respected optical component maker. He produced telescope mirrors and objective lenses for telescopes he manufactured, as well as for telescopes by other makers, including **H.N. Irving** and **Dudley Fuller**.

Willard & Co., J.W.
J.W. Willard & Co., [122], was a photographic and optical wholesaler in New York, USA. The company was established in 1857 by John W. Willard (1826-1901) as the J.W. Willard Manufacturing Co., and by 1870 the name had been changed to J.W. Willard & Co. Among their products, they manufactured Willard cameras and lenses. In 1866 they took over the business of **Charles F. Usener** and began to supply cameras made by him. In 1866 they advertised thus:

"Important notice to the photographic public. The undersigned have the pleasure to announce that they have purchased all the improved curves, gauges, tools and machinery used by Mr. Charles F. Usener in the manufacture of camera tubes and lenses in the City of New York. We have for some time past felt the importance in engaging in this important branch of manufacture; but, knowing that the reputation of a camera is of the highest importance, we have refrained from doing so until such time as we could introduce instruments of superior order of excellence. Our camera factory which we have purchased at a heavy expense, possesses all the advantages for producing cameras on the most extensive scale, and we engage in the manufacture of them under the most favourable circumstances. Mr. Usener is well known in the profession and trade generally to be the oldest and most talented optician in the country, and has manufactured cameras for **Holmes, Booth & Haydens'** for thirteen years. His valuable services, with those of all old hands who have been engaged with him for many years past, have been secured, and his talents in future will be exclusively directed to our interest. He will at once adopt some decidedly new improvements in the cameras which he manufactures for us bearing our name, and we guarantee to furnish instruments of the most superior quality and finish that have ever been offered to the public"

Willats, Richard
Richard Willats, (b.1819 fl.1850-1860) [61] [379] [438], was an optical and philosophical instrument maker of London, England. Addresses: 98 Cheapside (1850);

28 Ironmonger Lane (1851-1856); 2 Church Lane, Homerton (1857-1860). He was the son of Benjamin Willats, a chemist of Fore St., Cripplegate. He was an enthusiastic amateur photographer. He worked in partnership with his brother **Thomas Willats** until 1853, trading as **T.&R. Willats**, after which he traded under his own name.

Willats, T.&R.

Thomas & Richard Willats, (fl.c1845-1853) [**18**] [**61**] [**306**] [**322**] [**379**] [**438**], of London, England, were opticians, and optical, photographic, mathematical, and philosophical instrument makers, and suppliers of photographic equipment and chemicals. Addresses 98 Cheapside (1845-1853); 28 Ironmonger Lane (1851-1853). The business was a partnership between the two brothers, **Richard Willats** and **Thomas Willats**, and was terminated in 1853 [**35** #21453 p1864]. They published several books and manuals on photography, and are known to have made barometers, thermometers, and photographic apparatus. Instruments bearing their name also include microscopes [**113**], and electrostatic apparatus [**53**]. They exhibited at the 1851 Great Exhibition in London [**294**], where their catalogue entry listed "Portable photographic camera and stand. Registering thread counter, or linen prover."

Willats, Thomas

Thomas Willats, (b.1818 fl.1844-1857) [**18**] [**61**] [**379**], was an optical, mathematical, and philosophical instrument maker of London, England. Addresses: 98 Cheapside (1844-1845) and 28 Ironmonger Lane (1857). He was the son of Benjamin Willats, a chemist of Fore St., Cripplegate. He may have served an apprenticeship with Edward Palmer, a London chemist and philosophical instrument maker who, according to [**18**], had an apprentice named Thomas Willats. He traded under his own name prior to entering a partnership with his brother **Richard Willats**, trading as **T.&R. Willats** until 1853.

Willcox, Walton

Walton Willcox, (fl.1775-1781 d.c1783) [**18**] [**90**] [**385**], was a mathematical instrument maker of Hermitage, London, England. He served his apprenticeship with **John Urings**, beginning in 1765, and was freed of the **Joiners' Company** in 1772. From 1772 until 1777 he worked as a journeyman for John Urings. In 1775 **Richard Lekeux** was turned over to him as an apprentice by **Thomas Ripley**. **George Mills** began an apprenticeship with him in 1782, but was turned over to **William Spencer (I)** in 1783.

Willdey & Brandreth

Willdey & Brandreth, [**18**] [**53**] [**319**] [**322**] [**361**], optical instrument makers of London, England, was a partnership between **George Willdey** and **Timothy Brandreth** which began c1706, and continued until c1713. They initially traded at the *Archimedes & Globe* in Ludgate Street, the corner next St. Paul's, and at Exchange Alley, running out of Cornhill. Their address is also recorded as *The Great Toy, Spectacle, Chinaware and Print Shop*, next the Dog Tavern, corner of Ludgate Street, near St. Paul's, and *The Archimedes and Globe*, Spectacle and Toy Shop, Next the Dog Tavern in Ludgate Street. Their trade card listed spectacles, telescopes, microscopes, and other optical instruments. They also sold maps, curiosities, and toys, as well as cases of mathematical instruments made by **Edmund Culpeper**.

Willdey, George

George Willdey, (fl.1695-1733 d.1737) [**18**] [**274**] [**275** #498] [**276** #134] [**319**] [**442**], was an optical instrument maker, toyman, and map seller of London, England. Addresses: *Archimedes & Globe*, Ludgate Street (1706-1709); *The Great Toy, Spectacle, Chinaware and Print Shop*, next the Dog Tavern, corner of Ludgate Street, near St. Paul's (1709-1715). The West End of St. Paul's, London (1718-1737).

He served his apprenticeship with **John Yarwell**, beginning in 1694/5, and was freed of the **Spectacle Makers' Company** in 1702, at which time John Yarwell was working in the partnership **Sterrop & Yarwell**. He worked for a time as a journeyman instrument maker for **John Marshall**. In c1706 he began to work in partnership with **Timothy Brandreth**, an ex-apprentice of **Ralph Sterrop**, trading as **Willdey & Brandreth** until c1713. He also worked in partnership, from c1709 until c1713, with the hydrographer, engraver, and map and globe seller, Charles Price.

He was Master of the Spectacle Makers' Company from 1722 until 1733. He had at least 15 apprentices, among whom were his grand-daughter, Frances Wildey, and **Thomas Clark**. He was succeeded by his wife, **Judith Willdey**, who ran the business until his son, **Thomas Willdey**, gained his freedom in 1739, and took over the business.

Willdey, Judith

Judith Willdey, (fl.1737-1739) [**18**] [**319**], was an optical instrument maker and map seller, trading at the West End of St. Paul's, London, England. She was freed of the **Spectacle Makers' Company**, and took over the business of her husband, **George Willdey**, upon his death in 1737. She continued to run the business until her son, **Thomas**

Willdey, became a Freeman in 1739 and took over the business.

Willdey, Thomas

Thomas Willdey, (fl.1739-1747 d.c1748) [18] [274] [276 #319] [319], was an optical instrument maker, toyman, and map seller of London, England, who was the son of **George Willdey** and **Judith Willdey**. He began his apprenticeship in 1732, and was freed by patrimony of the **Spectacle Makers' Company** in 1739, upon which he took over the family business from Judith Willdey, trading at *Great Goldsmith, Toy, Spectacle, China and Print Shop*, at the corner of Ludgate Street, by St. Paul's. He advertised for sale perspective glasses, telescopes, globes, quadrants, and maps. Upon his death in 1748 his executors advertised for creditors to put forward their claims on his estate, [35 #8770 p3].

William Optics

William Optics Corp., [136], was founded in 1996 as Optical Technology Ltd. by brothers William and David Yang to make telescopes for use by the "more demanding" amateur astronomer and birder. The company is based in Taipei, Taiwan, with optical production of their telescopes by **Guan Sheng Optical**.

Willson & Dixey

Willson & Dixey, [18], was a partnership between **George Willson** and **George Dixey**, trading as opticians from 1802 until 1809 in London, England.

Addresses: Opposite St. James' Church, Piccadilly (1802-1809); 9 Wardrobe Place, Doctors' Commons (1802); 35 Piccadilly (1803).

Willson, George

George Willson, (fl.1798-1809) [18], was an optician and optical turner of London, England. Addresses: Sermon Lane, Doctors' Commons (1798); Wardrobe Place, Doctors' Commons (1799-1802). He served his apprenticeship with the mathematical instrument maker James Moulding, beginning in 1781, and was turned over to **John Mason** in 1797. He was freed of the **Stationers' Company** later that year. Among his apprentices was **George Dixey**, with whom he worked in partnership from 1802 until 1809, trading as **Willson & Dixey**.

Instruments bearing his name include: two-draw brass and leather "day and night" telescope; single-draw brass and mahogany "day and night" telescope. Note that his instruments are sometimes miss-attributed to **George Wilson**.

Willstrop, Dr. Roderick V.

Dr. Roderick V. Willstrop, [199], astronomer and optical designer, was a Fellow of the University of Cambridge's Institute of Astronomy. Among his contributions to optics, he has constructed several telescopes, and is responsible for the design of the Three Mirror Telescope, as well as other innovations in optical design. He designed a low-cost tabletop machine for producing batches of aspheric lenses. During his observations of pulsars, he designed a photometer with a filter wheel containing four filters to enable him to better determine the "colour" of the objects under investigation.

Wilson & Gillie

Wilson & Gillie were magnetic compass and nautical instrument makers of North Shields, Tyne & Wear, England. The company was founded by John Wilson Gillie in 1885. In 1913 the company merged with **John Lilley & Son**, and in 1943 the merged company was renamed **John Lilley & Gillie**.

Instruments bearing their name include: Brass sextant signed *Wilson & Gillie Quay North Shields*. Their trade label, with the address New Quay, North Shields, is also in the lid of the case for a sextant by **James Gilkerson & Co.** [16].

Wilson, Charles

Charles Wilson, (fl.1842-1851) [18] [276 #1744], was a navigational instrument maker, teacher, and map seller. He worked with **John William Norrie** in the partnership **J.W. Norrie & Co.** until it was terminated in 1840, after which he worked under his own name. He was freed by purchase of the **Spectacle Makers' Company** in 1842, and continued operation of the Navigation Warehouse at 157 Leadenhall St., London, England, until 1851.

He also worked in the partnership **Norrie & Wilson**, as evidenced by his trade label, "Charles Wilson, late J.W. Norrie & Wilson, Bookseller, Publisher, Chart & map seller at the Navigation Warehouse & Naval academy. 157 Leadenhall St. London, where may be had all the publications of Steel & Cº. late of Cornhill. Nautical instruments made, cleaned, & repaired."

Wilson, George

George Wilson, (fl.1805-1817), [18], was an optician and optical instrument maker of London, England. Addresses: 16 Charterhouse St., West Smithfield (1805); 44 Kirby Street, Hatton Garden (1809-1817); 115 Holborn Hill (1816-1817). His name is spelled Willson in [276 #1463], which may result from confusion with **George Willson**.

Wilson, James
James Wilson, (fl.1702-1710) [18] [264] [274] [275 #450], was an optical instrument maker of *The Willow Tree*, Cross St., Hatton Garden, London, England. The Wilson type microscope, a screw-barrel microscope, was named after him, but not invented by him. He did, however, publish his own design for it in the *Philosophical Transactions of the Royal Society of London* in 1702. He is known to have also made other kinds of microscope, including compass microscopes, as well as telescopes, prospects, camera obscuras, and magic lanterns. Instruments bearing his name include; screw-barrel microscopes; compass microscopes; compound microscope; refracting telescope.

Wilson Microscope
See microscope.

Wilson, Robert William
Robert William Wilson, (fl.1828-1830) [18], was an optical, mathematical, and philosophical instrument maker of 2 New Compton St., Soho, London, England.

Wilton, William
William Wilton, (fl.c1800-1847 d.1859) [18] [276 #1246] [361] [379], was an optical, mathematical, and philosophical instrument maker, and clock and watch maker, of Cornwall, England. Addresses: St. Day (1830-1851) and Market Place, Camboune (1852-1856).

According to his trade card, he sold miners & surveyors theodolites, dials and quadrants, levels, drawing instruments, and more. He exhibited at the Great Exhibition in 1851 in London [328]. His catalogue entry no. 402 lists a magnetic dip and intensity instrument, with an extensive description, and a miner's theodolite. He was succeeded in business by **E.T. Newton**.

Wing, Tycho
Tycho Wing, (1726-1776) [18] [90], was a mathematical instrument maker of London, England. Address: Near Exeter Exchange, Strand, London (1751). He served his apprenticeship with his father-in-law, **Thomas Heath**, and was freed of the **Grocers' Company** in 1751. In 1751 he began to work in partnership with Thomas Heath, trading as **Heath and Wing**. Among his apprentices were the mathematical and philosophical instrument maker, Thomas Newman, beginning in 1751, and **Charles Fairbone (I)**, beginning in 1753.

In 1764 Tycho Wing was one of the petitioners who attempted to revoke the patent obtained by **John Dollond (I)** for the achromatic lens, and which was enforced after Dollond's death by his son, Peter (See Appendix X).

The Heath & Wing partnership ended upon the death of Thomas Heath in 1773, at which time Tycho Wing retired. Thomas Newman succeeded Heath & Wing in business.

Winkel, Rudolf
Rudolf Winkel, (1827-1905) [374], of Göttingen, Germany, was a mechanical engineer noted for his production of microscopes. He served an apprenticeship with **Moritz Meyerstein**, and an apprenticeship with **Breithaupt** from 1848-1857. Upon completion in 1857, he started his own company. In 1858 he took on his first apprentice, **F.G. Voigt**, who went on to establish **Voigt & Hochgesang** in 1869. He began making microscopes in the 1860s.

By 1880 his three sons, Carl (1857-1908), Herman (1860-1935), and Albert (1863-1919), were working for the company, and by 1900 the company had about 30 employees. In 1907, after the death of Rudolf, his sons built a new factory, where they expanded their product range and began mass production. By 1911 **Zeiss** were the major shareholder, and the company was incorporated as R. Winkel GmbH. During WWII they marked their instruments with the **ordnance code** eaw. In 1957 the company became Winkel-Zeiss, part of Zeiss.

Winkel-Zeiss
See **Rudolf Winkel**.

Winspear, John Emanuel
John Emanuel Winspear, (1844-1906) [50] [264] [320], was a manufacturer of achromatic microscopes, and a philosophical instrument maker. He trained as an optician, and was working as an optician in Bradford, Yorkshire, England, by age 22. In 1870 he moved his business to Derrington Works, Hull. He acted as the sole agent for immersion and other objectives by **Ernst Gundlach** of Berlin, Germany, until sometime in 1871, at which time **R.&J. Beck** of London also became a distributor. In 1872 Gundlach emigrated to the USA, and Winspear's agency ceased. His business continued until liquidation proceedings began in 1877 [35 #24453 p2990], leading to bankruptcy in 1879 [35 #24730 p3773].

Winter, T.B.; Winter & Son, T.B.
T.B. Winter, (fl.1857-1878), of 55 Grey Street, Newcastle-Upon-Tyne, England, was a mathematical, nautical, philosophical, and optical instrument maker. In 1858 the business relocated to 21 and 27 Grey Street. 21 Grey Street was the same address as **John Cail** whom he succeeded in business at that address.

According to his trade card dated 1857-1858, [53], "T.B. Winter, 55 Grey Street, Corner of High Bridge,

Newcastle on Tyne. Manufacturer of Mathematical, Nautical, Philosophical and Optical instruments. Spectacles to suit all sights. Transit instruments, Theodolites, Levels and circumferentors, Sextants, Quadrants, Compasses and telescopes."

T.B. Winter was succeeded in business by T.B. Winter & Son. Instruments bearing this name include: sextant; barometer; surveyor's level; pocket compass; barograph; mining dial; opera glasses; surveying protractor; triple optic binocular made by **Lemaire**.

Winter, Thomas

Thomas Winter, (fl.1794-1849) [18] [53] [319] [322], was an optical and mathematical instrument maker of London, England. Addresses: 9 Wells St., Oxford St.; Mount St., Grosvenor Square; 9 New Bond St.; 4 Ebenezer Place, Commercial Rd., Limehouse; 33 Poland St., Oxford St.; 37 Brewer St., Golden Square (1800); 6 Brewer St., Golden Square (1805-1816); 5 Market St., Oxford St. (1839-1849); 24 Great Castle St., Regent St. (1849).

He was noted for his museum microscopes. He was subject to bankruptcy proceedings in 1803 [35 #15583 p558], for which the final dividend was paid in 1805 [35 #15785 p298].

Note: [276 #1060, #2048, #2049] has three separate entries for Thomas Winter, with different addresses, and [18] lists them as the same person.

Wirgin

Wirgin was a camera maker of Wiesbaden, Germany, founded in the 1920s by the brothers Heinrich, Max, Wolf, and Joseph Wirgin, who were Polish Jews, and had recently emigrated to Germany. Their company, Kamerawerk Gebr. Wirgin, manufactured cameras under the brand-names *Wirgin* and *Edixa*. Shortly before WWII the brothers emigrated to the USA, and during the war the Wirgin company was operated as a part of **ADOX**. Heinrich returned to Germany in 1948, by then using the name Henry, and re-acquired the Wirgin factory, which he continued to run successfully. In 1962 Wirgin acquired the camera maker **Franka**. Following Henry's retirement in 1968, the company continued under new ownership, trading as Edixa, until 1972. The Edixa name was subsequently re-used by an unrelated company.

With, George Henry

George Henry With, (d.1904) [14] [76], was an eminent mirror maker from Hereford, England. He was a science professor at local colleges, and an amateur mirror maker. He is believed to have made around 200 mirrors for reflecting telescopes, some of which were mounted by **John Browning (III)**.

Withering, William

Dr. William Withering, (1741-1799) [46] [274] [287] [319], was a medical doctor in Stafford, England, and relocated to Birmingham in 1775. During the 1770s he developed an interest in botany, and published a textbook on botany in 1776. He is noted for his discovery of the use of foxglove (Digitalis purpurea) for the treatment of certain cardiac conditions. He invented two designs of simple microscope: a botanical microscope, and a boxed botanical microscope that unfolds as the box is opened.

Witherspoon, Earl

Earl Witherspoon, [185], of Sumpter, California, USA, was a maker of fine objective lenses, ranging from 4 to 6 inches, for refracting telescopes, during the 1940s and 1950s. By the early 1960s he was no longer making lenses.

Woerle, Martin

See **Martin Wörle**.

Wöhler, Dr. F.A.

Dr. F.A. Wöhler of Kassel, Germany, was an optical instrument manufacturer. Examples of the company's products include riflescopes, artillery telescopes, and prism binoculars. During WWII they marked their optical instruments with the **ordnance code** clb. Their production of optical instruments ended in 1967.

Wollaston, Francis

Francis Wollaston, (1731-1815) [46] [276], was an English priest, author and astronomer. In 1758 he married Althea Hyde, the fifth daughter of John Hyde. They had ten daughters and seven sons, including **William Hyde Wollaston**, and the natural philosopher Francis John Hyde Wollaston. Francis Wollaston was a keen amateur astronomer, with a private observatory at his home in Chislehurst, Kent, and was elected a fellow of the Royal Society in April 1769. He commissioned an altazimuth telescope from **William Cary**, and produced a star catalogue that was used by **William Herschel**.

Wollaston, William Hyde

William Hyde Wollaston, (1766-1828), [37] [45] [46] [70] [155], son of **Francis Wollaston**, was a physicist, chemist and a physiologist. He began his career practising medicine, but gave this up in 1801 to set up a private laboratory in which he could pursue his scientific interests. He devised a method for producing metallic platinum, which resulted in considerable wealth for him, and about three quarters of his output of malleable platinum was marketed by **William Cary** [128]. Wollaston is also known for discovering the elements palladium and rhodium.

His interest in astronomy, optics and crystallography led him in other directions. He was the first to observe the dark lines in the Sun's spectrum, later studied in detail by **Fraunhofer**, but he believed them simply to be boundaries between colours. His inventions include: a process for making very fine wire, as thin as 0.0001 inch or less, for use as cross-hairs in telescope eyepieces; a method using total internal reflection for measuring the refractive index of crystals; the reflective goniometer; the camera lucida; the Wollaston Doublet; and the Wollaston Prism.

Wollaston Doublet
The Wollaston doublet, [70], an innovation by **William Hyde Wollaston**, replaced the conventional concavo-convex lens in a microscope objective with a plano-convex lens. This was easier to produce and, with the inclusion of a diaphragm between the lenses, reduced chromatic and spherical aberration. In 1829 he presented a paper, describing Wollaston's doublet lens, to the Royal Society [84].

Wollaston Prism
See prism.

Wood, Benjamin
Benjamin Wood, (fl.1810-1841) [3] [18], was a mathematical and nautical instrument maker of Liverpool, England. Addresses: 41 Wapping (1810); 52 Wapping (1811); 50 Wapping (1813-1832); 51 Wapping (1816); 49 Wapping (1818-1828); 6 Bath St. (1829); 21 Bath St. (1831-1834); 46 Wapping (1834-1835); 28 Bath St. (1835).

Instruments bearing his name include an octant with scale divided by **Edward Pritchard** (see Appendix XI). He was declared bankrupt in 1823 [35 #17906 p464].

He is believed to have been the same person as Benjamin Jasper Wood, (fl.1796-1809), who served his apprenticeship with **Michael Dancer**, the grandfather of **John Benjamin Dancer**, beginning in 1788, and was freed of the **Joiners' Company** in 1796. He traded at 10 Paved Buildings, Leadenhall Market, London (1796-1800) and 28 Rosomon St., Clerkenwell, London, (1802-1809), and is then believed to have moved to Liverpool, where he restarted his business at 41 Wapping in 1810.

His son, Benjamin Jasper Wood Jr., (fl.1828-1865), was an optician, mathematical instrument maker, and teacher of navigation, whose early Liverpool addresses corresponded with those of his father in Wapping and Bath St., in addition to: 23 Bath St. (1832-1837); 45 Wapping (1837); 7 Bath St. (1841-1853); 64 Chatsworth St. (1857-1860); 12 College Lane (1865).

In 1842 Benjamin Jasper Wood was declared bankrupt, [35 #20147 p2683], but it is not clear whether this was the father or the son. The absence of "Jr." in the notice may suggest that it was the father.

Wood, Edward George
Edward George Wood, (1811-1896) [50] [306], a mathematical and philosophical instrument maker and machinist, was the older brother of **George Smart Wood**. In 1844 he entered a partnership with Fallon Horne and William Thornthwaite, trading as **Horne, Thornthwaite and Wood**. He operated his own business at 74 Cheapside, London, England, trading as E.G. Wood from 1854 until his death in 1896. The company was continued by his grandson, L.A.S. Wood, as an electrical business, trading as E.G. Wood, and as Wood & Co.

Wood, George Smart
George Smart Wood, (1816-1882) [3] [18] [50], of London, England, younger brother of **Edward George Wood**, was an optical, mathematical and philosophical instrument maker. In 1849 Wood went into partnership in Liverpool with **A. Abraham** and Abraham's manager, **Charles West (II)**. The partnership traded as Abraham & Co., and was terminated in 1855 [35 #21840 p216]. The shop traded at the Liverpool addresses: 46 Prescot St. (1847); 52 Prescot St. (1849); 76 South Castle St. (1871); 20 Lord St. (1875-1896); 15 London Rd. (1875). George Wood continued the business as Abraham & Co. He signed his instruments *Wood Late Abraham*, and in the early 1880s he traded as Wood, George S. (late Abraham & Co.)

Following George's death in 1882 his son, John Wood (1849-1918), ran the business, largely making spectacles and binoculars. By 1900 he was trading as Wood-Abraham, and at about that time he opened a branch in 24 Market Street, Manchester. In about 1910 the Manchester branch was acquired by Richard Thomas, who operated it until 1941 when it closed.

Wooley, Sons & Co. Ltd., James
James Wooley, Sons & Co. Ltd., [53] [122] [320], of Manchester, England, was established in 1833 by James Wooley (1811-1858) as a chemist and druggist. They began to manufacture scientific apparatus and surgical instruments and, when James Wooley died in 1858, the company was taken over by his son George (c1837-1918). By 1872 George's brothers, Harold and Herman, were working with him, and the company was trading as James Wooley, Sons and Company. They expanded their products to include photographic equipment as well as microscopes, accessories, and prepared slides. They sold instruments by other manufacturers, as well as those

bearing their own name. In 1895 the company was incorporated as a limited company. During the early C20th they ceased selling photographic equipment. When they stopped selling microscopes and accessories is unclear, but the company continued as a drug and chemical supplier.

Worgan, John

John Worgan (fl.1682-1700), was a mathematical instrument maker of London, England, specialising in surveying instruments and dials. He served his apprenticeship with Nathaniel Anderton, and was freed of the **Grocers' Company** in 1682. There is, however, some confusion in the Grocers' Company records, and it is possible that he was turned over to Walter Hayes to complete his apprenticeship, prior to being freed in 1682. Records place him working in George Alley, Fleet Ditch in 1685, then *Under the Dial of St. Dunstan's Church*, Fleet Street, between 1686 and at least 1700, as well as Fetter Lane in 1693. He is known to have had five apprentices, two of whom were turned over to other masters during or after 1700. A Gunther's quadrant made by John Worgan is in the Science Museum, London. Other examples of his work include compass dials and circumferentors. He is also known to have sold alidades, plane tables, and sectors. According [**234**], his premises may have been taken over by **John Rowley**. However, while this is confirmed in [**90**] [**442**], some doubt is cast by [**18**] and [**275** #451] since, according to a print dated 1737, there were six or seven shops Under St. Dunstan's Church, of which the westernmost shop was immediately under the Dial.

Wörle, Martin

Martin Wörle (Woerle), (fl.1811-1851) [**174**] [**374**], was an optical instrument maker of Munich, Germany. He initially worked for **Josef Ritter von Fraunhofer**, from whom he learned how to produce and test achromatic lenses. At around the time of Fraunhofer's death in 1826, Martin Wörle set up his own business in Munich. **Michael Baader** began to work for him in 1838. Baader subsequently married Wörle's daughter, and continued to work for him until Wörle's death in 1851.

Worthington & Allan

Worthington & Allan, [**2**] [**18**], of 196 Piccadilly, London, England, was a partnership between **Nathaniel Worthington** and **James Allan (II)**, which appeared in the trade directories from 1822 until it was terminated in 1832 [**35** #18951 p1561]. Both partners were previously employees of **Jesse Ramsden**. Surviving instruments include sextants, hand-held telescopes and barometers.

Worthington, Nathaniel

Nathaniel Worthington, (c1790-1853) [**2**] [**18**] [**53**], mathematical and optical instrument maker of Gloucester, England, was the only registered apprentice of **Matthew Berge**, beginning his seven-year apprenticeship in 1804. In 1819 he succeeded Berge in his business at 196 Piccadilly, London, and retained **Ramsden**'s brass standard yard and one of Ramsden's circular dividing engines. He formed a partnership with **James Allan (II)**, trading as **Worthington & Allan**. The partnership appeared in the trade directories from 1822 until James Allan (II) died in 1832, after which Nathaniel Worthington traded under his own name until 1852. Upon Worthington's retirement the Ramsden circular dividing engine was acquired by **Knox & Shain** of Philadelphia, Pennsylvania, USA. A 15½-inch altazimuth circle bearing Worthington's signature is at Palmero Observatory.

Wratten, Frederick

Frederick Charles Luther Wratten, (1840-1926) [**37**] [**61**] [**71**], was an English inventor and manufacturer of photographic plates. His interest in photography arose while working in London as a clerk for Joseph Solomon's Photographic & Optical Warehouse. He began experimenting with the new gelatine dry-plate process, and came up with his own innovative process for producing the plates. In 1877 he founded **Wratten & Wainwright** in partnership with Henry Wainwright. He continued to work for the company after it was acquired by **Kodak** in 1912.

Wratten filter

Sometime between 1909 and 1912, **Wratten & Wainwright** introduced an extensive set of colour photographic filters, known as *Wratten filters*, each being numbered as an index of its colour and degree of light attenuation. In 1912, after **Kodak** acquired Wratten & Wainwright, they continued production of the filters.

The numbering system subsequently became an industry standard for coloured filters. The filter range includes *neutral density* filters, which attenuate light to varying degrees without altering colour. Wratten filters are also used in astronomy applications, for observation and imaging. When Wratten filters are used in photographic imaging, the attenuation of light must be compensated by adjustment of the focal ratio or shutter speed.

Wratten & Wainwright

Wratten & Wainwright, [**37**] [**61**] [**70**] [**71**] [**306**], of 38 Great Queen Street, Long Acre, London, England, were primarily makers of photographic plates. The company was founded in 1877 by **Frederick Wratten** and Henry Wainwright (d.1882). In 1878 they began to sell gelatine silver bromide dry plates. These plates had improved sensitivity, and the company exported them in large numbers to continental Europe. In 1887 they started their Photographic Apparatus Department, and began to market cameras bearing their name. They continued to sell cameras until the mid-1890s.

The company relocated to Canterbury Road, Croydon in 1890, and was incorporated in 1896 with owners including Frederick's son, Sidney Herbert Wratten (1871-1944), and a recent graduate from London University, C.E. Kenneth Mees (1882-1960). Shortly after this, they introduced the first truly panchromatic plates. The company developed a series of colour photographic filters, known as *Wratten filters*, and in 1912 the company was taken over by **Kodak**, who continued production of the filters. Both Frederick Wratten and C.E. Kenneth Mees went on to work for Kodak.

Wray Optical Works Ltd.

Wray Optical Works Ltd., of Ashgrove Road, Bromley Hill, Kent, England, [29] [44] [124] [306], was founded in 1850 by William Wray, a solicitor and amateur astronomer who made his own telescopes and polished his own lenses. The company built a strong reputation making photographic lenses and cameras. In 1908, suffering a decline in business, Wray was saved by a merger with **Aitchison & Co.** During WWI, Wray manufactured photographic lenses and binoculars under their own name, and also made binoculars for Aitchison.

Wray designed and manufactured optics for **Cintel** following their formation in 1927. During WWII Wray manufactured military prism binoculars, photographic lenses, and reflector gunsights. In 1948 Wray was acquired by **British Optical and Precision Engineers**, a subsidiary of **Rank**, and continued to make photographic lenses, camaras and binoculars branded Wray. Increasingly the competition from imported Japanese brands impacted their business. Their response was to attempt to reduce costs and thus offer better competition, and the *Wrayvu* model of binoculars, introduced in 1957, is an example of the result.

Wray sought out other markets, and in the mid-1960s they were approached by Elliott Automation (then the parent company of **Elliott Brothers**), for whom they completed the optical design for a head-up display. Following this Wray took on further contracts for HUD optics. In 1971 Rank closed the Wray Optical Works Ltd. factory in Bromley, and moved the plant to the **Kershaw** factory in Leeds. Following this move Kershaw never delivered under their HUD optics contracts, and lost the business.

Wrench, Edward

Edward Wrench, (fl.1822-1853) [18] [53] [276 #2280] [322], was an optical, mathematical, philosophical, and surveying instrument maker of London, England. Addresses: 57 Red Lion St., Holborn; Red Lion St., Holborn (1825); 6 Gray's Inn Terrace (1839-1852). In 1825 he attended the London Mechanics' Institute. Instruments bearing his name include: telescope; theodolite; surveyor's compass; celestial globe; miniature terrestrial globe; pantograph; wheel barometer; marine barometer.

Wright, David

David Wright, (fl.1821-1849) [3] [18] [276 #2054], was an optician and optical instrument maker of Sheffield, England. Addresses: Corn Hill (1821-2822); 58 Campo Lane (1830); 94 Fargate (1833-1837); 30 Fargate (1839-1841); 72 Fargate (1845-1846); 90 Norfolk Road (1849). From 1818 until 1825 he worked in partnership with **William Chadburn**, trading as **Chadburn & Wright**.

Wright, Gabriel

Gabriel Wright, (fl.1779-1803) [16] [18], was an optical and mathematical instrument maker of 36 Little Britain, London, England, and 148 Leadenhall Street, London. He worked for **Benjamin Martin** from 1764 until 1782. He was freed of the **Girdlers' Company** in 1783 by purchase. He patented index-glass adjustment in a sextant in 1779 [95]. This was noted as an improvement to the Hadley quadrant. The index-glass is mounted on a platform on the index arm, and may be rotated around the pivot by means of screws for parallel adjustment. He took **Benjamin Hooke** as an apprentice in 1786.

He engaged in several partnerships trading at *The Navigation Warehouse*, 148 Leadenhall St., London. In 1782 he entered a partnership with **Henry Gregory**, trading as **Gregory & Wright**. Subsequently **William Gilbert (I)**, joined the partnership, and they began to trade as **Gregory, Gilbert & Wright**. This partnership was terminated in 1789, [35 #13124 p560], after which they conducted their business as **Gilbert & Wright** until 1792. In 1794 he formed a new partnership, this time with William Gilbert (I) and Benjamin Hooke, who had recently completed his apprenticeship, and they traded as **Gilbert, Wright & Hooke** until 1801.

Wright, John

John Wright, (fl.1750-1760) [3] [18], from London, England, was a mathematical, philosophical and optical instrument maker, trading at the *Sphere and Hadley's Quadrant*, Near St. Stephen's Church, Bristol. He served his apprenticeship with **Benjamin Cole (I)** beginning in

1750, and was freed of the **Merchant Taylors' Company** in 1760.

The business address of the *Sphere and Hadley's Quadrant*, Near St. Stephen's Church, Bristol, was also listed as the *Sphere and Hadley's Quadrant*, in St. Stephen's Lane, and is almost certainly the same address as *Hadley's Quadrant*, in St. Stephen's Lane, Bristol, subsequently used for the business of **Joshua Springer**, who succeeded John Wright in business.

According to John Wright's advertisement in *Felix Farley's Bristol Journal* in 1756: "Accurately makes, according to the best and latest improvements, in silver, brass, ivory, wood, &c. all kinds of Mathematical, Philosophical and Optical instruments; where gentlemen, upon signifying the plate and figure of any instrument in *Desauliers*, *Graevsand*, or other authors, may be served therewith, or have any model or instrument made according to their own contrivance.—Makes orreries and spheres of different sizes.—Measuring wheels, and coach or chaise way-wisers for measuring the roads, &c., Theodolites, plain tables, circumferentors, and levels of various sorts.—Hadley's sea quadrant made with the utmost accuracy with glasses, whose planes are truly parallel; also Davis's and other quadrants.—Azimuth, amplitude, and other compasses, either for cabin, steerage, or pocket: Artificial magnets, made according to Dr. Knight's improvements, which are particularly useful for touching compass needles.—Gunners quadrants, and all instruments for fortification, &c.,—Gauging instruments, and rules of all sorts.—Variety of curious pocket cases of drawing instruments in silver, brass, &c., Also magazine cases with variety of useful instruments, proper for gentlemen who travel.—Elliptical, proportional, and triangular compasses, &c.,—Sun dials, horizontal and other sorts, for any latitude; Also universal, and variety of other portable ones, with new improvements.—Air pumps, either double or single barrel, to demonstrate the curious experiments depending on the pressure, and spring or elasticity of the air, &c., with all their apparatus; Also hydrostatical balances, carefully adjusted for determining the specifick gravity of solids and fluids.—Barometers, either standard, diagonal, or portable, with or without thermometers; Also the famed mercurial thermometer, truly adjusted and made to any scale.—Reflecting telescopes, either Newtonian or Gregorian, made with the utmost accuracy.—Refracting telescopes with the late improvement, for sea or land.—Microscopes of various kinds, either double or single, to be used with or without the solar apparatus.—Camera obscura for drawing in perspective in which the external objects are represented in their just proportions and proper colours.—Prisms for demonstrating the theory of light and colours.—Diagonal mirrors for viewing perspective prints.—Spectacles, either crown, white, or true Venetian green glass, ground on brass tools, as approved of by the Royal Society, set in variety of commodious frames; Also reading glasses, fitted in silver, metal, tortoiseshell, or horn, &c.,—Convex and concave mirrors, Opera glasses, and magic lanthorns, with a variety of other instruments, made and sold wholesale or retail (with books of their use) as cheap as in London.

"Gentlemen may depend upon being served with the above, and all other instruments, made according to the latest discoveries, John Wright being late an apprentice to Mr. Cole, successor to Mr. Thomas Wright, instrument maker to his Majesty."

Wright, Lewis
Lewis Wright (1838-1905), [44] [109], was an English scientific instrument maker and author of several books on the subject of microscopy, as well as a journalist and editor. He was born in Bristol. By 1855 he had designed a projection lantern microscope for educational purposes, which was manufactured by **Newton & Co.** He continued to work with **Herbert C. Newton** of Newton & Co. on the design of optical instruments. His son was **Russel S. Wright**.

Wright, Russel S.
Russel S. Wright, (1876-1961) [44] [109], son of **Lewis Wright** was a scientific instrument maker and x-ray engineer. He began his career working with Herbert C. Newton of **Newton and Co.** producing optical instruments. In 1911 he went into partnership with H.C. Newton, trading as **Newton & Wright**.

Wright, Samuel
Samuel Wright, (fl.1764) [364], of Bedford Street, London, England, was one of the petitioners who attempted, in 1764, to revoke the patent obtained by **John Dollond (I)** for the achromatic lens, and which was enforced after Dollond's death by his son, Peter (See Appendix X). As a petitioner, Samuel Wright is likely to have been a telescope maker.

Wright, Thomas (I)
Thomas Wright (I), (b.c1686, retired 1748) [14] [18] [234] [276 #143] [396], was a mathematical instrument maker of Fleet Street, London, England. He served his apprenticeship with **John Rowley**, beginning in 1707, and was freed of the **Broderers' Company** in 1715. He succeeded Rowley in business in 1718, trading at the *Orrery and Globe*, Fleet Street, which is believed to have been the same address as Rowley. His main products were globes and orreries, and he may have been assisted in the

construction of his fine orreries by the clock-maker George Graham (1673-1751), who invented the orrery, and shared his address for a period around 1720.

Thomas Wright (I) also made other instruments, including pocket dials, Gunter's quadrants, surveyor's perambulators, circumferentors, plane tables, sun dials, sextants, magnetic compasses, reflecting telescopes and compound microscopes. In 1727 he was appointed Mathematical Instrument Maker in Ordinary to King George II. In about 1740 he entered a partnership with the mathematical instrument maker William Wyeth, and the partnership ended when Wyeth died in 1741. By 1747 Wright's trade address was known as the *Orrery & Globe, next the Globe and Marlborough Head Tavern*, Fleet Street. Wright retired in 1748, and was succeeded in his business by his former employee, **Benjamin Cole (I)**.

Wright, Thomas (II)

Thomas Wright (II), (1711-1786) [14] [18] [46] [276 #321], was a mathematical instrument maker, inventor, astronomer and landscape gardener. He was born in Byers Green, Durham, England, the son of John (a yeoman and carpenter) and Margaret Wright. In 1730 he spent some time working for instrument makers **Thomas Heath** and **Jonathan Sisson**. At the age of twenty he set up a school to teach navigation, and sell instruments at the *Sign of ye Creation*, Sunderland from 1730 to 1734, then at St James's, London from 1735 to 1748, and in Byers Green, Durham in 1762. He also gave private tuition in maths and astronomy, and in 1750 he published *An Original Theory of the Universe*.

Wright, Thomas (III)

Thomas Wright (III), (fl.1826-1842) [18] [276 #2055], of London, England, was an optician, optical and mathematical instrument maker, drawing instrument maker, and dealer in cabinet work and ivory goods. Addresses: 27 City Terrace, City Rd.; 28 City Terrace, City Rd.; 15 St. Mary Street Hill; 42 Allen St., Goswell St.; 1 Worship Square, Worship St., Hoxton (1839-1841); 25 City Rd. (1841).

Würdemann, William

William Würdemann, (1811-1900) [189] [208], was a German born instrument maker who moved to Washington DC, USA, to work for the US Coast Survey. They recruited him in order to end the reliance on European instrument makers and suppliers. From 1849 to 1881 he had his own workshop as a mathematical and optical instrument maker, primarily producing surveying instruments for the US Coast Survey and other customers. He had access to the Coast Survey workshops and instruments, and used their dividing engine, made by **Troughton & Simms** in 1847. However, he eventually built his own dividing engine, together with **Gustav Heyde** of Dresden, Germany, and with it he continued to make fine instruments including portable transit and zenith instruments. He recruited the German optical instrument makers **George Saegmuller** and **Camill Fauth**, who settled in Washington DC and worked for him for about four years before starting their own firm, **Fauth & Co**.

Wyck, Johan van der

Johan van der Wyck, (1623-1679) [419] [420] [421], was a military engineer and optical instrument maker. He was born in Germany to an aristocratic family, and in 1650 he matriculated from the military and civil training college, Collegium Auriacum in Breda in the Netherlands, where he is believed to have learned practical optics from a mathematics professor. By this time van der Wijk was already making optical instruments, and in 1654 **Christian Huygens** and his brother Constantijn, ordered two telescope lenses from him, which he delivered later that year. Over his career, he acquired a strong reputation for grinding lenses for telescopes.

He joined the army of the Calvinist Dutch Republic and, from 1654 until 1657 Van der Wyck was a military engineer in Delft. In 1657 he entered military service for king Carl X Gustav of Sweden, and in 1658 he was captured by Danish forces, and remained a prisoner-of-war until 1659. Following this, he remained in military service, during which time he is believed to have continued to make optical instruments. In 1674, under the influence of a controversial spiritual leader, he left military service, and returned to the Dutch Republic.

He is known to have made: microscopes; telescopes; camera obscuras, some of which would have been used by the eminent artists of Delft at the time.

Wye-Level

Another name for a Y level.

WZFO

See **Warsaw Photo-Optical Works**.

X

Xibei Optical Instrument Factory

The Xibei Optical Instrument Factory was founded in Xi'an in the Shaanxi province in China in 1957. An example of their camera models was the *Huashan DF-S* budget SLR, produced in the 1980s.

X-Ray Optics

X-rays, [7], are high-energy photons, which are generated in one of two ways. The first is when they are emitted by an atom or molecule during the transition of an inner, tightly bound electron. The second is the case of *bremsstrahlung* (German for braking radiation), in which they are emitted when a high-energy charged particle, often an electron, is slowed or deflected from its path. X-rays are in the frequency range 3×10^{17} Hz to 5×10^{19} Hz, and were discovered in 1895 by the German engineer and physicist Wilhelm Conrad Röntgen.

X-ray optics is the branch of optics that uses x-ray photons rather than those in the visible or other parts of the electromagnetic spectrum. The optical properties of x-rays are employed in areas such as medical imaging, x-ray astronomy, x-ray microscopy, x-ray interferometry, and others. The high energy of x-ray photons means that the optical components required are very different from those used in visible optics, and as such are outside the scope of this book. See Appendix VIII.

Y

Yamamoto Seisakusyo

Yamamoto Seisakusyo (Yamamoto Co. Ltd.) was a small Japanese company established by Mr. Yamamoto, probably in the late-1960s, to design and construct telescopes. The company designed and assembled telescopes to Mr. Yamamoto's designs, having the parts manufactured by other companies. Yamamoto telescopes bore the maker's mark illustrated in Appendix IV, and retailed under various brand-names including **Busch** (Germany), Sky Master (Italy), Perl (France), Jupiter (France), **YOSCO** (Australia), Sport Master (New Zealand), Milo, Palomar, and Satellite.

Yarwell, John

John Yarwell, (1648-1712) [2] [14] [18] [24] [53] [275 #343], was an optician and optical instrument maker of London, England.

Addresses: *Archimedes and Three Golden Prospects*, near the great North Door in St. Paul's Church Yard; *Ye Archimedes & Spectacles* in St. Paul's Church Yard (1671-1692); North Side of St. Paul's (1672); St. Paul's Church Yard (1676-1696); *Archimedes & 3 pair of Golden Spectacles* in Ludgate St., the Shop next Ludgate (1692); *Archimedes and Three Golden Prospects*, St. Paul's Church Yard (1694); *Sign of Archimedes* in Ludgate St., first Spectacle Shop in Ludgate (1697); *Archimedes & Crown* (1698-1712).

He was one of the two leading microscope makers at the time in London, and his business was next door to that of his rival, **John Marshall**. He served his apprenticeship with the spectacle maker Richard Edwards of Fenchurch Street, London, and became a skilled glass-grinder. He was freed of the **Spectacle Makers' Company** in 1669. Among his apprentices was **George Willdey**, beginning in 1694/5.

In 1697 he advertised a new microscope which was capable of observing the circulation of blood in, for example, fish tails. According to his trade card, in addition to spectacles and reading glasses he made telescopes, microscopes, magnifying glasses, multiplying glasses, triangular prisms, ear trumpets, and "all other sorts of glasses, both concave and convex."

In 1697 John Yarwell began to work in partnership with **Ralph Sterrop**, trading as **Sterrop & Yarwell**. This partnership continued until John Yarwell's retirement in 1708, after which Ralph Sterrop continued the business under his own name.

Yawman & Erbe

Yawman & Erbe, [113] [201], was a furniture maker founded by Gustav Erbe and Philip H. Yawman, ex-employees of **Bausch & Lomb**, in Rochester, New York, USA. For a brief period in the early 1880s they made microscope stands with optics made by **Ernst Gundlach**. It is said that Yawman & Erbe discontinued making microscopes when Bausch & Lomb threatened to begin manufacturing furniture. They also made the first film-roll holders for **Kodak**, and assembled the wooden bodies of Kodak cameras up to 1895. **Frank Brownell** served an apprenticeship with Yawman & Erbe prior to setting up his own company in the 1880s.

Yashica

Yashica was a Japanese manufacturing company, founded in 1945 as Yashima Seiki Seisakusho. They were incorporated in 1949, and in 1953 they first began production of cameras. In the same year they changed their name to Yashima Kōgaku Seiki K.K. In 1957 they opened a subsidiary in the USA, Yashica Inc., and in 1958 the company name was changed to Yashica Corporation. By the late-1960s, and perhaps before then, Yashica were producing binoculars as well as cameras and lenses. They continued to expand their successful product range and, in 1973, **Zeiss Ikon** licensed the name *Contax* to Yashica who used it for their *Contax* range of cameras with interchangeable Contax/Yashica or Zeiss lenses. In 1982 Yashica was acquired by **Kyocera**, who continued to produce *Contax* cameras and lenses until 2005. In 2008 the Yashica brand was sold to the MF Jebsen Group, a manufacturing company.

Yeates & Son

Yeates & Son of Dublin, Ireland, [53] [304], was a partnership between **Samuel Yeates** and his son **George Yeates**, trading at 2 Grafton Street from 1832 until 1839. Yeates & Son then continued as a partnership between George Yeates and his son Stephen from 1840 until 1887, also at 2 Grafton Street. The business is also recorded at 9 Nassau Street, Dublin in 1839, but this may be the same address, since 2 Grafton Street is on the corner adjoining Nassau Street. Yeates & Son were, by appointment, Mathematical Instrument Makers to the University, (Trinity College, Dublin).

Examples of works from Yeates & Son include: equatorial mount on iron stand with three levelling screws, declination circle and hour circles; equatorial refractor with brass declination and hour circles, with finder scope, on an oak base; telescope on a brass stand with a rotatable optical wheel which may be for polarizing; three-inch refractor on a pillar and tripod stand, signed *Yeates & Son, Dublin*.

Yeates, Andrew
Andrew Yeates, (1800-1876) [14] [18] [34] [53] [304], was the third son of **Samuel Yeates**. He moved to London, England in the early 1820s to work with **Edward Troughton**, and under Troughton's supervision he repaired instruments for the Greenwich Observatory. He ran his own business in London from 1837 until 1873. His business card reads "A Yeates, Mathematical instrument maker, 12 Brighton Place, New Kent Road, London."

Yeates, George
George Yeates, (1796-1882) [34] [53] [304], was an optical instrument maker in Grafton Street, Dublin, Ireland. He was the second son of **Samuel Yeates**, and he continued his father's business, designing and making surveying instruments, including clinometers, theodolites, rangefinders and levels. He traded as **Yeates & Son** in partnership with his father, at 2 Grafton Street, from 1832 until 1839, then from 1840 he traded as Yeates & Son in partnership with his son Stephen at the same address. This partnership continued until 1887. The family business continued at 2 Grafton Street until 1922, and is recorded at 9 Nassau Street, Dublin in 1839, although this may be the same address, since 2 Grafton Street is on the corner adjoining Nassau Street.

Yeates, Samuel
Samuel Yeates, (1762-1834) [18] [34] [53] [304], was an apprentice of **Seacombe Mason**, but it is not clear whether he took his freedom of the Glovers' & Skinners' guild of Dublin. In about 1790 he set up his own business as an optician in Grafton Street, Dublin, Ireland. His second son, **George Yeates** continued the family business, and his third son, **Andrew Yeates** left Ireland to work with **Edward Troughton** in London. Samuel traded in partnership with George as **Yeates & Son**, at 2 Grafton Street, from 1832 until 1839. The family business continued in Grafton Street until 1922.

Y Level
A Y level, or *Wye-level*, is a form of surveyor's level in which the telescope is mounted in Y-shaped supporting frames. The telescope may be rotated on its axis or removed from the frame and reversed end-for-end in order to test for errors in adjustment. The dumpy level is an improvement on the Y Level.

YOSCO
YOSCO was the York Optical and Scientific Company of Melbourne, Australia. The company was established in 1950 by Arthur Geddes and Martin Gertler, and they sold imported optical instruments under their own brand-name. They sold a variety of optical instruments as well as mirror making kits and spectacles. The company was sold by the original owners in 1990 and was closed by the new owner. An example of their products is a YOSCO branded 108mm refractor with focal length 1600mm manufactured by **Yamamoto Seisakusyo** bearing their maker's mark (see Appendix IV).

Youle, William
William Youle (I), (fl.1822-1866) [18] [276 #1752], of 22 Fieldgate Street, Whitechapel, London, England, was an optical, mathematical, and philosophical instrument maker. He was freed of the **Founders' Company** in 1836, and in 1837 he took his son, William Youle (II), as an apprentice [195]. From 1840 his business was located at 79 Leadenhall Street. In 1843, when William (II) would have completed his apprenticeship, records show them trading as Youle & Son at 79 Leadenhall Street. From 1845 until 1866 William (I) was trading at 83 Leadenhall Street. Examples of his work include a cased simple botanical microscope [53], and a single-draw brass and mahogany telescope, 25½ inches long. He is also known to have sold levels and theodolites.

Young & Sons
See **William J. Young**.

Young, Thomas
Thomas Young, (1773-1829) [7] [46] [155] [276], was an English physician and natural philosopher. He was born to a Quaker family in Somerset, although later became a member of the Church of England, and he pursued a career as a physician. He was a scholar of the classics, and a successful Egyptologist. He studied many subjects, including natural philosophy, and took a particular interest in optics. In 1801 he was appointed professor of natural philosophy by the Royal Institution, but this did not end his career as a physician. His lecture notes, published later, included his definition of the modulus of elasticity relating longitudinal stress and strain, subsequently known as *Young's modulus*. In 1804 he was appointed foreign secretary of the Royal Society, a post in which he served for the rest of his life. In 1808 he became a fellow of the Royal College of Physicians. In 1818 he was appointed secretary of the Board of Longitude, and superintendent of the Nautical Almanac. In 1827 he was elected a member of the Paris Académie des Sciences.

Young was also interested in acoustics, and drew comparisons between light and sound. Like his contemporary, **Augustin Fresnel**, he was a proponent of the wave theory of light, believing it to be a longitudinal vibration in a luminiferous ether, contrary to the particle theory of light supported by many of his contemporaries. He subsequently suggested that light is a transverse wave, thus providing a means to explain polarization. He researched the interference of light, leading to his experiments with what came to be called *Young's slits* (see interferometer: *Young's double slit*). Although the luminiferous ether was later proved not to exist, he contributed significantly to the development of the wave theory of light. In explaining colour vision, he derived the three-colour theory, or *trichromatic colour theory*, further developed by **Hermann von Helmholtz**, to become known as the *Young-Helmholtz theory*. Young also researched the physiology of the eye, and is generally credited with the invention of the ophthalmometer and the keratometer.

Young, William J.

William J. Young, (1800-1870) [58], was a mathematical and surveying instrument maker in Philadelphia, USA. He served an apprenticeship with the mathematical instrument maker Thomas Whitney, and started his own business in about 1820. He generally had about ten apprentices or journeymen in his workshop, and produced high quality hand-made mathematical and surveying instruments. Young's foreman, Joseph Knox, and a former apprentice of Young, Charles J. Shain, formed the company **Knox & Shain** in 1850.

In 1866 Charles S. Heller and Thomas N. Watson became partners in the business, and they began to trade as William J. Young & Co. Upon William Young's death in 1870, the partnership was terminated, and Charles Heller formed his company **Heller & Brightly**. Alfred Young, (who may have been William's son), continued the business as Young & Sons. The company incorporated in 1917, and became a part of **Keuffel & Esser** in 1918, where they continued to operate as Young & Sons.

Yukon Advanced Optics Worldwide

Yukon Advanced Optics Worldwide was founded in 1998. The company's trading branch is in Texas, USA. Their products include monoculars, binoculars, spotting scopes, and riflescopes, in both day vision and night vision options. Their manufacturing plant, **Beltex Optics**, is in Belarus, and began producing spotting scopes in 1991.

Z

Zaalberg van Zelst, J.J.
Johannes Jacobus Zaalberg van Zelst, (1827-1903) [374] [378], was an optical instrument maker of Leiden and Amsterdam, Netherlands. He served his apprenticeship with the scientific instrument maker Biense Rienks (1830-1871) at the Breestraat in Leiden. In 1865, he exhibited a compound microscope, with lenses he had made himself, at the exhibition of physical and mathematical instruments, held in Leiden, and was awarded a medal for his work.

Zagorsk Optical and Mechanical Plant (ZOMZ)
Zagorsky Optiko-Mekhanichesky Zavod (ZOMZ) (Загорский оптико-механический завод). The Zagorsk Optical and Mechanical Plant is a Russian manufacturer of optical components and optical instruments. They enjoy a good reputation for optical instruments such as camera lenses, microscopes, monoculars and binoculars. They make (made) binoculars for the **Kronos** brand, and binoculars and cameras using the **Zenith** brand (although their use of the *Zenith* brand-name is not exclusive). The mark *ЗОМЗ* indicates ZOMZ. For Russian and Soviet makers' marks, see Appendix V.

Zambra, Joseph
Joseph Warren Zambra, (1822-1897) [18] [247] [276 #1755] [379] [438], was a barometer maker and meteorological instrument maker of London, England. Addresses: 11 Brook St. (1843-1846); 9 Manchester St., Argyle Square (1848-1851). He was born in London, and his father, Joseph Cesare Zambra, was a barometer maker and travelling barometer salesman. Joseph served an apprenticeship as a glassblower, and attended the London Mechanics' Institute for three periods between 1843 and 1851. He worked in partnership with **John Tagliabue**, trading as **Tagliabue & Zambra**, philosophical instrument makers, until the partnership was dissolved in 1850, [35 #21084 p1005]. In 1850 he began to work in partnership with **Henry Negretti**, trading as **Negretti & Zambra**. He continued to work in this partnership until his death in 1897. He was succeeded as a partner in the company by his son M.W. Zambra, and the company continued to operate until 1981.

Zapp, Walter
Walter Zapp, (1905-2003), was a Latvian born inventor and engineer. He invented and designed the *VEF Minox Riga* sub-miniature camera that was made from 1937 until 1944 by the radio manufacturer Valsts Electrotechniska Fabrika in Riga, Latvia. After WWII Walter Zapp took his designs to West Germany, where they were manufactured by **Minox** GmbH, beginning in 1948, and were the basis of a successful range of cameras. He left the company in 1950, but was hired back as a consultant engineer in 1988.

Zeiss, Carl
Carl Zeiss, (1816-1888) [24] [29] [33] [44], was a German optician commonly known for the company he founded, Carl Zeiss AG.

Carl Zeiss founded his company in 1846 in Jena, Germany, as a precision mechanics and optics workshop. He began to produce simple microscopes in 1847, and compound microscopes in 1857. In 1866 Zeiss hired **Ernst Abbe** as research director of the Zeiss optical works. Based on Abbe's contribution, from 1872, the optics of Zeiss microscopes were built on the basis of scientific calculations, resulting in considerably improved optical properties, and giving Zeiss a technological lead. In 1876 Zeiss appointed Abbe to be his partner and successor. Zeiss and Abbe began to collaborate with **Otto Schott** in the production of improved optical glass, resulting in the establishment in 1884 of the Schott glassworks. In 1888, upon the death of Zeiss, Abbe founded the Carl-Zeiss-Stiftung (Carl Zeiss Foundation), located in Heidenheim an der Brenz and Jena, Germany, to be the sole shareholder of the two companies Carl Zeiss AG and **Schott AG**.

Among their expanding range of optical instruments, Zeiss made microscopes, refractometers, spectrometers, industrial and metallurgical measuring instruments, photographic lenses, binoculars, periscopes, terrestrial and astronomical telescopes, eyeglasses, ophthalmological testing instruments, and geodetic instruments.

In 1893, the first representative of Carl Zeiss AG outside Germany opened in London, England. The first production site outside Jena opened in Vienna, Austria in 1906.

In 1908, Zeiss formed a commercial agreement with **Bausch & Lomb** and, shortly after this, the Bausch & Lomb, Zeiss, and **Saegmuller** subsidiaries merged to form the Bausch & Lomb Optical Company, which became known as the **Triple Alliance**, and lasted until WWI.

In 1910 Zeiss acquired a stake in the company **AGC**, and in 1931 they took ownership.

WWI was a setback to the company's internationalization endeavours. Carl Zeiss (London) began manufacturing binoculars in the UK in 1909 using parts imported from Jena. It was, however, closed down in 1917 under the Trading with the Enemy Act. The factory at Mill Hill was sold, and **Ross Ltd.** acquired all their assets.

In 1921, Zeiss founded the Nederlandse Instrumenten Compagnie, **Nedinsco**, which was acquired by the entrepreneur G. Beusker in 1953.

In 1926, **Zeiss Ikon** was formed as a part of the Carl Zeiss Foundation.

In 1927, Zeiss became the majority shareholder in Emil Busch AG Optische Industrie (see **Emil Busch**).

Between WWI and WWII, Zeiss increasingly produced military optical ordnance for use with aircraft, submarines, tanks, and guns. During WWII, aided by the use of forced labour, Carl Zeiss AG manufactured binoculars, rangefinders, and other optical munitions for the war effort. These were marked with the **ordnance codes** blc, emq, lmq or rln. At the end of WWII, with the partitioning of Germany, the company was split into two:

In East Germany, VEB Carl Zeiss Jena operated as a separate nationalised company. In 1966, VEB Rathenower Optische Werke (see Emil Busch) became a part of VEB Carl Zeiss Jena. In 1985, VEB Pentacon (see **Pentacon**) became a part of VEB Carl Zeiss Jena.

In West Germany the company established facilities at Oberkochen as Opton Optische Werke Oberkochen GmbH, which in 1947 was renamed Zeiss-Opton Optische Werke Oberkochen GmbH. The name Carl Zeiss was registered in Heidenheim in 1951, and from 1953 the firm traded as Carl Zeiss.

In 1953 the West German Emil Busch company was renamed Emil Busch GmbH, Göttingen, which later became part of Carl Zeiss AG.

By 1954 Zeiss was a majority shareholder in M. Hensoldt & Soehne AG (see **Moritz Hensoldt**), and in 1968 Zeiss completed its acquisition of Hensoldt, which became a fully owned subsidiary. Zeiss production of sports optics is still located in Hensoldt.

In 1957 **R. Winkel** in Göttingen became Winkel-Zeiss, part of Carl Zeiss. Zeiss had been the majority shareholder in Winkel since about 1911. Zeiss production of microscopes and related products is still located in Göttingen.

Following a prolonged trademark dispute between the East German and West Germany companies, from 1971 they used the names Carl Zeiss Jena and Carl Zeiss Oberkochen, with an agreement on which global markets each could target.

After the reunification of Germany in 1990, the optical division of VEB Carl Zeiss Jena was privatised as Carl Zeiss Jena GmbH. This was acquired by Carl Zeiss Oberkochen. VEB Rathenower Optische Werke was separated from Zeiss to become **Rathenower Optische Werke GmbH**, and the remainder of VEB Carl Zeiss Jena continued as **Jenoptik** GmbH.

During the 1990s, rationalisation, workforce reduction, and elimination of competition between Carl Zeiss Oberkochen and Carl Zeiss Jena resulted in restructuring and reorganization of the two business as part of Carl Zeiss AG. With this, the company focussed on four key markets: semiconductors/microelectronics, life sciences, eye care, and industrial metrology.

In 1999 **Spectra Precision** entered a joint venture with Carl Zeiss AG to manage the Zeiss surveying instrument business in Jena. In 2001 **Trimble**, having acquired Spectra Precision, acquired the remaining interest in the Zeiss surveying instrument business, which continued as Trimble Jena GmbH, and retained its location at the Zeiss plant in Jena.

For information about Zeiss binocular serial numbers, see [**86**], and for information about Zeiss binocular models, see [**87**]. For information about Zeiss microscope models and serial numbers, see [**320**].

A proliferation of fake Carl Zeiss binoculars will be found on auction sites, many of which were made in Japan. For information on some of these, see [**255**].

Zeiss Ikon

Zeiss Ikon was a German company, formed as part of the Carl Zeiss Foundation in 1926, with finance from **Zeiss**, by the merger of the four camera makers **Contessa-Nettel**, **Ernemann**, **Goerz** and **ICA**. In 1927, they acquired **AG Hahn für Optik und Mechanik**. They introduced the **Contax** brand-name in 1932 with a rangefinder camera. During WWII Zeiss Ikon manufactured optical munitions for the war effort, and marked their instruments with the **ordnance codes** dpv, dpw, or dpx. At the end of the war, with the partitioning of Germany, Zeiss Ikon was divided into two separate companies:

In West Germany, Zeiss Ikon AG Stuttgart continued camera production. The Zeiss Ikon brand-name **Pentax** was derived from the words *pentaprism* and *Contax*, and was sold to **Asahi Optical Company** in 1957. In 1965 Zeiss Ikon AG Stuttgart merged with **Voigtländer**. Zeiss Ikon ceased camera production in 1972, with the *Contax* brand-name sold to **Yashica** in 1973. In the early C21st, Zeiss revived the Zeiss Ikon name with cameras manufactured by **Cosina**. Production continued until 2012.

In East Germany, VEB Zeiss Ikon Dresden was a nationalised company. Some of the camera production tooling was removed after the war to be used by the **Arsenal Factory** in Kiev. However, camera production soon resumed. In 1958 the company name was changed to VEB Kinowerke Dresden, and in 1959 they merged with several other companies, including VEB Kamera-Werke Niedersedlitz (see **Kamera Werkstätten**), VEB Welta Kamera Werke (see **Welta**), and VEB Kamera und Kinowerke Dresden. In 1964 the newly formed company became VEB Pentacon (see **Pentacon**).

Zenit
Zenit is a Russian brand of cameras, binoculars and microscopes manufactured by **Krasnogorskij Mechanicheskij Zavod (KMZ)**. Some Zenit products have been referred to as *Zenith* in marketing and sales literature, etc. However, see **Zenith**.

Zenith
Zenith is a brand-name that arises in at least four different contexts:

(1) Zenith is a Japanese brand of binoculars and monoculars, made from the 1950s to the late 1970s. They sometimes bear a **JB code** to identify the manufacturer, and an oval quality control sticker. Some Zenith binoculars are branded *Super Zenith*.

(2) Some **Zenit** products have been referred to as *Zenith* in marketing and sales literature.

(3) Zenith is a brand-name that has been used by **Zagorsky Optiko-Mekhanichesky Zavod (ZOMZ)** for some of their microscope and camera products.

(4) Zenith is a brand of microscopes available from various UK retailers, and manufactured in China.

Zenith Sector
A zenith sector is a variant of the zenith telescope that is hinged at the top so that it can move in an arc, aligned to the meridian. Thus, it may be used to measure distances from the zenith (measured as an angle).

Zenith Telescope
A zenith telescope is one that is designed to point directly to the zenith, and is used for accurate measurements of star positions, astronomical latitude, and the orientation of the poles. A variant of this is the zenith sector.

Zenobia Kōgaku
Zenobia, was a brand-name used by **Daiichi Kōgaku** for some of its cameras. Having failed to adapt quickly enough to the shift in demand from folding cameras to 35mm cameras, Daiichi Kōgaku closed in 1955. In 1956 the company re-emerged, this time under the name Zenobia Kōgaku, and continued in business until 1958.

Zenone & Butti
Zenone & Butti, (fl.1823) [18] [263], were carvers and guilders, and looking glass makers of 5 Calton St., Edinburgh, Scotland. The firm was a short-lived partnership between John Zenone and Louis Joseph Butti. They retailed instruments bearing their name, but almost certainly did not manufacture them. Instruments bearing their name include: 3-draw telescope with 1½-inch objective, brass with wooden barrel. Also, by Zenone: stick barometer; wheel barometers; telescope, and by Butti: wheel barometer.

Zentmayer, Joseph
Joseph Zentmayer, (1826-1888) [201], was a German born scientific instrument maker. He served his apprenticeship before working for instrument makers in Karlsruhe, Frankfurt, and Munich, and for **A. Repsold & Söhne** in Hamburg. He emigrated to the USA in 1948, where he worked for instrument makers in Cleveland, Baltimore and Washington DC before going to Philadelphia in 1851 to work for **William J. Young**, makers of surveying instruments.

In 1853 Zentmayer started his own business in Philadelphia, and in 1855 he began making microscopes, for which he gained a significant reputation. He produced a variety of microscope models. Upon his death in 1888 his sons continued to make microscopes, using the name Joseph Zentmayer until at least 1895.

Zimmer, J.G.
J.G. Zimmer, (fl.1741-1760) [53] [322], was an optical, mathematical, and philosophical instrument maker of Dresden, Germany. Along with **J.S. Merklein**, he served as mathematical and optical instrument maker to Count Hans Löser at Reinharz Castle in Wittenberg, where the two instrument makers worked together as **Merklein und Zimmer**. Instruments bearing his name include: pedometer; thermometers; sundial.

Zoetrope
A Zoetrope [45] is an early device for displaying moving images. It consists of a vertical drum with vertical slots cut at regular intervals around it. The inner surface of the drum has a sequence of still images printed onto it. The user sets the drum spinning, and the slots form a rudimentary shutter so that, when viewed from the side, one image at a time is seen as each slot passes the user's eyeline. An improved version of the Zoetrope is the Praxinoscope.

ZOMZ
See **Zagorsk Optical and Mechanical Plant**.

Zoogyriscope, Zoogyroscope
An earlier name used by **Eadweard Muybridge** for his invention, the Zoöpraxiscope.

Zoom Lens
See lens.

Zoöpraxiscope (Zoogyroscope)
The Zoöpraxiscope, [45] [159], was a lantern projector used for projecting enlarged motion picture images onto a screen from a sequence of photographs printed onto a rotating glass disk. It is, in essence, a projecting version of the Phénakistiscope. The Zoöpraxiscope was devised by **Eadweard Muybridge**, and used by him to illustrate his lectures on animal locomotion. It was an important precursor to the moving image projector that became the foundation of the early film industry. Prior to calling his invention the Zoöpraxiscope, Muybridge named it the Zoogyriscope.

Zorki
Zorki was a brand of rangefinder cameras produced after WWII by **Krasnogorskij Mechanicheskij Zavod (KMZ)**. They were initially produced for **Kharkov Machine-Building Plant (FED)**, using an *FED-Zorki* brand-name. After FED had recovered its production capability following the war, KMZ produced a number of camera models under the brand-name *Zorki*, almost all of which were rangefinders. Production of Zorki cameras continued until 1978.

Zschokke, Paul
Paul Zschokke, (1853-1932) [161] [219] [374] [422], was an optical instrument maker, born in Switzerland. He served his apprenticeship with **Steinheil & Söhne** in Germany, and rose to become a director of the company. In 1897 he became a partner in **Reinfelder & Hertel**.

In 1903 Paul Zschokke left Reinfelder & Hertel, and purchased **G.&S. Merz**, which was at that time operating as the two companies, **Optische Glasfabrik in Benediktbeuern, Sigmund Merz München**, and **Optische Werkstätte von Jakob Merz**. He continued the business, trading as Optisches Institut G.&S. Merz vormals Utzschneider und Fraunhofer in München. He used this name with permission from **Sigmund Merz**.

In 1907, Paul Zschokke moved the company to new premises in the district of Pasing in Munich, seeking cleaner air and a quieter location. He continued to sell his remaining stock of large Merz refractors, but concentrated his production on small telescopes and binoculars.

From 1908-1912 the firm traded as G.&S. Merz vormals Utzschneider und Fraunhofer in Pasing bei München. He signed his instruments *G.&S. Merz in München*, or *G.&S. Merz vorm. Utzschneider u. Fraunhofer München*.

From 1912, the company traded as G.&S. Merz in Pasing. In 1914, he relocated the firm to his home address, and in 1918, **Georg Tremel** began to work for him.

The company ceased trading in 1932 due to bankruptcy. The company, it's machines and tooling, were taken over by Georg Tremel in 1933, together with August Lösch, another employee of Merz.

Glossary

Atwood machine: An educational machine designed to demonstrate the laws of motion under constant acceleration. It was invented by the English mathematician, George Atwood, in 1784.

Azimuth: See Altitude-Azimuth (Alt-Az) Telescope Mount for a definition of the term.

Brevet or *breveté*: (French) patent.

Brevetti: (Italian) patent.

Bté. SGDG: (French) *Breveté Sans Garantie du Gouvernement* Patented Without State Guarantee.

B^{tee} Fr. & Etr., (French) *Brevetée France & Étranger* Patented in France and Overseas.

Cable: A unit of length equal to one tenth of a nautical mile. Approximately 185.2 meters.

Chandler: A retailer who sells merchandise for a specific trade. A ship chandler is a retailer who sells equipment, supplies, and provisions for ships.

Chondrometer: A weighing device designed to measure the bulk density of grain.

Coherent: (Relating to light) having the same phase, or a fixed phase difference.

Cometarium: A mechanical instrument which demonstrates the elliptical path of a comet around a heavenly body.

Concavo-convex: (Referring to a lens) concave on one side and convex on the other.

Declination: See Equatorial Mount for definition.

Déposé: (French) registered.

Dienstglas: (German) service glass.

Ehmals: (German) formally or previously.

Ellipsoidal: (Adjective: ellipsoidal; Noun: ellipsoid). A mirror or other optical surface that is described as ellipsoidal is the central part of a surface that is described by rotating an ellipse about its major axis. This is a special case of the more general ellipsoidal surface shape which is symmetrical about three axes, and its plane sections on those axes are ellipses or circles.

Endless screw: A worm screw meshing with, and acting at a tangent to the edge of a gear, providing endless motion. The worm screw is turned to accurately effect small rotations of the gear.

Erect: An erect image is one that is the right way up.

Figure: Relating to a lens or mirror. (n) The figure of a lens or mirror is its shape, so designed to yield the requisite optical properties. (v) To figure a lens or mirror is to work its optical surface(s) into the required shape.

Fire-damp: A flammable gas, mostly methane, found in coal mines.

Gebrüder: (German) brothers.

Geodetic: Of or relating to geodesy, which is the science of determining the exact location of geographical points. The term also refers to determining the exact shape and size of the earth.

Gyro-compass: A specially mounted, continuously driven gyroscope that is used as a compass. It is mounted so that its spin axis is parallel to the Earth's surface, and gimballed so that it may rotate in an axis normal to the Earth's surface. The combined effects of the gyro's rotation and the rotation of the Earth result in the gyro pointing toward the geographical north.

Hodometer: A surveyor's instrument used for measuring distance travelled.

Holosteric: Wholly constructed of solid materials, without any liquids.

Hyperboloidal: (Adjective: hyperboloidal; Noun: hyperboloid). A mirror or any other surface that is hyperboloidal is one whose shape is described by rotating a hyperbolic curve about its axis of symmetry.

Ion: An ion is an atom which is electrically charged as a result of the loss or gain of one or more electrons. An atom or group of atoms thus electrically charged is referred to as *ionised*.

Journeyman: A person who is qualified in their trade, and works for an employer.

Laser: A laser is a device that uses the physical properties of *stimulated emission of radiation* to produce a narrow beam of monochromatic, coherent light. The word is an acronym, which stands for *light amplification by stimulated emission of radiation*.

Lenticular: Of or relating to a lens or lenses.

Limelight: Limelight is a form of intense illumination. It is created by burning a mixture of oxygen and hydrogen (oxyhydrogen). The flame is directed at a cylinder of calcium oxide (quicklime), which is thus heated to incandescence.

Longitudinal Wave: A wave which propagates in the same direction as the direction of displacement of the medium in which it is transmitted. For example, sound. Compare with *transverse wave*.

Magnetometer: An instrument used to measure the intensity or direction of a magnetic field, such as the Earth's magnetic field.

Marque: (French) brand.

Marque Déposé: (French) trademark.

Marque de Fabrique: (French) trademark.

Maser: A maser is a device that uses the physical properties of *stimulated emission of radiation* to produce a narrow beam of monochromatic, coherent microwaves. The word is an acronym, which stands for *microwave amplification by stimulated emission of radiation*.

Mathematical Instrument: Mathematical instruments are those used in the study or practice of mathematics, and largely consist of instruments whose main purpose is to measure or compute lengths, areas and angles. However, in common usage, the terms mathematical, scientific, and philosophical instruments are less tightly defined, and there is much overlap in usage.

Meridian: A meridian is an imaginary line joining the north and south poles, crossing the equator at an angle of 90°. Meridian lines are designated by degrees of longitude, beginning with 0° at Greenwich.

Microscopist: A person who uses a microscope.

Modele Déposé: (French) registered design.

Monochromatic: (Relating to light) of one single wavelength, or a very small range of wavelengths.

Mural: A mural instrument is one which is mounted on a wall, such as a mural circle or a mural quadrant.

Natural Philosophy: A name historically used for the science now known as physics.

Nonius: A nonius consists of a series of concentric rings marked or engraved on a quadrant or other instrument, embellished with additional marks which enable accurate reading of an angle. The nonius fell into disuse following the invention of the vernier.

Optician: In modern usage the term optician is used to indicate a person whose role is to test people's eyesight, and to provide corrective glasses or contact lenses. Historically, however, the term also frequently indicated a maker of optical instruments.

Optoelectronics: Of or relating to optical components that, when stimulated by light in the visible or infrared spectrum, produce an electrical output, or conversely, when stimulated by an electrical input, produce light in the visible or infrared spectrum.

Optronics: Another word for optoelectronics.

Panchromatic: Sensitive to all colours of the visible spectrum.

Paraboloidal: (Adjective: paraboloidal; Noun: paraboloid). A mirror or any other optical surface that is described as paraboloidal is generally a circular paraboloid. This is one whose shape is described by rotating a parabolic curve about its axis of symmetry. Since a circle is a special case of an ellipse, a circular paraboloidal surface is a special case of an elliptical paraboloidal surface.

Petrography: A branch of geology concerning the study and classification of rocks, often using a microscope.

Petrology: A branch of geology concerning the origin, structure, and composition of rocks.

Phantasmagoria: An exhibition of optical illusions, usually projected using one or more magic lanterns, often to display preternatural phenomena. Frequently achieved using back projection. The first such exhibition was in London, England in 1801.

Phase: (Relating to an electromagnetic wave such as light) The phase of a periodic wave is a particular stage or point in the cycle in relation to a reference position or time, or in relation to another such wave. Two waves are *in phase* if the crests of the waves coincide. They are *out of phase* if the crests do not coincide. The phase difference between two sinusoidal waves is enumerated as an angle.

Philosophical Instrument: A philosophical instrument is one whose purpose is to investigate aspects of nature that cannot be measured directly, such as magnetism, pressure, gravity, and temperature. Hence, this class of instruments includes compasses, barometers, gravity pendulums, thermometers, and others. However, in common usage, the terms mathematical, scientific, and philosophical instruments are less tightly defined, and there is much overlap in usage.

Plano-convex: (Relating to a lens) plane (flat) on one side and convex on the other.

Prolate: Having a polar diameter of greater length than the equatorial diameter.

Reticule: An arrangement of fine lines or wires, frequently in the shape of a cross or grid, set in the focal plane of an optical instrument. This is usually used to assist in positioning or aiming the instrument, or to measure the size, position, or movement of an observed object.

Right Ascension: See Equatorial Mount for definition.

Scientific Instrument: Scientific instruments are those whose main purpose is to investigate or demonstrate the principles of physics, generally in scientific investigation or education. However, in common usage, the terms mathematical, scientific, and philosophical instruments are less tightly defined, and there is much overlap in usage.

Sic: sic is Latin for *so* or *thus*. It is used to indicate that wording which may appear questionable or incorrect is written as intended.

Sidereal: Of or relating to the stars. A sidereal clock measures the time of rotation of the Earth in relation to the stars. Thus, one sidereal day is the time between successive meridian crossings of a star. A sidereal day is 23 hours 56 minutes, compared with a solar day, measured between meridian crossings of the Sun, which is 24 hours.

Sympiesometer: (1) A type of barometer in which has, above the liquid, a confined gas instead of vacuum. (2) An instrument used to measure the pressure or velocity of a current of liquid.

Société: French for company.

Spherical: When a lens or mirror is described as spherical this means that the surface has a curve of constant radius, i.e. it is spherically curved. It does not mean that the lens or mirror has the form of a full sphere.

Tellurian: A working model for showing the movement of the earth around the sun.

Thunder house: An C18th gadget designed to demonstrate the effects of static electricity, particularly in relation to the action of a lightning conductor.

Transverse Wave: A wave which propagates in a direction perpendicular to the direction of displacement of the medium in which it is transmitted. For example, ripples on water; electromagnetic waves; light. Compare with *longitudinal wave*.

Vector: A variable quantity, such as velocity, that has both magnitude and direction.

Vernier: A vernier, or vernier scale, is a short, graduated scale that slides alongside an instrument's main graduated scale. It is used for fractional readings of the divisions on the main scale.

Vormals: (German) formally.

Vorm: (German) Abbreviation for vormals.

Zenith: The zenith is the point vertically above the observer. It is at an altitude of 90° above the horizon.

Abbreviations

It is assumed that the reader is familiar with common abbreviations for: Nations, such as UK and USA, etc.; States in the USA, such as NY, PA, etc.; Metric and imperial measures such as mm and in, etc; Eras, such as BCE, AD.

abs: Acrylonitrile butadiene styrene, (a thermoplastic polymer).
AG: Aktiengesellschaft. (German) public limited company.
Alt-Az: Altitude-Azimuth. See Altitude-Azimuth (Alt-Az) Telescope Mount.
APO: apochromat or apochromatic.
b.: (With a date) born; date of birth.
BAA: British Astronomical Association.
bap.: Bapitsed (baptism date).
B.O.T.: Board of Trade.
c: Lower-case c used in conjunction with a date stands for the Latin *circa*, and means *approximately*. It is used when exact dates are not known, but approximate dates are.
C19th: Nineteenth century. Similarly, other centuries.
CA: Chromatic Aberration.
CCD: Charge-coupled device - an imaging sensor technology.
CCTV: Closed-circuit television.
CEO: Chief Executive Officer (of a company).
Ch.: (in a reference) chapter.
CMO: Common Main Objective (in reference to a microscope).
CMOS: Complementary metal-oxide-semiconductor - an imaging sensor technology.
d.: (With a date) died; date of death.
EO: Electro-optical.
est.: (with a date) established: date of establishment.
f/n: (where n is a number). Focal ratio. For example, f/5 indicates an aperture of 1/5 of the focal length.
fl: Flourished. This is used to show dates when a person's career flourished. It indicates the date or date range when the person is known to have been working, and is not necessarily indicative of the full span of their career.
f/l: Focal length.
fr: (with a date) freed of a guild.
FRAS: Fellow of the Royal Astronomical Society.
Gebr: (German) brothers. Abbreviation for *Gebrüder*.
GmbH: Gesellschaft mit beschränkter Haftung. (German) limited liability company. *Ges.m.b.H.* or *GesmbH* is the abbreviation used in Austria.
GPS: Global Positioning System.
HM: Her (or His) Majesty.
HMS: Her (or His) Majesty's Ship.
HRH: Her (or His) Royal Highness.
HTML: Hypertext markup language.
HUD: Head-up display.
IR: Infrared.
Jr.: Junior.
KG: Kommanditgesellschaft. (German) limited partnership business.
KK: Kabushikikaisha (Kabushiki kaisha or kabushiki gaisha). (Japanese) share company. It is sometimes translated as corporation, and sometimes as Co. Ltd.
LED: Light emitting diode.
MG: Ministère de la Guerre, which was the French Ministry of War until shortly after WWII.
M.O.T.: Ministry of Transport.
nr: Near.
oem: Original equipment manufacturer. The term is used to refer to a manufacturer who makes and supplies items or parts to the company who will sell the final product under their own name.

OHG: (German) company that operates as a general partnership.
OU: Open University.
p: Page.
pp: Pages.
PSS: Photographic Society of Scotland.
RAF: Royal Air Force.
RAS: Royal Astronomical Society.
Ret.: (with a date) retired.
RMS: Royal Microscopical Society.
SA: Société anonyme, a kind of French corporation.
S.G.D.G.: Sans Garantie Du Gouvernement. (French) without government guarantee. The marking *Breveté S.G.D.G.* means *patent without Government guarantee.*
SLR: Single lens reflex.
SpA: Società per azioni. (Italian) public limited company.
TLR: Twin lens reflex.
TPI: Threads per inch.
UC: University of California.
VEB: Volkseigener Betrieb. (German) This was a designation used for a publicly owned enterprise in East Germany following the end of WWII.
VP: Vice-President (of a company).
WC: (In a London address) West Central.
WWI: World war 1.
WWII: World war 2.

Appendix I: Barr & Stroud Binocular Models

The information about Barr and Stroud binocular models shown here is based on information in References [**19**], [**26**], [**64**], [**65**], [**66**], [**233**], and information from inspection of extant binoculars. Reliable information has not thus far been available, and hence is not included, for the following models: CA1, CA5, CA6, CF11, CF13, CF14, CF16, CF17, CF21, CF22, CF32-CF35, CF48, CF51, CF52, CF54, CF55, CF57-CF59.

Galilean Binoculars

Barr & Stroud made Galilean binoculars, models CA1 to CA7. Those for which information is available are detailed below.

The photo to the left shows typical markings on Barr and Stroud Galilean binoculars. This shows the letters B and S on the upper bar and the inscription C.A. 4 and 63.5mm on the lower bar. These are CA4 binoculars with an interpupillary distance (IPD) of 63.5mm.

The *CA2* (3½x55.6), *CA3* (3½x50.8), and *CA4* (3½x44.5), were introduced in 1919, and have a field of view of 112 yards at 1000 yards. The CA2 has an IPD of 65mm, and the CA3 and CA4 were each available in two models with IPD of 62mm or 63mm. The CA2 was discontinued in 1929/30.

The *CA7* was introduced in 1929/30, and is 4x55.6mm with a field of view of 104 yards at 1000 yards. This model differs from the CA2, 3 and 4 in that it is largely made of Bakelite. It was available with or without eye-shields, and has no lens hoods. The interpupillary distance is not marked on the binocular, but is measured as 65mm. Unlike the earlier Galilean binoculars, the maker's name and the model number are moulded in relief on the upper bridge.

Porro Prism Binoculars

The following are Barr & Stroud hand-held Porro prism binocular models that were produced prior to the company's acquisition by the Thales Group in 2000. During this period Barr and Stroud did not produce any roof prism binoculars:

6x Models

The CF1, CF2, CF3, and CF4 were Porro-I prismatic binoculars introduced in 1919/20 and discontinued in 1929/30. They have a field of view of 138 yards at 1000 yards

The *CF1*, 6x0.9 inch (23mm), was the central focus version of the independent focusing *CF3* model, and was referred to on internal documentation as standard army pattern. This was Barr and Stroud's first prismatic binocular model. The CF1 bears the early Barr and Stroud logo (shown) comprising a B intertwined with an S, enclosed in a circle.

The *CF2*, 6x1.2 inch (30.5mm), was the central focus version of the independent focusing *CF4* model, and was referred to in internal documentation as standard navy pattern. The CF4 also bears the early Barr and Stroud logo (shown) comprising a B intertwined with an S, enclosed in a circle. In 1928-29 the CF2 was replaced by the *CF10*, and the CF4 was replaced by the *CF20*. The specifications of these new models were the same as shown above for the CF2 and CF4.

At least three other 6x models were produced, all for military use. The *CF12* was an independent focusing Porro-II 6x50 model with a field of view of 121 yards at 1000 yards, and was designed in 1919 as for night use by the Royal Navy. The *CF45s* was an independent focusing Porro-I 6x30.5 model built to the same specification as the British Army Binocular No. 2 Mk I and Mk II. The CF45s had coated optics and a graticule on the right-hand side.

The *GK5* was an independent focussing Porro-II 6x42 deck-mounted *Director* binocular, meaning it was used primarily to acquire bearings to targets. It was a heavy, rugged binocular, used by the British Admiralty, during and after WWII, as Admiralty Pattern 1947a. It had selectable yellow, grey, and dark green filters, operated by a knob on each prism housing.

Two CF5 Models

The *CF5* model designation was used for two distinctly different models of Porro-I prismatic binoculars. The *CF5* model, 8x0.9 inch (23mm), introduced in 1919 was the central focus version of the independent focusing *CF7*, introduced in 1920. The *CF5* model, 6x24mm, was introduced in 1929/30, and was the central focus version of the independent focusing *CF19*, with a field of view of 138 yards at 1000 yards. The later CF5 and the CF19 were marked as discontinued in a *Leaflet B.60* in 1962.

7x Models

All documented B&S 7x models were of Porro-II design, and were either 7x42, 7x50 or 7x50.8.

The story of the iconic *CF41* 7x50 binocular, widely used by the British Navy, begins with the *CF15*. This was an independent focus 7x50 model with a field of view of

121 yards at 1000 yards. It was adopted for use by the British Admiralty in 1931 as Admiralty Pattern 1900 (AP1900) for night use. The *CF23* was a modified CF15 with coloured filters. The *CF30*, replaced the CF15 as AP1900 in late-1932. This was 7x50.8 with the same field of view as the CF15, and was the independent focusing version of the *CF31*. The *CF40*, also 7x50.8, improved on the CF30, introducing colour filter glasses behind the field lens. This was adopted as AP1900A in 1934. This was not found to be the optimum position for the colour filters, and in 1935 the *CF41*, 7x50, was adopted as AP1900A with colour filters on the objective side of the prisms. Earlier models have click-stops on the oculars, and some have extendable lens-hoods. At some point in production desiccator vents were added to the CF41. A hand-held desiccator pump was attached to the desiccator vents to replace the air in the binoculars with dry air from the pump. The *CF42* was a CF41 fitted with a graticule on the right-hand side. This was adopted as AP1907A by mid-1936, and a fully sealed version was given store code AC2032.

At least two more 7x50 models were produced, both Porro-II. The *CF44* was introduced around 1946 or 1947 with a new focusing mechanism. The *CF60* was Barr & Stroud's last hand-held binocular design, and had a field of view of 121 yards at 1000 yards.

The *CF25*, *CF28* and *CF29* are all 7x42 models with a field of view of 127 yards at 1000 yards. The CF25 was an individual focusing Admiralty Pattern 1949 (AP1949), and at least some of these were fitted with extendable lens-hoods. The CF28 was the civilian version of the CF25, and the CF29 was the central focusing version.

8x Models

The CF6, CF8, CF18, CF24, CF27 and CF38 models are all Porro-I 8x1.2 inch (30.5mm) binoculars.

The *CF6*, is the central focusing version of the independent focusing *CF8*, with a field of view of 130 yards at 1000 yards. Production of the CF6 and CF8 began in 1920. The *CF18* is the central focusing version of the *CF38*, with a field of view of 127 yards at 1000 yards; production began on or before 1934. The *CF24* is the central focusing version of the *CF27*. This is a "wide angle" binocular with a field of view of 150 yards at 1000 yards.

The *CF50* was a central focusing 8x30 binocular with a field of view of 8°. This was a close-range model introduced in the mid-1960s, with a close focus of 24 inches (61mm). Achieving this close focus necessitated a complex design, resulting in a very expensive binocular with an unusual appearance.

10x Models

The CF26, CF36, CF37 and CF47 were Porro-II 10x50 models. The *CF26*, with a field of view of 87 yards at 1000 yards, was the independent focusing version of the *CF36*. The CF26 was replaced in 1947 by the *CF47*, with a field of view of 112 yards at 1000 yards. The *CF37* was the central focusing version of the CF47.

The *CF43* is the central focusing version of the independent focusing *CF53*. They are 10x42 Porro-I model with a field of view of 121 yards at 1000 yards. These two models were introduced in 1947 or 1948.

12x Models

The *CF39* is the central focusing version of the independent focusing *CF49*, and is a 12x50 Porro-II model with a field of view of 104 yards at 1000 yards. These two models were introduced in 1947 or 1948.

15x Models

The *CF46* is the central focusing version of the independent focusing *CF56*, and is a 15x60 Porro-II model with a field of view of 79 yards at 1000 yards. These two models were introduced in 1949.

Opera Glasses

The *CF9*, 4x0.6 inch (15mm), is a central focusing Porro-I opera glass, introduced in 1920. The shape of the prism housings is unusual for Barr & Stroud, and only a small number were ever made.

Serial Numbers

Barr & Stroud binoculars were allocated serial numbers in order of production, and can be dated using the guide to serial numbers with approximate dates given in [**125**].

Binoculars by OVL

For information on Barr & Stroud binocular models since the brand-name was taken over by OVL in 1998, please refer to **Optical Vision Ltd.**

Appendix II: Eyepieces

Overview
An eyepiece is the lens or combination of lenses, next to the observer's eye, used in microscopes, telescopes, binoculars, and other optical instruments to examine the image formed at the focus of the objective lens. Some eyepieces are interchangeable, and some are fixed, such as in a conventional pair of binoculars. Telescopes and microscopes, however, commonly have interchangeable eyepieces.

Telescope Eyepieces
Telescope eyepiece barrel sizes generally conform to one of the following standards:

0.965" (24.5mm) This is an obsolete barrel size found on older telescopes. It was the standard eyepiece size of Carl Zeiss, Jena from 1897 until 1995, and the standard was used by Japanese companies in the 1930s and after WWII.

1.25" (31.75mm) This is the most common modern consumer telescope eyepiece barrel size. However, with long focal lengths the field of view becomes limited.

Some very old telescopes use a threaded barrel 1 1/4" in diameter with 16 threads to the inch; such eyepieces are referred to as *RAS thread* eyepieces, as the standard which they follow was established by the Royal Astronomical Society in Britain.

2" (50.8mm) This is also a common modern barrel size. This allows for a wider field of view with longer focal length.

3" (76.2mm) Used in observatories for very large focal length and field of view up to 120°.

Other barrel sizes exist, but are less common.

Microscope Eyepieces
The two main kinds of eyepiece found on older microscopes are the Huygenian and the Ramsden [142]. These are improved by the Kellner design which utilises a cemented achromatic doublet for the eye-lens. More complex eyepieces may be of the Periplan design. Microscope eyepieces may be designed to optically match the microscope objective, and hence be less interchangeable than telescope eyepieces. This will depend on the correction for field curvature and/or chromatic aberration that is built into the objective, necessitating a matching compensating eyepiece. Most microscope eyepieces have an integral field stop. Modern microscope eyepiece barrel sizes are commonly 23.2mm, 30mm or 30.5mm.

A compensating eyepiece is one which is designed to correct for aberrations in the objective of an optical instrument such as a microscope. They primarily correct for chromatic difference of magnification and field curvature, and sometimes other aberrations. A compensating eyepiece is normally specifically designed to match a particular objective.

Binocular Eyepieces
The most common eyepiece design for older binoculars is the three-element Kellner design. The Reversed Kellner is becoming more common, with the advantages of a slightly larger field of view, longer eye-relief, and working better with the short focal ratios of binocular objectives. Wide-field binoculars usually use modifications of the Erfle eyepiece, which has a relatively wide field of view. However, with these modified Erfle designs, eye-relief tends to suffer when the field of view is greater than about 60°.

Eyepiece Designs
There are many designs of eyepiece, [7] [22] [39] [142], and new or improved designs are frequently produced. Eyepiece designs vary in terms of their field of view (or apparent field of view), eye-relief, and the degree to which they suffer from the various kinds of optical aberrations and distortions. It is not possible to list every eyepiece design here. However, the following are some of the more commonly encountered designs: (Note: eye relief is given as a multiple of the eyepiece focal length F_e. Both eye relief and field of view are dependent on the detailed design of a given eyepiece and the optical glass used, so the figures given below are only guidelines).

Bertele
Ludwig Jakob Bertele designed the Bertele eyepiece for **Steinheil**. It is a four-element design for use in military binoculars. The Bertele eyepiece has a high degree of orthoscopy.

Field 70°, eye-relief $0.8F_e$

Erfle
A wide-angle lens designed in 1917 by **Heinrich Erfle** for military use. The Erfle eyepiece is a five-element design in three principal forms:

Erfle I: field 60°, eye-relief $0.3F_e$
Erfle II: field 70°, eye-relief $0.6F_e$
Erfle III: field 55°, eye-relief $0.32F_e$

Ehrlich
A microscope eyepiece with a variable diaphragm, used to estimate the number of red and white corpuscles in a sample of blood. The eyepiece was made by **Leitz** under direction from Paul Ehrlich (1854-1915), who was an eminent physician and scientist working in the fields of haematology, immunology, and antimicrobial chemotherapy.

Huygenian
The Huygenian eyepiece, designed by **Christian Huygens** in 1703, consists of two plano-convex lenses, separated by some distance, with their convex surfaces facing away from the eye. The lens farther from the eye is a field lens, positioned such that the real image is formed between it and the eye-lens. The two lenses are separated by a distance of half the sum of their focal lengths.

The purpose of the Huygenian eyepiece design was to minimise lateral chromatic aberration. This is to some extent achieved, but longitudinal chromatic aberration and spherical aberration, together with coma and distortion, are noticeable at focal ratios faster than f/12.

Field 40°, eye-relief less than $0.3F_e$

Kaspereit
A six-element eyepiece developed by **Otto Karl Kaspereit** as a modification of the Erfle design. This design suffers from lateral chromatic aberration, distortion, and astigmatism.

Field greater than 68°, eye-relief $0.3F_e$

Kellner
Designed by **Carl Kellner**, this is a modification of the Ramsden eyepiece, and consists of an eye-lens, which is a cemented achromatic doublet with its plane, flint glass side to the eye, and a field lens which is a singlet plano-convex lens, with its plane surface away from the eye. The field and eye lenses are of equal focal length, and are separated by a distance equal to the focal length of either. The real image is formed at the plane surface of the field lens.

Field 40-45°, eye-relief 0.4-$0.45F_e$

König
A four-element eyepiece designed by **Albert König** for **Zeiss**. The König eyepiece is a variation of the Kellner eyepiece, and as such was originally referred to as the Kellner II. König improved on this with the four-element König II.

König: field 55°, eye-relief $0.67F_e$
König II: field 55°, eye-relief $0.92F_e$

Nagler
Al Nagler, the founder of **Tele Vue**, patented a seven-element eyepiece design with a wide 82-degree apparent field of view. The optical design uses a Smythe achromatic field flattener to widen the field of view, but this component makes the eyepiece bulky and heavy. The design suffers from a degree of spherical aberration, and the eight-element Nagler II design, which supersedes it, corrects this to some extent.

Nagler I: field 52-82°, eye-relief $1.2F_e$
Nagler II: field 52-82°, eye-relief $1.2F_e$

Orthoscopic
An orthoscopic eyepiece is, in principle, corrected for distortion. The concept was introduced by **Ernst Abbe** for **Zeiss** in 1880, for use in microscopes. It is a four-element design in which the field lens is an overcorrected triplet whose middle component is biconcave (negative), and a plano-convex eye lens with its curved surface almost in contact with the field lens. Many variants on this design have since been devised.

Field 30-64°, eye-relief 0.66-$0.91F_e$.

Periplan
The Periplan eyepiece is a seven-element design for microscopy. The lens elements are arranged as a doublet, a triplet, and two individual lenses. The design results in reduced chromatic aberration, and a flatter field.

Plössl
Developed in 1860 by **G.S. Plössl** as an improved version of the Ramsden eyepiece, resulting in a class of orthoscopic, achromatic, wide-field eyepieces of three types; the Symmetrical, the Dial-Sight, and the Plössl.

field 40-45°, eye-relief 0.7-$0.8F_e$.

Ramsden
Designed by **Jesse Ramsden**, this eyepiece consists of two plano-convex lenses of the same kind of glass, and of equal focal length, but with their curved faces turned towards each other. These two lenses, the field lens and the eye-lens, are of equal focal length and separated by a distance equal to two-thirds of the focal length of either. This eyepiece design suffers from lateral chromatic aberration.

Field 35°, eye-relief $0F_e$

Reversed Kellner Eyepiece (RKE)
The Reversed Kellner Eyepiece (RKE) is a modified Kellner, designed for **Edmund Scientific** by David Rank. In this design, the achromatic doublet is the field lens, and the singlet is the eye-lens.

Field 50°, eye-relief $0.9F_e$

Appendix III: German Wartime Makers' Ordnance Codes for Optical Products

During WWII, military optical instruments made in Germany and occupied territories were not marked with the name of the manufacturer for security reasons. Instead, they were marked with a three-letter code that represented the maker's name. The following list is incomplete, and is compiled from a variety of sources. This system was common across all military ordnance manufacturing, including optical products. (A more complete list of military ordnance codes may be found in [168]).

Code	Maker
bck	Brüninghaus & Co, Leder und Metallwarenfabrik, Versmold, Germany
beh	**Ernst Leitz GmbH**, Wetzlar, Germany
bek	**Hensoldt-Werk fuer Optik und Mechanik**, Herborn, Germany
blc	**Carl Zeiss**, Military Division, Jena, Germany
bmh	K Jirasek Fabrik für Feinmechanik und Optik, Prague, Czechoslovakia
bmj	**Hensoldt & Söhne, Mechanisch-Optische Werke AG**, Wetzlar, Germany
bmk	**Srb & Stys**, Fabrik Präziser Messinstrumente, Prague, Czechoslovakia
bmt	**C.A. Steinheil & Söhne GmbH**, Optische Werke, Munich, Germany
bpd	**C.P. Goerz GmbH**, Vienna, Austria
bvf	**C. Reichert Optische Werke AG**, Vienna, Austria
bvu	**Franz Kuhlmann**, Wilhelmshaven, Germany
bwt	**Gustav Heyde**, Dresden, Germany
bxx	**Askania Werke AG**, Berlin, Germany
byg	Johann Wyksen, Optische & Feinmaschinen, Katowitz, Poland
bzz	**IG Farbenindustrie AG** Kamerawerk, Munich, Germany
cad	**Karl Kahles Optiker**, Vienna, Austria
cag	**D. Swarowski, Glasfabrik & Tyrolit-Schleifmittel-Werke**, Wattens, Tyrol, Austria
cau	**Kodak AG**, Stuttgart, Germany
ccx	**Hugo Meyer & Co.**, Optische & Feinmaschinenwerke, Görlitz, Germany
clb	**Dr. F.A. Wöhler**, Kassel, Germany
clk	**F.W. Breithaupt & Söhne** Fabrik Geodätischer Instrumente, Kassel, Germany
cll	August Baumgart Feinmechanik und Präzisionsmaschinenbau, Rathenow, Germany
cln	**E Sprenger** Optische-Mechanische Werkstatten, Berlin, Germany
crh	**Franz Schmidt & Haensch**, Berlin, Germany
crj	**Otto Fennel** Söhne KG, Instrumente, Kassel, Germany
crn	Friedrichs & Co., Hamburg, Germany
cro	**R. Fuess**, Optische Industrie, Berlin, Germany
ctn	Fredricks & Co, Hanseatische Werkstätten für Feinmechanik und Optik, Hamburg, Germany
cwu	Georg Kremp, Wetzlar, Germany
cxn	**Emil Busch AG**, Optische Industrie, Rathenow, Germany
czn	**Emil Busch AG**, Optische Industrie, Rathenow, Germany
dag	Oculus GmbH, Berlin, Germany
ddv	Oculus GmbH, Berlin, Germany
ddx	**Voigtländer & Sohn AG**, Braunschweig, Germany
dhq	**J.D. Möller GmbH**, Wedel, Germany
dkl	**Josef Schneider**, Kreuznach, Germany
doq	**Deutsche Spiegelglas AG**, Leine, Germany
dow	Waffenwerke Bruenn, Prerau plant, Czechoslovakia (Renamed in 1943 *Opticotechna GmbH*, Werk Prerau, Protectorate of Bohemia and Moravia)
dpg	**ADOX Kamerawerk GmbH**, Wiesbaden, Germany
dpv	**Zeiss-Ikon**, Dresden, Germany
dpw	**Zeiss-Ikon**, Goerz plant, Berlin-Zehlendorf, Germany

dpx	**Zeiss-Ikon**, Stuttgart, Germany
dqc	Eugen Ising, Metallwarenfabrik, Bergneustadt, Germany
dym	Runge & Kaulfuss, Rathenow, Germany
dys	Heinrich Zeiss, Berlin, Germany
dzl	**Optische Anstalt Oigee**, Berlin-Schöeneberg, Germany
eaf	Aude & Reipe, Mechanoptik-Gesellschaft für Präzisionstechnik, Babelsberg, Germany
eaw	**R. Winkel GmbH** Optische Industrie, Göttingen, Germany
eed	**Kürbi & Niggeloh**, Radevormwald, Germany
emq	**Carl Zeiss**, Jena, Germany
emv	**Hertel & Reuss**, Kassel, Germany
erv	Fritz Hofmann, Erlangen, Germany
eso	**G. Rodenstock**, Optische Werke, Munich, Germany
esu	**Steinheil & Söhne GmbH**, Astronom Industrie, Munich, Germany
eug	Optische Präzisionswerke GmbH, Warsaw, Poland
fco	Sendlinger Optische Glaswerke GmbH, Berlin-Zehlendorf, Germany
fjt	Photogrammetrie GmbH, Berlin, Germany
fln	Franz Rapsch Optische Fabriken AG, Rathenow, Germany
fvs	**Spindler & Hoyer**, Göttingen, Germany
fvx	**Christoph Beck & Söhne** (CBS), Kassel, Germany
fwq	Saalfelder Apparatebau GmbH, Saalfeld, Germany
fwr	Optische Anstalt Saalfeld GmbH, Saalfeld, Germany
fxp	Hans Kollmorgen, Optische Anstalt, Berlin, Germany
gag	F. Mollenkopf Optische Industrieanstalt, Stuttgart, Germany
gcg	H. Maihak AG, Hamburg, Germany
gkp	Schütz Ruf & Co., Kassel, Germany
gug	Ungarische Optische Werke AG, Budapest, Hungary
guj	Werner D. Kuhn, Optische Industrie, Berlin-Steglitz, Germany
gwr	**Dennert & Pape**, Hamburg, Germany
gwv	Ernst Plank Fabrik Optische und Mechanischer Waren, Nürnberg, Germany
gxh	**Nitsche & Günther Optische Werke KG**, Rathenow, Germany
gxl	**Franke & Heidecke**, Braunschweig, Germany
gxp	Homrich & Sohn, Hamburg, Germany
hap	Max Kohl AG, Chemnitz, Germany
hdv	Optische Werk Osterode GmbH, Osterode am Harz, Germany
hfo	**Valentin Linhof OHG**, Munich, Germany
hkm	**Karl Braun AG**, Optische Industrie, Nürnberg, Germany
hna	Korelle Werke Brandtmann & Co, Dresden, Germany
hrw	Hoh & Hahne, Leipzig, Germany
hwt	**Ihagee Kamerawerk Steenbergen & co.**, Dresden, Germany
hxh	**A. Krüss**, Optisch-Mechanische Werkstätten, Hamburg, Germany
jfn	Tetanal Photowerk, Berlin, Germany
jfp	Dr. Karl Leiss, Optische Mechanische Instrumente, Berlin-Steglitz, Germany
jnh	**Hensoldt Werke für Optik & Mechanik**, Herborn, Dillkreis, Germany
jon	Voigtlander-Gavaert, Berlin, Germany
jux	**Nedinsco**, Nederlandsche Instrumenten Compagnie, Venlo, Netherlands
jve	Ernst Ludwig Optische Werke, Weixdorf, Germany
jxn	Helmut Korth, Feinmechanik und Optik, Berlin, Germany
kce	**Schneider & Co.**, Le Creusot, France
khc	**Otto Himmler**, Mikroskope, Berlin, Germany
kjj	**Askania Werke AG**, Berlin, Germany
kln	**Ernst & Wilhelm Bertram**, Munich, Germany
kna	Berning & Co. KG, Photographische Apparate und Bedarfsartikel, Düsseldorf, Germany

kov	**Bénard et Turenne, Établissement Barbier**, Paris, France
kqc	**Jos. Schneider & Co. Optische Werke KG**, Göttingen, Germany
krq	**Emil Busch AG**, Optische Werke, Rathenow, Brandenburg, Germany
kwc	Gamma Feinmechanische & Optische Werke, Budapest, Hungary
kxv	**A. Jackenkroll Optische Anstalt GmbH**, Berlin, Germany
lae	Heinrich Zeiss, Gostingen, Luxembourg
lfn	C. Richter Kamarafabrik, Tharandt, Germany
lmq	**Carl Zeiss**, Jena, Germany
lwg	Optische Werke Osterode GmbH, Freiheit, near Osterode am Harz, Germany
lww	**Huet & Cie**, Paris, France
lwx	**Société Optique et Précision de Levallois S.A. (OPL)**, Levallois, Paris
lwy	**Societe d'Optique et Mechanique de Haute Precision (SOM)**, Paris
mbv	**I.G. Farbenindustrie AG**, Berlin, Germany
mca	**Dr. C. Schleussner Fotowerke GmbH**, Frankfurt, Germany
mlr	**Società Scientifica Radio Brevetti Ducati (SSRD)**, Bologna, Italy
mtq	Roland Risse GmbH, Photochemische Fabrik, Florsheim, Germany
mtr	**Voigtländer & Sohn AG**, Berlin, Germany
mtu	A. Lorenz Feinmechanik-Optik, Dresden, Germany
mtv	A. Lorenz Feinmechanik-Optik, Dresden, Germany
mtw	A. Lorenz Feinmechanik-Optik, Dresden, Germany
nms	Richard Holz, Optischmechan, Berlin, Germany
nxt	S.A.I. Ottico Mechanica e Rilevamenti Aerofotogrammetrici, Rome, Italy
ocp	Aktophot GmbH, Sabechtlitz, Prague, Czechoslovakia
ocv	W. Klazer Feinmechanik und Fotobedarf, Prague, Czechoslovakia
okc	Hauff AG, Stuttgart, Germany
phq	**Möller**, Wedel, Germany
ppx	**Askania Werke AG**, Berlin, Germany
pvf	**C. Reichert, Optische Werke AG**, Vienna, Austria
pwf	**C. Reichert, Optische Werke AG**, Vienna, Austria
qhg	**Möller**, Wedel, Germany
rln	**Carl Zeiss**, Jena, Germany
xa	Busch-Jager, Lüdenscheider Metallwerke, Lüdenscheid, Germany

Appendix IV: Japanese Symbols and Codes for Optical Products

The post-war Japanese optical industry was huge – but unfortunately little reliable documentation seems to exist for it. This is an attempt to gather together some of the available information, even though much of it is not verifiable.

Makers' Marks

The following is drawn from many sources, including references [67], [68] and [94], other internet sources, extant optical instruments, and numerous forum posts and discussions. A complete list would be pretty much impossible to compile, and would be encyclopaedic. This is an incomplete list of some of the more common Japanese makers' marks:

Asahi Optical Company.

Apollo Labs.

Astro Optical Industries Co. Ltd.

Also used by Astro Optical Industries.

Also used by Astro Optical Industries.

Also used by Astro Optical Industries.

Used on early **Swift Sport Optics** telescopes manufactured by **Takahashi**.

Bushnell Optical Laboratory Inc.

Carton Optical Industries Ltd.

Gailand Optical Co.

Hiyoshi Optical Company.

The Hiyoshi Optical Company is also believed to have used this logo.

Kenko.

Koyu Company Ltd.

MIZAR Optical Japan.

Nishimura.

Shimadzu.

Yamamoto Seisakuyso.

Towa (but not exclusively).

(single or double circle) **Vixen**.

Eikow.

Tanzutsu.

JB codes

For about three decades following 1959 all optical products exported from Japan were subject to rigorous quality checks by the Japan Telescopes Inspection Institute (JTII). Only products that passed these quality checks were authorised for export. A silvered oval label, marked JTII, indicated that the product had passed the quality tests.

Binoculars produced during this era were marked with two codes:

The first, commonly referred to as the JB code, indicates the manufacturer of the finished product. This consists of a symbol which looks like a J with an extra limb horizontally to the right. It is a combination of the letters J and L, standing for "Light machinery of Japan". This is followed by B and a number of up to three digits which designates the manufacturer of the finished product.

The second mark is commonly referred to as the JE code, and comprises the same J L combination followed by E and a number of up to three digits which designates the manufacturer of the metalwork.

For more information, see [68], [88], [89]. The following is an incomplete list of manufacturers by JB code (reproduced with kind permission, see **acknowledgements**):

JB 1 Toa Kogaku Co.Ltd., Tokyo
JB 2 Katsuma Kogaku Kikai Co.Ltd.
JB 3 Toei Kogaku Seisakujo Co.Ltd.
JB 4 Toei Kogaku Co. Ltd., Hatogaya-Shi
JB 5 Meiji Seiko Co. Ltd.
JB 6 **Asahi Kogaku Kogyo Co. Ltd.**, Tokyo.
JB 7 **Nippon Kogaku Kogyo Co. Ltd.**
JB 8 Fuji Sbashin Koki Co. Ltd.
JB 9 Sato Kogaku Kogyo Co. Ltd.
JB 10 Toko Seiki Co. Ltd.
JB 11 Omiya Kogaku Kikai Seisalcujo, Tokyo
JB 12 Orora Kogaku Co. Ltd. - Aurora Kogaku Co. Ltd.
JB 14 Ueta Seiki Co. Ltd.
JB 15 Tokyo Oputikaru Co. Ltd.- Tokyo Optical Co. Ltd.
JB 16 **Sankei Koki Seisakujo Inc.**
JB 17 Otake Kogaku Kogyo Co. Ltd., Tokyo
JB 18 **Tokyo Kogaku Kikai Co. Ltd.**
JB 19 FujiKogeisha Co. Ltd.
JB 20 Mitsui Koki Seisakujo Co. Ltd.
JB 21 Kokisha Co. Ltd., Tokyo
JB 22 Itabashi Kogaku Kikai Seisakujo Inc.,Tokyo
JB 23 Ishii Kogaku Co. Ltd.,Yokohama
JB 24 Ichikawa Kogaku Kogyo Co. Ltd.,Tokyo
JB 25 Zuiho Kogaku Seiki Co. Ltd., Tokyo
JB 26 Futaba Kogaku Kogyo Co. Ltd.
JB 27 Sanyo Koki Co. Ltd.
JB 28 Fuji Seinñtsukiki Seisakujo. Inc.
JB 29 Meikosha Inc.
JB 30 Kofu Kogaku Kogyo Co. Ltd.
JB 31 Muraki Koko Co. Ltd.
JB 32 Miyako Seiki Co. Ltd.
JB 33 Teito Koki Co. Ltd.
JB 34 Musashi Kogaku Co. Ltd.
JB 35 Raito Koki Seisakujo Co. Ltd. - Lite Koki Seisakujo Co. Ltd.
JB 36 Jiyama Seiko Co.
JB 37 Yoshinon Kogaku Kikai Co. Ltd.
JB 38 Nakabishi Kogaku Inc.

JB 39 Josei Koki Inc.
JB 40 Mutsu Koki Inc.
JB 41 Shinsei Kogaku Seiki Co. Ltd.
JB 42 Nippon Garasu Kogyo Co. Ltd. Takinokawa Syuchojo
JB 43 Tozaki Kogaku Kogyo Co. Ltd.
JB 44 likura Kogaku Seisakujo Inc.
JB 45 **Taisei Kogaku Kogyo Co; Ltd.**
JB 45 **Tamron Co. Ltd.**, Tokyo
JB 46 Otsuka Kogaku Co. Ltd., Tokyo
JB 47 Tokuhiro Koki Seisakusho Inc., Tokyo
JB 48 Kazusa Koki Seisakujo Inc.
JB 49 Sankyo Kogaku Kogyo Co. Ltd.
JB 50 Tanaka Koki Seisakujo. Inc.
JB 51 Yoshimoto Kogaku Co. Ltd.
JB 52 Kanto Kogaku Kogyo Co. Ltd.
JB 53 Inoue Koki Seisakujo Inc.
JB 54 Suzuki Kogaku Seiki Co. Ltd.
JB 55 Enshu Kogaku Seiki Co. Ltd.,Tokyo.
JB 56 Hiyoshi Kogaku Co. Ltd.
JB 57 Oji Kogaku Kikai Co. Ltd.
JB 58 Ryuko Seisakujo
JB 59 Mitsui Kogaku Seisakujo
JB 60 Akebono Optical, Tokyo
JB 60 Wakaba Koki Seisakujo Inc.
JB 61 Meiho Kogaku Seisakujo Inc.
JB 62 Oshiro Kogaku Co. Ltd.
JB 63 Ofuna Kogaku Kogyo Co. Ltd.
JB 64 Kobayashi Kogaku Scisakitjo. Inc.
JB 67 Esaka Kogaku
JB 68 Sono Kogaku Kikai Co. Ltd.
JB 69 Akebono Kogaku Kogyo Co
JB 70 Sugamo Kogaku Seisakujo
JB 71 Toho Koki Co. Ltd., Tokyo
JB 72 Rubina Koki Co. Ltd.
JB 73 Tsuchida Kogaku Seisakujo.
JB 74 Omori Sogo Kogaku Kogyo
JB 75 Seki Kogaku Kikai Co. Ltd.

JB 76 Izumi Seiki Seisakujo. Inc.
JB 77 Koronbia Kogaku Co. Ltd. – Columbia Kogaku Co.
JB 78 Kuribayashi Kogaku Seisakujo
JB 79 Furukawa Kogaku Seisakujo
JB 80 Sansei Kogaku Kogyo Co. Ltd
JB 81 Copitar Co. Ltd., Tokyo
JB 81 Takahisa Kogaku Kogyo Co. Ltd.
JB 82 Sanwa Kogaku Co. Ltd., Tokyo
JB 83 Nakamura Kogaku Kogyo Seisakujo Inc.
JB 84 Oei Kogaku Co. Ltd.
JB 85 Kawashima Kogaku Seisakujo
JB 86 Niigaki Kogaku Seiki Seisakujo
JB 87 Yachiyo Kogaku Co. Ltd
JB 88 Kofuku Sangyo Co. Ltd. Kowa Koki Seisakujo
JB 89 Sekiguchi Kogaku Seisakujo. Inc., Tokyo
JB 90 Someno Koki Seisakujo
JB 92 Hayashi Kokisha
JB 93 Seiwa Kogaku Co. Ltd., Tokyo
JB 94 Ibuki Kogaku Co. Ltd.
JB 96 Fuji Koki Seisakujo Inc.
JB 96 Nikkei Optical Co. Ltd., Tokyo
JB 97 Soneda Kogaku Kenkyujo Inc.
JB 98 Seiwa Seiki Seisakujo Inc.
JB 99 Tokushu Kogaku Seisakujo Co. Ltd.
JB 100 Nippon Koki Co. Ltd.
JB 101 Hifumi Kogaku Kikai Co. Ltd.
JB 102 Hoya Kogaku Co. Ltd.
JB 103 Teihoku Kogaku Kogyo Co. Ltd., Tokyo (Tohoku)
JB 104 Kyanon Kamara Co. Ltd. - Canon Camera Co. Ltd.
JB 105 Toyoshima Kogaku Kogyo Co. Ltd
JB 106 Sanko Kogaku Seisakujo Inc.
JB 107 Jyonan Kogaku Kikai Seisakujo. Inc.
JB 108 Kenkosha Inc.
JB 109 Chiyoda Kogaku Seiko Co. Ltd.
JB 110 Taiyo Kogaku Seisakujo Inc.
JB 110 Tsukuba Kogaku Co. Ltd.
JB 111 Tosco Co. Ltd., Tokyo
JB 112 Keizan Kogaku Co. Ltd.
JB 113 Ito Kogaku Kikai Seisakujo. Inc.
JB 114 Arai Kogaku
JB 114 Hoya Optical Co.,Ltd.,Tokyo
JB 115 Kanagawa Kogaku Kogyo Co. Ltd, Kamakura-Shi
JB 116 Hattori Koki Seisakujo
JB 116 Nichiryo Co. Ltd., Tokyo
JB 117 Nikken Kogaku
JB 118 Yoshikawa Koki Co. Ltd.
JB 119 Nisshin Kogaku Kogyo Co. Ltd., Tokyo
JB 120 Noppon Koki

JB 121 Oizumi Kogaku Kogyo Co. Ltd.
JB 122 Imai Kogaku Seisakujo
JB 123 Kansai Kogaku Seisakujo
JB 124 Oda Kogaku Seiki Co. Ltd.
JB 125 Inoue Kogaku Kogyo Co. Ltd.
JB 126 Yabe Kogaku Kiki Seisakujo Co. Ltd.
JB 127 Koei Seiki Co. Ltd., Tokyo
JB 128 Taishin Kogyo Co. Ltd.
JB 129 Narimasu Koki Seisakujo
JB 130 KomiyaKogakuSangyo Co. Ltd.
JB 131 Yamanoi Kogaku Co. Ltd
JB 132 Taiei Kogaku Co. Ltd.
JB 133 Kamakura Koki Co. Ltd., Warabi-Shi
JB 134 Kozan Kogaku Seisakujo Co. Ltd.
JB 135 Eikosha Sagyojo Co. Ltd.
JB 136 Araki Kogaku Seiki Seisakujo
JB 137 Katsuno Koki Seisakujo
JB 138 The Oriental Trading Co. Ltd., Tokyo
JB 138 Toyo Jitsugyo Co. Ltd.
JB 139 Busho Kogaku Seisakujo. Inc. (Bushu), Saitama.
JB 140 Toyo Kogaku Kogyo Co. Ltd
JB 141 Tonan Kogaku Co. Ltd., Tokyo
JB 142 Subaru Kogaku Kikai Co. Ltd.
JB 143 Nagashima Kogaku Co. Ltd.
JB 144 Tokuei Seiki Kogyojo Co. Ltd.
JB 145 Kimura Kogaku Seisakujo
JB 146 Warabi Kokisha
JB 147 Sankaku Kogaku Kogyo Co. Ltd.
JB 148 Misuzu Kogaku Kogyo Co. Ltd
JB 149 Urawa Kogaku Seiki Seisakujo Co. Ltd.
JB 150 Kuramochi Kogyo Co. Ltd.
JB 151 Daito Kogaku Kogyo Co. Ltd.
JB 152 Taito Kogaku Co. Ltd.
JB 153 Kitano Koki Seisakujo Co. Ltd.
JB 154 Sanei Kagaku Kenkyojo
JB 155 Ichihara Kogaku Renzu Seisakujo
JB 156 Nippon Sogan Co. Ltd.
JB 157 Kuroki Kogaku Kogyo Co. Ltd.
JB 158 Towa Koeki Co. Ltd.
JB 159 Myoko Kogaku
JB 160 Myoko Kogaku
JB 161 Hakko Seiki Co. Ltd.
JB 162 Akebono Kogaku Seiki Seisakuju
JB 163 Tokyo Koki Seisakujo
JB 164 Toyoshima Koki Seisakujo
JB 165 Yabuki Kogaku Kogyo Co. Ltd.
JB 166 Komiya Kogaku Kenkyuio
JB 167 Ikkosha
JB 168 Hirabayashi Kogaku Seisakujo
JB 169 Ota Kogaku Kogyo Co. Ltd.
JB 170 Koseiki Seisakujo. Inc.
JB 171 Noguchi Kogaku Kogyo Co. Ltd.

JB 172 **Showa Koki Seizo Co. Ltd.**
JB 173 Okaya Kogaku Kikai Co. Ltd.
JB 174 Chitose Kogaku Co. Ltd.
JB 175 Toyo Kogaku Kenkxuio.
JB 176 Sanyo Koki Co. Ltd., Tokyo
JB 176 Toyo Koki Kogyo Co.
JB 177 Keihin Seiko Co. Ltd.
JB 178 Shinyo Koki Seisakujo
JB 179 Otaki Kogaku Seiki Seisakujo
JB 180 Tsukumo Seisakujo
JB 181 Yamagami Kogaku Seisakujo
JB 182 Yarnato Koki Seisakujo
JB 183 Izumi Kogaku Co. Ltd.
JB 184 Akatsuki Kogaku Kogyo Co. Ltd.
JB 185 Fuji Kogaku Co. Ltd.
JB 186 Nikko Seisakujo Co. Ltd.
JB 187 Kurumada Kogaku Kogyojo
JB 188 Daiichi Seiko Co. Ltd.
JB 189 Sawama Kogaku Seisakujo
JB 190 Nitto Koki Co. Ltd.
JB 191 **Seiwa Optical Co. Ltd.**, Wako-Shi
JB 191 Yoko Sangyo Co. Ltd
JB 192 Hoja Koki Seisakujo. Inc.
JB 193 Atorasu Koki Seisakujo Inc. (Atlas Koki Seisakujo)
JB 194 Miyagaki Kogakusha
JB 195 Kawashima Seisakujo
JB 196 Kokoku Seiki Seisakujo Co. Ltd.
JB 197 Sankyo Kogaku Co. Ltd.
JB 198 Tenwa Seiki Kogyo Co. Ltd.
JB 199 Keihoku Kogaku
JB 200 Nansei Koki Seisakujo. Inc.
JB 201 Takeuchi Shokai Seisakujo
JB 202 Jyohoku Kogaku Kogyo Kyodo Kumiai Sogankyo Chosei Daiichi Jigyojo
JB 203 Fujita Kogaku Kogyo Co. I.td.
JB 204 Chuo Koki Seisakujo
JB 205 Ileiwa Koki Co. Ltd.
JB 206 Fujita Koki Seisakuio. Inc.,Tokyo.
JB 207 Aporon Koki Seisakujo
JB 207 Hitachi Optical Co. Ltd., Tokyo
JB 207 **Hiyoshi Optical Co.,Ltd.,Tokyo**
JB 208 Katon Kogaku Seisakujo Inc. (Carton Kogaku Seisakujo)
JB 210 Oishi Kogaku Kogyosho. Inc.
JB 211 Toho Seiki Co. Ltd.

JB 212 Sanwa Seiki Co. Ltd.
JB 213 Daito Kikai Seisakujo Co. Ltd.
JB 214 Miyama Kogaku Co. Ltd.
JB 215 Jyosei Koki Seisakujo
JB 216 Noguchi Koki Seisakujo
JB 217 Kawashima Seisakujo
JB 219 Tokoki
JB 220 Ato Kogaku Co. Ltd. - Art Kogaku Co. Ltd.
JB 221 Akabori Kogaku Kikai Seisakujo
JB 222 Kanda Koki
JB 223 Zaika Co. Ltd.
JB 224 Ikko Seiki
JB 225 Fuji Kogaku Kenkyujo
JB 226 Otake Kogaku Kogyo Co. Ltd.
JB 227 Jya Seikosha
JB 228 Kanagawa Kogaku Kogyo Co. Ltd.
JB 229 Kanto Seimitsu Sogankyo Chosei Gijutsu Kyodo Kumiai
JB 230 Higashi Nippon Kogaku Kikai Kyodo Kumiai
JB 230 Palus Optical, Tokyo
JB 231 Hiroyuki Tochihara
JB 231 Tochihara Optical Co. Ltd., Tokyo
JB 232 Akira Ishii
JB 233 Shuzaburo Ishikawa (Ishikawa Optical Ind.),Tokyo
JB 234 Yoshitada Matsumaru
JB 235 Takeo Saito
JB 236 Shuzaburo Wakabayashi
JB 237 Motooka Yoshikimi?
JB 238 Toshio Maruyama
JB 239 Ogoshi (?)
JB 240 Teiji Hirose
JB 241 Kazuo Tsuchihashi
JB 242 Tornoaki Ogawa
JB 243 Suwa Koki Co. Ltd.
JB 244 Tosei Kogaku
JB 248 Opal Optical, Tokyo
JB 250 Fujimori Optical Co. Ltd., Tokyo
JB 251 Toho Optical Mfg. Co. Ltd., Tokyo
JB 256 Shinsei Optical Co. Ltd., Tokyo
JB 257 Kyoshi Optical Co. Ltd., Tokyo
JB 272 Atsugi Optical, Tokyo.
JB 274 Kohiyama Seisakusho, Tokyo
JB 277 Orient Co., Saitama
JB 345 Kobayashi Co. Ltd., Kyoto

Appendix V: Russian and Soviet Makers' Marks

A Note on Soviet Era Marks

During the Soviet era the USSR makers' marks were placed adjacent to (above, below or beside) a hammer and sickle depiction, illustrated here. Normally the makers' mark is simply reproduced adjacent to the hammer and sickle. However, in some cases the original makers' mark is altered to go with the hammer and sickle.

As an example, the NPZ pentaprism, (see below), is slightly elongated, and reproduced twice at the base of the hammer and sickle, as illustrated here.

Makers' Marks

This information is drawn from many sources, including references [181] [182] [183], other internet sources, and extant optical instruments.

The makers' marks shown below do not constitute an exhaustive list.

Arsenal: **Arsenal Factory**, Kiev.

FED: **Kharkov Machine-Building Plant**.

GOMZ: **Gosudarstvennyi Optiko-Mechanicheskiy Zavod**.

KMZ: **Krasnogorskij Mechanicheskij Zavod**. This design depicts a Dove prism, and was used from 1942, with the hammer and sickle sign above it (see above). The following two illustrations show how the design evolved:

From 1946 KMZ used this design of a Dove prism with a beam of light traversing it.

The modern KMZ design shows a Dove prism with a bold beam through it, but no arrow head.

KOMZ: **Kazan Optical and Mechanical Plant**.

LOMO: **Leningrad Optical Mechanical Association**.
This mark appears with or without the small circle at the top.

LZOS: **Lytkarino Optical Glass Factory**.

 NPZ: **Novosibirsk Instrument-Making Plant**. The Russian НПЗ translates as NPZ.

 NPZ also used the central "pentaprism" shape of the above image, as illustrated here, and this was modified as described above in the Soviet era symbol with the hammer and sickle image.

 Progress: **Progress Optics**. The company used a depiction of a plano-convex lens, but with an open contour, illustrated here.

 Progress Optics also used this depiction of a plano-convex lens. In this case the contour is complete.

 ZOMZ: **Zagorsk Optical and Mechanical Plant**. This depiction of three lenses joined together was used until 1962, after which they used the symbol shown below:

 ZOMZ used this design after 1962. It is a stylised depiction of an eye, and reflects the company's increased involvement in optical instruments for ophthalmology.

Appendix VI: Optical Glass

Properties of Optical Glass
The pertinent properties of optical glass, [7] [21] [22], are:

Refractive Index: The *absolute index of refraction*, n, of a material is the ratio of the speed of an electromagnetic wave in vacuum to that in the material. The refractive index n is given by:

$$n = c/v$$

where c is the speed of light in a vacuum, and v is the speed of light in the material [7]. It is a dimensionless quantity which defines how fast light travels in the material. Tolerance and homogeneity of the refractive index of optical glass affect its quality, and a measure for these values can be found in the specifications from glass manufacturers.

Dispersion: Named after **Ernst Abbe**, the *Abbe number*, or V-number is a measure of the material's dispersion – the variation of refractive index with wavelength. The Abbe number, V_d is defined as:

$$(n_d - 1) / (n_f - n_c)$$

where where n_d, n_f and n_c are the refractive indices of the material at the wavelengths of the Fraunhofer spectral lines: d (i.e. D_3 = Yellow: 589.3nm), f (= Blue: 486.1nm) and c (= Red: 656.3nm). A higher Abbe number indicates a lower dispersion.

Precision: The precision to which the optical surfaces have been ground. There is rarely any empirical measure of this precision given in the specification of an optical device such as a telescope or a pair of binoculars. Cost is often an indicator.

Transmittance: No glass is perfectly transparent. Incident light on the surface of an optical component can be reflected, absorbed or transmitted. Reflection is dealt with below in the discussion of coatings. The degree of absorption depends on the wavelength of light. For example, a very thick piece of the purest glass will show a blue or green tint.

Homogeneity: The main factors affecting optical homogeneity are striae, bubbles and inclusions. Striae are bands of deviation in refractive index, typically on a scale of tenths of a mm to several mm.

Stress birefringence: Birefringence is an anisotropy in the refractive index. The material displays two different indices of refraction. In optical glass this can result from inherent stresses that depend on the annealing conditions, the glass type, and the dimensions.

Types of Optical Glass
There are hundreds of optical glass types, and a large number of glass types used in telescopes, binoculars and other instruments. The glass type alone is not a suitable indicator of quality. Criteria for evaluating the suitability of a given glass type for any particular optical instrument is beyond the scope of this book. See also Pyrex.

Crown Glass
Crown glass is made without any lead or iron, originally hand-blown in circular sheets, and made for use as window glass. It is used as an optical glass because of its low refractive index and low dispersion.

Flint Glass
Flint glass, originally made from ground flint, contains lead oxide, and is also known as *crystal* or *lead crystal* glass. It is used as an optical glass because of its relatively high refractive index and low dispersion.

Other Glass Compositions
The international glass type designation is a six-digit number. Schott augment these digits with a further three of their own. The first three digits of the international designation indicate the refractive index, n_d. The second set of three digits indicate the dispersion V_d (see Abbe Number, above). The third set of three digits used by Schott indicates the density. For example, the Schott BaK4 glass has the designation 569560.305. This example has n_d=569, V_d=560, and international designation 569560.

Optical Coatings
Where lens coatings are referred to in the specifications of instruments such as binoculars or telescopes, they are generally referred to as one of the following:

Coated. This simply means that at least one of the optical surfaces has a minimum of a single coating.

Fully Coated. This indicates that all optical surfaces have at least one coating.

Multi-Coated. This means that at least one of the optical surfaces is multi-coated – however, it says nothing about the quality or design of the multi-coating, nor how many layers it comprises, or whether more than one surface is coated.

Fully Multi-Coated. This means that all optical surfaces have a multi-coating – however, it says nothing about the quality or design of the multi-coating, nor how many layers it comprises.

Poorly designed optical coatings, including many of the "ruby" coatings found on inexpensive binoculars, are not desirable. These "ruby" coatings work by blocking part of the optical spectrum. Any poorly designed coating, "ruby" or otherwise, will have a detrimental effect on image quality.

Optical Glass Makers

The following optical glass makers are featured in this book:

Barr and Stroud
Benoist Berthiot
British Crown Glass Co.
CDGM
Chance Brothers and Co.
Chance-Pilkington Optical Works
Choisy-le-Roi Glass-Works
Corning
Cosina
Derby Crown Glass Company
Deutsche Spiegelglas AG
Glass Perfection
Pierre Guinand
Hoya
Lytkarino Optical Glass Factory (LZOS)
Mathematical-Mechanical Institute Reichenbach, Utzschneider and Liebherr
Merz
Nikon
Nippon Kōgaku Kōgyō Kabushikigaisha
Ohara
Parra Mantois et Cie
Parsons Optical Glass Company
Petrograd porcelain works
Pilkington
Rodenstock
Schott
Swarovski Optik
UQG Optics

See also the 1951 *Directory for the British Glass Industry*, [**327**].

Bonding Optical Glasses

Canada balsam is a turpentine made by distillation of the resin of the Balsam Fir tree (Abies Balsamea) of North America. Its refractive index when dried is similar to that of crown glass. When suitably purified and filtered, it is optically clear. Because of these two properties, Canada Balsam was historically used as an adhesive for optical glasses. After WWII manufacturers favoured *polyester*, *epoxy* and *urethane* single and two-part adhesives. Most modern optical bonding is carried out using *UV-cured adhesives* [**146**].

Appendix VII: Charts and Timelines

This appendix shows the following:

Charts of business, family, and apprenticeship relationships for:

    Cooke, Troughton & Simms

    Spencer, Browning, and Rust

Charts of mergers, acquisitions, and business successions for:

    Leitz/Cambridge Instrument Co./Zeiss, and associated companies (sheet 1)

    Leitz/Cambridge Instrument Co./Zeiss, and associated companies (sheet 2)

A timeline for:

    Reichenbach/Fraunhofer/Merz, successive company names, proprietors, and locations

These charts and timelines provide additional clarification to the information in this book, relating to these companies.

Brass and Glass: Optical Instruments and their Makers

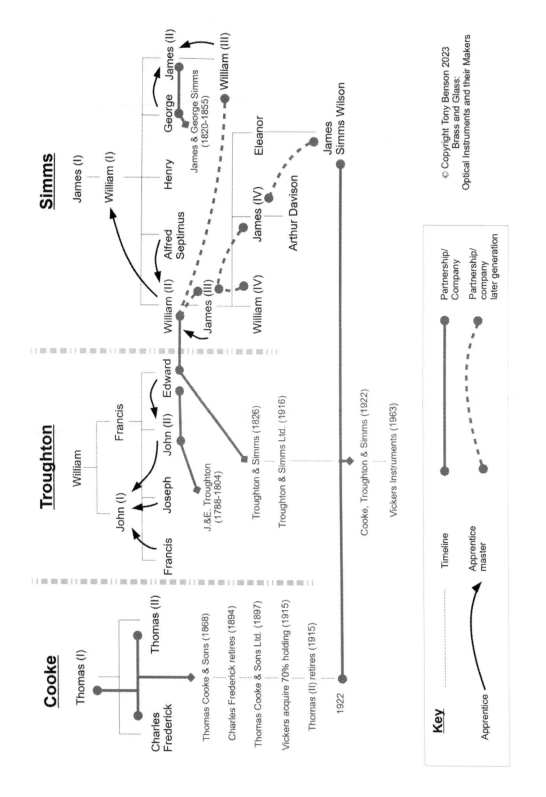

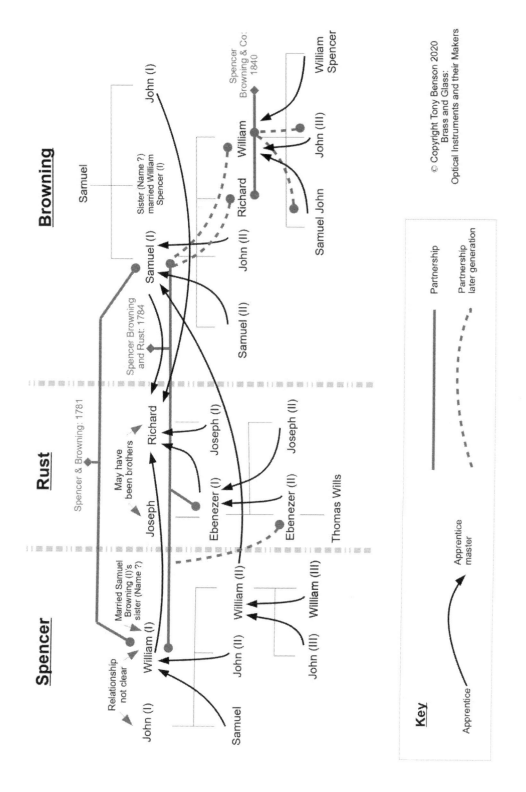

Brass and Glass: Optical Instruments and their Makers

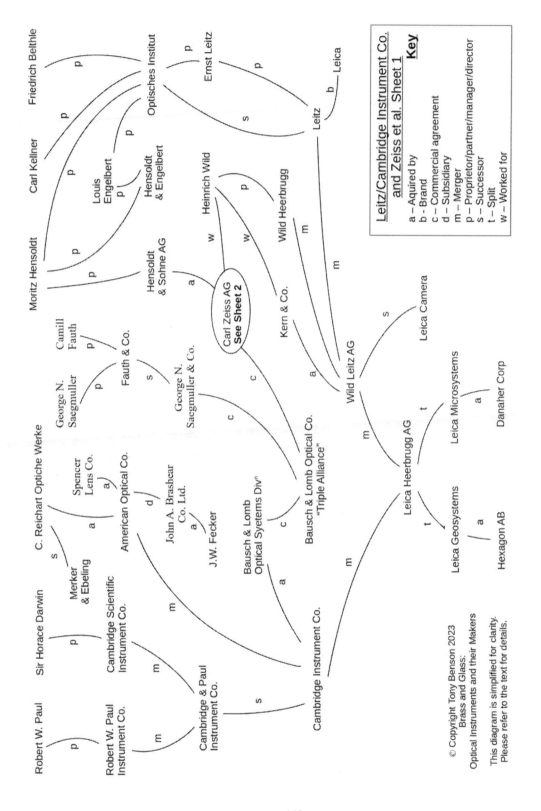

Brass and Glass: Optical Instruments and their Makers

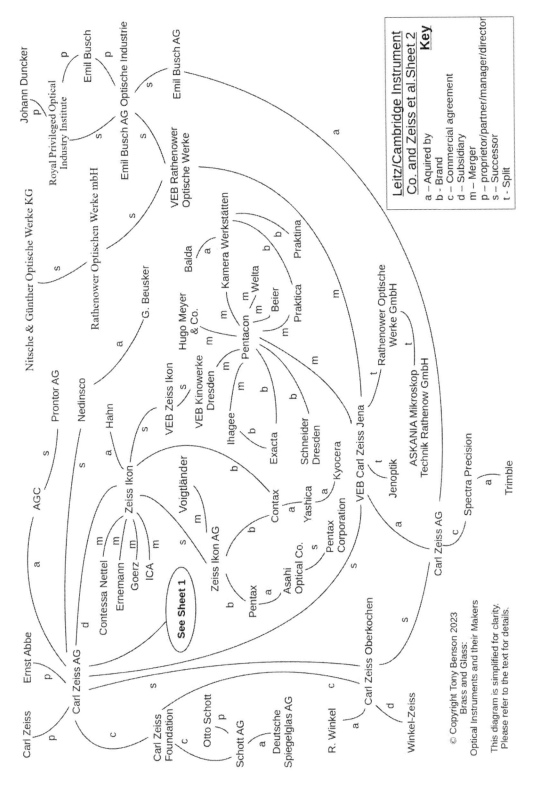

443

Reichenbach/Fraunhofer/Merz Company Timeline

Date	Proprietor(s)	Trading as	Location
1802-1804	Georg Friedrich von Reichenbach, Joseph Liebherr	Mathematisch-mechanisches Institut von Reichenbach & Liebherr	Munich
1804-1809	Georg Friedrich von Reichenbach, Joseph von Utzschneider, Joseph Liebherr	Mathematisch-mechanisches Institut von Utzschneider, Reichenbach & Liebherr	Munich & Benediktbeuern from 1806
1809-1814	Joseph von Utzschneider, Georg Friedrich von Reichenbach, Joseph von Fraunhofer	Optisches Institut von Utzschneider, Reichenbach & Fraunhofer in Benediktbeuern	Munich & Benediktbeuern
1814-1819	Joseph von Utzschneider, Joseph von Fraunhofer	Optisches Institut von Utzschneider und Fraunhofer in Benediktbeuern	Munich & Benediktbeuern
1819-1826	Joseph von Utzschneider, Joseph von Fraunhofer	Optisches Institut von Utzschneider und Fraunhofer in München	Munich & Benediktbeuern
1826-1838	Joseph von Utzschneider	Optisches Institut von Utzschneider und Fraunhofer in München	Munich & Benediktbeuern
1838-1839	Joseph von Utzschneider, Georg Merz, Joseph Mahler	Optisches Institut von Utzschneider und Fraunhofer in München	Munich & Benediktbeuern
1839-1845	Georg Merz, Joseph Mahler	Optisches Institut von Utzschneider und Fraunhofer in München	Munich & Benediktbeuern
1845-1847	Georg Merz, Sigmund Merz	G. Merz & Söhne in München	Munich & Benediktbeuern
1847-1858	Georg Merz, Ludwig Merz, Sigmund Merz	Merz, Utzschneider & Fraunhofer in München	Munich & Benediktbeuern
1858-1867	Georg Merz, Sigmund Merz	G.&S. Merz in München	Munich & Benediktbeuern
1867-1883	Sigmund Merz	G.&S. Merz vormals Utzschneider & Fraunhofer in München	Munich & Benediktbeuern
1882-1883	Jakob Merz, Matthias Merz Feb 25 1882- June 21 1883	Optische Werkstätte der Gebrüder Matthias und Jakob Merz	Munich
1883-1889	Sigmund Merz	Optische Glasfabrik in Benediktbeuern, Sigmund Merz München	Benediktbeuern
1884-1903	Jakob Merz	Optische Werkstätte von Jakob Merz	Munich
1903-1908	Paul Zschokke	G.&S. Merz vormals Utzschneider & Fraunhofer in München	Munich
1908-1912	Paul Zschokke	G.&S. Merz vormals Utzschneider & Fraunhofer in Pasing bei München	Munich
1912-1932	Paul Zschokke	G.&S. Merz GmbH in Pasing	Munich
1933-1976	Georg Tremel	Georg Tremel	Munich

Appendix VIII: The Electromagnetic Spectrum

The electromagnetic spectrum is the range of frequencies/wavelengths of electromagnetic radiation, ranging from the longest wavelength (lowest frequency) radio waves to the shortest wavelength (highest frequency) gamma rays. This is illustrated in the diagram overleaf.

Electromagnetic (EM) radiation takes the form of electromagnetic waves propagated through space or through matter. An electromagnetic wave is a periodic fluctuation in a magnetic field coupled with, and at right-angles to a periodic fluctuation in an electric field at the same frequency. The relationship between the fluctuating magnetic and electric fields results from a physical phenomenon, in which a moving electric field induces a moving magnetic field, and vice-versa. These two fields may thus be considered two aspects of the electromagnetic field. The fluctuating fields form an electromagnetic wave which travels from its source, independently of it, in a direction perpendicular to both the electric and magnetic fluctuations.

Visible light is electromagnetic radiation, and is a part of the wider EM spectrum. Most consumer optical instruments operate in visible light. However, some operate outside of this part of the spectrum, for example in infrared or ultraviolet light. Research in physics, and notably in astronomy and astrophysics, uses imaging techniques in all parts of the EM spectrum from the longest radio waves to gamma rays. Microscopy is carried out beyond the visible spectrum, utilising wavelengths ranging from microwave to the shortest gamma rays. It should be noted that imaging outside of the visible spectrum, of necessity, uses false colour for human-viewable display.

Radio Frequency (RF) imaging techniques are used in various applications. In medicine, MRI uses RF to image in strong magnetic fields. RADAR, which is *radio detection and ranging*, is used in air traffic, marine, and *ground penetrating RADAR*, as well as other applications. RF imaging is also used in radio astronomy.

Microwave imaging is used in medical applications such as detection of breast cancer. It is also used in civil engineering in non-destructive testing for applications such as imaging reinforcing bars through concrete. Other applications include the detection of concealed weapons for security purposes, through the wall imaging, landmine detection, and archaeological investigation.

Infrared imaging, otherwise known as thermal imaging, has a vast number of applications, ranging from low-light sensing, firefighting, and thermal mapping, to detecting heat leaks, and measuring the efficiency of heating and air-conditioning systems.

Ultra-violet imaging is used for applications such as document examination, glass identification, forensic examination and detection of latent fingerprints, detection of fine surface scratches, and detecting surface contamination.

X-rays are used in various medical imaging techniques, including computed tomography (CT) scanning, projection radiography, and fluoroscopy. These same techniques are also used in industrial and other inspection and investigation processes.

Gamma rays are used in medical imaging techniques, including positron emission tomography (PET) imaging.

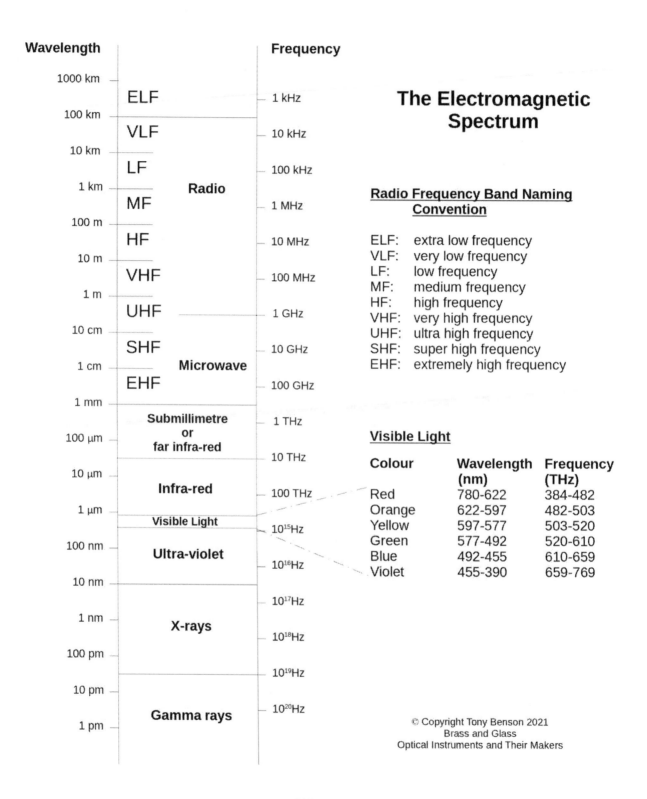

Appendix IX: The London Craft Guilds

[90] [195] [355] [385] [389] [390] [401] [414] [415]

Overview

The craft guilds in England first began to emerge during the C13th and C14th in many towns and cities. Their purpose was to bring together craftsmen of a particular trade, and protect their livelihood from competition from those from other towns or other countries, and from workmen who were not skilled or properly trained in the craft. The craft guilds were known as Companies, and in order to carry out their trade, craftsmen were required to be Freemen of a Company. With time, some guilds began to distinguish themselves as Livery Companies, so named because of the livery, or clothing they wore for special occasions. Not only did the guilds regulate the activities of tradespeople in the city, they also provided social community for members, and the guild's benevolent fund cared for widows and orphans of deceased members, and provided a pension to retired members who needed it.

In 1562 the Statute of Artificers was enacted, enforcing tight control over prices, wages, training, and the freedoms of workers. The act required every practitioner of a craft to have served an apprenticeship of at least seven years. With this act, the guilds gained significant control over the trades, although the authority of the craft guilds of London did not extend beyond the city walls.

The accuracy and usefulness of the London guild records vary, and many were lost in the Great Fire of 1666 (although the records of the Grocers' Company were happily not lost). In general, from that time, records were starting to be well enough kept and preserved to be useful to future historians. Following the repeal of the Statute of Artificers in 1814, the quality of record keeping rapidly declined. The records maintained during the intervening years provide a valuable historical record.

The usual route for a tradesperson to become a Freeman was to complete an apprenticeship with an established Freeman of a Company, thus enabling them to be freed of the same Company. Having acquired this status, a Freeman was permitted to conduct the trade or craft of their own, or another Company. Freedom came with an associated cost, and hence not all apprentices took their freedom, and some took their freedom at a later date. There were, however, several ways to obtain the freedom of a Company:

Apprenticeship: An apprenticeship was usually started at age fourteen, when a person was bound to serve his apprentice master (NOTE: In this text, apprentices are referred to with male pronouns for convenience only. For mathematical and optical instrument makers, the apprentice was usually a boy, but not always. Daniel Thomas, mathematical instrument maker of the Clockmakers' Company took on three girl apprentices, and **George Willdey**, optical instrument maker of the Spectacle Makers' Company, took on seven girl apprentices. [**385** p331, footnotes 40, 41]. One of George Willdey's apprentices, Susanna Passavant, obtained her freedom of the Spectacle Makers' Company in 1735/1736). The rights and responsibilities of both the apprentice and his master were defined in the terms of the indenture. The master was usually paid a fee for taking on the apprentice (although the fee varied considerably), and agreed to provide food, lodging, clothes, and tuition in the skills of his trade. If the binding fee was waived for a member of the apprentice master's family, it was referred to as "Love and affection". Generally, the apprentice lived in the Master's house as one of the family. In return the apprentice agreed to abide by strict rules of lifestyle and work. The apprenticeship usually lasted seven years.

In some cases, the apprentice was subject to a convenience binding, in that he was bound apprentice to a Freeman of the Company he wished to join, and immediately turned over to an apprentice master of a different Company, with whom he served his apprenticeship. As an example, **Henry Hemsley (I)** began his apprenticeship with **Henry Raynes Shuttleworth** of the Spectacle Makers' Company, beginning in 1774, and was immediately turned over to John Graves of the Fishmongers' Company, with whom he learned his trade as an optician. He was freed of the Spectacle Makers' Company in 1786.

Upon completion of his apprenticeship the candidate was expected to work as a journeyman for a Freeman of the Company, usually for two years, after which he was normally expected to pay a fee to obtain freedom of the Company. However, not all who served apprenticeships took their freedom, and many who did so took their freedom sometime, even many years, later. In these cases, they might work for a Freeman as a journeyman, and not seek to gain their freedom unless/until they wanted to start their own business, take apprentices of their own, or if sufficient pressure was applied by the officials of the guild.

Patrimony: Another route to guild membership was freedom by patrimony. In this case a Freeman of the guild, usually a parent, paid a patronage fee to purchase the candidate's freedom.

Redemption/Purchase: A person could be freed by redemption, also referred to as purchase. The candidate had to satisfy the officers of the guild that they were suitable for the position, and would be freed upon payment of a fee.

Honoris causa: Freedom could be conferred on a distinguished person as an honour, and would be granted upon payment of a fee.

According to [385], a *Brother* was a person who was already free of one guild, and wanted to become a member of another. They could not become a Freeman, since it was not allowed for a person to be a Freeman of more than one guild. If a person was a Freeman of another London guild, or a provincial guild, they were known as a foreigner. If they paid a suitable fee to become a member of a second guild, they were referred to in the second guild as a *Foreign Brother*. An example is **Thomas Whitford**, who was admitted to the Spectacle Makers' Company in 1743 as a Foreign Brother. The meaning of the term Foreign Brother may, however, have varied from guild to guild. According to [401], the term Foreign Brother applied to a person who had served an apprenticeship but not taken their freedom. Such a person was not allowed to make or sell their goods within the city walls. A person who learned their trade in a foreign country was known as an alien stranger. When admitted after paying the required fee, they were known as an *Alien Brother*. A Brother was lower in rank than a full Freeman and, in some Companies, they could only take on apprentices by turn-over from a Freeman.

Widow: A widow who had worked for her husband for at least seven years was deemed to have served an apprenticeship, and would be awarded their freedom upon payment of a fee, [385 p331, footnote 39] [390]. **Jane Sterrop** is an example of this. Having succeeded her late husband in business, she ran the company for eighteen years after his death, and trained seven apprentices.

In some guilds the term *Yeoman* was used instead of Freeman, although in the earliest days of the guilds a Yeoman was a rank between that of a Freeman and a Liveryman (see below). In some cases, an apprentice was turned over to complete his apprenticeship with a new master. This could be for one of several reasons: the death or incapacity of the master, incompatibility between apprentice and master, to give the apprentice wider experience, convenience bindings (mentioned above), etc. Freemen could, by payment of an additional fee, gain the freedom of the city.

Members of the guilds paid quarterly fees, known as quarterage, to uphold their membership. The guilds had strict regulations, which applied to all members, on matters such as the payment of fees, who could take on apprentices, how many apprentices they could take on, and the progress of a member through the ranks of the Company. It appears that guild members generally abided by these rules, but not always. Those who did not were frequently subject to fines and punishment, but in many cases the guild officers turned a blind eye to the transgression. The fees associated with being a senior officer, and the other expenses they were expected to pay, ensured that the government of the guild was exclusively carried out by the wealthier members.

The guild hierarchy and structure varied from Company to Company. However, it generally consisted of various ranks, the lowest being apprentice, then Brother, and Freeman. Next was the Liveryman, who was required to wear the guild livery, and contribute both money and time to the activities of the guild. Senior to the Liverymen were the Assistants, who were members of the Court of Assistants, the ruling court of the guild, with significant powers. Assistants retained that office for life, and could be called upon to serve as Warden or Master of the Company. There were generally at least two Wardens, a Renter Warden (treasurer), and an Upper Warden. The most senior rank of all was the Master. Each of these ranks had clearly defined responsibilities and fees payable to the Company. If a member was called upon to serve as a Liveryman, Assistant, Warden, or Master, they were required to comply. To decline the role would incur a large fine, payable to the Company. There were two officers who carried out the day-to-day business of the guild. The Clerk maintained the minutes and guild records, and the Beadle carried out general management duties.

Of the 110 Livery Companies of the City of London, a small number emerged as leaders during the C16[th], mainly due to them being the wealthiest, and hence the most powerful. These were known as the Great Twelve. They were: Mercers, Grocers, Drapers, Fishmongers, Goldsmiths, Merchant Taylors, Skinners, Haberdashers, Salters, Ironmongers, Vintners, and Clothworkers.

The London Livery Companies had names such as the Worshipful Company of Grocers, and the Worshipful Company of Spectacle Makers, etc. These are today frequently referred to simply as the Grocers' Company, and the Spectacle Makers' Company, etc. This convention is used in the body of this book, and in this appendix.

In the C17[th] the trades of mathematical, scientific, philosophical, and optical instrument maker were relatively new, and hence no craft guild existed that specialised in these trades. Thus, the instrument makers who wished to operate within the City of London became members of a variety of different Companies, most notably the Grocers' Company and the Spectacle Makers' Company.

In the early C19[th], a dispute arose between the Clockmakers' Company and the Spectacle Makers'

Company, concerning which of the two guilds should have control of the mathematical instrument makers. This centred around a test-case concerning **Robert Brettell Bate**, who was summonsed in 1813 by the Clockmakers' Company to take up his freedom, since he was carrying on his business, but not a member of a guild [355]. He did not respond to the summons and, instead, enrolled with the Spectacle Makers' Company. The dispute was resolved in 1817, when the two companies agreed that those capable of making clocks, watches, and sundials, should be members of the Clockmakers' Company, while those who only retailed these items may be members of the Spectacle Makers' Company. The extent to which efforts were made to enforce this ruling is unclear.

During the C17th and C18th, as the London suburbs outside the city walls grew, so did the number of skilled craftsmen who were not members of any guild. Instrument makers such as **Jesse Ramsden** and **John Bird** never joined a guild, since they operated their businesses outside the city walls. Thus, increasing numbers of craftsmen took their business outside of the influence of the guilds. During this same period, an increase in the admission of Freemen by purchase or by patrimony meant that increasing numbers of Freemen were not skilled practitioners of their guild's core trade. These two factors led to a gradual decline in the power exerted by the guilds over their trades.

The decline in the relevance of the guilds after the repeal of the Statute of Artificers in 1814 was significant. The objective of the guilds had always been to establish a monopoly over each guild's represented trade. Nonetheless, with mathematical and optical instrument makers, the need for apprenticeships to hand down and disseminate knowledge of the craft was highly significant. Very few were self-taught, and learning the craft skills from a master, in a formal setting, was generally a necessary preliminary to becoming a competent instrument maker. Despite this, industrialisation, evolving working practices, lack of guild authority, and the trend for manufacturing companies to replace individual traders in instrument making, led the decline in apprentice numbers over the course of the C19th.

Example: the text of an apprenticeship indenture document

The following text is transcribed from a scan of the original indenture document dated 1738, binding **Samuel Johnson** to serve a seven-year apprenticeship with **James Mann** of the Spectacle Makers' Company:

This Indenture witnesseth that Samuel Johnson son of Abraham Johnson of Westerham in the county of Kent, carpenter, doth put himself apprentice to James Mann, Citizen and Spectaclemaker of London, to learn his art and with him (in the manner of an Apprentice) to serve from the date hereof unto the full end term of seven years, from thence next following to be fully complete and ended. During which term the said Apprentice his said Master faithfully shall serve, his secrets keep, his lawful commandments every where gladly do. He shall do no damage to his said Master nor see to be done of others but that he to his power shall let or forthwith give warning to his said Master of the same. He shall not waste the goods of his said Master, nor lend them unlawfully to any. He shall not commit fornication nor contract matrimony within the said term. He shall not play at cards, dice, tables, or any other unlawful games, whereby his said Master may have any loss with his own goods or others during the said term. Without license of his said Master he shall neither buy nor sell. He shall not haunt taverns or play-houses nor absent himself from his said Master's service day nor night unlawfully: But in all things as a faithful Apprentice he shall behave himself towards his said Master and all his during the said term. And the said Master in consideration of twenty pounds being ye money given with his said Apprentice, his said Apprentice in ye same Art and Mystery which he useth by the best means that he can, shall teach and instruct or cause to be taught and instructed finding unto his said Apprentice meat, drink, apparel, lodging and all other necessaries according to the custom of London during the said term. And for the true performance of all and every the said covenants and agreements, either of the said parties bindeth himself unto the other by these presents. In witness whereof the parties above nam'd do these Indentures interchangeably have put their hands and seals the twelfth day of June in the twelfth year of the reign of our Sovereign Lord George the Second of Great Britain, France, and Ireland, King, Defender of the Faith and so forth and in the year of our Lord one thousand seven hundred and thirty eight. Sealed and delivered in the presence of <there follow three illegible signatures, one of which is the Warden>

A full list of the London Guilds and Freemen mentioned in this book follows:

Armourers and Braziers' Company
Grant Preston.

Blacksmiths' Company
Michael Adams, John Symonds Marratt, Samuel Whitford.

Broderers' Company
See [**385**].
Joseph Howe, John Rowley, Thomas Wright (I).

Butchers' Company
William Simms (I).

Coachmakers' Company
William Elliott.
Freedom not recorded or not known: Charles Alfred Elliott.

Cordwainers' Company
Freedom not recorded or not known: Daniel Scatliff (II).

Clockmakers' Company
See [**389**].
William Harris (I), who was Master of the Clockmakers' Company from 1830-1832.
Freedom not recorded or not known: Charles Manning.

Clothworkers' Company
One of the Great Twelve Livery Companies.
Thomas Dainteth, Francis Arthur Darton, James Simms (II).
Freedom not recorded or not known: George Simms, William Simms (III).

Coopers' Company
Frederick Cox.

Drapers' Company
One of the Great Twelve Livery Companies.
William Bruce, Robert Gogerty, Francis Hauksbee, John Mason, George William Stanton, John Stanton, John Alfred Stanton, Robert Stanton.

Fanmakers' Company
Joseph Long (I).
Freedom not recorded or not known: William Hobcraft (II).

Farriers' Company
William Clarke (III).

Fishmongers' Company
John Jones, Richard Lekeux.

Fletchers' Company
Charles Lincoln, Charles Silberrad.

Founders' Company
William Youle (I).
Freedom not recorded or not known: William Youle (II).

Girdlers' Company
George Casson, Albert Alfred Dixey, Charles Anderson Dixey, Charles Wastell Dixey, Edward Dixey, Benjamin Hooke, Gabriel Wright.

Goldsmiths' Company
One of the Great Twelve Livery Companies.
Joseph Beck, George Cary, John Cary (I), John Cary (II), Joseph Hitch, John Kimbell (I), John Kimbell (II), John Leverton, Benjamin Martin, William Matthews (I), William Simms (II), Thomas Smithies Taylor.
Freedom not recorded or not known: Conrad Beck, David Jones, James Simms (III).

Grocers' Company
The Worshipful Company of Grocers is one of the Great Twelve Livery Companies. Established in 1345, the Worshipful Company of Grocers was responsible for such matters as weights and measures. The historian Joyce Brown, [90], has discovered and documented a whole dynasty of instrument makers in the Worshipful Company of Grocers.

George Adams (I)	J.J. Dallaway	Thomas Heath (I)	Ebenezer Rust (I)
George Adams (II)	George Dollond (I)	Henry Hemsley (II)	Ebenezer Rust (II)
Dudley Adams	George Dollond (II)	Thomas Hemsley (I)	Joseph Rust (I)
John Bailey	Charles Fairbone (I)	John Holyman	Joseph Rust (II)
John Browning (I)	Charles Fairbone (II)	John Huggins (II)	Richard Rust
John Browning (II)	Joseph Fairey (I)	Joseph Jackson	Benjamin Scott
John Browning (III)	Richard Fairey (I)	James Martin	William Spencer (I)
Richard Browning	John Gilbert (I)	James Parsons	William Spencer (II)
Samuel Browning (I)	John Gilbert (II)	William Parsons (II)	Edward Troughton
William Browning	John Gilbert (III)	William Parsons (III)	John Troughton (I)
Henry Crawford	Thomas Gilbert	Charles Potter (I)	John Troughton (II)
Edmund Culpeper (I)	William Gilbert (I)	James Ripley	Tycho Wing
Edmund Culpeper (II)	William Gilbert (II)	Thomas Ripley	John Worgan

Freedom not recorded or not known: George Adams (III), Samuel Browning (II), Samuel John Browning, William Dollond, Henry Gilbert, Charles Gilbert, Thomas Heath (II), Nathaniel Hill, John Pallant, Charles Potter (II), John Thomas Schmalcalder, Samuel Spencer, John Spencer (II).

Joiners' Company
See [385].
Michael Dancer, William Green (I), George Hearne, Ebenezer Hoppe, George Mills, Francis Morgan, John Morgan, Thomas Parnell (I), William Parnell, Robert Tangate, John Urings, Walton Willcox, Benjamin Wood.
Freedom not recorded or not known: Thomas Pickering.

Loriners' Company
Thomas Harris (II).

Masons' Company
William Cook, Henry Gregory, Henry Hughes, Joseph Hughes (I).
Freedom not recorded or not known: Thomas Cole, George Heath, Joseph Hughes (II), William Littlewort.

Merchant Taylors' Company
One of the Great Twelve Livery Companies.
Edward Chester, Benjamin Cole (I), Benjamin Cole (II), John Dobson (II), Nathaniel Hill, John Lilley (I), John Lilley (II), Henry Macrae, Michael Weeden, William John Weeden, John Wright.
Freedom not recorded or not known: John Snart, Charles Cartwright Weeden, Charles Fox Weeden (I), Daniel Weeden, Michael Weeden.

Needlemakers' Company
Oliver Combs

Pattenmakers' Company
Freedom not recorded or not known: John Radford (II).

Shipwrights' Company
Thomas Lorkin.

Spectacle Makers' Company
The Worshipful Company of Spectacle Makers, [401], was established in 1629 by a group of spectacle makers led by Robert Alt. The first Master of the Company was a spectacle maker named Edward Gregorie, and Robert Alt is believed to have been his Upper Warden. The records of the Company were lost during the Great Fire of 1666. Optical instrument makers who were Freemen of the Company, featured in this book, are:

Nathaniel Adams	David Deane	James Jameson	George Richardson
Robert Alt,	George Dollond (I)	Samuel Johnson	Matthew Richardson
Joseph Philip Amadio	George Dollond (II)	David Jones	Edward Scarlett (I)
Herbert Francis Angus	John Dollond (II)	John Jones	Edward Scarlett (II)
James Ayscough	Peter Dollond	Samuel Jones	John Scatliff
Thomas Barnett	William Dowling	William Jones	Samuel Scatliff
George Bass	David Drakeford	Charles Lincoln	Henry Raynes Shuttleworth
Bartholomew Bate	John Dunnell	Thomas Lincoln	Addison Smith
Robert Brettell Bate	William Eastland (I)	Elijah Linnell	Joseph Smith (II)
John Berge	William Eastland (II)	George Linnell	Avis Sterrop
John Bleuler	William Eden	Joseph Linnell	George Sterrop
Edward Blunt	John Eglington	Robert Linnell	Jane Sterrop
Thomas Blunt (I)	Peter Eglington	Matthew Loft	Mary Sterrop
Thomas Blunt (II)	John Field	James Long	Ralph Sterrop
William Blunt	Willliam Ford	William Longland	Thomas Sterrop (I)
Timothy Brandreth	Thomas Gay	James Mann (I)	Thomas Sterrop (II)
William Edmund Callaghan	William Gilbert (I)	James Mann (II)	James Tomlinson
Alfred Chislett	William Gilbert (II)	John Mann	Francis Walker
Thomas Clark	William Gofton	John Margas	Francis Watkins (I)
James Cleare	Henry Hemsley (I)	Damaine Middlefell	Charles Robert West
Christopher Cock	Thomas Hemsley (II)	William Moon	Francis West
Oliver Combs	Thomas Hemsley (III)	Edward Nairne	George Whitbread
Matthew Cooberow	Thomas Hails	Frederick Newton	Thomas Whitford
James Cox	Thomas Harris (I)	Andrew Pritchard	George Willdey
John Cox (II)	William Harris (II)	Henry Pyefinch	Judith Willdey
William Cox (II)	John William Holmes	John Radford	Thomas Willdey
John Cuff	John Howe	William Radford	Charles Wilson
John Davis (III)	Joseph Howe	George Ribright	John Yarwell
Francis Day	Alexander Hughes	Thomas Ribright (I)	
John Day	Henry Hughes	Thomas Ribright (II)	

Freedom not recorded or not known: John Huggins (I), Thomas McIntosh, Francis Linsell West, Henry West, John George West.

Optical instrument makers who served as Master of the Worshipful Company of Spectacle Makers included:

John Radford 1669-1670, and 1687	David Deane 1760
Joseph Howe 1672-1674, and 1689-1690	Peter Eglington 1761-1762
William Longland 1686-1687, and 1694-1695	James Tomlinson 1764-1768
Thomas Sterrop (I) 1701	Edward Nairne 1769-1773, and 1795-1796
Ralph Sterrop 1702-1708	Peter Dollond 1774-1881, 1797-1798, and 1801-1802
James Mann (I) 1716	Henry Shuttleworth 1782-1786
Thomas Gay 1718-1720	Charles Lincoln 1787-1789
Edward Scarlett (I) 1720-1721	John Dollond (II) 1790-1791
James Mann (II) 1735-1737	Thomas Blunt (I) 1792-1793, and 1815-1816
Matthew Loft 1744-1745	John Field 1799-1801, and 1820-1822
Edward Scarlett (II) 1745	James Long 1805-1807
Thomas Lincoln 1746	William Gilbert (I) 1807-1808
George Bass 1747, and 1754-1757	John Bleuler 1794, and 1809-1810
John Cuff 1748	George Dollond (I) 1811-1812
George Sterrop 1750	William Harris 1824-1825
Samuel Scatliff 1751	Robert Brettell Bate 1828-1829
James Ayscough 1752	George Dollond (II) 1862
Thomas Ribright (I) 1758-1759	

Stationers' Company
William Archer, John Bennett (I), William Bush, James Champneys, James Chapman (I), John Cooke, John Cuthbertson, George Dixey, John Johnson Evans, William Heather, William Hobcraft (I), John Munford, William Price, James Search, James Simons, Christopher Stedman, Samuel Street (I), Walter Watkins, Alexander Wellington, George Willson.

Freedom not recorded or not known: James Chapman (II), Samuel Street (II), Samuel Wells.

Tinplate Workers' Company
Alexender Mackenzie.

Turners' Company
James Haly Burton, Christopher Cock, Joseph Cox, William Cox (I), John Dunnell, John Gray (II), James Mann (II), John Marshall, John Smith (I), John Ustonson, Charles Fox Weeden (I), Daniel Weeden.

Vintners' Company
One of the Great Twelve Livery Companies.
James Long, George Stebbing, George James Stebbing, Francis Watkins (II).
Freedom not recorded or not known: Frederick George Stebbing, Richard William Stebbing.

Wheelwrights' Company
Stephen Bithray.

Worshipful Company of Makers of Playing Cards
John Crichton, Benjamin Messer, John Newman.
Freedom not recorded or not known: Anthony James Lear, Robert Murray.

Appendix X: The Dollond Patent Dispute and the Petition of 1764

[1] [2] [24] [177] [364] [365] [366] [388] [400]

Background

Sir Isaac Newton carried out considerable experimentation on the dispersion of light in a refracting medium such as a lens or a prism. He concluded in 1672 that refraction is always accompanied by dispersion, and included this result in his book *Opticks*, first published in 1704. This research, and the conclusions he drew, led to the enduring, widely held belief among optical instrument makers that chromatic aberration could not be corrected in a refracting telescope.

Nonetheless, in 1695, David Gregory, the nephew of **James Gregory**, had already observed that, since the human eye does not suffer from the colour-fringing consequences of dispersion, it must be possible to learn something from a study of the eye. In his words:

"Perhaps it would be of service to make the object-lens of a different medium, as we see done in the fabric of the eye, where the crystalline humour (whose power of refracting rays of light differs very little from that of glass) is by nature, who never does anything in vain, joined with the aqueous and vitreous humours (not differing from water as to their power of refraction) in order that the image may be painted as distinct as possible upon the bottom of the eye."

Subsequent research has shown that the human eye is not achromatic. Nonetheless this idea informed some of the later research, including the theories of the eminent Swiss mathematician and physicist, Leonhard Euler (see below), and some of the consequent experiments made by **John Dollond (I)**.

c1730: The Invention

Chester Moor Hall was a barrister of the Inner Temple, London, England, who is generally credited with inventing the achromat. His hobby was optics, and in 1729-1730, based on his own experiments he devised a new theory: "that of correcting the refrangibility of the rays of light in the construction of object glasses for refracting telescopes." His idea was to use two lenses of different refractive index which, when combined, would serve the function of an objective lens, and at the same time reduce the colour dispersion in the image. He described his invention to **Addison Smith**, who later wrote that Hall "…told me that about the year 1729 he had made some prismatical experiments in order to illustrate a new theory in optics, that of correcting the refrangibility of the rays of light in the construction of object glasses for refracting telescopes, and the success attending his experiments had fully convinced him that his theory was well founded…"

Chester Moor Hall went to two different opticians, **Edward Scarlett (I)** and **James Mann (II)** and commissioned each to grind one of the two optical elements, the crown and flint glasses. Both of these opticians coincidentally subcontracted the work to the same person, **George Bass**, a grinder and polisher in Bridewell. This is believed to have been the first ever compound object glass, and had a diameter of 2½ inches, and a focal length of 20 inches. Chester Moor Hall, with this invention, successfully reduced chromatic aberration in a telescope to a level where it was considered tolerable. Having thus achieved his goal, he never publicised his work in this field, and made no attempt to seek a patent for it. He did, however, give written instructions on how to make such a lens to **John Bird** and **James Ayscough**.

In 1747, Leonhard Euler submitted a memoir to the Berlin Academy of Sciences in which he claimed to show that both chromatic aberration and spherical aberration could be corrected by means of an arrangement of two meniscus lenses separated by water. John Dollond (I) was made aware of this memoir by **James Short (I)**, and informed Euler, with reference to Newton's results, that he was in error. He also wrote to Short, saying "It is therefore, Sir, somewhat strange that anybody now-a-days should attempt to do that, which so long ago has been demonstrated impossible." Nonetheless, following correspondence with Euler, John Dollond (I) carried out experiments of his own, but failed to reproduce a workable result. Euler, however, was convinced that his method would significantly reduce chromatic aberration, but conceded that it did suffer from spherical aberration.

James Ayscough, was James Mann (II)'s partner and ex-apprentice. According to later testimony by **William Eastland**, in 1752 he had received a subcontract from Ayscough for another order from Chester Moor Hall. This was accompanied by written instructions on how to make the lens. He completed the work for Hall and, at this time, both Eastland and Ayscough began to make telescopes using Hall's invention.

In 1754, **Samuel Klingenstierna** was the first to publish a comprehensive theory of optical systems with no colour dispersion or spherical aberration. He showed that

Sir Isaac Newton had made an experimental mistake, and that his results applied only to prisms whose apex angle is small. He corresponded with **John Dollond (I)**, sharing these results, and encouraging Dollond's experimentation. John Dollond (I) experimented using crown glass prisms, with water as a second medium of different refractive index. The results, however, were unsatisfactory. In order to overcome both chromatic aberration and spherical aberration, he needed to replace the water with a material of higher dispersion.

According to a statement made much later by **Jesse Ramsden**, it was James Ayscough who first brought Chester Moor Hall's invention to the attention of John Dollond (I), although other accounts give this credit to **Robert Rew**. In 1757, John Dollond (I) had a conversation with George Bass in his workshop, and understood that Chester Moor Hall had used a flint glass optical element, together with a crown glass element, to overcome chromatic aberration. This gave Dollond the second optical medium, with higher dispersion, that he needed.

1758: The Patent, and the Watkins Partnership

By 1758, John Dollond (I) had satisfied himself that he had a workable solution to the problem of chromatic aberration in a telescope, in the form of a crown and flint compound objective lens. He published his findings in 1758, in the *Philosophical Transactions of the Royal Society of London*, with no acknowledgement of Hall's or Klingenstierna's contribution to his work, with the result that his relationship with Klingenstierna faltered. Encouraged by his son, **Peter Dollond**, and with financial assistance from **Francis Watkins (I)**, he petitioned for, and was awarded a 14-year patent for the invention.

The patent was announced in the *London Gazette* on 3rd June 1758 [35 #9797 p2]: "By His MAJESTY'S Patent, A new-invented refracting telescope, by which objects are seen much clearer and distincter than by any before made, the object glass being made of mediums of different refractive qualities are so adapted as to correct the errors arising from the different refrangibility of light, as well as to those which proceed from the spherical surfaces of the glasses. These telescopes, will therefore be found of infinite advantage to the publick in general, but more particularly to the navy, where good refracting telescopes are of the utmost importance. To be had of Mr. Francis Watkins, optician, at Charing Cross; and of Mr. John Dollond, in Denmark Court in the Strand. Mr. Watkins has also reflecting telescopes of all sizes, finished to the greatest perfection."

By this time a partnership was already in place between John Dollond (I) and Francis Watkins (I). This was agreed by Dollond in recognition of the financial contribution Watkins had made to obtaining the patent. The agreement enabled them to both trade in, and share the profits from the sale of telescopes with the new compound achromatic objective. The partnership, dated 29 May 1758, was set to continue for the 14-year term of the patent. Almost immediately after entering this agreement, Francis Watkins learned about Chester Moor Hall's priority in the invention, and that Dollond was not the true inventor.

In 1761 John Dollond (I) was elected a Fellow of the Royal Society. He was awarded the Copley Medal, largely for his contribution to the development of the achromatic telescope lens. The term *achromatic* was coined by the amateur astronomer, John Bevis.

During the remaining years of his life, John Dollond (I) made no attempt to enforce his patent, and many London optical instrument makers benefited from it while paying no fees to him. Following his death in 1761, his son, Peter Dollond, fell into dispute with Francis Watkins (I) and, with a payment of £200, terminated the agreement. However, Watkins continued to make and sell telescopes in violation of the patent, and Peter Dollond successfully sued him. The judge asserted Chester Moor Hall's priority as inventor, but in 1764 the court ordered Watkins to cease production and sale of these telescopes. By this time, considerable resentment was building among the London optical and mathematical instrument makers. Thanks to George Bass, Edward Scarlett (I), James Mann (II), James Ayscough, and William Eastland, they were aware of the earlier invention by Chester Moor Hall. Indeed, some had been making achromatic lenses based on Hall's design prior to the granting of the Dollond patent. They viewed Peter Dollond's aggressive enforcement of the patent he had inherited from his father, for an invention he didn't make, as an unreasonable impairment of their ability to carry out their trade.

1764: The Petition

In 1763 Francis Watkins became Master of the **Spectacle Makers' Company**. In 1764 he organised a number of mathematical and optical instrument makers of London to submit a petition to the Privy Council, requesting that they revoke the patent. The 35 petitioners, in the order listed in [364], are as follows:

Addison Smith, Charing Cross
John Bird, Strand
Tycho Wing, Strand
Samuel Scatliff, St. Paul's Churchyard
Joseph Linnell, Ludgate Street
William Eastland, Clerkenwell
George Ribright, Poultry

Jas. Champneys, Cornhill
John Eglington, Hatton Garden
David Deane, Smithfield
Benjamin Martin, Fleet Street
John Bennett (I), Crown Court, Soho
John Troughton (I), Surrey Street (*Note 1*)
Nathaniel Hill, Chancery Lane
John Cuff, Strand
Joshua Bostock, Drury Lane
Samuel Wright, Bedford Street
James Jameson, Saffron Hill
Jos. Hitch, Eagle Court, St. John's Lane
Peter Eglington, Strand
John Cox (II), St. John's Court, Cow Lane
William Cole, Strand
Francis Morgan, Cary Street
J. Cleare, Fleet Street (*Note 2*)
J. Burton, Fleet Street (*Note 3*)
George Bass (Bast), Fleet Ditch
David Drakeford, Fleet Ditch
John Cooke, Snow Hill
Robert Featley, Fleet Street
Robert Rew, Coldbath Fields
William Ford, Cannon Street
John Davies, Charing Cross
J. Burton, Johnson's Court, Fleet Street (*Note 3*)
J. Cleare, Mitre Court, Fleet Street (*Note 2*)
J. Clack, Saffron Hill

Note 1: [364] states that the John Troughton listed here is the brother of **Edward Troughton** – i.e., **John Troughton (II)** as listed in this volume. However, [18] states that it was **John Troughton (I)**. The 1764 date suggests it was more likely to be John (I), who was trading at Surrey Street at the time, while John (II) didn't gain his freedom until 1764.

Note 2: Two entries for J. Cleare of Fleet Street: [18] lists **James Cleare** of Mitre Court, Fleet Street, who was a spectacle maker and telescope maker. His father was John Cleare, for whom no occupation is listed. It is possible that the two entries in the list for J. Cleare may have been father and son.

Note 3: Two entries for J. Burton of Fleet Street: **John Burton**, optical instrument maker of Johnson Court, Fleet Street was one of the petitioners. Additionally, [18] lists **James Haly Burton**, mathematical instrument maker of Johnson's Court, Fleet Street, who may also have been one of the petitioners, and would account for the two entries in the list.

The basis of the petition was:
- The method of producing an achromatic lens was not new at the time the patent was granted.
- John Dollond (I) was not the first and true inventor.
- Objective lenses constructed to the method described in Dollond's patent were being made and publicly sold prior to the granting of the patent.
- John Dollond (I) had not in his lifetime made any attempt to enforce the patent.

The Spectacle Makers' Company contributed £20 towards the cost of the petition, despite the fact that Peter Dollond was a member, as his father, John Dollond (I), had been during his lifetime. Records of a ruling in the case, if one was made, are not known to have survived. Subsequent events, however, make it clear that the patent remained in force.

After the Petition

In 1765 Peter Dollond initiated legal action against William Eastland, **Christopher Stedman** and **James Champneys** for violation of his patent. The case against Champneys was tried in February 1766 in the Common Pleas (a common law court), before Lord Camden, and that judgement was delivered on 20th February 1766 in favour of Peter Dollond.

The judge is reputed to have said: "It was not the person who locked up his invention in his scrutoire that ought to profit by a patent for such invention, but he who brought it forth for the benefit of the public."

The judge thereby reasserted Lord Camden's ruling, effectively enforcing Dollond's right to demand a royalty from anyone who produced achromatic telescopes until the expiration of the patent in 1772. This enabled Dollond to increase the market price for achromatic telescopes, and substantially increase his profits. He used two glass elements for his telescope objectives, and three glass elements for his opera glass objectives.

In 1766 Peter Dollond took action against Francis Watkins (I) and Addison Smith, who were at that time working in partnership, as joint defendants.

Some sources state that in 1766, when John Dollond (I)'s daughter, Sarah, was married to Jesse Ramsden, her dowry consisted of half of the patent. If this is true, the dowry would have been given by Peter Dollond. No record has been found of any involvement of Ramsden in the patent dispute or the ensuing lawsuits [2].

In 1768, Peter Dollond took action against **Henry Pyefinch**. This was the last legal action taken by Peter Dollond in enforcement of his patent.

By the end of 1768, William Eastland and James Champneys both moved to Holland, where they settled, Eastland in The Hague, and Champneys in Amsterdam. James Champneys also brought with him his former apprentice, the mathematical and philosophical instrument maker, **John Cuthbertson**, who by then was also his son-in-law.

In 1772 the patent expired. The bad feelings engendered by Peter Dollond's actions, however, rumbled on. Peter Dollond repeatedly justified his enforcement actions, but never once mentioned Chester Moor Hall. In 1790 a letter was published in the *Gentleman's Magazine*, signed by "Veritas", mentioning Mr Euler and Mr Klingenstierna, but clearly attributing the invention to Hall. Part of the letter read:

"Mr Hall used to employ the working opticians to grind his lenses; at the same time he furnished them with the radii of the surfaces, not only to correct the different refrangibility of rays but also the aberration arising from the spherical figures of lenses. Old Mr Bass, who at that time lived in Bridewell Precinct, was one of these working opticians, from whom Mr Hall's invention seems to have been obtained."

He also says: "That Mr Ayscough, optician on Ludgate Hill, was in possession of one of Mr Hall's achromatic telescopes in 1754, is a fact which at this time will not be disputed."

Veritas is generally believed to have been Jesse Ramsden.

Appendix XI: Scale Dividers' Marks

Many astronomical, navigation, surveying, and other instruments have a curved scale, marked in degrees and fractions, to assist the user in measuring angles. With the advent of the dividing engine in the C18th, and the resulting increase in accuracy, makers who used them sought to distinguish their work from the less accurate scales marked without the use of a dividing engine. The mark used by a scale divider sometimes indicated the dividing engine used, rather than the person who used it to divide the scale. Instrument makers who were not in possession of a dividing engine took their instruments to a maker who did have one. Such instruments would normally have the maker's name on them, as well as a divider's mark on the scale. Most instruments known with these scale dividers' marks are navigation instruments including octants, some sextants and a quintant. [16] [374] [381]. However, they are not exclusively navigation instruments, as illustrated by a theodolite, made by Nairne & Blunt, with a scale divided on Jesse Ramsden's dividing engine.

The foul anchor symbol was used by some makers (see entry for foul anchor). However, this was not the only anchor representation used in scale dividers' marks, and nor were the marks exclusively anchors. In cases where the mark has no initials, it is difficult to be sure who the scale divider was.

Anchor Symbols

It is believed that this mark may have been used by **James Allen (II)**. It appears on an octant signed by **R.B. Bate**.

This mark was used by **Dring & Fage** on instruments bearing their name, e.g. octant dated c1800.

The mark with initials GN is found on two unsigned octants. The scale divider is not known.

The mark with initials GW is found on an unsigned octant dated c1790. The scale divider is not known.

The mark with the initials ID is found on an octant by **Thomas Blunt** dated c1815, and on an unsigned octant dated c1790. The scale divider is not known.

This mark is found on an octant by **Richard Patten**. The scale divider is not known.

This mark is found on an octant by **Dring & Fage** with the address 248 Tooley St., which dates it to sometime between 1800 and 1822. The scale divider is not known.

The anchor with no initials is thought to have been used by **John Troughton (II)** and **Edward Troughton**, both of whom built their own dividing engines, and this symbol may also have been used by other makers. It is found on octants, dated from c1800 to c1840, bearing the names: Robert Banks (**Robert Bancks**); **Richard Barry**; **Isaac/George Bradford**; **Benjamin Brown**; **William Cary**; **William Cook**; **William Harris (I)**; **William Heather**; **G.W. Peter & Zoon**; M. Wardell & Sons;

Foul Anchor Symbols

The foul anchor with initials EP was used by **Edward Pritchard** when engraving marine instruments with the small dividing engine he inherited from **Jesse Ramsden**. It is found on octants bearing the names **William Cary**; **A. Chevalier**; **Dring & Fage**; **Thomas Jones**; **Jesse Ramsden**; **George Richardson**; **Benjamin Wood**; as well as unsigned octants.

The foul anchor symbol with the letters I R was used by **Jesse Ramsden** to identify a scale as having been engraved on his dividing engine. He used this mark at the centre of the scales on marine sextants and octants. The use of this symbol continued to be associated with his dividing engine after it came into the possession of **Matthew Berge** and, later, **Nathaniel Worthington**. It is fond on:

Sextants bearing the names **George Adams (II)**; **Nairne & Blunt**; Nathaniel Worthington; and unsigned.

Octants bearing the names: George Adams (II); Matthew Berge; **William Cary**; J.&L. Hardy; **Benjamin Martin**; **Joshua Lover Martin**; **John H. Moore**; Nairne & Blunt; Jesse Ramsden; **George Stebbing**; and unsigned.

A quintant bearing the name Nathaniel Worthington.

A theodolite bearing the name Nairne & Blunt.

The foul anchor with initials JA is thought to have been used by **James Allan (I)**, when engraving instruments using the dividing engine he built some time before 1809. It is found on octants bearing the names: **Robert Brettell Bate**; **Joseph Fairey**; **Richard Patten**.

The foul anchor symbol is also found with other initials and marks:
GN: on unsigned English octants
RG: on an octant by **Dring & Fage**.
HD: on an octant by **Benjamin Messer**.
B&J: on an octant by **Blachford & Imray**.
WH: on an unsigned English octant
SP: on an unsigned English octant.
Fleur-de-lis: on an octant by **Thomas Ripley & Son**.

The foul anchor with no initials is found on:
Two unsigned English sextants.
Octants bearing the names: **Richard Barry**; **Isaac/George Bradford**; **Gilbert & Co.**; **Gilkerson & Co.**; **J.&L. Hardy**; **William Heather**; **William Holliwell**; **G. Hulst van Keulen**; **Thomas Jones**; **Thomas Ripley & Son**; **Alexander Wellington**; Wilkinson & Sons (London).

Other Symbols

B&J — This is the initials of **Blachford & Imray**, and is found on an octant bearing their name. It is believed they may have acquired a dividing engine second-hand. They are also known to have used a foul anchor symbol with the initials B&J (see above).

FW — Scale divider not known. These initials are found on:
  A sextant bearing the name **James Grant**, and an unsigned one.
  Octants bearing the names: Brokovski (Liverpool); **George Christian**; **Henry Hughes**; **J.W. Norie & Co.**; **Somalvico & Co.**; **Benjamin Wood**;

SBR — This is the initials for **Spencer, Browning & Rust**. The mark is found on all octants and wooden framed sextants made by them. It is also found octants by **Alexander** of Yarmouth, **Charles Jones**, **Alexander Adie**, **David Heron**, **William Parnell**, **Henry Duren**, and **Joseph Fairey**, as well as numerous unsigned octants.

IID — Scale divider unknown. This mark is found on an octant by **Benjamin Messer**.

References

Bibiliography and references

[1] *Eyes Right: The Story of Dollond & Aitchison Opticians, 1750-1985*. Hugh Barty-King, 1986.
[2] *Jesse Ramsden (1735-1800): London's Leading Scientific Instrument Maker*. Anita McConnell, 2007.
[3] *Making Scientific Instruments in the Industrial Revolution*. A.D. Morrison-Low.
[4] *Admiralty Manual of Navigation. Volume I*. HMSO 1954.
[5] *Admiralty Manual of Navigation. Volume II*. HMSO 1960.
[6] *Admiralty Manual of Navigation. Volume III*. HMSO 1954.
[7] *Optics*. Hecht & Zajac. 1974.
[8] *Cyclopaedia of Telescope Makers Part 1 (A-F)*. A.D. Andrews. Irish Astronomical Journal, 1992, vol. 20(3).
[9] *Cyclopaedia of Telescope Makers Part 2 (G-J)*. A.D. Andrews. Irish Astronomical Journal, 1993, vol. 21(1).
[10] *Cyclopaedia of Telescope Makers Part 3 (K-N)*. A.D. Andrews. Irish Astronomical Journal, 1994, vol. 21(3/4).
[11] *Cyclopaedia of Telescope Makers Part 4 (O-R)*. A.D. Andrews. Irish Astronomical Journal, 1995, vol. 22(1).
[12] *Cyclopaedia of Telescope Makers Part 5 (Sae-Sim)*. A.D. Andrews. Irish Astronomical Journal, 1996, vol. 23(1).
[13] *Cyclopaedia of Telescope Makers Part 6 (Sin-Syk)*. A.D. Andrews. Irish Astronomical Journal, 1996, vol. 23(2).
[14] *Cyclopaedia of Telescope Makers Part 7 (T-Z)*. A.D. Andrews. Irish Astronomical Journal, 1997, vol. 24(2).
[15] *Appendix to the Cyclopaedia of Telescope Makers*. A.D. Andrews. Irish Astronomical Journal, 1998, vol. 25(1).
[16] *Sextants at Greenwich: A Catalogue of the Mariner's Quadrants, Mariner's Astrolabes Cross-Staffs, Backstaffs, Octants, Sextants, Quintants, Reflecting Circles and Artificial Horizons in the National Maritime Museum, Greenwich*. W.F.J. Morzer Bruyns, Dr. Richard Dunn. Oxford University Press, 2009.
[17] *Nineteenth-century Scientific Instruments*. Gerard L'Estrange Turner.
[18] *Directory of British Scientific Instrument Makers 1550-1851*. Gloria Clifton. 1995.
[19] *We're certainly not afraid of Zeiss: Barr and Stroud Binoculars and the Royal Navy*. William Reid. 2001, ISBN 1-901663-66-3.
[20] *The Optical Munitions Industry in Great Britain, 1888–1923*. Stephen C. Sambrook.
[21] *Optical Glass - Description of Properties. Pocket Catalogue*. Schott. 2000 and 2016.
[22] *Basic Optics and Optical Instruments*. The Bureau of Navy Personnel (US). Dover, 1969.
[23] *Russian Defense Business Directory: St. Petersburg And Leningrad Oblast*. Edited by Franklin J. Carvalho. US Department of Commerce, 1996.
[24] *The History of the Telescope*. Henry C. King. Charles Griffin & Co. Ltd., 1955. Dover Reprint, 1979 and 2003.
[25] *Directions for making the best composition for the metals for reflecting telescopes, together with a description of the process for grinding, polishing, and giving the great speculum the true parabolic curve*. John Mudge. Philosophical Transactions of the Royal Society of London, Vol 67, 1777. P296.
[26] *Range and Vision: The First Hundred Years of Barr & Stroud*. Michael Moss and Iain Russell. 1988.
[27] *From Tsarism to the New Economic Policy. Studies in Soviet History and Society*. Edited by R.W. Davies (University of Birmingham). 1990.
[28] *Blackmail Sabotage: Attacks on French industries during World War Two*. Bernard O'Connor.
[29] *Notes on British Makers of Prismatic Binoculars During World War 1*. Terence Wayland.
[30] *Dictionary of British Scientific Instruments*. Issued by the British Optical Instrument Manufacturers' Association. Pub: London Constable and Company Ltd. 1921 (Reprinted without changes by Interbook International BV, Schiedam, 1976).
[31] *From Camera to Cinemascope: Photography Was Born and Raised in France*. Pierre G. Harmant. Published in "France Actuelle", vol. VII, No 6, March 15, 1958.
[32] *The Littorio Class: Italy's Last and Largest Battleships*. Ermingo Bagnasco, 2011.
[33] *The Zeiss works and the Carl-Zeiss stiftung in Jena; Their scientific, technical and sociological development and importance*. Felix Auerbach, 1903. Translated from the German second edition by Siegfried F. Paul and Frederick J Cheshire. Pub Marshall, Brookes & Chalkley Ltd., London. 1904.
[34] *The Telescope in Ireland: Obscure makers & marks. Irish telescope makers and Irish signatures on telescopes*. Peter Abrahams.
[35] *London Gazette*.

[36] *Voigtländer & Sohn: Die Firmengeschichte von 1756 bis 1914 [Voigtländer & Son: The company history from 1756 to 1914].* Grabenhorst, Carsten. Braunschweig: Museum für Photographie - Appelhans Verlag, 2002.
[37] *Biographical Dictionary of the History of Technology.* Lance Day (Editor), Ian McNeil (Editor). Routledge, London, 1998.
[38] *Scientific Instruments 1500-1900: An Introduction.* Gerard L'Estrange Turner. Philip Wilson Publishers, 1998.
[39] *Evolution of the Astronomical Eyepiece.* C.J.R. Lord. Brayebrook Observatory, 1996.
[40] *A History of the Photographic Lens.* Rudolf Kingslake.
[41] *The Cambridge Illustrated History of the World's Science.* Colin A. Ronan. Cambridge University Press. 1983.
[42] *Classic Telescopes: A Guide to Collecting, Restoring and Using Telescopes of Yesteryear.* Neil English. Springer New York. 2013.
[43] *Binocular Astronomy.* Stephen Tonkin. Springer. 2014.
[44] *Grace's Guide to British Industrial History.* Edited by Andrew Ian Tweedie. (https://gracesguide.co.uk).
[45] *Encyclopædia Britannica.* Published by Encyclopædia Britannica Inc.
[46] *Oxford Dictionary of National Biography.* Oxford University Press.
[47] *Deutsche Biographie.* Duncker & Humblot, Berlin.
[48] *Officine Galileo: 150 Years of History and Technology.* Museo Galileo. Istituto e Museo di Storia della Scienza. Italy.
[49] *The Daguerreotype in Liverpool in 1839: Ste Croix, A. Abraham and J.B. Dancer.* R. Derek Wood
[50] *Historical Makers of Microscopes and Microscope Slides.* Brian Stevenson. (http://microscopist.net).
[51] *Australian Dictionary of Biography.* Australian National University.
[52] *Museum of Science and Industry.* Liverpool Road, Manchester. Collections database.
[53] *Science Museum.* Exhibition Road, South Kensington, London. Collections database.
[54] *Royal Museums Greenwich.* National Maritime Museum; Cutty Sark; Queen's House; Royal Observatory.
[55] *Binocular and Cine Collection.* Anna and Terry Vacani. (https://binoculars-cinecollectors.com/).
[56] *John Henry Barton. A Biography.* Terence Wayland.
[57] *Beck Microscopes.* Catalogue by R.J. Beck Ltd., London, England.
[58] *National Museum of American History.* Smithsonian; Kenneth E. Behring Center.
[59] *John Brashear: Biography.* Glenn A. Walsh. 2001.
[60] *Gore's General Advertiser.* 8 May 1800, p3.
[61] *Early Photography.* (http://earlyphotography.co.uk).
[62] *The Dutch Virtual Magic Lantern Museum.* Henc R.A. de Roo, Huizen, the Netherlands. (https://luikerwaal.com).
[63] *Astronomical Optics: Part 4 - Optical Aberrations.* Black Oak Observatory.
[64] *Binoculars and Telescopes for Field Work.* J.R. Hebditch. British Trust for Ornithology Field Guide Number Two. 1967.
[65] *Binoculars.* J.R. Hebditch. 1959.
[66] *Choosing a Binocular.* Booklet C. 6 by Barr and Stroud (undated).
[67] *Guide Book of Japanese Optical and Precision Instruments.* The Japan Optical and Precision Instruments Manufacturers' Association. Edition Dates: 1953-1974.
[68] *The History of the Telescope and the Binocular.* Peter Abrahams, Portland, Oregon.
[69] *Number Code on Reflecting Telescopes by Nairne and Blunt.* George L'E Turner. Journal for the History of Astronomy, Vol. 10, P. 177, 1979.
[70] *The History of Photography from the camera obscura to the beginning of the modern era.* Helmut Gernsheim in collaboration with Alison Gernsheim. Oxford University Press, 1955.
[71] *History of Photography.* Josef Maria Eder, translated by Edward Epstean. Dover Publications, 1972.
[72] *Evolution of the Focault-Secretan Reflecting Telescope.* William Tobin. Journal of Astronomical History and Heritage, 19(2), 106–184 (2016).
[73] *Venetian Makers of Optical Instruments of the 18th-19th Centuries. Part 1.* Alberto Lualdi. Bulletin of the Scientific Instrument Society. No. 76, 2003.
[74] *Venetian Makers of Optical Instruments of the 18th-19th Centuries. Part 2.* Alberto Lualdi. Bulletin of the Scientific Instrument Society. No. 77, 2003.
[75] *Venetian Makers of Optical Instruments of the 18th-19th Centuries. Part 3.* Alberto Lualdi. Bulletin of the Scientific Instrument Society. No. 78, 2003.

[76] *George Calver - East Anglian Telescope Maker*. H.E. Dall. Journal of the British Astronomical Association. Vol 86, p. 49-52. 1975.

[77] *Photographic chart of the heavens to Argelander's scale $1^0=20mm$*. J. Franklin-Adams. Monthly Notices of the Royal Astronomical Society Vol. 64, p.608. 1904.

[78] *Binoculars by Moritz Hensoldt & Sons, of Wetzlar, Germany*. Compiled by Peter Abrahams.

[79] *Obituary: Ronald Nicholas Irving, 1915-2005*. C. Lord. Journal of the British Astronomical Association, Vol. 116, No. 3, p.146. 2006.

[80] *Ross Military Binoculars*. Terence Wayland

[81] *Fellows deceased :- William Charles Cox*. Monthly Notices of the Royal Astronomical Society, Vol. 35, p.167. 1875.

[82] *Ross of London, a Chronology*. Terence Wayland

[83] *Obituary notice: Henry Wildey, 1913-2003*. Daniels, D. & Wildey, H. Journal of the British Astronomical Association, vol.114, no.1, p.48-49. 2004.

[84] *A Description of a Microscopic Doublet*. William Hyde Wollaston. Philosophical Transactions of the Royal Society of London, Vol. 119, pp.9-13. 1829.

[85] *Binoculars by Moritz Hensoldt & Sons, of Wetzlar, Germany*. Peter Abrahams.

[86] *Zeiss Binocular Production Numbers, Military and Civilian*. Peter Abrahams, based on production and sales statistics from Zeiss archives BACZ 8401, 11809, 17119, 19607, 23209-23270.

[87] *List of Zeiss Binoculars From 1894-1950 (Jena) and 1954-1972 (Oberkochen & Wetzlar)*. Wolfgang Kind and Peter Hudemann; Translated and extended by Jack Kelly, Fred Schwartzman, and Peter Abrahams.

[88] *Outline of Japanese binocular production*. Peter Abrahams.

[89] *Japanese Manufacturers' Codes for Optical Products*. Peter Abrahams, based on a list obtained by Bill Beacom.

[90] *Mathematical Instrument-Makers in the Grocers' Company 1688-1800*. Joyce Brown. Science Museum. 1979.

[91] *Henry Barrow, Instrument Maker*. John T. Stock. Bulletin of the Scientific Instrument Society. No. 9 1986.

[92] *Ross Binoculars Code Words and Model Names*. Terence Wayland.

[93] *Serial Numbers of Ross Prismatic Binoculars*. Terence Wayland.

[94] *Vintage Miniature Binoculars: Virtual Museum*. Mark Ohno. (https://miniaturebinoculars.com).

[95] *The Description and Construction of a new constructed octant, sextant and quintant. By His Majesty's Patent, Being an Improvement on the Hadley's Quadrant*. Gabriel Wright. London, 1779.

[96] *The English equatorial mounting and the history of the Fletcher Telescope*. Wayne Orchiston. Journal of Astronomical History and Heritage, Vol. 4, No. 1, p. 29-42 2001.

[97] *The Hale Telescope on Palomar Mountain*. Ron Maddison. Journal of the Antique Telescope Society. Vol. 13, pp.4-14. 1997.

[98] *J.H. Steward: A Family Dynasty*. David G. Rance. Slide Rule Gazette, Issue 12, Autumn, 2011.

[99] *The Corfield Story*. Bev Parker. (https://corfield.org/camera/corfield.htm).

[100] *Obituary Notices: Fellows:- Franklin-Adams, John*. Monthly Notices of the Royal Astronomical Society, Vol. 73, p.210. 1913.

[101] *Telescope: 6-inch Franklin Adams wide-angle Star Camera (1898)*. The Royal Observatory, Greenwich.

[102] *A Time-Telling Telescope*. Richard Dunn and Lavinia Maddaluno. Antiquarian Horological Society. Antiquarian Horology. March 2012.

[103] *Surveying instrument makers of Central Canada*. Randall C. Brooks and William J. Daniels. Canadian Journal of Civil Engineering. 20, pp.1037-1046 1993.

[104] *Rates of chronometers and watches on trial at the Observatory, 1766–1915*. The Royal Observatory, Greenwich.

[105] *History of Kodak*. (http://archive.org).

[106] *A Short History of BBT Krauss*. Binocular and Cine Collection – Anna and Terry Vacani. (https://binoculars-cinecollectors.com).

[107] *Lieberman & Gortz Short History*. Binocular and Cine Collection – Anna and Terry Vacani. (https://binoculars-cinecollectors.com).

[108] *Microfilm – A Brief History*. University of California Southern Regional Library Facility.

[109] *LUCERNA. The Magic Lantern Web Resource*. (https://lucerna.exeter.ac.uk).

[110] *Whitaker's Red Book of Commerce or Who's Who in Business*. J. Whitaker & Sons Ltd. London. 1914.

[111] *The Story of Pathéscope Apparatus*. Graham L. Newnham. 2017. Based on 3 articles in *Amateur Cine World* magazine in 1960.
[112] *Japan's Hoya to sell Pentax camera business to Ricoh*. Reuters. July 21, 2011.
[113] *Antique Microscopes*. Allan Wissner. (http://antique-microscopes.com).
[114] *History of C. Plath*. Sperry Marine.
[115] *The history of PZO - or "Polish people have also something to boast of..."*. Szymon Starczewski, Part 1, 2 Nov 2011; Part 2, 22 Nov 2011.
[116] *Guide to the Riggs & Brother Company Records*. The Independence Seaport Museum, Pennsylvania.
[117] *Riggs & Brother Records*. The Joseph Downs Collection of Manuscripts and Printed Ephemera. The Winterthur Library, Delaware.
[118] *Das Ende einer Berliner Marke*. Berliner Morgenpost, 22nd Oct 2002.
[119] *Bristol, High Street And The Blitz 1940 - a Memory of Bristol*. The Francis Frith Collection.
[120] *William Banks, 30 Corporation Street, Bolton, Lancashire*. Monthly Notices of the Royal Astronomical Society, Volume 56, Issue 8, May 1896, Pages 407-408.
[121] *The invention of spectacles. How and where glasses may have begun*. The College of Optometrists.
[122] *Historic Camera*. (http://historiccamera.com).
[123] *Special report 2003: Henry Wildey (1913-2003)*. Hampstead Scientific Society - Astronomy Section. 2003.
[124] Rochester Avionic Archives.
[125] *Barr & Stroud Binoculars*. University of Glasgow Archive Services. 2014.
[126] *Scientific Instrument Trade in Provincial England During the Industrial Revolution, 1760-1851*. A.D. Morrison-Low. PhD Thesis, University of York, 1999.
[127] *John Benjamin Dancer: Manchester Instrument Maker*. Jenny Wetton. Bulletin of the Scientific Instrument Society No. 29 (1991).
[128] *William Cary and His Association with William Hyde Wollaston. The Marketing of Malleable Platinum in Britain from 1805 to 1824*. John A. Chaldecott, The Science Museum, London. Platinum Metals Review, 1979, 23, (3).
[129] *Jesse Ramsden: the craftsman who believed that big is beautiful*. Anita McConnel. Antiquarian Astronomer, 2013, Issue 7, p. 41-53.
[130] *A Description of a new Construction of Eye-glasses for such Telescopes as may be applied to Mathematical Instruments*. Jesse Ramsden. Philosophical Transactions of the Royal Society of London. Vol XV, 1809.
[131] *Antique Medical Instruments*. Elisabeth Bennion. Sotheby Park Bernet; University of California Press. 1979.
[132] *Thomas Harvey (1812-1884). Anti-slavery campaigner and philanthropist*. The Thoresby Society. The Historical Society for Leeds and District. 2013.
[133] *A. Kershaw and Sons. The Early Years*. Peter Kelley. The Oakwood and District Historical Society. Oak Leaves, Part 7. 2007.
[134] *The Times*. March 10th 1948.
[135] *The Times*. October 1st 1947.
[136] *Company Seven*. (http://company7.com/wo/index.html).
[137] *A history of WZFO – or "Polish people have also something to boast of..." Part III: Warsaw Photo-optical Works – a history of Polish photographic cameras*. Szymon Starczewski. 16th Dec. 2011.
[138] *Scientific Instrument Making in Manchester 1870-1940. I: Setting the Scene*. Jenny Wetton. Bulletin of the Scientific Instrument Society. No. 51, 1996.
[139] *Scientific Instrument Making in Manchester 1870-1940. II: Thomas Armstrong & Brother, and G. Cussons & Company*. Jenny Wetton. Bulletin of the Scientific Instrument Society. No. 52, 1997.
[140] *Scientific Instrument Making in Manchester 1870-1940. III: Flatters and Garnet Limited, and Fowler & Company*. Jenny Wetton. Bulletin of the Scientific Instrument Society. No. 53, 1997.
[141] *Scientific Instrument Making in Manchester 1870-1940. IV: Joseph Halden & Company, and A.G. Thornton Limited*. Jenny Wetton. Bulletin of the Scientific Instrument Society. No. 54, 1997.
[142] *Optical Microscopy Primer. Anatomy of a Microscope*. Florida State University. (https://micro.magnet.fsu.edu/primer/index.html).
[143] *The Dictionary of Irish Biography*. A project of the Royal Irish Academy. Cambridge University Press.
[144] *GENi*. Genealogy website. (https://www.geni.com).

[145] *The British Museum.*
[146] *The Bonding of Optical Elements Techniques and Troubleshooting.* Summers Optical, A. Division of EMS Acquisition, Inc.
[147] *The 300-Year Quest for Binoculars.* John E. Greivenkamp and David Steed. The International Society for Optics and Photonics, SPIE. 2010.
[148] *Ignazio Porro.* Pioneers in Optics. Florida State University.
[149] *Professional papers of the Corps of Royal Engineers.* Edited by Captain A.T. Moore, R.E. Occasional Papers Vol XXVIII. 1902. Published by The Royal Engineers Institute, Chatham. 1903.
[150] *Book Review: Den Himmel Fest im Blick: eine Wissenschaftliche Biografie uber dem Astro-Optiker Bernhard Schmidt. Franz Stein Verlag, 2002.* by Osterbrock, D.E. Journal for the History of Astronomy, vol. 36, pt. 1, no. 122, p. 118 (2005).
[151] *Coudé Systems.* A.E. Whitford. The Construction of Large Telescopes, Proceedings from Symposium no. 27 held in Tucson, Arizona, Pasadena and Mount Hamilton, California, U.S.A., 5-12 April 1965.
[152] *Heliostats, Siderostats, and Coelostats: A Review of Practical Instruments for Astronomical Applications.* A.A. Mills. Journal of the British Astronomical Association, Vol. 95, NO.3/APR, P. 89, 1985.
[153] *19th Century French Scientific Instrument Makers. Part III. Lerebours et Secretan.* Paolo Brenni. Bulletin of the Scientific Instrument Society. No. 40. 1994.
[154] *19th Century French Scientific Instrument Makers. Part XI. The Brunners and Paul Gautier.* Paolo Brenni. Bulletin of the Scientific Instrument Society. No. 49. 1996.
[155] *A Dictionary of Scientists.* Oxford University Press. 1999.
[156] *Le Centaire de Pierre-Louis Guinand.* D.P. L'Astronomie, vol. 39, pp.177-197.
[157] *Joseph Fraunhofer (1787-1826).* Howard-Duff, I. Journal of the British Astronomical Association, vol.97, no.6, p.339-347.
[158] *Out of the Inkwell: Max Fleischer and the Animation Revolution.* Richard Fleischer. University Press of Kentucky, 2005.
[159] *The Muybridge Collection.* Kingston Upon Thames Museum.
[160] *Encyclopedia of Early Cinema.* Edited by Richard Abel. Routledge. 2005.
[161] *Merz Telescopes. A global heritage worth preserving.* Editors: Chinnici, Ileana (Ed.) Springer. 2017.
[162] *Essays on the History of Mechanical Engineering.* Editors: Francesco Sorge, Giuseppe Genchi. Springer, 24 Nov 2015.
[163] *T. Ertel & Sohn G.m.b.H. Mathematical Mechanical Institute for Geodetic Military Scientific Instruments in Munich, Germany, 1802-1984.* J.B. te Pas. Bulletin of the Scientific Instrument Society. No. 56, 1998.
[164] *Jean Texereau, Master Optician.* By Roger W. Sinnott. Sky and Telescope, February 11, 2014.
[165] *Obituary Notices: Richey, George Willis.* F.J. Hargreaves. Monthly Notices of the Royal Astronomical Society, Vol. 107, p.36.
[166] *Henri Chrétien (1879-1956).* R., S. L'Astronomie, Vol. 71, p.257.
[167] *Unitron History Project.* Dave Komar. (https://www.unitronhistory.com).
[168] *German WWII Ordnance Codes.* (http://oldmilitarymarkings.com/codes_full.html).
[169] *Jewel Lenses - A Historical Curiosity.* Dr. Robert Nutall & Arthur Frank. New Scientist, 13th January 1972.
[170] *The Edinburgh Gazette.*
[171] *Memoir of the Late Mr. William Blackie.* John Coldstream M.D., Leith. Transactions of the Royal Scottish Society of Arts, Volume 2, 1844, p315-320.
[172] *The Microscope: Its Construction and Management. Including Technique, Photo-micrography, and the Past and Future of the Microscope.* Henri van Heurck. 1891. English translation. Crosby, Lockwood & Son, London. 1893.
[173] *Advertisement: Richardson Adie & Co., Cutlers and Opticians.* Fifeshire Advertiser Fife, Scotland 30 Nov 1912, page 5.
[174] *Baader Planetarium.* (https://www.baader-planetarium.com)
[175] *Geometrical and Graphical Essays containing a general description of the mathematical instruments used in geometry, civil and military surveying, levelling and perspective.* George Adams, Corrected and enlarged by William Jones. Second edition, 1797. Description of a new pocket box sextant, and an artificial horizon by the editor.
[176] *Never at Rest: A Biography of Isaac Newton.* by Richard S. Westfall. Cambridge University Press. 1980.

[177] *Opticks: A Treatise of the Reflections, Refractions, Inflections & Colours of Light.* Sir Isaac Newton. Dover Publications, New York. 1979.
[178] *Dictionary of National Biography.* 1885-1900 Edition.
[179] *John Wall biography.* (https://web.archive.org/web/20120306233913/crayfordjohn.webs.com/index.htm).
[180] *The Retrofocally Corrected Apochromatic Dialyte Telescope.* John Wall and Peter Wise. Journal of the British Astronomical Association, vol.117, no.1, p.29-34.
[181] *Logos and Factories.* Alfred's Camera Page. (http://cameras.alfredklomp.com/logos).
[182] *Guide to Russian Optics Manufacturers.* Forum post by user Badger on the Milsurps forum. (http://www.milsurps.com/content.php?r=264-Guide-to-Russian-Optics-Manufacturers).
[183] *Logos of enterprises of the optical industry of the USSR.* Zenit Camera: The Russian Cameras Memorial Site. (http://www.zenitcamera.com/qa/qa-logos.html).
[184] *Henry Fitz. American Telescope Maker.* Peter Abrahams. Journal of the Antique Telescope Society, Volume 6, Summer 1994. Revised 1995, 2000.
[185] *Bailey, Boyle, Bray, Byrne, Carroll, Clacey, Fecker, Ferson, Gardam, Gaertner, Howard, Jaegers, Martz, Petitdidier, Sellew, Spencer, Thomson, Tinsley, Tolles, Usner, Witherspoon.* Peter Abrahams.
[186] *On Telescopes of Short Focal Length.* H.L. Smith. Sidereal Messenger, vol. 1, January 1883. pp.239-243.
[187] *Catchers In The Light. Volume I: Catching Space.* Stefan Hughes.
[188] *Astronomy and Astrophysics. Volume XI. 1892.* Advertisement John Byrne's Astronomical and Terrestrial Telescopes. Gall & Lembke.
[189] *Public skies: telescopes and the popularization of astronomy in the twentieth century.* by Gary Leonard Cameron. Iowa State University. PhD Dissertation. 2010.
[190] *H. Dennis Taylor, Optical Designer for T. Cooke & Sons.* Peter Abrahams.
[191] *Les Observatoires du Clain.* Blog posts tagged Ždánice Observatory. (http://debeerst.ning.com/profiles/blogs).
[192] *Berthiot's Large Format Anastigmats.* Daniel W. Fromm.
[193] *Boller and Chivens – "Where Precision is a Way of Life".* Archive history project. (http://bollerandchivens.com).
[194] *The Hubble Space Telescope Optical Systems Failure Report.* NASA. November 1990.
[195] *Records of London's Livery Companies Online: Apprentices and Freemen 1400-1900.* (https://londonroll.org/home).
[196] *Warner & Swasey Co.* Encyclopedia of Cleveland History. Case Western Reserve University.
[197] *Bardou, Brunner, Cassegrain, Cauchoix, Chevalier, Gambey, Gautier, Krauss, Lerebours et Secretan, Mailhat, Vion.* Peter Abrahams.
[198] *19th Century French Scientific Instrument Makers. Part I. H.P. Gambey.* Paolo Brenni. Bulletin of the Scientific Instrument Society. No. 38. 1993.
[199] *English telescope makers & designers of the twentieth century.* Peter Abrahams.
[200] *John Johnson Collection of Printed Ephemera.* Bodlean Libraries. University of Oxford. (Case 10, nos. 245-281).
[201] *A Short History of the Early American Microscopes.* Donald L. Padgitt. Microscope Publications Ltd. 1975.
[202] *About George A. Carroll.* Stony Ridge Observatory. (http://stony-ridge.org).
[203] *John Clacey—Optician.* Popular Astronomy, Vol. 38, 1930. pp.472-477.
[204] *The Chronodeik.* Astronomical register, vol. 20, pp.138-138. 1882.
[205] *Seth Carlo Chandler.* A. Searle, Popular Astronomy, vol. 22, pp.271-274.
[206] *John Dobson - A Brief Biography.* The Sidewalk Astronomers. (https://www.sidewalkastronomers.us).
[207] *The Story of My Life. To My Descendants to Imitate Where I was Right And to Profit by My Mistakes.* George N. Saegmuller. Ed. Dave Burns.
[208] *The German-American Connection: William Würdemann, Camill Fauth and George N. Saegmuller.* Bart Fried. Journal of the Antique Telescope Society, Vol. 7, 1994, pp.13-15.
[209] *James Walter Fecker, 1891-1945.* C. Fisher. Popular Astronomy, Vol. 54, p.17-19.
[210] *100 Obscure American Telescope Makers & Designers.* Peter Abrahams.
[211] *William Gaertner, 1864-1948.* R.T. O'Connor. Popular Astronomy, Vol. 57, 1949, p.65.
[212] *Amasa Holcombe.* Peter Abrahams.

[213] *The challenges and frustrations of a veteran astronomical optician: Robert Lundin, 1880-1962.* J.W. Briggs & D.E. Osterbrock. Journal of Astronomical History and Heritage, Vol. 1, No. 2, p. 93-103. (1998).
[214] *Tectron Telescopes.* by Tom Clark.
[215] *Irish national inventory of historic scientific instruments.* R. Charles Mollan. Samton Ltd., Dublin, 1995.
[216] *On a new Form of Ghost Micrometer.* Chas. E. Burton & Howard Grubb. Monthly Notices of the Royal Astronomical Society, Vol. 41, p.59-62. 1880.
[217] *Dennert & Pape and Aristo Slide Rules 1872–1978.* Hans Dennert. The Journal of the Oughtred Society, Volume 6, No. 1, p4-14. March, 1997.
[218] *Land Surveying in Eighteenth and Early Nineteenth-Century Dublin.* Finnian O'Cionnaith. PhD Thesis, National University of Ireland Maynooth. 2011.
[219] *Bamberg/Askania, Busch, Erfle, Ertel, Fritsch, Merz, Reinfelder & Hertel, Repsold, Steinheil, Tremel.* Peter Abrahams.
[220] *91 Telescope Makers From the Netherlands.* Peter Abrahams.
[221] *The Start Of It All.* John Victor Thomson. Optical Surfaces.
[222] *F.J. Hargreaves, 1891 February 10 - 1970 September 4.* E.J. Hysom. Journal British Astronomy Association, Vol. 82, p.43-44, 1971.
[223] *Obituary Notices: John Henry Hindle.* William Porthouse. Monthly Notices of the Royal Astronomical Society, Vol. 103, p.66-67. 1943.
[224] *Three-Mirror Telescope.* Institute of Astronomy. University of Cambridge.
[225] *The Handbook to the manufacturers & exporters of Great Britain.* Ed G.T. Wright. 1870.
[226] *Clavé Paris - History of a French perspective.* by Astropleiades.
[227] *Introduction To KMZ.* Alfred's Camera Page. (http://cameras.alfredklomp.com/kmzintro).
[228] *Instrument Makers to the World. A History of Cooke, Troughton & Simms.* Anita McConnell. William Sessions, 1992.
[229] *The Hamblin Trust.* (https://www.thehamblinvision.org.uk).
[230] *Portraits et Histoire des Hommes Utiles: Jecker.* Arthur Barker, edited by A. Jarry de Mancy, Société Montyon et Franklin, Paris. 1837. pages 363-366.
[231] *The Cambridge History of Science: Volume 4, Eighteenth-Century Science.* Ed. Roy Porter. Cambridge University Press, 2003.
[232] *Binoculars in the Air.* William Reid. Bulletin of the Scientific Instrument Society. No. 70, 2001.
[233] *'The Admirable Barr & Stroud 7 x 50': Admiralty Pattern 1900A Binoculars.* by William Reid. Bulletin of the Scientific Instrument Society. No. 54, 1997.
[234] *At the Sign of the Orrery: The Origins of the Firm of Cooke, Troughton & Simms Ltd.* From material collected by E. Wilfred Taylor and J. Simms Wilson, and brought up to date by P.D. Scott Maxwell. c1960.
[235] *Charles Darwin and Robert Brown – their microscopes and the microscopic image.* Brian J. Ford. InFocus Issue 15, September 2009.
[236] *Greenwich Time and the Discovery of the Longitude.* Derek Howse. Oxford University Press. 1980.
[237] *Scientific Trade Catalogues in Smithsonian Collections: Instruments for Science 1800-1914.* Smithsonian Libraries. Digital collection.
[238] *Surveying and Drawing Instruments.* C.F. Casella Ltd. 1911 Catalogue.
[239] *Colonial Optical Company.* 1962 catalogue.
[240] *Daguerreobase.* Collective cataloguing tool for daguerreotypes. (http://daguerreobase.org).
[241] *Hearn & Harrison Illustrated Catalogue of Magic Lanterns, Dissolving View Apparatus and Lantern Sliders.*
[242] *Mathematical, Philosophical, and Optical Instruments Made and Sold by Heath and Wing. Near Exeter-Exchange, in the Strand, London. Instruments for Geometry-Drawing, &c..* Catalogue of 1765.
[243] *Colonial Optical Company.* Catalogue of 1962
[244] *The Meopta Story.* (https://www.meopta.com/en/history).
[245] *About Al Nagler.* (https://televue.com).
[246] *Negretti & Zambra's encyclopædic illustrated and descriptive reference catalogue of optical, mathematical, physical, photographic and standard meteorological instruments, manufactured and sold by them.* Catalogue of c1887.
[247] *History of the Firm Negretti and Zambra.* W.J. Read. Bulletin of the Scientific Instrument Society. No. 5, 1985.

[248] *A compleat system of opticks in four books, viz. a popular, a mathematical, a mechanical, and a philosophical treatise. To which are added remarks upon the whole.* Robert Smith. 1738.
[249] *Takahashi historical catalogues.* (https://www.takahashijapan.com/ct-products/bnclog.html).
[250] *Telescope: Ramage's 25-foot telescope (c.1820).* The Royal Observatory, Greenwich.
[251] *On the Adjustment and Testing of Telescope Objectives.* H.D. Taylor.
[252] *Description of an engine for dividing mathematical instruments.* Jesse Ramsden. 1777.
[253] *Catalogue of microscopes and accessories.* W. Watson & Sons, Ltd.
[254] *Virtual Archive of Wild Heerbrugg.* (https://wild-heerbrugg.ch).
[255] *Jenoptem Fakes: Based on information by Claudio Manetti.* (http://holgermerlitz.de/jenoptem.html).
[256] *Scientific Instruments of the 17th and 18th Centuries and Their Makers.* Maurice Daumas. Portman Books, London. 1989.
[257] *Russian Defense Business Directory.* US Department of Commerce Bureau of Export Administration. 4th edition, May 1995.
[258] *The International Exhibition of 1862. The illustrated Catalogue of the Industrial Department.* Volume 2: British Division 2.
[259] *The Record of the International Exhibition, 1862* William Mackenzie: Glasgow, Edinburgh, London.
[260] *Thomas Armstrong and Brother* Museum of Science and Industry, Manchester, 2006.
[261] *Imperial War Museum* London, Manchester and Duxford, UK.
[262] *National Museums Scotland.* Collections database.
[263] *Brass & Glass. Scientific Instrument Making Workshops in Scotland as Illustrated by Instruments from the Arthur Frank Collection of the Royal Museum of Scotland.* T.N. Clarke; A.D. Morrison-Low; A.D.C. Simpson. National Museums of Scotland. 1989.
[264] *The Great Age of the Microscope. The Collection of the Royal Microscopical Society through 150 Years.* Gerard L'E Turner. Adam Hilger, Bristol and New York. 1989.
[265] *Galignani's New Paris Guide.* 1839
[266] *Galignani's New Paris Guide.* 1845
[267] *Galignani's New Paris Guide.* 1884
[268] *The Evolution of the Microscope.* S. Bradbury. Pergamon Press. 1967.
[269] *The Australian National Maritime Museum.* Collections database.
[270] *History of Science Museum.* University of Oxford. Collections database.
[271] *Scottish Scientific Instrument-Makers 1600-1900.* D.J. Bryden. Royal Scottish Museum Information Series. Edinburgh, 1972.
[272] *Collection Appereils.* (http://collection-appareils.fr).
[273] *Eighteenth Century Spectacles. Eyewear in the 1700s.* The College of Optometrists.
[274] *The History of the Microscope.* Reginald S. Clay and Thomas H. Court. Charles Griffin, London 1932.
[275] *The Mathematical Practitioners of Tudor & Stuart England 1485-1714.* E.G.R. Taylor. Cambridge. 1954.
[276] *The Mathematical Practitioners of Hanoverian England 1714-1840.* E.G.R. Taylor. Cambridge. 1966.
[277] *Telescope: 4-inch Danjon Prismatic Astrolabe OPL. No. 9 (1956).* The Royal Observatory, Greenwich.
[278] *The Astrolabe.* James E. Morrison. Janus. 2007.
[279] *Marine Works of Art.* Auction Catalogue. Bonhams, Knightsbridge, London, England. 25th Feb 2004.
[280] *Essays on the History of the Microscope.* Gerard L'E Turner. Senecio, Oxford. 1980.
[281] *Company: John Barker & Co. Ltd.* House of Fraser Archive. (https://housefraserarchive.ac.uk/company/?id=c0537).
[282] *Literary Panorama.* Volume 14. August 1813
[283] *Ulrich Schenk, a forgotten Swiss instrument-maker.* Anthony Turner. Bulletin of the Scientific Instrument Society. No. 78. 2003.
[284] *The Philosophical Transactions of the Royal Society of London, from Their Commencement, in 1665, to the Year 1800: 1785-1790.* Vol XVI. 1809.
[285] *Microscopy-UK, Micscape Magazine, Microscopy Library.* (http://www.microscopy-uk.org.uk).
[286] Manuscript letter from J.T. Slugg to customer. 1872.
[287] *Antique Microscopes and Other Antique Scientific Instruments.* Dr. Barry J. Sobel and Dr. Jurriaan de Groot. (https://microscope-antiques.com).

[288] *The Quest for Perfect Vision. Chérubin d'Orléans's optical instruments and the development of theories of binocular perception in late seventeenth-century France.* Antoine Gallay, 2013. MPhil Dissertation, University of Cambridge.
[289] *Benedictus Spinoza.* Peter Abrahams.
[290] *The John A. McAllister Collection: The McAllister Family. Biographical essay.* The Library Company of Philadelphia. 2007.
[291] *Klönne & Müller Arbeitsmikroskop für Apotheker (1880's).* Giovanni Maga. L'Istituto di Genetica Molecolare Luigi Luca Cavalli-Sforza (IGM).
[292] *Chemical News and Journal of Industrial Science, Volume 1.* Sir William Crookes, 1860.
[293] *The Ninth Londoniad: (complete in Itself) Giving a Full Description of the Principal Establishments in the Capital of England, etc.* London. 1863.
[294] *Official Catalogue of the Great Exhibition of the Works of Industry of All Nations, 1851.* London. 1851.
[295] *International Exhibition 1862 Official Catalogue. Industrial Department.* London. 1862.
[296] *Catalogue Général No 89 des Appareils et Accessoires usités en projection: Quatrième Partie Cinématographie.* Radiguet & Massiot. c1910. Paris, France.
[297] *Histoire des Projections Lumineuses.* Patrice Guerin. (http://diaprojection.fr/).
[298] *Museum Optischer Instrumente.* Karlsurhe, Germany.
[299] *19th Century French Scientific Instrument Makers. Part XIII: Soleil, Duboscq, and Their Successors.* Paolo Brenni. Bulletin of the Scientific Instrument Society. No. 51. 1996.
[300] *Francois Soleil, Andrew Ross and William Cookson: the Fresnel Lens Applied.* Allen Simpson. Bulletin of the Scientific Instrument Society. No. 41. 1994.
[301] *Murray's Handbook for Modern London; or London As It Is.* John Murray. London. 1851.
[302] *19th Century French Scientific Instrument Makers. Part IX: Louis Joseph Deleuil (1795-1862) and his son Jean Adrien Deleuil (1825-1894).* Paolo Brenni. Bulletin of the Scientific Instrument Society. No. 47. 1995.
[303] *John Bird 1709-1776 Mathematical Instrument Maker in the Strand.* C.D. Hellman. ISIS. 1932.
[304] *Vulgar & Mechanick. The Scientific Instrument Trade in Ireland, 1650-1921.* J.E. Burnett, A.D. Morrison-Low. National Museums of Scotland & The Royal Dublin Society
[305] *Robson's London Street Directory* Robson & Co., London. 1832.
[306] *British Camera Makers, an A-Z Guide to Companies and Makers.* Norman Channing and Mike Dunn.
[307] *John Charles Dennis, Esq., 122 Bishopsgate Street, was balloted for, and duly elected a Fellow of the Society.* G.B. Airy, Esq. Monthly Notices of the Royal Astronomical Society, Volume 10, Issue 1, November 1849, Pages 1–2. 09 November 1849.
[308] *The Athenæum Journal of Literature, Science, and the Fine Arts. July to December 1862.* John Francis. London. 1862.
[309] *Rev. D.B. Marsh. Obituary.* Journal of the Royal Astronomical Society of Canada. Vol 27. 1993. pp.373-374.
[310] *The Oxford Companion to Ships and the Sea.* Ed. Peter Kemp. Paladin Granada, 1979.
[311] *Descriptive Account of a Variety of Intellectual Toys, Made for the Instruction and Amusement of Youth, of Both Sexes, as Recommended by Dr. Beddoes, Miss Edgeworth. Illustrated by 30 Engravings of the Various Instruments.* Francis West. London. 1836.
[312] *The Craig Telescope. The Story of London's Lost Leviathan.* (http://www.craig-telescope.co.uk).
[313] *Folded-Optics Microscopes 1 of 2. The Nm1 (Newton Microscopes): Their Heritage.* R. Jordan Kreindler. Miscape Magazine. Issue 213, July 2013.
[314] *The Nm1 (Newton Microscopes): Part 2 of 2. An in-depth examination and comparison to other folded-optics designs.* R. Jordan Kreindler. Miscape Magazine. Issue 218, Dec 2013.
[315] *Companies House.* (https://find-and-update.company-information.service.gov.uk).
[316] *The TWX-1 Folded-Optics Microscope: Monarch of the Folded-Optics Kingdom – Folded-Optics Microscopes: Part I with an introduction to early field microscopes.* R. Jordan Kreindler and Yuval Goren. Miscape Magazine. Issue 193, Nov 2011.
[317] *Musées en Franche-Compté.* Digitised Books
[318] *The Nobel Prize.* (https://www.nobelprize.org).
[319] Manuscript r*esearch notes.* Generously shared with the author by Steve Gill of the Quekett Microscopical Club.
[320] *Notes on Modern Microscope Manufacturers.* Brian Bracegirdle. Queckett Microscopical Club. 1996.

[321] *Obituary Notices - Conrady, Alexander Eugen*. R. Kingslake. Monthly Notices of the Royal Astronomical Society, Vol 105, p67.
[322] *Webster Signature Database*. (http://historydb.adlerplanetarium.org/signatures)
[323] *Pascall Atkey*. Website, now defunct. (http://web.archive.org/web/20120130231757/http://www.pascallatkey.co.uk).
[324] *19th Century French Scientific Instrument Makers. Part X: The Richard Family*. Paolo Brenni. Bulletin of the Scientific Instrument Society. No. 48. 1996.
[325] *Proctor & Beilby Part I: Early 19th Century Instrument Making in the English Midlands*. A.D. Morrison-Low. Bulletin of the Scientific Instrument Society. No. 41. 1994.
[326] *Proctor & Beilby Part II: Proctor & Beilby's Sheffield*. A.D. Morrison-Low. Bulletin of the Scientific Instrument Society. No. 42. 1994.
[327] *Directory for the British Glass Industry*. 1951.
[328] *Great Exhibition 1851. Official Descriptive and Illustrated Catalogue. Part II. Classes V. to X. Machinary*. Class X. Philosophical, Musical, Horological and Surgical Instruments.
[329] *British Patents Relating to Small Telescopes and Binoculars*. Terrance Wayland.
[330] *Monthly Bulletin of the Scientific Instrument Manufacturers' Association of Great Britain Ltd*. September 1948.
[331] *GAMBICA History*. (https://www.gambica.org.uk/about-us/gambica-history.html).
[332] *WorldCat Identities*. (https://www.oclc.org/research/areas/data-science/identities.html).
[333] *Surveying Instruments*. Fritz Deumlich. Wulter de Gruyter. 1982
[334] *Little Imp Suite of Books and Catalogues on Microscopes/Microscopy*. Courtesy of Steve Gill, and hosted on Microscopy-UK.org.uk.
[335] *Who's Who of Victorian Cinema*. Edited by Stephen Herbert and Luke McKernan. London: British Film Institute, 1996.
[336] *Elizabethan Instrument Makers. The Origins of the London Trade in Precision Instrument Making*. Gerard L'E Turner. Oxford University Press. 2000.
[337] *A History of British Pathé*. (https://web.archive.org/web/20210227032028/britishpathe.com/pages/history/#18961910)
[338] *20th Century Stereo Viewers*. (https://viewmaster.co.uk).
[339] *Stereo's missing link*. John Dennis. Stereo World magazine, Vol. 7, No. 3, July-August 1980.
[340] *Flavelle Bros. & Co, Opticians & Jewellers, Sydney, New South Wales*. Museums Victoria.
[341] *Antique Optics*. (https://antiqueoptics.eu).
[342] *Horace Dall*. British Astronomical Association. Newsletter of the Instruments and Imaging Section. 9 May 2013.
[343] *Obituary of Dr. L.D. McIntosh*. American Monthly Microscopical Journal. 1892.
[344] *Sartory Instruments Ltd – A Very British Enterprise*. Carel Sartory. Quekett Journal of Microscopy, 2011, 41.
[345] *Louis Joblot and His Microscopes*. Hubert Lechevalier. American Society for Microbiology. Bacteriological Reviews. Vol. 40 No. 1, Mar. 1976.
[346] *Bolex Collector*. (http://bolexcollector.com).
[347] *Beier Cameras*. Thomas Knorre (http://beier-kamera.de).
[348] *Bencini and Italian Cameras - A Guide for the Collector*. (http://bencinistory.altervista.org).
[349] *Collecting and Using Classic Cameras*. Ivor Matanle. Thames & Hudson. 1992.
[350] *Collecting and Using Classic SLRs*. Ivor Matanle. Thames & Hudson. 1997.
[351] *McKeown's Price Guide to Antique and Classic Cameras 1995-1996*. James M. McKeown. Centennial Photo Service, U.S. 1994
[352] *The scientific value of the Carte du Ciel*. Derek Jones, FRAS. Astronomy & Geophysics, Volume 41, Issue 5, October 2000, Pages 5.16–5.20.
[355] *R B Bate of The Poultry*. A McConnell. Scientific Instrument Society Monograph, 1993
[356] *The Messer Family of London*. Family website. (https://sites.google.com/site/themesserfamilyoflondon/Home)
[357] *Davis Derby – A History of Engineering*. David Hind
[358] *Science Preserved. A Directory of Scientific Instruments in Collections in the United Kingdom and Eire*. Mary Holbrook. Trustees of the Science Museum. 1992. HMSO Publications.
[359] *Archives Commerciales de la France*. (https://gallica.bnf.fr/ark:/12148/bpt6k5592100n).

[360] *The Planimetrica Collection.* (https://planimetrica.jimdofree.com)
[361] *Scientific Trade Cards in the Science Museum Collection.* H.R. Calvert M.A. HMSO London, 1971.
[362] *Martinus Van Marum Life and Work. Volume IV. Van Marum's Scientific Instruments in Teyler's Museum.* G. L'E. Turner and T.H. Levere. Noordhoff International Publishing, Leiden. 1973.
[363] *Abraham Sharp 1653-1742.* Elizabeth Connor. Publications of the Astronomical Society of the Pacific, Vol. 54, No. 321, p.237.
[364] *The Invention of the Achromatic Lens.* R.B. Prosser. The Observatory, Vol. 40, pp.297-301. August 1917.
[365] *Note with respect to the Invention of the Achromatic Telescope.* A.C. Ranyard. Monthly Notices of the Royal Astronomical Society. Vol. 46, Issue 8, pp.460–462. June 1886.
[366] *The Evasion of Dollond's Notorious Patent on the Achromatic Telescope by the Move to the Dutch Republic of the Instrument Makers Eastland and Champneys.* Huib J. Zuidervaart. Bulletin of the Scientific Instrument Society No. 128. 2016.
[367] *J.H. Onderdewijngaart Canzius, Instrument Manufacturer and Museum Director.* Peter de Clercq. Bulletin of the Scientific Instrument Society. No. 49, 1996.
[368] *The Makers of Surveying Instruments in America Since 1700.* Charles E. Smart. Regal Art Press. 1962.
[369] *The Billings Microscope Collection of the Medical Museum, Armed Forces Institute of Pathology.* Various authors. Armed Forces Institute of Pathology, Washington DC. 1974. Reprinted 1987.
[370] *Memoirs of Literature Containing a Large Account of Many Valuable Books, Letters and Dissertations Upon Several Subjects, Miscellaneous Observations, &c.* Vol. IV. London. 1722.
[371] *A Trade Brochure by François Baillou.* Alberto Lualdi. Bulletin of the Scientific Instrument Society. No. 40, 1994.
[372] *Pietro Patroni, an 18th-century Milanese Optician.* Alberto Lualdi. Bulletin of the Scientific Instrument Society. No. 47, 1995.
[373] Monthly Notices of the Royal Astronomical Society. Volume 21, Issue 4.
[374] *Nineteenth-Century Scientific Instruments and their Makers.* Edited by P.R. de Clercq. Communication 221 of the National Museum for the History of Science and Medicine "Museum Boerhaave". Presented at the Fourth Scientific Instrument Symposium, Amsterdam 23-26 October 1984. Pub: Leiden 1985.
[375] *New Dutch Biographical Dictionary, Part 5.* P.J. Blok & P.C. Molhuysen. A.W. Sijthoff, Leiden 1921.
[376] *Telescopes from Leiden Observatory and other collections 1656-1859. A Descriptive Catalogue.* Huib J. Zuidervaart. Museum Boerhaave Communication 320. Leiden 2007.
[377] *Navigational Instruments in the Netherlands during the 19th century: Production, Distribution and Use.* Bulletin of the Scientific Instrument Society. No. 6, 1985.
[378] *Digital Web Centre for the History of Science in the Low Countries.*
(https://dwc.knaw.nl/biografie/biografisch-apparaat/scientific-instrument-makers)
[379] *Barometer Makers and Retailers 1660-1900.* Edwin Banfield. Baros Books, 1991.
[380] Book Reviews, *In de Gekroonde Lootsman. Het kaarten-, boekuitgevers en instrumentenmakershuis Van Keulen te Amsterdam 1680-1885.* Bulletin of the Scientific Instrument Society. No. 25. 1990.
[381] *The Navigating Instruments of the National Maritime Museum Greenwich: Research in Progress.* Willem F.J. Mörzer Bruyns. Bulletin of the Scientific Instrument Society. No. 92. 2007.
[382] *Collection of Historical Scientific Instruments.* Harvard University. (http://waywiser.fas.harvard.edu/collections)
[383] *Evidence from Trade Cards for the Scientific Instrument Industry.* M.A. Crawforth. Annals of Science, 42 (1985), pp453-554.
[384] *Scientific Instrument Making in Manchester 1790-1870.* Jenny Wetton. Memoirs and Proceedings of the Manchester Literary and Philosophical Society, Vol 130, 1990-91. Also published in the Proceedings of the Eleventh International Scientific Symposium, Bologna University, Sept 1991.
[385] *Instrument Makers of the London Guilds.* M.A. Crawforth. Annals of Science. Vol. 44, 1987, pp319-377.
[386] *A Note on Henry Edgeworth, Mathematical, Philosophical and Optical Instrument Maker.* Terrance J. Bryant. Bulletin of the Scientific Instrument Society. No. 61. 1999.
[387] *Van der Bildt Telescope in Teyler's Museum.* Gerard Turner. Bulletin of the Scientific Instrument Society. No. 9. 1986.
[388] *A History of the Development of the Telescope from about 1675 to 1830 based on Documents in the Court Collection.* Thomas H Court and Moritz von Rohr. Published under licence by IOP Publishing Ltd. Transactions of the Optical Society, Vol 30, Issue 5, June 1929, pp207-260.

[389] *Guild Organisation and the Instrument-Making Trade, 15-1830: The Grocers' and Clockmakers' Companies.* Joyce Brown. Annals of Science, Vol 36, Issue 1, 1979, pp1-34.

[390] *Instrument Makers in the London Livery Companies.* An Overview. Gloria Clifton. Bulletin of the Scientific Instrument Society. No. 88. 2006.

[391] *Krüss. The History of a Hamburg Based Family.* A. Krüss Optronic GmbH. 2021.

[392] *Poor Humphrey's calendar, wherein are given prophecies concerning things to come in 1829.* Poor Humphrey. 1828.

[393] *Triplet for a Microscope, &c.* J. Holland. Transactions of the Society, Instituted at London, for the Encouragement of Arts, Manufactures, and Commerce. Vol. 49, Part I (1831-1832), pp. 120-126

[394] *The Description of Two New Micrometers.* Jesse Ramsden. Philosophical Transactions of the Royal Society of London. Vol. 69, Dec. 1779.

[395] *Makers of Jewel Lenses in Scotland in the Early Nineteenth Century.* R.H. Nuttall and A. Frank. Annals of Science. 30. 1973, pp407-416.

[396] *English Instrument Making in the 18th Century.* Reginald S. Clay & Thomas H. Court. Transactions of the Newcomen Society. Volume 16, 1935 - Issue 1, pp45-54.

[397] *From Craft Workshop to Big Business – The London Scientific Instrument Trade's Response to Increasing Demand, 1750–1820.* Anita McConnell. The London Journal Volume 19, 1994 - Issue 1.

[398] *The Burgesses & Guild Brethren of Glasgow, 1751-1846.* Scottish Record Society. Edited by James R. Anderson.

[399] *Automating Time Determination: The Photographic Zenith Tube (PZT) of the Neuchatel Observatory.* Julien Gressot. Bulletin of the Scientific Instrument Society. No. 156. 2023.

[400] *Francis Watkins and the Dollond Telescope Patent Dispute. Dr. B. Gee's Take on the Subject.* A.D. Morrison-Low. Bulletin of the Scientific Instrument Society. No. 122. 2014.

[401] *Contributions to the History of the Worshipful Company of Spectaclemakers.* T.H. Court & M. von Rohr. Transactions of the Optical Society. Vol. XXXI. 1929-30. No. 2.

[402] *C.D. AHRENS (1837-1918) Optical Prism Worker of Every Description: Inventor and Maker of the Best Polarizing Prisms & Maker of the Largest Nicol Prisms in Existence.* Stuart Talbot, FRAS. Bulletin of the Scientific Instrument Society. No. 121. 2014.

[403] *Patent Agents and the Newtons in 19th-century London.* John R. Millburn, Bulletin of the Scientific instrument Society No. 20. 1989

[404] *The Newtons of Chancery Lane and Fleet Street Revisited. Part I: A Question of Establishment.* Brian Gee. Bulletin of the Scientific Instrument Society No. 35. 1992

[405] *The Newtons of Chancery Lane and Fleet Street Revisited. Part II: The Fleet Street Business and Other Genealogy.* Brian Gee. Bulletin of the Scientific Instrument Society No. 36. 1993

[406] *The Manufacture of Precision Brass Tubing.* Allan Mills. Bulletin of the Scientific Instrument Society. No. 27. 1990.

[407] *Broadhurst Clarkson & Co., Ltd. of London: Their 1971 Catalogue.* Stuart Talbot, FRAS. Bulletin of the Scientific Instrument Society. No. 112. 2012.

[408] *19th Century French Scientific Instrument Makers II: The Chevalier Dynasty.* Paolo Brenni. Bulletin of the Scientific Instrument Society. No. 39. 1993.

[409] *Manuscript research notes.* Generously shared with the author by Dominic de Mattos.

[410] *19th Century French Scientific Instrument Makers V: Jules Carpentier (1851-1921).* Paulo Brenni. Bulletin of the Scientific Instrument Sockety. No. 43. 1994.

[411] *Telescope Manufacturing and Testing at J. Hammersley in Early 20th Century.* John K. Bradley. Bulletin of the Scientific Instrument Society. No. 113. 2012.

[412] *Thomas Crickmore of Ipswich (1781-1822): 'an artist of distinguished talent as an optical instrument maker'.* D.J. Bryden. Bulletin of the Scientific Instrument Society. No. 134. 2017.

[413] *The History and Description of the Town and Borough of Ipswich.* G.R. Clarke. 1830. p459.

[414] *The Development of London Livery Companies: an historical essay and select bibliography.* William F. Kahl. Baker Library, Harvard Graduate School of Business Administration, Boston, Massachusetts. 1960.

[415] *The Gilds and Companies of London.* George Unwin. Meuthen & Co. London. 1908.

[416] *GONICHON & PARIS: Three Generations of Opticians in Paris from 1730 to 1830.* Michel Morizet. Bulletin of the Scientific Instrument Society. No. 136. 2018.

[417] *A Magical Astronomy Find.* Martin Beech. Bulletin of the Scientific Instrument Society. No. 144. 2020.

[418] *Charles Potter, Optician and Instrument Maker.* J.A. Smith. Journal of the Royal Astronomical Society of Canada, Vol. 87, No. 1. 1993.

[419] *The Remarkable Career of a 'Most Rare Workman': Johan van der Wyck (1623-1679), a Dutch-educated Military Engineer and Optical Practitioner. PART 1: In the Service of the Dutch Republic.* Huib J. Zuidervaart. Bulletin of the Scientific Instrument Society. No. 138. 2018.

[420] *The Remarkable Career of a 'Most Remarkable Workman': Johan van der Wyck (1623-1679). A Dutch-educated Military Engineer and Optical Practitioner. PART 2: In the service of the Swedish King and the Duke of Schleswig-Holstein-Gottorp.* Huib J. Zuidervaart. Bulletin of the Scientific Instrument Society. No. 139. 2018.

[421] *'Most rare workmen': optical practitioners in early seventeenth-century Delft.* Huib J. Zuidervaart & Marlise Rijks. The British Journal for the History of Science. March 2014.

[422] *Wissenschaftlicher Instrumentenbau der Firma Merz in München (1838-1932).* Dissertation zur Erlangung des Doktorgrades an der Fakultät für Mathematik, Informatik und Naturwissenschaften Fachbereich Geowissenschaften der Universität Hamburg vorgelegt von Jürgen Kost. Hamburg, 2014.

[423] *Dumotiez and Pixii: The Transformation of French Philosophical Instruments.* Paolo Brenni. Bulletin of the Scientific Instrument Society. No. 89. 2006.

[424] *Historical microbiology – using a Van Musschenbroek microscope.* Koen D. Quint 1,2 and Lesley A. Robertson. Microbiology Letters, Volume 366, Issue 16. August 2019.

[425] *A Remarkable Family Piece: A Hand-Held Telescope from the Musschenbroek Workshop.* Peter de Clercq and Charles de Mooij. Bulletin of the Scientific Instrument Society. No. 66. 2000.

[426] *Do You Know the Ménards? A Dynasty of Parisian Opticians in the 17th and 18th Centuries.* Michel Morizet. Bulletin of the Scientific Instrument Society. No. 106. 2010.

[427] *Papers in Mechanics.* Transactions of the Society, Instituted at London, for the Encouragement of Arts, Manufactures, and Commerce. Vol. 28, 1810.

[428] *Papers in Mechanics.* Transactions of the Society, Instituted at London, for the Encouragement of Arts, Manufactures, and Commerce. Vol. 29, 1811.

[429] *Papers in Mechanics.* Transactions of the Society, Instituted at London, for the Encouragement of Arts, Manufactures, and Commerce. Vol. 34, 1816.

[430] *Supplement.* Transactions of the Society, Instituted at London, for the Encouragement of Arts, Manufactures, and Commerce, Vol. 38, 1820.

[431] *Obituary. James Allan.* Inverness Courier. 13th September 1821.

[432] *Information About Telescopes; Containing Two Addresses to the Public on Cheap Telescopes, and Remarks by a Clergyman on Astronomical Observations,* With an Enlarged Catalogue and Testimonials. J.T. Slugg. 1861.

[433] *The Stuff of Legends.* David Gardner. Photographica World No 108.

[434] *P. & J. Dollond Catalogue: A Trade Handbill of c. 1780.* Stuart Talbot, FRAS. Bulletin of the Scientific Instrument Society. No. 100. 2009.

[435] *An Introduction to the History of Elliott Brothers up to 1900.* Gloria Clifton. Bulletin of the Scientific Instrument Society No. 36. 1993.

[436] *The Gerard Turner Medal Lecture, 2013: British Scientific Instrument Makers 1851-1914.* Gloria C. Clifton. Bulletin of the Scientific Instrument Society 120. 2014.

[437] *Central-Zeitung für Optik und Mechanik, Elektrotechnik und verwandte Berufszweige. Erstes, reich illustrirtes, fachwissenschaftliches Organ unter Mktwirkung bedeutender Fachgelehrten herausgegeben und redigirt von Dr. Oscar Schneider.* Berlin 1901. Druk von Rosenthal & Co.

[438] *English Barometers 1680-1860. A History of Domestic Barometers and Their Makers and Retailers.* Nicholas Goodison. Antique Collectors' Club Ltd., Woodbridge, Suffolk. 1977.

[439] *Dick and Swift's Patent Petrological Microscope.* Mr. A. Dick. Journal of the Royal Microscopical Society. 1889 part 1. pp432-436.

[440] *James Short, F.R.S., and His Contribution to the Construction of Reflecting Telescopes.* G. L'E. Turner. Notes and Records of the Royal Society, London. Vol. 24, Issue 1. June 1969. pp91-108.

[441] *The Office of Ordnance and the Instrument-Making Trade in the Mid-Eighteenth Century.* J.R. Millburn. Annals of Science 45, 1988. pp221-293.

[442] *Evidence from Advertising for Mathematical Instrument Making in London*, 1556-1714. D.J. Bryden. Annals of Science 49, 1992. pp 301-336.
[443] *Dividing the Circle: Development of Critical Angular Measurement in Astronomy, 1500-1850*. Allan Chapman. Wiley-Praxis. 1995.
[444] *The Precision Makers: A History of the Instruments Industry in Britain and France 1870-1939*. Mari E.W. Williams. Routledge. 1994.
[445] *Adams of Fleet Street, Instrument Makers to King George III*. John R. Millburn. Ashgate. 2000.
[446] *Cooke Optics. A Tale of Technical Excellence and Endurance*. Dudley Darby. ZERB (The journal of the Guild of Television Camera Professionals). Autumn 2011.
[447] *Catalogue of the British Industries Fair, 1922*.
[448] *Catalogue of the British Industries Fair, 1929*.
[449] *Catalogue of the British Industries Fair, 1937*.
[450] *Catalogue of the British Industries Fair, 1947*.
[451] *The Hull Packet*. Friday, September 25, 1835.
[452] *AG Hahn für Optik und Mechanik, Kassel*. Larry Gubas. Zeiss Historica - Journal of the Zeiss Historica Society. Vol 34. No 2. 2012.
[453] *IHAGEE: Its history until 1945*. Hugo D. Ruys. Originally pubilished in Photohistorisch Tijdschrift (the journal of the Dutch Society of Photographica Collectors). 1984. Subsequently updated.
[454] *R.W. Paul - A Versatile Instrument Maker*. IEE Review Volume 42, Issue 6, November 1996

NOTES

NOTES

Made in the USA
Middletown, DE
06 October 2024